Neutrino Physics
and Astrophysics

Neutrino Physics and Astrophysics

Proceedings of the XVIII International Conference on Neutrino Physics and Astrophysics, Takayama, Japan, 4-9 June 1998

Edited by

Y. Suzuki
Kamioka Observatory, Institute for Cosmic Ray Research,
Higashi Mozumi, Kamioka Gifu, 506-12 Japan

Y. Totsuka
Institute for Cosmic Ray Research, University of Tokyo,
3-2-1, Midori-cho, Tanashi, Tokyo 188, Japan

1999

ELSEVIER
Amsterdam - Lausanne - New York - Oxford - Shannon - Singapore - Tokyo

Elsevier Science B.V.
Sara Burgerhartstraat 25
P.O. Box 211, 1000 AE Amsterdam, The Netherlands

Reprinted from: Nuclear Physics B, Volume 77

Library of Congress Cataloging in Publication Data
A catalog record from the Library of Congress has been applied for.

ISBN: 0 444 50289 0

Transferred to digital printing 2006

FOREWORD

The 18th International Conference on Neutrino Physics and Astrophysics (Neutrino'98) was held at the Public Cultural Hall in Takayama from June 4th to 9th, 1998. Takayama is 250 km from Tokyo and located about 50 km south from the site of Super-Kamiokande.

The scientific program was arranged to cover most of the neutrino physics. But more than half of the total 70 presentations were related to the neutrino mass and oscillations including atmospheric and solar neutrino studies, which shows the rapid growth of interests stimulated by the interesting new results from the field.

Neutrino mass and oscillations may imply the existence of a mass scale many orders of magnitude higher than the current physics and will guide us to physics beyond the standard model of particle physics.

We prepared a poster session this time, which gave an opportunity for young physicists to present their results and made it easier for them to obtain some funds for attending the conference. The dedicated poster session was supplied with wine and cheese that, we hope, made discussions very active.

About 350 participants attended Neutrino '98 from 24 countries. We are grateful to all the participants of the conference, to the session chairmen and to all the speakers whose excellent presentations contributed to the scientific success of the conference. We are indebted to the members of the International Advisory Committee who made the scientific program interesting and valuable.

The conference was sponsored by the International Union of Pure and Applied Physics and Science Council of Japan. It was supported by Gifu Prefecture and Hida District Administrative Office by which many of the social programs were planned and operated. Those volunteers preparing the coffee break and interpreters were also arranged by the Hida Office.

We are grateful to many private companies for their generous contributions.

Special thanks go to the young physicists and secretaries at the Kamioka Observatory and other universities including those from US institutions who spent many hours in preparing the conference. This conference would not have been successful without their efforts.

Yoji Totsuka
Chairman of the Conference

INTERNATIONAL NEUTRINO CONFERENCE COMMITTEE

CONFERENCE ORGANIZATION

Organizing Institution

Kamioka Observatory, Institute for Cosmic Ray Research, The University of Tokyo

International Advisory Committee

J. Bahcall	IAS	P. Monacelli	Gran Sasso
B. Barish	Caltech	A. Morales	Zaragoza
F. Boehm	Caltech	L. Okun	Moscow
J. Cronin	Chicago	F. Reines	Irvine
A. Dar	Technion	M. Roos	Helsinki
J. Ellis	CERN	C. Rubbia	CERN
E. Fiorini	INFN, Milan	B. Sadoulet	Berkeley
S. Glashow	Harvard	J. Schneps	Tufts
F. Halzen	Wisconsin	J. Steinberger	CERN
T. Kirsten	Heidelberg	F. Vannucci	Paris
M. Koshiba	Tokyo	S. Weinberg	Austin
L. Lederman	Fermilab	K. Winter	CERN
G. Marx	Budapest	S. Wojcicki	Stanford
A. McDonald	Queens	G. Zatsepin	Moscow

Local Organizing Committee

J. Arafune	ICRR, Tokyo	K. Sato	Tokyo
H. Ejiri	RCNP, Osaka	H. Sobel	Irvine
T. Kajita	ICRR, Tokyo	Y. Suzuki (Secretary)	ICRR, Tokyo
M. Nakahata	ICRR, Tokyo	S. Tasaka	Gifu
K. Nakamura	KEK	Y. Totsuka (Chairman)	ICRR, Tokyo
K. Niwa	Nagoya	T. Yanagida	Tokyo
I. Sanda	Nagoya		

Sponsors

International Union of Pure and Applied Physics
Gifu Prefecture
Science Council of Japan
Hida District Administrative Office

CONTENTS

(Abstracted/Indexed in: Current Contents: Physical, Chemical & Earth Sciences/INSPEC)

*Paper presented at the conference, but not published in these proceedings.

Part 8. Neutrino and Particle Physics

Chairman: A. Halprin

Chairman: M. Goldhaber

Part 9. Direct Search for Neutrino Mass

Chairman: I.R. Barabanov

Part 10. Double Beta Decay

Chairman: E. Takasugi

Part 1

Opening Lecture

ELSEVIER Nuclear Physics B (Proc. Suppl.) 77 (1999) 3–9

NUCLEAR PHYSICS B
PROCEEDINGS
SUPPLEMENTS

Neutrinos: A Glimpse Beyond the Standard Model

P. Ramond[a]

[a]Institute for Fundamental Theory
Department of Physics
University of Florida
Gainesville, Fl 32611

Dedicated to the Memory of Dick Slansky

1. A Short History of Neutrinos

Neutrinos are awesome: of all elementary particles, only neutrinos (not even quarks!) have their own conferences, this year Neutrino-98, on a par with Susy, Strings, Lattices, and the like.

It is sobering to remind ourselves that all weak interaction experiments start out wrong, even when performed by the greatest experimentalists of their times. In 1911-1912, using a magnetic spectrometer and photographic plates, O. Von Bayer, O. Hahn, and L. Meitner [1,2] were the first to measure the spectrum of electrons in β radioactivity. Their conclusion: like α radioactivity, the spectrum of the decay product is discrete!

In 1914, Chadwick [3], performed similar measurements in Geiger's laboratory in Berlin and came out with a different conclusion, that the spectrum of β electrons is continuous. The Great War interrupted the discourse, and the next step in the story were measurements by C. D. Ellis [4] who showed that the discrete lines found earlier were due to internal conversion. Finally in 1927, C.D. Ellis and W. A. Wooster [5] found that the mean energy liberated in β decay accounted for only 1/3 of the allowed energy. By that time even Lise Meitner agreed that the electron spectrum was continuous, setting the stage for W. Pauli's famous letter.

In a December 1930 letter that starts with typical panache, "*Dear Radioactive Ladies and Gentlemen...*", W. Pauli proposes a "*desperate*" way out: there is a companion particle to the β electron. Undetected, it must be electrically neutral, and in order to balance the $N - Li^6$ statistics, it carries spin 1/2. He calls it the *neutron*. It is clear from the letter that Pauli saw no reason why this new particle could not be massive.

In 1933, E. Fermi in his formulation of the theory of β decay gave it its final name, the little neutron or *neutrino*, as it is clearly much lighter than Chadwick's neutron which had been discovered since Pauli's letter.

The next step in our story is in 1945, when B. Pontecorvo [6] puts forward the idea that neutrinos can be detected. It is based on the following observation: an electron neutrino can hit a ^{37}Cl atom and transform it into ^{37}Ar. While the Chlorine atoms are plentiful, as in cleaning fluid C_2Cl_4, Argon is an inert gas that does not interact much; furthermore it is radioactive and sticks around just long enough to be detectable through its decay: its abundance can be monitored by patient and careful experimentalists. Pontecorvo did not publish the report, perhaps because of its secret classification, or perhaps because he showed it to Fermi who thought the idea ingenious but not immediately achievable.

In 1953, Cowan and Reines [7] proposed a different technique to detect neutrinos, by using a liquid scintilator.

In 1954, Davis [8] uses Pontecorvo's original proposal, by setting up outside a nuclear reactor, and then using radio-chemical techniques to detect the Argon atoms.

In 1956, Cowan and Reines [9] announced they

PII S0920-5632(99)00382-5

had detected $\overline{\nu}_e$'s through the reaction $\overline{\nu}_e + p \to e^+ + n$. Cowan passed away before 1995, the year Fred Reines was awarded the Nobel Prize for their discovery. There emerge two lessons in neutrino physics: not only is patience required but also longevity: it took 26 years from birth to detection and then another 39 for the Nobel Committee to recognize the achievement!

In 1956, motivated by rumors that Davis had found evidence for antineutrinos coming from a pile, Pontecorvo [10] reasoned, in analogy to Gell-Mann and Pais, who had just shown how a K-meson could oscillate into its antiparticle, that it could be due to a similar effect: an electron neutrino produced in the Savannah reactor could oscillate into its own antiparticle and be detected by Davis. The rumor went away, but the idea of neutrino oscillations was born; it has remained with us ever since, and proven the most potent tool in hunting for neutrino masses.

Having detected the neutrino, there remained to determine its spin and mass. Its helicity was measured in 1958 by M. Goldhaber [11], but convincing evidence for its mass has, up to this meeting, eluded experimentalists.

In 1957, Lee and Yang propose that weak interactions violate parity, and the neutrino is again at the center of the action. Unlike the charged elementary particles which have both left- and right-handed components, neutrinos are purely left-handed (antineutrinos are right-handed), which means that lepton-number is chiral.

In 1962, a second neutrino, the muon neutrino is detected [12], (long anticipated by theorists Inouë and Sakata in 1943 [13]). This time things went a bit faster as it took only 19 years from theory (1943) to discovery (1962) and 26 years to Nobel recognition (1988).

That same year, Maki, Nakagawa and Sakata [14] introduce two crucial ideas; one is that these two neutrinos can mix, and the second is that this mixing can cause one type of neutrino to oscillate into the other (called today flavor oscillation). This is possible only if the two neutrino flavors have different masses.

In 1963, the Astrophysics group at Caltech, Bahcall, Fowler, Iben and Sears [15] puts forward the most accurate of neutrino fluxes from the Sun. Their calculations included the all important Boron decay spectrum, which produces neutrinos with the right energy range for the Chlorine experiment.

In 1964, using Bahcall's result [16] of an enhanced capture rate of ^8B neutrinos through an excited state of ^{37}Ar, Davis [17] proposes to search for ^8B solar neutrinos using a $100,000$ gallon tank of cleaning fluid deep underground. Soon after, R. Davis starts his epochal experiment at the Homestake mine, marking the beginning of the solar neutrino watch which continues to this day. In 1968, Davis et al reported [18] a deficit in the solar neutrino flux, a result that has withstood scrutiny to this day, and stands as a truly remarkable experimental *tour de force*. Shortly after, Gribov and Pontecorvo [19] interpreted the deficit as evidence for neutrino oscillations.

2. Standard Model Neutrinos

The standard model of electro-weak and strong interactions contains three left-handed neutrinos. The three neutrinos are represented by two-components Weyl spinors, ν_i, $i = e, \mu, \tau$, each describing a left-handed fermion (right-handed antifermion). As the upper components of weak isodoublets L_i, they have $I_{3W} = 1/2$, and a unit of the global ith lepton number.

These standard model neutrinos are strictly massless. The only Lorentz scalar made out of these neutrinos is the Majorana mass, of the form $\nu_i^t \nu_j$; it has the quantum numbers of a weak isotriplet, with third component $I_{3W} = 1$, as well as two units of total lepton number. Thus to generate a Majorana mass term at tree-level, one needs a Higgs isotriplet with two units of lepton number. Since the standard model Higgs is a weak isodoublet Higgs, there are no tree-level neutrino masses.

What about quantum corrections? Their effects are not limited to renormalizable couplings, and it is easy to make a weak isotriplet out of two isodoublets, yielding the $SU(2) \times U(1)$ invariant $L_i^t \vec{\tau} L_j \cdot H^t \vec{\tau} H$, where H is the Higgs doublet. As this term is not invariant under lepton number, it is not be generated in perturbation theory. Thus the important conclusion: *The standard model neutrinos are kept massless by global chiral lepton number symmetry.* The detection of non-zero neutrino masses is a tangible indication of physics beyond the standard model.

3. Neutrino Mass Models

The present experimental limits on neutrino masses are quite impressive, $m_{\nu_e} < 10$ eV, $m_{\nu_\mu} <$ 170 keV, $m_{\nu_\tau} < 18$ MeV [20]. Any model that generates neutrino masses must contain a natural mechanism that explains their small value, relative to that of their charged counterparts.

To generate neutrino masses without new fermions, we must break lepton number. This requires adding to the standard model Higgs fields which carry lepton number, as one can arrange to break lepton number explicitly or spontaneously through their interactions.

To impart Higgs with lepton number, they must be coupled to standard model leptons. From invariance requirements, we see that there can be only three such fields with two units of lepton number: An isotriplet Higgs, $\vec{\mathbf{T}}$, and two isosinglets, one positively charged, S^+, the other doubly charged, S^{--}, with renormalizable couplings

$$\vec{\mathbf{T}} \cdot L_{(i}\vec{\tau}L_{j)} \; ; \qquad S^+ L_{[i}\vec{\tau}L_{j]} \; ; \qquad S^{--}\overline{e}_{(i}\overline{e}_{j)} \; . \tag{1}$$

The curvy brackets denote flavor-symmetrization, the square ones flavor-antisymmetrization.

With these fields we can construct three types of cubic interactions that break lepton number: $H\vec{\tau}H \cdot \vec{\mathbf{T}}$, $S^+S^+S^{--}$, and $\vec{\mathbf{T}} \cdot \vec{\mathbf{T}}S^{--}$, which introduce through their couplings an unknown scale at which lepton number is violated. There are no quartic interactions that violate lepton number.

The Higgs isotriplet has a neutral compent; it can be arranged to get a vacuum value, breaking lepton number spontaneously. This leads to a Nambu-Goldstone boson, called the Majoron. Since it is part of an isotriplet, it couples to the Z boson, whose measured width rules out isotriplet breaking of lepton number. One needs electroweak singlet scalars with lepton number to devise Majoron models that are not in manifest conflict with experiment.

Perhaps the simplest way to give neutrinos masses is to introduce for each one an electroweak singlet Dirac partner, $\overline{N}_i$. These appear naturally in the Grand Unified group $SO(10)$. Neutrino Dirac masses are generated by the couplings $L_i\overline{N}_jH$ after electroweak breaking. Unfortunately, these Yukawa couplings yield masses which are too big: they are along the electroweak breaking parameter, of the same order of magnitude as the masses of the charged elementary particles $m \sim \Delta I_w = 1/2$. The situation is remedied by introducing Majorana mass terms $\overline{N}_i\overline{N}_j$ for the right-handed neutrinos. The masses of these new degrees of freedom is arbitrary, since it has no electroweak quantum numbers, $M \sim \Delta I_w = 0$. If it is much larger than the electroweak scale, the neutrino masses are suppressed relative to that of their charged counterparts by the ratio of the electroweak scale to that new scale: the mass matrix (in 3×3 block form) is

$$\begin{pmatrix} 0 & m \\ m & M \end{pmatrix} , \tag{2}$$

leading to one small and one large eigenvalue

$$m_\nu \sim m \cdot \frac{m}{M} \sim \left(\Delta I_w = \frac{1}{2}\right) \cdot \left(\frac{\Delta I_w = \frac{1}{2}}{\Delta I_w = 0}\right) . \tag{3}$$

This seesaw mechanism [21] provides a natural explanation for the smallness of the neutrino masses as long as lepton number is broken at a large scale M. With M around the energy at which the gauge couplings unify, this yields neutrino masses at or below the eV region.

The flavor mixing comes from two different parts, the diagonalization of the charged lepton Yukawa couplings, and that of the neutrino masses. From the charged lepton Yukawas, we obtain $\mathcal{U}_e$, the unitary matrix that rotates the lepton doublets L_i. From the neutrino Majorana matrix, we obtain $\mathcal{U}_\nu$, the matrix that diagonalizes the Majorana mass matrix. The 6×6 seesaw Majorana matrix can be written in 3×3 block form

$$\mathcal{M} = \mathcal{V}_\nu^t \, \mathcal{D}\mathcal{V}_\nu \sim \begin{pmatrix} \mathcal{U}_\nu & \epsilon\mathcal{U}_{N\nu} \\ \epsilon\mathcal{U}_{N\nu}^t & \mathcal{U}_N \end{pmatrix} , \tag{4}$$

where ϵ is the tiny rastio of the electroweak to lepton number violating scales, and $\mathcal{D} = \mathrm{diag}(\epsilon^2\mathcal{D}_\nu, \mathcal{D}_N)$, is a diagonal matrix. $\mathcal{D}_\nu$ contains the three neutrino masses, and ϵ^2 is the seesaw suppression. The weak charged current is then given by

$$j_\mu^+ = e_i^\dagger \sigma_\mu \mathcal{U}_{MNS}^{ij} \nu_j \ , \tag{5}$$

where

$$\mathcal{U}_{MNS} = \mathcal{U}_e \mathcal{U}_\nu^\dagger \ , \tag{6}$$

is the matrix first introduced in ref [14], the analog of the CKM matrix in the quark sector.

In the seesaw-augmented standard model, this mixing matrix is totally arbitrary. It contains, as does the CKM matrix, three rotation angles, and one CP-violating phase, and also two additional CP-violating phases which cannot be absorbed in a redefinition of the neutrino fields, because of their Majorana masses (these extra phases can be measured only in $\Delta\mathcal{L} = 2$ processes). All are additional parameters of the seesaw-augmented standard model, to be determined by experiment.

Their prediction, as for for the quark hierarchies and mixings, necessitates further theoretical assumptions. Below we present such a framework, which predicts maximal mixing between ν_μ and ν_τ [22] and a thrice Cabibbo suppression of ν_e into $\nu_{\mu,\tau}$.

4. A Neutrino Mixing Model

This model [23] follows from the Cabibbo suppresions of the Yukawa couplings of the standard model. Using the well-known Cabibbo suppressions in the quark sector, we identify family symmetries on the quarks that reproduce the patterns. We generalize this symmetry to the leptons, using grand-unified groups in a very simple way, and then use the lepton assignments to produce Cabibbo suppressions in the lepton sectors. Using special properties of the seesaw mechanism, we find a unique lepton mixing matrix, with the properties already described.

We assume that the Cabibbo supression comes about because of extra family symmetries in the standard model. A standard model invariant operator, such as $\mathbf{Q}_i\overline{\mathbf{d}}_j H_d$, if not invariant under the additional symmetry, cannot be present at tree-level. Assuming the existence of an electroweak singlet field θ, which serves as the order paramameter for this new symmetry, the interaction

$$\mathbf{Q}_i\overline{\mathbf{d}}_j H_d \left(\frac{\theta}{\Lambda}\right)^{n_{ij}} \tag{7}$$

can appear in the potential as long as the family charges balance under the new symmetry. When θ acquires a *vev*, this leads to a suppression of the Yukawa couplings of the order of $\lambda^{n_{ij}}$ for each matrix element, where $\lambda = \theta/\Lambda$ is assumed to be like the Cabibbo angle, and Λ is the natural cut-off of the theory. This is a natural mechanism in the context of an effective low energy theory with cut-off Λ. As a consequence of the charge balance equation

$$X_{if}^{[d]} + n_{ij} X_\theta = 0 \ , \tag{8}$$

the exponents of the suppression is related to the charge of the standard model invariant operator. That charge is the sum of the charges of the fields that make up the invariant. Let us now apply this mechanism to the invariants in the seesaw mechanism.

We start with the charged lepton Yukawa couplings of the form $L_i\overline{N}_j H_u$, with charges $X_{L_i} + X_{Nj} + X_H$, which gives the Cabibbo suppression of the ij matrix element. It follows that we can write the orders of magnitude of these couplings in the form

$$\begin{pmatrix} \lambda^{l_1} & 0 & 0 \\ 0 & \lambda^{l_2} & 0 \\ 0 & 0 & \lambda^{l_3} \end{pmatrix} \mathcal{Y} \begin{pmatrix} \lambda^{p_1} & 0 & 0 \\ 0 & \lambda^{p_2} & 0 \\ 0 & 0 & \lambda^{p_3} \end{pmatrix} \ , \tag{9}$$

where $\mathcal{Y}$ is a Yukawa matrix with no Cabibbo suppressions, $l_i = X_{L_i}/X_\theta$, $p_i = X_{N_i}/X_\theta$. The first matrix will form the first half of the MNS matrix in the charged lepton current.

Similarly, the mass matrix for the right-handed neutrinos, $\overline{N}_i\overline{N}_j$ will be written in the form

$$\begin{pmatrix} \lambda^{p_1} & 0 & 0 \\ 0 & \lambda^{p_2} & 0 \\ 0 & 0 & \lambda^{p_3} \end{pmatrix} \mathcal{M} \begin{pmatrix} \lambda^{p_1} & 0 & 0 \\ 0 & \lambda^{p_2} & 0 \\ 0 & 0 & \lambda^{p_3} \end{pmatrix} . \tag{10}$$

The diagonalization of the seesaw matrix is of the form

$$L_i H_u \overline{N}_j \left(\frac{1}{\overline{N}\,\overline{N}}\right)_{jk} \overline{N}_k H_u L_l \ , \tag{11}$$

from which the Cabibbo suppression matrix from the $\overline{N}_i$ fields *cancels*, leaving us with

$$\begin{pmatrix} \lambda^{l_1} & 0 & 0 \\ 0 & \lambda^{l_2} & 0 \\ 0 & 0 & \lambda^{l_3} \end{pmatrix} \mathcal{M}' \begin{pmatrix} \lambda^{l_1} & 0 & 0 \\ 0 & \lambda^{l_2} & 0 \\ 0 & 0 & \lambda^{l_3} \end{pmatrix} , \quad (12)$$

where $\mathcal{M}'$ is a matrix with elements of order one. The Cabibbo structure of the seesaw neutrino matrix is determined solely by the charges of the lepton doublets! As a result, the Cabibbo structure of the MNS mixing matrix is also due entirely to the charges of the three lepton doublets. This general conclusion depends on the existence of at least one Abelian family symmetry, which we argue is implied by the observed structure in the quark sector.

The Wolfenstein parametrization of the CKM matrix [24],

$$\begin{pmatrix} 1 & \lambda & \lambda^3 \\ \lambda & 1 & \lambda^2 \\ \lambda^3 & \lambda^2 & 1 \end{pmatrix} , \quad (13)$$

and the Cabibbo structure of the quark mass ratios

$$\frac{m_u}{m_t} \sim \lambda^8 \quad \frac{m_c}{m_t} \sim \lambda^4 \quad ; \quad \frac{m_d}{m_b} \sim \lambda^4 \quad \frac{m_s}{m_b} \sim \lambda^2 , \quad (14)$$

are reproduced by a simple charge assignment on the three quark families, namely

$$X_{\mathbf{Q},\overline{\mathbf{u}},\overline{\mathbf{d}}} = \mathcal{B}(2,-1,-1) + \eta_{\mathbf{Q},\overline{\mathbf{u}},\overline{\mathbf{d}}}(1,0,-1) , \quad (15)$$

where $\mathcal{B}$ is baryon number, $\eta_{\overline{\mathbf{d}}} = 0$, and $\eta_{\mathbf{Q}} = \eta_{\overline{\mathbf{u}}} = 2$. Two noteworthy features emerge: the charges of the down quarks associated with the second and third families are the same, and the η values for both $\mathbf{Q}$ and $\overline{\mathbf{u}}$ are the same.

Theoretical prejudices based on grand unified quantum numbers determine for us the family charges of the leptons from those of the quarks. In grand unified extensions of the standard model, baryon number generalizes in $SO(10)$ to $\mathcal{B} - \mathcal{L}$, where $\mathcal{L}$ is total lepton number, and the standard model families split under $SU(5)$ as $\overline{\mathbf{5}} = \overline{\mathbf{d}} + L$, and $\mathbf{10} = \mathbf{Q} + \overline{\mathbf{u}} + \overline{e}$. Thus a natural assignment is to assign $\eta = 0$ to the lepton doublet L_i, and $\eta = 2$ to the electron singlet $\overline{e}_i$. In this way, the charges of the lepton doublets are simply $X_{L_i} = -1(2,-1,-1)$. As we have just argued, these charges determine the Cabibbo structure of the MNS mixing matrix to be

$$\mathcal{U}_{MNS} \sim \begin{pmatrix} \mathcal{O}(1) & \mathcal{O}(\lambda^3) & \mathcal{O}(\lambda^3) \\ \mathcal{O}(\lambda^3) & \mathcal{O}(1) & \mathcal{O}(1) \\ \mathcal{O}(\lambda^3) & \mathcal{O}(1) & \mathcal{O}(1) \end{pmatrix} . \quad (16)$$

We therefore expect no Cabibbo suppression in the mixing between ν_μ and ν_τ. This mixing scheme is consistent with the preliminary results of SuperKamiokande announced at the 1977 ITP workshop [25], and also consistent with the small angle MSW [26] solution to the solar neutrino deficit.

The determination of the mass values is more complicated, as it not only depends on the relative interfamily charge assignments but also on the overall intrafamily charges. Here we simply quote the results from a particular model [23]. The masses of the right-handed neutrinos are found to be of the following orders of magnitude

$$m_{\overline{N}_e} \sim \Lambda\lambda^{13} ; \qquad m_{\overline{N}_\mu} \sim m_{\overline{N}_\tau} \sim \Lambda\lambda^7 , \quad (17)$$

where Λ is the cut-off. The seesaw mass matrix for the three light neutrinos comes out to be

$$m_0 \begin{pmatrix} a\lambda^6 & b\lambda^3 & c\lambda^3 \\ b\lambda^3 & d & e \\ c\lambda^3 & e & f \end{pmatrix} , \quad (18)$$

where we have added for future reference the prefactors a, b, c, d, e, f, all of order one, and

$$m_0 = \frac{v_u^2}{\Lambda\lambda^3} , \quad (19)$$

where v_u is the *vev* of the Higgs doublet. This matrix has one light eigenvalue

$$m_{\nu_e} \sim m_0\lambda^6 . \quad (20)$$

Without a detailed analysis of the prefactors, the masses of the other two neutrinos come out to be both of order m_0. However, the mass difference inferred by the superKamiokande result [25] (up to this conference) can be reproduced, but only if the prefactors are carefully taken into account. The two heavier mass eigenstates and their mixing angle are written in terms of

$$x = \frac{df - e^2}{(d+f)^2} , \qquad y = \frac{d-f}{d+f} , \quad (21)$$

as

$$\frac{m_{\nu_2}}{m_{\nu_3}} = \frac{1-\sqrt{1-4x}}{1+\sqrt{1-4x}} , \qquad \sin^2 2\theta_{\mu\tau} = 1-\frac{y^2}{1-4x} . \quad (22)$$

If $4x \sim 1$, the two heaviest neutrinos are nearly degenerate. If $4x \ll 1$, a condition easy to achieve if d and f have the same sign, we can obtain an adequate split between the two mass eigenstates. For illustrative purposes, when $0.03 < x < 0.15$, we find

$$4.4 \times 10^{-6} \leq \Delta m^2_{\nu_e - \nu_\mu} \leq 10^{-5}\ rmeV^2 , \qquad (23)$$

which yields the correct non-adiabatic MSW effect, and

$$5 \times 10^{-4} \leq \Delta m^2_{\nu_\mu - \nu_\tau} \leq 5 \times 10^{-3}\ \mathrm{eV}^2 , \qquad (24)$$

for the atmospheric neutrino effect. These were calculated with a cut-off, 10^{16} GeV $< \Lambda < 4 \times 10^{17}$ GeV, and a mixing angle, $0.9 < \sin^2 2\theta_{\mu-\tau} < 1$. It is satisfying that these values are compatible not only with the data but also with the gauge unification scale, and the basic ideas of Grand Unification. With poetic justice, we note that Grand Unification with its prediction of proton decay motivated the building of large underground water Čerenkov counters. The serendipitous detection of neutrinos from SN1987A by the IMB, Kamiokande, and other collaborations, established these detectors as major tools for the discovery of neutrino properties.

5. Outlook

The present field of neutrino physics is being driven by many experimental findings that challenge theoretical expectations. All can be explained in terms of neutrino oscillations, implying neutrino masses and mixing angles, but one should be cautious as evidence for neutrino oscillations has often been reported, only to either be withdrawn or else contradicted by other experiments.

The reported anomalies associated with solar neutrinos [27], neutrinos produced in cosmic ray cascades [25], and also in low energy reactions [28], cannot all be correct without introducing a new type of neutrino which does not couple to the Z boson, a *sterile* neutrino.

Small neutrino masses are naturally generated by the seesaw mechanism, which works because of the weak interactions of the neutrinos. A similar mass suppression for sterile neutrinos involves new hitherto unknown interactions, resulting in substantial additions to the standard model, for which there is no independent evidence. Also, the case for a heavier cosmological neutrino in aiding structure formation may not be as pressing, in view of the measurements of a small cosmological constant.

To conclude, experimental neutrino physics is in a most exciting stage, as it provides in the near future the best opportunities for finding evidence of physics beyond the standard model.

REFERENCES

1. Much of the following follows the excellent article by P. Radvanyi, *From Becquerel to Pauli* in Neutrinos, Dark Matter and the Universe, T. Stolarcyk, J. Tran Thanh Van, F. Vanucci, editors (Editions Frontières, France).
2. O. Von Bayer, O. Hahn, and L. Meitner, Phys. Zeitschr. **12**, 273(1911); *ibid* **13**, 273(1911); **13** 264(1912).
3. J. Chadwick, Verh. d. D. Phys. Ges., **16**, 383(1914).
4. C. D. Ellis, Proc Royal Soc. **A99**, 261(1921).
5. C. D. Ellis and W. A. Wooster, Proc. Royal Soc. **A117**, 109(1927).
6. B. Pontecorvo, Chalk river Report PD-205, November 1946, unpublished.
7. C. L. Cowan and F. Reines, Phys.Rev. **90**, 492(1953).
8. Raymond Davis Jr., Phys Rev **97**, 766(1955).
9. C.L. Cowan, F. Reines, F.B. Harrison, H.W. Kruse, A.D. McGuire , Science **124**, 103(1956).
10. B. Pontecorvo, JETP (USSR) **34**, 247(1958).
11. M. Goldhaber, L. Grodzins, A.W. Sunyar, Phys.Rev. **109**, 1015(1958).
12. G. Danby, J.M. Gaillard, K. Goulianos, L.M. Lederman, N. Mistry, M. Schwartz, J. Steinberger, Phys.Rev.Lett. **9**, 36(1962).
13. S. Sakata and T. Inouë, Prog. Theo. Physics, **1**, 143(1946).
14. Z. Maki, M. Nakagawa and S. Sakata, Prog.

Theo. Physics, **28**, 247(1962). B. Pontecorvo, Zh. Eksp. Teor. Fiz. **53**, 1717(1967).
15. J. Bahcall, W. A. Fowler, I. Iben and R. L. Sears, Astrophysics. J. **137**, 344(1963).
16. J. Bahcall, Phys. Rev. Lett. **12**, 300(1964).
17. Raymond Davis Jr., Phys. Rev. Lett. **12**, 303(1964).
18. Raymond Davis Jr., D. Harmer and K. Hoffman, Phys. Rev. Lett. **20**, 1205(1968).
19. V. Gribov and B. Pontecorvo, Phys. Lett. **B28**, 493(1969).
20. Particle Data Group, R. M. Barnett *et al.*, Phys Rev **D54**, 1(1996).
21. M. Gell-Mann, P. Ramond, and R. Slansky in Sanibel Talk, CALT-68-709, Feb 1979 (unpublished), and in *Supergravity* (North Holland, Amsterdam 1979). T. Yanagida, in *Proceedings of the Workshop on Unified Theory and Baryon Number of the Universe*, KEK, Japan, 1979.
22. Maximal mixing was recognized long ago as a feature of the grand-unified group $SO(10)$: see J. A. Harvey, P. Ramond and D. B. Reiss, Nucl. Phys. **B199**, 223(1982)
23. N. Irges, S. Lavignac and P. Ramond, Phys. Rev. **D58**, 035003(1998).
24. L. Wolfenstein, Phys. Rev. Lett. **51**, 1945(1983).
25. E. Kearns, presented at the ITP conference on Solar Neutrinos: News about SNUs, Dec 1997.
26. L. Wolfenstein, Phys. Rev. D17, 2369 (1978); S. Mikheyev and A. Yu Smirnov, Nuovo Cim. **9C**, 17 (1986).
27. See N. Hata and P. Langacker, Phys. Rev. **D56**, 6107(1997).
28. C. Athanassopoulos *et al.*, Phys. Rev. Lett. **75**, 2560(1995); **77**, 3082(1996); nucl-ex/9706006.

Part 2

Solar Neutrinos

ELSEVIER

Nuclear Physics B (Proc. Suppl.) 77 (1999) 13–19

NUCLEAR PHYSICS B
PROCEEDINGS
SUPPLEMENTS

The Homestake Solar Neutrino Program

K. Lande, B.T. Cleveland, R. Davis, Jr., J. Distel, P. Wildenhain [a]
J. Abdurashitov, V.N. Gavrin, I. Mirmov, E. Veretenkin, V.E. Yants [b]
Yu. S. Khomyakov [c]

[a]Department of Physics and Astronomy, University of Pennsylvania, Philadelphia, PA, 19104, United States of America

[b]Institute for Nuclear Research of the Russian Academy of Sciences, Moscow, 117312, Russia

[c]Institute of Physics and Power Engineering, Obninsk, Russia

1. INTRODUCTION

It is now well established and accepted that the intensity and energy distribution of the solar neutrino flux observed by the four operating detectors does not match that predicted by the standard solar model or any of its physically viable variants [1,2].

Thus, the foci of solar neutrino observations now are:

1. Precise determination of the spectrum of electron neutrinos reaching the Earth from the Sun, and
2. Demonstration, by an "appearance" observation that non-electron neutrinos are being detected, if neutrino flavor transition is indeed the source of the difference between predicted and observed solar neutrino fluxes.

The ^{8}B electron neutrino flux will be directly measured by the rate of charged current interactions in the forthcoming SNO [3] detector. In order to determine the fluxes from the other electron neutrino generating reactions from the present data set, it is necessary to carry out a series of differences between the measured rates of various experiments. For example, in order to obtain the electron neutrino flux from the sources in the 1 MeV range, ^{7}Be, PeP and CNO, the ^{8}B electron neutrino flux as determined by SNO, is subtracted from the ^{37}Cl measurement [4]. This 1 MeV flux can then be subtracted from the gallium detector observations, GALLEX [5] and SAGE [6] to get the p-p electron neutrino flux.

A similar process can be utilized to detect or put limits on the flux of non-electron neutrinos reaching the Earth from the Sun. For example, the non-electron neutrino flux from ^{8}B can be determined by subtracting the charged current signal determined by SNO from that of Super Kamiokande. In the 1 MeV domain, the non-electron neutrino signal can be determined by subtracting the electron neutrino flux (as determined by the Chlorine-SNO difference described above) from the BOREXINO measurement [7].

The difference determinations, as described above, are very sensitive to the precision of each of the measurements in the data set. Thus, it is critical that each of the involved components in the data set be determined as precisely as possible and that the data set have as much redundancy as possible. In the ^{8}B domain redundancy will be provided by the combination of Kamiokande, Super Kamiokande and SNO. In the low energy domain, this redundancy is already provided by GALLEX and SAGE. In the 1 MeV region, we presently have only one observation, that of the Homestake chlorine detector.

With this in mind, we have set two future goals

0920-5632/99/$ – see front matter
PII S0920-5632(99)00383-7

for the Homestake program, the reduction of the uncertainties in the chlorine measurement and a second determination of the electron neutrino flux in the 1 MeV region using an ^{127}I detector.

In this paper we will (1) demonstrate the role that the chlorine and iodine detectors play in determining the electron neutrino flux in the 1 MeV region, (2) discuss the plans to reduce the uncertainty in the chlorine measurement, and (3) discuss the calibration, operation and supernova investigation potential of the iodine neutrino detector.

2. DETERMINATION OF THE ^{7}Be FLUX USING ^{37}Cl AND SUPER-KAMIOKANDE RESULTS

The threshold of the chlorine detector for the reaction, ^{37}Cl(ν_e,e$^-$)^{37}Ar is 814 keV. This threshold permits the electron neutrinos from ^{7}Be (862 keV), PeP (1.44 MeV), the CNO and ^{8}B (15 MeV endpoint) to drive this reaction. The total solar electron neutrino capture rate for this detector is 2.56 $\pm$ 0.16 (stat) $\pm$ 0.16 (syst) SNU.

Since we do not yet have the SNO measurement of the ^{8}B solar electron neutrino flux, we will use the SuperKamiokande flux measurement. The only concern here is that SK does not distinguish between electron and non-electron neutrino interactions. The present SK ^{8}B measurement is 2.44$\pm$ 0.10$\times 10^6$ neutrinos cm^{-2} sec^{-1}. Multiplying this flux by the cross section for ^{8}B neutrino on ^{37}Cl of 1.14$\pm$ 0.03 $\times 10^{-42}$ cm^2 [8,9] gives a production rate of 2.78 $\pm$ 0.14 SNU. Subtracting this SK derived ^{8}B flux from the ^{37}Cl measurement gives -0.22 $\pm$ 0.26 SNU for the electron neutrino flux in the 1 MeV region. This total absence of electron neutrinos in the 1 MeV region is the most startling result of the solar neutrino observations.

The implication of this result is that all the intermediate energy solar electron neutrinos have been transformed into non- electron neutrinos before they reach the terrestrial detectors. If the transformations have been into active neutrinos, then BOREXINO, which detects neutrinos by elastic scattering from electrons, will observe a signal that is 1/6 of that which would have occured had all these neutrinos been electron type. In Cl SNU terms, the resultant signal would be 0.3 rather than 1.8 SNU.

It should be noted that the above result is solar model independent since it uses only measured neutrino fluxes and cross sections. The only assumption is that there is not an appreciable distortion of the ^{8}B neutrino energy spectrum, a result consistent with the SK observations.

If there is an appreciable non-electron component in the observed ^{8}B neutrino signal at SK, then the ^{8}B electron neutrino flux to be subtracted from the Cl observations will be reduced and a non- zero 1 MeV electron neutrino signal will result. As an example, suppose that SNO observes an electron neutrino flux from ^{8}B of 1.9 $\times 10^6$ cm^{-2} sec^{-1}. Then the difference between SK and SNO of 0.54$\times 10^6$ cm^{-2} sec^{-1} implies a non-electron neutrino ^{8}B flux of 6 $\times$ 0.54 $\times 10^6$ cm^{-2} sec^{-1}, giving a total ^{8}B neutrino flux of (1.90 + 3.24) $\times 10^6$ cm^{-2} sec^{-1}, the flux prediction of BP98 [10]. Applying these results to the chlorine measurements would give an electron neutrino flux in the 1 MeV region of 0.39 SNU. Of course, the actual fluxes must await the measurements from SNO.

Unfortunately, the uncertainty of the ^{7}Be electron neutrino flux as obtained from the combination of the chlorine and SK measurements is comparable to the difference between that ^{7}Be electron neutrino flux and the anticipated BOREXINO measurement. The demonstration of the "appearance" of non-electron neutrinos in the 1 MeV region requires a clear, statistically significant difference between the flavor independent BOREXINO measurement and the ^{7}Be electron neutrino flux. There is clearly a critical need both to reduce the Cl result uncertainties and to provide redundancy for those results with a second electron neutrino detector that functions in the same energy window.

Since the present statistical and systematic errors in the chlorine measurement are equal, in order to reduce the overall uncertainty it will be necessary to reduce both of these errors. We begin with a discussion of systematic errors. There are three contributions, uncertainty in extraction efficiency, uncertainty in counting efficiency and

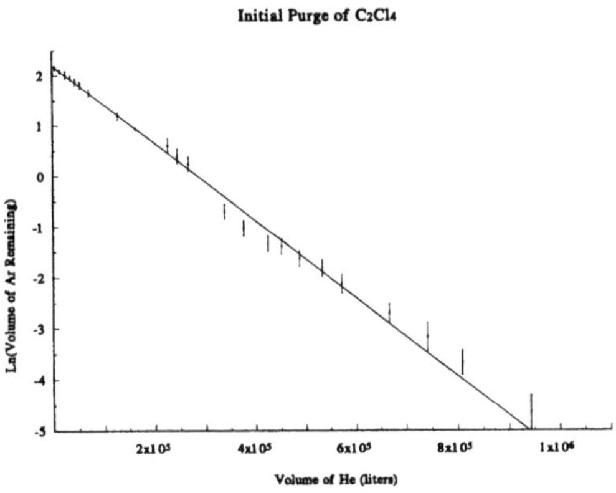

Figure 1. Initial purge of air argon from the Homestake tank. The extraction rate is measured to be linear for over three orders of magnitude in argon concentration.

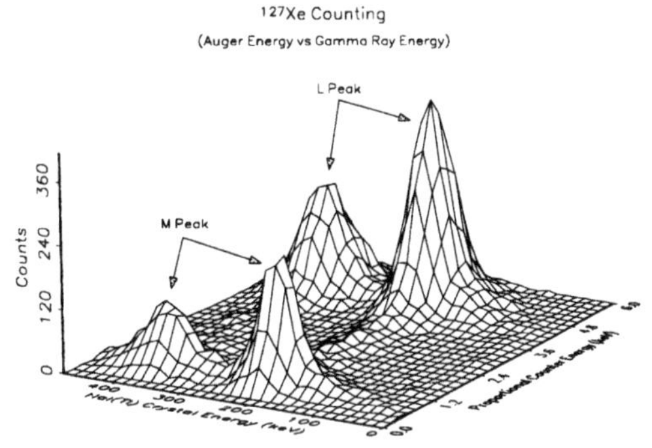

Figure 2. Auger electrons observed in coincidence with nuclear gamma rays in the decay of ^{127}Xe

uncertainty in background correction.

The uncertainty in extraction efficiency is now 1.3%. This is limited by the precision with which we can measure the fraction of isotopically labeled carrier that is recovered. In the initial extraction of air argon from the C_2Cl_4 it was demonstrated that the extraction of uniformly distributed argon from the Homestake detector can be described by a single exponent to the 99.9% extraction efficiency level. By increasing the extraction running time by a factor of 2, we can increase the recovery of argon from the present 95% to over 99.7% and reduce this uncertainty to less than 0.3%, negligible on the proposed precision scale.

The uncertainty in counting efficiency is controlled by the determination of fractional effective volume and by correction of counting background. In the past, each of these factors contributed almost 2%.

The volumetric efficiency of each counter was determined by filling the counter with an intense sample of ^{37}Ar and measuring the counting rate of this sample. This argon sample was then transferred into a counter of known efficiency and recounted. The volumetric efficiency of the test counter was thus the ratio of counts in the two counters. Uncertainties were introduced by the uncertainty of gas transfer precision and by absolute volumetric efficiency of the normalization counter. The calculated precision for this calibration procedure, 2%, agrees well with the scatter of efficiencies obtained by multiple independent calibrations of a given counter.

We have recently begun determining counter efficiency with counter fills of ^{127}Xe. ^{127}Xe decays into ^{127}I with the emission of Auger electrons in coincidence with either one (375 or 203 keV) or two (203 and 172 keV) nuclear deexcitation gamma rays. The two gamma ray events permit triple coincidence detection of the Auger electrons within the proportional counter in coincidence with the two gamma rays in two external NaI detectors. The ratio of this triple coincidence rate to the three double coincidence rates, proportional counter with either of the NaI detector and the two NaI detectors, provides the efficiency of the missing component, proportional counter or either NaI counter.

The volumetric efficiency of a given proportional counter determined with ^{127}Xe and compares very well with the ^{37}Ar calibration of that counter. In addition, repeated calibrations of a given proportional counter with ^{127}Xe gives volumetrical efficiencies that agree to about 0.5%, close to the counting statistics of the calibration.

We have also observed a very small, but identifiable cosmogenically induced background in our proportional counters. We suspect that this is induced in the iron cathodes before they are brought underground. Counters that do not use iron in their cathodes do not exhibit this background. We intend to eliminate this effect and the associated contribution to the unceertainty by making future counters with non-ferrous cathodes.

Because of decreased gain near the ends of the cathodes, a small but significant fraction of the ^{37}Ar decays in the counter have signal amplitudes less than our acceptance window. Recently constructed counters have fringing field correction plates that avoid this effect. In the future, we intend to use only fringe field corrected counters.

Finally, there is a gap between the cathode and the surrounding supersil envelope to allow for expansion of the cathode during counter bakeout. This gap constitutes a dead region in that decays that occur in this region are not observed. The volumetric efficiency measurement, of course, corrects for this loss. However, the counting statistics are degraded by this event loss. Again, we have recently introduced modifications in cathode construction that avoid this event loss.

The anticipated cumulative effect of all the above counter modifications is to both reduce the systematic uncertainties in counter efficiency to less than 1% and approximately double the effective counting efficiency. The increase in counting efficiency will permit us to go to a 35 day extraction cycle, 10 extractions per year, rather than the present 60 day extraction cycle, with more counts per extraction than before, thus appreciably improving the counting statistics per year.

The final systematic uncertainty to be attacked is that due to non-neutrino backgrounds. These are primarily associated with neutrons from the local rock and the remnant of cosmic rays that reach our depth. The present neutron correction is based on measurements made over 20 years ago. We hope to appreciably improve the precision of the neutron correction by use of a new neutron spectrometer similar to that built by Abdurashitov and Yants in Baksan. The intent is to reduce the uncertainty contribution to our SNU measurement from the neutron correction to 1%.

Combining the above effects, we hope to achieve a total statistical precision of better than 4.5% and a total systematic precision of less than 2.8% in an additional four year running period for an overall precision of 5.3% or better, approximately matching the present Cl SNU equivalent precision of the SK determination of the ^{8}B flux.

3. IODINE

Given the central role in the determination of the solar electron neutrino spectrum served by a neutrino detector with ^{7}Be sensitivity, it is extremely desirable to achieve some redundancy in this flux determination. About a decade ago, Haxton [11] pointed out that ^{127}I could play a role similar to that of ^{37}Cl in this energy domain. ^{127}I had not been previously considered because the transition to the ground state of ^{127}Xe is not allowed and thus the first transition of interest is to the 125 keV excited state of ^{127}Xe. This transition, which has a threshold of 789 keV, about 25 keV less than that from ^{37}Cl to ^{37}Ar, is sensitive to the same solar neutrinos as is ^{37}Cl. Thus, a solar neutrino measurement with an iodine detector would provide confirmation of the chlorine observations. In addition, if the iodine solar electron neutrino flux measurement can be made with a precision comparable to that of the chlorine detector, the combined results will provide an improved determination of the ^{7}Be results. Theoretical estimates of the cross section for solar neutrinos on ^{127}I [12,13] suggest that this cross section is much larger than that for ^{37}Cl and that it has a relatively enhanced for ^{7}Be neutrinos compared to that for ^{8}B.

The primary problem with an iodine detector is the need to determine the cross section for $^{127}\mathrm{I}(\nu_e,e^-)^{127}\mathrm{Xe}$ as a function of energy. This determination has been of upmost importance to us. We are engaged in a program of simultaneously obtaining this cross section from the B(G-T) measured in $^{127}\mathrm{I}(p,n)^{127}\mathrm{Xe}$ and by direct neutrino calibration over the entire solar neutrino energy range. The steps in this program are:

1. The measurement of B(G-T)

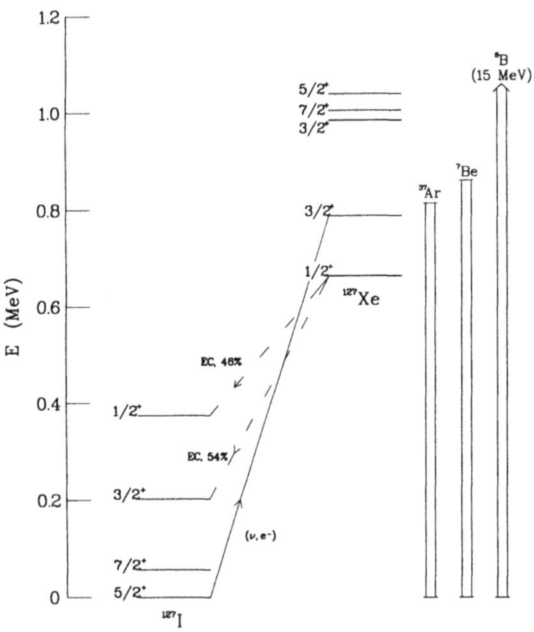

Figure 3. Nuclear level scheme for ^{127}I and ^{127}Xe. Ground state transitions are forbidden by spin considerations.

in ^{127}I(p,n)^{127}Xe reactions was carried out at the Indiana University Cyclotron Facility (IUCF) [14]. The results of this measurement are: σ [ν_e+ ^{8}B] = (4.3± 0.6) x 10^{-42} cm^2 and σ [ν_e+ ^{7}Be] = (1.2± 0.4) x 10^{-45} cm^2.

2. The measurement in a radiochemical detector of the integral of the cross section for ^{127}I(ν_e,e$^-$)^{127}Xe over the spectrum of electron neutrinos from the decay of stopped muons. This measurement was carried out at the Los Alamos Meson Physics Facility (LAMPF) [15]. Since these neutrinos range up to 53 MeV, this cross section does not directly apply to the needed ^{8}B spectrum measurement. However, the very good agreement between this cross section and that determined for this energy range from the IUCF (p,n) measurement is extremely useful in establishing the applicability of B(G-T) to neutrino cross sections. We intend to repeat this measurement with an electronic iodine detector, NaI crystal, in which we can measure the neutrino cross section as a function of energy and thus directly obtain σ [ν_e+ ^{8}B].

3. The direct measurement of σ [ν_e+ ^{7}Be] with an intense source of ^{37}Ar. ^{37}Ar decays by orbital electron capture producing monoenergetic neutrinos of 814 keV. These neutrinos are 25 keV above the threshold of the first excited state of ^{127}Xe, and 48 keV less than the neutrinos from the decay of ^{7}Be. Thus, ^{37}Ar provides an ideal source with which to directly determine the sensitivity of an ^{127}I detector to ^{7}Be neutrinos [16].

The best reaction for producing ^{37}Ar is ^{40}Ca(n,α)^{37}Ar. Since this reaction requires energetic neutrons, such a source is most reasonably produced in a fast neutron (breeder) reactor. The ideal reactor for this purpose is the BN-600 reactor in Russia. This calibration will be carried out by a collaboration of Russian and US groups with the source production at BN-600 and the exposure at the Baksan Neutrino Observatory [17,18]. The intention is to produce a preliminary source of 400 kilocuries and a final source of about 2 megacuries. The expected cross section precision of the preliminary calibration is about 18% and the final calibration is about 8%.

Although the above calibration program is somewhat lengthy, it will result in a well calibrated iodine detector and thus provide the essential input for a second, independent measurement of the electron neutrino flux from ^{7}Be in the Sun.

4. HOMESTAKE IODINE DETECTOR

The iodine detector has been designed to operate in the same mode as the chlorine detector, extract the neutrino interaction product from the detector liquid, put those product atoms into a proportional counter and determine the number of these atoms by observing their decays.

The detector consists of a series of modules, 1.2 m diameter by 11 m long, each of which contain 10 tonnes of ^{127}I in the form of NaI dissolved in water. Electron neutrinos interacting with ^{127}I produce ^{127}Xe. During the extraction phase, the xenon is swept out of the solution and onto a cryogenically cooled charcoal trap by a flow of helium. By dividing the detector into multiple modules, each with its own extraction system, we were able to greatly increase the extraction rate. The extraction time constant measured in our test module was 12 minutes, resulting in a 99.3% extraction efficiency in one hour.

The detector now under construction in the Homestake Mine consists of 10 modules with a total target of 100 tonnes of ^{127}I. Each of these modules has a separate liquid circulation pump and sweeping system. The sweeping helium then goes into a common gas manifold. The system has two charcoal traps with valving that permits us to direct the sweeping gas onto either of these traps. One planned mode of operation is to carry out two sweeps per day, one at 6 A.M. and another at 6 P.M. with the morning sweep directed to one trap and the evening sweep directed to the other trap. Neutrino flavor transitions during the passage of solar neutrinos through the core of the

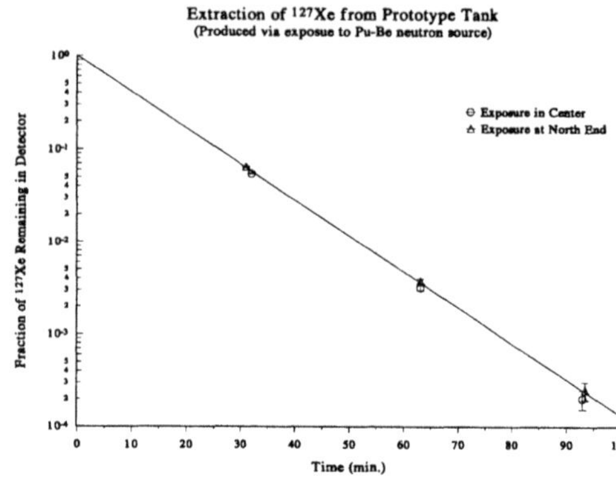

Figure 4. Extraction rate of ^{127}Xe for a neutron source exposure at one end and at the center of the prototype module. The extraction rates for the two locations are indistinguishable.

Earth would manifest themselves in a difference in the signal intensity at night as compared to that during the day, the "Day/Night Effect".

The counting of ^{127}Xe decays has already been described in the section on the calibration of proportional counters. The two and three fold coincidences, Auger electron in the proportional counter in coincidence with either one or two gamma rays in external NaI crystals, essentially eliminates uncorrelated counting backgrounds.

5. USE OF IODINE TO DETECT ELECTRON NEUTRINOS FROM SUPERNOVA

The detection of neutrino bursts from supernova are among the most spectacular events in astrophysics. These bursts although very brief, a few seconds, represent enormous instantaneous emission of energy, $\sim 10^{53}$ ergs. Since these bursts contain neutrinos of all flavors and originate in the core of the collapsar where densities are about 10^{13}, flavor transitions of all neutrinos will occur as these neutrinos traverse the collapsed star. Since neutrino pairs of all flavors are produced in the collapsar core, the flavor-energy relationship of the emitted neutrino provides a unique laboratory in which to investigate flavor transitions of electron, muon and tau neutrinos. The specific flavor sensitivity of inverse beta decay detectors to electron neutrinos permits a specific labeling of these emissions.

The iodine detector is very well suited to this supernova neutrino detection. The large cross section for electron neutrinos in the supernova energy range permits detection across the Galaxy. Frequent sweeping of the detector, i.e. Day/Night operation, insures that the detector contains no remnant ^{127}Xe atoms from background processes such as solar neutrinos. The detector would be swept into a third, reserve charcoal trap, whenever a supernova in the galaxy was detected, either by neutrinos or other means.

6. Acknowledgements

The authors would like to thank the National Science Foundation (PHY93-12480), the Civilian Research and Development Foundation (Grant RP1-213) and the University of Pennsylvania Research Foundation for their support of this work. We are deeply indebted to the Homestake Mining Company, both management and employees, for their continual assistance in our solar neutrino investigations.

REFERENCES

1. Bahcall, J. N., Krastev, P. I., Smirnov, A., Yu., Phys. Rev. D 58 (1998)
2. Hata, N., Langacker, P., Phys. Rev D 56, 6107 (1997)
3. McDonald, A. - Status of the SNO Project, these proceedings
4. Cleveland, B. T., Daily, T., Davis, Jr., R., Distel, J. R., Lande, K., Lee, C. K., Wildenhain, P., S., Ullman, J., Astrophys. J. 496, 505 (1998)
5. Kirsten, T. - Results from GALLEX and GNO, these proceedings
6. Gavrin, V. N. - Results from SAGE, these proceedings
7. Oberauer, L. - Status of BOREXINO, these proceedings
8. Aufderheide, M. B., Bloom, S. D., Resler, D., A., and Goodman, C. D., Phys. Rev. C 49, 678 (1994)
9. Bahcall, J., N. et. al. Phys. Rev. C 54, 411 (1996)
10. Bahcall, J. and Basu, S. Pinsonneault, M., Phys. Lett. B 433, 1 (1998)
11. Haxton, W., Phys. Rev. Lett. 60, 768 (1988)
12. Engel, J., Pittel, S., and Vogel, P, Phys. Rev. Lett. 67, 426 (1991)
13. Engel, J., Pittel, S., and Vogel, P, Phys. Rev. C 50, 1702 (1994)
14. Palarczyk, M., et. al., Phys. Rev. C. (1998), to be published
15. Distel, J., R. Thesis, University of Pennsylvania (1997)
16. Haxton, W. C., Phys. Rev. C 38, 3473 (1988)
17. Gavrin, V., N., Kachetkov, A. L., Kornoukhov, V. N., Kosarev, A. A., Yants, V. E., -"On The Possibility of an ^{37}Ar Artificial Neutrino Source" - INR preprint 777/92, August 1992
18. Khomuakov, Yu., S., Yevdokimov, V. P., Ievleva, J. I., Manturov, G. N., Matveenko, I., P., Rumyantsev, V. N., Tsyboulya, A. M., Chyorny, V. A. - "Reactor Technology for the Production of a Powerful ^{37}Ar Source" - IPPE Preprint, (1998), Obninsk

ELSEVIER

Nuclear Physics B (Proc. Suppl.) 77 (1999) 20–25

NUCLEAR PHYSICS B
PROCEEDINGS
SUPPLEMENTS

Solar neutrino results from SAGE*

V.N.Gavrin

Institute for Nuclear Research of the Russian Academy of Sciences Moscow 117312 Russia

For the SAGE Collaboration:
J.N.Abdurashitov[a], T.J.Bowles[b], M.L.Cherry[c], B.T.Cleveland[d], T.Daily[d], R.Davis Jr.[d], S.R.Elliott[e], V.N.Gavrin[a], S.V.Girin[a], V.V.Gorbachev[a], T.V.Ibragimova[a], A.V.Kalikhov[a], N.G.Khairnasov[a], T.V.Knodel[a], K.Lande[d], C.K.Lee[d], I.N.Mirmov[a], S.N.Nico[b], A.A.Shikhin[a], W.A.Teasdale[b], E.P.Veretenkin[a], V.M.Vermul[a], D.L.Wark[b], P.W.Wildenhain[d], J.F.Wilkerson[e], V.E.Yants[a], G.T.Zatsepin[a]

[a]Institute for Nuclear Research of the Russian Academy of Sciences Moscow 117312 Russia
[b]Los Alamos National Laboratory, Los Alamos, NM 87545 USA
[c]Louisiana State University, Baton Rouge, LA 70803 USA
[d]University of Pennsilvania, Philadelphia, PA 19104 USA
[e]University of Washington, Seattle, WA 98195 USA

We report the status of the Russian-American Gallium solar neutrino Experiment (SAGE). The solar neutrino result for SAGE III, 20 runs during the measuring period May 1995 through December 1997, is 56.7 +9.3/-8.7(stat.)+4.6/-4.8(syst.) SNU. The combined result for 57 measurements from 1990 through 1997 (SAGE I+II+III) is 66.9 +7.1/-6.8 (stat) +5.4/-5.7 (syst) SNU. The final result of the SAGE ^{51}Cr experiment to check the response of SAGE to low energy neutrinos is also presented.

1. INTRODUCTION

Elaborate analyses of the results of all operating solar neutrino experiments [1-5] leads one to the severe conclusion [see for example 6, 7, 8] that the astrophysical solution of the solar neutrino problem is almost or completely impossible. To explain what has been observed in all experiments - a significant deficit of the solar neutrino flux relative to the Standard Solar Model prediction [9] and the apparent absence of an appreciable flux of ^{7}Be neutrinos, a number of hypotheses have been developed based mainly on proposed new properties for the neutrino, particularly on the existence of neutrino oscillations. A significant test of these hypotheses is expected relatively soon from the measurement of the energy spectrum of ^{8}B neutrinos in Super-Kamiokande and from the meaurements of the neutral current to charged current ratio in SNO, as well as from BOREXINO data, results that are likely to appear not too long from now.

*The research described in this publication was made possible in part by Award No. RP2-159 of the U.S. Civilian Research & Development Foundation for the Independent States of the Former Soviet Union (CRDF), and in part by Russian Foundation for Fundamental Research Award No. 96-02-18399.

PII S0920-5632(99)00385-0

However among all of the operating solar neutrino experiments, and those planned to observe solar neutrinos in the near future, only the gallium experiments are able to provide information on the p-p neutrinos - the principal component of the solar neutrino spectrum.

Since the p-p flux is almost solely determined by the solar luminosity, and therefore is very reliably predicted, any experimentally measured deviation from the theoretical expectation could provide a direct confirmation that neutrinos have mass and oscillate, either in vacuum or in the matter of the Sun. Detailed information about the low energy part of the solar neutrino spectrum is thus greatly needed, and it should be measured by the HELLAZ and HERON experiments and the recently proposed LENCSE project. It is unlikely, however, that the results from these experiments will appear in the nearest years.

To obtain now the most precise information possible about the flux of p-p neutrinos, it is very important that the gallium experiments measure the solar neutrino flux with sufficient precision to test the theoretical prediction of the lowest possible solar neutrino capture rate in the gallium experiment of 79.5 ± 2 SNU (Bahcall 1997).

A separate and very important question is the possible existence of time variations of the solar neutrino flux, which the results of some experiments may indicate, and which should not be left out of consideration. If time variations exist, it is essential that the gallium, Borexino, SNO, and Super-Kamiokande experiments all take data at the same time. Only by simultaneous measurement of the p-p, 7Be, and 8B fluxes can the origin of any observed time variations be unfolded. Time variations have the great advantage that most of the systematic and all of the theoretical uncertainties can be ignored, since it is the relative change in the flux that is of interest. To maximize the sensitivity to time variations it is essential that the gallium experiments have the greatest possible mass..

The Russian American Gallium Experiment SAGE continues to operate and to carry out regular measurements of the integral flux of solar neutrinos. We report here new SAGE results for the period from May 1995 through December 1997. SAGE results prior to May 1995 were presented in [10].

2. THE BAKSAN GALLIUM EXPERIMENT

The Gallium Germanium Neutrino Telescope used in SAGE is situated in an underground laboratory at the Baksan Neutrino Observatory in the northern Caucasus mountains in southern Russia. The SAGE laboratory , with an overhead shielding of 4700 meters of water equivalent, is the second deepest laboratory in the world. This great depth makes the background from cosmic rays in SAGE measurements practically negligible.

SAGE has been measuring the capture rate of solar neutrinos with a target of gallium metal in the liquid state via the reaction $^{71}Ga(\nu_e, e^-)^{71}Ge$ since 1990. Metallic gallium has the advantage that its sensitivity to background from internal and external radioactivity is much less than in aqueous Ga solutions. The ^{71}Ge atoms are chemically extracted from the target at the end of each exposure period (earlier typically 1 month, now 1.5 months) and their decay is observed in low background proportional counters typically for 5-6 months of counting. The experimental layout and the chemical and counting procedures have been described elsewhere [3, 10]. In the future, we intend to introduce a number of changes in the process of germanium extraction, as well as its concentration in solution, which should allow us to reach a total efficiency of 94-95% instead of 82-84% that SAGE has now. This will improve the accuracy of measurements in an individual run by almost 7%.

Table 1.
Exposure periods of SAGE data.

Designations	Exposure periods	Number of runs	Best fit values and 1-σ uncertainties
SAGE I	Jan 1990 - May 1992	16	81+20/-18
SAGE II*	Sep 1992 - Dec 1994	21	78+13/-13
SAGE III	Mar 1994 - Dec 1997	20	57+ 9/-9
SAGE	Jan 1990 - Dec 1997	57	67+ 7/-7

*To eliminate possible bias of the SAGE results, ten runs of SAGE II during the period from November 1993 through June 1994 (when it is known that normal operations of SAGE were violated [10, 12]) are not included in the analysis.

Table 2.
Summary of systematic uncertainties for the combined SAGE data set.

Origin of uncertainty	1 -σ uncertainty in SNU
Chemical Extraction Efficiency	± 3.2
Saturation Factor	± 0.8
Counting Efficiency	+4.3/-3.2
Residual Rn after Cuts	- 3.4
Internal Radioactivity/Fast Neutrons	- 0.3
Cosmic Rays Muon Background	- 0.4
TOTAL	+5.4/-5.7

The mass of Ga now used in SAGE is slightly more than 50 tons. During each extraction approximately 0.1% of the gallium is converted to $GaCl_3$ and removed from the gallium target. The $GaCl_3$ that has accumulated during SAGE running could be converted back to metal, and thus returned for use in measurements. If SAGE will obtain funding for this convertion (about $500K), the active mass of gallium in the target could reach 58-59 tons. This would lead to a reduction of the statistical uncertainty in an individual SAGE run of about 8%.

3. EXTRACTIONS AND RESULTS

The data from SAGE naturally divides itself into several periods which are deffirenciated by various experimental conditions.

In SAGE I, monthly extractions to measure the solar neutrino flux were made from January 1990 through May 1992 (30 tons of gallium at the beginning and 57 tons after September 1991). These measurements used a hardware determination of the pulse rise time. The experiment was upgraded to SAGE II with a new counting system that included a transient digitizer with lower background and noise. Monthly extractions were made in SAGE II from September 1992 through December 1994. At that time, solar neutrino extractions were stopped and a ^{51}Cr calibration experiment was carried out. SAGE III began operation with a solar neutrino extraction in March 1995. Beginning in

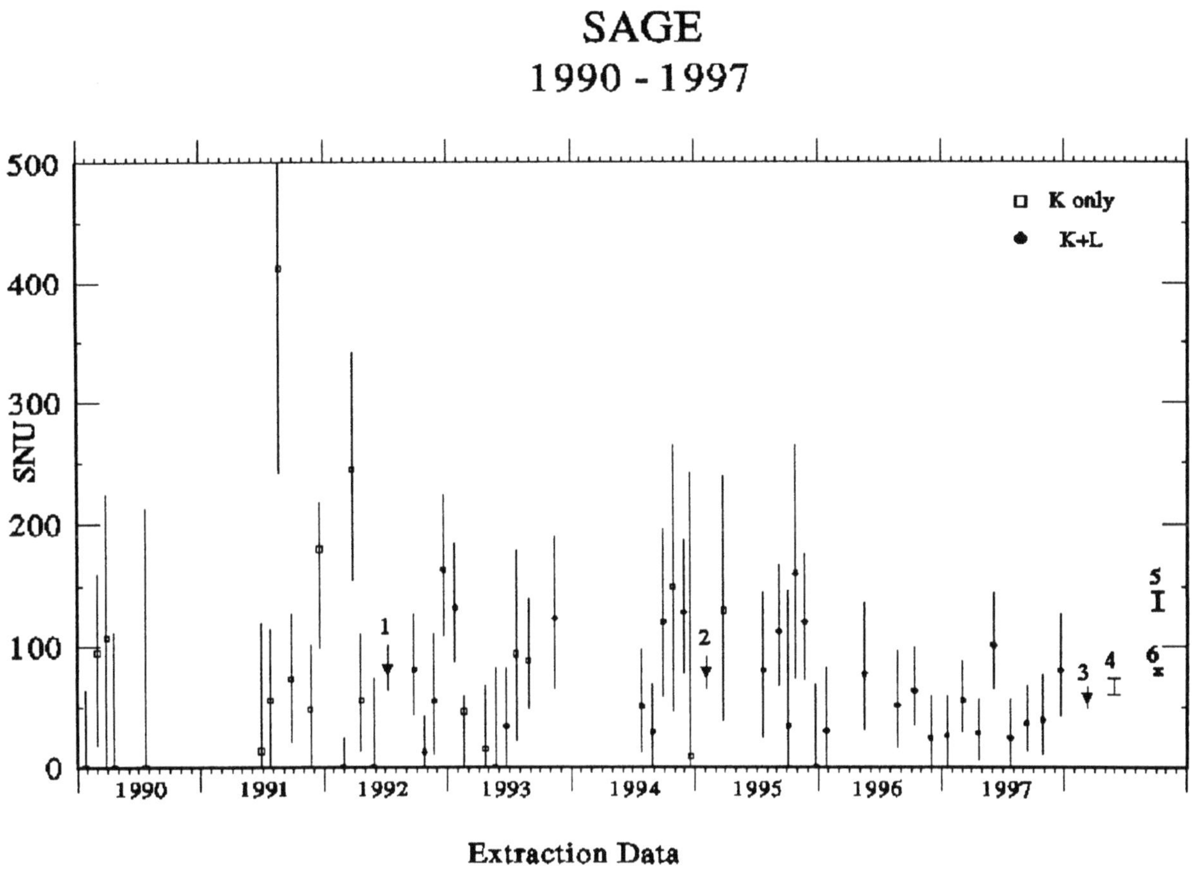

Figure 1. Best fit values and 1-σ uncertainties for each SAGE run in the period of Jan 1990 - Dec 1997. Ten runs made during Nov 1993 though June 1994 are not included [10, 12]. (1, 2, 3, 4) - Combined results for SAGE I, SAGE II, SAGE III, and the total SAGE exposure period; 5 - SSM prediction of Bahcall; 6 - Theoretically possible minimum 79.5±2 SNU.

1996, the extraction schedule was changed so that runs took place about every 45 days. This increased exposure period does not substantially deteriorate the statistical accuracy, but significantly reduces the level of effort and the amount of chemical reagents needed for the extraction.

Data from all the solar neutrino runs made between January 1990 and December 1997 are presented here except those runs in which either the extraction sample was accidentally lost or the electronics or proportional counter failed during the counting of the extraction sample. Figure 1 shows individual run results along with the combined results for each exposure period and for all measurements from 1990-1997. The time period, the number of runs, and the combined results for each SAGE exposure period are presented in Table 1. The same standard analysis procedure was used in SAGE III as in SAGE II [10, 11].

The systematic uncertainties determined for the full SAGE data set are summarized in Table 2. SAGE is continuing to work on reducing the systematic uncertainties. One can see preliminary improvement of systematic uncertainties by comparison of the values quoted here with those quoted earlier [3, 10]. It is expected that we will be able to further reduce some of the systematic uncertainties in the future.

The final statistical result of a combined fit of the data from January 1990 through December 1997 data is 66.9 +7.1/-6.8 SNU with a value of the Cramer-von Mises fit criterion Nw^2 of 0.056 corresponding to a goodness of fit probability of 66%. The best fit value for the half-life of the observed decay signal of ^{71}Ge is 11.1+2.3/-1.8 days, in good agreement with the known ^{71}Ge half-life of 11.4 days. Including the systematic uncertainties, the SAGE I + II + III result is 66.9 +7.1/-6.8 (stat) +5.4/-5.7 (syst)SNU.

4. ^{51}Cr EXPERIMENT FINAL RESULTS

A calibration experiment using a 517 kCi ^{51}Cr source irradiating 13 tons of Ga was carried out from December 1994 through May 1995. The measured production rate of ^{71}Ge from ^{51}Cr source was 14.0 ± 1.5(stat.) ± 0.8(syst) atoms per day, equivalent to about 3500 SNU, 50 times higher than the rate from solar neutrinos [13]. This is the largest production rate ever measured with a low-energy neutrino source. The neutrino capture cross section can be inferred from this measuring production rate and compared to theoretically calculated cross sections of Bahcall [14] and Haxton [15]

$$R = \sigma\,(\text{measured})/\sigma\,(\text{theoretical}) = \begin{cases} 0.95 \pm 0.12(\text{exp.}) +0.035/-0.027\ (\text{theor})\ (\text{Bahcall}) \\ 0.87 \pm 0.11(\text{exp.}) \quad \pm 0.09 \quad (\text{theor})\ (\text{Haxton}) \end{cases}$$

We present here the final result of ^{51}Cr calibration experiment as the ratio (R) of the measured cross section to the value calculated from theory. With either of these theoretical cross sections, (R) is consistent with 1.0, which implies that the total efficiency of the SAGE experiment to the neutrinos from ^{51}Cr is close to 100%.

5. CONCLUSIONS AND PLANS

SAGE has carried out measurements of the integral flux of solar neutrinos from 1990 to the present. The SAGE observation of only 52 ± 7% of the Standard Solar Model prediction of Bahcall (129 +8/-6 SNU) reported in this proceedings constitutes a 6-σ effect. The SAGE extraction efficiency measured in the ^{51}Cr calibration experiment demonstrates with a high confidence level that the observed deficit reflects a significant reduction in the integral flux of solar neutrinos compared to the Standard Solar Model predictions. SAGE intends to continue long-term measurements whose a main goals are to improve the accuracy of determination of the suppression of the solar neutrino flux and to clarify the question about the existence of time variations of the flux. Even though the present experimental data are statistically compatible with a constant flux, time variations can not be excluded. In the investigation of this question it would be useful to join the efforts of both gallium experiments by cooperatively running SAGE and GNO.

6. ACKNOWLEDGMENTS

First of all we would like to express our deep appreciation for the support of all our colleagues from the USA and other countries who do their best to help us to preserve the SAGE experiment.

SAGE wishes to thank E.N.Alexeyev, J.N.Bahcall, M.Baldo-Ceolin, V.A.Kuz'min, L.B.Bezrukov, S.Brice, A.E.Chudakov, G.T.Garvey, W.Haxton, V.A.Matveev, R.G.H.Robertson,V.A.Rubakov, A.Yu.Smirnov, A.N.Tavkhelidze and many members of

GALLEX for their continued interest and for stimulating discussions. We acknowledge support from the Russian Academy of Sciences, the Institute for Nuclear Research of the Russian Academy of Sciences, the Ministry of Science and Technology of Russian Federation, the Division of Nuclear Physics of the Department of Energy, and the National Science Foundation.

REFERENCES

1. R.Davis Prog. Part. Nucl. Phys. 32, 13 (1994); B.T.Cleveland et al., Astrophys.J. 496, (1998) 505; K.Lande, Proc. of this Conference.
2. Y.Suzuki et al., Nucl. Phys.B 38 (1995) 54.
3. J.N.Abdurashitov et al., Phys. Lett.B 328 (1994) 234
4. P.Anselmann et al., Phys. Lett.B 342 (1995) 440; T.A.Kirsten, Proc.of this Conference.
5. Y.Suzuki, Results from Super-Kamiokande & Kamiokande, Proc.of this Conference.
6. A.Yu.Smirnov, Proc. of the 17th Intern. Conf. on Neutrino Physics and Astrophysics, ed. by Kari Enqvist, Katri Huitu, Jukka Maalampi, Helsinki, Finland (1996) 38.
7. V.Berezinsky, Invited Lecture, Proc 25th Intern. Cosmic Ray Conference, Durban (1997), in press.
8. J.N.Bahcall, Proc. 4th Intern. Solar Neutrino Conf. ed by W.Hampel, Heidelberg, Germany, (1997) 3.
9. J.N.Bahcall, Phys. Lett. 338 B (1994) 276; S.Turck -Chieze and I.Lopez, Astrophys. J. 408 (1993) 347.
10. T.J.Bowles et al. (SAGE Collaboration), Proc. 4th Intern. Solar Neutrino Conf., ed. by W.Hampel, Heidelberg, Germany (1997) 109.
11. B.T.Cleveland, Nucl. Instrum. Methods Phys. Res. 214 (1983) 451; S.R.Elliott, Nucl. Instrum. Methods Phys. Res. Sect. A 290 (1990) 158.
12. V.N.Gavrin et al. (SAGE Collaboration), Proc. 17th Intern. Conf. on Neutrino Physics and Astrophysics, ed. by Kari Enqvist, Katri Huitu, Jukka Maalampi, Helsinki, Finland (1996) 14.
13. J.N.Abdurashitov et al. (SAGE Collaboration) Phys. Rev.C (1998), submitted.
14. J.N.Bahcall, Phys. Rev.C 56 (1997) 3391.
15. W.Haxton, Preprint nucl-th/9804011, Phys. Lett. B , in press.

ELSEVIER

Nuclear Physics B (Proc. Suppl.) 77 (1999) 26–34

NUCLEAR PHYSICS B
PROCEEDINGS
SUPPLEMENTS

GALLEX solar neutrino results and status of GNO

T. A. Kirsten

Max - Planck - Institut für Kernphysik, POB 103 980, D-69029, Heidelberg, Germany

for the GALLEX* and GNO collaborations†

We describe the results of the GALLEX solar neutrino experiment after completion of the project. In particular, we summarize the results for GALLEX IV (12 solar runs) and for our ^{71}As - spiking experiments. The integral GALLEX result for all 65 solar runs (GALLEX I-IV) is (78 ± 8) SNU (1σ). This is only slightly more than half of the expected rate. The significance of this deficit is assured

- directly by the results from our two ^{51}Cr neutrino source experiments (at the 10% - level), and
- indirectly by means of experiments in which ^{71}Ge is generated within the target solution from beta-decaying ^{71}As (at the percent level). GALLEX at Gran Sasso is now succeeded by GNO (Gallium Neutrino Observatory). The GNO status as of October 1998 is reported.

1. INTRODUCTION

The GALLEX detector at the Gran Sasso Underground Laboratory (LNGS) has observed since May 1991 a flux of solar neutrinos in sufficient quantity to account for the solar luminosity as reflected in the flux of low - energy pp-neutrinos from the primary hydrogen fusion reactions occurring in the solar core [1-5]. However, in assigning the measured signal to pp - neutrinos, little or no signal is left to account for the other neutrino fluxes also expected from the Standard Solar Model (SSM), in particular the ^{7}Be neutrinos ('^{7}Be - neutrino problem' [4-7]). The significance of this observation has developed with our continuing reduction of the statistical errors and with the demonstration of reliable detector performance in the ^{51}Cr neutrino source experiment. Recently we have aimed at an additional performance demonstration by spiking the gallium detector with ^{71}As.

The present article is organized as follows: In sect. 2 we report on the results from GALLEX IV and summarize the GALLEX solar data altogether. In sect. 3 we present the results from the ^{71}As spiking experiments and discuss the consequences, also in context with the results from our ^{51}Cr neutrino source experiments. Sect. 4 is a status report of the Gallium Neutrino Observatory ('GNO'), the project that succeeded GALLEX at the Gran Sasso underground laboratories.

2. GALLEX SOLAR DATA

The principal aim of our radiochemical GALLEX detector is the detection of solar pp - neutrinos via $\nu_e^- + {}^{71}Ga \rightarrow {}^{71}Ge + e^-$ ($T_{½}$=11.43 d, E_{thr}=233 keV).

* GALLEX collaboration: W. Hampel, J.Handt, G. Heusser, J. Kiko, T. Kirsten (spokesperson), M. Laubenstein, E. Pernicka, W.Rau, U. Rönn, M. Wojcik, Y. Zakharov (MPIK Heidelberg); R.v.Ammon, K.H. Ebert, T. Fritsch, D. Heidt, E. Henrich, L. Stieglitz, F. Weirich (FZK Karlsruhe); M. Balata, M.Sann, F.X.Hartmann (LNGS Gran Sasso); E. Bellotti, C. Cattadori, O. Cremonesi, N. Ferrari, E. Fiorini, L. Zanotti (Milano INFN); M. Altmann, F. v. Feilitzsch, R. Mößbauer, S.Wänninger (TUM München); G. Berthomieu, E. Schatzman (Nice Observatoire); I. Carmi, I. Dostrovsky (Weizmann Institute); C. Bacci, P. Belli, R. Bernabei, S. d'Angelo, L. Paoluzi (Roma INFN); M. Cribier, J. Rich, M. Spiro, C. Tao, D. Vignaud (CE Saclay); J.Boger, R. L. Hahn, J. K. Rowley, R. W. Stoenner, J. Weneser (Brookhaven Nat.Lab).

† GNO collaboration: E. Bellotti (spokesperson), C. Cattadori, N. Ferrari, L. Zanotti (Milano INFN); E. Burkert, W. Hampel, J.Handt, G. Heusser, J. Kiko, T. Kirsten, M. Laubenstein, W. Rau, H. Richter, M. Wojcik, Y. Zakharov (MPIK Heidelberg); K.H. Ebert, E. Henrich (FZK Karlsruhe); M. Balata, F. X. Hartmann (LNGS Gran Sasso); M. Altmann, F. v. Feilitzsch (TUM München); E. Schatzman (Nice Observatoire); P. Belli, R. Bernabei, S. d'Angelo (Roma INFN); M. Cribier (CE Saclay).

PII S0920-5632(99)00389-8

The target is 30.3 tons of gallium (12 tons of ^{71}Ga) contained in 100 tons of concentrated gallium-chloride solution. This chemical form was chosen to facilitate the extraction of the product, ^{71}Ge [8]. Extremely low counter backgrounds (order 1 count per month) are required in order to significantly detect the minute neutrino induced activity. Typically only a handful of ^{71}Ge decays are observed per run [9]. Energy and pulse shape analysis serve to select the candidate events. Then, the maximum likelihood method is used to partition the counts into the signal from ^{71}Ge decay and the background, constant in time. Counting of a run lasted typically for about 6 months in order to fully characterize the very low background after the decay of ^{71}Ge. Data taking started in May 1991. Since that date, about monthly runs were performed till January 1997 (counting lasted till June 19, 1997). This covers a total of 65 runs comprising 1594 net days of exposure. The GALLEX periods were separated by a scheduled change of the target tanks (after GALLEX I) and by two Cr - source exposure periods (between GX II/GX III and GX III/GX IV, respectively). Table 1 is a matrix of all solar runs performed within GALLEX I - IV.

Table 1
GALLEX solar run matrix. Statistical and systematic errors are combined in quadrature (1σ). Small contributions from side reactions are already subtracted (for details see the quoted references).

Date of data release	Ref.	Solar runs G I	G II	GIII	GIV	Total	Blank runs (total)	Result (SNU) (cumulative)
May 1992	[1]	14				14	5	83 ± 21 SNU
June 1993	[2]	15	6			21	11	87 ± 16 SNU
Febr 1994	[3]	15	15			30	19	79 ± 12 SNU
June 1995	[4]	15	24			39	27	77 ± 10 SNU
July 1996	[5]	15	24	14		53	31	70 ± 8 SNU
Oct 1998	[6]	15	24	14	12	65	36	78 ± 8 SNU

In addition to the solar runs, blank runs are also listed. Blanks were frequently performed in order to verify the absence of any target related artifacts. They are in every respect identical to solar runs except that the exposure time is reduced to one day, the minimum time required to mimic a real run (the blank data given are corrected for the neutrino production during that one day). The result from all 36 blank runs is (- 4 ± 5) SNU (1σ), consistent with a null result.

The individual run results for GALLEX IV are given in Table 2. They are also contained in Figure 1, where all 65 solar run results are shown. The data are corrected for side reactions. The magnitude of these corrections is summarized in Table 3 (see [2] for details).

Evidently, a single run result has little meaning because the error is large (typically only about four ^{71}Ge counts are recorded per solar run). However, the statistics assembled during more than 5 years of data taking allowed us to reduce the statistical error

Table 2 :
Results from individual solar runs in GALLEX IV (standard rise time cut), after corrections for side effects (see Table 3). 1σ statistical errors.

Run #	Run type	Exposure dates	Duration	Result [SNU]
SR 54	A146	02/13/96-03/05/96	21d	140^{+75}_{-63}
SR 55	A148	03/06/96-03/28/96	22d	57^{+54}_{-40}
SR 56	A149	03/28/96-04/16/96	19d	118^{+71}_{-56}
SR 57	A151	04/17/96-05/07/96	20d	155^{+73}_{-59}
SR 58	A157	06/26/96-07/16/96	20d	108^{+74}_{-61}
SR 59	A158	07/16/96-08/06/96	21d	151^{+73}_{-59}
SR 60	A161	08/28/96-09/17/96	20d	-78^{+62}_{-52}
SR 61	A162	09/17/96-10/09/96	22d	106^{+57}_{-46}
SR 62	A163	10/09/96-11/19/96	41d	186^{+63}_{-54}
SR 63	A165	11/20/96-12/10/96	20d	70^{+54}_{-41}
SR 64	A166	12/10/96-01/08/97	29d	184^{+71}_{-60}
SR 65	A167	01/08/97-01/21/97	13d	101^{+72}_{-54}

SR = solar run

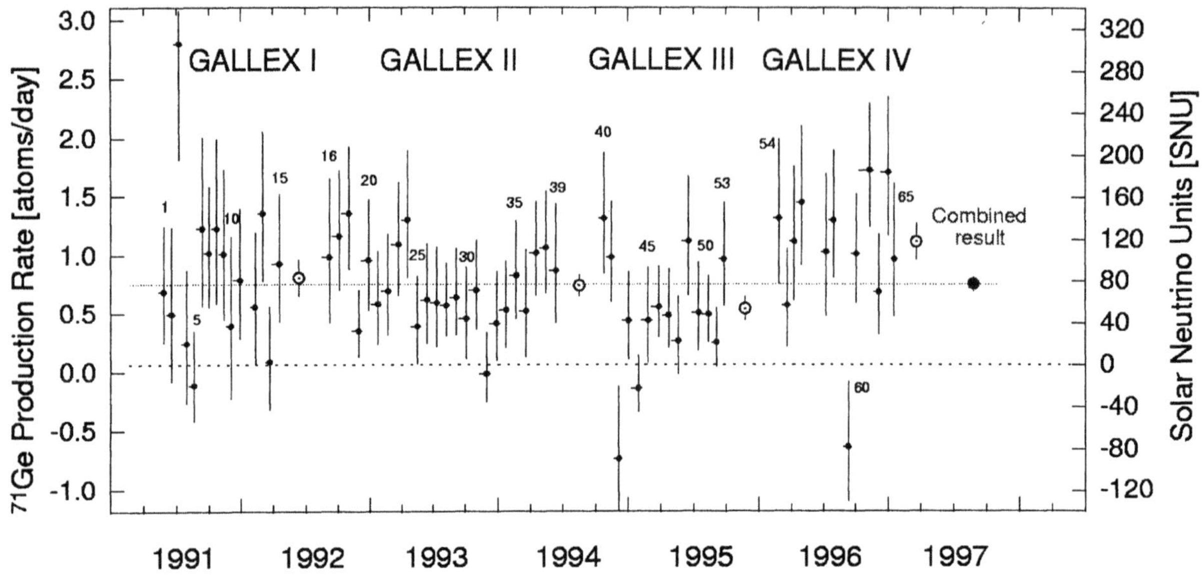

Figure 1. GALLEX single run result overview. The left hand scale is the measured ^{71}Ge production rate; the right hand scale, the net solar neutrino production rate (SNU) after subtraction of side reaction contributions. Error bars are ± 1σ, statistical only. The label 'combined' applies to the mean global value for the total of all 65 runs. Horizontal bars represent run duration; their asymmetry reflects the 'mean age' of the ^{71}Ge produced. See Table 1 for references (source of data).

Table 3
Contributions from side reactions and systematic errors in GALLEX. For details, see [2].

	GALLEX IV	**GALLEX (I-IV)**
SIDE REACTIONS		
muon induced background	2.8 ± 0.6 SNU	2.8 ± 0.6 SNU
fast neutrons	0.15 ± 0.1 SNU	0.15 ± 0.1 SNU
^{69}Ge from muons and ^{8}B-ν that is falsely attributed to ^{71}Ge	1.0 ± 1.0 SNU	1.0 ± 1.0 SNU
Rn outside the counters	0.3 ± 0.3 SNU	0.3 ± 0.3 SNU
Subtotal	*4.3 ± 1.2 SNU*	*4.3 ± 1.2 SNU*
Rn-cut inefficiency	2.2 ± 1.7 SNU	2.2 ± 1.2 SNU
TOTAL TO SUBTRACT	**6.5 ± 2.1 SNU**	**6.5 ± 1.7 SNU**
SYSTEMATIC ERRORS		
counting efficiency (incl. Energy- and rise-time cuts)	± 4.7 %	± 4.5 %
target size and chemical yield	± 2.2 %	± 2.2 %
^{68}Ge correction error	$\pm^{0.0}_{0.1}$ %	$\pm^{0.9}_{2.6}$ %
side reaction subtraction error	± 1.8 %	± 2.2 %
TOTAL SYSTEMATIC ERROR	**± 5.5 %**	$\pm^{\mathbf{5.6}}_{\mathbf{6.1}}$ **%**

Table 4
Results from solar exposure periods. Statistical and systematic errors are combined in quadrature.

	Time period	Net exposure days	Number of runs	Result [SNU]
GALLEX I	05/14/91-04/29/92	324	15	$83.4 \pm^{18.5}_{19.5}$
GALLEX II	08/19/92-06/22/94	649	24	$75.9 \pm^{10.5}_{10.7}$
GALLEX III	10/12/94-10/04/95	353	14	53.9 ± 11.0
GALLEX IV	02/14/96-01/23/97	268	12	118.5 ± 19.0 *
GALLEX (I-IV)	05/14/91-01/23/97	1594	65	$\mathbf{77.5} \pm^{7.6}_{7.8}$

to ≈ 6 % under the assumption of a production rate constant in time. Altogether about 320 ^{71}Ge atoms produced by solar neutrinos have actually been seen to decay.

The *differential* results for the four measuring periods are given in Table 4. The development in time of the *cumulative* GALLEX result is shown in the last column of Table 1. The data are self-compatible since the beginning and the errors shrunk in a consistent fashion. The joint result (for GALLEX I-IV) is $77.5 \pm^{7.6}_{7.8}$ SNU (1σ). Major sources for systematic errors are listed in Table 3. The errors quoted in Tables 1 and 4 contain the systematic errors after quadratic addition to the statistical errors.

As evident from Figure 1 and Table 4, the results for the four measuring periods GALLEX I, II, III and IV show appreciable scatter. We note that GALLEX III is 2.2 σ below, GALLEX IV 2.2 σ above the mean value. Obviously this is no compelling evidence for any time variability of the solar neutrino flux, but for a better apprehension of the situation we have analyzed the statistics of these low rate data by Monte-Carlo simulations. First we investigate whether the scatter of the 65 single run results is compatible with a Monte-Carlo generated distribution of 65,000 single run results (1,000 full GALLEX experiments) for a constant production rate of 84 SNU (signal 77.5 SNU + 6.5 SNU for side reactions) under the conditions of normal single runs, using the actual conditions of GALLEX I, II, III, and IV in appropriate proportions. The result is positive. As shown in Figure 2, the fit of the two distributions is very good. Hence, if anything, it is the time sequence of high and low results, not the results themselves, which display larger statistical departures from the mean.

The χ^2-test for the results of the four data taking periods to be compatible with the mean yields only a 1.5 % probability ($\chi^2 = 10.4$ with 3 d.o.f.). However,

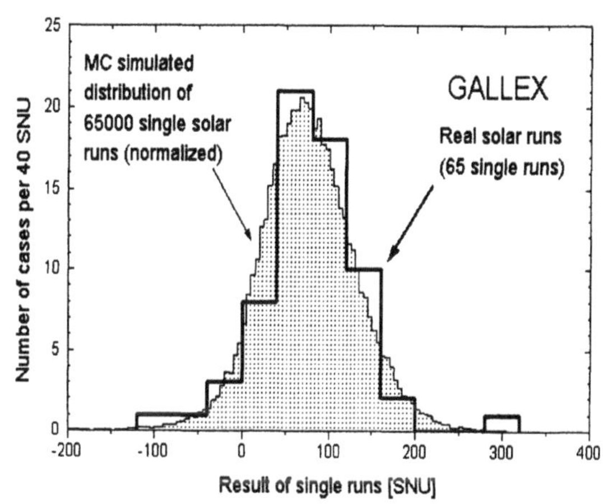

Figure 2.
Monte-Carlo distribution (1,000 full GALLEX experiments) vs. distribution of 65 experimental single results (bold histogram).

* With the alternative pulse shape analysis described in [10] the GALLEX IV result is 99.4± 15.5 (stat.) SNU, and the combined result for GALLEX I - IV is 74.9 ± 5.9 (stat.) SNU (see [11] for the proof that this result is consistent with our pulse shape analysis).

there is nothing peculiar in this way of grouping the runs. For example, if we divide all 65 solar runs into four quarters (subsequent run numbers 1-17, 18-33, 34-49, 50-65), the respective probability is 26.7 % ($\chi^2 = 3.9$ with 3 d.o.f.), other groupings tend also to give probabilities >10%. In this respect, it is just the GALLEX grouping (I-IV) which behaves unexpected in case of a production rate constant in time.

3. ^{71}As AND ^{51}Cr PERFORMANCE TESTS AND THE CONSEQUENCES FOR DATA INTERPRETATION

As all radiochemical solar neutrino experiments, GALLEX is a very complex experiment aiming for the detection of just a few atoms. To demonstrate credibility of any results obtained, convincing performance tests are required. This concerns in particular (but not exclusively) the chemical behavior of the neutrino produced ^{71}Ge in the gallium chloride target solution and the procedures of its separation and recovery in the course of the experiment. Simple labeling with inactive germanium spikes (as is routinely done by us with milligram carriers in every run for yield monitoring) is not sufficient for this proof if one wants to definitely exclude withholding mechanisms caused e.g. by recoil excitation which could alter the chemical behaviour of the neutrinogenic ^{71}Ge in statu nascendi (Szillard-Chalmer reactions, 'hot atom chemistry').

Previously we have exposed the GALLEX target to the strongest man-made (Megacurie) neutrino source ever made [12-14]. Even though we have outnumbered the solar neutrino production with the ^{51}Cr source by up to a factor of 10, still only a few hundred atoms could be produced. This has limited the result of this successful experiment to the 10% level.

A complementary approach is to add, carrier free, ^{71}As activity to the target solution where it beta-decays (β+, EC) with 2.72 day half-time into ^{71}Ge ($T_{½}$ = 11.43 d, Figure 3) [15,16]. This in-situ production mode for ^{71}Ge is relevant because the recoil energy range covers the respective ranges for solar neutrinos (Table 5). The advantage is that it is no problem to produce as many ^{71}Ge atoms as desired and to perform tests at the percent level.

Such experiments were prohibited during regular solar neutrino recording phases, when the addition of ^{71}Ge (at any level) is strictly prohibited by low-level background considerations. It became possible,

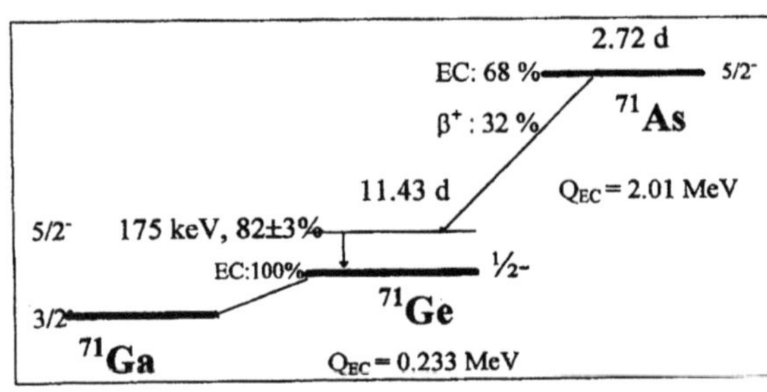

Figure 3. Decay scheme for ^{71}As.

Table 5. Recoil kinematics for solar neutrinos and for ^{71}As-decay in the GALLEX target

^{71}Ge-production process	energy of emitted particles [MeV]	recoil energy E_R of ^{71}Ge-atom (or nucleus)
Solar neutrino capture:		
$\nu_e + {}^{71}Ga \rightarrow {}^{71}Ge + e^-$		
pp-neutrinos (~57 % of expected rate)	0 - 0.19	≤ 1.5 eV
^{7}Be-neutrinos (~26 % of expected rate)	0.63 (90%)	7.9 eV
	0.15 (10%)	1.3 eV
^{8}B-neutrinos (~9 % of expected rate)	$<\sigma\cdot\phi>_{max}$ at ≈ 10	→ 830 eV
Arsenic in-situ decay:		
electron capture (68%):		
$^{71}As + e^- \rightarrow {}^{71}Ge^* + \nu$	1.838	25.6 eV
positron decay (32%):		
$^{71}As \rightarrow {}^{71}Ge^* + e^+ + \nu$	0.813	≤ 11.3 eV
γ-emission (82 %):		
$^{71}Ge^* \rightarrow {}^{71}Ge + \gamma$	0.175	0.23 eV

Table 6. Summary of ^{71}As tank experiments and results

As experiment #	1 (A1t)	2 (A2t)	3a (B3tS)	3b (B3tL)
tank residence time	long	long	short	long
description of the experiment	addition of mixture of As and Ge-carrier	carrier free	short exposure (3d)	long exposure (22d)
result (recovery in %)	100.7 ± 1.6 %	99.6 ± 1.4 %	99.7 ± 1.4 %	99.4 ± 1.8 %

however, in the scheduled maintenance period between the end of GALLEX solar observations and the start of GNO (see sect. 4). Dealing with quantities of order 10^5 ^{71}As atoms, all had decayed end of 1997, eight months after the last spiking of the tank (April 1997). Thus, the As-experiment does not present a background problem for GNO (this is true only since our procedure did avoid the involuntary introduction of any long lived ^{68}Ge) [16].

We have performed a number of ^{71}As - controlled experiments in which we have varied the mixing and extraction conditions, the standing time, and the quantity and time of eventual stable germanium carrier additions. For production of ^{71}As we used the reaction ^{69}Ga (^{3}He , n)^{71}As at the Heidelberg Tandem. The use of isotope separated ^{69}Ga (as oxide) reduces the production of long-lived side products such as ^{73,74}As. Irradiation conditions were 200 minutes @ 200 nA $^3He^{++}$ at 13.4 MeV, yielding ≈ 2 x 10^9 ^{71}As-atoms, sufficient to supply the desired O(10^6 atoms) batches for about 6 weeks. After chemical treatment an acidic master solution was prepared. Immediately (1-2 hours) before use at Gran Sasso, all ^{71}Ge is expelled by purging the solution with nitrogen in the presence of NaOCl and concentrated HCl. The expulsion yield was monitored by AAS and turned out to be quantitative in all cases.

For each of our 3 major spiking experiments we have prepared 3 samples (few ml scale) in precisely determined weight proportions.

(1) a sample to determine the absolute ^{71}As - activity by Ge(HP) spectrometry at Gran Sasso via the 175 keV gamma transition to the ^{71}Ge ground state (branching [82± 3]%). With O(10^6 atoms) sample sizes, the statistical errors ranged from 0.55 % to 0.62 %. The ± 4 % absolute error of these determinations is dominated by the gamma-branching uncertainty. Relative errors for the samples from different experiments are < 0.8 %.

(2) the actual spike sample added to the GALLEX tank. Sample sizes have been O(10^5 atoms).

(3) another sample that served as a reference for comparison. It is treated all the way from sample splitting through Ge-synthesis, counter filling, and counting just like the real spike, except that it bypasses the target tank, that is, the As-decay occurs not in the $GaCl_3$-target tank but rather in an external vial, avoiding any effects which have to do with the chemistry in gallium chloride solutions.

Table 6 is a summary of the results from all our spiking experiments. They were designed to test all potential withholding scenarios which have been suggested or speculated about. This includes testing purposely improper admixture of stable germanium carrier, eventual saturation of withholding sites in the presence of carrier germanium, time-dependent hiding- and release processes, and conditions of Ge-extraction. The major variables of the experiments can be seen in the Table. The results (last line) are very satisfactory, no indications exist for any withholding mechanisms > 1.8 % (the largest uncertainty in any of our experiments) even under the most 'critical' conditions (no carrier, long exposure, soft mixing) [16]. This statement is based on ratio - comparisons between the tank spike samples and the reference samples which have not even seen the target solution and for which crosswise the same individual counters were used to eliminate absolute efficiency calibration errors. We have also very good absolute agreement of our proportional counting results with the gamma-counted samples, however as mentioned, this is only within ± 4% (parallel) because of the uncertainty of the branching ratio. The fit would be ideal for a branching ratio of 82.8 % , instead of the nominal 82 ± 3 %, one may even consider this as a measurement of the γ-branching ratio.

The results of the Cr source experiments and now of the arsenic experiments validate the GALLEX solar neutrino data. Any systematic bias is now excluded or limited to at most a few percent.

The major result of GALLEX has been to establish the 'third solar neutrino problem' [4,17-20]. It consists in the apparent absence of most or all ^{7}Be neutrinos, the second largest expected contributor to the Ga-signal. As errors shrunk, the ^{7}Be deficit became more and more significant. After 65 solar runs the GALLEX solar result is 78 ± 8 SNU. This is substantially below the predictions of the various standard solar models (≈ 130 SNU [21-23]). The latest update for the result of the SAGE experiment (67 ± 7) SNU (1σ) [24] agrees well with our result.

The measured rate in GALLEX is almost exactly what is expected from the PPI cycle alone, but in addition a sizable contribution (≈35 SNU) from ^{7}Be neutrinos is also to be expected (^{8}B neutrinos are not distinct in the Ga detector signal, their direct contribution is rather small).

In view of the appropriateness of the standard solar model (as proved by recent helioseismological evidence, [25]) and in view of the Homestake [26] and Kamiokande [27] detection of a part of the expected ^{8}B-neutrinos, which demands the precursor ^{7}Be to be present in the solar core, it is now, after the exclusion of systematic bias, inescapable to invoke neutrino mass from our data. Then, the observed deficit can be explained by the reduction of ^{7}Be neutrinos through matter mediated neutrino oscillations (MSW-effect) or through vacuum oscillations. If pp-neutrinos rather than ^{7}Be-neutrinos were reduced instead, this would be even stronger evidence for non-zero neutrino mass.

A consistent MSW-solution for the data from all solar neutrino experiments exists and is centered at $\Delta m^2 \approx 4 \cdot 10^{-6}$ (eV/c^2)^2 and $\sin^2 2\theta \approx 8 \cdot 10^{-3}$. This is the 'small angle solution' in the (Δm^2, θ) parameter space for squared mass difference and mixing angle [28]. The second ('large angle') solution is centered at similar Δm^2 but near maximum mixing. It became somewhat disfavored with shrinking errors in GALLEX. In any case, Δm^2 is about the same for both solutions. If assigned to ν_e - ν_μ oscillations in a seesaw scenario, the mass of the muon neutrino would be ≈ 2.2 meV, while m(ν_e) is well below one micro-electronvolt.

4. STATUS OF GNO

GNO (Gallium Neutrino Observatory) is the successor project of GALLEX with newly defined motives and goals [29,30].

The end of the GALLEX observations in early 1997 was followed by a break till early 1998. During this period, a major overhaul and modernization of the experimental set-up (which has been in continuous operation since 1990) took place. Early 1998, all ^{71}Ge activity imported into Gran Sasso with the arsenic experiments has decayed to less than one atom and solar observations were resumed within the frame of GNO.

A non-physical 'zero-cross section' solar model can define an absolute minimum neutrino flux by setting *ad hoc* and against better knowledge all cross sections for PPII, PPIII, and for the CNO-cycle to zero. The only requirement is to sustain the solar luminosity. For this one expects 73 SNU from PPI plus some extra 7 SNU in order to compensate with PPI for the luminosity that is lost in turning off the other branches. Hence, a minimum requirement is ≈ 80 SNU [31]. If the CNO - neutrinos were preserved, the respective minimum for the rate in a gallium experiment would be ≈ 88 SNU. Suppose the experimental error of GALLEX could be further reduced. Then it could become possible to exclude standard (massless) neutrinos without any reference to solar models. This is the major goal for the GNO experiment.

Apart from the above motivation, GNO is intended to provide a long time record of low energy neutrino observations from 1998 onwards. Gallium detectors remain to be the only way to register pp-neutrinos during 1998 - 2002. Up-scaling to 100 tons of gallium is envisioned for a reduction of the total error to ≈ 4 SNU. This will also allow a close examination of the time constancy of the pp neutrino flux during a whole solar cycle.

To reduce the statistical error, it is planned to increase the target size in 2 steps from 30 tons of Ga to 60 t (GNO60) and 100 t (GNO 100). To make this pay, the systematic error must also be reduced further. This is attempted by developing completely new low-level counter types. In particular, we follow two approaches:

(i) Reduction of the error of the counting efficiency, presently the largest single contribution to the systematic error (see Table 3). This may be achieved either with a counter more uniform in shape, or with an 'ex and hopp' counter design for one-time use, thus allowing active internal calibration with ^{71}Ge after completion of the solar run counting.

(ii) About 40 % increase in ^{71}Ge counting efficiency (from 70 % to ≈100 %) by using a 4π low temperature calorimeter, operated at ≈50 mK. The Ge to be counted is thermally deposited on a sapphire absorber substrate in which the energy deposition from a decay event leads to a temperature increase, causing the transition from superconducting to the normal conducting state. The resistance change is read out with an attached superconducting phase transition thermometer (STP). Early tests (still in 2π-geometry) are promising. Figure 4 shows a spectrum obtained from ^{71}Ge that was deposited in a micron-thick metallic mirror on the sapphire surface by thermal decomposition of activated GeH_4.

The past and anticipated future development of the experimental errors in GNO is shown in Fig. 5, contingent to additional gallium.

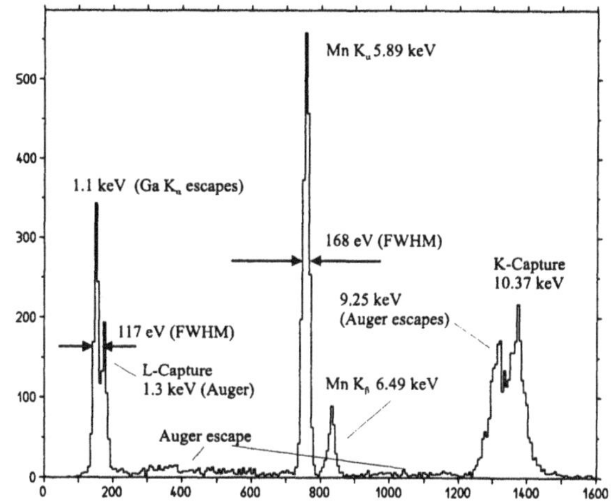

Figure 4. ^{71}Ge decay spectrum using a 2π low-temperature calorimeter with ≈1Bq ^{71}Ge deposited as a thin Ge-film on sapphire. The Mn-peaks are from a ^{55}Fe calibration source. Note the superior energy resolution, resolving the composite L- and K- peaks of Ge.

Actual GNO - data taking in regular monthly runs started in April 1998. Meanwhile, the seventh GNO run has been extracted on 20 Oct 1998.

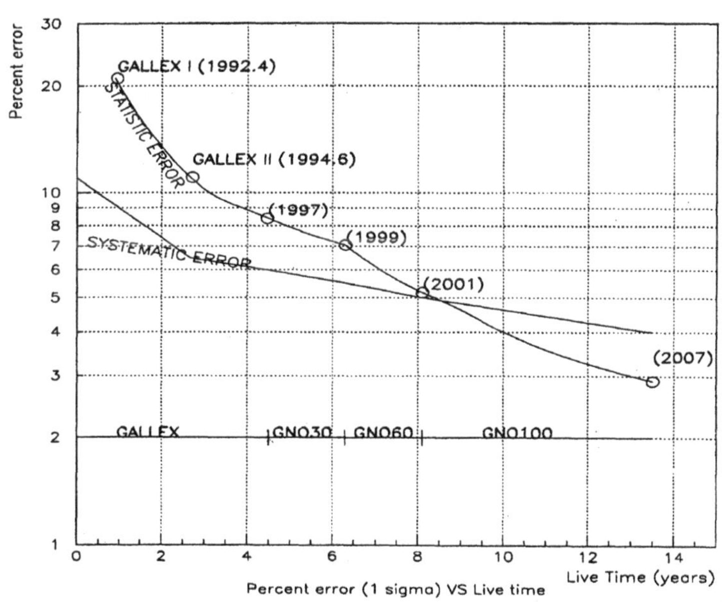

Figure 5. Time evolution of the errors in GNO. The time schedule of additional Ga acquisition is assumed to be as described in [29]. To be conservative, we have assumed 90% 'on-time' (duty factor).

REFERENCES

1. GALLEX Collaboration, P. Anselmann et. al., Phys. Lett. B285 (1992) 376.
2. GALLEX Collaboration, P. Anselmann et. al., Phys. Lett. B314 (1993) 445.
3. GALLEX Collaboration, P. Anselmann et. al., Phys. Lett. B327 (1994) 377.
4. GALLEX Collaboration, P. Anselmann et. al., Phys. Lett. B357 (1995) 237.
5. GALLEX Collaboration, W. Hampel et al., Phys. Lett. B388 (1996) 384.
6. GALLEX Collaboration, W. Hampel et al., 'GALLEX solar neutrino observations: Results for GALLEX IV' , submitted to Phys. Lett. B (Oct.1998).
7. T. Kirsten, Ann. N. Y. Acad. Sci. 759 (1995) 1
8. E. Henrich and K. H. Ebert, Angew. Chem., Int. Ed. (engl.) 31 (1992)1283.
9. T. Kirsten, AIP Conf. Proc. (7th RIS, Bernkastel-Kues 1994) 329 (1995)15.
10. M. Altmann, F. v. Feilitzsch, U. Schanda, Nucl. Instr. Meth. A 381 (1996) 398
11. M. Altmann, GALLEX internal note GX-124a (Aug. 1998).
12. M. Cribier et al., Nucl. Instr. Meth. A378 (1996) 233.
13. GALLEX Collaboration, P. Anselmann et. al., Phys. Lett. B342 (1995) 440.
14. GALLEX Collaboration, W. Hampel et al., Phys. Lett. B420 (1998) 114.
15. T. Kirsten, Progr. In Particle and Nucl. Phys. 40 (1998) 85.
16. GALLEX Collaboration, W. Hampel et al., Phys. Lett. B436 (1998) 158.
17. GALLEX Collaboration, P. Anselmann et. al., Phys. Lett. B285 (1992) 390.
18. J. N. Bahcall, Phys. Lett. B338 (1994) 276.
19. X. Shi, D. Schramm, and D. Dearborn, Phys. Rev. D50 (1994) 2414.
20. T. Kirsten, Nuovo Cimento C19 (1996) 821.
21. J. N .Bahcall, S. Basu, and M.H. Pinsonneault, Phys. Lett. B433 (1998) 1.
22. V. Castellani et al., Phys. Rep. 281 (1997) 309.
23. A. S. Bruns, S. Turck-Chieze and P. Morel, astro-ph/9806272.
24. Sage Collaboration, J.N. Abdurashitov et al., these Proceedings (1998).
25. S. Basu et al., Month. Notes Royal Soc. 292 (1997) 234.
26. B. Cleveland et al., Proc. 4th Intern. Solar Neutrino Conf., W. Hampel (edit.), MPI Kernphysik, Heidelberg, Germany (publ.), 85.
27. Y. Suzuki, these Proceedings (1998).
28. J. N. Bahcall, P. I. Krastev, and A. Y. Smirnov, hep-ph/ 9807216.
29. E. Bellotti, Proc. 4th Intern. Solar Neutrino Conf., W. Hampel (edit.), MPI Kernphysik, Heidelberg, Germany (publ.), 173.
30. E. Bellotti et al., GNO-Proposal, Gran Sasso Lab. (LNGS) report INFN/AE-96/27, also from http:// kosmopc.mpi-hd. mpg.de/gallex. html.
31. J. N. Bahcall, Phys. Rev. C56 (1997) 2839.

ELSEVIER

Nuclear Physics B (Proc. Suppl.) 77 (1999) 35–42

NUCLEAR PHYSICS B
PROCEEDINGS
SUPPLEMENTS

Solar Neutrino Results from Super-Kamiokande

Y. Suzuki[a]
(for the Super-Kamiokamde Collaboration)

[a]Kamioka Observatory, Institute for Cosmic Ray Research, The University of Tokyo, Higashi-Mozumi, Kamioka, Gifu 506-1205 Japan: suzuki@suklow12.icrr.u-tokyo.ac.jp

The recent results from the solar neutrino observation in Super-Kamiokande is presented. From the 504 days of data, taken between the 31st of May, 1996 and the 25th of March, 1998, we have obtained the ^8B–solar neutrino flux ratio to the prediction of the standard solar model of BP98 to be $0.474^{+0.010}_{-0.009}\,^{+0.017}_{-0.014}$ ($0.506^{+0.011}_{-0.010}\,^{+0.018}_{-0.015}$ for TC98). We found no significant difference between day and nighttime flux. The spectrum shape is compared with the expected from the ^{8}B-neutrino spectrum, the neutrino interaction and the detector response, and we have obtained χ^2 of 25.3 (15 d.o.f), corresponding to a 4.6% C.L. Implications of these measurements to neutrino oscillations are discussed.

1. Introduction

Goal of the Super-Kamiokande solar neutrino measurement is to find an definitive evidence for neutrino oscillations independent of the absolute flux calculations. The neutrino oscillations are phenomena which depend on the distance between the source and the detector and the neutrino energy. The matter effect also strongly depend on the neutrino energy. Therefore the discovery of the energy-dependent phenomena is a definite evidence of the neutrino oscillations. Both the vacuum oscillation and the MSW effect may have large observable energy dependence. The day/night flux difference—another model-independent evidence for neutrino oscillations—can be caused by the regeneration of electron neutrinos through the earth. Time variations (seasonal or semi-annual) are related not only to the vacuum oscillations, but also to the neutrino magnetic moment. The correlation to the solar activity is also important for the magnetic moment of neutrinos.

Experimentally, the measurements of the day/night flux variations and other time variations are relatively easy, but require high statistics and therefore long experimental run time. On the other hand, in order to obtain the shape of the energy spectrum, very careful determination of the energy scale of the detector is needed . For this purpose we have used the electron LINAC[1]. This report is organized in the following way. After the short introduction, we will briefly describe about the energy calibration, and the data analysis. The results on the day and nighttime flux measurements and the spectrum measurements will be shown. Implication of the day-night results and the E_e-shape measurement will then be discussed. We conclude this report by pointing the direction of the future of the solar neutrino measurement at the Super-Kamiokande detector.

2. Detector

Super-Kamiokande, located at Kamioka Observatory, Institute for Cosmic Ray Research, The University of Tokyo, is a 50,000tons imaging water Cherenkov detector placed underground, 1,000m water equivalent, 137.32 degree east longitude and 36.43 degree north latitude. The inner detector consists of 32ktons of water, viewed by 11,146 50 cm diameter PMTs, is used to detect neutrinos. The areas of the PMT photocathode covers about 40% of the inner surface. The outside of the inner detector is an outer detector, ~2.7m thick water layers, which serves as an active shield against incoming radiations, like γ-rays and neutrons. The active part of the anti-detector is viewed by out-facing 1,881 8" PMTs attached with the wave length shifter plates. The

0920-5632/99/$ – see front matter
PII S0920-5632(99)00393-X

details on the detector will be found in ref [2]

2.1. Operation

Super-Kamiokande has been operated since April, 1996. The cumulative live-time is nearly 90%. Most of the dead time is due to the time for calibrations by LINAC and other methods. The number of dead inner PMTs, as of June, 1998 (about 2 years after the beginning of the experiment), is about 110 (~1%) and that for the outer detector is about 150.

2.2. Detection of solar neutrinos

Solar neutrinos are detected through ν_e+e→ ν_e+e interactions. If solar neutrinos oscillate into $\nu_{\mu,\tau}$, then those $\nu_{\mu,\tau}$ are detected through $\nu_{\mu,\tau}$+e→ $\nu_{\mu,\tau}$+e interactions. The total cross sections of those interactions are

$$\sigma(\nu_e + e \rightarrow \nu_e + e) = 0.920 \times 10^{-43}(E_\nu/10MeV)cm^2,$$

$$\sigma(\nu_{\mu,\tau} + e \rightarrow \nu_{\mu,\tau} + e) = 0.157 \times 10^{-43}(E_\nu/10MeV)cm^2,$$

for $\sin^2\theta_W$=0.225.

The ν_e+e→ ν_e+e interactions has a good characteristics for the solar neutrino detection. Since the recoil electrons are emitted with the constraint of $E_e\ \theta^2 \leq 2m_e$, they keep the incoming direction of neutrinos, where E_e is the energy of the electrons, θ is the scattering angle and m_e is the electron mass.

The recoil electron energy is given by:

$$\frac{d\sigma}{dE_e} = \frac{2G_F^2 m_e}{\pi}\left[c_L^2 + c_R^2\left(1 - \frac{E_e'}{E_\nu}\right)^2 + c_L c_R \frac{m_e}{E_\nu}\right],$$

where $c_L = 1/2 + sin^2\theta_W$ and $c_R=\sin^2\theta_W$ for ν_ee scattering, and $c_L = -1/2 + sin^2\theta_W$ and $c_R=\sin^2\theta_W$ for ν_μe scattering. The ν_ee interaction is mediated by both charged and neutral currents, and the ν_μe interaction is mediated only by the neutral current.

These recoil electron energies are used to to study the spectrum distortion.

2.3. Calibration

The primarily energy calibration was performed by an electron LINAC, located near the Super-Kamiokande water tank. The energy of the LINAC is variable, covering 5 to 15 MeV, which exactly matches to the solar neutrino spectrum. The diameter of the beam is 6mmϕ at the end of the beam pipe. The electron LINAC can be used not only for the energy calibration, but also used for obtaining the angular and position resolution. We have also performed the various tests on the reconstruction programs by using the LINAC data. The goal of the energy-scale error is less than 1% which contributes to the flux error of ~2% at 6.5 MeV.

The LINAC beam energy is calibrated by the Ge detector which is calibrated by the monochromatic electrons from the internal conversion of ^{207}Bi through a magnetic spectrometer. The uncertainty of the beam energy is ±0.6% at 6 MeV and ±0.3% at 10MeV.

The LINAC data are taken at 8 different positions almost uniformly distributed. By using those LINAC data, a Monte Carlo simulation program was tuned to reproduce the data. The evaluation of the energy scale and the various resolutions was done by the tuned MC simulation. The energy resolution over the volume was estimated by the MC simulation to be 14.4% for 10MeV electrons. The vertex resolution averaged over 22.5kton volume is obtained to be 71.4cm for 10 MeV electrons and the angular resolutions is estimated to be 26.7deg by the MC simulation. The MC simulation is also used to evaluate the systematic errors of the energy scale and its position dependence and energy dependence.

The additional information has also been obtained by the spallation products, ^{16}N, electrons from the stopping muon decay and γ rays from the Ni-Cf sources utilizing the reaction Ni(n,γ).

The muons penetrating the detector produces the spallation products. About 600 events/day of those products, very high statistics, were used to evaluate the angular dependence of the energy scale, and used for testing the stability of the detector. ^{16}N, of which the energy is precisely known to be better than 0.1%, but with small statistics, are used to check the absolute scale of energy and check the angular dependence of the energy scale. The electrons from the muon decay— 1500 events/day, are used to trace the change of the water transparency. Those γ-rays produced through the Ni(n,γ)Ni reactions are used for testing the stability of the detector.

The total systematic errors on the energy scale thus obtained is 0.9% at 6 MeV and 0.7% at 10MeV. Those errors are from 0.53% at 6MeV and 0.26% at 10MeV for LINAC beam energy, 0.1% for tuning precision, 0.5% for position dependence, 0.2% for the stability, 0.5% for the directional dependence, and 0.3% at 6 MeV and 0.1% at 10MeV.

3. Data Reduction

The results presented here are based on the data obtained from May 31, 1996 to March 25, 1998, the effective 504 days. Starting from the 11 Hz of raw data, then by removing the cosmic rays entering outside of the detector and by selecting low energy contained events by applying simple computer programs, we reduce the data down to 6.2 Hz. Then the vertex position, direction and the energy of the events are reconstructed by using the timing of the hit PMTs and the pattern of the event. We have then applied crude energy and fiducial volume cuts requiring 5.5MeV and 1.5m from the wall, resulting 17,000 events per day. Those reduced events consist of the solar neutrinos, γ rays from the rocks surrounding the detector, spallation products and the electrons and γ from the daughter of Rn contaminated in the detector water. We then have applied the spallation cuts, further tight cuts on the fiducial volume (2m from the wall) and the 6.5 MeV (6.0 MeV kinetic energy) minimum energy requirement. Finally we have applied the remaining gamma cut, which yielded the 173 events per day—the final sample. The details of those cuts were explained in ref [2].

4. Solar neutrino signal

From the final data sample of 504 days of data, the directional distribution towards the sun, $\cos\theta_{sun}$, is made as shown in Fig.1. From Figure 1, by using the maximum likelihood method[2], we have obtained the corresponding total solar neutrino signal of

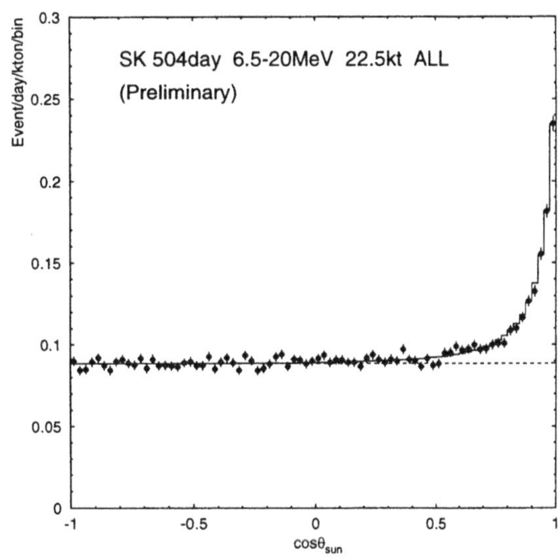

Figure 1. The directional distribution of the final sample towards the sun.

6823^{+148}_{-130}(stat.)$^{+239}_{-198}$(syst.) events—13.5 events per day.

The resulting total ^{8}B solar neutrino flux is $2.44 \pm 0.05^{+0.09}_{-0.07}$ $\times 10^6 \text{cm}^{-2}$ in the 22.5kton fiducial volume. The ratio to the expected flux for the standard solar models are: Data/SSM = $0.473^{+0.010}_{-0.009}$(stat.)$^{+0.017}_{-0.014}$(*syst.*) for BP98[3], Data/SSM = $0.368^{+0.008}_{-0.007}$(*stat.*) $^{+0.013}_{-0.011}$(*syst.*) for BP95[4] and Data/SSM = $0.506^{+0.011}_{-0.010}$(stat.) $^{+0.018}_{-0.015}$(syst.) for TC98(Brun et al.)[5].

4.1. Seasonal variations

The data are divided into 1.5 months time periods, which is useful to check the vacuum oscillation solutions of the solar neutrino problem. In Fig. 2 the one year variations in these run time divisions are shown. The expected variations due to the eccentricity of the earth is shown by the solid line. The χ^2 (for the solid line) is 8.78 for 14 d.o.f. (84.4% C.L.). The agreement is quite well and we see no seasonal variations within our experimental sensitivity.

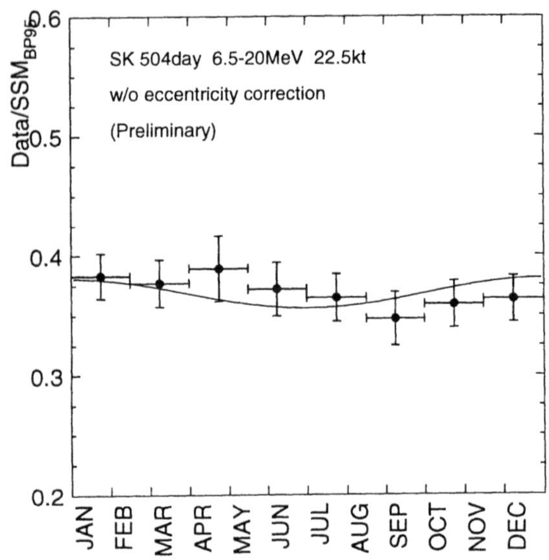

Figure 2. The solar neutrino flux in every 1.5 months periods.

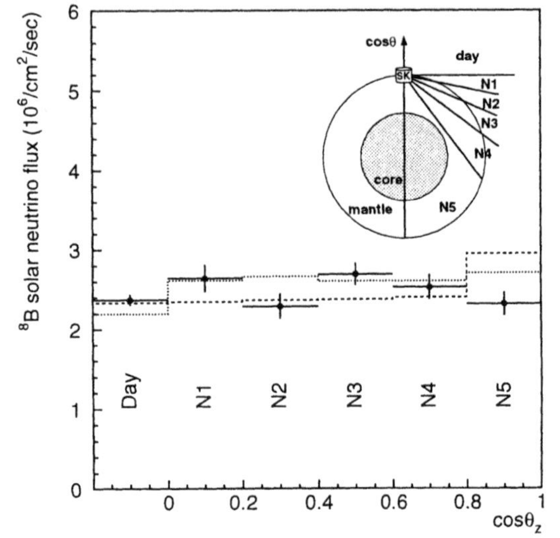

Figure 3. Solar neutrino fluxes as a function of the nadir of the Sun. Night data is divided into 5 bins with an interval of $\Delta\cos\theta_Z$=0.2. Dotted histogram is the expected variation of a typical large mixing angle solution ($\sin^2 2\theta = 0.56, \Delta m^2 = 1.2 \times 10^{-5} eV^2$) and dashed histogram is that of a typical small mixing angle solution ($\sin^2 2\theta = 0.01, \Delta m^2 = 6.3\times 10^{-6} eV^2$).

4.2. Day/Night flux difference

The separately measured day and nighttime fluxes are important to check the matter effect through the earth. We expect, for the most of the case, higher flux in nighttime flux. The flux difference of day and nighttime determines the neutrino oscillation parameters independent of the absolute flux calculations of Standard Solar Models.

The measured day-night fluxes corrected for the eccentricity of the earth's orbit at 1 AU, are: 2.37±0.07(stat)$^{+0.08}_{-0.07}$(syst) $\times 10^6$/cm^2/sec for the daytime, and 2.48 $^{+0.07}_{-0.06}$(stat) $^{+0.08}_{-0.07}$(syst) $\times$ 10^6/cm^2/sec for the nighttime.

The relative systematic errors becomes smaller to be $^{+0.6}_{-0.5}$% if we take a ratio. The major contribution comes from the relative energy scale (0.5%), background subtraction (0.4%). The flux ratio is:

$$\frac{N}{D} - 1 = 0.047 \pm 0.042(stat) \pm 0.008(syst).$$

There is no significant difference within the experimental errors.

Figure 3 shows the day-time flux and the nighttime fluxes which are divided into 5 bins with the bin width of $\Delta\cos\theta = 0.2$.

At the Super-Kamiokande site, the sun covers the nadir angle within con$\theta_z \leq 0.974$, therefore Super-Kamiokande does not see neutrinos which pass through the inner core (0.981<cosθ_z<1.) The boundary of the outer core and the mantle is cosθ_z= 0.838, where the density of the earth jumps from $\leq$5.5g/cm^2 to 10∼12g/ cm^2. For the data with the zenith angle greater than

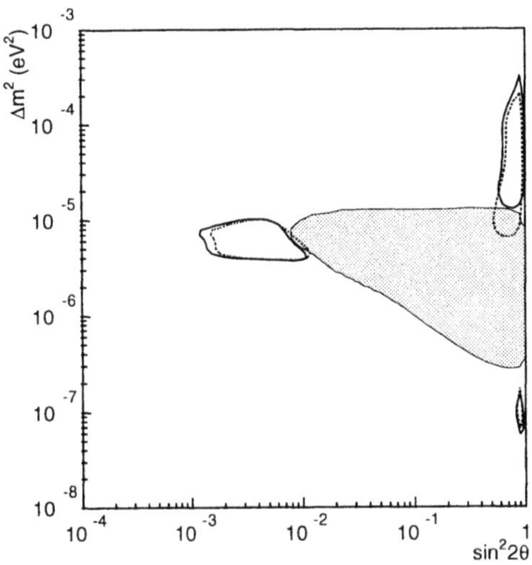

Figure 4. Excluded region by SK day/night flux measurement for $\nu_e \to \nu_{\mu,\tau}$ oscillations. The shaded shows the excluded region with more than 99% C.L.. The region shown by the dotted lines are allowed at the 99% C.L. from the combined analysis of Homestake, SAGE, Gallex and SK-flux assuming the BP98[3] flux. The thick solid lines shows the allowed region at the 99% C.L. from the combined rate analysis of the rates and the SK D/N variation.

0.838 (N5), those neutrinos detected in the Super-Kamiokande has passed through the earth's outer core. Those neutrinos are expected to be enhanced for some parameter of the small mixing angle solutions [6].

The observed flux for N_5 is:

$$\frac{N5}{\langle D, N1 \cdots N4 \rangle} - 1 = -0.055 \pm 0.063(stat) \pm 0.013(syst).$$

No significant excess in N5 is seen in the data.

In Fig. 3, the typical expected variations for the assumed neutrino oscillations are also shown.

The relative difference of the day and nighttime flux are used for the flux independent neutirno oscillation analysis, treating the flux normalization factor α as a free parameter in the χ^2 definition:

$$\chi^2_{DN} = \sum_{i=D,N1,\cdots,N5} \left\{ \frac{\phi_i - \alpha \times \phi_i^{osci}(\sin^2 2\theta, \Delta m^2)}{\sqrt{\sigma_i^2 + \sigma_{sys,i}^2}} \right\}^2,$$

where ϕ_i is the measured flux, $\phi_i^{osci}(\sin^2 2\theta, \Delta m^2)$ is the effective flux for a given set of oscillation parameters derived from the ratio of the expected number of events with and without oscillations. σ_i and $\sigma_{sys,i}$ are the statistical and systematic errors of ith-bin.

The χ^2 value for the case of no oscillations is 7.4 with 5 degrees of freedom (dof), which corresponds to a 19% probability.

The confidence level of the typical the oscillation parameters are 0.011% for ($\Delta m^2 = 7.9 \times 10^{-6}$, $\sin^2 2\theta = 7.9 \times 10^{-3}$) and 12.2% for ($\Delta m^2 = 7.1 \times 10^{-6}$, $\sin^2 2\theta = 2.5 \times 10^{-3}$). Minimum χ^2 of 5.2 was found at ($\sin^2 2\theta = 3.2 \times 10^{-2}, \Delta m^2 = 1.6 \times 10^{-6} eV^2$) for the $\nu_e \to \nu_{\mu,\tau}$.

The shaded region in Figure 4 shows the region where the exclusion probabilities are more than 99% ($\chi^2 > 15.09$ for 5 dof), using the expected zenith angle dependence. Absolute flux information was not used in these calculations, therefore the results are solar model independent.

4.3. Energy spectrum

It is crucial to understand the energy scale and energy resolution of the detector in order to make a precise measurement of the spectrum. For this purpose we have used an electron LINAC to calibrate the energy response of the detector as mentioned above.

The solar neutrino signal in each bin was obtained by the maximum likelihood method, as is explained in ref. [2]. The angular resolution used for the signal extraction was tested against the LINAC data and the small difference between the data and the Monte Carlo calculation was treated as a systematic error (2.2% error for the total flux measurement). The contribution from the non-flat component of the background was evaluated to be less than 0.1%. The resultant energy spectrum is shown in Fig. 5, where the ratio to the expected ^8B spectrum shape [7] was taken.

The last three data points are slightly higher than the rest of the data points. If those excess is cased by the error of the energy scale or the energy resolution, then those must be shifted by 3.6% (for the scale error) or 20% (for the resolution error), where those systematic errors are 0.8% and 2%, respectively. Therefore it is un-

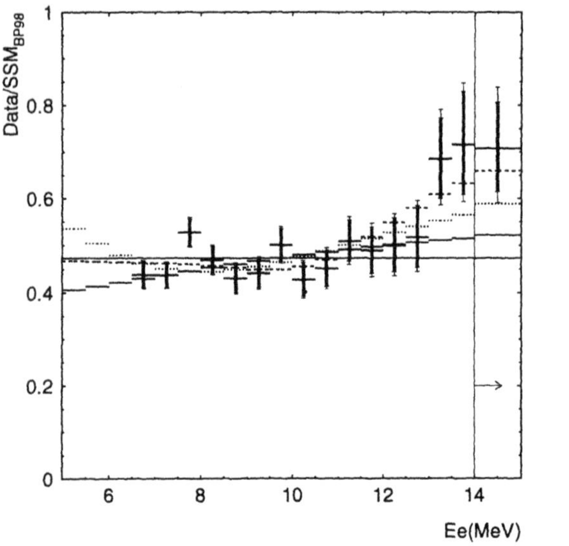

Figure 5. Observed electron energy spectrum in the ratio to the expectation from the SSM. The inner thick and outer thin error bars show the statistical and systematic errors. The expected spectrum distortion due to the neutrino oscillations are shown for the typical oscillation parameters: solid line for a small mixing angle solution (Δm^2=7.1×10^{-6}eV^2, $\sin^2 2\theta$=5×10^{-3}); dotted line for a vacuum oscillation (Δm^2=7.5×10^{-11}eV^2, $\sin^2 2\theta$=0.83); dashed line for the best fit parameters (Δm^2=4.2×10^{-10}eV^2, $\sin^2 2\theta$=0.93).

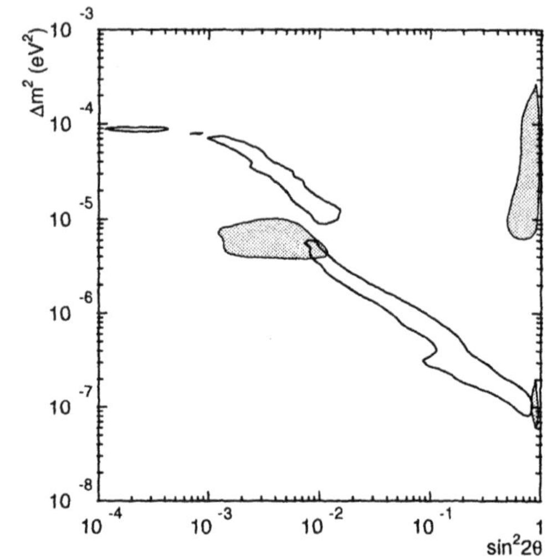

Figure 6. The 99% C.L. allowed region obtained by the day-time and the nighttime spectrum data (solid line). The allowed region obtaained by the global fit using the flux results of all the solar neutrino experiments (including the flux measurement of Super-Kamiokande) are shown by the shaded region.

likely that those higher data points are caused by the improper understanding of the response of the detector.

It might be suspected that those excess events are coming from the Hep-neutrino (^{3}He + p → ^{4}He + e^+ + ν_e) contribution[8]. The Hep-neutrino contributions for the standard flux of ϕ_{Hep}^{BP98} is 1.1% and 3.0% for the energy bins between 13 and 14 MeV and between 14 and 20 MeV, respectively. An unexpected contribution of about 20× ϕ_{Hep}^{BP98} is needed to explain those higher data points. However there is no convincing reason to increase the flux by an arbitrary factor.

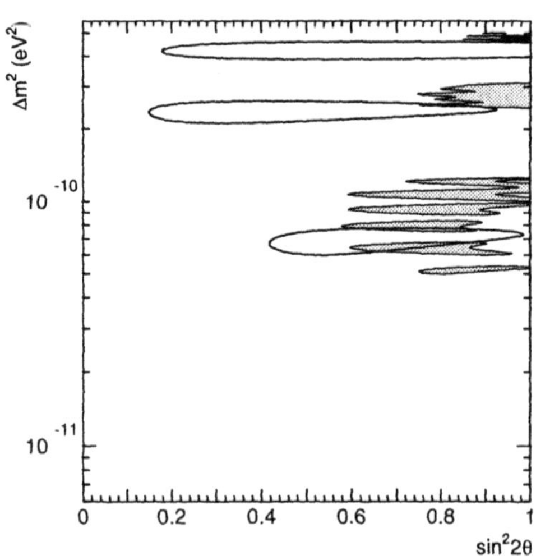

Figure 7. Same as Fig. 6, but for the vacuum oscillation region.

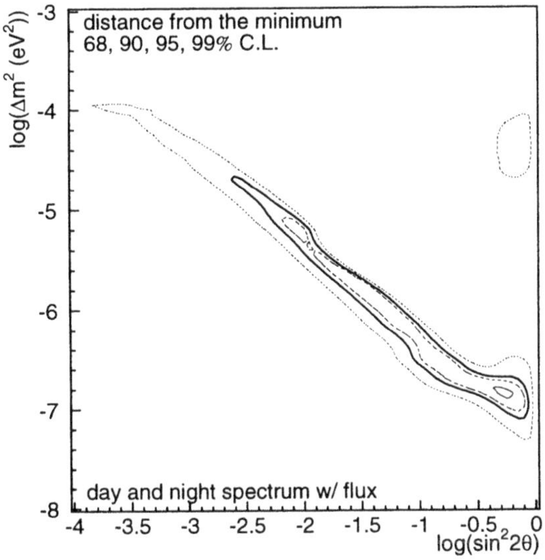

Figure 8. Same as Fig. 6, but the additional constraint from the absolute flux is demanded.

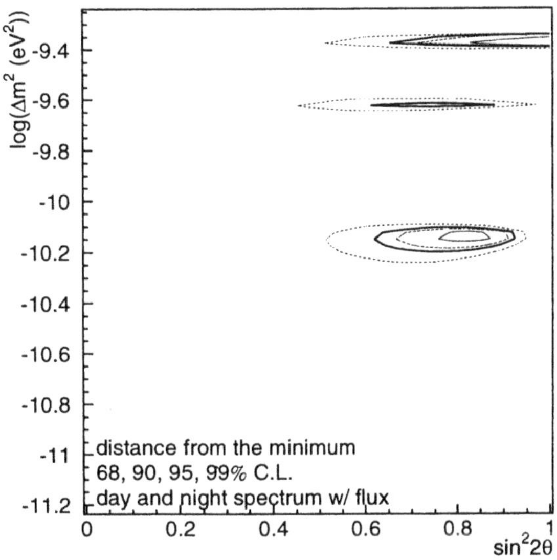

Figure 9. Same as Fig. 8, but for the vacuum oscillation region.

The observed energy spectrum is examined using the following χ^2 :

$$\begin{aligned}\chi^2 &= \sum_{i=1}^{16or32} \{((\frac{data}{SSM})_i \\ &- \alpha/((1+\delta_{i,exp}\times\beta)(1+\delta_{i,cal}\times\gamma)))/\sigma_i\}^2 \\ &+ \beta^2+\gamma^2,\end{aligned}$$

where $\delta_{i,exp}$, $\delta_{i,cal}$ and σ_i are the correlated experimental error, the uncertainty of the spectrum calculation[7] and errors associated with each energy bin defined as a sum of statistical and uncorrelated errors added in quadrature. And α is a free parameter which normalizes the measured flux relative to the expected flux. β and γ are also free parameters used for constraining the variation of correlated systematic errors.

The χ^2 for the case of no oscillation is obtained to be 25.3 with 15 degrees of freedom at α=0.449, β=-1.49 and γ=-0.93. This corresponds to an agreement with the expected energy shape at the 4.6 % C.L. or, conversely, to a deviation from the expected energy shape with a significance of 95.4 %.

In order to in cooperate day/night effect, we have used day spectrum and night spectrum separately.

The χ^2 for the flat distribution (no oscillation) for this treatment is 50.2 for 31dof(1.6% C.L.).

The typical expected spectrum distortions from neutrino oscillation are also shown in Fig. 5.

The best fit is observed in the vacuum oscillation region for the oscillation parameters, Δm^2=4.2×10^{-10}eV^2 and $\sin^2 2\theta$=0.93 with the χ^2 of 38.7 (31 dof).

In Fig. 6 and Fig. 7, we have shown the allowed parameter regions which are obtained by the combination of the day-time spectrum and nighttime spectrum. Those allowed regions are independent from the flux calculations. In Fig. 8 and Fig.9, the absolute flux constraint is used for the χ^2 calculations.

5. Conclusion

The solar ^8B-flux is measured. The obtained flux value is

$$2.44 \pm 0.05(stat)^{+0.09}_{-0.07}(syst) \times 10^6/cm^2/s$$

and the flux ratio to the BP98 is $0.474^{+0.010}_{-0.009}$ $^{+0.017}_{-0.014}$. No seasonal variations are found by using the data above 6.5 MeV. The day-night flux

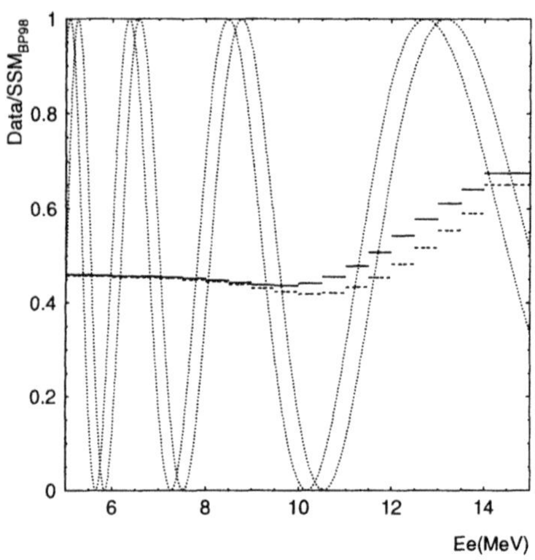

Figure 10. The electron energy spectrum for a vacuum oscillation. The electron spectrum at peripheron (solid) and at apherion (dashed) is shown. The dotted lines are the expected "neutrino" spectrum for those cases.

ratio is

$$\frac{N}{D} - 1 = 0.047 \pm 0.042(stat) \pm 0.008(syst).$$

No core enhancement was found. The excluded region by the non-observation of the day/night effect is extended beyond the edge of the small mixing angle region. Using the day/night and E_e-shape data with the assumption of the standard contribution of Hep-neutrino, the best fit parameters were found in a vacuum oscillation region. The allowed region independent of the absolute flux calculation is consistent with those constrained by the absolute flux.

6. Future prospect

As shown in Fig. 5, low energy data below the current energy threshold of Super-K has an ability to distinguish solutions which is complimentary the high energy data. We will be able to reduce the threshold down, hopefully to 5.5MeV (total energy) by the end of 1998.

We certainly need more statistics and have to make systematics error smaller, since the high energy data points are still poor statistic.

If the true oscillation parameters are in the vacuum oscillation regions , then we may see seasonal variations. In Fig. 10, the expected flux for the time when the earth is at the periheron and at the apherion. The best fit oscillation parameters are used. The effect is only visible in high energy bins where those electrons are coming from the near-monochromatic neutrinos, since they are close to the end point energy. The magnitude of the expected seasonal variations for the events between 11.5 and 20 MeV is about 10% on top of the 7% variation due to the eccentricity of the earth. Since the event rates are rapidly decrease as the energy of the event goes up, we need to accumulate high statistic data and hopefully see that in a few years' data.

REFERENCES

1. Nakahata et al., Nucl. Instr. Methods, to be published.
2. Y. Fukuda et al., Phys. Rev. Lett.
3. J.N.Bahcall, S.Basu and M.Pinsonneault, astro-ph/9805135
4. J.N.Bahcall and M.Pinsonneault, Rev. Mod. Phys. 67,781(1995).
5. A.S.Brun,S.Turck-Chieze and and P.Morel Ap.J, October,1998; astro-ph/9806272.
6. S.Petcov, in these proceedings; A.Smirnov, in these proceedings.
7. J.N.Bahcall et al., Phys. Rev. C54, 411 (1996).
8. R.Escribano, hep-ph/9805238.

ELSEVIER

Nuclear Physics B (Proc. Suppl.) 77 (1999) 43–47

NUCLEAR PHYSICS B
PROCEEDINGS
SUPPLEMENTS

THE SUDBURY NEUTRINO OBSERVATORY PROJECT

A.B. McDonald, Quenn's University
Kingston, Ontario, Canada
For the SNO Collaboration*

The construction of the Sudbury Neutrino Observatory (SNO) project was completed in April, 1998 and the detector is presently in operation as it is being filled with light and heavy water. The SNO detector is a 1,000 tonne heavy water Cerenkov detector situated 2,000 meters underground in INCO's Creighton Mine near Sudbury, Ontario, Canada.[1] The project is a Canadian, US and UK collaboration. Through the use of heavy water SNO will be able to detect a number of neutrino reactions, including one sensitive specifically to electron neutrinos and another one which is sensitive to all neutrino types. With these two reactions the detector will be used to search for solar neutrino flavour change without the requirement of electron neutrino flux normalization by solar model calculations. It will also provide unusual sensitivity for other measurements of solar neutrino properties, atmospheric neutrinos and supernova neutrinos.

1. DETECTION REACTIONS

The SNO detector will use heavy water to observe neutrinos via four principal reactions.

1) Charge Current (CC) reaction: $d + \nu_e \rightarrow p + p + e^-$ which is sensitive exclusively to electron neutrinos.
2) Neutral Current (NC) reaction: $\nu_x + d \rightarrow n + p + \nu_x$, which is sensitive to all non-sterile neutrino types equally.
3) Elastic Scattering (ES) reaction: $\nu_x + e^- \rightarrow e^- + \nu_x$, which is predominantly sensitive to electron neutrinos; they have about six times greater cross-section than other neutrino types.
4) $\nu_e + d \rightarrow n + n + e^+$, which is a sensitive reaction for electron anti-neutrinos.

2. DETECTOR DESCRIPTION

A schematic picture of the Sudbury Neutrino Observatory is provided in Figure 1. The detector consists of 1,000 tonnes of heavy water contained within a spherical acrylic vessel of 12m diameter and 5cm thickness. This vessel is surrounded by about 9500 20-cm-diameter photomultiplier tubes (PMT's), mounted on a geodesic structure 18 m in diameter. These photomultipliers are equipped with light collectors which increases the total geometric efficiency to about 70% of 4π. The barrel shaped cavity is about 22m in diameter at the midpoint and is about 34m high. It is lined with a 8-mm thick polyurethane barrier impermeable to water and radon. The whole cavity outside the acrylic vessel will be filled with ultra pure light water.

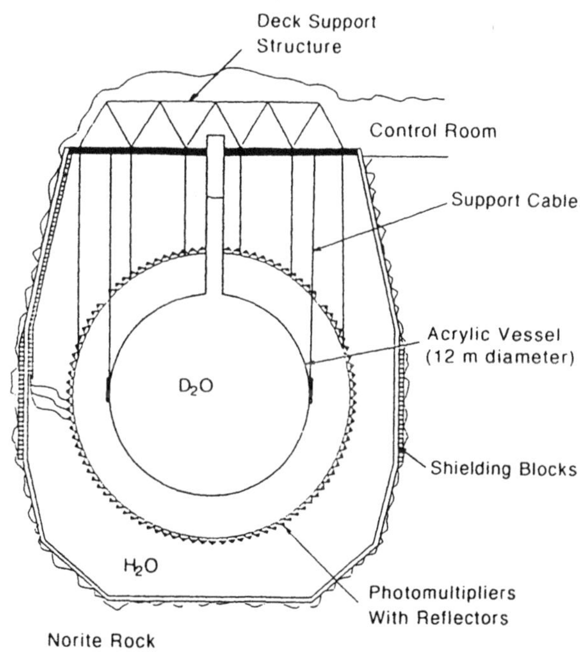

Figure 1

The SNO detector is designed to provide very low levels of radioactive background for the reactions to be used to detect neutrinos. The entire detector is constructed from materials which have been carefully selected for low ^{238}U and ^{232}Th content. The levels of uranium and thorium in the surrounding norite rock are

0920-5632/99/$ – see front matter
PII S0920-5632(99)00395-3

about 10^{-6} grams per gram. By careful selection, the levels in the photomultiplier tubes and support structure are about 10^{-8} grams per gram and in the acrylic material less than about 10^{-12} grams per gram. In addition, the detector was constructed and is being operated under ultra-clean conditions. The air is carefully filtered and the workers wear clean-room clothing to maintain an air quality on the order of Class 2000. An extensive water purification system is used to reduce the equivalent concentrations of ^{238}U and ^{232}Th to less than 1 x 10^{-14} gram per gram of water. Preliminary measurements of light water in the detector outer cavity and measurements of heavy water in the underground storage tanks are near the design objectives.

The photomultiplier structure is 99.9% impermeable to water, enabling the water flow to be directed outwards from the region between the acrylic vessel and the photomultipliers into the external region where it is removed, purified, stripped of radon and reinjected in the inner region. This maintains a lower level of radioactivity in the light water in the inner region than is required in the outer region.

The construction of the detector was carried out as follows. The cavity was excavated and lined with the polyurethane waterproof material. The top half of the photomultiplier support structure containing about 5,000 PMT's and light collectors was constructed and suspended in the cavity. All construction was accomplished using detector parts which were sized to fit within the roughly 3m x 3m dimensions of the mine hoist. The acrylic vessel was constructed by bonding together carefully-machined panels with these dimensions, using liquid acrylic as the bonding agent. The top half of the acrylic vessel was constructed on a construction platform at the bottom of the cavity. It was then lifted into place and suspended from low-radioactivity Vectran ropes. Construction then proceeded with the bonding of the lower half of the acrylic vessel, addition of the lower half of the geodesic structure and remaining photomultipliers, final cleanup of the detector cavity and start of water fill. The bonding of the acrylic panels took considerably longer than originally planned. About one kilometer of bonding was accomplished with no flaws, but about 0.5% of the bonds showed small flaws which required time-consuming removal in situ and repeated rebonding for success. The completed vessel met all engineering and scientific requirements. Figure 2 is a picture of the completed vessel and the surrounding photomultipliers taken with a wide angle lens inserted at the bottom of the photomultiplier structure.

To enhance the sensitivity to the neutral current (NC) reaction, one technique will be the addition of $MgCl_2$ to the heavy water. Neutron capture in ^{35}Cl results in gamma rays with energy summing to 8.6 MeV, which are detectable from subsequent Cerenkov light production. A reverse osmosis system has been developed for the removal of $MgCl_2$ from the detector. To identify the contribution from capture in the chlorine, the detector will be operated for some period with no $MgCl_2$ and other periods with about 0.25 % $MgCl_2$ dissolved in the heavy water. Simulations have indicated that it may also be possible to distinguish CC events from NC events through the event topology. The distribution of light from CC events arising from a single energetic electron will be different from the NC event arising from a gamma ray cascade.

An additional technique for the measurement of neutrons from the NC reaction will be the introduction of an array of over 100 ^{3}He gas-filled proportional counters. These counters are being constructed using ultra-pure nickel tubing fabricated by a chemical vapour deposition process. The neutrons will be detected through energetic protons and tritons produced by the neutron capture on ^{3}He. The distinctive time evolution of these events will enable them to be distinguished from alpha particle radioactivity from the walls of the ultra-pure nickel tubes used as the bodies of the proportional counters. A decision as to which technique will be used first to observe the NC reaction will be made after the detector characteristics and radioactive backgrounds have been defined during the running with pure heavy water.

3. PHYSICS OBJECTIVES

A primary objective of the SNO detector will be the observation of solar neutrinos via the charged current and neutral current reactions. The threshold for the CC reaction will be about 5 MeV in electron energy, corresponding to about 6.4 MeV in neutrino energy. The threshold for the NC reaction will be 2.2 MeV. Therefore both of these reactions will be sensitive only to the neutrinos from ^{8}B decay in the sun. However a comparison of the flux measured by these two

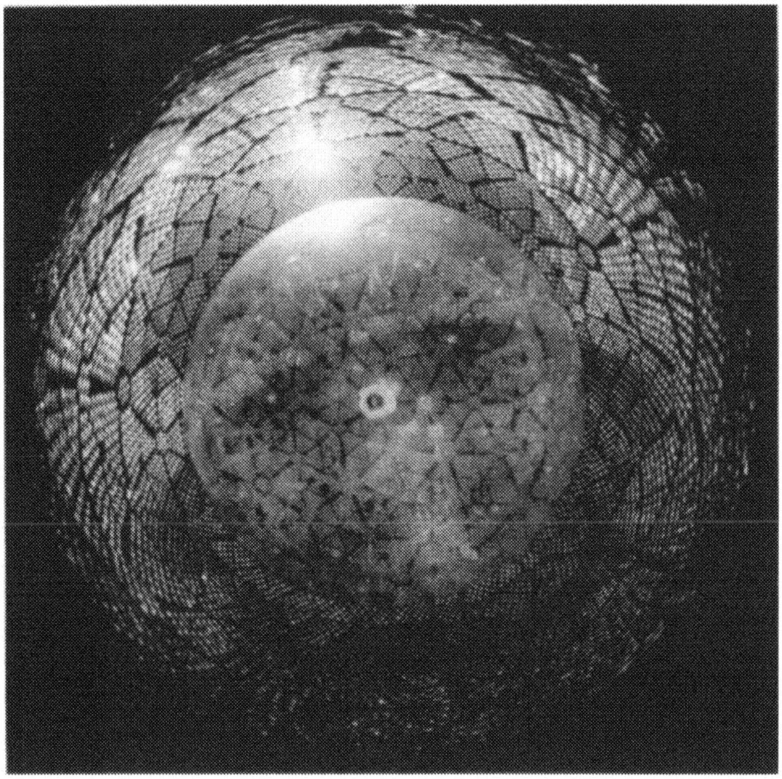

Figure 2

reactions should provide a determination of whether the ^{8}B neutrinos are changing from electron neutrinos to another non-sterile type.

The CC reaction will provide an accurate measure of the shape of the ^{8}B neutrino energy spectrum. For this reaction, the outgoing electron carries almost all the energy of the incoming neutrino less the Q-value of 1.4 MeV. Therefore the ^{8}B spectrum can be observed with an energy resolution of better than about 20%. This is an advantage compared to the ES reaction where the incident neutrino energy is shared between the outgoing electron and a scattered neutrino with comparable energy. The CC reaction has a relatively large cross-section, so the SNO detector will observe about 3,600 counts per year for a flux corresponding to about 40% of the flux calculated by typical standard solar models [2].

The NC reaction would produce about 3,000 detected events per year for a full standard solar model flux. The NC reaction will be detected by two different techniques as described above. The two techniques have different sensitivities to systematic effects and provide somewhat independent measurements of the NC flux.

With information obtained from the NC reaction, the CC reaction, including the ^{8}B spectral shape, and temporal information on the event rates, it will be possible to examine a region of oscillation parameters for electron neutrinos ranging from

$\Delta m^2 = 10^{-4} eV^2$ to $10^{-11} eV^2$,

and values of $\sin^2 2\theta$ ranging from 1 down to 10^{-4}.

With this set of information it will be possible to determine whether electron neutrinos are making transitions to sterile or non-sterile neutrinos for each of the possible situations allowed by the existing set of solar neutrino data. As the regions allowed by the existing data have been described in other papers in this conference and in many other publications, we will proceed to discuss the characteristic data which would be observed by SNO for the different regions.

Non-Adiabatic MSW:

$\Delta m^2 \cong 5\times10^{-6}eV^2$, $\sin^2 2\theta \cong 10^{-3}$

For the Non-Adiabatic MSW region allowed by existing experiments, the results would be as follows: for electron neutrinos converting to non-sterile neutrinos there would be a distortion of the ^8B spectrum at low energies, and the ratio of CC/NC would be smaller than the normalized ratio expected for a flux containing purely electron neutrinos. For ^8B electron neutrinos making a transition to sterile neutrinos, the CC spectrum would still be distorted on the lower energy side but the normalized CC/NC ratio would be 1.0.

Large Angle MSW:

$\Delta m^2 \cong 5\times10^{-6}\ eV^2$, $\sin^2 2\theta \cong 1$

This region of transition for electron neutrinos to sterile neutrinos may be ruled out by considerations of Big Bang nucleosynthesis. For a coupling constant this large, the effect is similar to the addition of a fourth neutrino species during the Big Bang. Although this region has been ruled out in previous calculations, recent uncertainties in measurements of nuclear abundances makes a restriction based on the Big Bang less certain at present. For electron neutrinos making a transition to non-sterile neutrinos the signature would be a normalized CC/NC ratio smaller than 1 and no distortion of the ^8B spectrum as observed by the CC reaction.

Region Sensitive to Earth Regeneration:

$\Delta m^2 \cong 5\times10^{-6}eV^2$, $\sin^2 2\theta$ from 10^{-2} to 1.0

For electron neutrino transitions to non-sterile neutrinos, regeneration effects in the earth can result in day-night effects which vary according to season and also show some variation in magnitude according to time of night. This distinctive temporal pattern would be indicative of a neutrino flavour change within this parameter region.

Vacuum Oscillations:

$\Delta m^2 \cong 5\times10^{-10}eV^2$, $\sin^2 2\theta \cong 1.0$

One additional remaining solution which agrees with the existing solar neutrino measurements is vacuum oscillations, where the distance from the earth to the sun is an appropriate value to result in the strong suppression of flux for some of the existing experiments and less significant suppression for the others.

The peak to peak variations in the seasonal counting rates observed for the charge current and neutral current reactions will be about 7% due to the change in the solid angle from the small eccentricity of the earth's orbit. With electron neutrino flavour change to either sterile or non-sterile neutrinos, there will be significant spectral distortion observable in the CC reaction and a different seasonal variation in the integrated CC rate above threshold. For some parameter values the spectral distortion can be very large. Observation of these seasonal effects would be an indication of such vacuum oscillations and the NC reaction can then be used to determine whether the transitions are to sterile or non-sterile neutrinos.

Experimental Effects

Background radioactivity in the detector must be minimized and the residual radioactivity must be measured accurately. The threshold for the CC reaction will be determined by beta and gamma rays from the uranium and thorium chain elements in the components in the detector. The minimization of this radioactivity is the motivation for constructing all components from ultra pure materials and controlling cleanliness during construction. In addition, it is particularly important to restrict and determine the gamma flux in the heavy water with energies greater than 2.2 MeV, the threshold for photo-disintegration of deuterium. These gammas can produce a free neutron in the D_2O and therefore can simulate the NC reaction induced by neutrinos.

In order to deal with this background, extensive water purification equipment has been developed for the light water, heavy water and $MgCl_2$. Sensitive techniques have been developed for the measurement of the principal sources of such gamma rays: uranium and thorium chain decay products in the heavy water. These techniques will include the measurement of such elements in samples of the water taken during recirculation, including ^{222}Ra obtained by degassing, as well as techniques for the observation of radioactive decay patterns from the Cerenkov light observed in the detector. The materials used in the ^3He detectors have also been carefully selected to have very low levels of uranium and thorium and their daughters. Techniques have been developed for the measurement of inherent radioactivity in these detectors which might lead to photodisintegration of deuterium.

Calibration of the detector is being accomplished with the use of laser light sources with varying wavelengths and a diffuser ball which can be positioned almost anywhere within the D_2O volume. Light-emitting diodes are also mounted on the photomultiplier support structure and can be actuated at any time for measurements of detector stability. For absolute energy calibration, a series of gamma ray sources will be used, including a thorium source at 2.62 MeV, ^{16}N at about 6 MeV and gammas from the (p,t) reaction at about 20 Mev. A sono-luminescent source (sonoball) will be used to provide short pulses for time calibration. A ^8Li source has been developed to provide electrons forming an energy spectrum with a shape directly related to the ^8B decay spectrum. The ^{16}N source and the ^8Li source are produced by nuclear reactions using a compact neutron generator and will be transported to the center of the SNO heavy water volume via capillary tubing. They are attached to the manipulator mechanism which enables them to be positioned within the D_2O volume.

To date, the laser-ball, ^{16}N, sonoball and light-emitting diode sources have been used for initial calibration of the detector, with light-water levels below the photomultiplier support structure. The ^{16}N source was used for studies of gamma rays interacting in the acrylic sphere. Comparison of these measurements with preliminary measurements of radioactive background levels observed in the empty detector indicate that fluxes of gammas from the rock walls are near the levels assumed in previous simulations of detector performance.

4. CONCLUSIONS

From this discussion it is apparent that the detection abilities of the SNO detector provide a wide ranging sensitivity. The full set of measurements could provide definitive answers to the remaining questions of whether neutrino flavour change takes place for solar neutrinos and with what parameters, including transitions to sterile or non-sterile neutrinos. The detector capabilities provide the opportunity to determine the total flux of ^8B neutrinos from the sun even if neutrino flavour change has taken place and so can contribute significant astrophysical information as well.

REFERENCES

1. G. Ewan, Nucl. Instrum. and Meth. **A314** (1992) 373; SNO Collaboration, Physics in Canada **48**, 112, 1992; SNO Proposal, SNO-87-12, October, 1987.
2. For example, J.N. Bahcall and M.H. Pinsonneault, Rev. Mod. Phys. **64**, 885, (1994) and other references in this conference.

*The SNO Collaboration includes participants from Queen's University, Centre for Research in Particle Physics at Carleton University, University of Guelph, Laurentian University, University of British Columbia, University of Pennsylvania, Los Alamos National Laboratory, Lawrence Berkeley National Laboratory, University of Washington, Brookhaven National Laboratory, University of Oxford.

ELSEVIER

Nuclear Physics B (Proc. Suppl.) 77 (1999) 48–54

NUCLEAR PHYSICS B
PROCEEDINGS
SUPPLEMENTS

Status of the BOREXINO Solar Neutrino Experiment

L. Oberauer[a]

[a]Technische Universität München, James-Franck Str., D-85747 Garching
SFB 375 Astro- Particlephysics

Aim of BOREXINO is to measure in real time solar neutrinos with an energy threshold of about 250 keV via pure leptonic neutrino electron scattering $\nu + e \rightarrow \nu + e$. Special interest in BOREXINO is the first determination of the low energy ^7Be solar neutrino flux. In the MSW-scenario a strong depletion of this flux should be found. In case of vacuum oscillations a clear time fluctuation in the measured flux is to be expected. In addition the large mixing angle MSW solution can be probed independently by measuring the $\bar{\nu}_e$ flux of european power reactors in a long baseline oscillation experiment.

1. Introduction

The solar neutrino problem originates from the discrepancy between the ν_e expectations of the solar neutrino flux, as calculated by the Solar Standard Model (see e.g. [1]), and the experimental results. Information from helioseismology confirms the solar standard model to an accuracy within <1%. In addition, results from the Luna experiment [2] at low energies disfavour explanations based on discrepancics in this nuclear cross section. At last, the overall performance of both gallium experiments has been tested by terrestial neutrino sources, and no discrepancy from the expected values has been found [3].

Data analysis of the existing experiments leads to the assumption of severe suppression of the charged current solar ^7Be-branch. Fig. 1 shows a model independent constraint on this flux [4]. The only inputs are the integral flux of the SuperKamiokande experiment, which is taken as the value for the ^8B–neutrinos, and the known luminosity of the sun, whereas contributions from other branches of the pp-cycle and the CNO-cycle are even neglected. Within the 90% confidence limits a flux of only 20% of the expected charged current ^7Be neutrinos is consistent with the data, revealing a strong hint for physics beyond the standard model.

If neutrinos have non-zero masses and if they mix in analogy to the quark sector, neutrino oscillations arise. It can be shown that certain parameter on neutrino masses and mixing angles provide a solution to the result of all solar neutrino experiments. There exist three parameter areas: the MSW[1] small mixing angle solution with values typically for neutrino mass differences $\Delta m^2 \approx 10^{-6} \rightarrow 10^{-5}\ eV^2$ and mixing strength in the region $10^{-6} < sin^2 2\theta < 2 \cdot 10^{-5}$, the MSW large mixing angle solution for the same mass region but almost full mixing strength, and the vacuum oscillation solution with strong mixing at $\Delta m^2 \approx 10^{-10}$ eV^2.

Fig. 2 shows the deformation of the solar neutrino spectrum as expected in case of the MSW solution. The ^7Be neutrino line at 861 keV is almost fully suppressed in the small mixing angle solution as it is indicated by experimental data. The spectral shape of the high energy ^8B neutrino branch is deformed. The measurement of this shape is aim of SuperKamiokande and SNO, whereas BOREXINO will deliver direct information about the ^7Be neutrino flux.

2. Physics with BOREXINO

BOREXINO is going to be build up in the underground laboratory at Gran Sasso, Italy. The collaboration consists of several european and american institutes and is listed at the end of this article.

The aim of BOREXINO is to measure in real

[1]Mikhejev, Smirnov, Wolfenstein effect: resonant flavour conversion inside the sun, e.g. ref [5].

PII S0920-5632(99)00397-7

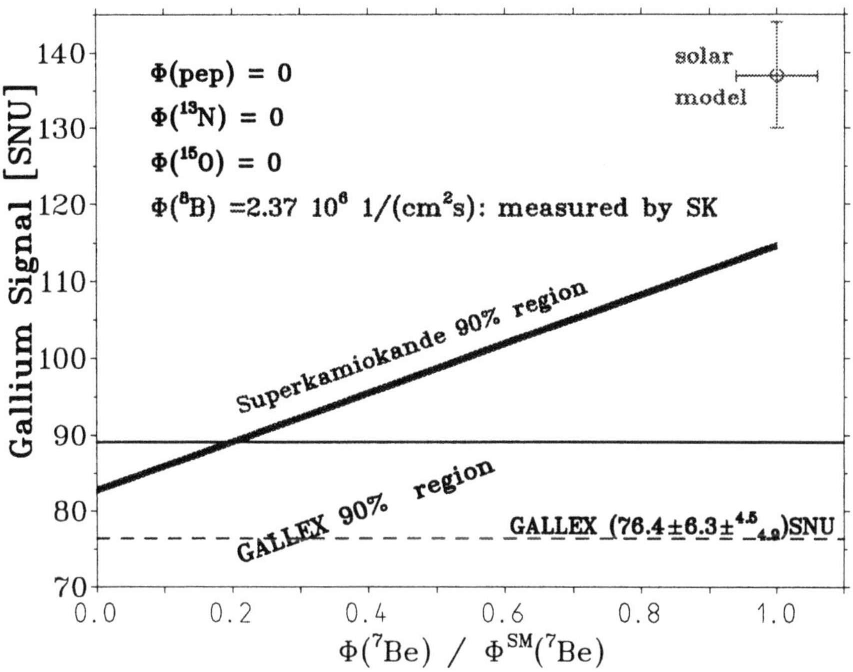

Figure 1. Model independent constraint on the charged current solar 7-Be neutrino flux derived by the experimental results from SuperKamiokande, Gallex, and the value of the solar luminosity. The signal strength (1 SNU = 1 capture in 10^{36} target atoms per second) of a gallium experiment is depicted as function of the 7-Be neutrino flux, resulting in the black line. The width of the line represents the experimental uncertainty of SuperKamiokande. Within 90% CL only about 20% of the expected 7-Be neutrino flux is compatible with experimental results.

time the solar neutrino flux at a low energy threshold with high statistics and energy resolving via pure leptonic neutrino electron scattering $\nu + e \rightarrow \nu + e$. Liquid scintillator (300t total mass) will be target and detector material. Monoenergetic ^7Be-neutrinos give rise to a compton like recoil spectrum in BOREXINO with a edge at 660 keV. Assuming validity of the standard model a counting rate for ^7Be-neutrinos, which would consist in this case purely as ν_e, of roughly 55/day in BOREXINO is expected. Here charged current as well as neutral current interaction in the neutrino electron scattering takes place.

In scenarios of total neutrino flavour conversion, i.e. for neutrino mass differences $\Delta m^2 \approx 10^{-6} \rightarrow 10^{-5}\ eV^2$, a reduced flux of approximately 12/day would be measured due to the lower cross section of $\nu_{\mu,\tau}$ scattering, which occurs only via neutral current interaction.

In case of vacuum oscillations, i.e. for neutrino mass differences $\Delta m^2 \approx 10^{-10}\ eV^2$, BOREXINO would see a distinct time dependent periodical neutrino signal due to the seasonal eccentricity of the earths orbit around the sun. In fig. 3 the counting rate of the 7-Be solar neutrino flux in BOREXINO is depicted in case of vacuum oscillations for parameters favoured by the recent SuperKamiokande results. A very distinct seasonal dependence should be observed. As a dashed line the oscillation probability is indicated with the

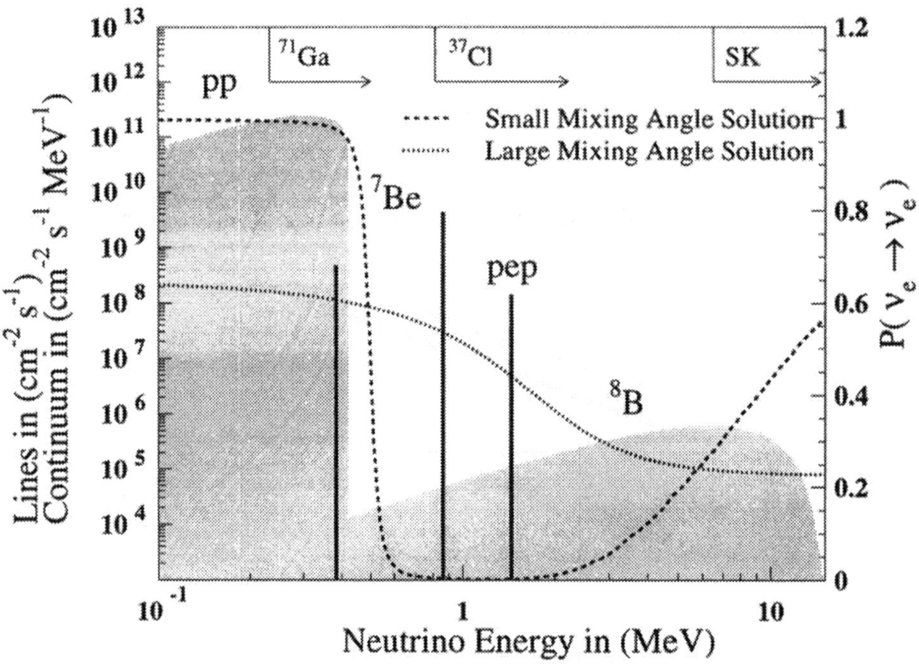

Figure 2. The solar neutrino spectrum as calculated in the standard model and the electron neutrino survival curve due to resonant neutrino flavour transversion. The 7-Be neutrino line at 861 keV is almost completely suppressed in the small mixing angle solution, as it is indicated by experimental data. The spectral shape of the high energy 8-B neutrino branch should be deformed. The measurement of this shape is aim of SuperKamiokande and SNO, BOREXINO will deliver information about the 7-Be neutrino flux.

scale on the right side. But also for other parameters in the range of $\Delta m^2 \approx 10^{-10}\ eV^2$ a very clear seasonal fluctuation should be expected.

For neutrino mass differences in the range of $\Delta m^2 \approx 10^{-7}\ eV^2$ and for large mixing BOREXINO should see a 'day/night' effect due to electron neutrino recovery during the path through the earth.

BOREXINO also can serve for additional projects in neutrino physics. Search for a magnetic moment can be performed by means of terrestial neutrino sources by investigating the electron recoil shape at low momentum transfer. Via the inverse beta-decay $\bar{\nu_e} + p \rightarrow e^+ + n$ BOREXINO can look for signals from geophysical neutrinos [6] as well as for neutrinos emitted by european nuclear power plants [7]. The latter would serve as a long baseline neutrino oscillation experiment probing the large mixing angle solution for the solar neutrino problem.

2.1. The BOREXINO Detector

BOREXINO is placed in hall C of the underground laboratory at Gran Sasso, Italy. An overburden of about 3500 m.w.e. suppresses the cosmic muon flux to 1.1 $/h\,m^2$. The detector is shielded successively from outer radioactivity.

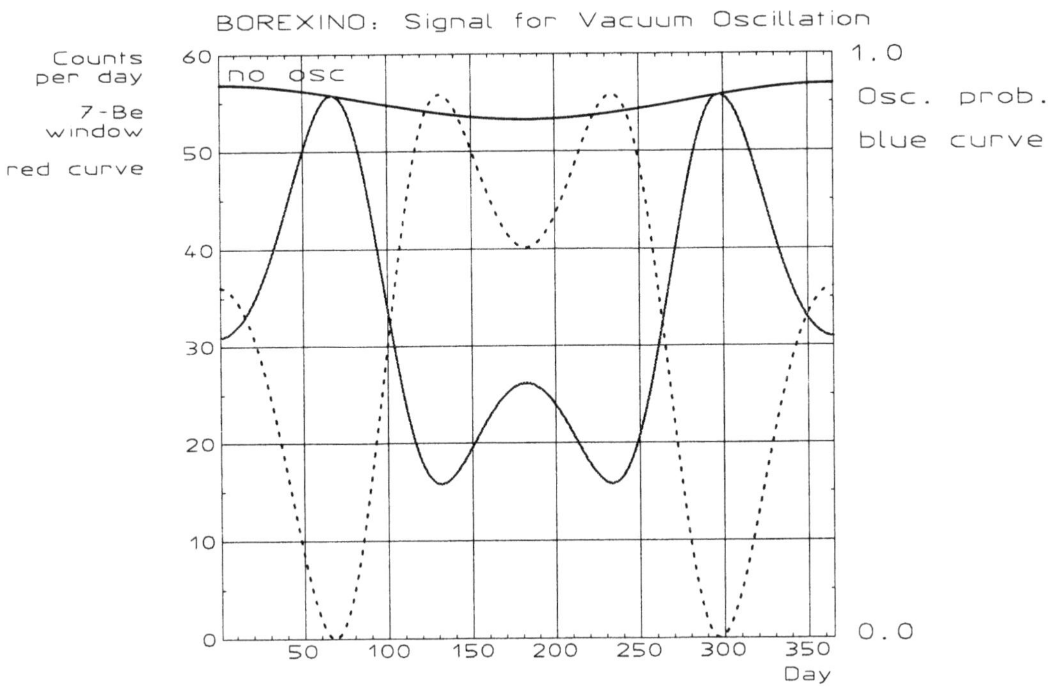

Figure 3. Counting rate of the 7-Be solar neutrino flux in BOREXINO in case of vacuum oscillations for parameters favoured by the recent SuperKamiokande results. A very distinct seasonal dependence should be observed. As a dashed line the oscillation probability is indicated with the scale on the right side.

The adjacent inner layer serves as shielding and has to provide an increased purity in terms of internal radioactivity. Fig. 4 shows the experimental setup.

Inside the 'external' tank a stainless steel sphere will support about 2200 phototubes on the inside and 200 tubes at the outside. Latter provide the muon veto Cherenkov system. Tubes inside the sphere will be equipped with light guides in order to increase the geometrical coverage and hence the energy resolution.

The steel sphere will be filled with a transparent, high purity buffer liquid which itself holds a nylon sphere, filled with organic scintillator. Between the steel sphere and the scintillator region an additional nylon shroud hinders radon convection from the outer area towards the critical region. The active scintillator mass will be around 300t. As our first option serves pseudocumol (PC) with a wavelengthshifter at concentrations of about 0.15%.

Time information in each channel provides reconstruction of the event position. A fiducial volume of about 100t for solar neutrino interaction will be defined, establishing a counting rate of 55 neutrinos per day according to the standard solar model. The outer part of the scintillator sphere serves as an active additional shielding against external background.

3. Results of CTF and NAA

The demands on purity in terms of radioactivity in BOREXINO, especially for the scintil-

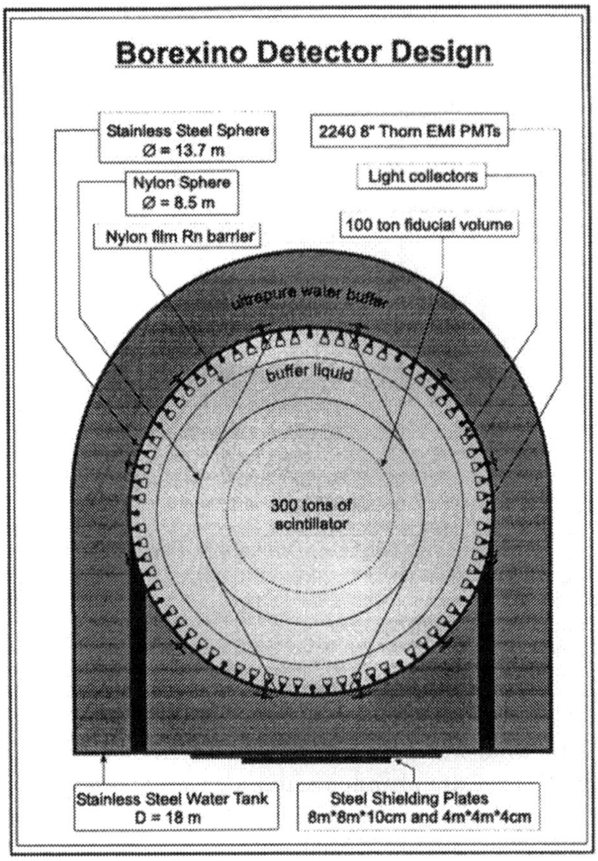

Figure 4. Schematic view of BOREXINO. The ultrapure scintillator (300t) inside a nylon vessel is shielded differentially by means of a liquid buffer (1040t), a steel sphere on which the tubes are mounted, and the outer water buffer which is contained in an external steel tank with dimensions of about 18m. A transparent nylon shroud hinders radon convection in the liquid buffer region. Additional tubes mounted on the outer surface of the steel sphere allow detection of penetrating muons via the Cherenkov effect.

lator itself, are challenging. In order to be able to extract a clear signal from background events also in case of total flavour conversion, an intrinsic concentration in Uranium and Thorium of about 10^{-16} should not be exceeded significantly. The amount of $^{14}C/^{12}C$ must not be higher than $\approx 10^{-18}$. In order to test scintillating materials a large Counting Test Facility (CTF) has been built up in hall C of the underground laboratory at Gran Sasso, which resembles a small prototype (ca. 5t of scintillator) of BOREXINO. In addition high sensitive neutron activation analysis (NAA) has been developped at the Technical University Munich in Garching.

3.1. CTF results

From beginning of 1995 until summer 1997 several CTF-tests about the feasibility of BOREXINO have been performed. This includes procedures to maintain or to improve the purity of the scintillator and encouraging results have been obtained: $^{14}C/^{12}C = 1.94 \cdot 10^{-18}$, $^{238}U = (3.5 \pm 1.3) \cdot 10^{-16}\ g/g$, $^{232}Th = (4.4 \pm 1.5) \cdot 10^{-16}\ g/g$. A complete discussion of the CTF results including experimental techniques for further background suppression are given in ref. [8] and ref. [9]. Details about the experimental setup of the CTF can be found in ref. [10].

Due to the three-dimensional array of phototubes the position of events in CTF could be reconstructed. Careful data analysis including source tests demonstrated, that the observed single rate was dominated by external background, mainly by radon present in the water shielding.

Several purification tests on the scintillator compounds have been performed, including N_2-bubbling, water extraction, distillation as well as column separation. Removal of radioactive gases present in the environment like Krypton and improvements in the concentration levels of ^{210}Pb-daughters could be demonstrated [8]. Column separation has been tested with an alternative scintillator and very promising results have been obtained.

The CTF response on intersecting cosmic muons has been studied and the necessity of an outer veto system in BOREXINO became clear. According to the requirements of the experiment a design for the outer detector has been finished. Measurements done at the high energy muon beam at SPS in CERN allowed for the careful study of cross sections of the generation of cosmogenic radionucleides in the scintillator. Data analysis is on-going and results can be expected in summer 1998.

Experience with the CTF helped to understand technical problems. Some of them concern deterioration of detector materials, the sealing of the tubes, radon diffusion through wet nylon, and radon tightness of liquid handling systems.

3.2. NAA results

Highly developped neutron activation analysis (NAA) of scintillation samples performed in Garching provides an independent test on the radiopurity and allows important tests on the secular equilibrium of the decay chains. With NAA an upper limit for uranium in PC/PPO of $^{238}U < 2 \cdot 10^{-16}\ g/g$ (90% CL) has been obtained. In addition concentration values or limits have been measured by this method for various isotopes, including man-made nuclei for different detector materials. For details, see ref. [11].

In Garching a new low background laboratory with shielding against the cosmic hadronic and soft electromagnetic component has been built up and completed in 1998. First measurements with neutron activation at the research reactor FRM-I in Garching, chemical extraction in a radiochemical laboratory, and final data taking in the low background laboratory show, that sensitivities in Uranium and Thorium far below the 10^{-16} limit can be reached now.

4. Conclusion

BOREXINO is dedicated to detect in real time low energy solar neutrinos. The demands on the purity of detector materials are challanging. However, results from the CTF and NAA show very encouraging results on the radiopurity of liquid scintillators. The CTF will be reinstalled and finds further use for testing purposes. NAA in Garching has been developped to concentration levels far below the 10^{-16} range. In hall C of the underground laboratory at Gran Sasso the external tank of BOREXINO has been finished. Start of the filling procedure of BOREXINO is expected in 2000.

5. Collaboration list

Germany: Max-Planck-Institut für Kernphysik Heidelberg: B. Freudiger, W. Hampel, J.Handt, G. Heusser, J. Kiko, T. Kirsten, H. Neder, W. Rau, M.Wojcik, Y. Zakharov. Technische Universität München: F. von Feilitzsch, C. Hagner, T. Hagner, R. von Hentig, G. Korschinek, L. Oberauer, J. Jochum, S. Schönert, K.H. Schuhbeck.

Hungary: KFKI-RMKI Budapest: L. Cser, D. Kiss, I. Manno, G. Marx.

Italy: Universita e INFN di Genova: F. Gatti, V. Lagomarsino, G. Manuzio, P. Musico, A. Nostro, A. Razeto, E. Resconi, C. Salvo, G. Testera, S. Vitale. LNGS, Gran Sasso: C. Arpesella, M. Balata, A. Falgiani, A. Goretti, A. Ianni, M. Laubenstein, M. Neff, S. Nisi, R. Tartaglia. Universita e INFN di Milano: G. Alimonti, G. Bellini, S. Bonetti, A. Brigatti, B. Caccianiga, R. Dossi, C. Galbiati, A. Garagiola, M.G. Giammarchi, D. Giugni, A. Golubchikov, F.X. Hartmann, G. Korga, P. Lombardi, S. Magni, S. Malvezzi, J. Maneira, E. Meroni, L. Perasso, G. Pieri, G. Ranucci, P. Saggese, R. Scardaoni. Universita e INFN di Pavia: G. Cecchet, A. De Bari, A. Perotti, G. Sau. Universita e INFN di Perugia: F. Elisei, F. Masetti, U. Mazzucato.

Russia: J.I.N.R. Dubna: O. Smirnov, A. Sotnikov, O. Zaimidoroga.

USA: AT&T Bell Laboratories: R.S. Raghavan. Massachusetts Institute of Technology: M. Deutsch. Princeton University: J. Benziger, M. Johnson, L. Cadonati, F. Calaprice, M. Chen, R. Eisenstein, R. Fernholz, F. Loeser, R. Parsells, R.B. Vogelaar, R. Walls.

REFERENCES

1. J.N. Bahcall, M. Pinsonneault, Rev. Mod. Phys. 67, (1995) 781.
2. M. Junker et al., Nucl. Phys. B 70, (Proc. Suppl.), (1999), 382.
3. Gallex collaboration, Phys. Lett. B 388, (1996), 384.
4. M. Altmann, Naturwissenschaften 84, (1997), 105.
5. S.P. Mikheyev, A. Yu. Smirnov, Sov. J. Nucl. Phys. 42, (1985), 913. L. Wolfenstein, Phys. Rev. D20, (1979), 2634.
6. R.S.Raghavan, S. Schönert, S. Enomoto, J. Shirai, F. Suekane and A. Suzuki, Phys. Rev. Letters 80, (1998) 635
7. S. Schönert, Nucl. Phys. B 70, (Proc. Suppl.), (1999), 195.
8. G. Alimonti et al., BOREXINO collaboration, Astr. Part. Phys. 8 (1998), 141.
9. G.Alimonti et al., BOREXINO collaboration, Phys. Lett. B 422, (1998), 349.
10. G.Alimonti et al., BOREXINO collaboration, Nucl. Instr. Meth. (1998), accepted for publication.
11. T.Goldbrunner et al., Journ. of Rad. Nucl. Chem. 216, (1997) 293.

NUCLEAR PHYSICS B PROCEEDINGS SUPPLEMENTS

ELSEVIER

Nuclear Physics B (Proc. Suppl.) 77 (1999) 55–63

Future Solar Neutrino Projects

Robert E. Lanou, Jr.*
Department of Physics, Brown University, Providence, RI 02912

In the program to determine the intrinsic properties of the neutrinos and their role in physics and astrophysics, solar neutrino experiments are playing a fundamental part. The "first" and "second" generation detectors in the field have been enormously successful and promise to provide still greater contributions. The challenge to create a "third generation" of solar neutrino detectors arises principally from interest in making detailed investigations of the portion of the neutrino spectrum from a few keV to 1.5 MeV. That portion contains fluxes from the p-p and CNO continuum as well as the ^{7}Be and p-e-p lines. The need to overcome the experimental difficulties presented by working at these low energies have given rise to new ideas for detector technologies. The range of technologies is impressive in its variety and reflects choices of emphasis using real-time and radiochemical methods. In this paper, I present a review of the status of several R & D efforts known to me which are making progress toward providing detectors suitable to meet the various challenges of this low energy region.

1. INTRODUCTION

The difficulties of creating effective detectors for solar neutrino research are well known involving, in varying degree, radioactive backgrounds in the target and its surroundings, extraction of signal from massive targets, energy resolution and choices of detection reaction mechanisms. As lower energy neutrinos are to be detected, most of these difficulties increase or choices are narrowed especially for real-time detectors, although the higher flux ameliorates somewhat the massiveness of the targets. As the field of solar neutrino research has matured and moves toward its goal of achieving a measurement of the shape and neutrino-flavor composition of the flux over the full spectrum, the need for a "third generation" of detectors able to carry out detailed experiments in the region from a few keV to $\sim$ 1.5 MeV has been apparent. This is the region containing the p-p and CNO continuum as well as the ^{7}Be and p-e-p lines. From the "first generation" radiochemical experiments (Homestake, GALLEX and SAGE) [1] we already know some important things about this energy region. For example, that the integrated fluxes of ν_e are $\leq \frac{1}{2}$ of those expected from the standard solar model (SSM) [2] and that, when taken together in a global analysis of all solar experiments, suggest that the ^{7}Be ν_e flux may be much diminished. From the present status it is not possible to know how this flux reduction is shared among the different contributing fusion reactions nor whether there are neutrino flavors other then ν_e present. The upcoming, "second generation" BOREXINO [3] liquid scintillator experiment will directly address the ^{7}Be question in real-time via the (nearly) flavor-independent elastic scattering reaction on electrons. This represents an important next step; however, many questions will still remain, among them whether or not the full p-p flux is seen as ν_e, to what extent is the CNO flux affected, what is the flavor composition of the flux at all energies and whether the p-e-p and ^{7}Be lines can be resolved. The experimental technology to attack these questions in this energy regime has not existed but in recent years there have emerged — and there are still emerging — some new approaches in various stages of research and development which are showing significant promise. No single technique is likely to emerge which will be able to address all questions in a single detector and the diversity of the projects I will discuss reflect choices, on the experimenters' part, to focus on the different physics opportunities. In this talk I will try more strongly to emphasize how some of these new techniques work, some of their recent R & D results, what still needs to

*This work supported in part by DoE DE-FG02-88ER40452 & NSF PHY-9420744.

PII S0920-5632(99)00398-9

be done and what their physics foci are rather than what a finished detector might look like. By and large they are still R & D projects and have not yet shown full feasibility for implementation as a neutrino detector. Some of the projects are known explicitly by their target material such as Lithium or Gallium Arsenide and others with the target identity subsumed into acronyms such as HERON, HELLAZ, LENCSE and GNO.

2. HELIUM AS A TARGET

These projects, HERON and HELLAZ, are both based upon the use of helium as the target medium but they have radically different approaches and somewhat complementary goals. Both will utilize the elastic reaction, $\nu_{e,\mu,\tau} + e^- \rightarrow \nu_{e,\mu,\tau} + e^-$, for real-time detection in the energy region dominated by the p-p and ^{7}Be neutrinos. They will both measure the energy of the recoil electron and the overall rate. The similarities and contrasting differences are best seen by looking at the specifics of each project in turn.

2.1. HERON:

This project uses ^{4}He in its superfluid state [4] and is a new particle detection technique. For purity from radioactivity, superfluid helium ^{4}He is ideal for several reasons. It has no long-lived isotopes, its first excited nuclear state is at 20 MeV, it is self-cleaning in that all other atomic species freeze out and ^{3}He (which has a large neutron cross section) is easily reduced to negligible amounts in the superfluid during the detector filling. It has a high density of 0.14 g/cc, it is inexpensive and standard commercial methods exist for handling the liquid in large volumes. The mechanism for event identification and energy extraction from the target utilizes the high multiplicity of carriers ($\sim 10^7$ rotons/phonons and $\sim 10^3$ uv photons for a 50 keV e^-) generated in the liquid by the recoil. An array of low mass silicon or sapphire wafer calorimeters external to the liquid serve to detect pulses created by the direct uv photons as well as the delayed pulse generated by the rotons/phonons through quantum evaporation at the free surface. Event position in the detector would be found by observing the spatial distribution of hit wafers and relative times of arrival of the photon and evaporated atom pulses. The recoil electron energy would be found by the total pulse height on all hit wafers. Backgrounds are primarily Compton electrons from residual activity in the cryostat external to the liquid and would be subtracted by topological cuts and reference to fiducial and non-fiducial volumes. A detector large enough to detect 18 p-p and 7 ^{7}Be events/day (SSM) would have 10 tons of helium in its fiducial volume and measure ~5x5x5 meters overall. There would be $\sim$ 1000 wafer readout channels and both liquid and wafers held at 30mK. Refrigeration on this scale is commensurate with that for cryogenic gravity wave detectors.

Through earlier experiments detecting 3 to 5 MeV alpha particles the group had established the validity of the basic physics principles of the particle detection method in superfluid helium combined with the use of wafer calorimeters. Among other things it was seen that both the prompt uv photons and delayed roton/phonon evaporation signal were easily detected on the same silicon or sapphire wafer. Many of the details of the evaporation process were established including the existence of the so-called critical angle ($\pm 17^o$ for alphas) within which rotons/phonons were able to initiate evaporation. It was expected that this critical angle could play a useful role in event position location. Nonetheless, it still needed to be established that low energy electrons could also be detected and with what roton and photon characteristics. Improvements in wafer sensitivity by adoption of superconducting thin film thermometers and SQUID technology have been made to do this. A threshold of 300 eV was reached for energy deposition into an individual wafer. In a recent series of experiments, 364 keV single, mono-energetic electrons (from ^{113}Sn radioactive sources movable in the test cell) have been successfully detected in 3 liters of superfluid. The measured spectrum is shown in Figure 1; the peak at 31 keV is for the electron in this particular experimental geometry. For calibration purposes x-rays of 5.9 and 25 keV from ^{55}Fe and ^{113}Sn sources, respectively, are deposited directly into the wafer and

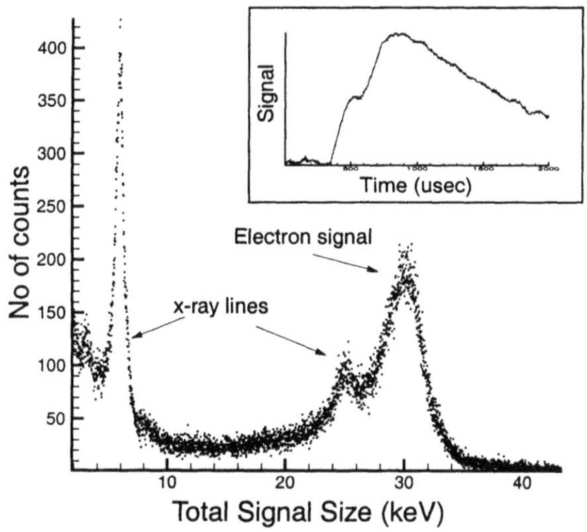

Figure 1. Energy spectrum of electrons from ^{113}Sn in superfluid ^{4}He. (Inset: Pulse for single, 364 keV electron with structure due to early arrival of uv photons.

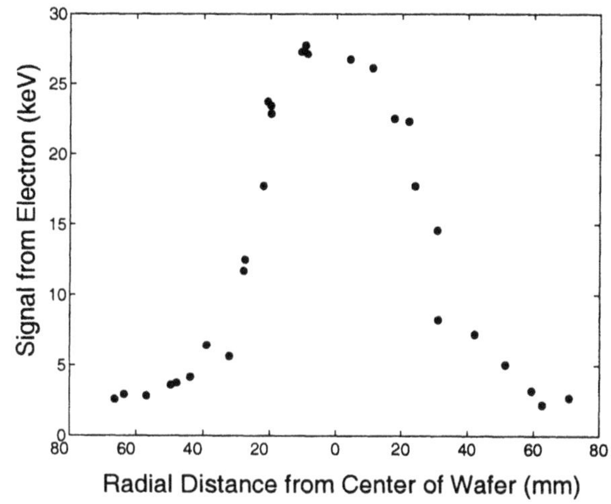

Figure 2. Horizontal distribution of electron signal at 2 mm above superfluid surface.

can also be seen. As the electron source is moved horizontally below a single wafer the effect of the critical angle can be seen (Figure 2) in the position dependence of the combined evaporation and photon signal and suggests a larger critical angle ($\geq 30^o$) for the electrons. Particular attention was paid to studying the total amount of energy detectable and the resulting partition of that energy between the evaporation signal and the uv photon signal. In the inset of Figure 1, the leading edge shows the prompt photon signal followed 250 μs later by the evaporation signal. While both alphas and electrons yield roughly the same total energy fraction ($\sim$ 10%) they share that energy quite differently between uv radiation and evaporation. The electron signal relative to the alphas is nearly 3 times that in the uv photons and about $\frac{1}{2}$ for the evaporation pulse. This corresponds, for an electron, to 14 photons/keV and 2x10^5 (rotons/phonons)/keV . The uv photons have a wavelength of $\sim$ 80 nm; a wavelength for which helium is self-transparent. The group believes it has completed its study of the physics processes underlying the proposed particle detection technique and has a good understanding of the mechanisms involved. The results are consistent with a microscopic model in which the initial ionization density along the track governs the rate of formation and collision relaxation of helium dimers and therefore the relative portions of energy detectable as rotons/phonons and uv photons.

Among the future work toward demonstrating feasibility for solar neutrino detection are several items. Foremost is the need to improve the sensitivity of the wafers for use in large volumes by a factor $\sim$ 20 . This is closely related, in part, to trying to capitalize on the increased uv radiation found to be emitted by the low energy electrons and incorporating it most effectively with the evaporation signal. A particularly promising avenue being pursued for increased sensitivity is based on incorporating new developments in magnetic calorimetry. Further work with superconducting thin film thermometers is also in progress. A multi-wafer cell is under development to test various features of energy and position determination as well as implementation of cryogenics for such a system.

2.2. HELLAZ:

This project [5] will utilize helium (mixed with $\sim$ 1% CH_4) in gaseous form under pressure and reduced temperature (5 atmos. and 77K) yielding a 0.003 g/cc density. It seeks to use an existing particle detection technique and scale it up in size (2000 m^3) and into the high pressure and low temperature regime. The mechanism for event identification and energy extraction makes use of the drift of the ionization electrons ($\sim$4x10^3 for 100 keV recoil e^-) in a large time projection chamber (TPC). The event signature consists of full e^- recoil track reconstruction. The recoil energy is to be determined by counting individual secondary, ionization electrons as they drift to the x-y imaging plane. Use of the first arriving secondary electron to set a t_0 and the time delays of the subsequent secondaries on the earliest part of the track should allow determination of the track orientation in the TPC. Detection of the Bragg peak should distinguish front from rear track end. Combination of this information with the contemporaneous position of the Sun would be used to deduce the energy of the incident neutrino. Backgrounds here are of two types: a) Compton electrons from the TPC and the pressure vessel and b) from beta decay of ^{14}C in the gas mixture. Backgrounds would be subtracted using a sample of events constructed assuming the Sun to be located 180° from its true position. A detector of 6 tons would be expected to yield 11 p-p and 4 ^7Be events/day (SSM); it would be 25 meters long and 10 meters in diameter. With x-y detector planes at both ends the recoil electron track would be drifted up to 10 meters. If experimental hall dimensions require, it could in principle by split into two portions.

In earlier work the group had extensively tested various gas mixtures to find one suitable for TPC operation under the conditions envisioned for the solar neutrino application. The He-CH_4 mixture was selected as the most promising but it imposes severe requirements on the speed ($\leq$ 20 ns) and gain ($\sim 10^6$) needed in the x-y readout for single secondary electron counting which is to provide the energy and space orientation of the recoil track. However, recently they have made important progress on that front by testing a new

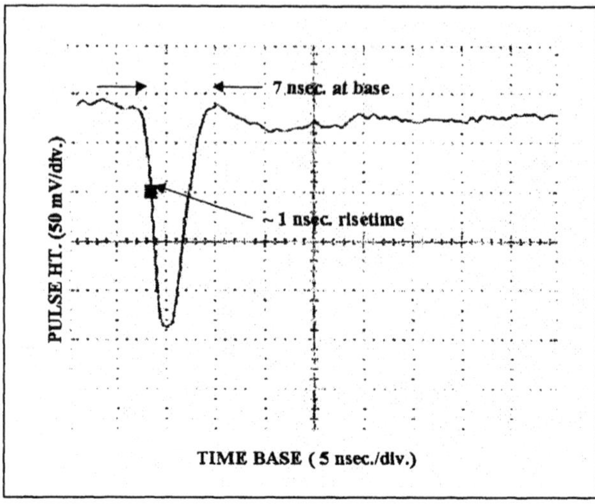

Figure 3. Single photoelectron pulse in "micromegas" chamber.

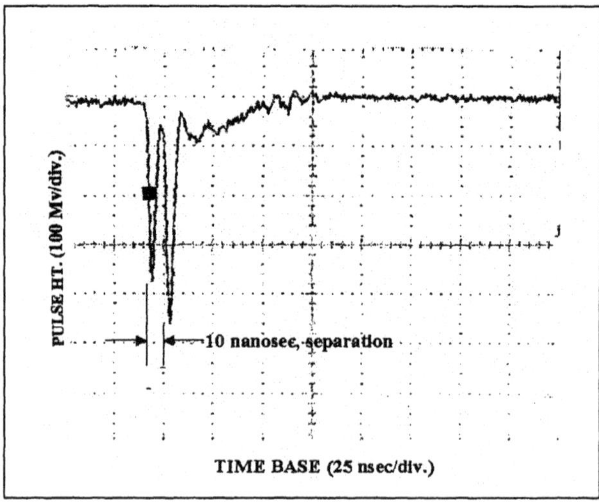

Figure 4. Two, successive photoelectron pulses in "micromegas" chamber.

wire chamber technology — the "Micromegas" [6] chamber. In that test a small chamber was used which was filled with a 1 atmosphere helium and 6% isobutane mixture and constructed of fine mesh forming a 100 μm gap above a one-dimensional strip read. An x-ray source of known strength produced single photoelectrons in the chamber. These were detected with a rise time of $\sim$ 1 ns and a gain of 2x10^6. Details of a single photoelectron pulse are shown in Figure 3. The successful detection of two, successive photoelectrons separated by only 10 ns is illustrated in Figure 4.

Future work is planned on several fronts. Of particular importance is the need to adapt this counting success to the large scale TPC environment : e.g., extend to two dimensional capability, find a gas and operational compatibility with the track drift and manage the potential, large increase in readout channels. Several ideas for achieving these points are under active consideration among them, incorporating an electrostatic lens to reduce the effective area of the 100 m^2 endcaps. Nearing completion is a 5 liter TPC capable of 5 bar and 77K operation to be used in tests to produce and detect Compton recoils from 511 gammas. They are also preparing to initiate tests for low activity materials as well as to carry out further Monte Carlo studies related to recognition of electron track direction and energy determination.

3. LENCSE:

A recent proposal [7] has been made to utilize ^{176}Yb in a real-time, liquid scintillator detector. By detecting ν_e through inverse beta decay it would serve as a complement to those detectors using the flavor independent elastic scattering from electrons. This project will be discussed in a separate presentation by Dr. Raghavan.

4. CRYOGENICS IN RADIOCHEMICAL EXPERIMENTS

A new approach to counting the back-decay of atoms collected in radiochemical experiments is being carried out by two groups (GNO and Lithium). Although this new technology has not yet been perfected for use in an operating solar neutrino detector, the groups are making impressive progress. The idea is to replace the classical method of small proportional counters by cryogenic micro-calorimeters. Micro-calorimeters are small (few micro- or milli-gram) crystals at milli-Kelvin temperatures to which the extracted radioactive sample has been applied. Because of the exceptionally low heat capacity of the calorimeter, and if 4π coverage is achieved, then the Auger electrons and x-rays following the electron capture decay are contained hermetically and a temperature pulse results. This is a technique which is developing rapidly in other areas (e.g., for x-ray astronomy, double beta decay or dark matter searches) where resolutions less than 100 eV have been obtained. Although the primary motivations for the two groups discussed here to employ the cryogenic method are somewhat different, the technique holds out promise in several areas important to both: lower thresholds, improved energy resolution, full energy deposit and increased counting efficiency. A particular challenge common to both experiments will be to devise a method which ensures that a fully efficient transfer of the precious sample, extracted from the chemical reactor, is made onto the micro-calorimeter crystal.

4.1. Lithium:

For some time, ^{7}Li has attracted interest as a radiochemical target with special relevance for the p-e-p and CNO neutrinos because the strength of the ground and first excited states can be accurately inferred from laboratory experiments [8]. The reaction is $\nu_e + {}^7\text{Li} \rightarrow e^- + {}^7\text{Be}$ with a threshold of 862 keV. A particular challenge to the realization of a lithium-based detector has been the counting of the electron capture decays ($\tau_{1/2}$ = 53d) of the extracted ^{7}Be in which 90% go to the ^{7}Li ground state but produce an Auger electron of only 55 eV together with a 57 eV nuclear recoil and is therefore not amenable to the usual proportional counter methods. Consequently, interest had centered on utilizing the decay to the first excited state by detecting the subsequent 474 keV gamma but, with only a 10%

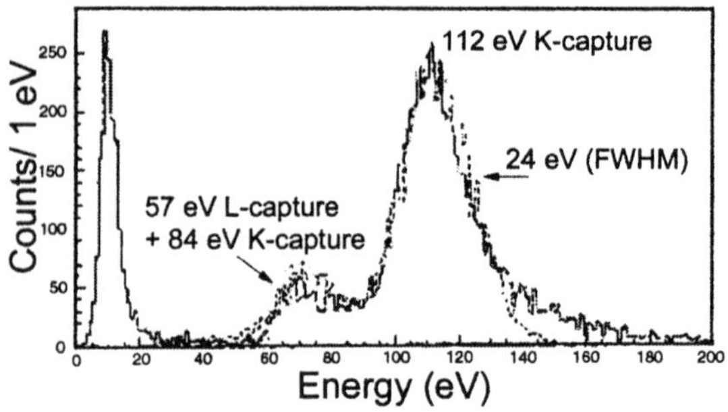

Figure 5. Spectrum of $^7Be \rightarrow {}^7Li$ and $^7Be \rightarrow {}^7Li^*$ in micro-calorimeter.

branching ratio, a 100 ton detector was required to achieve a rate of 0.5/day. Presently, a joint Russian-Italian project [9] has made interesting progress on two fronts: a) first, by capitalizing on the use of the 90% channel by developing the micro-calorimeter method thus making conceivable a detector of only 10 tons and b) second, by developing a large scale prototype for the lithium reactor itself to test the difficulties of the lithium handling and sample extraction.

In tests using cryogenic micro-calorimeter techniques with accelerator produced samples of Be and BeO, they have detected ^{7}Be decays via the summed energy deposited by the Auger electron (55 eV) and the recoiling nucleus (57 eV) giving a well resolved peak at 112 eV with a 24 eV FWHM at $\sim$ 80% efficiency (Figure 5). The microcalorimeter consisted of a 100x200x200 μm NTD thermistor at 45mK glued to the few μg sample.

While further work along these lines is needed, this result seems to establish feasibility of the method. Among the future projects are plans to study other Be compounds for improved phonon coupling to the thermistor and minimization of delayed counting effects from quasi-particle trapping in the crystal (see the peak at very low energies).

The mechanical work on a prototype reactor vessel to hold 300 kg of liquid lithium has been completed at INR(Troitsk). The principal R&D work to be carried out with the prototype are studies of the complete lithium handling method and the Be extraction process into forms most suitable for the micro-calorimetry counting method. In order to carry out this important series of tests with the prototype, funds are urgently needed to move it to another laboratory and to operate it there. Funding has been applied for and a decision enabling them to go forward is awaited.

4.2. GNO:

The Gallium Neutrino Observatory [10] is designed to improve and extend the very successful gallium radiochemical technique. Details of GNO's goals and present operating status can be found in Dr. Kirsten's presentation. My emphasis here is on the potential, cryogenic innovations which may have broad application. The events to be counted in GNO are the electron capture decays of ^{71}Ge. In the recently completed GALLEX experiment, 70% of the contribution to the systematic error was from the energy acceptance window. Consequently, the much improved energy resolution which the micro-calorimeters promise is one reason arguing for attempting to exploit them. Another reason is the potential for increased counting efficiency; the proportional counter efficiencies are typically 70% due to several factors including dead volume and missed x-rays. Every improvement in efficiency allows a similar reduction in gallium mass for the same statistics. Improved systematics and statistics are among the principle goals of GNO. K-capture constitutes 88% of the decays ($\sim$ 41% each into the 10.37 keV Auger electron or a 9.35 keV K_α with a 1.12 keV e^- and 5.3% into a 10.26 keV K_β with a 0.11 keV e^-). The L-capture is 10.3% of the rate giving only a 1.30 keV Auger electron. Hermetic detection of the x-ray would prevent the "contamination" of the L-peak by missed K_α x-rays. An attractive bonus of an additional 1.7% of the rate could be gained if the threshold can be kept below 150 eV since M-capture would be accessible.

The Munich group has made significant progress in experiments for cryogenic detection of ^{71}Ge. Two types of experiments are of particular interest. In the first, 1 mm^3 crystal of ^{nat}Ge was neutron activated thus producing ^{71}Ge whose de-

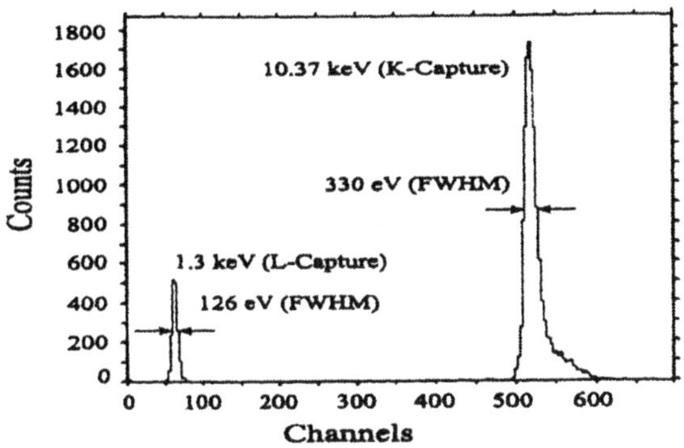

Figure 6. Spectrum of $^{71}Ge \rightarrow {}^{71}Ga$ in microcalorimeter.

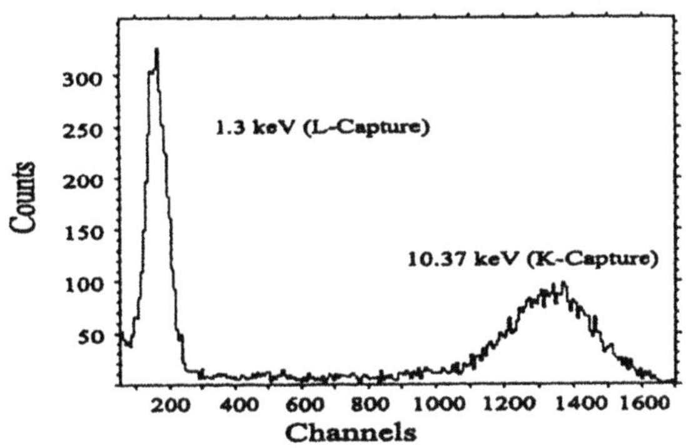

Figure 7. Spectrum of $^{71}Ge \rightarrow {}^{71}Ga$ in GALLEX proportional counter.

cays had 4π coverage. Attached to the crystal was a tungsten superconducting thin film transition edge thermometer (at 15 mK) with which the temperature pulses were measured. The resulting spectrum is shown in Figure 6 and for comparison Figure 7 is a typical GALLEX proportional counter result for the same decay. The effect of 4π acceptance and improved resolution is clearly evident. In a solar neutrino experiment it is Ge from GeH_4 which must be attached to a crystal. As a first step in using GeH_4 a thermally decomposed sample of GeH_4 was deposited onto a 1μg sapphire crystal equipped with an iridium-gold transition edge thermometer. Due to the 2π nature of the decay coverage the detection was not hermetic; however good resolution (160 eV on the 5.89 keV Mn K_β calibration line) was obtained and the features of the ^{71}Ge decay were clearly observed. They have also done some initial work on ^{37}Ar.

This work is continuing and among the immediate projects are improvements in the electronics to obtain a threshold low enough for seeing the M-line, developing a two crystal "sandwich" as a 4π hermetic device with readout on both sides for energy summing, reducing the contributions from deposition in the thermometer glue (see the high-side tail in Fig. 6) and, crucially, finding a high efficiency method for placing the Ge from GeH_4 on the crystals.

5. GALLIUM ARSENIDE:

This is perhaps the most ambitious project [11] and while it does not aim to be the "complete" solar neutrino detector it does give a measure of the task to realize a single detector more inclusive of reaction types and range of energies. Its principal aims are to make a model independent test of flavor non-conservation (including sterile ν's), to determine the precise energies and widths of the lines of ^{7}Be (862 keV), p-e-p (1.4 MeV) and the end point of the p-p continuum using the three channels of elastic, charged current and neutral current scattering of ν's. Further, in the case of any observed flavor non-conservation, they wish to determine the MSW parameters precisely if that is the mechanism involved. To carry out such a program, a massive detector with $\leq$ 2 keV energy resolution is required. Their idea is to create a GaAs based, electronic, real-time detector. Gallium and arsenic are chosen because they have no long-lived isotopes or (n,γ) daughters and for the possibility of good resolution due to the high multiplicity of e-hole pairs and low noise when cooled. However, they estimate this would require a 125 tonne (60 t of Ga) in 40,000 hyperpure 3.2 kg segments. Among the primary reactions to be exploited are $\nu_{e,\mu,\tau} + e^- \rightarrow \nu_{e,\mu,\tau} + e^-$, $\nu_e + {}^{71}Ga \rightarrow e^- + {}^{71}Ge$

and $\nu_{e,\mu,\tau} + {}^{71}Ga \rightarrow \nu_{e,\mu,\tau} + ({}^{71}Ga)^*$. Good resolution of the p-e-p and upper ^{7}Be line would be an important advance; however, a successful data analysis to achieve fully the above goals will be a very challenging one in that it relies upon use of the shape of the elastic differential cross-section to separate ν_e and $\nu_{\mu,\tau}$ as well as requiring a separation of the charged current p-p events. Extreme purity (from U, Th, ^{40}K & ^{14}C) of the GaAs as well as its electronic performance for devices as large as 3.2kg must be tested and assured.
R&D on this project has recently begun. It centers primarily on three areas: a) production of testable, small GaAs devices made from sizable boules, b) tests of the electronic properties of these small devices and c), based on results from these tests, investigation of new crystal growing methods to produce improved quality and larger boules. They report that in Russia twenty, high quality 1 kg GaAs ingots have been grown for them. And that from these ingots, several 400 μm thick working detectors have been made and their electrical properties measured. They find that most properties for these small devices are equal in quality to the best in the field but still fall short of what would be needed in the full-sized solar neutrino device. They find high electron mobility but need significant improvement in charge collection. They are preparing to take the next step to increase the thickness of the electrical test devices to 1 mm thickness and they are studying new crystal growing methods adaptable to their existing furnaces.
There is still a very, very long way to go on this very bold project but if it can be shown to be feasible it may be the closest we will ever get to a "universal" solar neutrino detector.

6. COMMENTS & CONCLUSIONS:

The field of solar neutrino physics has certainly come of age as the exciting results so far are showing. From Super-Kamiokande and the other second generation experiments, SNO, BOREXINO and GNO, just coming on line we can expect new crucial tests of neutrino properties. However, it is well to keep in mind that these experiments and their predecessors would not have been possible without the invention of new (or by pushing to the limit) experimental techniques. We are still only part of the way to having the experimental capabilities we need over the full solar spectrum. In order to go beyond what we now can do, we need new techniques.
I have tried to show by the projects described here that there is an active world-wide R & D effort making important progress on several fronts toward these goals; but we should keep in mind that we are not fully there yet and there is always room for more, new ideas.

7. ACKNOWLEDGMENTS:

I am very grateful to J. Adams, M. Altmann, T. Bowles, V. Gavrin, P. Gorodetzky, T. Kirsten, A. Kopylov, R. Raghavan, S. Vitale, and T. Ypsilantis for very helpful information and discussion concerning the specifics of the projects in which they are involved.

REFERENCES

1. Homestake: B.T. Cleavland et al, *Ap.J.* **496**, 505 (1998); GALLEX: P. Anselmann et al, *Phys. Lett.* **B342**, 440 (1995); SAGE: J.N. Abdurashitov et al, *Phys. Lett.* **B328**, 234 (1994). See also talks by K. Lande, T. Kirsten and V. Gavrin in these proceedings.
2. J.N.Bachall and M. Pinnsoneault, *Rev. Mod. Phys.* **67**, 781 (1995); S. Turck-Chieze and I. Lopes, *Ap. J.* **408**, 347 (1993).
3. C. Arpesella et al, *BOREXINO: Proposal for a Real Time Detector of Solar Neutrinos* Vol. I & II, INFN Preprint (1991); see also talk by L. Oberauer in these proceedings.
4. Work at Brown University: S. R. Bandler et al, *PRL* **74**, 3169 (1997); J.S. Adams et al, *PL* **B341**, 431 (1995); S. R. Bandler et al, *J. Low Temp. Phys.* **93**, 785 (1993); S.R. Bandler et al, *NIM* **A370**, 578 (1996).
5. Work at College de France with Florida State and Wayne State Univ.: private communications P. Gorodetzky and T. Ypsilantis; F. Arzarello et al, LPC-94-28; J. Seguinot et al LPC-92-31; G. Bonvicini *NP (Proc.Supp.)* **438** (1994); C. Tao in 4th *Int'l Solar Neu-*

trino Conference, Editor: W. Hampel, Max-Planck-Institut fur Kernphysik Press (Heidelberg), p.238 (1997).
6. Y. Giomataris et al, *NIM* **A376**, 29 (1996) and private communication P. Gorodetzky.
7. Work at Lucent Technology's Bell Laboratories. R.S. Raghavan, *PRL* **78**, 3618 (1997); see also talk by R.S. Raghavan in this volume.
8. J.N. Bahcall, *Neutrino Astrophysics*, 372ff (and refs. therein),Cambridge University Press (1989).
9. Cryogenic work at INFN (Genoa) and lithium reactor work at INR (Troitsk): Private communications by S. Vitale and A.V. Kopylov; M. Galeazzi et al, *Phys. Lett.* **B398**, 187 (1997); A.V. Kopylov in 4th *Int'l Solar Neutrino Conference*, Editor: W. Hampel, Max-Planck-Institut fur Kernphysik Press (Heidelberg), p.263 (1997).
10. GNO is at Gran Sasso Laboratory (see talk by T. Kirsten in this volume and proposal available at http://kosmopc.mpi-hd.mpg.de/gallex/gallex.htm). Cryogenic work at Technical University of Munich: Private communication M. Altmann; also J. Hoene et al, in *Proceedings of Low Temperature Detector Workshop ("LTD7")*, Editor: S. Cooper, Max Planck Institute of Physics (Munich) p.184 (1997).
11. Work at INR (Troitsk) and Los Alamos. Private communication T.J. Bowles and V.N.Gavrin.

ELSEVIER

Nuclear Physics B (Proc. Suppl.) 77 (1999) 64–72

NUCLEAR PHYSICS B
PROCEEDINGS
SUPPLEMENTS

Standard Solar Models

John N. Bahcall[a]

[a]Institute for Advanced Study, Princeton, NJ 08540

I review recent developments that affect standard solar model predictions of solar neutrino fluxes.

1. INTRODUCTION

A lot of progress has been made in understanding the robustness of solar model predictions since Neutrino 96 [1]. In this talk, I will first summarize the new ingredients and then give the current best-estimates. Then I will discuss the uncertainties in the predictions.

Many of the results given here are adopted from the recent BP98 paper [2]. As we shall see in Section 3, there is excellent agreement between standard solar models calculated by different groups with different codes.

If you want to obtain the numerical data that are discussed in this talk, you can copy them from my Web site: http://www.sns.ias.edu/~jnb .

2. NEW INGREDIENTS

In this section, I will first summarize the new and relevant results on nuclear fusion reactions and on the screening of nuclear reactions and then summarize the situation with respect to neutrino cross sections. Finally, I will mention a few miscellaneous improvements that have been made since Neutrino 96.

2.1. Nuclear reaction cross sections

In January, 1997, the Institute for Nuclear Theory (INT) hosted a workshop devoted to determining the best estimates and the uncertainties in the most important solar fusion reactions. Thirty-nine experts in low energy nuclear experiments and theory, representing many different research groups and points of view, participated in the workshop and evaluated the existing experimental data and theoretical calculations. Their conclusions have been summarized in a detailed article authored jointly by the participants and published by the Reviews of Modern Physics [3]. In general outline, the conclusions of the INT workshop paper confirmed and strengthened previous standard analyses of nuclear fusion rates, although in a few important cases (for the ^{3}He$(\alpha,\gamma)^7$Be, ^{7}Be$(p,\gamma)^8$B, and ^{14}N$(p,\gamma)^{15}$O reactions) the estimated uncertainties were determined to be larger than previously believed.

The largest change from what was used in the results presented at Neutrino 96 is the lower ^{7}Be$(p,\gamma)^8$B cross section adopted by Adelberger et al. [3]. Previously, most authors constructing standard solar models used the Caltech (CIT) value for the ^{8}B production cross section [4]. The difference between the INT and the CIT estimates of the ^{8}B production cross section is due almost entirely to the decision by the INT group to base their estimate on only one (the best documented) of the six experiments analyzed by the CIT collaboration.

As we go alone, I will indicate how the principal predictions of solar models depend upon the assumed ^{8}B production cross section.

2.2. Screening of nuclear reactions

In one respect, the calculation of neutrino fluxes has simplified from Neutrino 96 to Neutrino 98. The rather complicated expressions in the literature for the screening of nuclear fusion reactions by electrons and ions have been replaced by a simple analytic expression that was originally derived by Salpeter [5] for the case of "weak screening" only.

Gruzinov and Bahcall [6] employed a mean field formalism to calculate the electron density of the screening cloud using the appropriate density matrix equation of quantum statistical mechanics. Because of well understood physical effects that

PII S0920-5632(99)00399-0

are included for the first time in this treatment, the calculated enhancement of reaction rates does not agree with the frequently used interpolation formulae. For the sun, screening effects cause only small uncertainties in the predicted neutrino fluxes if the appropriate Salpeter formula is used.

2.3. Neutrino cross sections and energy spectra

Improved cross sections for neutrino absorption by gallium and by chlorine are now available as well as somewhat more precise standard (undistorted) neutrino energy spectra. I summarize the results here.

A number of authors do not use the best available data for the neutrino cross sections and energy spectra and some authors even give event rates (or SNU values) but do not say which cross sections and energy spectra they use so that one cannot interpret their results precisely.

I have calculated [7] neutrino absorption cross sections for ^{71}Ga for all solar neutrino sources with standard energy spectra, and for laboratory sources of ^{51}Cr and ^{37}Ar; the calculations including, where appropriate, the thermal energy of fusing solar ions and use improved nuclear and atomic data. The ratio, R, of measured (in GALLEX and SAGE) to calculated ^{51}Cr capture rate is $R = 0.95 \pm 0.07$ (exp) $+\ ^{+0.04}_{-0.03}$ (theory) and was discussed extensively at Neutrino 98 by Gavrin and by Kirsten.

I also calculated cross sections for specific neutrino energies chosen so that a spline fit determines accurately the event rates in a gallium detector even if new physics changes the energy spectrum of solar neutrinos. In order to make possible more precise analyses of event rates for neutrino scenarios which change the shape of the neutrino energy spectra from individual neutrino sources, I evaluated and presented, for the first time, theoretical uncertainties for absorption cross sections at specific energies, as well as for the standard (undistorted) neutrino energy spectra. Also for use by people doing neutrino oscillation calculations, I calculated standard energy spectra for pp and CNO neutrino sources and presented the results in Appendices and on my Web site.

I note in passing that neutrino fluxes predicted by standard solar models, corrected for diffusion, have been in the range 120 SNU to 141 SNU since 1968 [7].

A group of us have recently redetermined the standard shape for the ^{8}B neutrino energy spectrum [8]. The available data all seem to be consistent with each other within rather small uncertainties, so we were able to determine not only a best-fit energy spectrum but also two extreme spectra that are different by what we estimate is effectively $\pm 3\sigma$. The uncertainties include estimates of the radiative and forbidden corrections. This improved spectrum yields a slightly different ^{8}B absorption cross section [8].

I am somewhat nervous about the standard (undistorted) ^{8}Be neutrino energy spectrum since it does depend upon rather old data [8]. It would be very good if the α-particle energy spectrum from ^{8}Be decay could be remeasured accurately in a new laboratory experiment.

The situation for neutrino-electron scattering is good; cross sections are available [9] that include electroweak radiative corrections.

2.4. Miscellaneous improvements

The standard solar model, BP98, that is discussed in this report includes somewhat improved radiative opacities calculated by the Livermore National Laboratory group, the so-called OPAL96 opacities [10], and the improved OPAL equation of state [11].

One improvement that I am rather proud of is new publicly available software that I have made available on my Web site (under Neutrino Software and Data) is a program to calculate solar neutrino rates and uncertainties. The problem of calculating the uncertainties in the predicted fluxes is somewhat complicated, especially since the uncertainties are asymmetric for some of the important input parameters. I decided to make this software publicly available (and therefore polished it significantly) so that people could see explicitly what uncertainties were included for each parameter and how the uncertainties were combined.

I have also polished my nuclear energy generation code. The changes made in this code, al-

Table 1
Standard Model Predictions (BP98): solar neutrino fluxes and neutrino capture rates, with 1σ uncertainties from all sources (combined quadratically).

Source	Flux (10^{10} cm^{-2}s^{-1})	Cl (SNU)	Ga (SNU)
pp	$5.94\,(1.00^{+0.01}_{-0.01})$	0.0	69.6
pep	$1.39\times10^{-2}\,(1.00^{+0.01}_{-0.01})$	0.2	2.8
hep	2.10×10^{-7}	0.0	0.0
^{7}Be	$4.80\times10^{-1}\,(1.00^{+0.09}_{-0.09})$	1.15	34.4
^{8}B	$5.15\times10^{-4}\,(1.00^{+0.19}_{-0.14})$	5.9	12.4
^{13}N	$6.05\times10^{-2}\,(1.00^{+0.19}_{-0.13})$	0.1	3.7
^{15}O	$5.32\times10^{-2}\,(1.00^{+0.22}_{-0.15})$	0.4	6.0
^{17}F	$6.33\times10^{-4}\,(1.00^{+0.12}_{-0.11})$	0.0	0.1
Total		$7.7^{+1.2}_{-1.0}$	129^{+8}_{-6}

though they probably took me altogether a few weeks of programming and debugging time, ended up not changing neutrino flux predictions by more than a percent.

3. BEST-ESTIMATE FLUXES AND EVENT RATES

Table 1 gives the neutrino fluxes and their uncertainties for our best standard solar model, hereafter BP98. As discussed in the previous section, the solar model makes use of the INT nuclear reaction rates, recent (1996) Livermore OPAL radiative opacities, the OPAL equation of state, and electron and ion screening as determined by the recent density matrix calculation.

Figure 1 displays the calculated ^{7}Be and ^{8}B neutrino fluxes for all 19 standard solar models with which we are familiar which have been published in the last 10 years in refereed science journals. The fluxes are normalized by dividing each published value by the flux from the BP98 solar model [2]; the abscissa is the normalized ^{8}B flux and the ordinate is the normalized ^{7}Be neutrino flux. The rectangular box shows the estimated 3σ uncertainties in the predictions of the BP98 solar model. The abbreviations, which indicate references to individual models, are identified in the caption of Figure 1.

All of the solar model results from different groups fall within the estimated 3σ uncertainties in the BP98 analysis (with the exception of the Dar-Shaviv model whose results have not been reproduced by other groups). This agreement demonstrates the robustness of the predictions since the calculations use different computer codes (which achieve varying degrees of precision) and involve a variety of choices for the nuclear parameters, the equation of state, the stellar radiative opacity, the initial heavy element abundances, and the physical processes that are included.

The largest contributions to the dispersion in values in Figure 1 are due to the choice of the normalization for S_{17} (the production cross-section factor for ^{8}B neutrinos) and the inclusion, or non-inclusion, of element diffusion in the stellar evolution codes. The effect in the plane of Fig. 1 of the normalization of S_{17} is shown by the difference between the point for BP98 (1.0,1.0), which was computed using the INT normalization, and the point at (1.18,1.0) which corresponds to the BP98 result with the CIT normalization.

Helioseismological observations have shown [1, 13] that diffusion is occurring and must be included in solar models, so that the most recent models shown in Fig. 1 now all include helium and heavy element diffusion. By comparing a large number of earlier models, it was shown that all published standard solar models give the same results for solar neutrino fluxes to an accuracy of better than 10% if the same input parameters and physical processes are included [14,15].

The theoretical predictions in Table 1 disagree with the observed neutrino event rates, which are, see Ref. [16] and the results presented at this conference by Lande, Gavrin, Kirsten, and Suzuki: 2.56 ± 0.23 SNU (chlorine), 72.2 ± 5.6 SNU (GALLEX and SAGE gallium experiments), and $(2.44 \pm 0.10) \times 10^6$cm^{-2}s^{-1} (^{8}B flux from SuperKamiokande).

Bahcall, Krastev, and Smirnov [17] have compared the observed rates with the calculated, standard model values, combining quadratically

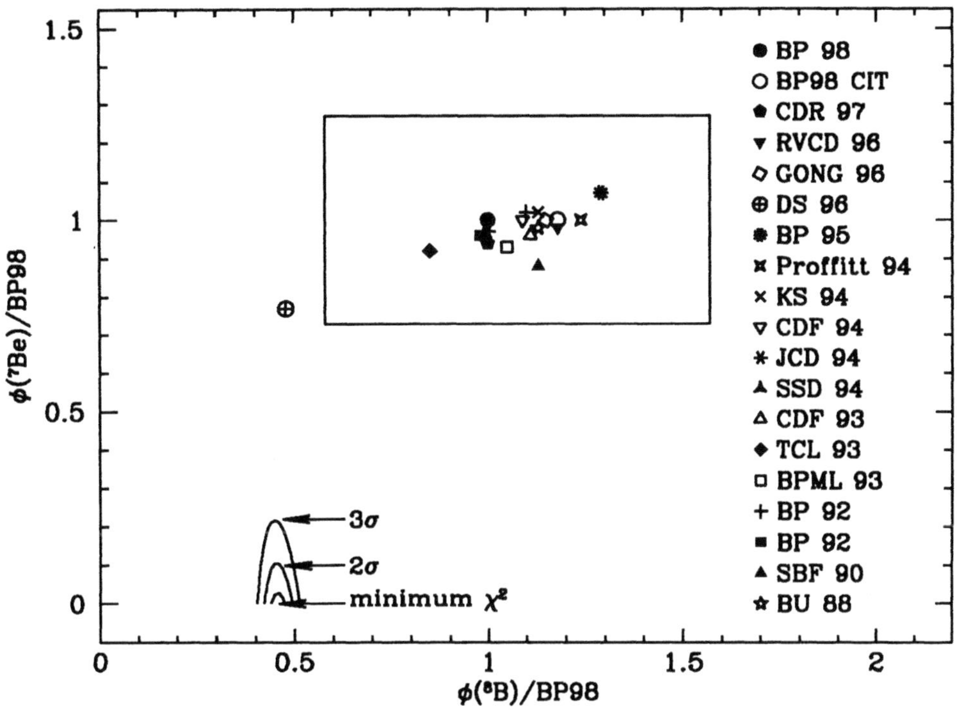

Figure 1. Predictions of standard solar models since 1988. The figure shows the predictions of 19 standard solar models in the plane defined by the ^{7}Be and ^{8}B neutrino fluxes. The abbreviations that are used in the figure to identify different solar models are defined in the bibliographical item, Ref. [12]. We include all standard solar models with which we are familiar that were published in refereed journals in the decade 1988-1998. All of the fluxes are normalized to the predictions of the Bahcall-Pinsonneault 98 solar model, BP98 [2]. The rectangular error box defines the 3σ error range of the BP98 fluxes. The best-fit ^{7}Be neutrino flux is negative. At the 99% C.L., there is no solution with all positive neutrino fluxes if the fluxes of CNO neutrinos are arbitrarily set equal to zero. There is no solution at the 99.9% C.L. if the CNO neutrinos are fixed at their standard solar model values. All of the standard model solutions lie far from the best-fit solution, even far from the 3σ contour.

the theoretical solar model and experimental uncertainties, as well as the uncertainties in the neutrino cross sections. Since the GALLEX and SAGE experiments measure the same quantity, we treat the weighted average rate in gallium as one experimental number. We adopt the SuperKamiokande measurement as the most precise direct determination of the higher-energy ^{8}B neutrino flux.

Using the predicted fluxes from the BP98 model, the χ^2 for the fit to the three experimental rates (chlorine, gallium, and SuperKamiokande) is

$$\chi^2_{\rm SSM}(3 \text{ experimental rates}) = 61 \ . \qquad (1)$$

The result given in Eq. (1), which is approximately equivalent to a 20σ discrepancy, is a quantitative expression of the fact that the standard model predictions do not fit the observed solar neutrino measurements.

The principal differences between the results shown in Table 1 and the results presented in our last systematic publication of calculated solar neutrino fluxes [15] is a 1.3σ decrease in the ^{8}B neutrino flux and 1.1σ decreases in the ^{37}Cl and ^{71}Ga capture rates. These decreases are due mainly to the lower ^{7}Be$(p,\gamma)^8$B cross section adopted by Adelberger et al. [3]. If we use, as in our recent previous publications, the Caltech (CIT) value for the ^{8}B production cross section [4], then the ^{8}B flux is $\phi\,(^8\mathrm{B,CIT}) = 6.1^{+1.1}_{-0.9} \times 10^6\ \mathrm{cm^{-2}s^{-1}}$, $\Sigma\,(\phi\sigma)_i\big|_{\mathrm{Cl,\,CIT}} = 8.8^{+1.4}_{-1.1}$ SNU, and $\Sigma\,(\phi\sigma)_i\big|_{\mathrm{Gallium}} = 131^{+9}_{-7}$ SNU, all of which are within ten percent of the Bahcall-Pinsonneault 1995 best-estimates.

4. UNCERTAINTIES

In this section, I will first discuss the formal uncertainties in the solar model flux calculations and then review the strong constraints that helioseismology places on perturbations of the standard solar model.

4.1. Uncertainties in the flux calculations

Table 2 summarizes the uncertainties in the most important solar neutrino fluxes and in the Cl and Ga event rates due to different nuclear fusion reactions (the first four entries), the heavy element to hydrogen mass ratio (Z/X), the radiative opacity, the solar luminosity, the assumed solar age, and the helium and heavy element diffusion coefficients. The ^{14}N $+ p$ reaction causes a 0.2% uncertainty in the predicted pp flux and a 0.1 SNU uncertainty in the Cl (Ga) event rates.

The predicted event rates for the chlorine and gallium experiments use recent improved calculations of neutrino absorption cross sections [7,8]. The uncertainty in the prediction for the gallium rate is dominated by uncertainties in the neutrino absorption cross sections, +6.7 SNU (7% of the predicted rate) and −3.8 SNU (3% of the predicted rate). The uncertainties in the chlorine absorption cross sections cause an error, ±0.2 SNU (3% of the predicted rate), that is relatively small compared to other uncertainties in predicting the rate for this experiment. For non-standard neutrino energy spectra that result from new neutrino physics, the uncertainties in the predictions for currently favored solutions (which reduce the contributions from the least well-determined ^{8}B neutrinos) will in general be less than the values quoted here for standard spectra and must be calculated using the appropriate cross section uncertainty for each neutrino energy [7,8].

The nuclear fusion uncertainties in Table 2 were taken from Adelberger et al. [3], the neutrino cross section uncertainties from [7,8], the heavy element uncertainty was taken from helioseismological measurements [19], the luminosity and age uncertainties were adopted from BP95 [15], the 1σ fractional uncertainty in the diffusion rate was taken to be 15% [20], which is supported by helioseismological evidence [13], and the opacity uncertainty was determined by comparing the results of fluxes computed using the older Los Alamos opacities with fluxes computed using the modern Livermore opacities [14]. To include the effects of asymmetric errors, the now public-available code for calculating rates and uncertainties (see discussion in previous section) was run with different input uncertainties and the results averaged. The software contains a description of how each of the uncertainties listed in Table 2 were determined and used.

The low energy cross section of the ^{7}Be $+ p$ reaction is the most important quantity that must be determined more accurately in order to decrease the error in the predicted event rates in solar neutrino experiments. The ^{8}B neutrino flux that is measured by the Kamiokande [16], Super-Kamiokande [21], and SNO [22] experiments is, in all standard solar model calculations, directly proportional to the ^{7}Be$+p$ cross section. If the 1σ uncertainty in this cross section can be reduced by a factor of two to 5%, then it will no longer be the limiting uncertainty in predicting the crucial ^{8}B neutrino flux (cf. Table 2).

4.2. How large an uncertainty does helioseismology suggest?

Could the solar model calculations be wrong by enough to explain the discrepancies between predictions and measurements for solar neutrino experiments? Helioseismology, which confirms predictions of the standard solar model to high precision, suggests that the answer is probably "No."

Figure 2 shows the fractional differences between the most accurate available sound speeds measured by helioseismology [23] and sound

Table 2
Average uncertainties in neutrino fluxes and event rates due to different input data. The flux uncertainties are expressed in fractions of the total flux and the event rate uncertainties are expressed in SNU. The ^{7}Be electron capture rate causes an uncertainty of $\pm 2\%$ [18] that affects only the ^{7}Be neutrino flux. The average fractional uncertainties for individual parameters are shown.

<Fractional uncertainty>	pp	^{3}He^{3}He	^{3}He^{4}He	$^{7}\mathrm{Be}+p$	Z/X	opac	lum	age	diffuse
	0.017	0.060	0.094	0.106	0.033		0.004	0.004	
Flux									
pp	0.002	0.002	0.005	0.000	0.002	0.003	0.003	0.0	0.003
^{7}Be	0.0155	0.023	0.080	0.000	0.019	0.028	0.014	0.003	0.018
^{8}B	0.040	0.021	0.075	0.105	0.042	0.052	0.028	0.006	0.040
SNUs									
Cl	0.3	0.2	0.5	0.6	0.3	0.4	0.2	0.04	0.3
Ga	1.3	0.9	3.3	1.3	1.6	1.8	1.3	0.20	1.5

speeds calculated with our best solar model (with no free parameters). The horizontal line corresponds to the hypothetical case in which the model predictions exactly match the observed values. The rms fractional difference between the calculated and the measured sound speeds is 1.1×10^{-3} for the entire region over which the sound speeds are measured, $0.05R_{\odot} < R < 0.95R_{\odot}$. In the solar core, $0.05R_{\odot} < R < 0.25R_{\odot}$ (in which about 95% of the solar energy and neutrino flux is produced in a standard model), the rms fractional difference between measured and calculated sound speeds is 0.7×10^{-3}.

Helioseismological measurements also determine two other parameters that help characterize the outer part of the sun (far from the inner region in which neutrinos are produced): the depth of the solar convective zone (CZ), the region in the outer part of the sun that is fully convective, and the present-day surface abundance by mass of helium ($Y_{\rm surf}$). The measured values, $R_{\rm CZ} = (0.713 \pm 0.001)R_{\odot}$ [24], and $Y_{\rm surf} = 0.249 \pm 0.003$ [19], are in satisfactory agreement with the values predicted by the solar model BP98, namely, $R_{\rm CZ} = 0.714R_{\odot}$, and $Y_{\rm surf} = 0.243$. However, we shall see below that precision measurements of the sound speed near the transition between the radiative interior (in which energy is transported by radiation) and the outer convective zone (in which energy is transported by convection) reveal small discrepancies between the model predictions and the observations in this region.

If solar physics were responsible for the solar neutrino problems, how large would one expect the discrepancies to be between solar model predictions and helioseismological observations? The characteristic size of the discrepancies can be estimated using the results of the neutrino experiments and scaling laws for neutrino fluxes and sound speeds.

All recently published solar models predict essentially the same fluxes from the fundamental pp and pep reactions (amounting to 72.4 SNU in gallium experiments, cf. Table 1), which are closely related to the solar luminosity. Comparing the measured gallium rates (reported at Neutrino 98) and the standard predicted rate for the gallium experiments, the ^{7}Be flux must be reduced by a factor N if the disagreement is not to exceed n standard deviations, where N and n satisfy $72.4 + (34.4)/N = 72.2 + n\sigma$. For a 1σ (3σ) disagreement, $N = 6.1(2.05)$. Sound speeds scale like the square root of the local temperature divided by the mean molecular weight and the ^{7}Be neutrino flux scales approximately as the 10th power of the temperature [25]. Assuming that the temperature changes are dominant, agreement to within 1σ would require fractional changes of or-

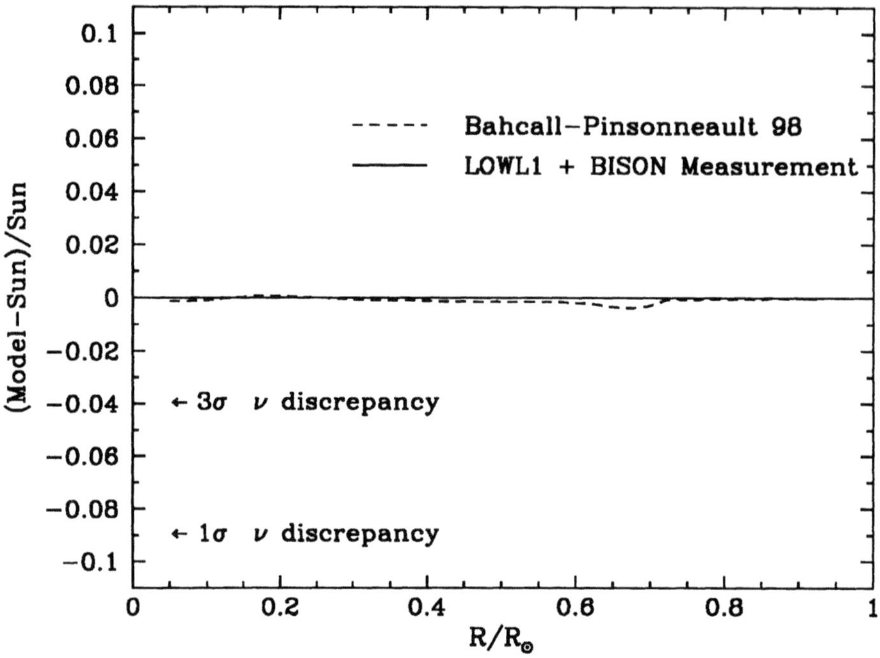

Figure 2. Predicted versus Measured Sound Speeds. This figure shows the excellent agreement between the calculated (solar model BP98, Model) and the measured (Sun) sound speeds, a fractional difference of 0.001 rms for all speeds measured between $0.05R_{\odot}$ and $0.95R_{\odot}$. The vertical scale is chosen so as to emphasize that the fractional error is much smaller than generic changes in the model, 0.03 to 0.08, that might significantly affect the solar neutrino predictions.

der 0.09 in sound speeds (3σ could be reached with 0.04 changes), if all model changes were in the temperature[1]. This argument is conservative because it ignores the contributions from the ^{8}B and CNO neutrinos which contribute to the observed counting rate (cf. Table 1) and which, if included, would require an even larger reduction of the ^{7}Be flux.

I have chosen the vertical scale in Fig. 1 to be appropriate for fractional differences between measured and predicted sound speeds that are of order 0.04 to 0.09 and that might therefore affect solar neutrino calculations. Fig. 1 shows that the characteristic agreement between solar model predictions and helioseismological measurements is more than a factor of 30 better than would be expected if there were a solar model explanation of the solar neutrino problems.

[1] I have used in this calculation the GALLEX and SAGE measured rates reported by Kirsten and Gavrin at Neutrino 98. The experimental rates used in BP98 were not as precise and therefore resulted in slightly less stringent constraints than those imposed here. In BP98, we found that agreement to within 1σ with the then available experimental numbers would require fractional changes of order 0.08 in sound speeds (3σ could be reached with 0.03 changes.)

5. DISCUSSION AND CONCLUSION

Three decades of refining the input data and the solar model calculations has led to a predicted standard model event rate for the chlorine experiment, 7.7 SNU, which is very close to the best-estimate value obtained in 1968 [26], which was 7.5 SNU. The situation regarding solar neutrinos is, however, completely different now, thirty years later. Four experiments have confirmed the detection of solar neutrinos. Helioseismological measurements show (cf. Fig. 1) that hypothetical deviations from the standard solar model that seem to be required by simple scaling laws to fit just the gallium solar neutrino results are at least a factor of 40 larger than the rms disagreement between the standard solar model predictions and the helioseismological observations. This conclusion does not make use of the strong evidence which points in the same direction from the chlorine, Kamiokande, and SuperKamiokande experiments.

The improvement in helioseismological measurements over the past two years, from Neutrino 96 to Neutrino 98 (cf. Figure 2 of the Neutrino 96 talk [1] with Figure 2 of this talk), has resulted in a five-fold improvement in the agreement with the calculated standard solar model sound speeds and the measured solar velocities! I believe that this improved agreement is yet another reason to believe that standard solar models reliably predict solar neutrino fluxes.

I am grateful to Y. Suzuki for special efforts that made possible my attendance at Neutrino 98 and which enriched the scientific experience of this extraordinarily exciting conference. I acknowledge support from NSF grant #PHY95-13835.

REFERENCES

1. J.N. Bahcall and M.H. Pinsonneault, in Neutrino '96, Proceedings of the 17th International Conference on Neutrino Physics and Astrophysics (Helsinki) K. Enqvist, K. Huitu and J. Maalampi (eds.) (World Scientific, Singapore, 1997), p. 56.
2. J. N. Bahcall, S. Basu and M. H. Pinsonneault, Phys. Lett. B 433 (1998) 1.
3. E. Adelberger et al., Rev. Mod. Phys. (accepted, Oct. 1998) astro-ph/9805121.
4. C.W. Johnson, E. Kolbe, S.E. Koonin and K. Langanke, Astrophys. J. 392 (1992) 320.
5. E.E. Salpeter, , Australian J. Phys. 7 (1954) 373.
6. A.V. Gruzinov and J.N. Bahcall, Astrophys. J. 504 (1998, to be published).
7. J.N. Bahcall, Phys. Rev. C 56 (1997) 3391.
8. J.N. Bahcall, E. Lisi, D.E. Alburger, L. De Braeckeleer, S.J. Freedman and J. Napolitano, Phys. Rev. C 54 (1996) 411.
9. J.N. Bahcall, M. Kamionkowski and A. Sirlin, Phys. Rev. D 51 (1995) 6146.
10. C.A. Iglesias and F.J. Rogers, Astrophys. J. 464 (1996) 943; D.R. Alexander and J.W. Ferguson, Astrophys. J. 437 (1994) 879. These references describe the different versions of the OPAL opacities.
11. F.J. Rogers, F.J. Swenson and C.A. Iglesias, Astrophys. J. 456 (1996) 902.
12. (GONG) J. Christensen-Dalsgaard et al., GONG Collaboration, Science 272 (1996) 1286; (BP95) J. N. Bahcall and M. H. Pinsonneault, Rev. Mod. Phys. 67 (1995) 781; (KS94) A. Kovetz and G. Shaviv, Astrophys. J. 426 (1994) 787; (CDF94) V. Castellani, S. Degl'Innocenti, G. Fiorentini, L.M. Lissia and B. Ricci, Phys. Lett. B 324 (1994) 425; (JCD94) J. Christensen-Dalsgaard, Europhys. News 25 (1994) 71; (SSD94) X. Shi, D.N. Schramm and D.S.P. Dearborn, Phys. Rev. D 50 (1994) 2414; (DS96) A. Dar and G. Shaviv, Astrophys. J. 468 (1996) 933; (CDF93) V. Castellani, S. Degl'Innocenti and G. Fiorentini, Astron. Astrophys. 271 (1993) 601; (TCL93) S. Turck-Chièze and I. Lopes, Astrophys. J. 408 (1993) 347; (BPML93) G. Berthomieu, J. Provost, P. Morel and Y. Lebreton, Astron. Astrophys. 268 (1993) 775; (BP92) J.N. Bahcall and M.H. Pinsonneault, Rev. Mod. Phys. 64 (1992) 885; (SBF90) I.-J. Sackman, A.I. Boothroyd and W.A. Fowler, Astrophys. J. 360 (1990) 727; (BU88) J.N. Bahcall and R.K. Ulrich, Rev. Mod. Phys. 60 (1988) 297; (RVCD96) O. Richard, S. Vauclair, C. Charbonnel and W.A. Dziembowski, Astron. Astrophys. 312 (1996) 1000; (CDR97) F. Ciacio, S. Degl'Innocenti and B. Ricci, Astron. Astrophys. Suppl. Ser. 123 (1997) 449.
13. J.N. Bahcall, M.H. Pinsonneault, S. Basu and J. Christensen-Dalsgaard, Phys. Rev. Lett. 78 (1997) 171.
14. J.N. Bahcall and M.H. Pinsonneault, Rev. Mod. Phys. 64 (1992) 885.
15. J.N. Bahcall and M.H. Pinsonneault, Rev. Mod. Phys. 67 (1995) 781.

16. R. Davis, Jr., Prog. Part. Nucl. Phys. 32 (1994) 13; B.T. Cleveland, T. Daily, R. Davis, Jr., J.R. Distel, K. Lande, C.K. Lee, P.S. Wildenhain and J. Ullman, Astrophys. J. 496 (1998) 505; GALLEX Collaboration, P. Anselmann et al., Phys. Lett. B 342 (1995) 440; GALLEX Collaboration, W. Hampel et al., Phys. Lett. B 388 (1996) 364; SAGE Collaboration, V. Gavrin et al., in Neutrino '96, Proceedings of the 17th International Conference on Neutrino Physics and Astrophysics (Helsinki) K. Enqvist, K. Huitu and J. Maalampi (eds.) (World Scientific, Singapore, 1997), p. 14; KAMIOKANDE Collaboration, Y. Fukuda et al., Phys. Rev. Lett. 77 (1996) 1683.
17. J.N. Bahcall, P.I. Krastev and A.Yu. Smirnov, submitted to Phys. Rev. D (1998).
18. A.V. Gruzinov and J.N. Bahcall, Astrophys. J. 490 (1997) 437.
19. S. Basu and H.M. Antia, Mon. Not. R. Astron. Soc. 287 (1997) 189.
20. A.A. Thoul, J.N. Bahcall and A. Loeb, Astrophys. J. 421 (1994) 828.
21. Y. Totsuka, in the Proceedings of the 18th Texas Symposium on Relativistic Astrophysics and Cosmology, December 15–20, 1996, Chicago, Illinois, A. Olinto, J. Frieman and D. Schramm (eds.) (World Scientific, Singapore, 1998) p. 114.
22. A.B. McDonald, in the Proceedings of the 9th Lake Louise Winter Institute, A. Astbury et al. (eds.) (World Scientific, Singapore, 1994), p. 1.
23. S. Basu et al., Mon. Not. R. Astron. Soc. 292 (1997) 234.
24. S. Basu and H.M. Antia, Mon. Not. R. Astron. Soc. 276 (1995) 1402.
25. J.N. Bahcall and A. Ulmer, Phys. Rev. D 53 (1996) 4202.
26. J.N. Bahcall, N.A. Bahcall and G. Shaviv, Phys. Rev. Lett. 20 (1968) 1209.

ELSEVIER

Nuclear Physics B (Proc. Suppl.) 77 (1999) 73–80

NUCLEAR PHYSICS B
PROCEEDINGS
SUPPLEMENTS

Uncertainties in the Solar Neutrino Flux

W.C. Haxton[a]

[a]Institute for Nuclear Theory, Box 351550, and Department of Physics, Box 351560
University of Washington, Seattle, WA 98195, USA

I discuss three issues relevant to solar neutrino flux measurements: cross section uncertainties in pp chain reactions, uncertainties in the GALLEX/SAGE response to ^{7}Be and ^{51}Cr neutrinos, and the implications of helioseismology for nonstandard suns with mixed cores. A few comments are also offered on $\nu_e \leftrightarrow \nu_\tau$ oscillations, cosmologically interesting neutrino masses, and recent proposals for supernova neutrino observatories.

1. INTRODUCTION

It is a pleasure to be present for this historic meeting hosted by the Superkamiokande collaboration. In this talk I will address three issues affecting the solar neutrino flux and one connected with future detectors for supernova neutrinos.

1.1. Nuclear Physics of the pp Chain

One of the crucial inputs into the solar model is the network of nuclear reactions comprising the pp chain (and CNO cycle). This network involves nonresonant charged particle reactions occurring at center-of-mass energies well below the height of the Coulomb barrier. As the solar core temperature $T_c \sim 1.5 \cdot 10^7 K$, the typical kinetic energy for a nucleus in the core is $< E > \sim 2$ keV. The competition between the Coulomb barrier and Boltzman distribution leads to a typical energy for reacting nuclei of $\langle E_{\rm reacting} \rangle \sim 10$ keV, a value that is generally lower than that where such reactions can be measured in the laboratory. Thus the task for nuclear physicists is to measure such reactions as accurately as possible over the accessible range of laboratory energies, then extrapolate these measurements to the energies relevant for the sun.

In the case of the driving reaction of the pp chain

$$p + p \rightarrow {}^2{\rm H} + {\rm e}^+ + \nu_e \; , \qquad (1)$$

the cross section is not measurable in the laboratory. Thus we must rely on theory. Fortunately deuterium is the simplest nucleus, and its properties are very well reproduced by NN potentials (Bonn, Paris, Argonne v18, etc.) carefully fit to phase shifts. The calculated cross section depends on the accuracy with which the axial vector coupling g_A is known and on two-body corrections to the space-like component of the axial current, which are fortunately of order $({\rm v/c})^2 \sim 1$ %, where v is a typical bound nucleon velocity.

The other major reactions of the pp chain are measureable, but generally not at the low energies relevant to our sun. The necessary extrapolation of the cross section $\sigma(E)$ to lower energies is accomplished via the S factor

$$\sigma(E) \equiv \frac{S(E)}{E} e^{-2\pi Z_1 Z_2 \alpha/\beta} \qquad (2)$$

where E is the center-of-mass energy, Z_1 and Z_2 are the charges of the interacting nuclei, and β is the relative velocity. The introduction of $S(E)$ removes the s-wave Coulomb interaction of point particles and thus provides a much smoother quantity for use in extrapolating data. $S(E)$ depends on a number of physical effects - nuclear finite size, atomic screening corrections, higher particle waves, etc. - that the theorist must evaluate before this extrapolation can be done.

Because the solar neutrino problem is at a crucial juncture, a group of about 40 experts recently met at the Institute for Nuclear Theory, Seattle, to discuss the nuclear physics of the pp chain and CNO cycle. The questions addressed included the best current values for cross sections, critiques of assigned uncertainties, and recommendations for future experimental and theoretical work that could further improve our understanding of the

PII S0920-5632(99)00400-4

nuclear physics. The summary of this workshop will appear in Reviews of Modern Physics (October, 1998) and is also available on the LANL preprint archive [1].

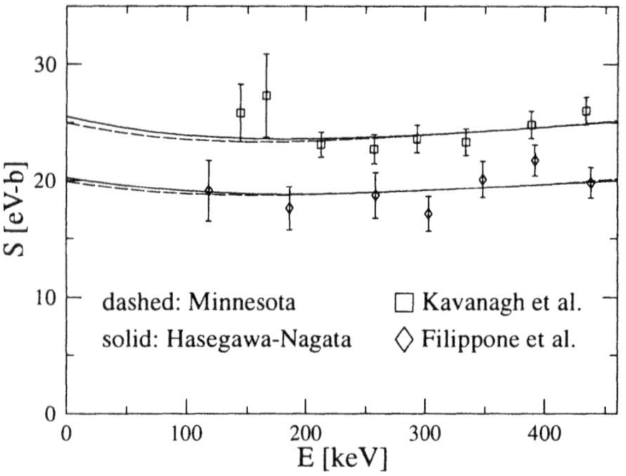

Figure 1. The $^7Be(p,\gamma)^8B$ S-factor as measured by Filippone et al. [3] and by Kavanagh et al. [4]. For each data set two theoretical extrapolations to S(0), reflecting different choices for the strong potential, are shown [2].

While I cannot give an adequate summary here, I will mention two of the reactions where significant changes were recommended. The first of these is $^7Be(p,\gamma)^8B$, where the standard $S_{17}(0)\sim$ 22.4 eVb is that given by Johnson et al [2]. Measurements of $S_{17}(E)$ are complicated by the need to use radioactive targets and thus to determine the areal density of the ^7Be target nuclei. Two techniques have been employed, measuring the rate of 478 keV photons from ^7Be decay or counting the daughter ^7Li nuclei via the reaction ^7Li (d,p)^8Li. The low-energy data sets [3,4] for $S_{17}(E)$ disagree by 25%, a systematic effect apparent in Fig. 1. Each data set is consistent with theory in its dependence on E: this dependence is simple in the illustrated low energy region as it is determined by the asymptotic nuclear wave function.

The Seattle working group on $S_{17}(E)$ found that only one low-energy data set, that of Filippone et al. [3], was described in the published literature in sufficient detail to be evaluated. The target activity in that experiment had been measured by both 478 keV gamma rays and by the (d,p) reaction, with consistent results. The resulting recommended value was thus based on this measurement, yielding

$$S_{17}(0) = 19^{+4}_{-2} eV\ b\ ,\ 1\sigma\ . \qquad (3)$$

Since the workshop, two developments have occurred. The Orsay/Bordeaux/Paris-Sud/USTHB group published [5] a new measurement of S_{17} (0) = 18.5 ± 1.0 eV b, while a preliminary value from the Weizmann/Troitzk/Mainz/Isolde group of $S_{17}(E_p = 1.2$ MeV) = 22.5 ± 2.5 eV b has been announced [6].

The $^3He(\alpha,\gamma)^7Be$ reaction has been measured by two techniques, by counting the capture γ rays and by detecting the resulting ^7Be activity. While the two techniques have been used by several groups and have yielded separately consistent results, the capture γ ray value $S_{17}(0) = 0.507 \pm 0.016$ keV b is not in good agreement with the ^7Be activity value 0.572 ± 0.026 keV-b. The Seattle working group concluded that the evidence for a systematic discrepancy of unknown origin was reasonably strong and recommended that standard procedures be used in assigning a suitably expanded error. The recommended value S_{34} (0) is 0.53 ± 0.05.

These and other recommended values were recently incorporated into the Bahcall and Pinnsoneault (BP98) solar model calculation [7]. While the workshop's recommended values involve no qualitative changes, there is some broadening of error bars and a downward shift in $S_{17}(0)$, leading to the lower BP98 ^8B flux. The workshop's Reviews of Modern Physics article summarizes a substantial amount of work on topics not discussed here: screening effects, weak radiative corrections to and exchange current effects on p+p, the atomic physics of $^7Be + e^-$, etc. Much of this discussion was useful in evaluating possible uncertainties in solar microphysics, and in identifying opportunities for reducing those uncertainities.

1.2. The Nuclear Physics of the GALLEX/SAGE ^{51}Cr Calibrations

The ^{51}Cr neutrino source experiments provide an important check on the overall gallium detector operations under few atom, hot chemistry conditions. The issue discussed here, and which was mentioned in the earlier experimental talks, is the potential complication due to contributions of uncertain strength to the $5/2^-$ and $3/2^-$ ^{71}Ge excited states (see Fig. 2).

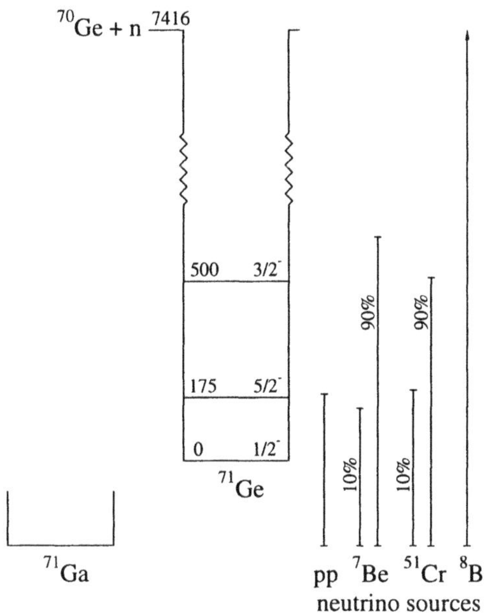

Figure 2. Level scheme for ^{71}Ge showing the excited states that contribute to absorption of pp, ^{7}Be, ^{51}Cr, and ^{8}B neutrinos.

The results of the source experiments can be normalized to the known ^{71}Ge ground state contribution [8], yielding

$$R_o = E\left[1 + 0.67\frac{\mathrm{BGT}(5/2^-)}{\mathrm{BGT(gs)}} + 0.22\frac{\mathrm{BGT}(3/2^-)}{\mathrm{BGT(gs)}}\right]$$

$$= \begin{cases} 0.98 \pm 0.08, & \text{GALLEX [9]} \\ 1.00 \pm 0.13, & \text{SAGE [10]} \end{cases} \tag{4}$$

where E represents any departure of the efficiency from the value determined from tracers and used by the experimentalists in their analyses. The dependence of the results on the unknown BGT values is explicit. Clearly the conclusion $E \sim 1$ requires an independent determination that the unknown BGT values are much smaller than the ground state value BGT(gs).

It had been assumed that forward-angle (p,n) charge exchange measurements determine the unknown BGT values

$$\frac{\mathrm{BGT}(5/2^-)}{\mathrm{BGT(gs)}} \lesssim 0.06$$

$$\frac{\mathrm{BGT}(3/2^-)}{\mathrm{BGT(gs)}} = 0.12 \pm 0.02 \tag{5}$$

But extensive investigations [11,12] of the proportionality between (p,n) cross sections and known weak interaction BGT values have shown that the relationship is a complicated one. Existing discrepancies can be removed by the assumption that forward-angle (p,n) effective operator contains a spin-tensor contribution of relative strength $\delta \sim 0.1$, in addition to the Gamow-Teller operator,

$$\hat{O}_{(p,n)} = \sigma(i)\tau_+(i) +$$

$$\delta\sqrt{8\pi}[Y_2(\Omega_i) \otimes \sigma(i)]_{J=1}\tau_+(i). \tag{6}$$

Simple considerations of the nuclear structure of ^{71}Ga and ^{71}Ge suggest that the tensor operator might be particularly troublesome for the ^{71}Ga$(3/2^-) \rightarrow ^{71}Ge(5/2^-)$ transition to the first excited state. The naive description of this transition is

$$1f_{5/2}(n) \rightarrow 2p_{3/2}(p), \tag{7}$$

an ℓ-forbidden amplitude that generates an enormous spin-tensor and vanishing Gamow-Teller contributions. Well-known transitions of a similar character, e.g. to the first excited state in ^{39}K(p,n) ^{39}Ca, have produced discrepancies between (p,n) and weak interaction transition probabilities of factors of ~100. The conclusion [8] is that it might be unwise to use the ^{71}Ga (p,n) results as a reliable independent measurement of BGT $(5/2^-)$.

To explore this further, I did a large-basis shell model calculation [8] of the ^{71}Ga $\rightarrow ^{71}$Ge weak and (p,n) transitions. The results agree reasonably with what is known experimentally: the calculated BGT (g.s.) = 0.051, compared to the experimental value 0.087, while the calculated (p,n) BGT ($5/2^-$), corresponding to the operator in Eq. (6), is 0.0006, in agreement with the experimental bound <0.005. However the latter result stemmed from a cancellation between the GT and spin-tensor operators comprising $O_{(p,n)}$,

$$\langle 5/2^- || O_{p,n} || 3/2^- \rangle = 0.264 - \delta 2.23. \quad (8)$$

The cancellation between the second term - an enormous spin-tensor amplitude - and the GT amplitude leads to a much larger beta decay BGT($5/2^-$) than would be allowed by in a more naive interpretation of the (p,n) BGT value. Thus this is an explicit demonstration that previous bounds on BGT ($5/2^-$) are too aggressive.

It is argued in Ref. [8] that, by relying on the shell model calculation of the very strong tensor amplitude in Eq. (8), a reasonable range of beta decay BGT ($5/2^-$) can be extracted from the (p,n) measurements. This results in a corresponding change in the ^{51}Cr cross section from the previous standard value

$$\sigma(^{51}\mathrm{Cr}) = (5.81^{+0.21}_{-0.16}) \cdot 10^{-45}\mathrm{cm}^2 \quad (9a)$$

to

$$\sigma(^{51}\mathrm{Cr}) = (6.39 \pm 0.68) \cdot 10^{-45}\mathrm{cm}^2 \quad (9b)$$

where the error in Eq. (9b) represents that due to excited state uncertainties only. If this value is used in Eq. (4), one finds

$$E = \left\{ \begin{array}{ll} 0.86 \pm 0.07 \pm 0.09, & \mathrm{GALLEX} \\ 0.875 \pm 0.11 \pm 0.09, & \mathrm{SAGE} \end{array} \right. \quad (10)$$

where the first uncertainty in the source experiment error while the second corresponds to the ^{51}Cr cross section. Note that E $\sim$ 1 is allowed, though it is certainly not demanded. It is important to note that the difference between (9a) and (9b) is one of an extended error range: all of the range in (9a) that is attributable to excited state uncertainties is allowed in (9b). It is also notable that the cross section uncertainty in Eq. (10) is comparable to the source experiment uncertainty.

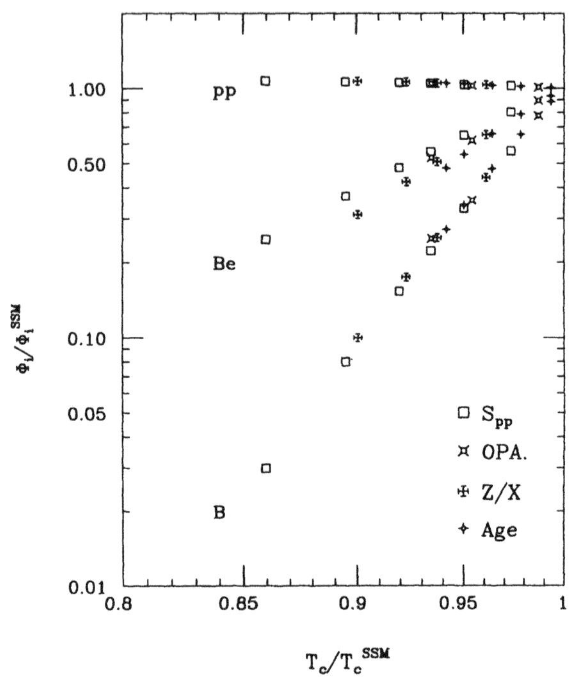

Figure 3. The response of the pp, Be, and B neutrino fluxes to the indicated variations in solar model input parameters, displayed as a function of the resulting central temperature T_c. (From Castellani et al. [13].)

The conclusion is that the source experiments have become ^{7}Be neutrino cross section measurements. Indeed, one can express the GALLEX/SAGE responses to the ^{7}Be neutrinos in terms of R_o, independent of almost all nuclear physics uncertainties. This means of course that other tests of E$\sim$ 1 under few-atom, hot chemistry conditions take on added importance. Thus the GALLEX ^{71}As test discussed by Prof. Kirsten at this meeting is crucial. The GALLEX/SAGE results are central to the conclusions of global analyses of the solar neutrino experiments that yield $\phi(^7\mathrm{Be}) \lesssim 0$.

1.3. Solar Core Mixing of ^{3}He and Helioseismology

The crux of the solar neutrino problem can be captured in two experimental quantities. First the ^{8}B neutrino flux $\phi(^8\mathrm{B})$, which varies approximately as T_c^{18} where T_c is the solar core temperature, is known to be reduced by about a factor of 1/3 (0.47 using the new BP98 results). Naively this result requires a cooler sun,

$$T_c \sim 0.96 T_c^{\mathrm{SSM}},$$

where the superscript SSM denotes the standard solar model result. However the flux ratio $\phi(^7\mathrm{Be})/\phi(^8\mathrm{B})$, which varies as T_c^{-10}, also appears to be reduced relative to the standard model. This then requires

$$T_c > T_c^{\mathrm{SSM}}$$

with the extent of the increase depending on how strongly one wants to suppress this ratio. ($\phi(^7\mathrm{Be}) \sim 0$ provides the best fit to the ^{37}Cl, GALLEX/SAGE, and Superkamiokande results.) It appears that the experimental results on $\phi(^8\mathrm{B})$ and $\phi(^7\mathrm{Be})/\phi(^8\mathrm{B})$ are thus in conflict, with the first requiring a cooler sun and the second a hotter one.

These arguments depend on the assumption that neutrino fluxes will track T_c as described above, but this appears to hold remarkably well. Figure 3, from Castellani et al. [13], illustrates this: changes due to modified nuclear cross sections, opacity, lowered metallicity, and the solar age produce neutrino fluxes that track the resulting T_c quite accurately. This led some in the field to argue that no nonstandard solar model could produce the observed pattern of neutrino fluxes.

Andrew Cumming and I decided to test this claim phenomenologically, under the assumption of a steady-state sun with conventional microphysics producing the correct luminosity. As our procedures are described elsewhere [14], I'll just state here our basic result. There appears to be only one possibility for constructing a steady state model with a neutrino flux pattern reasonably close to experiment: the solar core must mix on timescales of ^{3}He equilibration ($\sim$ few 10^6 years) in the "elevator convection" pattern illustrated in Fig. 4. This mixing produces

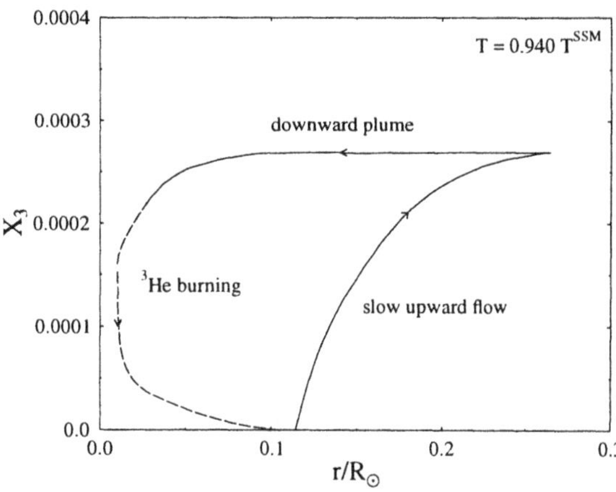

Figure 4. A phenomenologically derived core convection pattern that will suppress both $\phi(^8\mathrm{B})$ and $\phi(^7\mathrm{Be})/\phi(^8\mathrm{B})$ [14]. The downward flow is in plumes, rapid and localized, requiring $\sim$ few $\cdot\ 10^6$ years. This leads to out-of-equilibrium burning of ^{3}He at small r. The slow, broad, upward flow allows the cycle to replenish the ^{3}He. Typical upward times are $\sim$ few $\cdot\ 10^7$ years.

the desired flux pattern because it modifies the ppI/(ppII + ppIII) and ppII/ppIII branching ratios in the proper way. The former is enhanced because the resulting ^{3}He enrichment of the core favors the ^{3}He + ^{3}He reaction. The latter is reduced because the fraction of ^{3}He that burns by ^{3}He + ^{4}He produces ^{7}Be deep in the core, where the higher temperatures favor ppIII over ppII.

This exercise indicated that arguments against nonstandard solar model solutions based on how neutrino fluxes scale with T_c are not completely general. While the pattern of core mixing was derived phenomenologically, and not on physical grounds, it nevertheless has some physical appeal. The possibility of core mixing generated by the standard solar model overstability in the ^{3}He gradient was first discussed by Dilke and Gough. Roxburgh discussed a persistent convective core as a possible consequence of the growth of the SSM ^{3}He gradient during our sun's early convective stage.

Several astrophysical consequences of such mixing were discussed, with one, helioseismology, appearing problematic. Bahcall et al. [15] and Fiorentini et al. [16] evaluated the helioseismological consequences of replacing the SSM core molecular weight profile by a constant one, such as would occur for a continuously mixed model. This yielded 8% deviations in core sound speeds, far outside allowable bounds. My summary of this work is that it might be viewed as an attempt to estimate the natural scale of expected helioseismology changes. However, it does not convincingly settle the issue because such a modification of the SSM produces a "model" that fails to satisfy the equations of stellar evolution. One can envision that it might be more difficult to change the sound speed profile c(r) in a dynamically consistent model where the pressure and density profiles are coupled through the condition of hydrostatic equilibrium.

A stronger argument, which arose from discussions at last December's ITP conference on solar neutrinos, is that existing helioseismology determinations of c(r) coupled with a constant molecular weight profile in the core would necessarily lead to an unphysical temperature profile T(r) that would increase away from r=0. It appears this conclusion is overstated: adiabatically mixed models lacking molecular weight gradients in the core generically have profiles in T(r) that are convex downward. [While this counterargument to the ITP discussions was originally made to me by Richard Epstein, similar remarks have been made by John Bahcall and by Doug Gough at this conference.]

My view is that such mixed models, while quite unlikely, are not yet definitively ruled out. The issue of their viability is an important one, given the argument that a core mixed on the ^{3}He equilibration timescale is the only possibility for producing acceptable neutrino fluxes in a steady state model. Plans are underway at Los Alamos to evolve a series of 1D models where mixing is included through mixing length theory, adjusting these in the usual way to produce the proper luminosity after 4.6 b.y. of the stellar burning. Helioseismology studies performed on these models should then settle the issue.

1.4. Supernova $\nu_e \leftrightarrow \nu_\tau$ Oscillations and Cosmologically Interesting Neutrino Masses

In a Type II supernova 99% of the energy released by the core collapse is carried off by neutrinos. The initial flavor equilibrium of neutrinos trapped within the core at densities $\rho \gtrsim 10^{12}$ g/cm^3, coupled with the flavor-dependent decoupling of neutrinos from the matter at the neutrinosphere, leads to an approximate equipartition of energy among the flavors and to a characteristic hierarchy of temperatures. The average energy of heavy flavor neutrinos $\langle E_{\nu_{\mathrm{HEAVY}}} \rangle \sim$ 25 MeV, while $\langle E_{\bar{\nu}_e} \rangle \sim$ 16 Mev and $\langle E_{\nu_e} \rangle \sim$ 11 MeV. The lower values for the ν_es and $\bar{\nu}_e$s reflects their stronger matter couplings due to charged current reactions with nucleons and to their greater scattering cross sections off electrons. The lower ν_e temperature, relative to $\bar{\nu}_e$, is due to the neutron richness of the matter near the neutrinosphere and resulting enhancement of $\nu_e + n \to p + e^-$. The neutrino energy hierarchy $\langle E_{\nu_{\mathrm{HEAVY}}} \rangle > \langle E_{\bar{\nu}_e} \rangle > \langle E_{\nu_e} \rangle$ appears to be a result independent of the details of supernova modelling, in contrast to the case of solar neutrinos where fluxes depend on nuclear reaction networks.

The spectrum of neutrinos is essentially fixed at the neutrinosphere, $\rho \sim 10^{12}$ g/cm^3, a density that corresponds to a neutrino oscillation level crossing for $\delta m^2 \sim 10^4$ eV^2. Furthermore, the density scale height at this density will produce an adiabatic neutrino level crossing for $\sin^2 2\theta \gtrsim 10^{-5}$. Thus the supernova neutrino spectrum provides a unique opportunity to probe neutrino oscillations. In particular, if the neutrino masses have a seesaw pattern

$$m_\nu \propto \frac{m_D^2}{M_R} \qquad m_D^2 \leftrightarrow m_u^2 : m_c^2 : m_t^2$$

one can fix M_R according to the solar neutrino small angle solution, assuming the solar neutrino puzzle is due to $\nu_e \to \nu_\mu$. Thus "m_{ν_μ}" $\sim$ few $\cdot 10^{-3}$ eV and "m_{ν_τ}" $\sim$ 1 eV, a value that would be interesting cosmologically and would induce $\nu_e \leftrightarrow \nu_\tau$ oscillations for supernova neutrinos. This is illustrated in Fig. 5. As a consequence of this crossing, supernova ν_es will emerge anoma-

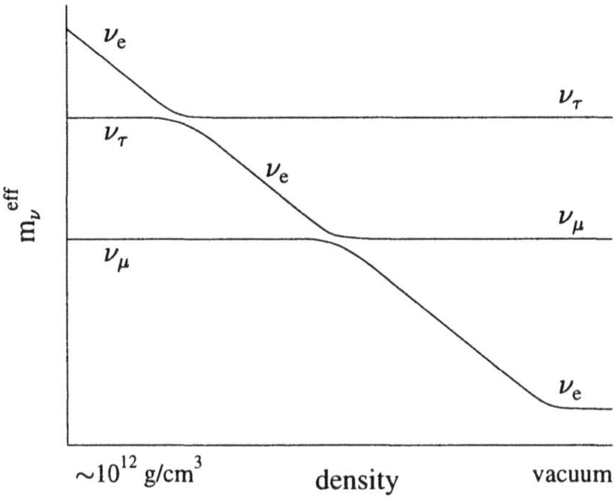

Figure 5. The three-flavor level crossing diagram showing two "crossings" that might be associated with matter enhanced oscillations of supernova neutrinos.

lously hot, with an $\langle E_{\nu_e}\rangle \sim 25$ MeV characteristic of heavy flavor neutrinos. An experimental demonstration that $\langle E_{\nu_e}\rangle > \langle E_{\bar{\nu}_e}\rangle$, for example, would provide strong evidence for oscillations and possibly provide information on massive tauon neutrinos.

Such an oscillation would have consequences for the supernova explosion, as it would enhance matter heating by the ν_es, and on nucleosynthesis, as the increased rate for $\nu_e + n \rightarrow e^- + p$ would drive the atmosphere above the protoneutron star proton rich, destroying any possibility for an r-process. But the aspect on which I would like to focus is the possibility of distinctive oscillation signals in terrestrial supernova neutrino observatories.

One detector of interest, despite being primarily sensitive to $\bar{\nu}_e$s, is Superkamokande. The usual ν_e signal, elastic scattering off electrons, will not be altered in total rate due to a $\nu_e \leftrightarrow \nu_\tau$ oscillation, since this rate is proportional to the luminosity, which is approximately independent of flavor. However there is some hardening of the spectrum of forward-scattered electrons: the question is whether this is enough of a signal, given uncertainties in the supernova ν_e and $\nu_\mu/\bar{\nu}_\mu/\nu_\tau/\bar{\nu}_\tau$ fluxes contributing to this forward scattering of electrons [17]. Perhaps more interesting is the reaction $\nu_e + ^{16}O \rightarrow ^{16}F + e^-$, which produces a backward-peaked distribution of electrons that would very likely be detectable above the $\bar{\nu}_e$ "background," given $\langle E_{\nu_e}\rangle \sim 25$ MeV [18]. There has also been a suggestion that the γ-ray cascades following ν-induced spallation reactions on ^{16}O might provide an attractive signal [19].

It seems to me that a detector of a different type - one flavor specific and economical to operate - might be useful in supernova watches, given that the characteristic time between galactic events might be $\sim$ 35 years. An attractive possibility is the 1 kiloton version of the iodine detector discussed by Lande, which would be able to view the entire galaxy with good statistics. The cross section averaged over the ν_e flux is predicted to increase by a factor of ~6.6 if there is a $\nu_e \leftrightarrow \nu_\tau$ oscillation. As the luminosity constraint leads to a smaller flux of (undistorted) ν_τs, this implies an increase in the iodine detector response of a factor of $\sim$ 2.9, a rather dramatic signal for oscillations.

Another possibility is the lead neutron spallation detector LAND proposed by Hargrove et al. [20]. The signal he discussed, single neutron emission, is not flavor specific nor is the cross section well determined. However, Fuller, McLaughlin, and I [21] recently found a rather attractive charge-current-specific signal in this detector, the emission of multiple neutrons. Due to details of the nuclear physics - the location of the giant resonances that are strongly excited by charged and neutral current scattering - this channel appears to "filter out" neutral current events, while leaving $\sim$ 70% of the charged current response. Thus if multiple neutron events are studied, LAND becomes flavor specific and, due to the high threshold for reaching the giant resonances, extraordinarily sensitive to the ν_e temperature. A $\nu_e \leftrightarrow \nu_\tau$ oscillation would produce approximately 40 times the number of events that would be measured in the absence of oscillations. The cross section for detecting $\langle E_{\nu_e}\rangle \sim 25$ MeV supernova neutrinos is a remarkable $4 \cdot 10^{39}$ cm^2. There is some work remaining to be done to verify these results, but the prospects appear quite promising.

This conference is a wonderful illustration of the power of experiments with astrophysical neutrinos. Supernova neutrinos differ from atmospheric and solar neutrinos in that they impact the earth only once or twice a century. But they provide unique windows on neutrino physics, such as ν_τ masses of cosmological interest. Thus it is probably important for us to prepare the right complement of neutrino observatories in anticipation of the next galactic supernova.

This work was supported in part by the U.S. Department of Energy.

REFERENCES

1. E.G. Adelberger, et al., to appear in Rev. Mod. Phys.
2. C.W. Johnson, E. Kolbe, S.E. Koonin, and K. Langanke, Ap. J, 392 (1992) 320.
3. B.W. Filippone, A. J. Elwyn, C. N. Davids, and D. D. Koetke, Phys. Rev. Lett., 50 (1983) 412 and Phys. Rev. C, 28 (1983) 2222.
4. R. W. Kavanagh, T. A. Tombrello, J. M. Mosher, and D. R. Goosman, Bull. Am. Phys. Soc., 14 (1969) 1209.
5. F. Hammache et al., Phys. Rev. Lett. 80 (1998) 928.
6. L. Weissman et al., Nucl. Phys. A, 630 (1998) 678; M. Hass et al., private communication.
7. J. N. Bahcall and M. H. Pinsonneault, Rev. Mod. Phys., 67 (1995) 781 and to be published (1998).
8. W. C. Haxton, Phys. Lett. B, 431 (1998) 110.
9. P. Anselmann, et al., Phys. Lett. B, 342 (1995) 440; W. Hampel et al., Phys. Lett. B, 388 (1996) 384.
10. D. N. Abdurashitov et al., Phys. Lett. B, 328 (1994) 234 and hep-ph/9803418 (March, 1998).
11. S.M. Austin, N. Anantaraman, and W. G. Love, Phys. Rev. Lett., 73 (1994) 30; J.W. Watson et al., Phys. Rev. Lett., 55 (1985) 1369.
12. N. Hata and W. C. Haxton, Phys. Lett. B, 353, (1995) 422.
13. V. Castellani, S. Degl'Innocenti, G. Fiorentini, M. Lissia, and B. Ricci, Phys. Rev. D, 50 (1994) 4749.
14. A. Cumming and W. C. Haxton, Phys. Rev. Lett., 77 (1996) 4286.
15. J.N. Bahcall, M. H. Pinsonneault, S. Basu, and J. Christensen-Dalsgaard, Phys. Rev. Lett., 78 (1997) 171.
16. G. Fiorentini, talk presented on the ITP Workshop on Solar Neutrinos, December 1997 (http://www.itp.ucsb.edu/online/snu/fiorentini).
17. H. Minakata, private communication.
18. W.C. Haxton, Phys. Rev. D, 36 (1987) 2283; Y. Z. Qian and G.M. Fuller, Phys. Rev. D, 49 (1996) 1762.
19. K. Langanke, P. Vogel, and E. Kolbe, Phys Rev. Lett., 76 (1997) 2629.
20. C.K. Hargrove et al., Astroparticle Physics, 5 (1996) 183.
21. G.M. Fuller, W. C. Haxton and G. C. McLaughlin, submitted to Phys. Rev. D.

ELSEVIER

Nuclear Physics B (Proc. Suppl.) 77 (1999) 81–88

NUCLEAR PHYSICS B
PROCEEDINGS
SUPPLEMENTS

Helioseismology and solar neutrinos

D.O. Gough

Institute of Astronomy and Department of Applied Mathematics and Theoretical Physics, Madingley Road, Cambridge, CB3 0HA, UK
and
W.W. Hansen Experimental Physics Laboratory, Stanford University, Stanford, CA 94305-4085, USA

The manner by which helioseismic date are obtained and analysed to diagnose the interior structure and rotation of the Sun is briefly described. The principal results of the analysis pertinent to solar neutrino production are presented; they have constrained standard theoretical models of the Sun in such a way as to prevent them from explaining the neutrino flux measurements without recourse to neutrino transitions. However, standard solar models do not represent the entire class of plausible models, and indeed they suffer the deficiency of being unstable. Models in which the products of the nuclear reactions are redistributed in the core may represent the Sun more closely, and should at least be considered and tested further in the process of developing a reliable model of the neutrino source.

1. WHAT IS HELIOSEISMOLOGY?

Helioseismology is the diagnostic study of the oscillations of the Sun. It shares many similarities with its parent discipline, geoseismology, and indeed most of the methods used to analyse the data were developed from geoseismological techniques. There are some very important differences however, most notably that, unlike in the Earth, shear waves do not exist. Moreover, rarely are waves seen to be emanating from a point as they do from earthquakes; the Sun is almost uniformly seismically active, the waves being generated by the turbulent convection that occupies the region immediately beneath the surface. However, these differences need not concern us here.

The Sun's seismicity is measured from observations of the motion and brightness variations of the surface. Figure 1 shows a typical observation: it is a Doppler image from which the mean of 45 similar images, taken at different times, has been subtracted in order to filter out the steady and slowly evolving components of the flow – principally rotation and convection, whose total velocity amplitude is about four times that of the oscillatory motion.

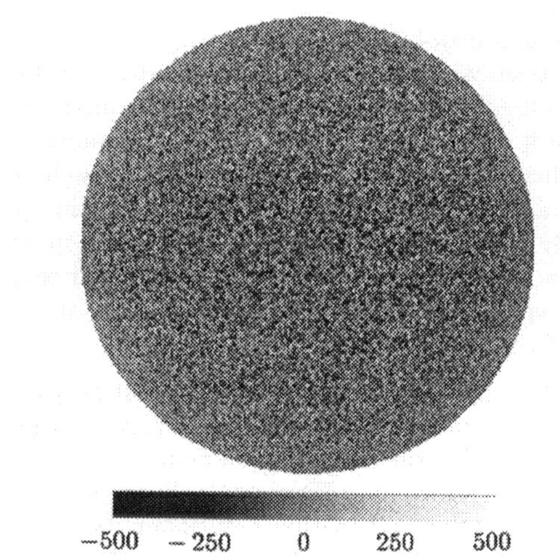

Fig. 1. SOI/MDI Dopplergram of the Sun from which the steady and slowly varying flow has been removed, leaving a grey-scale representation of the line-of-sight velocity of principally the acoustic modes.

The residual (almost vertical) motion can be regarded as the superposition of some 10^7 resonant modes of acoustic oscillation. The amplitude of a single mode is very small, typically about $\sqrt{10^{-7}} \times 500\ \mathrm{ms}^{-1} \simeq 15\ \mathrm{cm\,s}^{-1}$, which is only about 10^{-4} of the width of the spectrum line in which the Doppler shift is measured. Many observations must therefore be made in order to suppress contamination by noise. The image illustrated in the figure was taken by the Michelson Doppler Imager of the Solar Oscillations Investigation (SOI/MDI) on the spacecraft SOHO (Scherrer et al., 1997) which has viewed the Sun continuously for a little over two years.

The major objective of present-day helioseismology is to infer the internal structure and rotation of the Sun from the frequencies of as many of the modes as can be measured. Almost all the (cyclic) frequencies that have been measured pre-

PII S0920-5632(99)00401-6

cisely enough for structural diagnostic purposes lie in the range 2–4 mHz. Because the Sun is essentially steady on the timescale of an oscillation, the concept of oscillation frequency is a good one, and the equation governing the eigenfunction Ψ of an acoustic mode is linear (because the amplitudes are small) and essentially of Helmholtz type. Moreover, because the Sun is close to being spherically symmetrical, separable spherically harmonic solutions for Ψ with respect to spherical polar coordinates (r, θ, ϕ) can therefore be found:

$$\Psi = \Psi_{\boldsymbol{i}}(\mathbf{r}, t) = \psi_{nlm}(r) P_l^m(\cos\theta) e^{im\phi - i\omega t}, \qquad (1)$$

for the mode $\boldsymbol{i} = (n, l, m)$, where P_l^m is the associated Legendre function of the first kind. The principal quantum number n is called the *order* of the mode; the degree l of the Legendre function is call the *degree* of the mode, and the order m of the Legendre function is the *azimuthal order* of the mode. If the Sun were strictly spherically symmetrical, the eigenfrequencies $\omega = \omega_{\boldsymbol{i}}$ would be independent of m, because m defines the ϕ dependence which necessarily depends on the choice of the axis of the coordinate system, to which a spherically symmetrical Sun would be oblivious. However, slight deviations from spherical symmetry, due principally to rotation, cause the degeneracy to be split, measurement of which can be used to diagnose the symmetry-breaking agents. Most of what I shall discuss here concerns what we call the multiplet frequency, namely the uniformly weighted mean over m at fixed n and l, which can be shown to depend only on the spherically averaged component of the structure of the Sun.

Mode frequencies are obtained by first projecting the sequence of Doppler (or intensity) images onto spherical harmonics and then Fourier analysing the amplitudes in time. Because the harmonics are not orthogonal over the visible hemisphere (actually information from only about one-third of the Sun's surface is typically used because the signal-to-noise ratio is low near the limb), and also because the Doppler shift measures only the line-of-sight component of the velocity, the modes cannot be well separated by spatial projection alone. Nevertheless, by taking the various time dependences of the different modes into account one can achieve approximate isolation of modes with l up to about 150. Above that value the modes merge in the Fourier spectrum, an example of which is illustrated in Figure 2, and are best regarded as essentially continuous functions of l. Using long continuous observing sequences, the frequencies of some modes have been determined with a precision of about 1 part in 10^5. Modes of degree in excess of 1000 have been observed, although the precision is substantially lower.

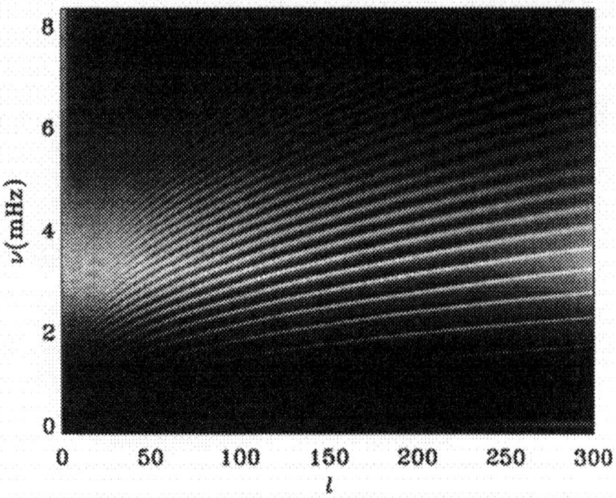

Fig. 2. Fourier power spectrum of solar observations, obtained from 360 days of continuous observation by SOI/MDI. White represents greatest power.

Each mode can be considered as the interference of an acoustic wave packet with itself as it propagates many times around the Sun. Figure 3 illustrates two such waves. (In a strictly spherical star the ray paths would lie in a plane.) In general, the ray paths are not closed, so the wave samples the whole of the Sun within the radius range accessible to it, a condition generically required for the eigenfrequencies to be discrete, as is the case in the similar theory of semi-classical quantization. All waves, except those with $l = 0$, penetrate to a nonzero lowest radius r_t, where they intersect, creating a caustic surface (i.e. the spherical surface of revolution of the caustic circle evident in Figure 3). They are reflected near the surface by the abrupt decline in density – the characteristic scale of variation of most physical quantities beneath the solar surface is roughly proportional to depth, and therefore is very small near the surface. Despite the fact that sound speed c increases with depth, the wave spends a comparatively long time near the caustic (the comparison here is with the region immediately above the caustic, not with the reflecting layers very near the surface where the smallness of the sound speed – c there is only 2 per cent of the value at the centre of the Sun – causes the wave to tarry longest). Not surprisingly, the extent to which the oscillation is influenced by the structure of the Sun is roughly proportional to the time spent; indeed, if the sound speed of the Sun were to change by

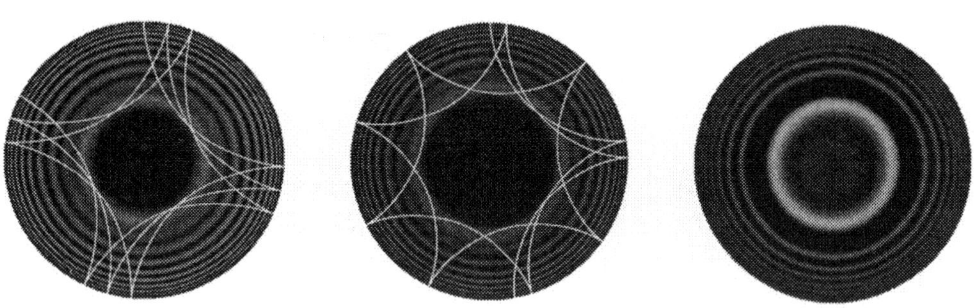

Fig. 3. (a) and (b) depict ray paths of two acoustic wave packets with slightly different $\omega/(l+1/2)$ which interfere to form normal modes of oscillation. The shading represents the extent to which the modes are sensitive to the structure of the Sun, lighter indicating greater sensitivity; it exhibits the inevitable wavelike dependence on radius, the envelope being the kernel K of equation (2). The high sensitivity near the inner caustic is evident; the even higher dominating sensitivity near the upper caustic is too close to the surface to be visible. (c) is the difference between the sensitivity functions of (a) and (b); notice the suppression of sensitivity in the outer regions (after Gough *et al.*, 1996).

$\delta c(r)$, the frequency of a mode would be altered by a relative amount $\delta\omega/\omega$ equal to the time-weighted average of the relative sound-speed change:

$$\frac{\delta\omega}{\omega} = \int K(r)\frac{\delta c}{c}dr \simeq \left(\int_{r_t}^{R} v_v^{-1}dr\right)^{-1}\int_{r_t}^{R} v_v^{-1}\frac{\delta c}{c}dr, \qquad (2)$$

where R is the radius of the upper reflecting surface (approximately the radius of the Sun) and $v_v(r;\omega,l)$ is the vertical component of the (group) velocity of the wave, which vanishes at the caustic. (Since the magnitude of the group velocity is essentially the sound speed c – there are actually small corrections due to the hydrostatic stratification under gravity – the dependence of v_v on ω and l arises predominantly from the geometry of the ray path, which actually depends just on the value of r_t.) Therefore, the frequency ω of the mode is influenced especially by conditions near the caustic.

The value of r_t is determined principally by the value of l (it is given approximately by the relation $r_t/c(r_t) \simeq (l+\frac{1}{2})/\omega$, simply the condition for a sound wave of frequency ω and horizontal wavenumber $(l+\frac{1}{2})/r$ to be travelling horizontally; but notice that the range of l is very much greater than the relative range of ω); the lower is l, the deeper is the penetration. Near the surface, however, the scale heights are too small for the waves to be influenced substantially by the comparatively gentle horizontal variation associated with l, and the influence of the structure of the Sun on the wave dynamics depends almost solely on ω. Consequently, if one were to subtract two of equations (2) corresponding to two modes of only slightly different values of r_t, and consequently with only slightly different values of $\delta\omega/\omega$, the kernels $K(r)$ would largely cancel except in the vicinity of the (neighbouring) caustics. The outcome is illustrated in Figure 3c. What one is then left with is a localized average of $\delta c/c$. By taking differences between modes with neighbouring caustics at different locations one can thus construct a (somewhat blurred) representation of $\delta c/c$.

In practice one nowadays works with differences $\delta\omega$ between the frequencies of the Sun and those of a theoretical model, in order to infer the sound-speed difference δc between the Sun and the model. Moreover, one takes rather more complicated linear combinations of those frequency differences than simple differences between two with neighbouring caustics. (Details of how those combinations are constructed can be found in the reviews by e.g. Gough and Thompson, 1991 or Thompson, 1995.) But my discussion above embodies the main essence of the more sophisticated procedures, and exhibits an important feature which is retained by them, namely that in order to obtain localized information one must take combinations of data $\delta\omega$ that almost cancel. Therefore the precision of the inferences is very much less than the precision of the seismic frequency data on which those inferences are based. This is particular true in the energy-generating core of the Sun, for the deeper the localization, the more severe is the cancellation, because v_v increases with depth and therefore the sensitivity of $\delta\omega/\omega$ to $\delta c/c$ decreases. The extent to which av-

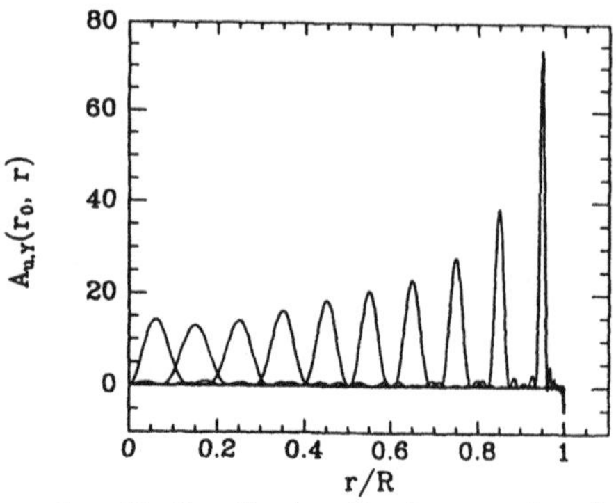

Fig. 4. Optimally localized averaging kernels $A(r; r_0)$ for sound speed centred about a selection of radii r_0.

eraging kernels A, which are linear combinations of the data kernels K, can be localized about different depths r_0 in the Sun is illustrated in Figure 4. The process of inferring properties such as localized averages of the structure (or of rotation or any other internal velocity) of the Sun from the frequencies of seismic oscillation is called *inversion*.

2. THE SOLAR NEUTRINO PROBLEM

Stated in its most elementary form, the solar neutrino problem has been the problem of finding a way of reconciling with observation the theoretical calculations of the fluxes of neutrinos from different components of the chain of nuclear reactions in the Sun. Initially it was only the ^{37}Cl capture rate that needed to be explained. Assuming for the sake of this part of my discussion that neutrino transitions do not occur, the difficulty experienced by solar physicists was that one could not construct a so-called standard solar model that reproduced the observed neutrino flux $^{37}F_\nu$ and at the same time satisfied other astrophysical prejudices. Stated briefly, a standard solar model is a spherically symmetrical model in hydrostatic and thermal equilibrium, in which the microphysics determining opacity, the equation of state, the nuclear reaction rates and the diffusion coefficients is treated as accurately as possible, yet the macrophysics of the fluid motion is ignored entirely, except in the convection envelope where solar modellers usually adopt a simple (mixing-length) turbulence theory, which yields typically a local relation with an undetermined scaling factor α between the temperature gradient and the fluxes of heat and (when it is included) momentum. Given the mass M of the Sun and the surface abundance ratio Z/X (where X is the relative abundance by mass of hydrogen and Z is the total abundance of all elements other than H and ^{4}He; the helium abundance is $Y = 1 - X - Z$), and assuming the age of the Sun to be known, there are two unknown parameters, α and the initial helium abundance Y_0, that can be adjusted to calibrate the model to reproduce the present solar radius R and luminosity L. One can then compute the neutrino fluxes.

In the old days (i.e. prior to helioseismology) a plethora of variants to the standard solar models were considered. Also considered, within the realm of standard (or almost standard) models, was the possibility that the value of Z/X was not as normally believed, but that perhaps instead the Sun had been formed with a low value of Z and the subsequent accretion of material from the interstellar medium masked the value in the interior (Joss, 1974), which is where Z really matters. The difficulty this idea presented is that whereas a standard solar model could then easily be constructed to reproduce the low observed value of $^{37}F_\nu$, it also had a low value of Y_0, about 0.14 or thereabouts, which is not only lower than any value measured in hot stars, but is substantially less than what cosmologists predict to have been created in the Big Bang.

This issue was one of the first to be addressed, and resolved, by helioseismology. To illustrate what needs to be achieved I show in Figure 5 the temperatures T and the squares of the sound speeds of two solar models, one with an initial helium abundance $Y_0 = 0.28$, which is more-or-less consistent with astronomical observations, the other with $Y_0 = 0.14$, which reproduces the neutrino flux $^{37}F_\nu$. Because for a perfect gas (the solar gas is nearly perfect, although in the core there is actually a non-negligible influence of partial electron degeneracy) $c^2 \propto T/\mu$, where μ is the mean molecular mass of the material, the plots of c^2 and T are very similar. The most striking difference is the dip in c^2 near the very centres of the evolved models, caused by the increased μ resulting from nuclear transmutations. For comparison I include c^2 in a zero-age model, which does not exhibit this property because the chemical composition is uniform. Included also in Figure 5b is an early seismological determination (for $r/R > 0.05$) of c^2; it differs from that of one of the models by no more than the thickness of the curve, but that difference varies with position, accounting for the variation in thickness of the apparently single curve.

The comparison demonstrates straight away that standard solar models really do present us with a neutrino problem and not a helium problem. That problem has recently been exacerbated by the neutrino flux measurements by Gallex and SAGE, and by Kamiokande and SuperKamiokande. Now the

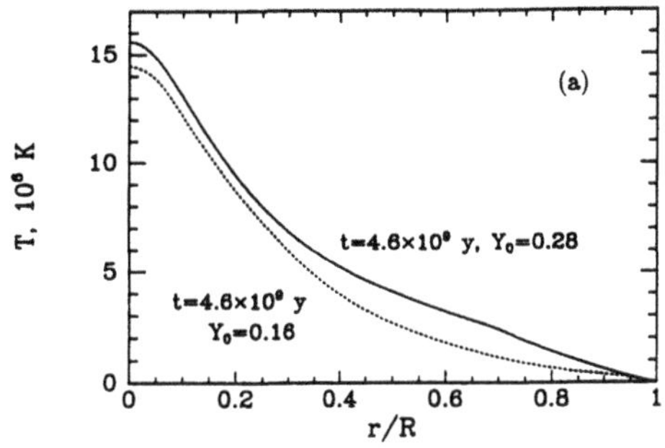

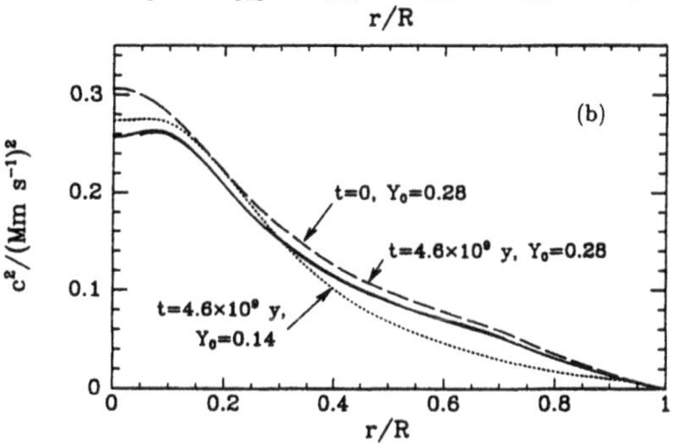

Fig. 5. (a) temperature $T(r)$ of two solar models with initial helium abundances $Y_0 = 0.28$ (solid curve) and $Y_0 = 0.14$ (dotted curve). (b) squares of sound speeds, $c^2(r)$, of the two models of (a); the dashed curve represents c^2 at zero age in the model with $Y_0 = 0.28$. A sound-speed inversion of solar data is included also as a continuous curve.

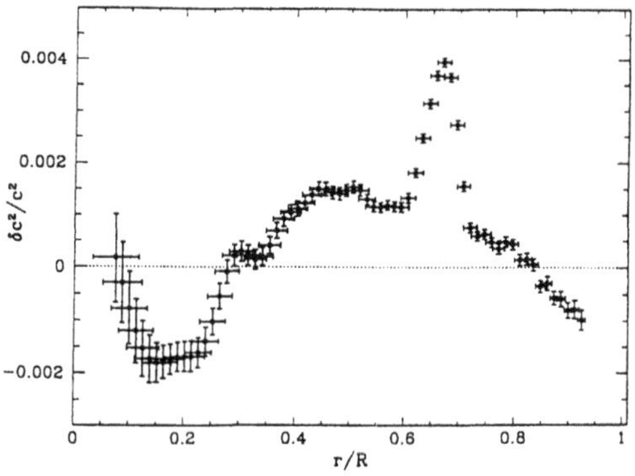

Fig. 6. Localized averages of the relative difference $\delta c^2/c^2$ between the squared sound speeds in the Sun and in the standard solar model of Christensen-Dalsgaard *et al.* (1996). The horizontal bars represent the characteristic widths of the averaging kernels A, such as are illustrated in Figure 4; the vertical bars are standard errors (which are correlated).

difficulty lies in explaining three different measurements, which, if taken at face value, cannot be reconciled by any simple adjustment of a standard solar model, because adjustments that bring one of the fluxes closer to observation tend to move other fluxes further away. I shall not describe the details, because they are amply discussed in these proceedings by Bahcall and by Haxton. However, before continuing I must point out, as does Bahcall, that the sound speed in modern standard models is quite close to the measured spherically averaged sound speed in the Sun. Figure 6 (from Kosovichev *et al.*, 1997) illustrates the discrepancy $\delta c^2/c^2$ with respect to one such model. The discrepancy is typically only about one part in 10^3, except for the hump centred at $r/R \simeq 0.67$, situated immediately beneath the base of the convection zone. Nevertheless, the discrepancy is certainly very significant, for in some regions it is greater than 10 formal standard errors. It is hardly out of the question that in the fullness of time minor revisions to the microphysics will be able to remove at least the large-scale component of the discrepancy, because different standard models computed today differ in magnitude by as much, if not more, than the $\delta c^2/c^2$ plotted in Figure 6 (cf. von Steiger, 1998).

But would merely removing the discrepancy be scientifically sufficient? Certainly not. Firstly, it is important to be sure that the model really does represent the spherically averaged structure of the Sun, which involves investigating more than just the sound speed. Secondly, we need also to investigate the asphericity, which, as Haxton has pointed out, may not be small in all respects.

3. A DIGRESSION ON ROTATION

Rotation splits the degeneracy of the eigenfrequencies of seismic modes with respect to azimuthal order m. Unlike the splitting due to all other symmetry-breaking agents (which cannot distinguish between east and west), rotationally induced frequency shifts are essentially odd functions of m, and can therefore readily be separated from any other component. Since harmonics of degree l with different values of m/l sample latitude differently (asymptotically for large l the mode is confined between latitudes $\pm\cos^{-1}[m/(l+\frac{1}{2})]$), a two-dimensional image of the angular velocity $\Omega(r,\theta)$ can be obtained by analysing the rotational splitting using a generalization of the procedure outlined in section 1. However, since the magnitude of this is small compared with the multipole frequencies – it is of order $m\Omega$ – the precision and the resolution of the inversion is substantially poorer than that for the spherically averaged structure.

In Figure 7 I illustrate the first rotational splitting inversion to have been published (Duvall *et al.*, 1984). It represents a (rather poorly defined) latitudinal average of the angular velocity of the

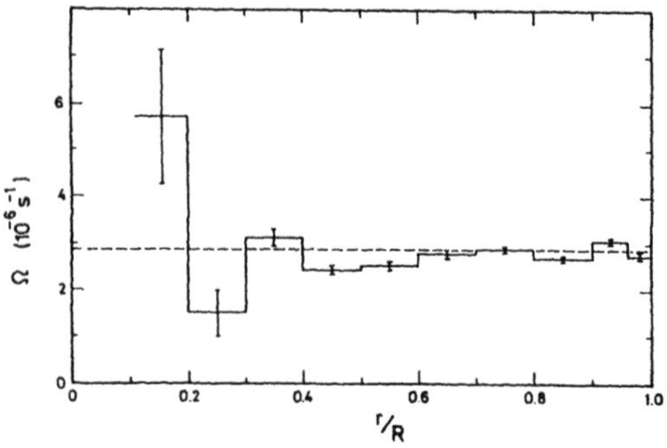

Fig.7. Fit of a piecewise-constant function to the angular velocity Ω in the Sun, averaged in the vicinity of the equatorial plane (after Duvall *et al.*, 1984). The vertical bars represent formal standard errors.

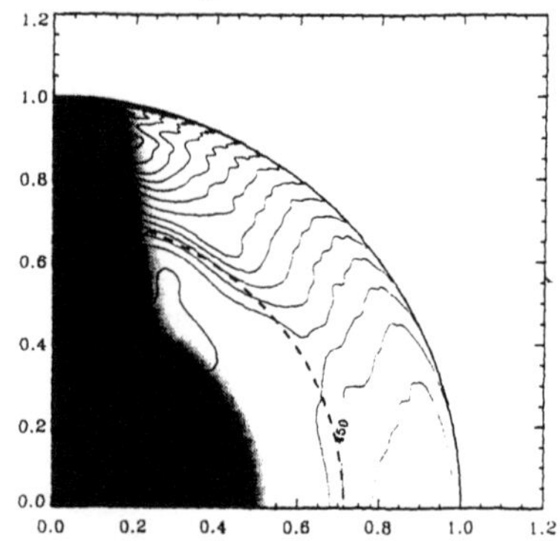

Fig. 8. Contours of constant (localized averages of the) angular velocity Ω of the Sun, plotted in a quadrant of a meridional plane. The dotted curve marks the location of the base of the convection zone (from Schou *et al.*, 1998).

Sun in the neighbourhood of the equatorial plane. Although it is certainly no longer the most reliable inversion available today, I present it because it exhibits an extremely interesting feature: that in a region surrounding the core the rotation appears to be slower than it is at the surface, whereas the very core appears to be rotating more rapidly. The latter is hardly significant, because the uncertainty in the frequency splitting of the lowest-degree modes on which the inference depends is statistically high, and it may also suffer from being biased. Indeed, those splittings have been contradicted by some (e.g. Elsworth *et al.*, 1995), although not all, subsequent measurements, to the extent that today the matter must be considered unresolved (e.g. Christensen-Dalsgaard *et al.*, 1998). Nevertheless, consideration of whether or not Figure 7 is more-or-less correct is pertinent to my theme. However, I postpone further discussion of it until the next section.

Figure 8 illustrates a modern two-dimensional angular-velocity inversion, using SOI/MDI data (Schou *et al.*, 1998). It does not extend as deeply as does the relatively crude equatorial averages plotted in Figure 7, because the authors regarded any inferences deeper than those plotted in Figure 8 to be too unreliable; not only does precision but also caution increase with experience. The feature to which I wish to draw attention here is that broadly speaking the latitudinal variation of Ω observed at the surface persists to the base of the convection zone, beneath which there is an abrupt transition to almost uniform rotation. The transition layer is called the tachocline, and is too thin to be properly resolved by the seismic data. (Recall that what are plotted in the figure are localized averages and not point values.) I do not wish to digress into a discussion of what agent holds the radiative interior rigid (at least in its outer layers – note that this figure is not significantly at variance with the averages plotted in Figure 7, which at greater depth imply a lesser rotation rate), except to point out that it seems to be most likely (Gough and McIntyre, 1998) – indeed at least possible – that the agent is a large-scale magnetic field. On the other hand, it is doubtless the case that the differential rotation of the convection zone is a consequence of a balance between the anisotropic Reynolds stresses of the turbulence and angular-momentum transport by large-scale material circulation. What happens in the tachocline?

Whenever a fluid experiences rotational shear, a lateral pressure gradient is established which drives a circulation. Perhaps the most commonly experienced example is a gently stirred cup of drink containing a sediment (e.g. tea leaves). The shear at the bottom of the rotating fluid produced by the viscous stress against the bottom of the (non-rotating) cup induces an inwardly directed flow, which causes the sediment to accumulate about the middle of the bottom of the cup. (Indeed, it is conservation of angular momentum by the outwardly directed return flow in the body of the drink, and not viscous stress, which reduces the angular velocity of the body of the fluid, and the angular momentum is eventually removed in the thin viscous shear layer at the bottom of the cup; therefore the timescale in which the drink stops rotating after stirring has ceased is substantially less than the characteristic viscous time $t_{\rm visc}$ based on the dimensions of the cup – about half an hour – and is instead the the geometric mean of $t_{\rm visc}$ and the period of stirring – say half a second: namely, half a minute.) The solar tachocline must behave likewise (Spiegel and Zahn, 1992; Gough and McIntyre,

1998). And, as in the cup, sediment is mixed. The sediment in this case is helium and other heavy elements that have been gradually settling under gravity since the Sun became a relatively quiescent star. The tachocline circulation homogenizes that sediment with the convection zone, leaving a hydrogen-rich (relative to current standard models that ignore the tachocline flow) layer, with a correspondingly augmented sound speed, immediately beneath the zone. That is presumably the phenomenon that is responsible for the sound-speed anomaly evident in Figure 6 centred at $r/R = 0.67$ to which I referred in the previous section. Moreover, one can calibrate standard solar models that do incorporate tachocline mixing to determine the tachocline thickness (Elliott and Gough, 1998).

I trust that now my digression is justified. Although at first sight one might have thought rotation to have little to do with the neutrino problem – the first angular-velocity inversion illustrated in Figure 7 made immediately evident that the centrifugal force, considered in the old days for providing a solution to the neutrino problem, is much too small to be relevant – it is now evident that that may not be so, for the associated secondary flow can redistribute the chemical composition and thereby affect the Sun's structure in an observable way. Even though the flow itself is much too slow to be detected by seismology directly, because it transports both material and angular momentum it thereby links the inferences one might draw from inversions for structure and rotation. Studying the tachocline via rotational shear has only a minor direct influence on the neutrino problem, however (through its small influence on the relation of the value of Z/X in the interior to that observed at the surface). More pertinent is the rotation of the core. I discuss that further in the next section.

4. THE SOLAR CORE

Here I add some further remarks to Haxton's discussion in these proceedings because of the extreme importance of his work to the study of solar neutrinos. As I mentioned earlier, and as Bahcall discusses in these proceedings, standard solar models, however they may be adjusted, appear to be unable to reproduce all the neutrino flux measurements simultaneously. This property has even led some to claim that the resolution of the solar neutrino problem cannot therefore lie in astrophysics, and to conclude that it must reside in nuclear or particle physics. Part of the importance of the discussion by Cumming and Haxton (1996) is to show that that conclusion is invalid. Of course in the light of the exciting announcement at this meeting of μ-neutrino transitions it seems not unlikely that the conclusion might nevertheless be correct. So is there likely to remain a role for solar physicists in this subject? I suggest that the answer is yes, and, moreover, that issues of the kind discussed by Cumming and Haxton might be central to that role.

The transitions whose detection has been announced do not involve the electron neutrinos, such as are emitted by the Sun. But it appears that almost everyone believes that if μ-neutrinos are associated with mass, then so too must be all the neutrinos of other flavours; to contemplate otherwise would entail an unjustifiably drastic modification to the standard model of particle physics. Consequently electron-neutrino transitions are almost inevitable, and indeed Suzuki offers evidence in these proceedings, albeit not yet convincingly significant, that such transitions do indeed occur. As more data are accumulated by SuperKamiokande, by the imminent Sudbury Neutrino Observatory, and by the various other new neutrino experiments, no doubt many of the parameters in the modified standard theory will be determined. But if it transpires that the electron-neutrino transition length is much greater than terrestrial dimensions, which the current SuperKamiokande measurements of atmospheric neutrinos suggest, then solar neutrinos are bound to be involved. And in that case it is evident that the quest of neutrino physicists would be aided substantially by having a realistic knowledge of the neutrino source. What is perhaps more important than that knowledge, however, is that we be not led astray by false information. And that is where the real importance of the Cumming-Haxton discussion becomes evident, for it makes us aware that there are possibilities beyond the confines of the standard theories of the Sun.

It has long been known that standard solar models are linearly unstable in their cores (e.g. Christensen-Dalsgaard, Dilke and Gough, 1974); but it is not known what the nonlinear development of that instability might be. I shall not discuss the possibilities here, but I raise the matter firstly to cast valid serious doubt on the standard models, and secondly to suggest that the existence of motion in the core of the genre of that discussed by Cumming and Haxton should hardly be surprising. Indeed it is superficially similar to that proposed earlier by Ghosal and Spiegel (1991). The motion is evidently too slow to be of dynamical significance: perturbations in nuclear energy generation induce (Lagrangian) temperature and composition fluctuations, causing any fluid element to move to a new location at which it is neutrally buoyant, in a manner analogous to the way in which a submarine (with its screws not turning) adjusts its depth. Therefore there is no horizontal variation in either density or pressure. Consequently, since the adiabatic exponent is nearly constant in the core, there is no horizontal variation in sound speed. The only modification to the structure comes from the advection of chemical species (there is only a negligible convective flux of heat), modifying the nuclear reaction rates and thereby changing the rate of energy generation, and thence the thermal stratification, both vertical and horizontal, and, of course, the neutrino production rates. It appears to be out of the question that the flow be so extensive as to homogenize the long-lived chemical elements H and He in the core, because if it did the mean molecular mass of the fluid would not increase towards the centre to cause the dip in c^2 evident in Figure 5: the disparity between the sound speeds in the model and the Sun would probably then be augmented by at least a factor ten above that of Figure 6 (cf. Bahcall *et al.*, 1997). But Cumming and Haxton (1996) insist that the model they present is for illustrative purposes only; the details should not be taken seriously. Indeed, in the light of this meeting's announcement of μ-neutrino transitions there will rightly be little incentive to explain the solar neutrino fluxes in the absence of e-neutrino transitions. However, motion of the kind discussed by Cumming and Haxton, doubtless less extensive than their original suggestion, perhaps confined to a shell surrounding the core as Ghosal and Spiegel (1991) have advocated, perhaps unsteady, should be contemplated further, because it may exert a significant influence on neutrino production.

If after further study such a flow were still to be plausible, how might one try to detect it? Evidently one cannot do so unambiguously from seismological inversions for sound speed: since the sound-speed variation is predicted to be spherically symmetrical, one cannot even hope to detect the consequences of the horizontal variation of the flow, and the cause of a minor radial variation cannot readily be distinguished from the effect of changes to the opacity or to nuclear reaction cross sections. So I return finally to the rotation. If the motion were axisymmetric about the axis of rotation, angular momentum would be conserved by the flow; regions closer to the axis would rotate faster than the mean and regions further out would rotate more slowly. Indeed, even if the motion were not axisymmetric there would still be a tendency towards conserving angular momentum, so the outcome would likely be similar. It is therefore intriguing that this behaviour is suggested by Figure 7. As I said when I first discussed the matter, our inference of the core's rotation is not yet significant. But surely it is now evident why it is important to strive to make it so.

I thank A.G. Kosovichev, P.H. Scherrer and J. Schou for help with producing the diagrams.

REFERENCES

1. Bahcall, J.N., Pinsonneault, M.H., Basu, S. and Christensen-Dalsgaard, J., 1997, *Phys. Rev. Lett.*, **78**, 131

2. Christensen-Dalsgaard, J., Dilke, F.W.W. and Gough, D.O., 1974, *Mon. Not. R. astr. Soc.*, **169**, 429

3. Christensen-Dalsgaard, J. *et al.*, 1996, *Science*, **272**, 1286

4. Chaplin, W. J., Christensen-Dalsgaard, J., Elsworth, Y., Howe, R., Isaak, G.R., Larsen, R.M., New, R., Schou, J., Thompson, M.J. & Tomczyk, S., 1998, *Mon. Not. R. astron. Soc.*, in preparation

5. Cumming, A. and Haxton, W.C., 1996, *Phys. Rev. Lett.*, **77**, 4286

6. Duvall Jr, T.L. *et al.*, 1984, *Nature*, **310**, 22

7. Elliott, J.R. and Gough, D.O., 1998, *Astrophys. J.*, submitted

8. Elsworth, Y., Howe, R., Isaak, G.R., McLeod, C.P., Miller, B.A., New R., Wheeler, S.J. and Gough, D.O., 1995, *Nature*, **376**, 669

9. Ghosal, S. and Spiegel, E.A., 1991, *Geophys. Astrophys. Fluid Dyn.*, **61**, 161

10. Gough, D.O. and McIntyre, M.E., 1998, *Nature*, **394**, 755

11. Gough, D. O. and Thompson, M. J., 1991, *Solar interior and atmosphere* (eds A. N. Cox, W. C. Livingston and M. S. Matthews, Univ. Arizona, Tucson) p. 519

12. Gough, D.O. *et al.*, 1996, *Science*, **272**, 1296

13. Joss, P.C., 1974, *ApJ*, **191**, 771

14. Kosovichev A.G., 1997, *et al.*, *Sol. Phys.*, **170**, 43

15. Scherrer, P.H. *et al.*, 1995, *Sol. Phys.*, **62**, 129

16. Schou, J. *et al.*, 1998, *Astrophys. J.*, **505**, 390

17. Spiegel, E.A. and Zahn, J-P., 1992, *Astron. Astrophys.*, **265**, 106

18. von Steiger, R, (ed.) 1998, *Solar Composition and its Evolution – from Core to Corona* (Kluwer, Dordrecht), *Space Sci. Rev.* in press

19. Thompson, M. J., 1995, *Inverse Prob.*, **11**, 709

ELSEVIER

Nuclear Physics B (Proc. Suppl.) 77 (1999) 89–92

NUCLEAR PHYSICS B
PROCEEDINGS
SUPPLEMENTS

Neutrino Magnetic Moment and Solar Neutrino Experiments

Ana M. Mourão [a] * and Anna Rossi[b] †

[a]CENTRA - Centro Multidisciplinar de Astrofísica and Dep. of Physics, I.S.T.
Avenida Rovisco Pais, 1, 1096 Lisboa Codex - Portugal

[b]CFIF - Centro de Física das Interacções Fundamentais - I.S.T.
Avenida Rovisco Pais, 1, 1096 Lisboa Codex - Portugal

We have studied the effect of a non-vanishing neutrino magnetic moment (μ_ν) on the ν_x (x=e, μ, τ) elastic scattering off electrons for the Super-Kamiokande detector. The bounds on the μ_ν we have obtained are comparable to that extracted from laboratory experiments. Furthemore, we outline the potential of the Borexino experiment which may be sensitive to neutrino magnetic moments $\lesssim 10^{-10}\mu_B$. In our analysis we have considered both cases of Majorana and Dirac neutrinos.

1. Introduction

The solar neutrino problem (SNP) is nowadays regarded as a direct evidence for physics beyond the standard electroweak model. This is due to the fact that the observed deficit of electron neutrinos in all solar neutrino experiments can only be explained assuming that non-zero neutrino masses and/or neutrino magnetic moments might lead to flavour, spin or spin-flavour neutrino oscillations of the solar left-handed neutrinos ν_{eL} [1]-[6]. At the same time a lot of work has being done also to understand the implications of the uncertainties in helioseismology [7] and nuclear physics [8] for the SNP.

As was already pointed out in the literature [9]-[13] a non-vanishing neutrino magnetic moment, μ_ν, can also affect the neutrino elastic scattering off electrons through which solar ν's are detected in Super-Kamiokande (SK). Therefore the expected signal in such a detector may depend also on the electromagnetic properties of the neutrinos.

In previous works [12,13] bounds have been obtained on μ_ν from Kamiokande and Super-Kamiokande data, taking into account the restrictions on the ν_e survival probability imposed by Homestake, Gallex and SAGE experiments.

In this work we present an updated analysis of the effect of a non-vanishing magnetic moment on the scattering off electrons for the Super-Kamiokande detector, similarly to the study carried on in[13]. We have considered both the cases of Dirac and Majorana neutrinos.

For definiteness we examine the two neutrino system $\nu_e - \nu_x$ (x=μ, τ) with a non-zero mass difference δm^2. In the solar interior the spin-flavour resonant conversion [5] $\nu_{eL} \to \nu_{xR}$, with probability P_R, occurs at higher matter density with respect to the usual MSW resonant conversion $\nu_{eL} \to \nu_{xL}$ [6], characterised by the probability P_L[1]. As a result of both conversions we have three different neutrino 'flavours' reaching the Earth:

$$\begin{aligned}\Phi_{\nu_{xR}} &= P_R\,\Phi_{SSM}\ ,\\ \Phi_{\nu_{xL}} &= P_L(1-P_R)\,\Phi_{SSM} \equiv P_{xL}\Phi_{SSM}\ , \qquad (1)\\ \Phi_{\nu_{eL}} &= (1-P_L)(1-P_R)\,\Phi_{SSM} \equiv P_{eL}\,\Phi_{SSM}\ ,\end{aligned}$$

here Φ_{SSM} is the standard solar model (SSM) prediction for a certain component of the neutrino flux [14]. In the above relations [2] there are only two independent quantities, *e.g.* P_{eL} and P_R. Hence $P_L = 1 - \frac{P_{eL}}{1-P_R}$ where $P_R \leq 1 - P_{eL}$. On the basis of this picture we have calculated the expected signal in SK experiment.

For the sake of simplicity, we assume that all the probabilities are energy independent. In par-

[1]We remind that the $\nu_{xL} \to \nu_{eR}$ spin-flavour resonance would occur at a much higher matter density, for given δm^2. We assume therefore that it does not take place in the solar interior.

[2]The picture envisaged from the eq. (1) could appear, for example, when the MSW and spin-flavour resonances lie far away from each other to be treated separately. This may be the case in the central region of the sun.

*This work was in part supported by JNICT projects PRAXIS/PCEX/P/FIS/4/96 and ESO/P/PRO/1127/96, Fundação Oriente and GTAE-Lisbon.

†Supported by a grant from GTAE - Lisbon.

PII S0920-5632(99)00402-8

ticular the present SK data on the ν_{eL} energy spectrum do not yet exclude that P_{eL} is energy independent for $E_\nu \geq 6.5$ MeV [15].

2. Signal in the Super-Kamiokande detector

The total signal in the Super-Kamiokande experiment, for the case of Dirac neutrinos, can be written as

$$\begin{aligned} R^{\rm total}_{SK} &= R^{\rm w+em}_{SK} \\ &= P_{e\rm L}\left[\langle\sigma^{\rm w}_{\nu_{e\rm L}}\rangle + \langle\sigma^{\rm em}_{\nu_{e\rm L}}\rangle\right] \\ &\quad + P_{\rm xL}\left[\langle\sigma^{\rm w}_{\nu_{\rm xL}}\rangle + \langle\sigma^{\rm em}_{\nu_{\rm xL}}\rangle\right] \\ &\quad + P_{\rm R}\langle\sigma^{\rm em}_{\nu_{\rm xR}}\rangle \end{aligned} \qquad (2)$$

where the averaged total $\nu - e$ cross sections are

$$< \sigma^i_\alpha > = \int dE_\nu \Phi^{^8B}_{\rm SSM}(E_\nu)\sigma^i_\alpha(E_\nu), \qquad (3)$$

where $\alpha = \nu_{eL}, \nu_{\rm xL,R}$ and i=em for electromagnetic and i=w for weak cross sections. $\Phi^{^8B}_{\rm SSM}(E_\nu)$ is the ^8B solar neutrino flux from BP98 [14]. In the calculation of the cross sections $\sigma^i_\alpha(E_\nu)$ we have taken into account the energy resolution of the detector [16]. The electromagnetic cross section can be taken e.g. from Kerimov *et al* [17]. Taking into account that $<\sigma^{\rm em}_\alpha> \propto \mu^2_\nu$ it is easy to see that for $P_{e\rm L} = 1, P_{\rm R} = P_{\rm xL} = 0$, and $\mu_{\nu_e} = 0$ we obtain the SSM expectation for the signal.

In our analysis we use the most recent SK data for the ^{8}B solar neutrino flux, $\Phi^{exp}_{\rm SK} = (2.44 \pm 0.05) \times 10^6$ cm^{2}s^{-1} [18], normalised to the SSM prediction - $\Phi^{th}_{\rm BP98} = (5.15 \pm 0.98) \times 10^6$ cm^{2}s^{-1} [14], namely

$$Z_{\rm K} = \frac{\Phi^{exp}_{\rm SK}}{\Phi^{th}_{\rm BP98}} = 0.47 \pm 0.09. \qquad (4)$$

As opposite to the Super-Kamiokande experiment, the Homestake detector is only sensitive to the $\nu_{e\rm L}$ component of the solar neutrino flux and the total rate is mainly due to ^{8}B neutrinos. Considering the experimental data [19] $Z_{\rm Cl} = 0.28 \pm 0.03$, we can assume that for the higher energy spectrum of ^{8}B neutrinos, $E_\nu \geq 6$ MeV, the neutrino survival probability is $P_{e\rm L} \sim 30\%$. Therefore this implies a total depletion of the intermediate energy ^{7}Be neutrinos as the present understanding of the SNP points to [1]-[3].

We can now find the values of the neutrino magnetic moment compatible with solar neutrino experiments, i.e needed to obtain the signal observed at the Superkamiokande.

3. Limits on neutrino magnetic moments

We have studied the impact of non-vanishing μ_ν in the SK signal taking into account the contribution of the several neutrino 'flavours' as shown in (1). In Fig. 1 we show the contour-plots for $Z_{\rm K}$ =0.47 in the parameter space $(P_{\rm R}, \mu_\nu)$, where $\mu_\nu \equiv \mu_{\nu_e} = \mu_{\nu_{\rm x}}$. We repeat our analysis with other values of $Z_{\rm K}$, just to understand implications of uncertainties in the SSM used in the evaluation of μ_ν. Notice that the larger the value of $P_{\rm R}$ the larger the value of μ_ν needed to saturate $Z_{\rm K}$ in order to compensate for the loss of the $\nu_{\rm xL}$ component (recall that $P_{e\rm L}$ is fixed at 0.3). The present experimental $Z_{\rm K}$ implies the bound $\mu_\nu \lesssim (2 \div 5) \times 10^{-10}\mu_{\rm B}$, almost independently of $P_{\rm R}$.

We have considered also the case with $\mu_\nu \equiv \mu_{\nu_{\rm x}}$ and $\mu_{\nu_e} = 0$. In this case we obtain a similar plot as that in Fig.1 and the limit $\mu_\nu \lesssim 5 \times 10^{-10}\mu_{\rm B}$.

For the sake of completeness we have also studied the case of Majorana neutrinos, for which the *antineutrino* state $\nu_{\rm xR} = \nu^c_{\rm xL} \equiv \tilde{\nu}_{\rm x}$, is 'active', having both electromagnetic and standard weak interactions. Hence in the eq. (2) one more term is to be added, *i.e.* $P_{\rm R}\langle\sigma^{\rm w}_{\tilde{\nu}_{\rm x}}\rangle$. We remember that for Dirac neutrinos both diagonal or transition μ_ν can be generated [20], while for Majorana neutrinos only transition (off-diagonal) magnetic moments are allowed [21].

In Fig. 2 we show our result in this scenario: the upper bound for Majorana neutrinos comes out to be similar, $\mu_\nu \lesssim (1.5 \div 4.0) \times 10^{-10}\mu_{\rm B}$ for the extreme case $P_{\rm R} = 0.7$. Note that for $P_{\rm R} <$ 0.7 , smaller values of μ_ν are in principle tolerated as the $\tilde{\nu}_{\rm x}$'s contribute to the weak cross section (compare with the Dirac case shown in Fig. 1). Our upper limit was obtained assuming vanishing vacuum mixing angle (and then $P_{\rm L} = 0$), thereby satisfying the experimental constraints on the $\tilde{\nu}_e$

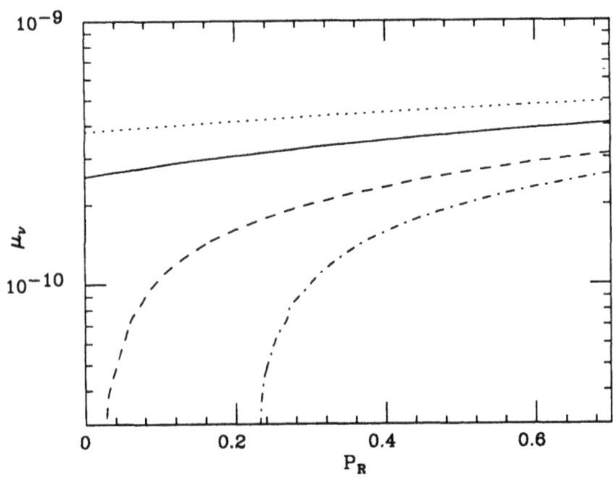

Figure 1. The contour plots of the expected signal versus the SSM prediction Z_K in Super-Kamiokande in the (P_R , μ_ν) parameter space in the case of Dirac neutrinos. The dotted, solid, dashed and dot-dashed curves correspond to $Z_K = 0.55, 0.47, 0.4$ and $Z_K = 0.37$, respectively. μ_ν is given in units of Bohr magneton.

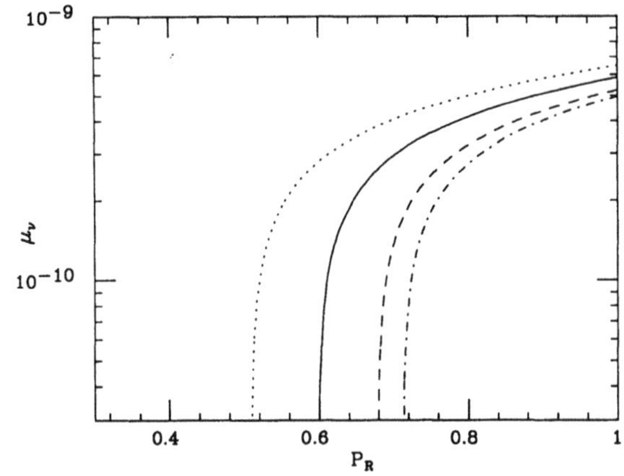

Figure 2. The same as in Fig.1 but for the Majorana case.

which could emerge from the subsequent vacuum oscillation $\tilde{\nu}_x \to \tilde{\nu}_e$[22].

We conclude that the bounds we have obtained are slightly more restrictive than those from accelerator experiments, namely in the case of μ_{ν_μ} it is $\mu_{\nu_\mu} < 7.4 \times 10^{-10} \mu_B$[25]. However our bounds are still not comparable to those from reactor experiments, - $\mu_{\nu_e} < 1.8 \times 10^{-10} \mu_B$. Needless to say that our results are more stringent in the case of μ_{ν_τ} for which $\mu_{\nu_\tau} < 5.4 \times 10^{-7} \mu_B$ [24].

Finally, we have discussed the potential of the future Borexino experiment [23] which will detect ^{7}Be neutrinos through $\nu - e$ elastic scattering. For $\mu_\nu = 10^{-10} \mu_B$, the cross section σ^{em} can be comparable to $\sigma^{W}_{\nu_e}$ for $E_\nu \leq 1$ MeV. Therefore we can expect a substantial signal in Borexino even in the case of complete conversion of the initial ^{7}Be-ν_e's into ν_{xR} or $\tilde{\nu}_x$. In Fig. 3 we have plotted the energy distribution of the events for Borexino in the case of complete conversion $\nu_e \to \tilde{\nu}_x$ (dotted line). We note that in this distribution we have taken into account the contributions from all solar neutrino components. However the ^{7}Be neutrinos contribute to more than 90% of the signal. Other important contribution is given by the *pep* neutrinos. For comparison we also shown the SSM distribution (solid line). We can expect $\sim$ 50% of the SSM prediction and a specific distortion of the spectrum. This is in contrast with the case of pure MSW conversion ($\mu_\nu = 0$) $\nu_e \to \nu_x$ that would imply a $(20 \div 25)$% reduction in the signal.

4. Conclusions

In this contribution we have updated the analysis on the effect of a non-vanishing neutrino magnetic moment on the $\nu - e$ cross section in Super-Kamiokande experiment. The limits we achieved -$\mu_\nu \lesssim (2 \div 5) \times 10^{-10} \mu_B$ - remain comparable to that extracted from the previous Kamiokande data as the electromagnetic cross section is smaller than the weak one for the energy range involved, $E \geq 6$ MeV. Therefore experiments with a much lower energy threshold - such as Borexino or Hellaz - could exhibit a much better sensitivity to a non-zero μ_ν and consequently provide a better testing of the spin-flavour resonant conversion itself as a solution to the SNP[23,26].

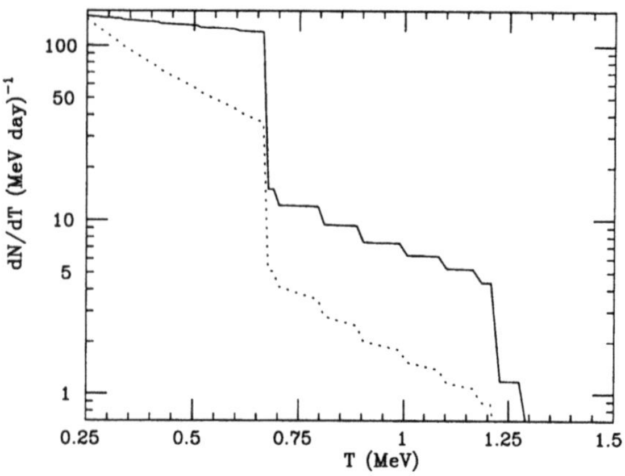

Figure 3. The energy distribution of the events in Borexino experiment. The dotted curve represents the expected signal in the case of complete $\nu_e \rightarrow \tilde{\nu}_x$ (Majorana case) with $\mu_\nu = 10^{-10}\mu_B$. The solid line corresponds to the SSM distribution.

Acknowledgments

We are grateful to E. Akhmedov, L. Bento, D. Gough and Y. Suzuki for very useful discussions. One of us (A.M.) would like to thank the organisers for the invitation to present this work and for the very stimulating ν98 Conference.

REFERENCES

1. V. Castellani, S. Degl'Innocenti and G. Fiorentini, Astron. Astrophys. 271 (1993) 601.
2. K. M. Heeger and R. G. H. Robertson, Phys. Rev. Lett. 77 (1996) 3720.
3. J. N. Bahcall, P. I. Krastev and A. Yu. Smirnov, Phys. Rev. D58 (1998) to appear.
4. A. Cisneros, Astrophys. Space Sci. 10 (1971) 87; M. B. Voloshin, M. I. Vysotsky, L. B. Okun, Sov. J. Nucl. Phys. 44 (1986) 440.
5. C. S. Lim and W. J. Marciano, Phys. Rev. D37 (1988) 1368; E. Kh. Akhmedov, Sov. J. Nucl. Phys. 48 (1988) 382; E. Kh Akhmedov, hep-ph/9705451; H. Minakata and H. Nunokawa Phys. Rev. Lett 63 (1989) 121.
6. L. Wolfenstein, Phys. Rev. D17 (1978) 2369; D20 (1979) 2364; S. P. Mikheyev, A. Smirnov, Sov. J. Nucl. Phys. 42 (1985)913.
7. D. Gough, in these Proceedings.
8. W. C. Haxton, in these Proceedings; E. G. Adelberger et al., astro-ph/9805121.
9. R. Barbieri and G. Fiorentini, Nucl. Phys. B304 (1988) 909; A. Suzuki et al., Phys. Rev. D43 (1991) 3557.
10. A. B. Balantekin, P. J. Hatchell and F. Loreti, Phys. Rev. D41 (1990) 3583.
11. K.S Babu, N. Mohapatra and I. Z. Rothstein, Phys. Rev. D44 (1991) 2265.
12. A. M. Mourão, J. Pulido, J. Ralston, Phys. Lett. B285 (1992) 364; Erratum-ibid. 288 (1992) 421.
13. J. Pulido and A. M. Mourão, Phys. Rev. D57 (1998) 1794.
14. J. N. Bahcall, S. Basu and M. H. Pinsonneault, Phys. Lett. B433 (1998) 1.
15. Y. Fukuda et al.- Super-Kamiokande Coll., Phys. Rev. Lett. 81 (1998) 1158.
16. J. N. Bahcall, Neutrino Astrophysics, CUP, 1993.
17. B. Kerimov, M. Ya. Safin, H. Nazih, Izvestyia Akademii Nauk SSSR, Ser. Fiz. 52 (1988) 126; P. Vogel, J. Engel, Phys. Rev. D39(1989) 3378.
18. Y. Suzuki, in these Proceedings.
19. B. Cleveland et al., Nucl. Phys. B38 (1995) 47.
20. K. Fujikawa and R. Shrock, Phys. Rev. Lett 45 (1980) 963.
21. J. Schechter and J. W. F. Valle, Phys. Rev. D24 (1981) 1883.
22. R. Barbieri et al., Phys. Lett. B259 (1991) 119; M. Aglietta et al., JETP Lett. 63 (1996) 791; A. A. Bykov et al., hep-ph/9808342.
23. C. Arpesella et al., (Borexino Collaboration), Proposal of BOREXINO (1991); Z. G. Berezhiani and A. Rossi Phys. Rev. D51 (1995) 5229.
24. Review of Particle Properties, C. Caso et al., EPJ C3 (1998) 1.
25. D. A. Krakauer et al. Phys. Lett. B 252 (1990) 177.
26. S. Pastor et al., hep-ph/9803378.

NUCLEAR PHYSICS B
PROCEEDINGS
SUPPLEMENTS

ELSEVIER

Nuclear Physics B (Proc. Suppl.) 77 (1999) 93–97

New Enhancement Mechanism of the Transitions in the Earth of the Solar and Atmospheric Neutrinos Crossing the Earth Core

S.T. Petcov[a b]

[a]Scuola Internazionale Superiore di Studi Avanzati, and INFN (Trieste), I–34014 Trieste, Italy.

[b]Inst. of Nuclear Research and Nuclear Energy, Bulgarian Academy of Sciences, 1784 Sofia, Bulgaria.

It is shown that the $\nu_2 \rightarrow \nu_e$ and $\nu_\mu \rightarrow \nu_e$ ($\nu_e \rightarrow \nu_{\mu(\tau)}$) transitions respectively of the solar and atmospheric neutrinos in the Earth in the case of $\nu_e - \nu_{\mu(\tau)}$ mixing in vacuum, are strongly enhanced by a new type of resonance when the neutrinos cross the Earth core. The resonance is operative at small mixing angles but differs from the MSW one. It is in many respects similar to the electron paramagnetic resonance taking place in a specific configuration of two magnetic fields. The conditions for existence of the new resonance include, in particular, specific constraints on the neutrino oscillation lengths in the Earth mantle and in the Earth core, thus the resonance is a "neutrino oscillation length resonance". It leads also to enhancement of the $\nu_2 \rightarrow \nu_e$ and $\nu_e \rightarrow \nu_s$ transitions in the case of $\nu_e - \nu_s$ mixing and of the $\bar{\nu}_\mu \rightarrow \bar{\nu}_s$ (or $\nu_\mu \rightarrow \nu_s$) transitions at small mixing angles. The presence of the neutrino oscillation length resonance in the transitions of solar and atmospheric neutrinos traversing the Earth core has important implications for current and future solar and atmospheric neutrino experiments, and more specifically, for the interpretation of the results of the Super-Kamiokande experiment.

1. Introduction

When the solar and atmospheric neutrinos traverse the Earth, the $\nu_2 \rightarrow \nu_e$ and $\nu_\mu \rightarrow \nu_e$ ($\nu_e \rightarrow \nu_{\mu(\tau)}$) transitions/oscillations they undergo due to small $\nu_\mu - \nu_e$ mixing in vacuum [1] can be dramatically enhanced by a new type of resonance which differs from the MSW one and takes place when the neutrinos cross the Earth core [1]. The resonance is present in the $\nu_2 \rightarrow \nu_e$ and $\nu_\mu \rightarrow \nu_e$ ($\nu_e \rightarrow \nu_{\mu(\tau)}$) transition probabilities, P_{e2} and $P(\nu_{\mu(e)} \rightarrow \nu_{e(\mu;\tau)})$, if the neutrino oscillation length (and mixing angles) in the Earth mantle and in the Earth core obey specific conditions [1]. When satisfied, these conditions ensure that the relevant oscillating factors in the probabilities P_{e2} and $P(\nu_{\mu(e)} \rightarrow \nu_{e(\mu;\tau)})$ are maximal [2] and that this produces a resonance maximum in P_{e2} and $P(\nu_{\mu(e)} \rightarrow \nu_{e(\mu;\tau)})$. Accordingly, the term "neutrino oscillation length resonance" or simply "oscillation length resonance" was used in [1] to denote the resonance of interest. There exists a beautiful analogy between the neutrino oscillation length resonance and the electron spin-flip resonance realized in a specific configuration of magnetic fields [3] (see [1] for further details).

At small mixing angles ($\sin^2 2\theta \lesssim 0.05$) the maxima due to the neutrino oscillation length resonance in P_{e2} and $P(\nu_{\mu(e)} \rightarrow \nu_{e(\mu;\tau)})$ are absolute maxima and dominate in P_{e2} and $P(\nu_{\mu(e)} \rightarrow \nu_{e(\mu;\tau)})$: the values of the probabilities at these maxima in the simplest case of two-neutrino mixing are considerably larger - by a factor of $\sim (2.5-4.0)$ ($\sim (3.0-7.0)$), than the values of P_{e2} and $P(\nu_\mu \rightarrow \nu_e) = P(\nu_e \rightarrow \nu_{\mu(\tau)})$ at the local maxima associated with the MSW effect taking place in the Earth core (mantle). The magnitude of the enhancement due to the oscillation length resonance depends on the neutrino trajectory through the Earth core: the enhancement is maximal for the center-crossing neutrinos [1,2]. Even at small mixing angles the resonance is rel-

[1]As is well-known, the $\nu_2 \rightarrow \nu_e$ transition probability accounts for the Earth effect in the solar neutrino survival probability in the case of the MSW two-neutrino $\nu_e \rightarrow \nu_{\mu(\tau)}$ and $\nu_e \rightarrow \nu_s$ transition solutions of the solar neutrino problem, ν_s being a sterile neutrino.

[2]Note that, in contrast, the MSW effect is a resonance amplifying the neutrino mixing.

[3]This analogy was brought to the attention of the author by L. Wolfenstein.

0920-5632/99/$ – see front matter
PII S0920-5632(99)00403-X

atively wide both in the neutrino energy (or resonance density) [1] - it is somewhat wider than the MSW resonance, and in the Nadir angle [2], h, specifying the neutrino trajectory in the Earth. It also exhibits strong energy dependence.

The presence of the oscillation length resonance in the transitions of solar and atmospheric neutrinos traversing the Earth has important implications [1–5] for the interpretation of the results, e.g., of the Super-Kamiokande experiment [6,7].

The Earth enhancement of the two-neutrino transitions of interest has been discussed rather extensively, see, e.g., refs. [5,8]. Some of the articles contain plots of the probabilities P_{e2} and/or $P(\nu_\mu \rightarrow \nu_e)$ or $P(\nu_e \rightarrow \nu_{\mu(\tau)})$ on which one can recognize now the dominating maximum due to the neutrino oscillation length resonance (see, e.g., [8]). However, this maximum was invariably interpreted to be due to the MSW effect in the Earth core before the appearance of [1].

2. The Neutrino Oscillation Length Resonance (NOLR)

All the interesting features of the solar and atmospheric neutrino transitions in the Earth, including those related to the neutrino oscillation length resonance, can be understood quantitatively in the framework of the two-layer model of the Earth density distribution [1,2,9,10]. The density profile of the Earth in the two-layer model is assumed to consist of two structures - the mantle and the core, having different constant densities, $\bar{\rho}_{man}$ and $\bar{\rho}_c$, and different constant electron fraction numbers, Y_e^{man} and Y_e^c [4]. The transitions of interest of the neutrinos traversing the Earth are essentially caused by two-neutrino oscillations taking place i) first in the mantle over a distance X' with a mixing angle θ'_m and oscillation length L_{man}, ii) then in the core over a distance X'' with different mixing angle θ''_m and oscillation length L_c, and iii) again in the mantle over a distance X' with θ'_m and L_{man}. Due to the matter effect $\theta'_m, \theta''_m \neq \theta$ and $L_{man,c} \neq L_{vac}$, L_{vac} being the oscillation length in vacuum (see, e.g., [13]). For fixed X' and X'' the neutrino oscillation length resonance occurs [1] if i) the relative phases acquired by the energy eigenstate neutrinos in the mantle and in the core, $\Delta E'X' = 2\pi X'/L_{man}$ and $\Delta E''X'' = 2\pi X''/L_c$, are correlated, being odd multiples of π, so that

$$\frac{X'}{L_{man}} = k + \frac{1}{2}, \quad \frac{X''}{L_c} = k' + \frac{1}{2}, \qquad (1)$$

where $k, k' = 0, 1, 2, ...$, and if ii) the inequality

$$\cos(2\theta''_m - 4\theta'_m + \theta) \;\; (\cos(2\theta''_m - 4\theta'_m)) \; < 0 \quad (2)$$

is fulfilled. Condition (2) is valid for the probability P_{e2} $(P(\nu_\mu \rightarrow \nu_e))$. When equalities (1) hold, (2) ensures that P_{e2} $(P(\nu_\mu \rightarrow \nu_e))$ has a maximum. In the region of the NOLR maximum where, e.g., $\Delta E'X' \cong \pi(2k+1)$, P_{e2} is given in the case of $\nu_e - \nu_\mu$ mixing by [1]:

$$P_{e2} \cong \sin^2\theta \; + \; \frac{1}{4}[1 - \cos\Delta E''X''] \times \left[\sin^2(2\theta''_m - 4\theta'_m + \theta) - \sin^2\theta\right]. \quad (3)$$

At the NOLR maximum P_{e2} takes the form [1]

$$P_{e2}^{max} = \sin^2(2\theta''_m - 4\theta'_m + \theta). \qquad (4)$$

The analogs of eqs. (3) - (4) for the probability $P(\nu_{\mu(e)} \rightarrow \nu_{e(\mu;\tau)})$ can be obtained by formally setting $\theta = 0$ while keeping $\theta'_m \neq 0$ and $\theta''_m \neq 0$ in (3) - (4). Note that one of the two NOLR requirements $\Delta E'X' = \Delta E''X'' = \pi$ is equivalent at $\sin^2 2\theta \lesssim 0.02$ to the physical condition [1]

$$\pi\,(\frac{1}{X'} + \frac{1}{X''}) \cong \sqrt{2}G_F(Y_e^c\bar{\rho}_c - Y_e^{man}\bar{\rho}_{man}). \quad (5)$$

Remarkably enough, for the $\nu_2 \rightarrow \nu_e$ and $\nu_{\mu(e)} \rightarrow \nu_{e(\mu;\tau)}$ transitions in the Earth, the NOLR conditions (1) with $k = k' = 0$ are approximately fulfilled at small mixing angles ($\sin^2 2\theta \lesssim 0.05$) in the regions where (2) holds [1]. The associated NOLR maxima in P_{e2} and $P(\nu_\mu \rightarrow \nu_e)$ are absolute maxima (Figs. 1 - 2).

Let us note that the study performed in [1] and discussed briefly above [5] differs substantially from

[4]The densities $\bar{\rho}_{man,c}$ should be considered as mean densities along the neutrino trajectories. In the Earth model [11] one has: $\bar{\rho}_{man} \cong (4-5)$ g/cm^3 and $\bar{\rho}_c \cong (11-12)$ g/cm^3. For Y_e one can use the standard values [11,12] (see also [3]) $Y_e^{man} = 0.49$ and $Y_e^c = 0.467$.

[5]For analysis of the NOLR effects in the $\nu_2 \rightarrow \nu_e$ and $\nu_e \rightarrow \nu_s$ transitions ($\nu_e - \nu_s$ mixing) and in the $\bar{\nu}_\mu \rightarrow \bar{\nu}_s$ (or $\nu_\mu \rightarrow \nu_s$) transitions at small mixing angles see [1,2].

the studies [14]. The authors of [14] considered the possibility of resonance enhancement of the $\nu_e \to \nu_{\mu(\tau)}$ transitions of neutrinos propagating in matter with density, varying *periodically* along the neutrino path (parametric resonance). It was found, in particular, that at small mixing angles strong enhancement is possible only if the neutrinos traverse at least 2 - 3 periods (in length) of the density oscillations. The density distribution in the Earth is not periodic [6]; and in order for the oscillation length resonance [7] to occur periodic variation of the density is not required.

In [16] the $\nu_\mu \to \nu_s$ transitions in the Earth were considered for $\sin^2 2\theta \cong 1$. It was noticed that in the region where $\sqrt{2}G_F N_n^{man,c} \gg \Delta m^2/E$, $N_n^{man,c}$ being the neutron number density, a new maximum in $P(\nu_\mu \to \nu_s)$ appears when $\sqrt{2}G_F N_n^{man(c)} X'^{('')} \cong 2\pi$, which was found to hold at $h \sim 28^0$. The height of the maximum is comparable to the heights of the other "ordinary" maxima present in $P(\nu_\mu \to \nu_s)$ for $\sin^2 2\theta = 1$. It is stated in [16] that the effect does not take place in the $\nu_{\mu(e)} \to \nu_{e(\mu;\tau)}$ transitions, which is incorrect both for $\sin^2 2\theta \ll 1$ and $\sin^2 2\theta \cong 1$ [1,2].

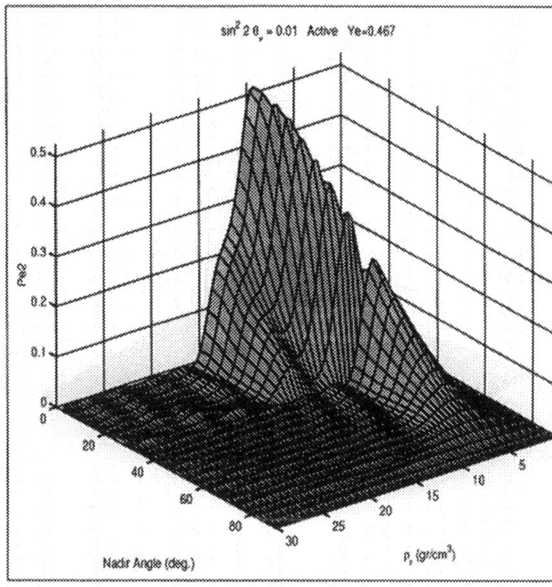

Figure 1. The probability P_{e2} as a function of h and the resonance density ρ_r for $\sin^2 2\theta = 0.01$ [2]. The prominent absolute maximum for $h \approx (0^0 - 28^0)$ at $\rho_r \sim (8 - 10)$ g/cm^3 is due to the NOLR. The "shoulder" at $\rho_r \sim 13$ g/cm^3 is caused by the MSW effect in the Earth core [1].

3. Implications of the Neutrino Oscillation Length Resonance

The implications of the oscillation length resonance enhancement of the probability P_{e2} for the Earth core crossing solar neutrinos, for the tests of the MSW $\nu_e \to \nu_{\mu(\tau)}$ and $\nu_e \to \nu_s$ solutions of the solar neutrino problem are discussed in refs. [1–4]. It is remarkable that for values of $\Delta m^2 \cong (4.0 - 8.0) \times 10^{-6}$ eV^2 from the small mixing angle (SMA) MSW solution region and the geographical latitudes at which the Super-Kamiokande, SNO and ICARUS detectors are located, the enhancement takes place in the $\nu_e \to \nu_{\mu(\tau)}$ case for values of the ^{8}B neutrino energy lying in the interval $\sim (5-12)$ MeV to which these detectors are sensitive. The resonance maximum in P_{e2} at $\sin^2 2\theta = 0.01$ for the trajectory with $h = 23^0$, for instance, is located at $E \cong 5.3$ (10.5) MeV if $\Delta m^2 = 4.0$ (8.0) $\times 10^{-6}$ eV^2. Accordingly, at small mixing angles the NOLR is predicted [3] to produce a much bigger - by a factor of ~ 6, day-night (D-N) asymmetry in the Super-Kamiokande sample of solar neutrino events, whose night fraction is due to the core-crossing neutrinos, in comparison with the asymmetry determined by using the *whole night* event sample. On the basis of these results it was concluded in [3] that it can be possible to test a substantial part of the MSW $\nu_e \to \nu_{\mu(\tau)}$ SMA solution region in the $\Delta m^2 - \sin^2 2\theta$ plane by performing *core* D-N asymmetry measurements. The Super-Kamiokande collaboration has already successfully applied this approach to the analysis of their solar neutrino data [6]: the limit the collaboration has obtained on the D-N asymmetry utilizing only the *core* event sample permitted

[6]The density change along the path of a neutrino crossing the Earth core is not periodic even in the two-layer model: it falls short of making even one and a half periods.

[7]Although this term may not be perfect, it underlines the physical essence of the new resonance. The objection to it raised in [15] is not convincing. The term "parametric resonance" used in [15], e.g., suggests incorrect analogies.

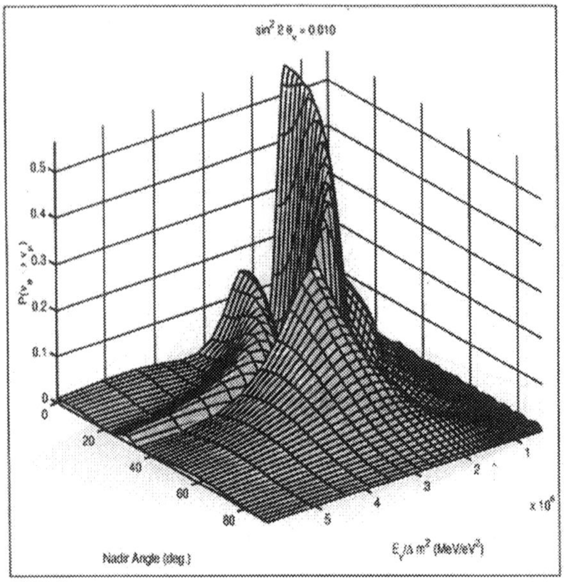

Figure 2. The probability $P(\nu_{e(\mu)} \to \nu_{\mu;\tau(e)})$ as a function of h and $E/\Delta m^2$ for $\sin^2 2\theta = 0.01$ [2]. The absolute maximum due to the NOLR for $h \approx (0^0 - 28^0)$ is clearly seen at $E/\Delta m^2 \sim (1.3 - 1.6) \times 10^6$ MeV/eV^2. The local maximum at $E/\Delta m^2 \sim (2.5 - 3.0) \times 10^6$ MeV/eV^2 is due to the MSW effect in the Earth mantle.

to exclude a part of the MSW SMA solution region located in the area $\sin^2 2\theta \cong (0.007 - 0.01)$, $\Delta m^2 \cong (0.5 - 1.0) \times 10^{-5}$ eV^2, which is allowed by the mean event rate data from all solar neutrino experiments (Homestake, GALLEX, SAGE, Kamiokande and Super-Kamiokande). In contrast, the current Super-Kamiokande upper limit on the *whole night* D-N asymmetry [6] does not permit to probe the SMA solution region: the predicted asymmetry is too small [3].

The strong NOLR enhancement of the $\nu_\mu \to \nu_e$ and $\nu_e \to \nu_{\mu(\tau)}$ transitions of atmospheric neutrinos crossing the Earth core can take place at small mixing angles practically for all neutrino trajectories through the core [1], e.g., for the trajectories with $h = (0^0 - 23^0)$ (Fig. 2). This is particularly relevant for the interpretation of the results of the atmospheric neutrino experiments and for the future studies of the oscillations/transitions of atmospheric neutrinos crossing the Earth. The Super-Kamiokande collaboration has reported at this Conference strong evidences for oscillations of the atmospheric ν_μ ($\bar{\nu}_\mu$) [7]. Assuming two-neutrino mixing, the data is best described in terms of $\nu_\mu(\bar{\nu}_\mu) \leftrightarrow \nu_\tau(\bar{\nu}_\tau)$ vacuum oscillations with parameters $\Delta m^2 \cong (0.5 - 6.0) \times 10^{-3})$ eV^2 and $\sin^2 2\theta \cong (0.8 - 1.0)$. The possibility of two-neutrino $\nu_\mu(\bar{\nu}_\mu) \leftrightarrow \nu_e(\bar{\nu}_e)$ large mixing oscillations is disfavored by the data [7]; at $\Delta m^2 \gtrsim 2 \times 10^{-3}$ eV^2 it is ruled out [17].

It is a remarkable coincidence that for $\Delta m^2 \sim (0.5 - 6.0) \times 10^{-3})$ eV^2 and small mixing, $\sin^2 2\theta \lesssim 0.10$, the oscillation length resonance in $P(\nu_\mu \to \nu_e) = P(\nu_e \to \nu_{\mu(\tau)})$ occurs [1] for values of the energy E of the atmospheric ν_e and ν_μ which contribute either to the sub-GeV or to the multi-GeV $e-$like and $\mu-$like Super-Kamiokande event samples [7]. For $\sin^2 2\theta = 0.01$, $\Delta m^2 = 5 \times 10^{-4}$; 10^{-3}; 5×10^{-3} eV^2, and $h = 0^0$ (Earth center crossing), for instance, the absolute maximum in $P(\nu_{\mu(e)} \to \nu_{e(\mu;\tau)})$ due to the NOLR takes place at $E \cong 0.75$; 1.50; 7.5 GeV. Thus, for values of Δm^2 from the region of the $\nu_\mu \leftrightarrow \nu_\tau$ oscillation solution of the atmospheric neutrino problem, the NOLR strongly enhances the $\nu_\mu \to \nu_e$ (and $\nu_e \to \nu_{\mu(\tau)}$) transitions of the atmospheric neutrinos crossing the Earth core, making the transitions detectable even at small mixing angles. It was suggested in [1] that the excess of e-like events in the region $-1 \leq \cos\theta_z \leq -0.6$, θ_z being the Zenith angle, either in the sub-GeV or in the multi-GeV sample, observed (in both samples) in the Super-Kamiokande experiment [7], is due to $\nu_\mu \to \nu_e$ small mixing angle transitions, $\sin^2 2\theta_{e\mu} \cong (0.01 - 0.10)$, with $\Delta m^2 \sim (0.5 - 1.0) \times 10^{-3})$ eV^2 or respectively $\Delta m^2 \sim (2 - 6) \times 10^{-3})$ eV^2, strongly enhanced by the NOLR [8]. The same resonantly enhanced transitions with $\Delta m^2 \sim (2 - 6) \times 10^{-3})$ eV^2 ($\Delta m^2 \sim (0.5 - 1.0) \times 10^{-3})$ eV^2) should produce at least part of the strong zenith angle dependence, exhibited by the $\mu-$like multi-GeV (sub-

[8] A more detailed investigation [2,19] performed within the indicated three-neutrino mixing scheme reveals, in particular, that the excess of e-like events in the Super-Kamiokande sub-GeV data at $-1 \leq \cos\theta_z \leq -0.6$ seems unlikely to be due to small mixing angle $\nu_\mu \to \nu_e$ transitions amplified by the oscillation length resonance.

GeV) Super-Kamiokande data [2].

The transitions of interest arise in a three-neutrino mixing scheme, in which the small mixing angle MSW $\nu_e \to \nu_\mu$ transitions with $\Delta m^2_{21} \sim (4-8) \times 10^{-6}$ eV^2, or large mixing angle $\nu_e \leftrightarrow \nu_\mu$ oscillations with $\Delta m^2_{21} \sim 10^{-10}$ eV^2, provide the solution of the solar neutrino problem and the atmospheric neutrino anomaly is due to $\nu_\mu \leftrightarrow \nu_\tau$ oscillations with $\Delta m^2_{31} \sim (0.5-6.0) \times 10^{-3}$ eV^2 [1]. For $\Delta m^2_{31} \gg \Delta m^2_{21}$ the three-neutrino $\nu_\mu \to \nu_e$ and $\nu_e \to \nu_{\mu(\tau)}$ transition probabilities reduce [18] to the two-neutrino transition probability $P(\nu_e \to \nu_\tau)$ (Fig. 2) with Δm^2_{31} and $\sin^2 2\theta_{e\mu} = 4|U_{e3}|^2(1-|U_{e3}|^2)$ playing the role of the two-neutrino oscillation parameters, where U_{e3} is the $e-\nu_3$ element of the lepton mixing matrix, ν_3 being the heaviest massive neutrino. The data [7,17] implies: $\sin^2 2\theta_{13} \lesssim 0.25$. Thus, searching for the $\nu_\mu \to \nu_e$ and $\nu_e \to \nu_{\mu(\tau)}$ transitions of atmospheric neutrinos, amplified by the oscillation length resonance, can provide also unique information about the magnitude of U_{e3} [19].

4. Conclusions

The neutrino oscillation length resonance should be present in the $\nu_2 \to \nu_e$ transitions taking place when the solar neutrinos cross the Earth core on the way to the detector, if the solar neutrino problem is due to small mixing angle MSW $\nu_e \to \nu_\mu$ transitions in the Sun. The same resonance should be operative also in the $\nu_\mu \to \nu_e$ ($\nu_e \to \nu_{\mu(\tau)}$) small mixing angle transitions of the atmospheric neutrinos crossing the Earth core if the atmospheric ν_μ and $\bar{\nu}_\mu$ indeed take part in large mixing vacuum $\nu_\mu(\bar{\nu}_\mu) \leftrightarrow \nu_\tau(\bar{\nu}_\tau)$, oscillations with $\Delta m^2 \sim (5\times10^{-4} - 6\times10^{-3})$ eV^2, as is strongly suggested by the Super-Kamiokande data [7], and if all three flavour neutrinos are mixed in vacuum. The existence of three-flavour-neutrino mixing in vacuum is a very natural possibility in view of the present experimental evidences for oscillations/transitions of the flavour neutrinos. In both cases the oscillation length resonance produces a strong enhancement of the corresponding transitions probabilities, making the effects of the transitions observable even at rather small mixing angles. Actually, the resonance may have already manifested itself in the excess of e-like events at $-1 \leq \cos\theta_z \leq -0.6$ observed in the Super-Kamiokande multi-GeV atmospheric neutrino data [1,2,19]. And it can be responsible for at least part of the strong zenith angle dependence present in the Super-Kamiokande multi-GeV and sub-GeV μ−like data [2,19].

REFERENCES

1. S.T. Petcov, Phys. Lett. **B434** (1998) 321.
2. M. Chizhov, M. Maris and S.T. Petcov, Report SISSA 53/98/EP.
3. M. Maris and S.T. Petcov, Phys. Rev. **D56** (1997) 7444.
4. M. Maris and S.T. Petcov, hep-ph/9803244.
5. Q.Y. Liu, M. Maris and S.T. Petcov, Phys. Rev. **D56** (1997) 5991.
6. Y. Suzuki, these Proceedings.
7. T. Kajita, these Proceedings.
8. S.P. Mikheyev and A.Yu. Smirnov, Proc. of the Moriond Workshop on Massive Neutrinos, 1986 (eds. O. Fackler and J. Tran Thanh Van Editions Frontières, France, 1986), p. 355; A. Dar et al., Phys. Rev. **D35** (1987) 3607; M. Cribier et al., Phys. Lett. **B182** (1986) 89; A.J. Baltz and J. Weneser, Phys. Rev. **D35** (1987) 528; J.M. Gelb, W. Kwong and S.P. Rosen, Phys. Rev. Lett. **78** (1997) 2296.
9. P.I. Krastev and S.T. Petcov, Phys. Lett. **B205** (1988) 84.
10. M. Maris, Q. Liu and S.T. Petcov, study performed in December of 1996 (unpublished).
11. F.D. Stacey, *Physics of the Earth, 2nd edition*, John Wiley and Sons, New York, 1977.
12. R. Jeanloz, Annu. Rev. Earth Planet. Sci. **18** (1990) 356.
13. S.T. Petcov, hep-ph/9806466.
14. V.K. Ermilova et al., Short Notices of the Lebedev Institute **5** (1986) 26; E.Kh. Akhmedov, Yad. Fiz. **47** (1988) 475; P.I. Krastev and A.Yu. Smirnov, Phys. Lett. **B226** (1989) 341.
15. A.Yu. Smirnov, these Proceedings.
16. Q. Liu and A.Yu. Smirnov, hep-ph/9712493.
17. M. Appolonio et al. (CHOOZ Collaboration), Phys. Lett. **B420** (1998) 397.
18. S.T. Petcov, Phys. Lett. **B214** (1988) 259.
19. S.T. Petcov, L. Wolfenstein and O. Yasuda, work in progress.

ELSEVIER

Nuclear Physics B (Proc. Suppl.) 77 (1999) 98–107

NUCLEAR PHYSICS B
PROCEEDINGS
SUPPLEMENTS

Towards the Solution of the Solar Neutrino Problem

A. Yu. Smirnov

The Abdus Salam International Center of Theoretical Physics, 34100 Trieste, Italy *

We discuss various aspects of the solar neutrino spectrum distortion and time variations of fluxes. (i) Oscillations of neutrinos which cross the mantle and the core of the Earth can be parametrically enhanced. The parametric effect gives correct physical interpretation of the calculated day-night asymmetry. (ii) Solution of the $\nu_\odot$-problem in schemes with three and more neutrinos which accommodate explanations of other neutrino anomalies, in particular, the atmospheric neutrino anomaly, can lead to complicated distortion of the boron neutrino spectrum. (iii) The study of correlations between time (seasonal or day-night) variations and spectrum distortion will help to identify the solution of the $\nu_\odot$-problem.

1. Introduction

Specific time variations of signals and distortion of the energy spectrum (along with the charged to neutral current events ratio) are the key signatures of the neutrino physics solutions of the solar neutrino problem. Preliminary SuperKamiokande (SK) data [1] indicate that the effects (if exist) are not strong: $(1-2)\sigma$, *i.e.* at the level of present sensitivity. Study of correlations between time variations and distortion of the spectrum strengthens a possibility of identification of the solution. In this connection, I will discuss some aspects of the time variations of signals (sect. II), distortion of the energy spectrum (sect. III) and correlation between time variations and spectrum distortion (sect. IV).

2. What Happens With Neutrinos Inside the Earth?

The matter of the Earth can modify properties of solar, atmospheric and supernova neutrinos. Numerical calculations have been performed in a number papers previously [2], however, physics of the effects has been understood only recently.

The density profile of the Earth has two main structures: the core and the mantle. Density changes slowly within the mantle and the core but it jumps sharply by a factor of two at their border. It is known for a long while that in the first approximation one can consider the mantle and the core as layers with constant density. Neutrinos arriving at the detector at zenith angle $\cos\Theta > -0.84$ cross the mantle only. For $\cos\Theta < -0.84$, neutrinos cross three layers: mantle, core and again mantle.

Let us introduce Φ_m and Φ_c – the oscillation phases acquired by neutrino in the mantle (one layer) and in the core of the Earth:

$$\Phi_i = 2\pi \int^{L_i} \frac{dL}{l_i} \approx \int^{L_i} dL\, \Delta H_i\ , \quad i = m, c, \quad (1)$$

where $l_i = 2\pi/\Delta H_i$ is the oscillation length in matter, and ΔH_i is the level splitting (difference of the eigenvalues of two neutrino states). In the layer with constant density: $\Phi_i = \Delta H_i L_i$.

In [3] it was realized that for neutrinos which cross both the mantle and the core of the Earth the equalities

$$\Phi_m \approx \Phi_c \approx \pi \quad (2)$$

can be approximately satisfied, and this leads to significant enhancement of oscillations. (The phases in both layers of mantle are obviously equal.) The transition probability can reach

$$P^{max} = \sin^2(4\theta_m - 2\theta_c), \quad (3)$$

where θ_m and θ_c are the mixing angles in the mantle and the core respectively. P^{max} can be much larger than $\sin^2 2\theta_m$ and $\sin^2 2\theta_c$ which correspond to maximal oscillation effect in one density layer.

*On leave of absence from INR RAN, Moscow

PII S0920-5632(99)00404-1

This is a kind of enhancement of oscillations which has been introduced by Ermilova et al., [4] and Akhmedov [5] (see also [6]) and called the parametric enhancement of neutrino oscillations. The parametric enhancement occurs when the parameter of system (the density in our case) changes periodically and the period, r_f, coincides with period of system.

The parametric enhancement of oscillations is due to certain synchronization of oscillation effects in the mantle and in the core. The frequencies of oscillations are different in the core and in the mantle. The enhancement occurs when the frequency change is synchronized with the frequency itself.

The condition (2) means that the size of the layer, L, (in mantle or core) coincides with half of the oscillation length: $L = l_M/2$.

In the approximation of constant densities in the mantle and the core the resonance condition for phases (2) can be written as

$$\Delta H_m L_m = \pi, \quad \Delta H_c L_c = \pi \ . \qquad (4)$$

(In general, the phase should be equal $\pi(2k+1)$, where $k = 0, 1, 2, ...$ fixes the order of resonance.)

In 1987 E. Akhmedov [5] has considered the case of the "castle wall" density profile when the period of perturbation consists of two layers with constant but different densities. The Earth realizes, in a sense, the case of "1.5 period".

The enhancement depends on number of periods (perturbations) and on the amplitude of perturbations which can be characterized by "swing" angle $\Delta\theta \equiv 2\theta_m - 2\theta_c$. For small perturbations, large transition probability can be achieved after many periods. In the Earth the perturbation is large $\Delta\theta \sim 2\theta_c$, and strong effect is realized even for "1.5 periods".

Physics of the effect can be well understood from the graphical representation [6] based on analogy of the neutrino evolution with behaviour of spin of the electron in the magnetic field. Indeed, a neutrino state can be described by vector

$$\vec{\nu} = \left(\text{Re}\psi_\mu^\dagger\psi_s, \ \text{Im}\psi_\mu^\dagger\psi_s, \ \psi_\mu^\dagger\psi_\mu - 1/2\right) \ , \qquad (5)$$

where ψ_i, $(i = \mu, s)$ are the neutrino wave functions. (The elements of this vector are nothing but components of the density matrix.) Introducing vector:

$$\vec{B} \equiv \frac{2\pi}{l_M}(\cos 2\theta_M, \ 0, \ \sin 2\theta_M) \qquad (6)$$

(θ_M is the mixing angle in medium) which corresponds to the magnetic field, one gets from the Schrödinger-like equation for ψ_i the evolution equation

$$\frac{d\vec{\nu}}{dt} = \left(\vec{B} \times \vec{\nu}\right) \ . \qquad (7)$$

In medium with constant density (θ_M = *const*), the evolution consists of $\vec{\nu}$- precession around $\vec{B}$: $\vec{\nu}$ is moves according to increase of the oscillation phase, Φ, on the surface of the cone with axis $\vec{B}$. The direction of the axis, $\vec{B}$, is determined uniquely by $2\theta_M$ (6). We will denote by $\vec{B}_m$ and $\vec{B}_c$ the axis in the mantle and in the core respectively. In fig. 1 we show a projection of the 3-dimensional picture on $(\text{Re}\psi_\mu^\dagger\psi_s, \ \psi_\mu^\dagger\psi_\mu - 1/2)$ plane [3].

The cone angle, θ_{cone} (the angle between $\vec{\nu}$ and $\vec{B}$) depends both on mixing angle and on the initial state. If an initial state coincides with ν_μ, the angle equals $\theta_{cone} = 2\theta_M$. The projection of $\vec{\nu}$ on the axis z, ν_z, gives the probability to find ν_μ in a state $\vec{\nu}$:

$$P \equiv \psi_\mu^\dagger\psi_\mu = \nu_z + \frac{1}{2} = \cos^2\frac{\theta_z}{2} \ . \qquad (8)$$

Here $\nu_z \equiv 0.5\cos\theta_z$, and θ_z is the angle between $\vec{\nu}$ and the axis z.

Let us consider an evolution of the neutrino which crosses the mantle, the core and then again the mantle and for which the resonance condition (2) is fulfilled. In the fig. 1, $2\theta_c < 2\theta_m < \pi/2$, so that both axes $\vec{B}_m$ and $\vec{B}_c$ are in the first quadrant. (Actually, such a situation corresponds to mixing above the resonance $2\theta_c > 2\theta_m > \pi/2$, when the axes are in the second quadrant. In fig. 1 for convenience of presentation we made redefinition $2\theta_c \to \pi - 2\theta_c$, $2\theta_m \to \pi - 2\theta_m$ which does not change result.) The initial state, $\vec{\nu}(1)$, coincides with flavor state, *e.g.*, ν_μ. (The picture corresponds to $\nu_\mu - \nu_s$ mixing considered in [3].) Neutrino first propagates in the mantle and this corresponds to $\vec{\nu}$ precession around

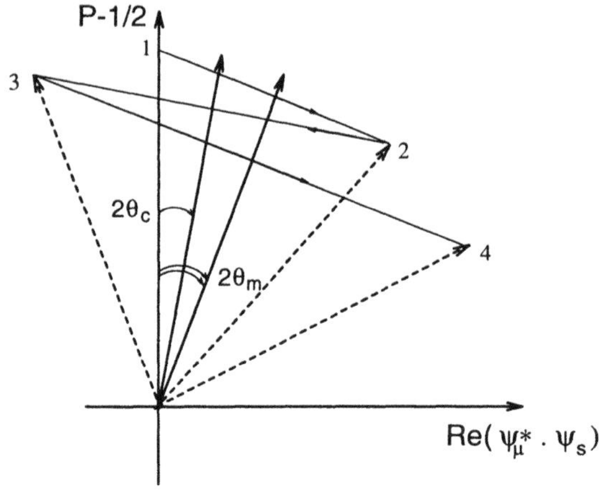

Figure 1. Parametric enhancement of the $\nu_\mu \leftrightarrow \nu_s$ oscillations inside the Earth. Graphical representation of evolution of the neutrino state; the case of parametric resonance. States of neutrino at the borders of the layers are shown by dashed vectors; the cone axes are shown by solid vectors.

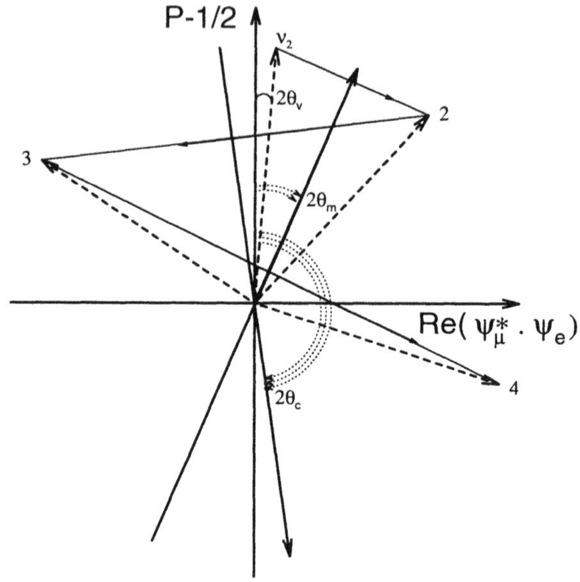

Figure 2. Parametric enhancement of the $\nu_2 \to \nu_e$ oscillations inside the Earth. Graphical representation of evolution of the neutrino state; the case of parametric resonance.

$\vec{B}_m = \vec{B}(2\theta_m)$. At the border between the mantle and the core the neutrino vector is in position $\vec{\nu}(2)$ (which corresponds to phase acquired in the mantle, $\Phi_m = \pi$). At the border the mixing angle changes suddenly: $\theta_m \to \theta_c$. In the core, $\vec{\nu}$ precesses around new position of axis, $\vec{B}_c \equiv \vec{B}(2\theta_c)$, with initial condition $\vec{\nu}(2)$. At the exit from the core, $\vec{\nu}$ will be in position $\vec{\nu}(3)$. When neutrino enters the mantle again, the value of mixing angle jumps back: $\theta_c \to \theta_m$. In the second layer of mantle, $\vec{\nu}$ precesses around $\vec{B}_m$ again. At the detector the neutrino vector will be in position $\vec{\nu}(4)$. After each jump of density the cone angle increases by the value of "swing" angle $\Delta\theta \equiv 2\theta_m - 2\theta_c$, thus enhancing the oscillations. According to fig. 1, a projection of $\vec{\nu}(4)$ on the axis z equals

$$\theta_z = 2\theta_m + 2\theta_m + 2\Delta\theta = 2(4\theta_m - 2\theta_c) \ .$$

Inserting this into (8) we get the survival probability $\cos^2(4\theta_m - 2\theta_c)$ which reproduces result in (3).

In [3] the parametric enhancement has been applied to $\nu_\mu \leftrightarrow \nu_s$ oscillations of atmospheric neutrinos.

It was realized by Petcov [7] that the conditions (2) is fulfilled for solar neutrinos leading to appearance of the peak in the regeneration probability. This allows one to get correct *interpretation* of the dependence of the probability on energy found in a number of papers before [2]. It gives correct understanding of the enhancement mechanism.

Notice that in [7] the condition (2) written in the form (4) has been *renamed* by "oscillation length resonance" and the enhancement due to the condition (4) is considered as a new effect which differs from that discussed in [4–6,3]. This change of the name is unjustified. Indeed, the eq. (4) is the condition on products of inverse oscillation length and width of the layer, that is, on the oscillation *phases*. The resonance associated with equality of phases is the parametric resonance. On the other hand, the MSW resonance can be considered as "the oscillation length resonance": in the MSW resonance the oscillation length coincides for small vacuum mixing with refraction length.

Detailed interpretation of the effect in terms of the parametric resonance has been given in [8].

In the case of solar neutrinos the survival probability (due to the averaging and lost of coherence) depends on the transition probability $\nu_2 \to \nu_e$ inside the Earth, where ν_2 is the heaviest mass eigenstate:

$$P \approx (1 - 2P_\odot)P_{2e} \ . \qquad (9)$$

Here $P_\odot$ is the ν_e survival probability inside the Sun.

Graphical representation of the evolution of the solar neutrinos inside the Earth in the case of parametric resonance is shown in fig. 2. Now $2\theta_c > \pi/2$ and $2\theta_m < \pi/2$, that is, the axis $\vec{B}_m$ is in the first and in the third quadrants, whereas $\vec{B}_c$ is in the second and in the fourth quadrants. Such a situation corresponds to neutrino energies between the MSW resonance energies in the core and in the mantle. (It is easy to show that when $2\theta_m < 2\theta_c < \pi/2$ the oscillations are suppressed.) The initial state is $\vec{\nu}(1) = \nu_2$. Neutrino vector $\vec{\nu}$ first precesses around $\vec{B}_m$ and at the border between the mantle will be in position $\vec{\nu}(2)$. Then in the core, $\vec{\nu}$ precesses around $\vec{B}_c$, with initial condition $\vec{\nu}(2)$, and at the exit from the core $\vec{\nu}$ turns out to be in position $\vec{\nu}(3)$. In the second layer of mantle, the vector $\vec{\nu}$ precesses around $\vec{B}_m$ with initial condition: $\vec{\nu} = \vec{\nu}(3)$, and at the detector it will be in position $\vec{\nu}(4)$. According to fig. 2, a projection of $\vec{\nu}(4)$ on the axis z equals

$$\theta_z = 2(4\theta_m - 2\theta_c) - 2\theta \ , \qquad (10)$$

and consequently, $P_{2e} = \sin^2(2\theta_m - \theta_c + \theta)$ [7], where the difference from (3) is related to difference in the initial state.

One can see from figs. 1 and 2 that enhancement considered in [3] for $\nu_\mu - \nu_s$ oscillations and the one in [7] for $\nu_2 - \nu_e$ are of the same nature: the swing of axes leads to an enhancement of oscillations. The difference is in the initial state and in inclination of the swing angle.

Maximal transition probability (3) can be achieved when the parametric resonance condition is fulfilled exactly. The oscillation phases are functions of the neutrino energy and the zenith angle Θ, and the two resonance conditions $\Phi_c(\Theta, E) = \pi$, $\Phi_m(\Theta, E) = \pi$ can be satisfied only for certain (resonance) values Θ_R and E_R. Deviations from Θ_R and E_R weaken the enhancement. Thus the parametric resonance leads to appearance of the peak (parametric peak) in the energy or/and zenith angle dependence of the transition probability. The width of the parametric peak is inversely proportional to number of periods of density perturbation: $\propto 1/n$ [6]. (The bigger the number of periods the sharper the synchronization condition.) In the case of the Earth the number of periods is small, $n \sim 1.5$, which means that the width of the peak is of the order one. Here the enhancement occurs even for significant detuning.

The probability P_{2e} (as well as $P(\nu_\mu \to \nu_e)$) [2] has rather complicated structure with three large peaks: two of them correspond to the MSW resonance enhancement of oscillations in the core and in the mantle. The third peak is between the MSW peaks and its height is bigger than $\sin^2 2\theta_m$ and $\sin^2 2\theta_c$ at the peak energy fig. 3. The appearance of this third peak associated with resonance condition (2) is the consequence of parametric enhancement. Notice that certain interplay of the oscillation effects in the mantle and in the core leads not only to appearance of the parametric peak but it also modifies the MSW peaks in the mantle and in the core. The MSW peaks become suppressed in comparison with peaks from only one layer (core or mantle).

Although the parametric enhancement can be rather strong: $P_{2e} \sim 1$, the regeneration effect turns out to be suppressed by factor $(1 - 2P_\odot)$ (9). Recent changes in the solar model predictions [9,10] indicate that the suppression can be even stronger than it was supposed before. Indeed, the predicted flux of the boron neutrinos is now smaller (due to smaller cross section of pBe reaction). This means that suppression of the boron neutrino flux due to oscillations should be weaker. We get $P_\odot \sim 0.5$ for the neutrino energy $E \sim 10$ MeV – in the center of the detectable region.

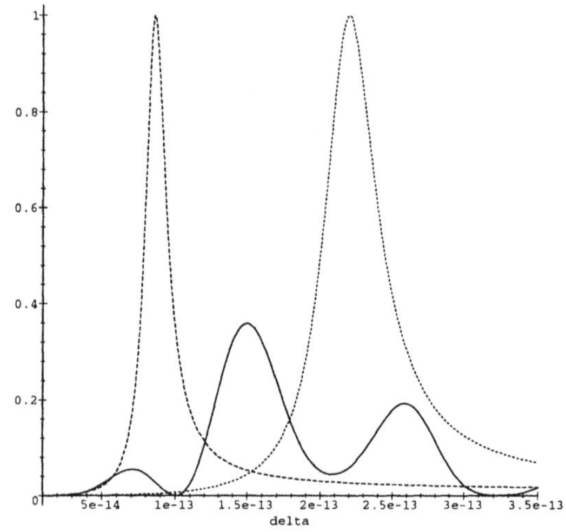

Figure 3. Transition probability for $\nu_e - \nu_\mu$ oscillations in the Earth (solid curve) as the function of *delta* $\equiv \Delta m^2/4E$. Also shown are $\sin^2 2\theta_c$ (dashed curve) and $\sin^2 2\theta_m$ (dotted curve); vacuum angle: $\sin^2 2\theta = 0.01$, the zenith angle $\cos\Theta = -0.88$. (From [8]).

3. Beyond the Solar Neutrino Problem

The solar neutrino problem should be considered in general particle physics context which allows one also to accommodate solutions of other neutrino anomalies, and first of all, the atmospheric neutrino anomaly whose oscillation interpretation has received strong confirmation [11]. In fact, the results on atmospheric neutrinos make even more plausible the solution of the solar neutrino problem in terms of neutrino mass and mixing.

Clearly, the same oscillation channel can not explain both the solar neutrino and the atmospheric neutrino anomalies. One should consider mixing of three or even more neutrino species. This can have some impact on solutions of the $\nu_\odot$-problem. In particular, one may expect additional modifications of the neutrino energy spectrum. On the other hand, the solution of the $\nu_\odot$-problem may shed some light on the origin of other neutrino anomalies.

The distortion can be characterized by a sole *slope parameter* s_e [12] defined as:

$$\frac{N_{osc}}{N_0} \approx R_0 + s_e T_e, \tag{11}$$

where N_{osc} and N_0 are the numbers of events with and without oscillations correspondingly, R_0 is a constant, T_e is the recoil electron energy in MeV, s_e is in the units MeV^{-1}. In fig. 4 we show the slope parameter predicted by different two neutrino solutions of the $\nu_\odot$-problem [13]. The dots correspond to the best fit points of the total rates. The ellipses show the experimental result. Clearly, at the moment it is impossible to make discrimination among solutions.

Let us describe some possibilities beyond simple two neutrino case.

1. In the three neutrino schemes which solve both the solar and the atmospheric neutrino problems there is the hierarchy: $\Delta m_{12}^2 \ll \Delta m_{13}^2$. In this case the heaviest state "decouples" from dynamics of the rest of system (leading to the averaged oscillation result) and the survival probability can be written as

$$P = \cos^4\theta_{e3} P_2 + \sin^4\theta_{e3}\ ,$$

where θ_{e3} describes the admixture of the ν_e in the heaviest state, and P_2 is the two neutrino survival probability which is characterized by Δm_{12}^2 and $\sin^2 2\theta_{12}$. For $\Delta m_{13}^2 > 2 \times 10^{-3}$ eV^2 the CHOOZ [15] and BUGEY [14] experiments give strong bounds on θ_{e3}, and therefore corrections due presence of the third neutrino are small. For $\Delta m_{13}^2 < 10^{-3}$ eV^2, the mixing can be large thus leading to strong modification of the probability. Notice, however, that these changes do not improve the fit of the solar neutrino data. For small θ_{e3}, the solutions of these two problems essentially decouple [16].

2. All three active neutrinos can be involved in the solar neutrino oscillations. This possibility can be naturally realized in the so called Grand

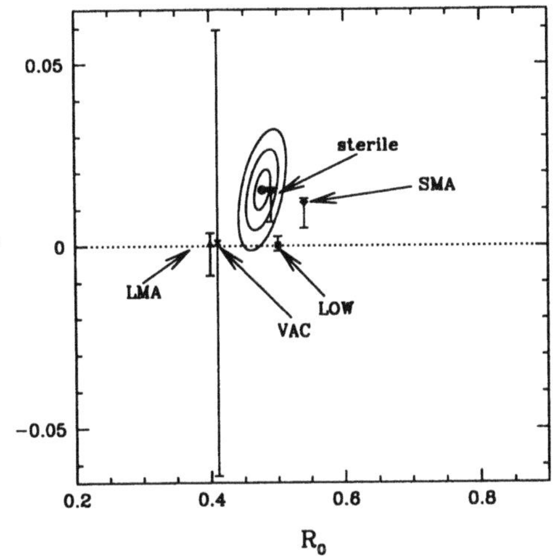

Figure 4. Deviation from an undistorted energy spectrum. The points with error bars show predictions from five possible 2ν - solutions: "SMA" stands for small mixing angle MSW conversion $\nu_e \to \nu_\mu$, "sterile" is the small mixing angle MSW conversion $\nu_e \to \nu_s$, VAC is the "just-so oscillations", LMA is the large mixing angle MSW solution and LOW is the large mixing angle MSW solution with low Δm^2. The points correspond to the best fit points of the total rates in four experiments. The ellipses show 1σ, 2σ and 3σ regions allowed by SK data. The errors in R_0 are large (not shown) so that all solution cross the ellipses in the horizontal scale. (From [13].)

Unification (GU) scenario [17]. Neutrino masses are generated by the see-saw mechanism; the neutrino Dirac mass matrix is similar to the mass matrix of the upper quarks at GU scale; the Majorana mass matrix of the RH neutrinos has weak mixing and linear mass hierarchy with the heaviest eigenvalue at the GU - scale. This scenario predicts naturally $\Delta m_{13}^2 \sim 10^{-5}$ eV^2 – in the range of the MSW solution of the solar neutrino problem and $\Delta m_{12}^2 \sim 10^{-10}$ eV^2 in the "just-so" oscillation region. It also leads to relatively large $\nu_e - \nu_\mu$ mixing. The solar neutrinos undergo both the $\nu_e - \nu_\tau$ resonance conversion and the $\nu_e - \nu_\mu$ oscillations on the way from the Sun to the Earth. The interplay of both effects results in a peculiar (oscillatory) distortion of the boron neutrino energy spectrum [18]. The corresponding distortion of the recoil energy spectrum is shown in fig. 5. Notice that the curve has a kink whose position depends on Δm^2. This may be relevant for interpretation of the SK data.

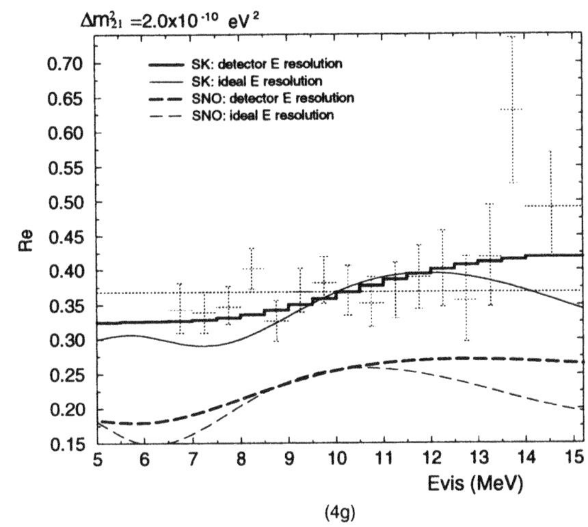

Figure 5. The expected distortion of the recoil electron energy spectrum in the SuperKamiokande (solid lines) and SNO (long dashed lines) experiments for hybrid solution of the $\nu_\odot$-problem with parameters: $\sin^2 2\theta_{e\mu} = 0.5$ $\sin^2 2\theta_{e\tau} = 6 \times 10^{-4}$, $\Delta m_{31}^2 = 8 \times 10^{-6}$ eV^2 and $\Delta m_{21}^2 = 2 \times 10^{-10}$ eV^2. (From [17].)

3. The atmospheric neutrino problem can be solved by oscillations $\nu_\mu \leftrightarrow \nu_s$ which involve the sterile neutrino. This opens a possibility to rescue small *flavor* mixing in lepton sector in analogy with quark mixing. Now inside the Sun the electron neutrino is converted into the mixture of the muon neutrino and sterile neutrino·

$\nu_2 = \cos\theta_{atm}\nu_\mu + \sin\theta_{atm}\nu_s$, where θ_{atm} is the angle responsible for deficit of the atmospheric neutrinos. Correspondingly, properties of this solution of the $\nu_\odot$- problem are intermediate between properties of solutions based on conversion into pure active and pure sterile states. In particular, a distortion of the spectrum is stronger than in pure active case but weaker than in pure sterile case [3] (fig. 6).

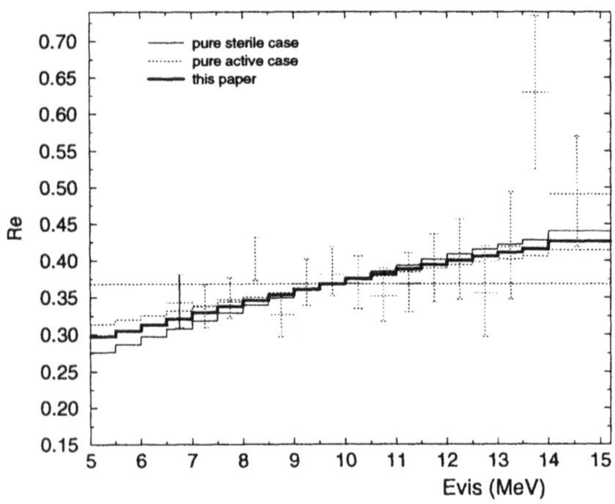

Figure 6. The expected distortion of the recoil electron energy spectrum in the SuperKamiokande experiment. The solid line corresponds to pure $\nu_e - \nu_s$ conversion, dotted line is for $\nu_e - \nu_\mu$, the bold solid line is for the mixed case $\nu_e - \nu_\mu, \nu_s$ with $\Delta m^2 = 5 \times 10^{-6}$ eV^2, $\sin^2\theta_{12} = 8.8 \times 10^{-3}$ and $\sin^2 2\theta_{atm} = 1$.

4. In the supergravity, the hidden sector and the observable sector communicate via the Planck scale $(1/M_P)$ suppressed interactions. In particular, a singlet field S from the hidden sector may have the coupling $(m_{3/2}/M_P)lHS$, where $m_{3/2} \sim 1$ TeV is the gravitino mass, H and l are the Higgs and the lepton doublets correspondingly. This interaction generates the $\nu - S$ mixing mass term

$$m_{es} = \frac{m_{3/2}}{M_P}\langle H\rangle \sim 10^{-4} \text{ eV}, \tag{12}$$

where $\langle H\rangle$ is the VEV of H [19].

Consequences of this mixing depend on the mass of the scalar, m_S. It turns out that for $m_S \sim m_{3/2}^2/M_P \sim 3 \times 10^{-3}$eV one gets $\Delta m^2 \sim 10^{-5}$ eV^2 and mixing angle $\sin^2 2\theta \sim 10^{-2}$, so that the $\nu_e \to S$ resonance conversion can solve the $\nu_\odot$- problem[19].

If m_S differs from the above value substantially, the other channel, *e.g.*, $\nu_e \to \nu_\mu$, can give a solution of the problem. In this case the $\nu_e - S$ mixing will modify the two neutrino effect. For $m_S > m_2$ ($m_2 \sim 3 \times 10^{-3}$eV), the $\nu_e - S$ mixing can lead to a dip in the non-adiabatic edge of the suppression pit at $E \sim (m_S/m_2)^2 E_a$, where $E_a \sim (0.5 - 0.7)$ MeV is the energy of the adiabatic edge. This will manifest as a dip in the recoil electron spectrum and can be relevant for explanation of the spectrum observed by the SuperKamiokande. Also flavor composition of the neutrino flux will depend on energy. The flux of the beryllium ν_e neutrinos is converted mainly to ν_μ, whereas boron neutrinos are transferred both to ν_μ and S. Correspondingly, an effect of the neutral currents is larger for low energies. Comparison of signals in BOREXINO and SNO experiments will check this effect.

4. Time Variations Versus Distortion

Existing solutions of the $\nu_\odot$-problem lead to specific correlations between time variations of signals and spectrum distortion. Therefore, using the data on spectrum distortion one can make predictions for time variations and *vice versa*. A study of these correlations strengthens the possibility to identify the solution.

1. For vacuum oscillation solution there is a strict correlation between a spectrum distortion and the amplitude of seasonal variations of neutrino flux [20]. The seasonal variations are due to ellipticity of the Earth orbit. The correlation originates from dependence of the oscillation probability P on the neutrino energy and distance to the Sun. Indeed, the phase of oscillations is

proportional to $\Phi \propto L/E$ which gives immediately

$$\frac{dP}{dL} = -\frac{dP}{dE} \cdot \frac{E}{L} \ . \tag{13}$$

Here $P^{-1}dP/dE$ is the slope of the neutrino spectrum distortion. According to (13), a positive slope, $dP/dE > 0$, is accompanied by decrease of probability with distance, so that the seasonal variations due to geometrical factor, L^{-2}, will be enhanced. In the case of negative slope, oscillations will suppress the seasonal variations due to geometrical factor.

The correlations can be expressed as correlations between the slope parameter for the energy spectrum of the recoil electrons (11) and the summer-winter asymmetry defined as

$$A_e \equiv 2\frac{N_W - N_S}{N_{SP} + N_A} . \tag{14}$$

Here N_W, N_S , N_{SP}, N_A are the numbers of events detected from November 20 to February 19, from May 22 to August 20, February 20 to May 21, from August 21 to November 19 respectively. It is convenient to describe the asymmetry due to oscillations by the parameter

$$r_e \equiv \frac{A_e}{A_e^0} - 1 \ , \tag{15}$$

where A_0 is the asymmetry related to the geometrical factor. Obviously, $r_e = 0$ in the no-oscillation case; $r_e > 0$ ($r_e < 0$) corresponds to enhancement (damping) of the geometrical effect. Fig. 7 shows the $s_e - r_e$ correlation. For the best fit value of the slope (fig. 4) we get $r_e \sim 0.4$, so that one expects an enhancement of asymmetry. This can be checked after 4 - 5 years of the SK operation.

2. In the case of the MSW solution there is a correlation between the day-night asymmetry and spectrum distortion. This helps do disentangle the large and small mixing solutions of the problem [21]. For large mixing solution one expects strong day-night asymmetry and weak distortion of the spectrum. In contrast, for small mixing solution stronger spectrum distortion is accompanied by weak day-night effect. In fig. 8 the distortion of spectrum is characterized by deviation of the average electron kinetic energy T_e from its standard value without oscillations. As follows from the figure the data favor a small mixing solution.

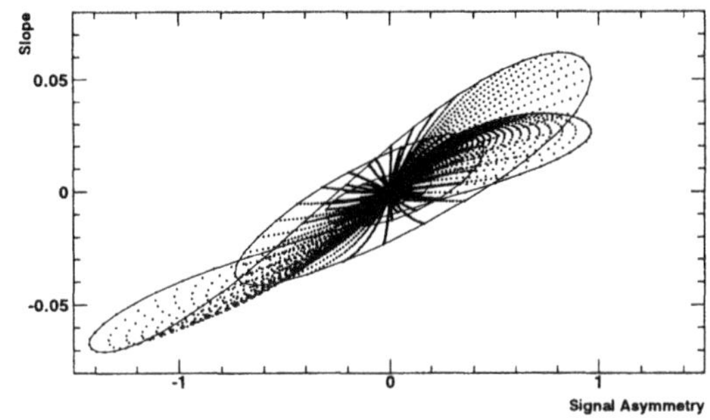

Figure 7. The slope - asymmetry plot. The points correspond to different values of Δm^2 between 10^{-11} and 10^{-9} eV^2, and $\sin^2 2\theta$ between 0.25 and 1.00. The solid line shows changes of the slope and asymmetry with Δm^2 for maximal mixing.

3. The correlation of the day-night effect and spectrum distortion allows one also to disentangle solutions based on conversion to active and to sterile neutrinos. Main difference comes from presence of the $\nu_\mu(\nu_\tau)$ contribution to νe-scattering in the case of active neutrino conversion. This contribution, being proportional to $(1 - P(E))$, leads to smearing of the spectrum distortion. Therefore for the same values of parameters the distortion is stronger in the sterile case. In contrast, the regeneration effect is weaker in the sterile neutrino case. This is related to the fact, that in the $\nu_e - \nu_s$ case the effective potential (which describes matter effect) is approximately two times smaller than in the $\nu_e - \nu_\mu$-case. Thus for $\nu_e - \nu_\mu$ conversion one expects larger day-night asymmetry and smaller slope, whereas $\nu_e - \nu_s$ conversion leads to larger slope but weaker asymmetry. In fig. 9 we show projection of the $(\Delta m^2, \sin^2 2\theta)$ regions of small mixing solutions onto D/N-asymmetry - slope plot which illustrates the correlation [22]. The correlation is solar model dependent. For the model BP95 [23]

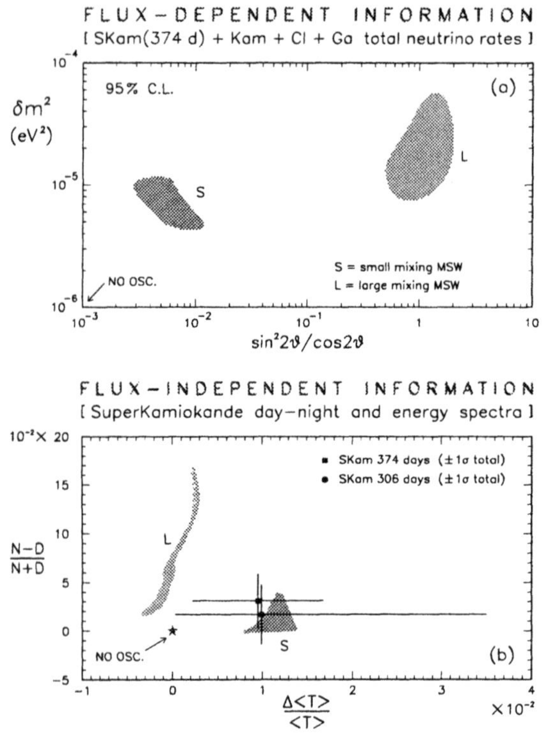

Figure 8. The day-night asymmetry - spectrum distortion plot. The distortion is characterized by the mean kinetic energy deviation. In panel (b) the regions show the map of the small (S) and large (L) mixing solutions at 95 % C. L. in the mass-mixing plane (panel(a)). (From [21].)

the regions corresponding to two channels of conversion are well separated. However in the models with smaller boron neutrino flux (see e.g. [9], [10]) both the slope and the D/N asymmetry become smaller and the two regions overlap. The identification of solutions (using this correlation) will be difficult. Notice that for small original boron neutrino flux the D/N asymmetry is negative in whole region of the $\nu_e - \nu_s$ solution and in part of the $\nu_e - \nu_\mu$ region. This is related to the fact, that for a small flux a required oscillation suppression should be weak, so that the survival probability $P_\odot > 1/2$ (see (9)). Moreover, due to additional contribution from ν_μ the $\nu_e - \nu_\mu$ solution requires stronger suppression. The two solutions can be also distinguished by measurements of the neutral current effect in SNO.

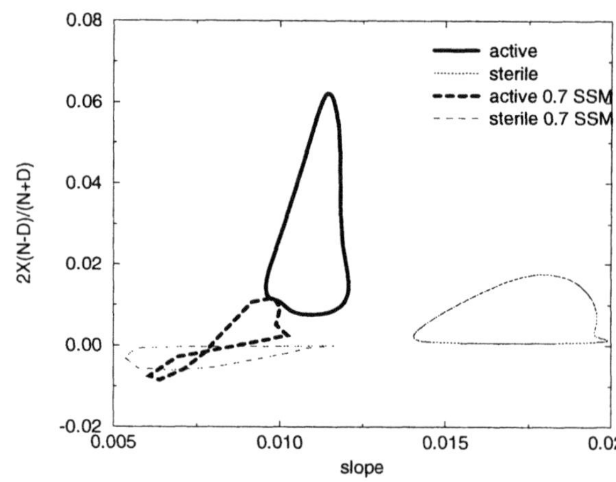

Figure 9. The slope - D/N-asymmetry plot [22]. The regions of predictions of small mixing MSW solutions: $\nu_e - \nu_\mu$ (bold lines), $\nu_e - \nu_s$ (thin lines). Solid lines correspond to solar model BP95, dashed lines are for BP95 model with diminished (by factor 0.7) boron neutrino flux.

5. Conclusion

Oscillations of neutrinos crossing the core of Earth can be parametrically enhanced. This leads to appearance of the parametric peak in the oscillation probability as function of neutrino energy. The parametric enhancement can be relevant for solar and atmospheric neutrinos as well as for neutrinos from supernova. Strong enhancement of the regeneration probability for solar neutrinos which cross the core is due to the parametric resonance.

Solution of the solar neutrino problem should be considered in wider particle physics context which allows one to explain, *e.g.*, the atmospheric neutrino problem. Under certain conditions the two problems "decouple" and the solution is still reduced to simple two neutrino case. However, in a number of schemes one gets modification of the simple two neutrino effect. This can manifest as complicated distortion of the neutrino (and the recoil electron) energy spectrum and also can lead to a peculiar change of the flavor composition of the solar neutrino flux with energy.

Precise measurements of spectrum can reveal physics "beyond the solar neutrino problem". One possibility is the Planck mass suppressed couplings of neutrinos with particles from the hidden sector.

Different solutions of the solar neutrino problem lead to specific correlations between the spectrum distortion and time variations of fluxes. This can be used to distinguish solutions.

Recent experimental data and new calculations of the fluxes require smaller oscillation effects (smaller mixing angles etc.), so that the identification of the solution becomes more difficult.

REFERENCES

1. Y. Suzuki, (these proceedings).
2. S. P Mikheyev, A. Yu. Smirnov, In *'86 Massive Neutrinos in Astrophysics and in Particle Physics*, Proc. of the 6th Moriond workshop, edited by O. Fackler and J. Tran Than Van, p 355. J. Bouchez *et al*, Z. Phys. C **32** (1986) 499; E. D. Carlson, Phys. Rev D **34**, 1454 (1986); M. Cribier, W. Hampel, J. Rich, and D. Vignaud, Phys. Lett. B **182**, 89 (1986); A. J. Baltz and J. Weneser, Phys. Rev. D **35**, 528 (1987); A. Dar, A. Mann, Y. Melina, and D. Zajfman, Phys. Rev. D **35**, 3607 (1988); G. Auriemma, M. Felcini, P. Lipari and J. L. Stone, Phys. Rev. D **37**, 665 (1988); A. Nicolaidis, Phys. Lett. B **200**, 553 (1988); P. I. Krastev and S. P. Petcov, Phys. Lett. **B205**, 84 (1988); J. M. LoSecco, Phys. Rev. D **47**, 2032 (1993); J. M. Gelb, W.-K. Kwong, and S. P. Rosen, Phys. Rev. Lett. **78**, 2296 (1997); Q. Y. Liu, M. Maris and S. T. Petcov, Phys. Rev. D **56**, 5991 (1997); M. Maris and S. T. Petcov, Phys. Rev. D **56**, 7444 (1997); M. Maris and S. T. Petcov, hep-ph/9803244.
3. Q. Y. Liu and A. Yu. Smirnov, Nucl. Phys. B524 (1998) 505, hep-ph/9712493; Q. Y. Liu, S. P. Mikheyev and A. Yu. Smirnov, hep-ph/9803415.
4. V. K. Ermilova, V. A. Tsarev and V. A. Chechin, Kr. Soob, Fiz. [Short Notices of the Lebedev Institute] **5**, 26 (1986).
5. E. Kh. Akhmedov, preprint IAE-4470/1, (1987); Yad. Fiz. **47**, 475 (1988) [Sov. J. Nucl. Phys. **47**, 301 (1988)].
6. P. I. Krastev and A. Yu. Smirnov, Phys. Lett. B **226**, 341 (1989).
7. S. T. Petcov, preprint SISSA 31/98/EP, hep-ph/9805262.
8. E. Kh. Akhmedov, hep-ph/9805272.
9. J. N. Bahcall, S. Basu, M.H. Pinsonneault, Phys. Lett. B **433** (1998), 1.
10. A. S. Brun, S. Turck-Chieze and P. Morel, astro-ph/9806272.
11. T. Kajita, these proceedings.
12. see *e.g.* W. Kwong, S. P. Rosen, Phys. Rev. D **51** (1995) 6159.
13. J. Bahcall, P. Krastev and A. Smirnov, hep-ph/9807216.
14. B Achkar et al., Nucl. Phys. B **434**, (1995) 503.
15. CHOOZ collaboration, M. Apollonio et al. hep-ex/9711002.
16. see *e.g.* C. Giunti, hep-ph/9802201.
17. K.S. Babu, Q.Y. Liu, A.Yu. Smirnov, Phys. Rev. D **57** (1998) 5825, hep-ph/9707457.
18. A. Yu. Smirnov, Proc. of the Int. symposium *Frontiers of Neutrino Astrophysics*, Takayama, October 1992 Ed. Y. Suzuki and K. Nakamura, (1992) p. 105; Q.Y. Liu, S.T. Petcov, Phys. Rev. D **56** (1997) 7392; B.C. Allanach, G.K. Leontaris, S.T. Petcov, Phys. Lett. B **431** (1998) 98.
19. K. Benakli, A. Yu. Smirnov, Phys. Rev. Lett. **79** (1997) 4314.
20. S.P. Mikheyev, and A.Yu. Smirnov, Phys. Lett. B **429** (1998) 343.
21. G.L. Fogli, E. Lisi, D. Montanino, hep-ph/9803309.
22. Q. Y. Liu and A. Yu. Smirnov (in preparation).
23. J. N. Bahcall and M. H. Pinsonneault, Rev. Mod. Phys. **67** (1995) 781.

Part 3

Atmospheric Neutrinos

NUCLEAR PHYSICS B
PROCEEDINGS
SUPPLEMENTS

ELSEVIER

Nuclear Physics B (Proc. Suppl.) 77 (1999) 111–116

Atmospheric Neutrino Studies in Soudan 2

Earl Peterson (for the Soudan 2 Collaboration) [a]

[a]School of Physics and Astronomy, University of Minnesota,
116 Church St. SE, Minneapolis MN, 55455 USA

We report a measurement of the atmospheric neutrino flavor ratio, R, using a sample of quasi-elastic neutrino interactions occurring in an iron calorimeter. The flavor ratio (tracks/showers) of atmospheric neutrinos in a 3.9 kiloton-year exposure of Soudan 2 is $0.64 \pm 0.11(stat.)\ ^{+0.06}_{-0.05}(syst.)$ of that expected. A preliminary look at a higher resolution sample, suitable for oscillation (L/E) analysis is also discussed.

1. Introduction

Six experiments have reported results on the flavor ratio of sub-GeV atmospheric neutrinos as measured in underground detectors [1–6]:

$$R = \frac{[(\nu_\mu + \overline{\nu}_\mu)/(\nu_e + \overline{\nu}_e)]_{Data}}{[(\nu_\mu + \overline{\nu}_\mu)/(\nu_e + \overline{\nu}_e)]_{MC}}. \quad (1)$$

These measurements suggest a value of R significantly lower than unity. The highest statistics on this measurement come from the water Cerenkov experiments Kamiokande, IMB, and SuperKamiokande. Three iron calorimeter experiments, NUSEX, Frejus, and Soudan 2, have reported results. Our previous result [6], $R=0.72 \pm 0.19(\text{stat.})^{+0.05}_{-0.07}(\text{syst.})$ was based on an exposure of 1.52 kton-years. The confirmation of the low atmospheric flavor ratio with good statistical significance in a calorimeter would provide additional evidence that there is no unknown source of systematic error in water detectors.

There are three stages involved in our analysis. First a sample of contained events is isolated. These are then classified for neutrino flavor. Finally a background subtraction is made and a value of R calculated. Each of these stages, particularly the flavor classification, could introduce bias into the flavor ratio measurement. We have therefore checked the procedure by using different analyses: they give consistent results and confirm the validity of our principal result.

Sections 3-4 describe our principal result, the flavor ratio from a sample of quasi-elastic interactions from a 3.9 fiducial kiloton-year exposure. By performing an analysis in which computer programs replace the scanning (an automated analysis), we have verified that the main procedure does not introduce biases due to subjectivity in the scanning. We have also checked our method of background subtraction and R calculation by an additional analysis in which an alternative method for background estimation is performed (a multivariate analysis). Section 5 is a preliminary discussion of oscillation analysis based on a high-resolution subsample of our data.

2. The Soudan 2 Detector

The Soudan 2 detector is a 963 ton fine-grained gas tracking calorimeter located in the Soudan Underground Mine State Park, Soudan, Minnesota. The detector currently operates with 90% live time and has been taking data since 1989. It consists of 224 1 meter x 1 meter x 2.7 meter iron modules weighing 4.3 tons each. Ionization deposited in the plastic tubes of a module drifts in an electric field to the faces of the module where it is detected by vertical anode wires and horizontal cathode strips. The third co-ordinate of the charge deposition is determined from the drift time in the module. The calorimeter modules operate in proportional mode; the measured pulse height is proportional to the ionization deposited in the tube. Pulse height measurements are used for particle identification. More details of the module construction and performance can be found in References [7,8].

The detector is surrounded by a 1700 m^2 ac-

PII S0920-5632(99)00405-3

tive shield mounted on the cavern walls. It has a measured efficiency of 95% for cosmic muons crossing a shield element. The complete shield covers about 97% of the total solid angle. Reference [9] contains more information about the shield.

3. Data Analysis

The data described in this paper come from a 3.9 kiloton-year exposure taken between April 1989 and January 1998. During this time some 100 million triggers were recorded.

3.1. Contained Event Selection

The initial stage of our data analysis consists of isolating a sample of contained events (in which all tracks and the main body of any showers are located within the fiducial volume, 20 cm inside the detector). Events are then scanned by physicists to finalize the containment. Monte Carlo events are interspersed with data so that the scanner does not know if he or she is scanning Monte Carlo or data. Shield information is not displayed so that the scanner does not know if the event is due to a neutrino or background. The contained event selection is fully described in References [6,10].

The Monte Carlo sample used in this analysis is 5.45 times the size of the expected neutrino sample and reproduces the actual performance of the Soudan 2 detector to a high degree of accuracy. The real detector geometry is simulated, as are local variations in the detector performance, particularly pulse height and drifting. Background hits in the detector are included by overlaying randomly initiated triggers onto Monte Carlo events. Data and MC events are analyzed identically at each stage of the data reduction.

3.2. Flavor Classification

During scanning, events are classified into one of three categories: single track, single shower, and multiprong. The single track category is further subdivided into mu-like tracks and protons as described in the following paragraph. The track and shower categories include primarily ν_μ and ν_e quasi-elastic scattering respectively; they are largely equivalent to the 'single ring' category in the water Cerenkov experiments. In addition to the lepton, events in these categories may contain recoil nucleons and/or small showers from muon decay at track endpoints. Events with two or more particles (other then recoil nucleons) emerging from the primary vertex, or single track events which are charged pions having visible scatters, are classified as multiprong. Some multiprong events whose flavor is clear are isolated (along with events with recoils) for the L/E analysis, discussed below.

Proton tracks can be identified because they are straight and highly ionizing. All tracks are fitted to a straight line trajectory and the track residual and average pulse height are calculated. Tracks with low fit residuals and high average pulse height are classified as protons. There is some overlap between protons and short, low energy muons where most of the observed track has $\beta << 1$. The separation algorithm is tuned to minimize the incorrect tagging of muons as protons. Muon tracks are incorrectly classified 4% of the time and 80% of protons are correctly identified.

4. The Flavor Ratio

4.1. Shield Classification

Contained events are a mixture of neutrino interactions and background processes. Neutral particles which originate with the interaction of cosmic ray muons in the rock surrounding the detector cavern are the principal source of background. These particles (neutrons and photons) can produce contained events if they travel into the fiducial volume of the detector before interacting. Such events are usually accompanied by large numbers of charged particles which strike the active shield located at the cavern walls. Shield activity therefore provides a tag for background events.

The shield information allows us to identify two separate event samples in our data. An event with zero shield hits is referred to as 'gold'; such an event is a neutrino candidate. Events with two or more shield hits are referred to as 'rock' events; they comprise a shield-tagged background sample. Table 1 gives the number of 'gold', 'rock',

Table 1
Raw numbers of gold, rock (shield-tagged background) and Monte Carlo events in each of the 4 categories.

Event Type	Track	Shower	Multi-prong	Proton
Gold	95	151	125	49
Rock	278	472	232	277
MC	749	729	711	82

and Monte Carlo events in each of the scanned categories. (Data events with one shield hit are a mixture of neutrino events and background and are excluded from our analysis.)

4.2. Background Corrections

Some muon interactions in the rock produce contained events *unaccompanied* by shield hits, due either to shield inefficiency or because the interaction did not produce any charged particles which entered the shield. The number of such interactions is determined by examining the distributions of event depths in the detector, where the event depth is defined as the minimum distance between the event vertex and the detector exterior, excluding the detector floor. These are shown in Figure 1.

We fit the depth distributions to determine the amount of background present in the gold sample. An extended maximum likelihood fit is performed which describes the data distributions as a sum of background (from the rock) and neutrino (from the MC) distributions. The fit determines that the track/shower ratio for background is 0.59 ± 0.04. We have previously shown that the track/shower ratio of the background does not vary as a function of shield hit multiplicity [6]. We therefore expect background present in the gold (zero shield hit) sample to occur in this same track/shower ratio and we include this expectation as a constraint in the fit. Note that the 'flavor ratio' in the background is very similar to that measured for neutrino events, hence the presence of background does not produce a large change in our measured ratio.

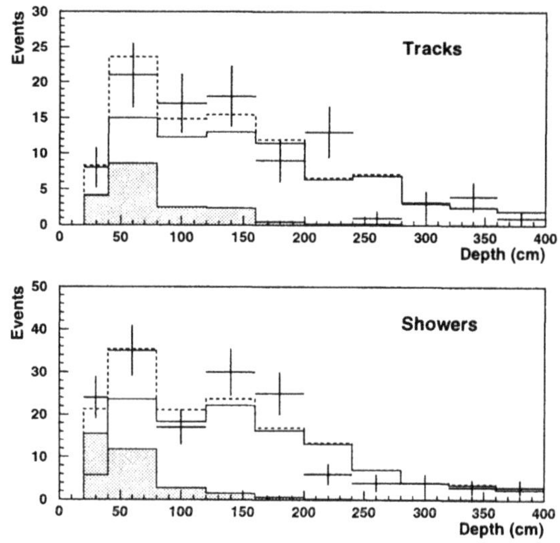

Figure 1. Depth distributions. Gold data are crosses. The rock distributions (shaded histograms) are normalized to the amount of background present in the gold sample as determined by the depth fit. The MC distributions (open histograms) are normalized to the number of neutrino events present in the gold data as determined by the depth fit. The dashed histogram shows the best fit to the data.

4.3. The Ratio

The results of the depth fits are that 76.9 ± 10.8 of the gold tracks and 116.3 ± 12.8 of the gold showers are due to neutrino interactions.

We use these numbers to calculate the background corrected atmospheric neutrino flavor ratio. Table 2 shows the flavor ratios with and without ('raw') the background subtraction.

The systematic error due to the background subtraction has two components.

1. The proton classification, which removes single protons from the track sample and places them in a separate category, serves as a background correction even before the depth fits are performed. An alternative approach to the one we have taken is to leave

Table 2
Data used in the calculation of the corrected flavor ratio. The Monte Carlo numbers in parentheses are normalized to the detector exposure. The error on the flavor ratio is statistical only.

Number of Gold Tracks	95
Number of Gold Showers	151
Number of MC Tracks	749 (137.4)
Number of MC Showers	729 (133.8)
Corrected Number of ν Tracks	76.9 ± 10.8
Corrected Number of ν Showers	116.3 ± 12.8
Raw Value of R	0.61 ± 0.09
Corrected Value of R	0.64 ± 0.11

the single protons in the track sample and determine the amount of background solely from the depth fits. The resulting value for the flavor ratio differs from our main value by $\delta R = +0.023$; the full difference is taken as a component of the systematic error.

2. Our method assumes that any background present in the gold sample behaves identically to the shield-tagged background of the rock sample. We have relaxed this assumption in various ways, and have determined the resulting uncertainty on R to be $\delta R = ^{+0.041}_{-0.030}$.

3. Systematic errors due to the uncertainty in the expected flavor ratio, Monte Carlo, and scanning procedure are calculated to be $\delta R = \pm 0.040$ [6].

The total error on R due to the background subtraction can be obtained by adding the contributions in quadrature: $\delta R = ^{+0.062}_{-0.054}$. Our primary measurement of the flavor ratio is therefore $R = 0.64 \pm 0.11(stat.)^{+0.06}_{-0.05}(syst.)$.

5. Neutrino Oscillation Analysis

Atmospheric neutrinos provide a laboratory for studying neutrino oscillations because they span a significant range in the important oscillation variable, L/E. The L/E dynamic range of atmospheric neutrinos in Soudan 2 is 10^1 (km/GeV) to 10^5 (km/GeV). Hence atmospheric neutrinos provide potential sensitivity to detecting neutrino oscillations signatures over four decades in Δm^2.

5.1. High-Resolution Data

In practice, the ability to identify an oscillation signature in an L/E distribution is mainly limited by the measurement of the incident neutrino direction. The neutrino directional measurement is smeared by detector resolution, target Fermi motion, and the failure to image all final state particles. We have found that by placing energy cuts on the data we can obtain a subsample of events which have the potential for good directional measurement, and hence better L/E determination [11]. The preliminary cuts that isolate this sample are:

- **Tracks and Showers**

  $P_{lept} > 150$ MeV/c if a recoil is present

  $P_{lept} > 600$ MeV/c if no recoil is present

- **Multiprongs**

  $E_{vis} > 700$ MeV

  $P_{vis} > 450$ MeV/c

  $P_{lept} > 250$ MeV/c .

Table 3 shows the effect of the high-resolution cuts on each data sample. Around 48% of the gold (neutrino candidate) events pass the high-resolution cuts. The high-resolution cuts are also very effective at eliminating background, which is predominantly low energy. As Table 3 shows, the amount of background present in the high-resolution sample (5.8 events) is an order of magnitude lower than that present in the total sample (56.2 events). This is important for an L/E analysis, since the background is mainly downwardgoing and may produce distortions in the measured L/E distribution. Also, at higher energy, the flavor misidentification is reduced (to 3%). For the high-resolution cuts, the average angle between the reconstructed neutrino direction and the true neutrino direction is 33.2° for the ν_μ CC sample and 21.3° for the ν_e CC sample.

Table 3
Numbers of events passing the high-resolution cuts. Numbers of MC events are normalized to 3.9 kty.

	CC Track	CC Shower	MP Track	MP Shower	Total ν_μ CC	Total ν_e CC
Gold	35	68	29	41	64	109
Background	2.8	2.1	0.4	0.5	3.2	2.6
MC (norm.)	78.3	60.6	33.2	40.6	111.5	101.2
			Preliminary			

5.2. Energy Distributions

After all identified particles in the event have been reconstructed, the neutrino energy and direction are determined. The neutrino momentum is simply the sum of the momenta of all particles in the event. For events with an identified recoil, the reconstructed neutrino energy is the total visible energy in the event. For events without an identified recoil, 111 MeV is added to the total visible energy in forming the neutrino energy. This number is determined from the Monte Carlo and is the average kinetic energy of the recoil in events where it is not visible. The energy resolution of the high-resolution sample is around 20% ($\Delta E/E$). The average reconstructed energy of ν_μ CC (ν_e CC) interactions is 1.0 (1.7) GeV.

5.3. L/E Distributions

The L/E distributions for the background subtracted preliminary data are compared to the Monte Carlo expectation in Figure 2. The most obvious feature is the overall deficit of ν_μ CC events compared with expectation. For the high resolution sample the flavor ratio (Data/MC) is R(high-res)=0.52 ± 0.09 (stat. only).

If we compare our data to Monte Carlo samples that include neutrino oscillations (of $\nu_\mu \rightarrow \nu_\tau$, where ν_τ CC interactions are ignored) preliminary indications are that $\sin^2(2\theta) = 1$ and $\Delta m^2 = 10^{-2}$ eV^2 provide a satisfactory fit. This is indicated in the figure by the abrupt deviation of the data just before log(L/E) of 2.0. The χ^2 values for these parameters is 12.7 for 12 degrees of freedom.

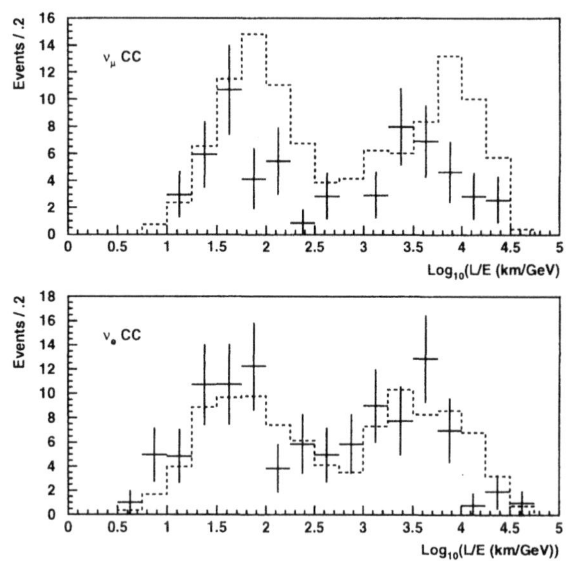

Figure 2. Preliminary L/E distributions for ν_μ CC and ν_e CC background subtracted data (crosses) and the Monte Carlo expectation (dashed histogram). The (unoscillated) MC is normalized to 3.9 kiloton-years data.

6. Conclusion

The flavor ratio of atmospheric neutrinos (data/MC) has been measured from a 3.9 kiloton-year exposure of the Soudan 2 detector to be $0.64 \pm 0.11(stat.)^{+0.06}_{-0.05}(syst.)$. This result is obtained after applying a background correction to a sample of 246 quasi-elastic neutrino candidates. Since our event acceptance and particle misidentification are different from those of the water Cerenkov experiments we would not necessarily expect to measure the same value of R. The probability of a statistical fluctuation from a true value of 1.0 to R=0.64 or below is less than 4×10^{-3}.

Two independent analyses have been carried out which check the contained event selection, flavor determination, and background correction procedures of our main analysis. This measurement is in good agreement with the previously published result from this experiment as well as the results from the water Cerenkov experiments.

A 'high-resolution' sample of events has been identified which has the capability of precision directional measurements. Preliminary indications from this sample indicate that a good fit to $\nu_\mu \rightarrow \nu_\tau$ can be obtained with $\Delta m^2 = 10^{-2}$ eV^2 and $\sin^2(2\theta) = 1$.

Acknowledgements

We acknowledge the support of the U.S. Department of Energy, the State and University of Minnesota and the U.K. Particle Physics and Astronomy Research Council. We would also like to thank: the Minnesota Department of Natural Resources for allowing us to use the Soudan Underground Mine State Park; the staff of the Park, particularly Park Managers D. Logan, P. Wannarka and J. Essig, for their day to day support; and Messrs B. Anderson, J. Beaty, G. Benson, D. Carlson, J. Eininger, J. Meier and W. Miller of the Soudan Mine Crew.

REFERENCES

1. Kamiokande Collaboration: K.S. Hirata *et al.*, Phys. Lett. B 205 (1988) 416. Kamiokande Collaboration: K.S. Hirata *et al.*, Phys. Lett. B 280 (1992) 146. Kamiokande Collaboration: Y. Fukuda *et al.*, Phys. Lett. B 335 (1994) 237.
2. IMB Collaboration: D. Casper *et al.*, Phys. Rev. Lett. 66 (1991) 2561. IMB Collaboration: R. Becker-Szendy *et al.*, Phys. Rev. D 46 (1992) 3720.
3. SuperKamiokande Collaboration: Y. Fukuda *et al.*, Phys. Rev. Lett. 81 (1998) 1562.
4. NUSEX Collaboration: M. Aglietta *et al.*, Europhys. Lett. 8 (1989) 611.
5. Frejus Collaboration: K. Daum *et al.*, Z. Phys. C 66 (1995) 417.
6. Soudan 2 Collaboration: W.W.M. Allison *et al.*, Phys. Lett. B 391 (1997) 491.
7. Soudan 2 Collaboration: W.W.M Allison *et al.*, Nucl. Instrum. Methods A 376 (1996) 36.
8. Soudan 2 Collaboration: W.W.M Allison *et al.*, Nucl. Instrum. Methods A 381 (1996) 385.
9. W.P. Oliver *et al.*, Nucl. Instrum. Methods A 276 (1989) 371.
10. H. Gallagher, Neutrino Oscillation Searches with the Soudan 2 Detector, PhD Thesis, University of Minnesota (1996).
11. H. Tom *et al.* 'Search for Neutrino Oscillation Effects Using Neutrino Zenith Angle and L/E Distributions in Soudan 2', Internal Memo PDK-699, March 1998.

NUCLEAR PHYSICS B
PROCEEDINGS
SUPPLEMENTS

ELSEVIER

Nuclear Physics B (Proc. Suppl.) 77 (1999) 117–122

Atmospheric neutrino induced muons in the MACRO detector

F. Ronga (for the MACRO collaboration)

INFN Laboratori Nazionali di Frascati,
P.O. Box 13 I-00044 Frascati Italy

A measurement of the flux of neutrino-induced muons using the MACRO detector is presented. Different event topologies, corresponding to different neutrino parent energies can be detected. The upward throughgoing muon sample is larger event sample. For this sample, produced by neutrinos having an average energy around 100 GeV, the ratio of the number of observed to expected events integrated over all zenith angles is $0.74 \pm 0.036_{stat} \pm 0.046_{syst} \pm 0.13_{theor}$. We have investigated whether the observed number of events and the shape of the zenith distribution can be explained by an hypothesis of $\nu_\mu \rightarrow \nu_\tau$ oscillation. The best probability (17%) is obtained for $\sin^2 2\theta \simeq 1.0$ and Δm^2 of a few times 10^{-3} eV^2, while the probability for the no oscillation hypothesis is 0.1 %. The other samples are due to the internally produced events and to upward-going stopping muons; the average parent neutrino energy is of the order of 4 GeV. The low energy data sets show a deficit of observed events similar to the one predicted by the oscillation model with maximum mixing suggested from the upward throughgoing muon sample.

1. Introduction

The interest in precise measurements of the flux of neutrinos produced in cosmic ray cascades in the atmosphere has been growing over the last years due to the anomaly in the ratio of contained muon neutrino to electron neutrino interactions. The observations of Kamiokande, IMB and Soudan 2 are now confirmed by those of SuperKamiokande with larger statistics and the anomaly finds explanation in the scenario of ν_μ oscillations [1].

The effects of neutrino oscillations have to appear also in higher energy ranges. The flux of muon neutrinos in the energy region from a few GeV up to a few TeV can be inferred from measurements of upward throughgoing muons [2]. As a consequence of oscillations, the flux of upward throughgoing muons should be affected both in the absolute number of events and in the shape of the zenith angle distribution, with relatively fewer observed events near the vertical than near the horizontal due to the longer path length of neutrinos from production to observation.

Here the measurement about the high energy muon neutrino flux is presented, together with the first results on low-energy neutrino events in MACRO.

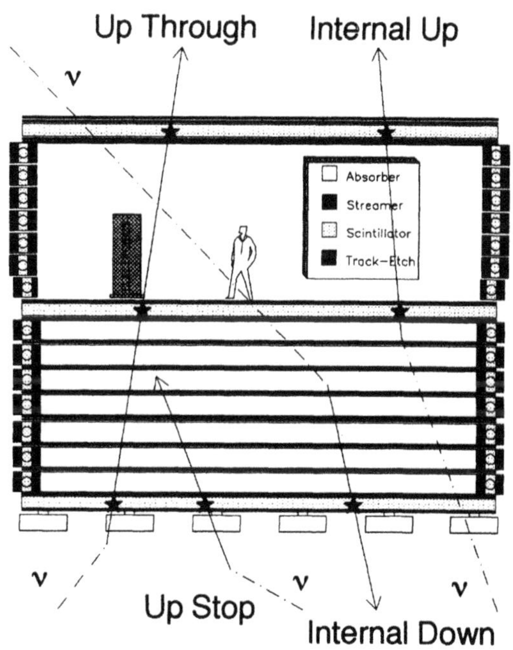

Figure 1. Sketch of different event topologies induced by neutrino interactions in or around MACRO (see text). In the figure, the stars represent the scintillator hits. The time of flight of the particle can be measured only for the *Internal Up* and *Up Through* events.

PII S0920-5632(99)00406-5

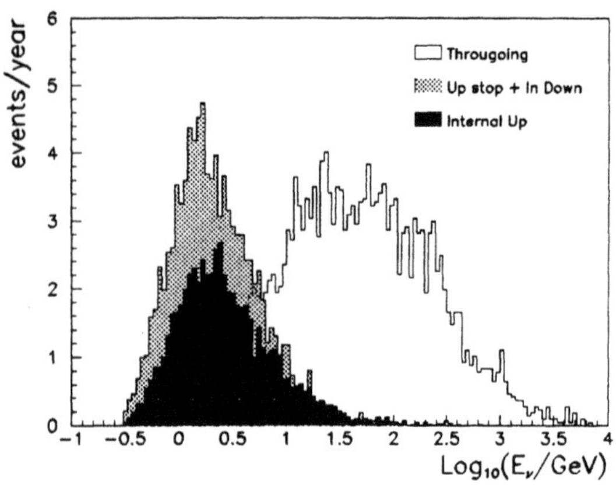

Figure 2. Distribution of the parent neutrino energy giving rise to the three different topologies of events (see Fig. 1), computed by Monte Carlo using the same cuts applied to the data. The distributions are normalized to one year of data taking. The average energies of the three samples are about 4 GeV, 4 GeV and 100 GeV, respectively.

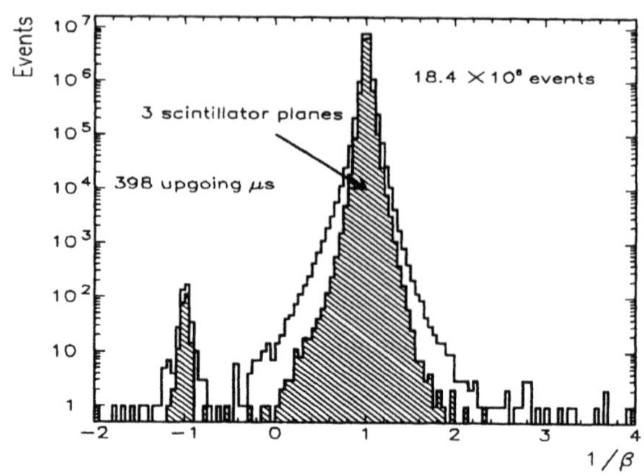

Figure 3. Distribution of $1/\beta$ for the full detector data set. A clear peak of upward muons is evident centered at $1/\beta = -1$. The widths of the distributions for upgoing and downgoing muons are consistent. The shaded part of the distribution is for the subset of events where three scintillator layers were hit.

2. MACRO as a neutrino detector

The MACRO detector provides an excellent tool for the study of upgoing muons. Its large area, fine tracking granularity, symmetric electronics with respect to upgoing versus downgoing muons and fully-automated analysis permit detailed studies of the detector acceptance and possible sources of backgrounds to the upgoing muon measurement. In addition, the overburden of the Gran Sasso Laboratory is significantly larger than that surrounding other experiments (Baksan and IMB), hence providing additional shielding against possible sources of background induced by down-going muons. In our first measurement of upgoing muons [2], we reported on a deficit in the total number of observed upgoing muons with respect to the expectation and also on an anomalous zenith angle distribution. In particular, too few muons were observed near the nadir. Here, we report on a much larger data set [3] which retains the same basic features as reported previously but with larger statistics.

The MACRO detector is located in Hall B of the Gran Sasso Laboratory, with a minimum rock overburden of 3150 hg/cm^2. It is a large rectangular box, 76.6 m × 12 m × 9.3 m, divided longitudinally in six similar supermodules and vertically in a lower part (4.8 m high) and an upper part (4.5 m high). The active detection elements are planes of streamer tubes for tracking and of liquid scintillation counters for fast timing. The lower half of the detector is filled with trays of crushed rock absorbers alternating with streamer tube planes, while the upper part is open and contains the electronics racks and work areas. There are 10 horizontal planes in the bottom half of the detector, and 4 planes on the top, made of wires and 27° stereo strip readouts. Six vertical planes of streamer tracking cover each side of the detector.

The scintillator system consists of three layers of horizontal boxes, with vertical boxes along the sides of the detector. The time (position) resolution for muons in a scintillator box in this analysis is about 500 ps ($\sim$ 11 cm).

Figure 1 shows a schematic plot of the three different topologies of neutrino events analyzed up to now: *Up Through* events, *Internal Up* events and *Internal Down* together with *Up Stop* events. Figure 2 shows the parent neutrino energy distribution for the three event topologies. The requirement of a reconstructed track selects events having a muon.

The *Up Through* tracks come from ν_μ interactions in the rock below MACRO. The muon crosses the whole detector ($E_\mu > 1$ GeV). The time information provided by scintillator counters permits to know the flight direction (time-of-flight method). Almost 50% of the tracks intercept 3 scintillator counters. The average neutrino energy for this kind of events is around 100 GeV. The data have been collected in three periods, with different detector configurations. In the first two periods (March 1989 - November 1991, December 1992 - June 1993) only the lower parts of MACRO were working. In the last period (April 1994 - November 1997) also the upper part of MACRO was in acquisition.

The *Internal Up* events come from ν interactions inside the apparatus. Since two scintillator layers are intercepted, the time-of-flight method is applied to identify the upward going events. The average neutrino energy for this kind of events is around 4 GeV. If the atmospheric neutrino anomalies are the results of ν_μ oscillations with maximum mixing and Δm^2 between 10^{-3} and 10^{-2} eV^2 it is expected a reduction in the flux of this kind of events of about a factor of two, without any distortion in the shape of the angular distribution. Only the data collected with the full MACRO (live-time around 3 years) have been used in this analysis.

The *Up Stop* and the *Internal Down* events are due to external interactions with upward-going tracks stopping in the detector (*Up Stop*) and to neutrino induced downgoing tracks with vertex in lower part of MACRO (*Internal Down*). These events are identified by means of topological criteria. The lack of time information prevents to distinguish the two sub samples. The data set used for this analysis is the same used for the *Internal Up* search. An almost equal number of *Up Stop* and *Internal Down* is expected if neutrinos do not oscillate. The average neutrino energy for this kind of events is around 4 GeV. In case of oscillations we expect a reduction in the flux of the *Up Stop* events similar to the one expected for the *Internal Up* events, while we do not expect any reduction of the *Internal Down* events (having path lengths of the order of 20 km).

3. Upward throughgoing muons (*Up Through*)

The direction that muons travel through MACRO is determined by the time-of-flight between two different layers of scintillator counters. The measured muon velocity is calculated with the convention that muons going down through the detector are expected to have $1/\beta$ near $+1$ while muons going up through the detector are expected to have $1/\beta$ near -1.

Several cuts are imposed to remove backgrounds caused by radioactivity in near coincidence with muons and showering events which may result in bad time reconstruction. The most important cut requires that the position of a muon hit in each scintillator as determined from the timing within the scintillator counter agrees within ± 70 cm with the position indicated by the streamer tube track.

When a muon hits 3 scintillator layers, there is redundancy in the time measurement and $1/\beta$ is calculated from a linear fit of the times as a function of the pathlength. Tracks with a poor fit are rejected. Other minor cuts are applied for the tracks with only two layers of scintillator hit.

It has been observed that downgoing muons which pass near or through MACRO may produce low-energy, upgoing particles. These could appear to be neutrino-induced upward throughgoing muons if the down-going muon misses the detector [4]. In order to reduce this background, we impose a cut requiring that each upgoing muon must cross at least 200 g/cm^2 of material in the bottom half of the detector. Finally, a large number of nearly horizontal ($\cos\theta > -0.1$), but upgoing muons have been observed coming from azimuth angles (in local coordinates) from -30° to 120°. This direction contains a cliff in the mountain where the overburden is insufficient to remove nearly horizontal, downgoing muons which

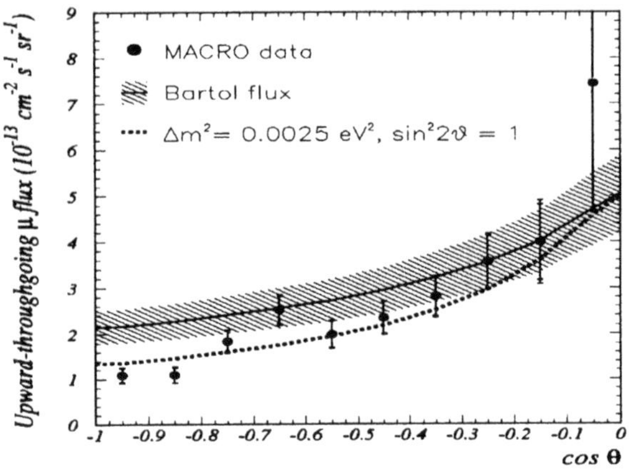

Figure 4. Zenith distribution of flux of upward throughgoing muons with energy greater than 1 GeV for data and Monte Carlo for the combined MACRO data. The solid curve shows the expectation for no oscillations and the shaded region shows the 17% uncertainty in the expectation. The dashed line shows the prediction for an oscillated flux with $\sin^2 2\theta = 1$ and $\Delta m^2 = 0.0025$ eV^2.

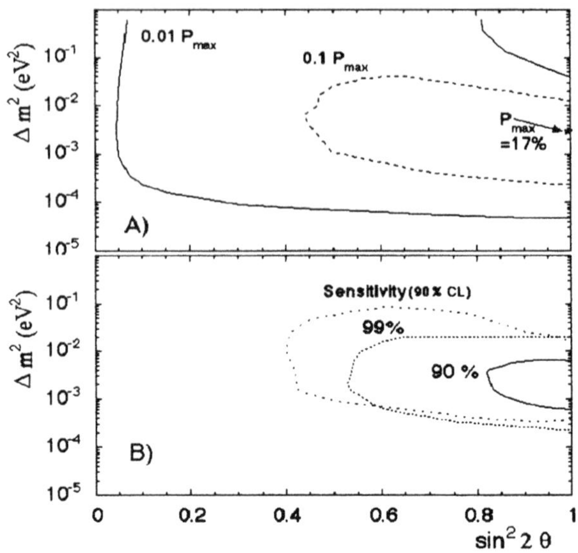

Figure 5. A) Probability contours for oscillation parameters for $\nu_\mu \to \nu_\tau$ oscillations based on the combined probabilities of zenith shape and number of events tests. The best probability in the physical regn is 17% and iso-probability contours are shown for 10% and 1% of this value (i.e. 1.7% and 0.17%). B) Confidence regions at the 90% and 99% levels calculated according to reference [5]. Since the best probability is outside the physical region the confidence intervals regions are smaller than the one expected from the sensitivity of the experiment.

have scattered in the mountain and appear as upgoing. We exclude this region from both our observation and Monte-Carlo calculation of the upgoing events.

Figure 3 shows the $1/\beta$ distribution for the upthroughgoing data from the full detector running. A clear peak of upgoing muons is evident centered on $1/\beta = -1$. There are 398 events in the range $-1.25 < 1/\beta < -0.75$ which we define as upgoing muons for this data set. We combine these data with the previously published data [2] (with 4 additional events due to an updated analysis) for a total of 479 upgoing events. Based on events outside the upgoing muon peak, we estimate there are 9 ± 5 background events in the total data set. In addition to these events, we estimate that there are 8 ± 3 events which result from upgoing charged particles produced by downgoing muons in the rock near MACRO. Finally, it is estimated that 11 ± 4 events are the result of interactions of neutrinos in the very bottom layer of MACRO scintillators. Hence, removing the backgrounds, the observed number of upgoing throughgoing muons integrated over all zenith angles is 451.

In the upgoing muon simulation we have used the neutrino flux computed by the Bartol group [6]. The cross-sections for the neutrino interactions have been calculated using the Morfin and Tung parton distributions set S_1 [7]. These parton distributions were chosen based on the good agreement of the resulting σ_T compared to the world average at E_ν = 100 GeV. The propagation of muons to the detector has been done using the energy loss calculation by Lohmann et al. [9] for standard rock. The total systematic uncertainty on the expected flux of muons adding the

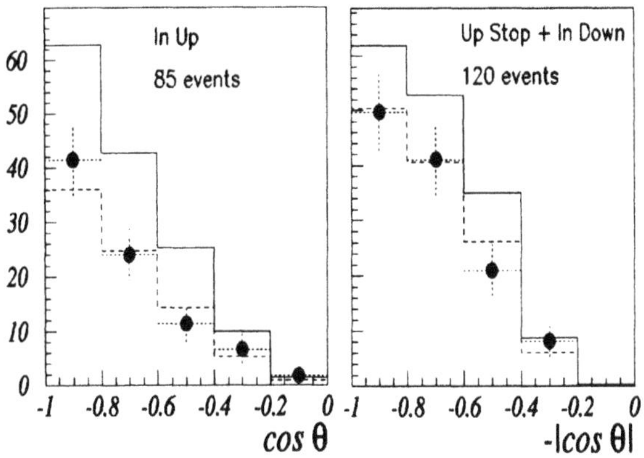

Figure 6. Comparison between measured and expected number of low energy events versus $cos(\theta)$. The dashed line is obtained assuming neutrino oscillation with the parameter suggested by the *Up Through* Sample. Note that in the second plot the flight direction is unknow

errors from neutrino flux, cross-section and muon propagation in quadrature is ±17%. This theoretical error in the prediction is mainly a scale error that doesn't change the shape of the angular distribution. The number of events expected integrated over all zenith angles is 612, giving a ratio of the observed number of events to the expectation of 0.74 ±0.036(stat) ±0.046(systematic) ±0.13(theoretical).

Figure 4 shows the zenith angle distribution of the measured flux of upgoing muons with energy greater than 1 GeV for all MACRO data compared to the Monte Carlo expectation for no oscillations and with a $\nu_\mu \to \nu_\tau$ oscillated flux with $\sin^2 2\theta = 1$ and $\Delta m^2 = 0.0025$ eV^2 (dashed line).

The shape of the angular distribution has been tested with the hypothesis of no oscillation excluding the last bin near the horizontal and normalizing data and predictions. The χ^2 is 26.1, for 8 degrees of freedom (probability of 0.1% for a shape at least this different from the expectation). We have considered also oscillations $\nu_\mu \to \nu_\tau$. The best χ^2 in the physical region of the oscillations parameters is 15.8 for Δm^2 around $0.0025 eV^2$ and maximum mixing (the best χ^2 is outside the physical region for mixing > 1).

To test oscillation hypothesis, we calculate the independent probability for obtaining the number of events observed and the angular distribution for various oscillation parameters. It is notable that the value of Δm^2 suggested from the shape of the angular distribution is similar to the value necessary in order to obtain the observed reduction in the total number of events in the hypothesis of maximum mixing. Figure 5 A) shows probability contours for oscillation parameters using the combination of probability for the number of events and χ^2 of the angular distribution. The maximum of the probability is 17%. The probability for no oscillation is 0.1%. Figure 5 B) shows the confidence regions at the 90% and 99% confidence levels based on application of the Monte Carlo prescription of reference [5]. We plot also the sensitivity of the experiment. The sensitivity is the 90% contour which would result from the preceding prescription if the data and Monte Carlo happened to be in perfect agreement at the best-fit point. The allowed regions are smaller than the one you could expect form the sensitivity of the experiment. This is because the best probability is outside the physical region.

The same procedure applied for sterile neutrino [10] gives 2% as maximum probability.

4. The Low Energy Events

The analysis of the *Internal Up* events is similar to the analysis of the *Up Through*. The main difference is due to the requirement that the interaction vertex should be inside the apparatus. About 87% of events are estimated to be ν_μ interactions. The preliminary uncertainty due to the acceptance and analysis cuts is 10%. After the background subtraction (3 events) 85 events are classified as *Internal Up* events

The *Internal Down* and the *Up Stop* events are identified via topological constraints. The main requirement is the presence of a reconstructed track crossing the bottom scintillator layer. All the track hits must be at least 1 m from the detector's edges. The criteria used to verify that the event vertex (or stop point) is inside the

Table 1
Event Summary. The predictions with oscillations are for maximum mixing and $\Delta m^2 = 0.0025 eV^2$

	Events detected	Predictions (Bartol neutrino flux)	
		No Oscillations	With oscillations
Up Through	451	$612 \pm 104_{theoret} \pm 37_{syst}$	$431 \pm 73_{theoret} \pm 26_{syst}$
Internal Up	85	$144 \pm 36_{theoret} \pm 14_{syst}$	$83 \pm 21_{theoret} \pm 8_{syst}$
In Down + Stop	120	$159 \pm 40_{theoret} \pm 16_{syst}$	$123 \pm 31_{theoret} \pm 12_{syst}$

detector are similar to those used for the *Internal Up* search. To reject ambiguous and/or wrongly tracked events which survived automated analysis cuts, real and simulated events were randomly merged and directly scanned with the MACRO Event Display. Three different events subsamples are considered according to the minimum number of streamer tube hits. We present here the sample with at least 3 streamer hits (corresponding roughly to 100 $gr\ cm^{-2}$). About 90% of the events are estimated to be ν_μ CC interactions. The main background for this search are the low energy particles produced by donwn-going muons [4]. After background subtraction (5 events) 120 events are classified as *Internal Down* and *Up Stop* events.

The Montecarlo simulation for the low energy events uses the Bartol neutrino flux [6] and the neutrino low energy cross sections reported in [8]. The simulation is performed in a large volume of rock (170 kton) around the MACRO detector (5.3 kton). The uncertainty on the expected muon flux is about 25%. The total number of events and the angular distributions are compared with the predictions in Table 1 and in Figure 6. The low energy samples show an uniform deficit of the measured number of events over the whole angular distribution with respect to the predictions, while there is a good agreement with the predictions based on neutrino oscillations.

5. Conclusions

The upgoing throughgoing muon data set is in favor of $\nu_\mu \to \nu_\tau$ oscillation with parameters similar to the those observed by Superkamiokande with a probability of 17% against the 0.1% for the no oscillation hypothesis. However the shape of the zenith distribution gives a maximum probability of only 4.6%. This could be due to a statistical fluctuation or to some hidden physics. We exclude effects due to the detector.

This oscillation hypothesis is also consistent with the MACRO low energy data. A combined statistical analysis of the three data samples will be performed in the future when more statistics will be available.

REFERENCES

1. Fukuda, Y., et al. (SuperKamiokande collaboration) : 1998, hep-ex/9807003 and T.Kajita in these proceedings.
   Hirata, K. S. et al. (Kamiokande collaboration) Phys. Lett. B280 (1992) 146 and Fukuda Y. et al. Phys. Lett. B335 (1994) 237.
   Becker-Szendy, R. et al (IMB collaboration): Phys. Rev.Lett. 66 (1991) 2561.
   Allison, W.W.M. et al (Soudan 2 Collaboration): Phys. Lett. B391,(1997) 491 and E. Peterson in these proceedings
2. S. Ahlen et al. (MACRO collabor.), Phys. Lett. B 357 (1995) 481.
3. M. Ambrosio et al.(MACRO collabor.) hep-ex/9807005 to be published on Phys Lett. B.
4. Ambrosio, M., et al. (MACRO collabor.) Astroparticle Physics 9 (1998) 105.
5. G. Feldman and R. Cousins, Phys. Rev. D57 (1998) 3873.
6. V. Agrawal, T.K. Gaisser, P. Lipari and T. Stanev, Phys. Rev. D53 (1996) 1314.
7. J. G. Morfin and W. K. Tung , Z. Phys. C52 (1991) 13.
8. P. Lipari, M. Lusignoli and F. Sartogo, Phys. Rev. Lett. 74 (1995) 4384.
9. W. Lohmann et al., CERN-EP/85-03 (1985).
10. Q.Y. Liu and A.Yu. Smirnov Nucl.Phys. B524 (1998) 505.

ELSEVIER Nuclear Physics B (Proc. Suppl.) 77 (1999) 123–132

NUCLEAR PHYSICS B
PROCEEDINGS
SUPPLEMENTS

Atmospheric neutrino results from Super-Kamiokande and Kamiokande – Evidence for ν_μ oscillations –

Takaaki Kajita[a]
for the Super-Kamiokande and Kamiokande collaborations

[a]Kamioka Observatory, Institute for Cosmic Ray Research, Univ. of Tokyo
Higashi-Mozumi, Kamioka-cho, Gifu, 506-1205, Japan

New atmospheric neutrino results from Super-Kamiokande are presented. Results from Kamiokande on upward going muons are also presented. All these data, together with the Kamiokande atmospheric neutrino data give evidence for neutrino oscillations. Two flavor $\nu_\mu \leftrightarrow \nu_\tau$ oscillations, with large $\sin^2 2\theta$ and Δm^2 in the region of 10^{-3} to 10^{-2} eV^2, explain all these data.

1. Introduction

Cosmic ray interactions in the atmosphere produce neutrinos. The prediction of the absolute flux has an uncertainty of $\pm 20\%$. However, the flavor ratio of the atmospheric neutrino flux, $(\nu_\mu + \bar{\nu}_\mu)/(\nu_e + \bar{\nu}_e)$, has been calculated to an accuracy of better than 5% in a broad energy range from 0.1 GeV to higher than 10 GeV. The calculated flux ratio has a value of about 2 for energies $\lesssim$1GeV and increases with increasing neutrino energy. For neutrino energies higher than a few GeV, the fluxes of upward and downward going neutrinos are expected to be nearly equal; the geomagnetic field effects on atmospheric neutrinos in this energy range are expected to be small because the primary cosmic rays that produce these neutrinos have rigidities exceeding the geomagnetic cutoff rigidity ($\sim 10\ GeV/Ze$).

The ν_μ/ν_e ratio has been measured in deep underground experiments by observing final-state leptons produced by charged-current interactions of neutrinos on nuclei, $\nu + N \rightarrow l + X$. The measurements are reported as $R \equiv (\mu/e)_{data}/(\mu/e)_{MC}$, where μ and e are the number of muon-like (μ-like) and electron-like (e-like) events observed in the detector for both data and Monte Carlo (MC) simulation. This ratio largely cancels experimental and theoretical uncertainties, especially the uncertainty in the absolute flux. The observed values of R by Kamiokande [1], [2] were significantly smaller than unity for both below and above 1 GeV energy ranges. Consistent results were obtained by IMB-3(<1.5GeV) [3] and Soudan-2 [4].

Neutrino oscillations have been suggested to explain the small values of R. For a two-neutrino oscillation hypothesis, the probability for a neutrino produced in a flavor state a to be observed in a flavor state b after traveling a distance L through a vacuum is:

$$P_{a\rightarrow b} = \sin^2 2\theta \sin^2 \left(\frac{1.27\Delta m^2(\mathrm{eV}^2)L(\mathrm{km})}{E_\nu(\mathrm{GeV})}\right), \quad (1)$$

where E_ν is the neutrino energy, θ is the mixing angle between the flavor eigenstates and the mass eigenstates, and Δm^2 is the mass-squared difference of the neutrino mass eigenstates. For detectors near the surface of the Earth, the neutrino flight distance, and thus the oscillation probability, is a function of the zenith angle of the neutrino direction. Vertically downward-going neutrinos travel about 15 km while vertically upward-going neutrinos travel about 13,000 km before interacting in the detector. The broad energy spectrum from a few hundred MeV to about 100 GeV and this range of neutrino flight distances makes measurements of atmospheric neutrinos sensitive to neutrino oscillations with Δm^2 down to 10^{-4} eV^2.

The zenith angle dependence of R measured by the Kamiokande experiment[2] at high energies, together with the small R values, has been cited as evidence for neutrino oscillations. Based

0920-5632/99/$ – see front matter
PII S0920-5632(99)00407-7

on these measurements, Kamiokande obtained the allowed parameter regions of neutrino oscillations. Because of the relatively small statistics, both $\nu_\mu \leftrightarrow \nu_e$ and $\nu_\mu \leftrightarrow \nu_\tau$ oscillations were allowed.

Recently, a long baseline reactor experiment, CHOOZ [5], excluded the $\nu_\mu \leftrightarrow \nu_e$ solution of the atmospheric neutrino problem.

The upward-going muons observed in underground detectors are the products of neutrino interactions in the rock. The mean neutrino energy of these events are of the order of 100(10) GeV for through-going(stopping) events. These events are used for the independent check of the neutrino oscillation analysis of the lower energy atmospheric neutrino data.

We present the analyses of atmospheric neutrino events from Super-Kamiokande. In addition to the events whose vertex positions are in the fiducial volume of the detector, we present the upward-going muon results from Kamiokande and Super-Kamiokande. We observed small values of R and a zenith angle dependent deficit of μ-like events. While no combination of known uncertainties in the experimental measurement or prediction of atmospheric neutrino fluxes is able to explain the data, a two-neutrino oscillation model of $\nu_\mu \leftrightarrow \nu_x$, where ν_x may be ν_τ or a new, non-interacting "sterile" neutrino, is consistent with the observed R values and zenith angle distributions. These data and the neutrino oscillation interpretation were further supported by a small (upward stopping muons)/(upward through-going muons) ratio and zenith-angle dependent deficit of upward through-going muon flux. From these measurements, we conclude that the atmospheric neutrino data give evidence for neutrino oscillations.

2. Super-Kamiokande detector

Super-Kamiokande is a cylindrical 50 kton water Cherenkov detector located at a depth of 2700 meters water equivalent in the Kamioka Observatory in Japan. The detector consists of an inner detector surrounded by an outer detector on all sides. The inner and outer detectors are optically separated by a pair of opaque sheets. 11146 50 cm ϕ photomultiplier tubes (PMTs), instrumented in all surfaces of the inner detector, detect Cherenkov photons radiated by relativistic charged particles. 1885 20 cm ϕ PMTs are instrumented in the outer detector. The outer detector is useful for identifying entering cosmic-ray muons and measuring exiting particles produced by neutrino interactions occurring in the inner detector. Pulse height and timing information from each PMT are recorded and used in the data analysis. The trigger threshold for electrons is 5.7 MeV/c at 50% efficiency.

For a description of the Kamiokande detector, see Ref.[6].

3. Fully and partially contained events

Super-Kamiokande observed a total of 4353 fully-contained (FC) events and 301 partially-contained (PC) events in a 33.0 kiloton-year exposure. FC events deposit all of their Cherenkov light in the inner detector while PC events have exiting tracks which deposit some Cherenkov light in the outer detector. For the present analyses, the neutrino interaction vertex was required to have been reconstructed within the 22.5 kiloton fiducial volume, defined to be $>$ 2 m from the PMT wall. The number of FC+PC events observed so far in Super-Kamiokande was about 4 times larger than that in Kamiokande.

FC events were separated into those with a single visible Cherenkov ring and those with multiple Cherenkov rings. For the analysis of FC events, only single-ring events were used. Single-ring events were identified as e-like or μ-like based on a likelihood analysis of light detected around the Cherenkov cone. The FC events were separated into "sub-GeV" ($E_{vis} < 1330$ MeV) and "multi-GeV" ($E_{vis} > 1330$ MeV) samples, where E_{vis} is defined to be the energy of an electron that would produce the observed amount of Cherenkov light. E_{vis} = 1330 MeV corresponds to ~1400 MeV/c for muons.

In a full-detector Monte Carlo simulation, 88% (96%) of the sub-GeV e-like (μ-like) events were ν_e (ν_μ) charged-current (CC) interactions and 84% (99%) of the multi-GeV e-like (μ-like) events were ν_e (ν_μ) CC interactions. PC events were es-

	Data	Monte Carlo
sub-GeV		
single-ring	2389	2622.6
e-like	1231	1049.1
μ-like	1158	1573.6
multi-ring	911	980.7
total	3300	3603.3
$R = 0.63 \pm 0.03(stat.) \pm 0.05(sys.)$		
multi-GeV		
FC events		
single-ring	520	531.7
e-like	290	236.0
μ-like	230	295.7
multi-ring	533	560.1
total	1053	1091.8
PC events		
total(=μ–like)	301	371.6
$R_{FC+PC} = 0.65 \pm 0.05(stat.) \pm 0.08(sys.)$		

Table 1
Summary of the sub-GeV, multi-GeV and PC event samples observed in 33 kiloton-year exposure of the Super-Kamiokande detector. The data are compared with the Monte Carlo prediction based on the neutrino flux calculation of Ref.[9].

timated to be 98% ν_μ CC interactions; hence, all PC events were classified as μ-like, and no single-ring and particle identification requirements were made. Table 1 summarizes the number of observed events for both data and Monte Carlo as well as the R values for the sub-GeV and multi-GeV samples. Further details of the detector, data selection and event reconstruction used in this analysis are given in Ref.[7,8].

Flavor ratio

Super – Kamiokande measured significantly small values of R in both the sub-GeV and multi-GeV samples. Several sources of systematic uncertainties in these measurements were considered[7,8]. For the sub-GeV sample, they were: 5% from uncertainty in the predicted ν_μ/ν_e flux ratio, 3.5% from the CC neutrino interaction cross sections and nuclear effects in the H_2O target, 3% from the neutral current cross sections, 2% from particle identification, 1% from the absolute energy calibration, 0.6% from the vertex fit and fiducial volume cut, less than 0.5% from non-neutrino background events and 1.5% from statistical uncertainty in the Monte Carlo. Adding these errors in quadrature, the total systematic uncertainty was 8%. The systematic uncertainty of R for the multi-GeV sample, obtained in a similar way, was 12%. Table 1 summarizes the measured R values in Super-Kamiokande. These results are consistent with the Kamiokande R values, which were $0.60\ ^{+0.06}_{-0.05}(stat.) \pm 0.05(sys.)$ for the sub-GeV data and $0.57\ ^{+0.08}_{-0.07}(stat.) \pm 0.07(sys.)$ for the multi-GeV data[2].

The Super-Kamiokande data have been analyzed independently by two groups, making the possibility of significant biases in data selection or event reconstruction algorithms remote[7,8]. Given the statistical error in R especially for the sub-GeV sample, statistical fluctuation can no longer explain the deviation of R from unity. Assuming that the systematic error has the gaussian form, we estimate the probability that the observed μ/e ratios could be due to statistical fluctuation is less than 0.001% and less than 1% for sub- and multi-GeV samples, respectively.

Zenith angle distribution

The μ-like data from Super-Kamiokande exhibit a strong up-down asymmetry in zenith angle (Θ) while no significant asymmetry is observed in the e-like data, see Figure 1. For further analyses, we define up-down ratio U/D where U is the number of upward-going events ($-1 < \cos\Theta < -0.2$) and D is the number of downward-going events ($0.2 < \cos\Theta < 1$). The ratio is expected to be near unity independent of flux model for $E_\nu > 1$ GeV, above which effects due to the Earth's magnetic field on cosmic rays are small. Based on a comparison of results from our Monte Carlo simulation using different flux models[10,9] as inputs, treatment of geomagnetic effects results in an uncertainty of roughly $\pm 0.02 \sim 0.03$ in the expected U/D values for e-like and μ-like sub-GeV events and less than ± 0.02 for multi-GeV events. These two flux calculations do not assume the existence

	Super-Kamiokande		Kamiokande
	MC	Data	Data
e-like			
Sub-GeV, <400MeV/c	1.00±0.04±0.03	$1.20^{+0.11}_{-0.10} \pm 0.03$	$1.29^{+0.27}_{-0.22}$
Sub-GeV, >400MeV/c	1.02±0.04±0.03	$1.10^{+0.11}_{-0.10} \pm 0.03$	$0.76^{+0.22}_{-0.18}$
Multi-GeV	1.01±0.06±0.03	$0.93^{+0.13}_{-0.12} \pm 0.02$	$1.38^{+0.39}_{-0.30}$
μ-like			
Sub-GeV, <400MeV/c	1.05±0.03±0.02	$1.03^{+0.11}_{-0.10} \pm 0.02$	$1.18^{+0.31}_{-0.24}$
Sub-GeV, >400MeV/c	1.00±0.03±0.02	$0.65^{+0.06}_{-0.05} \pm 0.01$	$1.09^{+0.22}_{-0.18}$
Multi-GeV (FC+PC)	0.98±0.03±0.02	$0.54^{+0.06}_{-0.05} \pm 0.01$	$0.58^{+0.13}_{-0.11}$

Table 2
Summary of the up/down ratio, U/D, for e-like and μ-like events from Super-Kamiokande and Kamiokande. The systematic errors of the Kamiokande experiment are not shown, but are similar to those of Super-Kamiokande.

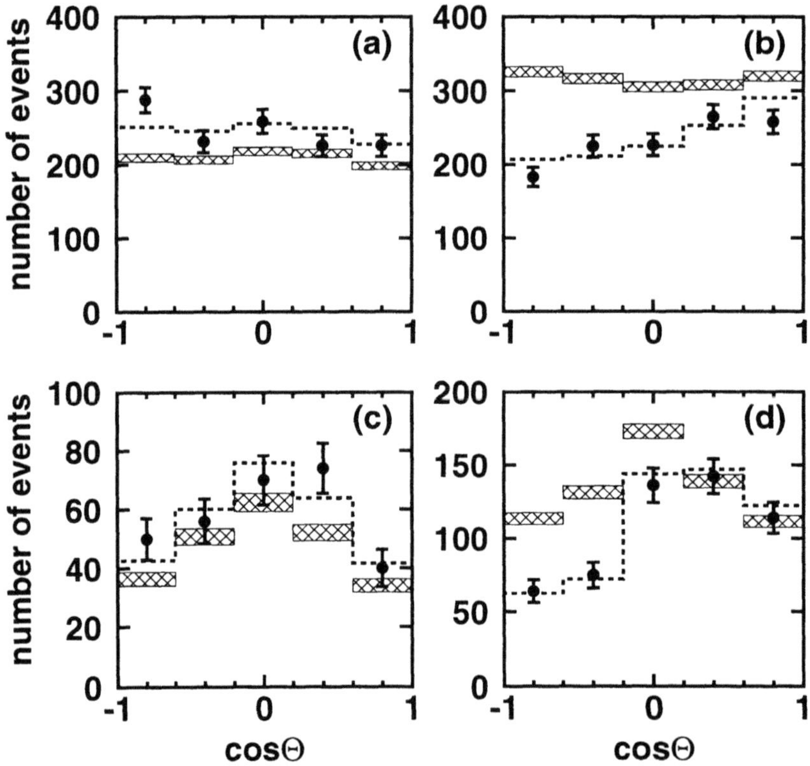

Figure 1. Zenith angle distributions observed in Super-Kamiokande for; (a)sub-GeV e-like, (b)sub-GeV μ-like, (c)multi-GeV e-like and (d) multi-GeV (FC+PC) μ-like events. CosΘ =1 means down-going particles. The histograms with shaded error bars show the MC prediction with their statistical errors for the no neutrino oscillation case. The dotted histograms shows the Monte Carlo prediction for $\nu_\mu \leftrightarrow \nu_\tau$ oscillations with $\sin^2 2\theta = 1$ and $\Delta m^2 = 2.2\times10^{-3}$ eV^2.

of 1 km mountain over the Super-Kamiokande detector. The rock reduces the neutrino flux because muons are stopped before they can decay in flight. We estimated the effect of the presence of rock on the flux. U/D changed about 2% for the multi-GeV events. The change was much smaller for the sub-GeV events.

Studies of decay electrons from stopping muons show at most a $\pm0.6\%$ gain difference, i.e., measured energy difference, between up-going and down-going particles. This gain difference caused 0.9%, and 0.7% uncertainty in U/D for the multi-GeV e-like and μ-like events. This uncertainty is much smaller for the sub-GeV sample. A contamination of non-neutrino background such as down-going cosmic ray muons could have directional correlation. The maximum contribution to the uncertainty in U/D from the contamination was estimated to be $\pm1.0\%$, $\pm0.1\%$, $\pm0.5\%$ and $\pm2.0\%$ for the sub-GeV e-like, μ-like, multi-GeV e-like and μ-like events, respectively. From these studies, the total systematic uncertainties in U/D for the data and MC are summarized in Table 2. In the present data, U/D for e-like events is consistent with expectations. U/D for high momentum μ-like events significantly deviates from unity, while U/D for low momentum μ-like events is consistent with unity. The average angle between the final state lepton direction and the incoming neutrino direction is 55° at $p = 400$ MeV/c and 20° at 1.5 GeV/c. At the momentum range below 400 MeV/c, the possible up-down asymmetry of the neutrino flux is largely washed out. We have found no detector bias differentiating e-like and μ-like events that could explain an asymmetry in μ-like events but not in e-like events [8].

The U/D value for the multi-GeV FC+PC μ-like events, $0.54\ ^{+0.06}_{-0.05} \pm 0.01$ deviates from unity by more than 6 standard deviations. This value is also consistent with the Kamiokande U/D value for the multi-GeV FC+PC μ-like events, $0.58\ ^{+0.13}_{-0.12}$. These numbers, which are close to 0.5, suggest a near maximal neutrino mixing.

Neutrino oscillation analysis

We have examined the hypotheses of two-flavor $\nu_\mu \leftrightarrow \nu_e$ and $\nu_\mu \leftrightarrow \nu_\tau$ oscillation models using a χ^2 comparison of the Super-Kamiokande data and Monte Carlo, allowing all important Monte Carlo parameters to vary weighted by their expected uncertainties[11]. The data were binned by particle type, momentum, and $\cos\Theta$. A χ^2 is defined as:

$$\chi^2 = \sum_{cos\Theta,p}^{70} \left(\frac{N_{Data} - N_{MC}(\sin^2 2\theta, \Delta m^2, \epsilon_j)}{\sigma} \right)^2 + \sum_j \left(\frac{\epsilon_j}{\sigma_j} \right)^2, \quad (2)$$

where the sum is over five bins equally spaced in $\cos\Theta$ and seven momentum bins for both e-like events and μ-like plus PC events (70 bins total). The statistical error, σ, accounts for both data statistics and the weighted Monte Carlo statistics. N_{Data} is the measured number of events in each bin. $N_{MC}(\sin^2 2\theta, \Delta m^2, \epsilon_j)$ is the expected number of Monte Carlo events and is a function of $\sin^2 2\theta$, Δm^2 and ϵ_j. ϵ_j are parameters which are related to the systematic uncertainties. The parameters (and their uncertainties) considered in this analysis are: overall normalization (25%, but this was fitted as a free parameter), E_ν spectral index (0.05), sub-GeV R (8%), multi-GeV R (12%), relative normalization of PC to FC (8%), L/E_ν (15%), sub-GeV (2.4%) and multi-GeV (2.7%) up-down ratios. See Ref.[11] for more details.

For $\nu_\mu \leftrightarrow \nu_e$, effects of matter on neutrino propagation through the Earth were included following Ref. [12]. Due to the small number of events expected from τ-production (15 to 20 events were expected in the present FC+PC sample), the effects of τ appearance and decay were neglected in simulations of $\nu_\mu \leftrightarrow \nu_\tau$. A global scan was made on a $(\sin^2 2\theta, \log\Delta m^2)$ grid minimizing χ^2 with respect to uncertainty parameters, ϵ_j, at each point.

The best-fit to $\nu_\mu \leftrightarrow \nu_\tau$ oscillations, $\chi^2_{min(phys)} = 65.2/67$ DOF, was obtained at $(\sin^2 2\theta, \Delta m^2) = (1.0,\ 2.2 \times 10^{-3}\ \text{eV}^2)$ inside the physical region ($0 \leq \sin^2 2\theta \leq 1$). The best-fit values of the Monte Carlo uncertainty parameters were all within their expected errors for this point. The global minimum occurred slightly outside the physical region at $(\sin^2 2\theta, \Delta m^2) = (1.05, 2.2 \times 10^{-3}\ \mathrm{eV}^2), \chi^2_{min(unphys)} = 64.8/67$ DOF. The 90% C.L. allowed region is located at $\chi^2_{min(phys)} + 5.0$, based on the minimum inside the physical region[13]. The allowed region is shown in Figure 6. In the region near χ^2 minimum, the χ^2 distribution is rather flat and has many local minima so that inside the 90% interval the best-fit Δm^2 is not well constrained. The χ^2 increases rapidly outside of the 90% C.L. region. We obtained $\chi^2 = 135/69$ DOF, when calculated at $\sin^2 2\theta = 0$, $\Delta m^2 = 0$ (i.e. assuming no oscillations).

The Kamiokande allowed region[2] obtained by the contained event analysis is also shown in Figure 6. The Super-Kamiokande allowed region favors lower Δm^2 than that of Kamiokande. However, the allowed regions from both experiments have a region of overlap. One of the reasons for the difference in the allowed region is due to the zenith angle distribution of the sub-GeV μ-like events. The U/D values of the μ-like events from Kamiokande and Super-Kamiokande are summarized in Table 2. It should be noted that Kamiokande U/D value was consistent with unity for μ-like events in the momentum range of sub-GeV, >400MeV/c but Super-Kamiokande observed smaller U/D value for the same sample; a 2.5σ difference in U/D for this sub-sample. This difference results in a difference in the favored Δm^2 region in these two experiments, since the energy region observing the small U/D is directly related to the determination of Δm^2,

For the test of $\nu_\mu \leftrightarrow \nu_e$ oscillations, the Super-Kamiokande data resulted in a relatively poor fit; $\chi^2_{min} = 87.8/67$ DOF, at $(\sin^2 2\theta, \Delta m^2) = (0.93,\ 3.2\times10^{-3}\ \mathrm{eV}^2)$. The expected U/D value of the multi-GeV e-like events for the best-fit $\nu_\mu \leftrightarrow \nu_e$ oscillation hypothesis, 1.52, differs from the measured value, $0.93\ ^{+0.13}_{-0.12} \pm 0.02$, by 3.4 standard deviations. We conclude that the $\nu_\mu \leftrightarrow \nu_e$ hypothesis is not favored.

The zenith angle distributions for the sub- and multi-GeV samples are shown in Figure 1. The data are compared to the Monte Carlo expectation (no oscillations, hatched region) and the best-fit expectation for $\nu_\mu \leftrightarrow \nu_\tau$ oscillations (dashed line). The oscillated Monte Carlo well reproduces the zenith angle distributions of the data.

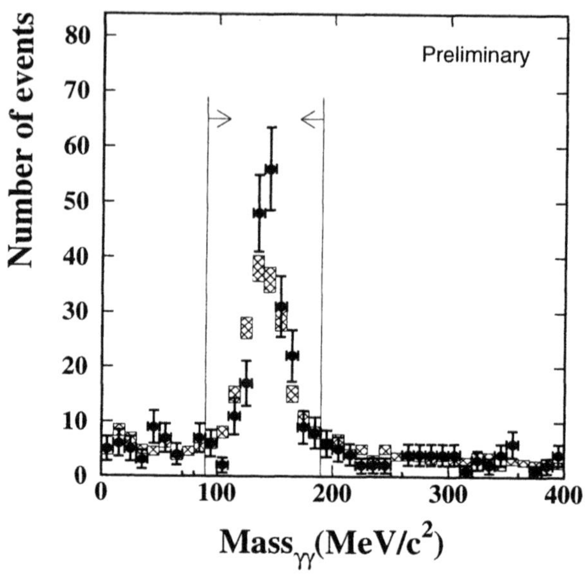

Figure 2. Invariant mass distribution observed in Super-Kamiokande for two ring events with "e-like, e-like" particle-identification results and with no μ-decay signal. The histogram with shaded error bars show the Monte Carlo prediction with their statistical errors. Cut region of π^0-mass is also shown.

In Super-Kamiokande, the oscillation parameters were also estimated by considering the R measurement and the zenith angle shape separately. The 90% C.L. allowed regions for each case overlapped at $1 \times 10^{-3} < \Delta m^2 < 4 \times 10^{-3}$ eV^2 for $\sin^2 2\theta = 1$.

As another check of the Super-Kamiokande data, we studied π^0 events. For $\nu_\mu \leftrightarrow \nu_\tau$ oscillations, the number of neutral-current (NC) events should be unchanged by neutrino oscillations. As a NC sample, we selected "π^0"-like events. The selection criteria were: 2 ring events, both rings should be e-like, no μ-decay signal and the invariant mass calculated from the charge and direction of the two rings assuming two γ's should be between 90 and 190MeV/c^2. We observed 210 π^0-like events. See Figure 2, which shows the invariant mass distribution for the two ring events. A clear excess at $\sim$140MeV/c^2 is seen. The fraction of NC events in this sample is estimated to be 82%. For $\nu_\mu \leftrightarrow \nu_\tau$ oscillations, the number of CC ν_e events ($\sim$ e-like events) should also be unchanged by neutrino oscillations. Therefore, the (π^0/e) ratio of the data should agree with the same ratio of the Monte Carlo without oscillations for the $\nu_\mu \leftrightarrow \nu_\tau$ case. We obtained: $(\pi^0/e)_{data}/(\pi^0/e)_{MC} = 0.93 \pm 0.07(stat) \pm 0.19(syst)$ (*Preliminary*). The result is consistent with the $\nu_\mu \leftrightarrow \nu_\tau$ interpretation of the data.

4. Upward going muons

Energetic atmospheric ν_μ's passing the Earth interact with rock surrounding the detector and produce muons via CC interactions. Because of the atmospheric muon background, it is difficult to select neutrino induced downward going muons. On the contrary, upward going muons are essentially neutrino origin.

Upward going muons can be categorized into two types. One is "upward through-going muons" which are the events which enter into the detector and exit, and the other is "upward stopping muons" which enter the detector and stop in the detector.

Upward through-going muons

The mean energy of neutrinos which produce upward through-going muons is about 100GeV. Kamiokande observed 372 upward going muons during 2456 detector live days. The selection criteria were; $\cos\Theta = -1 \sim -0.04$ and the minimum track length in the inner detector of 7 meters[14]. The minimum (mean) energy loss of these muons in the inner detector is 1.6(3.0)GeV. The average detection efficiency was 97%. With the requirement, $\cos\Theta < -0.04$, the background contamination was negligible. The observed flux of upward going muons was $1.94 \pm 0.10(stat.)^{+0.07}_{-0.06}(sys.) \times 10^{-13} cm^{-2} sec^{-1} sr^{-1}$. The expected flux based on the calculated flux of Ref.[10] was $2.46 \pm 0.54(theo.) \times 10^{-13} cm^{-2} sec^{-1} sr^{-1}$. Figure 3 shows the zenith angle distribution of the upward through-going muon flux observed in Kamiokande.

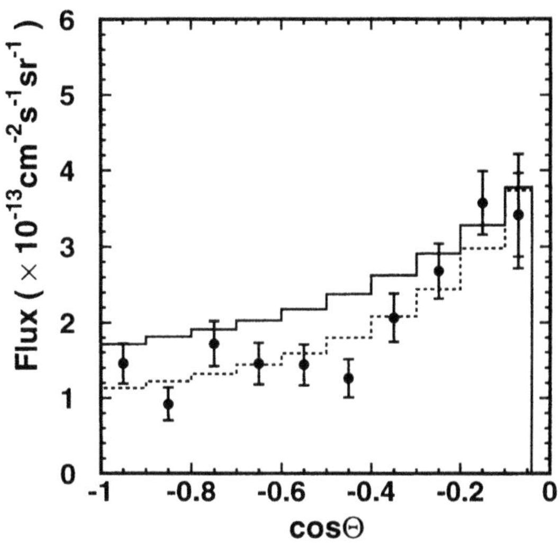

Figure 3. Zenith angle distribution of upward through-going muon flux observed in Kamiokande. Inner(outer) error bars show statistical (statistical+uncorrelated experimental systematic) errors. The solid histogram shows the expected flux for the null neutrino oscillation case. The dashed histogram shows the expected flux for the $\nu_\mu \leftrightarrow \nu_\tau$ oscillation case with $\sin^2 2\theta = 1.0$, $\Delta m^2 = 3.2\times10^{-3}$ eV^2 and $\alpha = 1.00$.

In Super-Kamiokande, 617 upward through-going muon events were observed during 537 detector live days[15]. The selection criteria were; $\cos\Theta < 0$, two outer detector cluster corresponding to the muon entrance and exit points and track length of a muon inside the inner detector should be longer than 7 m. The minimum (mean) energy loss of these muons in the inner detector is 1.6 (6) GeV. The average detection efficiency of these events was estimated to be >99%. The validity of this efficiency was tested using the real down-going cosmic-ray muons by assuming the up-down symmetry of the detector.

The number of background events, 4.6, was estimated by extrapolating the zenith-angle distribution of cosmic ray muons with $\cos\Theta = 0 \sim 0.08$. The background events are expected only in the $\cos\Theta = -0.1 \sim 0$ bin and are subtracted in

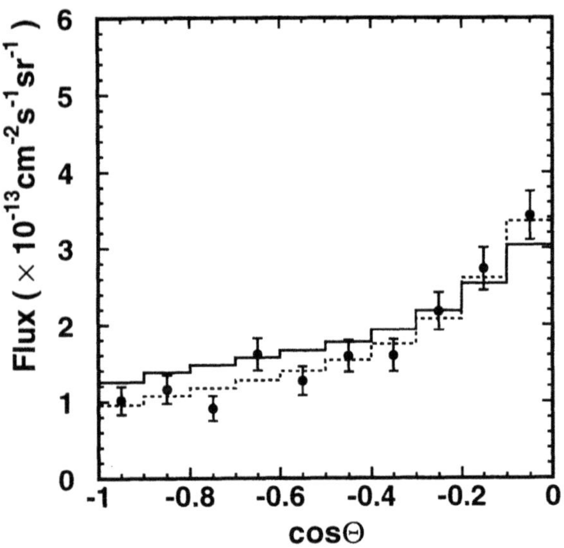

Figure 4. Zenith angle distribution of upward through-going muon flux observed in Super-Kamiokande. Error bars show statistical + uncorrelated experimental systematic errors. Estimated background is subtracted. The solid histogram shows the expected flux for the null neutrino oscillation case based on the calculated flux of Ref[9]. The dashed histogram shows the expected flux for the $\nu_\mu \leftrightarrow \nu_\tau$ oscillation case with $\sin^2 2\theta$ =1.0, Δm^2 =2.5×10^{-3} eV^2 and α =1.12.

further analyses. The observed flux of upward going muons was 1.75±0.07($stat.$)±0.09($sys.$) × $10^{-13} cm^{-2} sec^{-1} sr^{-1}$. The expected flux based on the calculated flux of Ref.[9]([10]) was 1.88±0.42($theo.$) × $10^{-13} cm^{-2} sec^{-1} sr^{-1}$ (2.01±0.45($theo.$) × $10^{-13} cm^{-2} sec^{-1} sr^{-1}$). Figure 4 shows the zenith-angle distribution of the upward through-going muon flux observed in Super-Kamiokande.

Because of the difference in the mean track length of muons, the observed and predicted fluxes are different between the two experiments. However, the observed/predicted flux ratios are consistent between the two experiments within the measurement errors. The measured zenith-angle distributions in these two experiments have similar shape; both experiment observed lower flux near the vertical direction compared with the predicted flux distributions. The χ^2 values of the comparison of the shape of the zenith angle distributions of the data and predictions are 21.3/9 DOF and 18.7/9 DOF for the Kamiokande and the Super-Kamiokande data, respectively.

Analyses of neutrino oscillations were carried out in these experiments. To test the oscillation hypothesis, a χ^2 is defined as:

$$\chi^2 = \sum_{cos\Theta}^{10} \left(\frac{\phi_{Data} - \alpha \cdot \phi_{MC}(\sin^2 2\theta, \Delta m^2)}{\sigma} \right)^2 + \left(\frac{\alpha - 1}{\sigma_\alpha} \right)^2, \qquad (3)$$

where σ is the statistical and systematic error in the observed flux (ϕ_{Data}) and α and σ_α are an absolute normalization factor of the expected flux($\phi_{MC}(\sin^2 2\theta, \Delta m^2)$) and its uncertainty, respectively. σ_α was taken to be ±22% by adding the theoretical uncertainty and correlated experimental errors. The uncertainty in the absolute neutrino flux (±20%) was the dominant source of σ_α. A minimum χ^2 was calculated by changing α for each ($\sin^2 2\theta$, Δm^2). Since the contained data prefer $\nu_\mu \leftrightarrow \nu_\tau$ oscillations and since the CHOOS experiment[5] excluded the $\nu_\mu \leftrightarrow \nu_e$ oscillation parameter region relevant to the atmospheric neutrino data, only $\nu_\mu \leftrightarrow \nu_\tau$ oscillations were tested.

The Kamiokande data had χ^2_{min} at $(\sin^2 2\theta, \Delta m^2) = (1.0, 3.2 \times 10^{-3}$ eV$^2)$ with α =1.00. The χ^2_{min} value was 12.8/8 DOF. χ^2_{min} for the Super-Kamiokande data occurred at $(\sin^2 2\theta, \Delta m^2) = (1.00, 2.5 \times 10^{-3}$ eV$^2)$ with α =1.12. The χ^2_{min} value was 7.3/8 DOF. The allowed regions did not change significantly for two assumptions of the neutrino flux[9],[10], suggesting that the zenith angle and energy dependences of the neutrino flux at high energies are understood well. Figure 6 shows the allowed regions obtained from these upward going muon data. The allowed regions are larger than those of contained events. But they overlap the allowed regions obtained by the contained data.

Upward stopping muons

Because of the large detector dimension of Super-Kamiokande, a substantial fraction of upward going muons stop in the detector. The mean neutrino energy of the upward stopping muons is about 10GeV, which is substantially lower than that of upward through-going muons. Therefore, in some neutrino oscillation parameters, it is expected that an observed (stopping/through-going) flux ratio of upward going muons is different from the calculated ratio.

Super-Kamiokande observed 137 upward stopping muons during 537 detector live days[15]. The selection criteria were similar to those of upward through-going muons except for a requirement of one outer detector cluster corresponding to the entrance point. The detection efficiency was estimated to be 99%. The estimated number of cosmic-ray background events was 13.2. It was estimated that these background events were mostly in the $\cos\Theta = -0.1 \sim 0$ bin. The background events were subtracted from this bin for further analyses. We then calculated the (stopping/through-going) flux ratio ($\equiv \Re$). The observed value was $\Re = 0.22 \pm 0.023(stat.) \pm 0.014(sys.)$, while the predicted value was $\Re = 0.39 \pm 0.05(theo.)$. The observed value was substantially smaller than the prediction. Figure 5 shows the zenith angle distribution of the (stopping/through-going) flux ratio. Also shown in the same figure are the predicted ratios with and without neutrino oscillations. Clearly, the predicted distribution with neutrino oscillations agrees well with the data within the systematic uncertainty (±14%).

The 90% C.L. allowed region of the neutrino oscillations was estimated by a χ^2 test:

$$\chi^2 = \sum_{cos\Theta}^{5} \left(\frac{\Re_{Data} - \beta \cdot \Re_{MC}(\sin^2 2\theta, \Delta m^2)}{\sigma} \right)^2 + \left(\frac{\beta - 1}{\sigma_\beta} \right)^2, \quad (4)$$

where σ is the experimental (mostly statistical) error in the observed ratio ($\Re_{Data}$) and β and σ_β are a uncertainty factor of the expected ratio ($\Re_{MC}(\sin^2 2\theta, \Delta m^2)$) and its 1 σ error, respectively. σ_α was taken to be ±14% by adding the theoretical uncertainty and correlated experimental systematic errors.

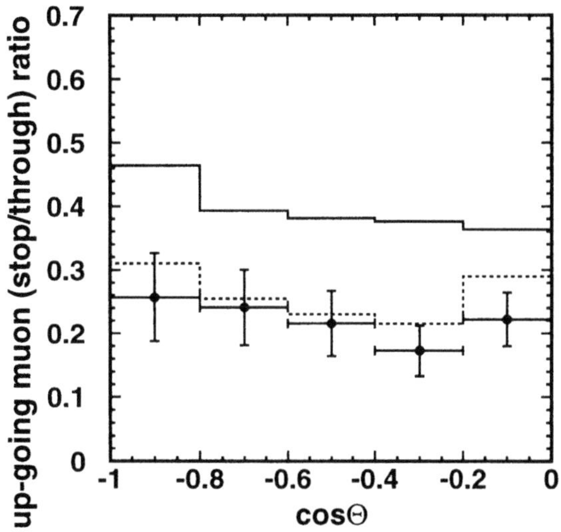

Figure 5. Zenith angle distribution of the (stopping/through-going) ratio of the upward going muon flux observed in Super-Kamiokande. Error bars show statistical + uncorrelated experimental systematic errors. The solid histogram shows the expected ratio for the null oscillation case. The dashed histogram shows the expected ratio for the $\nu_\mu \leftrightarrow \nu_\tau$ oscillation case with $\sin^2 2\theta = 1.0$ and $\Delta m^2 = 3 \times 10^{-3}$ eV^2. The expected ratio has ±14% correlated uncertainty.

χ^2_{min} occurred at $(\sin^2 2\theta, \Delta m^2) = (1.0, 3.4 \times 10^{-3}$ eV$^2)$. The χ^2_{min} value was 1.3/3 DOF. Figure 6 shows the allowed region obtained from this analysis. Again, the allowed region is larger than those of contained events. But it overlaps the allowed regions obtained by the contained and upward through-going muon data.

5. Conclusions

Both the zenith angle distribution of μ-like events and the (μ/e) values observed in Super-Kamiokande were significantly different from the predictions in the absence of neutrino oscillations. While uncertainties in the flux prediction, cross sections, and experimental biases are ruled out as explanations of the observations, the present data are in good agreement with two-flavor $\nu_\mu \leftrightarrow \nu_\tau$, or $\nu_\mu \leftrightarrow \nu_s$, oscillations. This

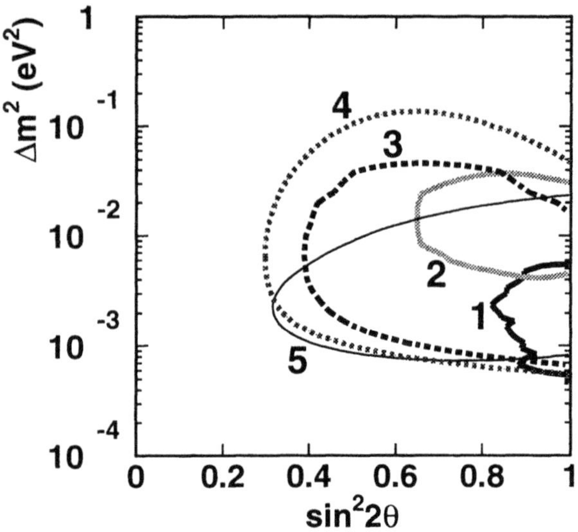

Figure 6. The allowed neutrino oscillation parameter regions obtained by Kamiokande and Super-Kamiokande. Inside of each curve is allowed at 90% C.L.. The (1)thick-black and (2)thick-gray curves show the regions obtained by contained event analyses from Super-Kamiokande and Kamiokande, respectively. The (3)black-dotted and (4)gray-dotted curves show the regions obtained by upward through-going muons from Super-Kamiokande and Kamiokande, respectively. The (5)thin-black curve shows the region obtained by the (stopping/trough-going) ratio analysis of upward going muons from Super-Kamiokande.

conclusion is consistent with the results from Kamiokande on contained data analysis and supported by the upward-going muon results from Super-Kamiokande and Kamiokande. Two experiments, Super-Kamiokande and Kamiokande, give consistent data and various techniques point to a common parameter region of neutrino oscillations: Δm^2 should be in the range around $10^{-3} \sim 10^{-2}$ eV^2 and $\sin^2 2\theta \gtrsim 0.8$. We conclude that the atmospheric neutrino data, especially from Super-Kamiokande, give evidence for neutrino oscillations.

The Super-Kamiokande experiment is supported by the Japanese Ministry of Education, Science, Sports and Culture and the United Status Department of Energy. The Kamiokande experiment was supported by the Japanese Ministry of Education, Science, Sports and Culture.

REFERENCES

1. K.S.Hirata *et al.*, Phys. Lett. B**205** (1988) 416; **280** (1992) 146.
2. Y.Fukuda *et al.*, Phys. Lett. B **335** (1994) 237.
3. D.Casper *et al.*, Phys. Rev. Lett. **66** (1991) 2561; R.Becker-Szendy *et al.*, Phys. Rev. D **46** (1992) 3720.
4. W.W.M.Allison *et al.*, Phys. Lett. B **391** (1997) 491; E.Peterson, for the Soudan-2 collaboration, in these Proceedings.
5. M.Apollonio *et al.*, Phys. Lett. B **420** (1998) 397.
6. K.Nakamura et al., in "Physics and Astrophysics of Neutrinos", Eds., M.Fukugita and A.Suzuki, Springer-Verlag (1994) p249.
7. Y.Fukuda, *et al.*, Phys. Lett. B **433** (1998) 9.
8. Super-Kamiokande Collaboration, Y.Fukuda, *et al.*, Phys. Lett. B (1998), accepted for publication, hep-ex/9805006
9. M.Honda *et al.*, Phys. Lett. **B248** (1990) 193; M.Honda *et al.*, Phys. Rev. **D52** (1995) 4985.
10. G.Barr *et al.*, Phys. Rev. **D39** (1989) 3532; V.Agrawal, *et al.*, Phys. Rev. **D53** (1996) 1313; T.K.Gaisser and T.Stanev, Proc. 24th Int. Cosmic Ray Conf. (Rome) Vol.1 (1995) p694.
11. Y.Fukuda, *et al.*, Phys. Rev. Lett. **81** (1998) 1562.
12. L.Wolfenstein, Phys. Rev. **D17** (1978) 2369; S.P.Mikheyev and A.Yu.Smirnov, Sov. J. Nucl. Phys. **42** (1985) 1441; S.P.Mikheyev and A.Yu.Smirnov, Nuovo Cim. **C9** (1986)17.
13. Based on a two-dimensional extension of the method in Review of Particle Properties, Section: Error and confidence intervals – Bounded physical region: R.M.Barnett *et al.*, Phys. Rev. **D54** (1996) 375.
14. S.Hatakeyama *et al.*, Phys. Rev. Lett. **81** (1998) 2016.
15. The Super-Kamiokande collaboration, drafts in preparation.

Nuclear Physics B (Proc. Suppl.) 77 (1999) 133–139

NUCLEAR PHYSICS B
PROCEEDINGS
SUPPLEMENTS

Fluxes of Atmospheric Neutrinos and Related Cosmic Rays

T.K. Gaisser[a]*

[a]Bartol Research Institute, University of Delaware,
Newark, DE 19716, USA

The atmospheric neutrino beam simultaneously spans a range of pathlengths from ten to ten thousand kilometers, which correspond respectively to downward- and upward-going neutrinos. As with any neutrino oscillation experiment, also in this case the interpretation of the data depends on a detailed knowledge of the neutrino beam. The ingredients are the primary spectrum of cosmic-ray nucleons, the geomagnetic fields in which the charged particles propagate and the properties of interactions of hadrons in the atmosphere. In this talk I review the status of calculations in light of the recent evidence for neutrino oscillations from Super-Kamiokande [1].

1. Introduction

When cosmic ray protons and nuclei interact in the atmosphere, the secondary cascades include neutrinos from decay of pions, muons and kaons. Production of these neutrinos depends on the local zenith angle because of the competition between decay and interaction of the parent mesons in the tenuous atmosphere. A simple geometric construction [2] shows that a trajectory from below with nadir angle θ has the same zenith angle θ on the other side of the earth. Therefore, since neutrinos with $E \ll 10^5$ GeV are virtually unattenuated by the earth, the flux of atmospheric neutrinos would be up-down symmetric in the absence of neutrino oscillations except to the extent that the isotropy of the primary cosmic rays is distorted by the geomagnetic field. Variation of the neutrino flux with azimuth is a consequence only of the geomagnetic field (the "East-West" effect) even in the presence of oscillations (because within a given band of zenith angle the distributions of neutrino pathlengths and energies are independent of azimuth). This fact allows [3] an important check of the systematics of the Super-K analysis [4], which I discuss in §2.

Typical altitudes of production of the neutrinos are between 10 and 20 kilometers, so the distribution of neutrino pathlengths ranges from 10 km for vertically downward neutrinos (neutrinos that originate directly overhead) to $\sim 10^4$ km for upward neutrinos from below. The range of neutrino energies for contained or partially contained events is from sub-GeV to multi-Gev. For neutrino-induced upward, throughgoing muons it extends to $\sim$ 1000 GeV. Thus the atmospheric neutrinos have a range of pathlength over neutrino energy $1 < L_{km} \,/\, E_{GeV} < 10^5$.

The $\pi \to \mu \to e$ decay chain is the predominant mode of production of atmospheric neutrinos in The sub-GeV to multi-GeV range. This leads to the basic prediction of

$$\frac{\nu_e + \bar{\nu}_e}{\nu_\mu + \bar{\nu}_\mu} \sim \frac{1}{2} \tag{1}$$

for $E_\nu \leq 1$ GeV. The ratio decreases as energy increases because muons are increasingly likely to reach the surface before decaying. Comparison of decay length to energy-loss length in the earth leads to the conclusion that virtually all muons that reach the ground stop before decay (or capture) occurs. Therefore muons that reach the ground do not contribute the neutrinos with energy high enough to contribute even to the sub-GeV sample ($p_e > 100$ MeV/c).

In contrast with the expectation of Eq. 1, several experiments [6–8,4] find

$$R = \frac{(\mu - like/e - like)_{data}}{(\mu - like/e - like)_{MC}} \approx 0.65, \tag{2}$$

which is equivalent to $(\nu_e + \bar{\nu}_e)/(\nu_\mu + \bar{\nu}_\mu) \approx 0.77$. Ingredients that enter the denominator of the ratio of ratios in Eq. 2 include the calculated neutrino flux, the cross sections for neutrinos to interact in the quasi-elastic and various multi-prong

*Research supported in part by the U.S. Department of Energy under Grant No. DE-FG02-91ER40626

0920-5632/99/$ – see front matter
PII S0920-5632(99)00408-9

channels and the detection and reconstruction efficiencies in the detector. The subject of this talk is the neutrino fluxes.

Three independent calculations [9–11] have been compared and analyzed [12] in order to identify and evaluate the sources of uncertainty in our knowledge of the atmospheric neutrino flux. Here I concentrate on comparison of the calculations of Honda *et al.* [10,13] with the "Bartol fluxes"[14,15], because these two have been used by the experimental groups [4,8,16] for analysis of their data. The calculations of Refs. [14] and [15] are extensions of the calculations of Ref. [9] respectively to high (>3 GeV) and to low (<200 MeV) energy. In addition, a more realistic treatment of the geomagnetic cutoffs is used [17], which reduces the calculated neutrino fluxes in the sub-GeV range by about 10 per cent. Honda *et al.* [13] also extended their calculation to high energy. At present the Bartol neutrino fluxes and the Honda *et al.* fluxes in the GeV range agree within 5% in magnitude as well as ratio. This level of agreement in magnitude is, however, smaller than the systematic uncertainties, as I will discuss in Sections 3 and 4.

In sections 3 and 4 I discuss the primary fluxes and the treatment of hadronic interactions, both of which influence the spectrum and shape of the neutrino spectra. In the Super-K analysis, the overall normalization is treated as a free parameter because of the large uncertainty in the normalization of the primary spectrum. There are recent measurements of the primary spectrum that should in principle allow one to reduce this source of uncertainty.

As emphasized by Perkins [18], muon fluxes high in the atmosphere are directly related to the neutrino fluxes, being produced by the same primary spectra and from the same interaction processes. In §5 I discuss how measurements of the flux of muons high in the atmosphere are being used to check the overall normalization and shape of the closely related neutrino flux. In the conclusion I list the various approximations common to the present calculations and how they might be expected to affect the results.

2. Geomagnetic effects

Propagation of a cosmic-ray nucleus through the geomagnetic field depends only on its gyroradius and hence on the magnetic rigidity,

$$R = A \times pc/(Ze) \tag{3}$$

Here A and Z are the mass and charge of a nucleus of momentum-per-nucleon p. Low energy particles at low geomagnetic latitudes cannot reach the atmosphere to produce secondaries. Since energy per nucleon is the important quantity for production of secondaries, nuclei become relatively more important compared to protons at low geomagnetic latitudes because of the factor $A/Z \approx 2$ in Eq. 3.

Both neutrino flux calculations [13,14] use geomagnetic cutoffs obtained by the standard method of backtracking antiprotons through the geomagnetic field to determine the cutoffs for a particular location. For example, in the calculation of Ref. [17], which is used in Ref. [14], antiparticles are injected at 20 km altitude on an outward trajectory. If the trajectory reaches $30\,R_\oplus$ before it travels $500\,R_\oplus$ and without intersecting the surface of the earth, then it is assumed that positive particles of the same rigidity can reach the atmosphere from that direction.

For the location of Super-K we have compared the cutoffs used in Ref. [13] with those of Ref. [17] used for the calculation of Refs. [14,15]. The cutoff maps are very similar, but with some noticeable differences toward the east, where the cutoffs are slightly higher in Ref. [17].

At low geomagnetic latitudes such as Kamioka, average cutoffs are higher locally (i.e. for cosmic rays entering the atmosphere above the detector) than for the opposite hemisphere (i.e. for the cosmic rays entering the atmosphere on the other side of the earth, which give rise to upward-going events). The opposite is the case for a detector at a high geomagnetic latitude, such as Soudan. There the local cutoffs are negligible in the sense that essentially all cosmic-rays from above with sufficient energy to produce pions and contribute to the flux of neutrinos can reach the atmosphere to interact. Upward events originate from the atmosphere over the entire hemisphere below each detector. Since the average over a full hemisphere is similar from any viewpoint, the upward/downward ratio should be greater than one at Kamioka but less than one at Soudan.

Fig. 1 illustrates the situation. The pair of curves labelled (A) shows the distribution of primary cosmic-ray energies that would contribute to the sub-GeV signal in Super-Kamiokande if there were no geomagnetic cutoff at all. The solid curve is for solar minimum and the dotted one for solar maximum. The middle pair (B) is the corresponding response from below, which would be similar if Super-K were moved to Soudan. The rightmost pair of curves (C) is the response for downward sub-GeV events at Super-Kamiokande. What is plotted is proportional to the event rate per logarithmic interval of primary energy, so in each case the area is proportional to the signal. Thus the upward/downward ratio at Super-K is $B/C > 1$. If Super-K were located at Soudan, the ratio would instead be $B/A < 1$. (The method used to simulate "sub-GeV" events is described in Ref. [3].)

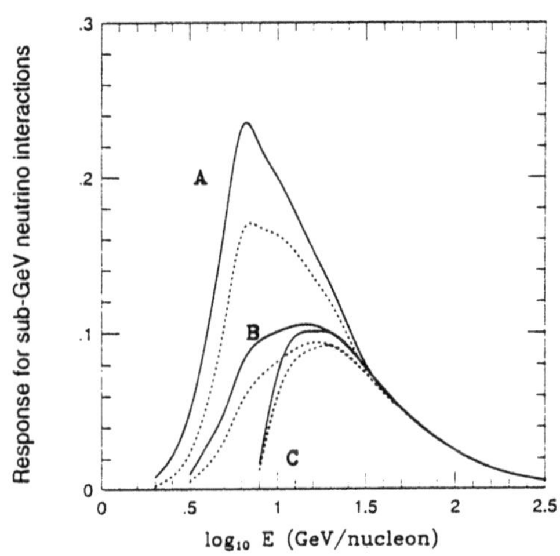

Figure 1. Response for sub-GeV muon-like events in Super-K.

As neutrino energy increases the up-down asymmetry from the geomagnetic effect diminishes. For this reason, the Super-Kamiokande group have emphasized the multi-GeV event sample in their search for neutrino oscillations [1,19]. On the other hand, the full data set contributes to the evidence for oscillations. Therefore it is important to note [3] that the geomagnetic effects themselves provide a way of testing the integrity of the entire analysis chain that is independent of whether or not there are oscillations.

At the low geomagnetic latitude of Kamioka there is a pronounced east-west effect on the cosmic radiation. Cutoffs are significantly lower for positive particles from the west than from the east. For example, the trajectory of a 20 GeV antiproton injected toward the east from above Super-K at 70° from the zenith would be bent down by the geomagnetic field and intersect the surface of the Earth, while the same antiproton injected toward the west would escape from the geomagnetic field. In other words, the cutoff for protons with zenith angle 70° from the east at Super-K is > 20 GeV. For directions closer to the horizon the cutoff from the east approaches 50 GV. In contrast, for directions above the horizon from the west the cutoff is 5 to 10 GV at Kamioka.

The excess of primary cosmic rays from the west at Kamioka produces a corresponding east-west asymmetry of the low-energy neutrino flux and hence of the sub-GeV event rate. There is a much smaller, but still non-negligible asymmetry for the multi-GeV event sample [3]. Since the east-west effect is an azimuthal asymmetry, it is independent of oscillations; oscillation effects depend on neutrino pathlengths, which vary with zenith angle but are independent of azimuth.

Figure 2 [20] compares the azimuthal dependence of the Super-K data ($0.4 < p_{lepton} < 2.0$ GeV/c, single ring events in 22.5 kton fiducial volume) with expectation. The solid line uses the neutrino fluxes of Ref. [13] and the dashed line the calculation of Ref. [14,15]. Although the fits are equally good ($\chi^2/d.o.f. \approx 1$ for all four comparisons), the geomagnetic effect is somewhat more pronounced with the Bartol neutrino flux [14,15] than with the flux of Honda *et al.* [13]. We have made some diagnostic tests to investigate the source of this difference and a similar difference between the two calculations that shows up in the zenith angle dependence of sub-GeV events. The difference arises in part from the difference in cutoffs mentioned above, but also from the difference in primary spectrum, as discussed below.

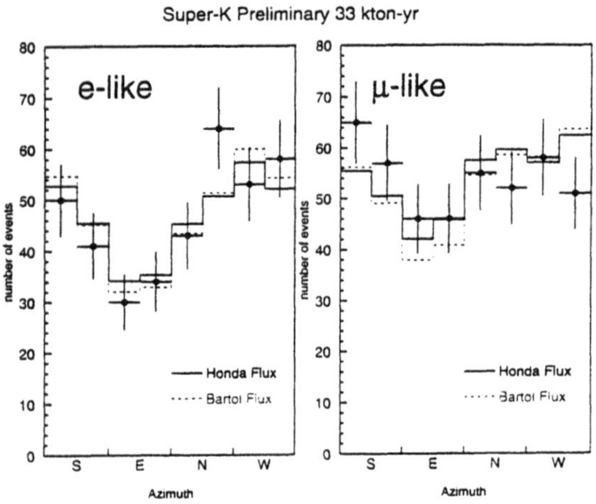

Figure 2. Azimuthal dependence: calculated and observed by Super-K.

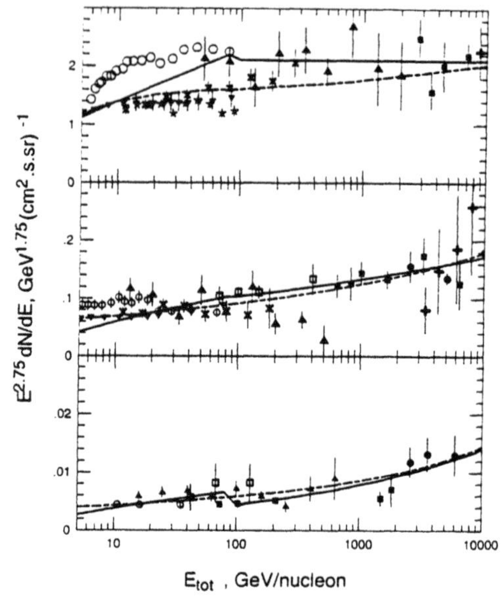

Figure 3. Summary of primary spectra.

3. Primary spectrum

Both the normalization and the shape of the assumed primary spectrum have important consequences for the calculation of the neutrino fluxes. The normalization propagates directly through to the event rate. The assumed spectral index affects the shape of the neutrino energy spectrum in an obvious way, but it also affects the angular dependence through its interaction with the geomagnetic effects. Thus, a softer spectrum will lead to more pronounced geomagnetic effects because a larger fraction of the event rate comes from lower energy primaries, which are most affected by the geomagnetic field.

Fig. 3 is a summary of measurements of spectra of protons, helium and the CNO group of nuclei, compared with the primary spectra of Honda *et al.* [13] (solid lines) and the spectra used in Ref. [14] (dashed lines). From Fig. 1, it is apparent that $5 < E < 50$ GeV/nucleon is the most important region of the primary spectrum for sub-GeV events. The harder spectrum of Ref. [13] in this energy region, coupled with the geomagnetic field, contributes significantly to the fact that the geomagnetic effects are somewhat smaller in the neutrino spectra of Ref. [13] than in Ref. [14], as mentioned in the previous section.

A marked feature of the plot for hydrogen is the fact that the data of Webber [21] (shown by the open circles in Fig. 3) are significantly higher than those from the LEAP experiment [22] and other more recent experiments in the same energy region. The difference is outside the error bars, indicating a systematic effect. The most recent result is from the BESS detector [23], and other recent experiments (Refs. [24,25]) are included in the BESS compilation. Generally (with the possible exception of the measurement of Ref. [26]), the interpretation of which is complicated by an unusually strong level of solar modulation) all the recent experiments are consistent with the LEAP results. In fits to their data the Super-K group have treated the overall normalization of their rates as a free parameter. The primary spectra and their potential consequences for interpretation of the data on neutrino interactions are discussed more fully in talks given at the Satellite Symposium [23,27,28].

4. Yields

It is important to note that the primary spectrum is not the only source of uncertainty in the normalization and shape of the energy spectrum of atmospheric neutrinos. Uncertainties in the yields of pions and kaons in interactions of hadrons with nuclei of the atmosphere are also important. Not all of phase space is covered in accelerator measurements with nuclear targets. For sub-GeV events the important range of beam energies is from a few GeV to several tens of GeV (see Fig. 1). In this energy range the atmospheric cascades are dominated by interactions of nucleons, and nearly all neutrinos are from the $\pi \to \mu \to e$ decay chain.

Existing measurements with beam energies around 20 GeV and light nuclear targets measure pions only above 3 or 4 GeV [29,30], and there are significant differences in how the lower energy pions are represented in the different neutrino flux calculations, as discussed in Ref. [12]. The pion multiplicities, and the momentum distributions as reflected by the spectrum-weighted moments for pion production, are highest in the calculation of [9,14,15]. This compensates to some extent for the higher assumed proton spectrum of Ref. [13] with the result that the calculated neutrino fluxes (comparing Refs. [14] and [13]) differ by less than either the primary spectrum or the yields.

Yields in a new calculation of Battistoni *et al.* [31] are intermediate between those of Refs. [9] and [10].

5. Muons

The same primary spectra and the same hadronic interactions determine both muon and neutrino fluxes. Therefore, comparison with measurements of muons high in the atmosphere offers a way to check directly the neutrino fluxes. The most important range of altitudes for pion decay is 10 to 25 kilometers, which corresponds to atmospheric depths of ~ 20 to ~ 200 g/cm^2.

Many of the same detectors referred to above in connection with recent measurements of the primary spectrum have also been used to measure the muon spectrum during ascent through the atmosphere and on the ground. The calculations of Refs. [14] (and [13]) compare reasonably well with the measurements of the MASS experiment [32], although there is a relative excess of muons below 1 GeV in the calculation. On the other hand, a recent comparison between [14] and the HEAT measurements of muons [33] showed better agreement in the shape of the spectrum but with an overall excess of the calculation relative to the data of as much as 50% in some bins.

Measurements on the ground and at float altitude necessarily have better statistics than data obtained during ascent. It is possible that some of the discrepancies referred to above could be a consequence of the short exposures during ascent. Both for MASS [32] and HEAT [33] there is a tendency for better agreement between calculation and measurement at float and at the ground than during ascent. This is an active area with further potential for reducing uncertainties in the flux of atmospheric neutrinos.

There are interesting possibilities with the muon measurements for probing details of the calculations. For one thing, muon fluxes at float altitude reflect directly the primary spectrum and the properties of pion production in single nucleon-nitrogen interactions with no intervening cascading.

A more interesting possibility arises from the fact that in some cases the same detector has been exposed at different locations with different geomagnetic cutoffs. The MASS experiment has been flown both in Northern Canada (essentially no cutoff) and from Ft. Sumner, NM where the vertical cutoff is ≈ 5 GV. The BESS detector has measured the muon charge ratio on the ground in northern Canada and in Japan. The low-energy behavior of the ratio is quite different in the two locations. Below ~ 1 GeV the μ^+/μ^- ratio decreases toward 1 in Japan, which can be understood as a consequence of the high local geomagnetic cutoff. There are two effects:
First, with a high cutoff, heavy primaries are relatively more important because they have a higher rigidity for a given energy per nucleon. Protons produce more positive than negative pions (and hence more μ^+) and *vice versa* for neutrons. Thus, enhancing the contribution from nuclei, which carry the neutrons, suppresses the muon charge ratio slightly. Secondly, vertical μ^+ at the ground have followed trajectories from slightly east of vertical (where the cutoffs are higher), whereas vertical μ^- will have come slightly from the west where more of the primary spectrum reaches the atmosphere to produce secondaries. This enhances the negative relative to the positive muons and hence reduces the μ^+/μ^- ratio preferentially at low energy where the bending is more significant.

6. Conclusion

Present calculations [13,14] include several approximations:

- They are one-dimensional; i.e. all neutrinos are assumed to follow the direction of the primary nucleon that produced them. This approximation has two effects:
    1. There should be some loss of particles that are produced at large angle. Given the momentum involved, as compared with the typical transverse momentum of produced pions, it is straightforward to check that this effect should be small for neutrino events in Super-K.
    2. Bending of charged particles in the atmosphere is not followed. This is perhaps the most important effect to check [34] because it is systematic. As explained above, the vertical muon charge ratio is reduced when the cutoffs are high. There is a corresponding decrease in the $\nu_e/\bar{\nu}_e$ ratio (and an increase in that part of the $\nu_\mu/\bar{\nu}_\mu$ ratio that comes from muon decay). Because $\sigma_\nu > \sigma_{\bar{\nu}}$, the calculated ratio of electron-like to muon-like events will decrease with respect to the one-dimensional calculation. This correction will therefore make the anomaly of Eq. (2) somewhat more pronounced.
- The superposition model has been used in Ref. [14] for interactions of nuclei. Within the framework of a standard multiple scattering picture, this approximation can be shown to give a good account of the distribution of first interactions of each nucleon. It will, however, lead to some overestimate of the multiplicity of pions in the target fragmentation region.
- The cascades are propagated to sea-level all over the globe. In particular, the exact terrain over the detector (i.e. the mountain in the case of Super-K) has been neglected [35]. This is negligible for muon neutrinos from pions, which decay high in the atmosphere. From the pathlength distributions of Ref. [5] it is possible to estimate the size of the effect of this approximation. To take an extreme case of a 4 km overburden (~NUSEX), the neutrinos from muon decay overhead are overestimated by about 10%, leading to a ~ 5% overestimate of the calculated $(\nu_e + \bar{\nu}_e)/(\nu_\mu + \bar{\nu}_\mu)$ ratio.
- The calculations are based on parametrizations of data in limited regions of phase space. Interpolations and extrapolations introduce some level of uncertainty. The yields of Ref. [14] are at the high end of a spectrum, with [11] the lowest and [13] and [31] in between.

At least two groups are embarking on three-dimensional calculations. The Italian group [31] has published a short account of their plan with a comparison of their one-dimensional results with those shown in Ref. [12]. They use FLUKA [36] with various hadronic interaction models at different energies. The authors of Refs. [9] and [14], together with Coutu, are also pursuing this goal. Although effects are generally expected to be small, in view of the importance of the experimental results, a greater level of detail in the calculations is warranted.

Acknowledgements. I am grateful to Ed Kearns for providing me with Fig. 2 [20] and to Todor Stanev for reading the manuscript and for collaboration on this work. I thank M. Goldhaber and M. Spiro for useful conversations, and M. Honda for very helpful exchanges of information about the calculations of Refs. [10,13].

REFERENCES

1. Y. Fukuda *et al.*, Phys. Rev. Letters 81 (1998) 1562.
2. D. Ayres *et al.*, Phys. Rev. D29 (1984) 902.
3. Paolo Lipari, Todor Stanev & T.K. Gaisser, Phys. Rev. D (to be published) astro-ph/9803093.
4. Y. Fukuda *et al.*, Phys. Lett. B433 (1998) 9 (sub-GeV) and hep-ex/9805006 (to be published) (multi-GeV).
5. T.K. Gaisser & Todor Stanev, Phys. Rev. D57 (1998) 1977.
6. R. Becker-Szendy *et al.* (IMB Collaboration), Phys. Rev. D46 (1992) 3720 and references therein.

7. Y. Fukuda *al.* (Kamiokande Collaboration) Phys. Lett. B335 (1994) 237 and references therein.
8. W.W.M. Allison *et al.* (Soudan Collaboration), Phys. Lett. B391 (1997) 491.
9. Giles Barr, T.K. Gaisser & Todor Stanev, Phys. Rev. D39 (1989) 3532.
10. M. Honda, K. Kasahara, K. Hidaka & S. Midorikawa, Phys. Lett. B248 (1990) 193.
11. E.V. Bugaev & V.A. Naumov, Phys. Lett. B232 (1989) 391.
12. T.K. Gaisser, M. Honda, K. Kasahara, H. Lee, S. Midorikawa, V. Naumov & Todor Stanev, Phys. Rev. D54 (1996) 5578.
13. M. Honda, T. Kajita, K. Kasahara & S. Midorikawa, Phys. Rev. D52 (1995) 4985.
14. Vivek Agrawal, T.K. Gaisser, Paolo Lipari & Todor Stanev, Phys. Rev. D53 (1996) 1314.
15. T.K. Gaisser & Todor Stanev, Proc. 24th Int. Cosmic Ray Conf. (Rome) vol. 1 (1995) 694.
16. M. Ambrosio *et al.* (MACRO Collaboration), hep-ex/9807005 (throughgoing muons) and M. Spurio, hep-ex/9808001 (stopping muons).
17. Paolo Lipari & Todor Stanev, Proc. 24th Int. Cosmic Ray Conf. (Rome) vol. 1 (1995) 516.
18. D.H. Perkins, Astroparticle Physics 2 (1994) 249.
19. T. Kajita, to appear in Proc. of Neutrino98.
20. C. McGrew (Super-Kamiokande Collaboration), to appear in Proc. Int. Conf. on High Energy Physics (Vancouver, 1998) and Super-K paper on azimuthal dependence (forthcoming).
21. W.R. Webber, R.L. Golden & S.A. Stephens, Proc. 20th Int. Cosmic Ray Conf. (Moscow) vol. 1 (1987) 325.
22. E.S. Seo *et al.* (LEAP) Ap.J. 378 (1991) 763.
23. S. Orito, T. Sanuki *et al.* (BESS), to appear in proceedings of the conference "New Era in Neutrino Physics", Tokyo, June 11/12 (1998).
24. W.R. Menn *et al.* (IMAX) Proc. 25th Int. Cosmic Ray Conf. (Durban) vol. 3 (1997) 409.
25. M. Boezio *et al.* (CAPRICE), Ap. J. (to be published).
26. W.R. Webber *et al.* (MASS), Ap.J. 380 (1991) 230.
27. M. Honda, to appear in proceedings of the conference "New Era in Neutrino Physics", Tokyo, June 11/12 (1998).
28. T.K. Gaisser, to appear in proceedings of the conference "New Era in Neutrino Physics", Tokyo, June 11/12 (1998).
29. T. Eichten *et al.*, Nucl. Phys. B44 (1972) 333.
30. J.V. Allaby *et al.*, CERN Yellow Report No. 70-12 (unpublished).
31. G. Battistoni *et al.*, Proc. 5th TAUP Conf., Gran Sasso (1997).
32. M. Circella, C.N. De Marzo, T.K. Gaisser & Todor Stanev, Proc. 25th Int. Cosmic Ray Conf. (Durban), vol. 7 (1997) 117.
33. Stephane Coutu (HEAT Collaboration) to appear in Proc. Int. Conf. on High Energy Physics (Vancouver, 1998).
34. Maurice Goldhaber has emphasized this point (private communication).
35. I thank M. Spiro for raising this question.
36. A. Fassò, A. Ferrari, J. Ranft and P.R. Sala, Proc. of the 3rd Workshop on Simulating Accelerator Radiation Environment, SARE-3, KEK-Tsukuba, May 7-9, 1997, (H. Hirayama, ed.) KEK Report 97-5, p. 32 (1997).

ELSEVIER

Nuclear Physics B (Proc. Suppl.) 77 (1999) 140–145

NUCLEAR PHYSICS B
PROCEEDINGS
SUPPLEMENTS

Uncertainty of the atmospheric neutrino fluxes

M.Honda[a]

[a]Institute for Cosmic Ray Research, University of Tokyo,
Tanashi 3-2-1, Tokyo 188 JAPAN

The uncertainty in the calculation of atmospheric neutrino fluxes is studied. The absolute value of atmospheric neutrino fluxes is sensitive to variation of the primary cosmic ray flux model and/or the interaction model. However, the ratios between different kind of neutrinos stay almost unchanged with these variations. It is unlikely that the anomalous ratio $(\nu_\mu/\nu_e)_{obs}/(\nu_\mu/\nu_e)_{MC}$ reported by Kamiokande and Super Kamiokande is caused by the uncertainty of predicted atmospheric neutrino fluxes.

1. Introduction

After the discovery of an anomalous ratio of ν_e/ν_μ in the atmospheric neutrinos by Kamiokande, refined calculations of atmospheric neutrino fluxes have been made intensively by several authors[1][2][3][4][5]. The differences between different authors have been compared in Ref. [6]. The major differences are in the primary cosmic ray flux and in the hadronic interaction model. These are also the main sources of uncertainty in the calculation of atmospheric neutrino fluxes.

There have been many measurements of the primary cosmic ray flux, but they do not agree with each other. We have to consider this disagreement as the uncertainty of the primary cosmic ray flux. There also have been many experimental studies of hadronic interactions. However, the most important piece of information, the energy spectrum of secondary particles in the projectile region, is not well known enough for a accurate calculation of atmospheric neutrino fluxes. This is partly because, as high energy physics experiments moved to the colliders, this measurement has become more difficult to make.

Adding to the above, the density structure of the atmosphere is a potential source of uncertainty. In many calculations, the US standard is often used as the 'standard' model. However, it is also known that the atmospheric density structure has a latitude dependence and seasonal variations.

In this paper, we study how these uncertainties affect the atmospheric neutrino calculation in the energy range of $\lesssim$10 GeV based on Ref. [5](HKKM). Note we use the one-dimensional approximation as HKKM throughout this paper.

2. Variation of calculation Model

2.1. Primary cosmic ray flux

In Fig. 1 are shown the observations of cosmic ray protons at $< 10^4$ GeV by different groups. We consider 3 flux models in this study: high, mid, and low shown in the figure for solar min. The flux models for primary cosmic rays agree in $\lesssim$5 GeV region, where the differences are small except for these caused by solar modulation.

Cosmic rays above a few 100 MeV originate in Galactic space. When they enter the solar sphere, they are pushed back by the solar wind. As this effect is more pronounced for lower energy cosmic rays, the energy spectrum of low energy cosmic rays varies with the strength of the solar wind, or with the solar activity. However, this modulation is expected to be around 5% from the minimum to the maximum of solar activity at 10 GeV. Above this energy, the effect of solar activity on the cosmic ray flux is very small.

The variation due to solar activity in the $\lesssim$10 GeV region agrees well with the expectation based on the Mid-flux model and conventional solar modulation formulae. The exception in Fig. 1 is the crosses which are the base data for the high

PII S0920-5632(99)00409-0

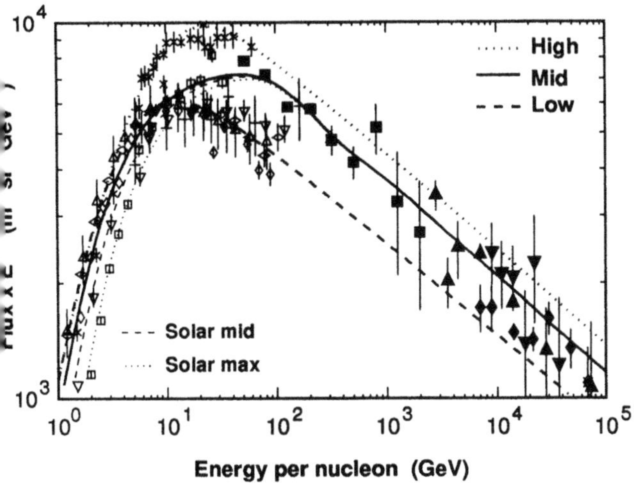

Figure 1. Observed cosmic ray proton flux. Crosses from Ref. [7], open upward triangles stand for LEAP[9], open squares for MASS[8], open downward triangles for IMAX[10], open vertical diamonds for CAPRICE[11], and open horizontal diamonds for BESS[12]. pluses, closed squares, closed vertical diamond, closed upward triangles, and closed downward triangles are from Refs[22],[26],[25],[34],[35],[36], and [38] of HKKM respectively.

flux model. We consider that the uncertainty in primary cosmic ray flux is small for $\lesssim$10 GeV.

Above 10 GeV, there are rather large differences among the different experiments. This may be caused by the inherent difficulties in making the measurements. Particularly, the problems in the estimation of instrumental efficiency and limited exposure time in balloon experiments. At this moment we have to assume that there are large uncertainties in the primary cosmic ray flux measurements in the $\gtrsim$10 GeV region. However, it is interesting that recent experimental values with super conductive magnetic spectrometers distribute rather around the Low flux model.

It is noted that around 20% of nucleons are carried by nuclei heavier than the proton. We do not show the details but for the heavier nucleus cosmic rays we use the flux value of HKKM.

Even using the one dimensional approximation, the geomagnetic field affects the calculation of atmospheric neutrino fluxes through the primary comsic ray flux due to the geomagnetic cutoff. Principally, however, the geomagnetic cutoff is not a source of uncertainty since it can be calculated to very good accuracy from the measured geomagnetic field.

2.2. Hadronic interaction

In the calculation of the atmospheric neutrino fluxes, hadronic interaction Monte Carlo code is one of the most important components. However, there are uncertainties in the hadronic interaction model, due to the lack of suitable experimental data. The interaction models used by different calculations are slightly different to each other.

A comparison of the secondary particle momentum spectrum is shown in Fig. 2 and in Fig. 3 for several Monte Carlo codes which have been used in the calculation of atmospheric neutrino fluxes. (This figure is taken from Ref. [6]). The variable x used in the figure is defined as $x = P_{2ndary}/P_{incident}$. It should be noted that the difference between the histograms (BGS[2] and HKHM[3]) and the smooth curve (BN[1]) is not so large as it appears. Since each value is multiplied by the variable x, their differences are emphasized in the low x region.

It is generally difficult to modify an established hadronic interaction Monte Carlo code due to the energy momentum conservation and to the correlation of secondary particles. However, in the case of the atmospheric neutrinos, the correlations are not important. We use an 'inclusive' hadronic interaction Monte Carlo code for the study of the variation of hadronic interactions.

The calculation of atmospheric neutrino fluxes with the inclusive interaction code is similar to the analytic calculation. In the case of the analytic calculation, however, it is difficult to treat the competition process between decay and interaction for hadronic particles. This is the most crucial point in the analytic calculation. Therefore, we make the Monte Carlo study with the inclusive hadronic interaction code in this paper.

Starting with the hadronic interaction model of FRITIOF 1.6 and Jetset 6.3, we consider 'mod-

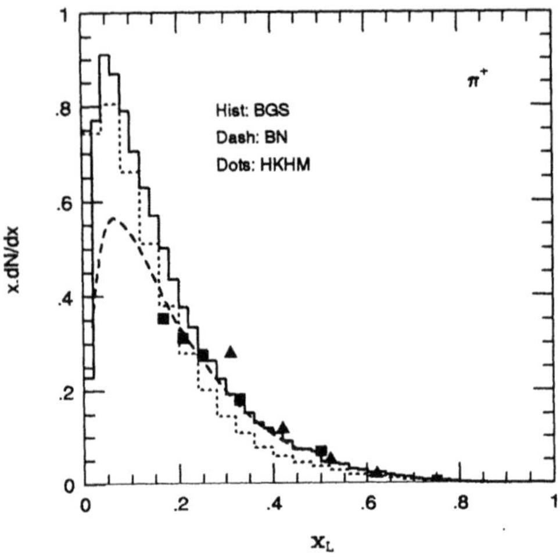

Figure 2. Secondary π^+ energy spectrum calculated using the interaction model used by various different authors (taken from Ref. [6]).

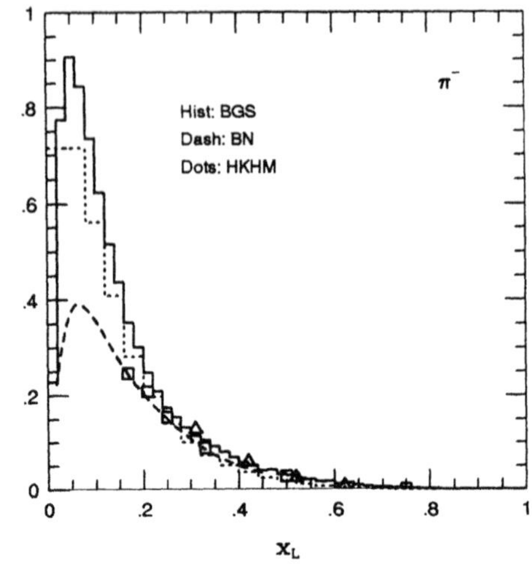

Figure 3. Secondary π^- energy spectrum calculated using the interaction model used by various different authors (taken from Ref. [6]).

erate' variations from that. The variation of secondary particle energy spectrum is created as follows: denoting a 'starting' energy spectrum of secondary particles as

$$\frac{dN}{dx} = f_{ptl}(x) ,$$

where we take $x = E_{2nd}/E_{inc}$ and E is the kinetic energy. Moderate variation to this secondary particle energy spectrum can be made with a parameter α as

$$\frac{dN}{dx} = A(\alpha) f_{ptl}(x^{1+\alpha}) ,$$

where $A(\alpha)$ is a factor introduced for energy conservation as:

$$A(\alpha) = \frac{\int x f_{ptl}(x)dx}{\int x f_{ptl}(x^{1+\alpha})dx} .$$

In other words, we request that the energy sum for each kind of particle is conserved. This modification method works efficiently to vary the slope of the secondary particle spectrum at $x \approx 0.3$.

This modification also changes the number of particles for each kind of secondary particle (= $\int x f_{ptl}(x)dx$). The variation of $\alpha = \pm 0.1$ roughly corresponds to ± 10 % variation at $x = 0.3$, and to a ± 20 % variation of the multiplicity. The variation of α from -0.2 to $+0.2$ is considered in this paper. The variation limit of α is set as 2 times of the possible uncertainty of multiplicity (± 10 %) in the hadronic interaction.

Examples of this modification are shown in Fig. 4 and Fig. 5 for $p + air \rightarrow \pi^{\pm}$ at $E_p = 32$ GeV with the secondary particle spectrum of BN at ≈ 20 GeV from Ref. [6] for comparison. One may notice that the secondary particle spectrum of BN is outside of our variation limit, even considering the difference of incident particle energy. To include BN secondary energy spectrum, probably a different starting point is needed.

In addition, variations of inelastic cross-sections from -20 % to $+20$ % and k/π-ratio from -40 % to $+40$ % are also considered. Note that we have assumed the same α for all kinds of secondary particles, although $f_{ptl}(x)$ is different for each particle, and σ_{inel} varies at the same ratio for all hadronic interacting particles. The variation range considered here is larger than the uncertainty normally considered[13].

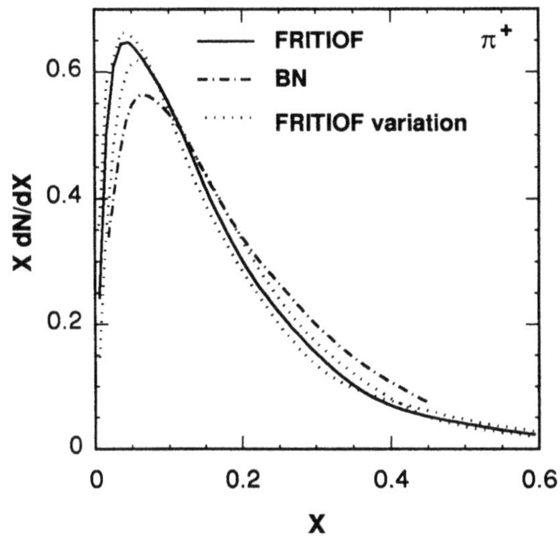

Figure 4. Secondary π^+ energy spectrum from FRITIOF 1.6 and Jetset 6.3 at 30 GeV with its variations for $\alpha = \pm 0.2$. For comparison, that of BN at $\approx$ 20 GeV is also shown.

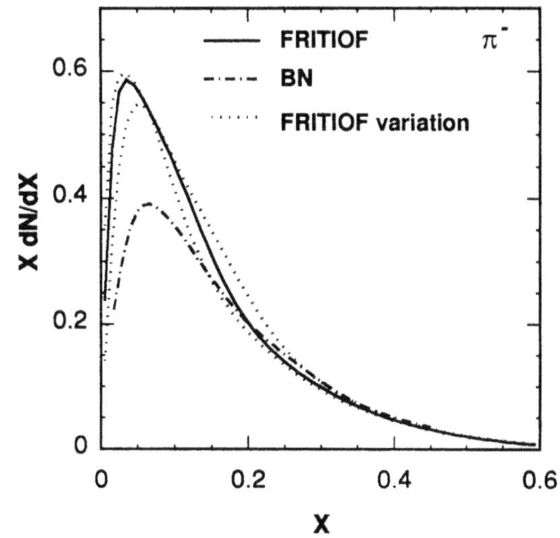

Figure 5. Secondary π^- energy spectrum from FRITIOF 1.6 and Jetset 6.3 at 30 GeV, with it's variations for $\alpha = \pm 0.2$. For comparison, that of BN at $\approx$ 20 GeV is also shown.

2.3. Density structure of the atmosphere

For the density structure of the atmosphere, the US-standard atmosphere model is often used. However, since we are interested in the effect of the variation of atmospheric structure, we introduce a simple model with a single scale height. The air density is expressed as a function of height as:

$$\rho = \frac{\rho_{col}}{h_0} \exp(-\frac{h}{h_0}) \ .$$

For the 'standard', we take $\rho_{col} = 1.231$ kg/m^3 and $h_0 = 8.4$ km, such that it agrees with the US-standard in global features.

We study the effect of variations from -10 % to $+10$ % both in the scale height and in the column density. However, when we study the variation of interaction model only, we use the US-standard atmosphere model.

3. Variation of neutrino fluxes

We have summarized the variation of atmospheric neutrino fluxes for the model variations of primary cosmic ray flux (Fig. 6), hadronic interaction(Fig. 7), and atmospheric density structure(Fig. 8).

From these figures, one can see that the variation of the primary cosmic ray flux and the secondary particle spectrum in the hadronic interactions have a large effect on the atmospheric neutrino fluxes. However, the variation of other hadronic interaction parameters and the density structure of atmosphere have far smaller effect on the atmospheric neutrino fluxes.

Shown in Fig. 9 is the variation region of the ratios among different kinds of neutrinos, with all the variations considered here. It is seen that the variation of the ratios is very small. All of the $(\bar{\nu}_e + \nu_e)/(\bar{\nu}_\mu + \nu_\mu)$ variations are inside of the boundary of ± 5 % at 1 GeV. However, the small statistics make the ratio variation larger in the $\gtrsim$3 GeV region due to Monte Carlo method.

We have also depicted the $(\bar{\nu}_e + \nu_e)/(\bar{\nu}_\mu + \nu_\mu)$ ratio by HKKM shifted by a factor $1.54 = 1/0.65$. Note that SuperKamiokande have observed the ratio $(\nu_\mu/\nu_e)_{obs}/(\nu_\mu/\nu_e)_{MC} \approx 0.65$ both for sub-

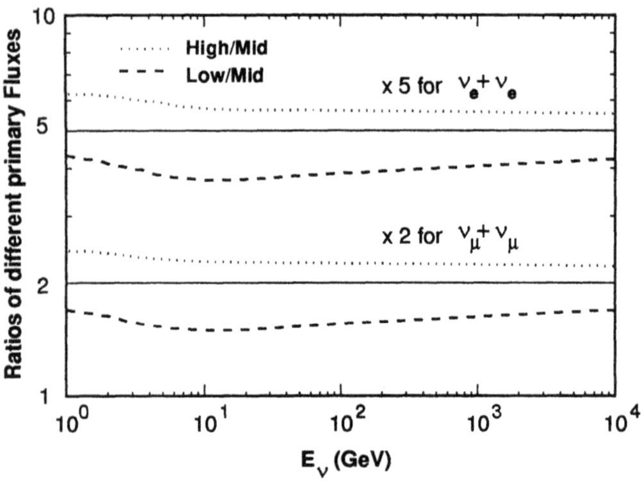

Figure 6. The variation of atmospheric neutrino fluxes due to the change of primary cosmic ray model. The ratio to that with Mid primary flux model (HKKM) is shown. For the flux models, see the text in section 2.1.

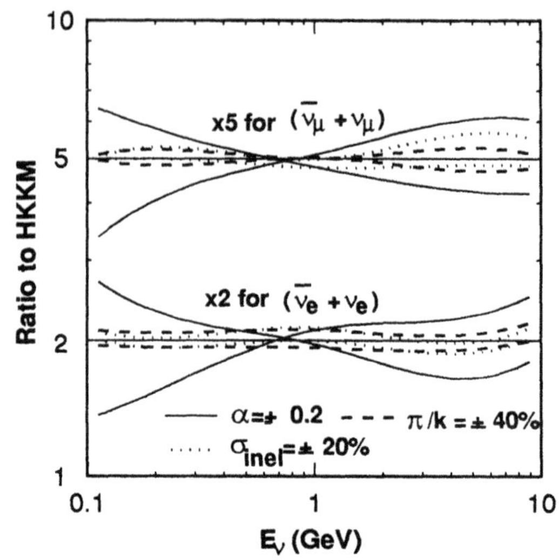

Figure 7. The variation of atmospheric neutrino fluxes due to the variation of interaction model. The ratio to the HKKM neutrino fluxes is shown. For the parameters used in the figure, see the text in section 2.2.

GeV and multi-GeV regions.

4. Summary and comments

We have studied the effect of variation in primary cosmic ray flux, hadronic interaction model, and density structure of atmosphere on the atmospheric neutrino fluxes. The variation of the primary cosmic ray flux is directly related to the absolute value of atmospheric neutrino fluxes. The secondary particle spectrum in the hadronic interactions also strongly affects the absolute value of the fluxes. Other variations considered here do not have a large effect on the atmospheric neutrino fluxes.

Variation of primary cosmic ray flux and/or interaction model does not cause a large change in the ratio of different kinds of neutrino. We have noted that the secondary particle spectrum of BN is outside of our variation limit. This may be the case for the interaction model used by different authors. However, the ratio $(\bar{\nu}_e+\nu_e)/(\bar{\nu}_\mu+\nu_\mu)$ does shows a good agreement among different authors (See the comparison in HKKM). Probably a variation from a different starting point would not give a very different answer. Thus, it is difficult to explain the ratio $(\nu_\mu/\nu_e)_{obs}/(\nu_\mu/\nu_e)_{MC}$ observed in Kamiokande and SuperKamiokande by the uncertainty in the calculation of atmospheric neutrino fluxes.

However, when applying the calculated atmospheric neutrino fluxes to the neutrino oscillation study, the absolute values of the fluxes cbecome important. The determination of cosmic ray flux to a high accuracy and the detailed study of hadronic interaction at $x = E_{2nd}/E_{inc} \approx 0.3$ are crucial.

The uncertainty for the arrival direction was not discussed so far in this paper. We shortly comment on the effect of the muon bending by geomagnetic field and transverse momentum in the hadronic interaction. For the muon in the geomagnetic field, we can calculate the average

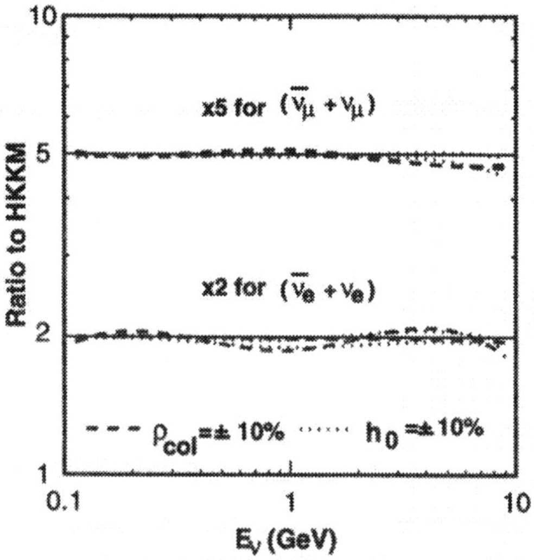

Figure 8. The variations of atmospheric neutrino fluxes due to the variation of atmospheric model. The ratio to the 'standard' model is shown. For the parameters used in the figure, see the text in section 2.3

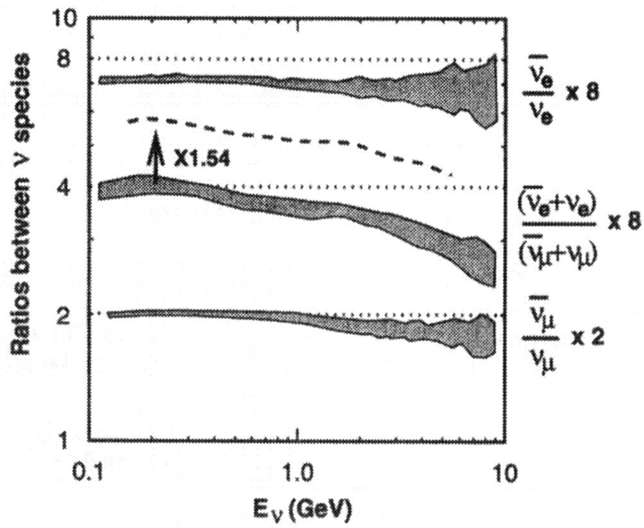

Figure 9. Variation of ratios among different kinds of neutrino with all the variation considered in sections 2.1–2.3 in all the combinations. The dashed line shows the ratio $(\bar{\nu}_e+\nu_e)/(\bar{\nu}_\mu+\nu_\mu)$ of HKKM shifted by a factor of 1.54 = (1/0.65).

bending angle before decay as:

$$\theta = c\gamma\tau_\mu/r_G = \frac{cE_\mu\tau_\mu}{m_\mu}/\frac{E_\mu}{c\cdot eB} = c^2\cdot\frac{eB}{m_\mu}\cdot\tau_\mu$$

$$= (3\times10^8(\mathrm{m/sec}))^2\frac{B(\mathrm{T})}{106\times10^6(\mathrm{eV})}2.2\times10^{-6}(\mathrm{sec})$$

$$= 0.19\times10^4 B(\mathrm{T})\ .$$

Therefore, the bending angle for the typical geomagnetic field ($\approx 0.5\times10^{-4}$ T) is around 0.1 radian or 5 degree. This is probably not an urgent problem at this moment.

We note that the typical value of the transverse momentum in a hadronic interaction is $\sim$ 0.3 GeV. Therefore, the transverse momentum is unimportant for $\gtrsim$ a few GeV.

REFERENCES

1. E. V. Bugaev, and V. A. Naumov, Phys. Lett. B **232** (1989) 391.
2. G. Barr, T. K. Gaisser, and T. Stanev, Phys. Rev. D **39** (1989) 3532.
3. M. Honda, K. Kasahara, K. Hidaka, and S. Midorikawa, Phys. Lett. B **248** (1990) 193.
4. H. Lee and Y. S. Koh, Nuovo Cimento B **105** (1990) 883.
5. M. Honda, T. Kajita, K. Kasahara, and S. Midorikawa, Phys. Rev. D **52** 4985 (HKKM).
6. T. K. Gaisser et. al., Phys. Rev. D **54** (1966) 5578.
7. Webber et. al., 20th ICRC, Moscow**1** 1987, vol 1, 325.
8. P. Pappini, et. al., 23rd ICRC, Calgary **1**, (1993) 579. (MASS)
9. W. S. Seo, et. al., ApJ, **378**, (1987) 763.
10. W. Menn, et. al., 25th ICRC, Durban, 1997, **3**, 409. (IMAX)
11. G. Barbiellini, et. al., 25th ICRC, Durban, 1997, **3**, 369. (CAPRICE)
12. S. Orito, private communications (1998) (BESS).
13. C. Caso et. al. The European Physical Journal C **3** no.1–4, (1998)

ELSEVIER

Nuclear Physics B (Proc. Suppl.) 77 (1999) 146–150

NUCLEAR PHYSICS B
PROCEEDINGS
SUPPLEMENTS

$\nu_\mu \leftrightarrow \nu_\tau$ vs $\nu_\mu \leftrightarrow \nu_s$ solutions for the atmospheric neutrino problem *

Osamu Yasuda

Department of Physics, Tokyo Metropolitan University
1-1 Minami-Osawa Hachioji, Tokyo 192-0397, Japan

The $\nu_\mu \leftrightarrow \nu_\tau$ and $\nu_\mu \leftrightarrow \nu_s$ solutions to the atmospheric neutrino problem are compared with Superkamiokande data. Both the solutions with a large mixing angle seem to be consistent with the data.

1. Introduction

Recent atmospheric neutrino data by Superkamiokande [1–3] provide strong evidence for neutrino oscillations. It has been shown [1–3] that atmospheric neutrino data favor $\nu_\mu \leftrightarrow \nu_\tau$ oscillations with maximal mixing, rather than $\nu_\mu \leftrightarrow \nu_e$. However, ν_μ disappearance alone does not imply uniquely a $\nu_\mu \leftrightarrow \nu_\tau$ solution and there is another solution $\nu_\mu \leftrightarrow \nu_s$, where ν_s denotes a sterile neutrino. In this talk some aspects of the $\nu_\mu \leftrightarrow \nu_\tau$ and $\nu_\mu \leftrightarrow \nu_s$ solutions are discussed.

In the past there has been a prejudice against the $\nu_\mu \leftrightarrow \nu_s$ solution to the atmospheric neutrino problem. The argument [4] was based on big bang nucleosynthesis which gives a condition $\Delta m^2 \sin^4 2\theta \lesssim 10^{-4}\text{eV}^2$ in order for sterile neutrinos not to be in thermal equilibrium. However, there was a loophole in this argument. Foot and Volkas [5] have shown that large lepton asymmetries will suppress $\nu_s \leftrightarrow \nu_x$ neutrino oscillations. Interestingly, given certain conditions, the required lepton asymmetries can actually be created by the oscillations themselves [5]. So there is no longer any obstruction to $\nu_\mu \leftrightarrow \nu_s$ as a solution to the atmospheric neutrino anomaly.

Many models [6,7] have been proposed which predict large or maximal active-sterile mixing. Among others, Foot and Volkas have been obsessed by exact parity symmetric models [7] and this was the main motivation of [8] in which $\nu_\mu \leftrightarrow \nu_s$ was examined in detail by fitting to the contained events of the Superkamiokande atmospheric neutrino data for 414 days.

2. Analysis of the Superkamiokande contained events for 414 days

The survival probability $P(\nu_\alpha \leftrightarrow \nu_\alpha)$ is obtained by solving the Schrödinger equation for neutrino evolution including matter effects. It is given by

$$i\frac{d}{dx}\begin{pmatrix}\nu_\mu(x)\\ \nu_{\tau,s}(x)\end{pmatrix} = M\begin{pmatrix}\nu_\mu(x)\\ \nu_{\tau,s}(x)\end{pmatrix}, \quad (1)$$

$$M \equiv U\text{diag}(0, \frac{\Delta m^2}{2E})U^{-1} + \text{diag}(0, A_{\tau,s}(x)),$$

where

$$U \equiv \begin{pmatrix}\cos\theta & \sin\theta\\ -\sin\theta & \cos\theta\end{pmatrix}$$

is the MNS mixing matrix [9], x is the distance traveled, Δm^2 the difference in squared masses, θ the vacuum mixing angle and $\nu_{\mu,\tau,s}(x)$ the wave-functions of the neutrinos. The quantities $A_{\tau,s}(x)$ are the effective potential differences generated through the matter effect [10]:

$$A_\tau(x) = 0$$

and, for electrically neutral terrestrial matter [11]

$$A_s(x) = \frac{1}{\sqrt{2}}G_F N_n(x),$$

where G_F is the Fermi constant, $N_n(x)$ is the number density of neutrons along the path of the neutrino. It is this matter effect $A_{\tau,s}$ that make a difference between the $\nu_\mu \leftrightarrow \nu_\tau$ and $\nu_\mu \leftrightarrow \nu_s$ oscillations. For antineutrinos the sign of A_s is reversed.

*Talk presented at "NEUTRINO 98", Takayama, Japan, June 4-9, 1998.

PII S0920-5632(99)00410-7

The way to obtain the numbers of events and evaluate χ^2 is described in [8], where two quantities have been introduced to perform a χ^2 analysis. One is the double ratio [12]

$$R \equiv \frac{(N_\mu/N_e)|_{\rm osc}}{(N_\mu/N_e)|_{\rm no-osc}}$$

where the quantities $N_{e,\mu}$ are the numbers of e-like and μ-like events. The numerator denotes numbers with oscillation probability obtained by (2), while the denominator the numbers expected with oscillations switched off. The other one is the quantity on up-down flux asymmetries for α-like (α=e,μ) events and is defined by

$$Y_\alpha \equiv \frac{(N_\alpha(\cos\Theta < -0.2)/N_\alpha(\cos\Theta > 0.2))|_{\rm osc}}{(N_\alpha(\cos\Theta < -0.2)/N_\alpha(\cos\Theta > 0.2))|_{\rm no-osc}},$$

where Θ is the zenith angle, $N_\alpha(\cos\Theta < -0.2)$ and $N_\alpha(\cos\Theta > 0.2)$ are the number of upward and downward going events, respectively. χ^2 with the double ratio R is defined by

$$\chi^2_{\rm atm}(R) = \sum_E \left(\frac{R^{SK} - R^{th}}{\delta R^{SK}}\right)^2,$$

and χ^2 with the up-down asymmetry Y_α is defined by

$$\chi^2_{\rm atm}(Y) = \sum_E \left[\left(\frac{Y_\mu^{SK} - Y_\mu^{th}}{\delta Y_\mu^{SK}}\right)^2 + \left(\frac{Y_e^{SK} - Y_e^{th}}{\delta Y_e^{SK}}\right)^2\right],$$

where the sum is over the sub-GeV and multi-GeV cases, the measured Superkamiokande values and errors are denoted by the superscript "SK" and the theoretical predictions for the quantities are labeled by "th".

The results of the χ^2 fits are displayed in Figs.1–8. In Figs. 1 and 2, χ^2 is plotted against Δm^2. For $\nu_\mu \leftrightarrow \nu_\tau$, χ^2 does not experience a deep minimum at the best fit point with respect to Δm^2 particularly when the R's are excluded from the fit. In general, for geometrical reasons, atmospheric neutrino analysis does not constrain Δm^2 very precisely. Note that the the situation is slightly different in case of $\nu_\mu \leftrightarrow \nu_s$. Figure 3 shows the allowed region of ($\sin^2 2\theta$, Δm^2) at

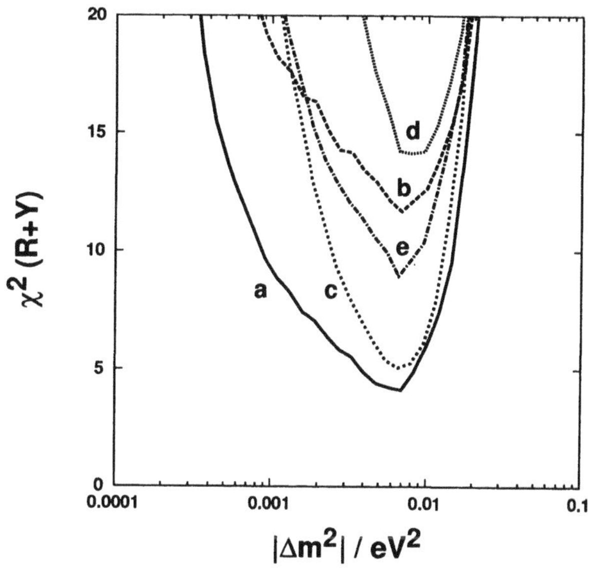

Figure 1. χ^2 as a function of Δm^2.
(a): $\nu_\mu \leftrightarrow \nu_\tau$, $\sin^2 2\theta = 1$;
(b): $\nu_\mu \leftrightarrow \nu_\tau$, $\sin^2 2\theta = 0.8$;
(c): $\nu_\mu \leftrightarrow \nu_s$, $\sin^2 2\theta = 1$;
(d): $\nu_\mu \leftrightarrow \nu_s$, $\sin^2 2\theta = 0.8$, $\Delta m^2 > 0$;
(e): $\nu_\mu \leftrightarrow \nu_s$, $\sin^2 2\theta = 0.8$, $\Delta m^2 < 0$.

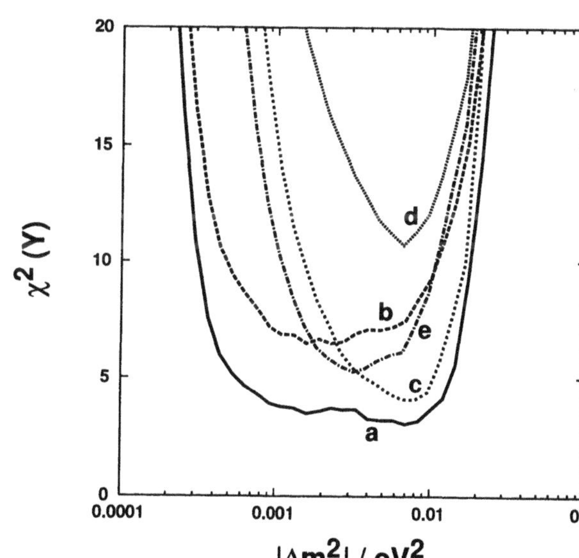

Figure 2. The same as Figure 1 but with R data excluded from the fit.

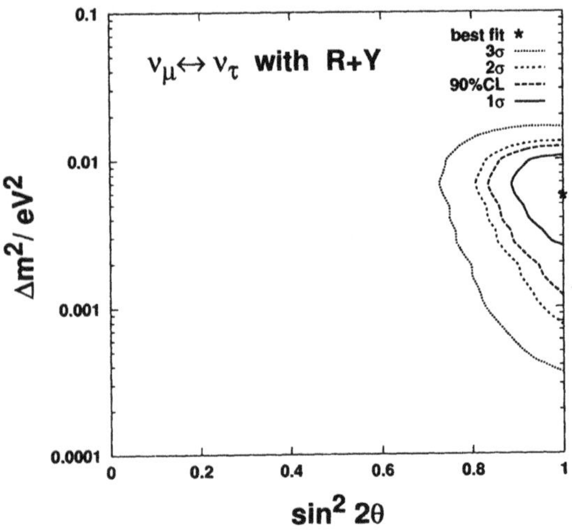

Figure 3. The allowed region in the $(\sin^2 2\theta,\ \Delta m^2)$ plane for the $\nu_\mu \leftrightarrow \nu_\tau$ scenario.

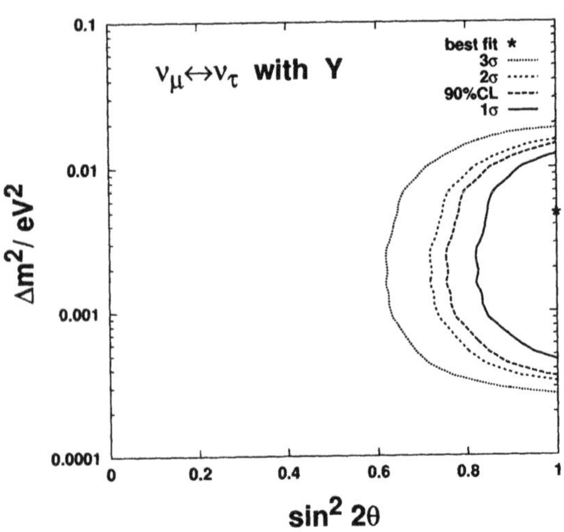

Figure 4. The same as Figure 3 but with R data excluded from the fit.

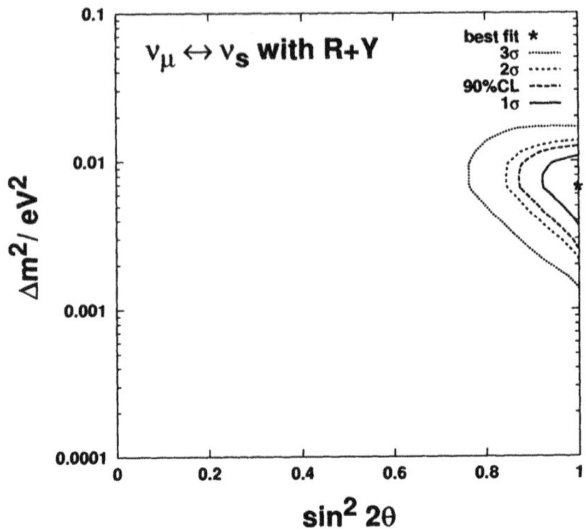

Figure 5. The allowed region in the $(\sin^2 2\theta,\ \Delta m^2)$ plane for the $\nu_\mu \leftrightarrow \nu_s$ scenario with $\Delta m^2 > 0$.

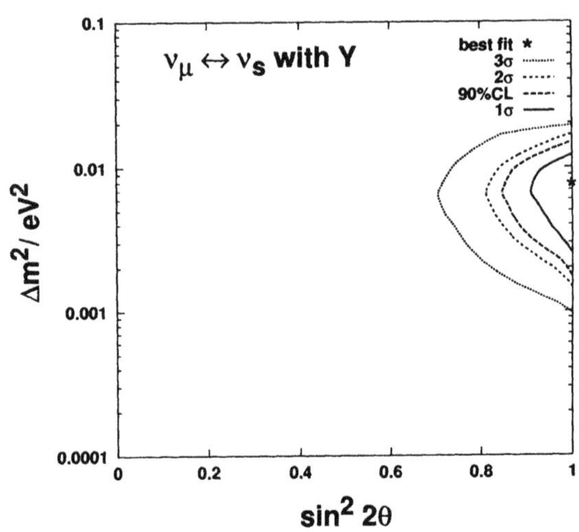

Figure 6. The same as Figure 5 but with R data excluded from the fit.

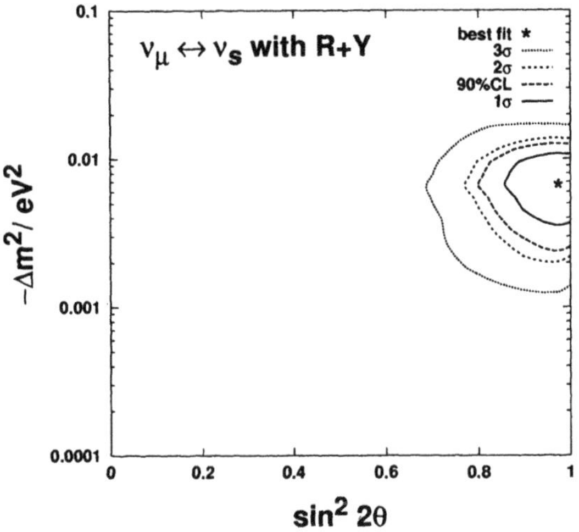

Figure 7. The allowed region in the $(\sin^2 2\theta,\ \Delta m^2)$ plane for the $\nu_\mu \leftrightarrow \nu_s$ scenario with $\Delta m^2 < 0$.

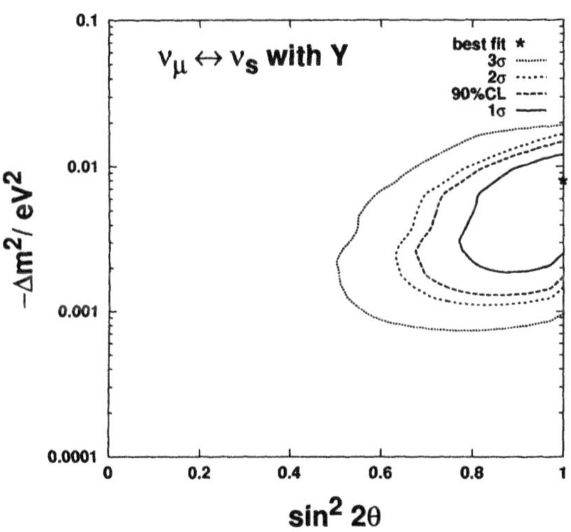

Figure 8. The same as Figure 7 but with R data excluded from the fit.

various confidence levels for the $\nu_\mu \leftrightarrow \nu_\tau$ scenario. Maximal mixing provides the best fit, and Δm^2 values in the 10^{-3} to 10^{-2} eV^2 range are favored. Note that the confidence levels are defined in the usual way by

$$\chi^2 = \chi^2_{\rm min} + \Delta\chi^2$$

where $\Delta\chi^2 = 2.3, 4.6, 6.2, 11.8$ for the 1σ, 90% C.L., 2σ and 3σ allowed region respectively. Our $\chi^2_{\rm min}$ for $\nu_\mu \leftrightarrow \nu_\tau$ oscillations is $\chi^2_{\rm min} = 4.5$ for 4 degrees of freedom. This is quite a good fit to the data (allowed at the 35% level). In Figure 4 we show the allowed region considering just the asymmetries instead of using both the asymmetries and the R ratios. Note that in this case there are 4 data points and 2 free parameters which gives 2 degrees of freedom.

Figures 5–8 show the corresponding results for the $\nu_\mu \leftrightarrow \nu_s$ scenario. If $\Delta m^2 > 0$, smaller values of Δm^2 are disfavored because the matter effect moves both R and Y away from the measured values, but if $\Delta m^2 < 0$, then smaller values of Δm^2 and $\sin^2 2\theta$ are permitted at the 90% confidence level. The value of $\chi^2_{\rm min}$ for the $\nu_\mu \leftrightarrow \nu_s$ scenario is $\chi^2_{\rm min} = 5.1$ for 4 degrees of freedom. This is similar to $\nu_\mu - \nu_\tau$ case and also represents quite a good fit (which is allowed at 28%).

To summarize, both the solutions $\nu_\mu \leftrightarrow \nu_\tau$ and $\nu_\mu \leftrightarrow \nu_s$ provide a good fit to the contained events of the Superkamiokande atmospheric neutrino data.

3. Other analyses

There have been several proposals to distinguish the $\nu_\mu \leftrightarrow \nu_\tau$ and $\nu_\mu \leftrightarrow \nu_s$ oscillations. Matter effects in $\nu_\mu \leftrightarrow \nu_s$ oscillations in upward going muon data were first analyzed by Akhmedov, Lipari and Lusignoli [13] and more recently by Lipari and Lusignoli [14]. It has been pointed out by Liu, Smirnov [15] and Liu, Mikheyev, Smirnov [16] that signatures due to parametric enhancement in $\nu_\mu \leftrightarrow \nu_s$ oscillations may be seen in upward going muon data. Vissani and Smirnov [17] proposed to look at the ratio (π^0-events)/(two ring events). Learned, Pakvasa and Stone [18] suggested that the up-down asymmetry (upward going π^0-events)/(downward going π^0-events) can tell a difference. Hall and Murayama [19] proposed a similar technique to use the up-down asymmetry in the multi-ring events. Kajita [2] mentioned the ratio (π^0-events)/(e-like events) which should in principle enable us to distinguish. All these analyses seem to be still inconclusive and we need more statistics and accurate knowledge on nuclear cross sections to draw a conclusion.

4. Conclusions

We have demonstrated that matter effects in the Earth have a significant role to play in comparing and contrasting the $\nu_\mu \leftrightarrow \nu_\tau$ and $\nu_\mu \leftrightarrow \nu_s$ solutions to the atmospheric neutrino anomaly with Superkamiokande data. So far both solutions provide a good fit to the data and we need more statistics to be conclusive. We hope that non-accelerator experiments such as Superkamiokande will distinguish them before future long baseline experiments with emulsion techniques [20] give direct evidence.

Acknowledgments

The author would like to thank R. Foot and R.R. Volkas for collaboration, discussions and comments on this manuscript. He also would like to thank T. Kajita and E. Kearns for useful communications. This research was supported in part by a Grant-in-Aid for Scientific Research of the Ministry of Education, Science and Culture, #09045036, #10140221, #10640280.

REFERENCES

1. Y. Fukuda et al., Phys. Lett. **B433** (1998) 9; hep-ex/9805006.
2. T. Kajita, these proceeding; http://www-sk.icrr.u-tokyo.ac.jp/nu98/scan/063/.
3. Y. Fukuda et al., hep-ex/9807003.
4. R. Barbieri and A. Dolgov, Phys. Lett. **B237** (1990) 440, Nucl. Phys. **B349** (1991) 743; K. Kainulainen, Phys. Lett. **B244** (1990) 191; K. Enqvist, K. Kainulainen and M. Thomson, Nucl. Phys. **B373** (1992) 498, Phys. Lett. **B288** (1992) 145; X. Shi, D.N. Schramm and B.D. Fields, Phys. Rev. **D48** (1993) 2563.
5. R. Foot and R. R. Volkas, Phys. Rev. **D55** (1997) 5147; Astropart. Phys. **7** (1997) 283; Phys. Rev. **D56** (1997) 6653.
6. M. Kobayashi, C. S. Lim and M. M. Nojiri, Phys. Rev. Lett. **67** (1991) 1685; C. Giunti, C. W. Kim and U. W. Lee, Phys. Rev. **D46** (1992) 3034; J. Bowes and R. R. Volkas, J. Phys. **G24** (1998) 1249; A. Geiser, preprint CERN-EP-98-056.
7. R. Foot, Mod. Phys. Lett. **A9** (1994) 169; R. Foot and R. R. Volkas, Phys. Rev. **D52** (1995) 6595.
8. R. Foot, R. R. Volkas and O. Yasuda, Phys. Rev. **D58** (1998) 13006.
9. Z. Maki, M. Nakagawa and S. Sakata, Prog. Theor. Phys. **28** (1962) 870.
10. L. Wolfenstein, Phys. Rev. **D17** (1978) 2369; S.P. Mikheyev and A.Yu. Smirnov, Yad. Fiz. **42** (1985) 1441 [Sov. J. Nucl. Phys. **42** (1985) 913]; Nuovo Cim. **C9** (1986) 17.
11. See for example, D. Notzold and G. Raffelt, Nucl. Phys. **B307** (1988) 924.
12. K.S. Hirata et al., Phys. Lett. **B280** (1992) 146; Y. Fukuda et al., Phys. Lett. **B335** (1994) 237.
13. E. Akhmedov, P. Lipari and M. Lusignoli, Phys. Lett. **B300** (1993) 128.
14. P. Lipari and M. Lusignoli, hep-ph/9803440.
15. Q.Y. Liu and A.Yu. Smirnov, hep-ph/9712493.
16. Q.Y. Liu, S.P. Mikheyev and A.Yu. Smirnov, hep-ph/9803415.
17. F. Vissani and A.Yu. Smirnov, Phys. Lett. **B432** (1998) 376; See also T. Kajita, Talk at *Topical Workshop on Neutrino Physics*, Institute for Theoretical Physics, The University of Adelaide, Nov. 1996.
18. J.G. Learned, S. Pakvasa and J.L. Stone, hep-ph/9805343. The author has learned from E. Kearns that this technique has been thought about for some time by the members of the Superkamiokande collaboration.
19. L. Hall and H. Murayama, hep-ph/9806218.
20. MINOS experiment, http://www.hep.anl.gov/NDK/HyperText/numi.html; OPERA experiment, http://www1.na.infn.it/wsubnucl/accel/neutrino/opera.html; ICARUS experiment, http://www.aquila.infn.it/icarus/.

NUCLEAR PHYSICS B
PROCEEDINGS
SUPPLEMENTS

ELSEVIER Nuclear Physics B (Proc. Suppl.) 77 (1999) 151–156

On the neutrino mass spectrum and neutrino mixing from oscillation data

S.M. Bilenky[a], C. Giunti[b] and W. Grimus[c]

[a]Joint Institute for Nuclear Research, Dubna, Russia, and Institut für Theoretische Physik, Technische Universität Munchen, D–85748 Garching, Germany

[b]INFN, Sezione di Torino, and Dipartimento di Fisica Teorica, Via P. Giuria 1, I–10125 Torino, Italy

[c]Institute for Theoretical Physics, University of Vienna, Boltzmanngasse 5, A–1090 Vienna, Austria

Two schemes of mixing of four massive neutrinos with two couples of close neutrino masses separated by a gap of the order of 1 eV can accommodate solar, atmospheric and LSND neutrino oscillation data. It is shown that long-baseline $\bar{\nu}_e \to \bar{\nu}_e$ and $\nu_\mu \to \nu_e$ transitions are strongly suppressed in these schemes. The scheme of mixing of three neutrino masses with a mass hierarchy that can describe solar and atmospheric neutrino data is also discussed. It is shown that in this scheme the effective Majorana mass $|\langle m \rangle|$ that characterizes the matrix element of neutrinoless double-β decay is less than $\sim 10^{-2}$ eV.

1. Introduction

The conference *Neutrino '98* is a very important event in neutrino physics. At this conference the Super-Kamiokande collaboration [1] presented the results of 535 days of measurement of the atmospheric neutrino fluxes which provide an impressive evidence for neutrino oscillations.

We discuss here which indications on the neutrino mass spectrum and on neutrino mixing can be obtained from the results of Super-Kamiokande and all other neutrino oscillation experiments. We will discuss also possible consequences for future experiments that can be inferred from the analysis of the existing data.

In accordance with the neutrino oscillation hypothesis (see [2]) the left-handed flavor neutrino fields ν_{eL}, $\nu_{\mu L}$ and $\nu_{\tau L}$ are linear combinations of the left-handed components of the (Dirac or Majorana) massive neutrino fields ν_i:

$$\nu_{\alpha L} = \sum_i U_{\alpha k} \, \nu_{iL} \,. \tag{1}$$

In the LEP experiments on the measurement of the invisible width of the Z-boson it was proved that only three flavor neutrinos exist in nature (see [3]). The number of massive neutrinos can be, however, bigger than three (see [2]).

If the total lepton number $L = L_e + L_\mu + L_\tau$ is conserved, the neutrinos ν_i are Dirac particles and the number of massive neutrinos is equal to three.

If the total lepton number is not conserved, the neutrinos ν_i are massive Majorana particles ($\nu_i = \nu_i^c \equiv \mathcal{C}\overline{\nu_i}^T$, where $\mathcal{C}$ is the charge-conjugation matrix). In the general case, the number of Majorana fields ν_i is $n = 3+m$, where m is the number of right-handed fields ν_{aR} that enter in the neutrino mass term. We have in this case

$$\nu_{\alpha L} = \sum_{i=1}^{n} U_{\alpha i} \, \nu_{iL} \,, \quad (\nu_{aR})^c = \sum_{i=1}^{n} U_{ai} \, \nu_{iL} \,, \tag{2}$$

where U is a $n{\times}n$ unitary mixing matrix.

Two possible options are usually discussed in the Majorana case:

1. The *see-saw* option [4]. If the total lepton number is violated by the right-handed Majorana mass term at an energy scale much larger than the electroweak scale, the Majorana mass spectrum is composed of three light masses m_i ($i = 1,2,3$) and three very heavy masses M_i ($i = 1,2,3$) that characterize the scale of lepton number violation. The light neutrino masses are given by the see-saw formula

$$m_i \sim \frac{(m_i^F)^2}{M_i} \ll m_i^F \quad (i = 1,2,3) \,. \tag{3}$$

PII S0920-5632(99)00411-9

where m_i^F is the mass of the charged lepton or up-quark in the ith generation. The see-saw mechanism provides a plausible explanation for the smallness of neutrino masses with respect to the masses of all other fundamental fermions.

2. The *sterile neutrino* option. If more than three Majorana mass terms are small, then there are light sterile neutrinos. In this case active neutrinos ν_e, ν_μ and ν_τ can transform into sterile states ν_a that are quanta of right-handed fields ν_{aR}. Notice that sterile neutrinos can be obtained in the framework of the see-saw mechanism with some additional assumptions ("singular see-saw" [5]).

From the analysis of the data of atmospheric neutrino experiments it follows that $\Delta m^2_{\text{atm}} \sim 2\times 10^{-3}\,\text{eV}^2$ [1], where Δm^2 is the difference between the squares of neutrino masses. Another scale of Δm^2 was obtained [6,7] from the analysis of the data of all solar neutrino experiments: $\Delta m^2_{\text{sun}} \sim 10^{-5}\,\text{eV}^2$ (MSW [8]) or $\Delta m^2_{\text{sun}} \sim 10^{-10}\,\text{eV}^2$ (vacuum oscillations). Finally, indications in favor of $\bar\nu_\mu \to \bar\nu_e$ and $\nu_\mu \to \nu_e$ oscillations with a third scale of Δm^2, $\Delta m^2_{\text{LSND}} \sim 1\,\text{eV}^2$, were obtained in the accelerator LSND experiment [9].

All these indications in favor of neutrino masses and mixing will be checked by future solar, long-baseline and short-baseline neutrino oscillation experiments (see these Proceedings).

We will consider two possible scenarios:

1. All three indications in favor of neutrino oscillations are confirmed.
2. Only the indications of solar and atmospheric neutrino experiments are confirmed.

2. Four massive neutrinos

At least four massive neutrinos are needed in order to have three different scales of Δm^2 [10–14,5]. The three types of neutrino mass spectra that can accommodate the solar, atmospheric and LSND scales of Δm^2 are shown in Fig. 1. In all these spectra there are two groups of close masses separated by a gap of the order of 1 eV which gives $\Delta m^2_{41} \equiv m_4^2 - m_1^2 \simeq \Delta m^2_{\text{LSND}}$.

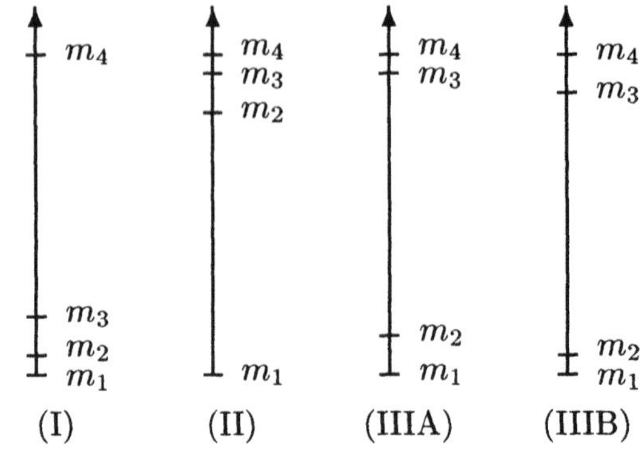

Figure 1

Only the largest mass-squared difference Δm^2_{41} is relevant for the oscillations in short-baseline (SBL) experiments and the SBL transition probabilities have the same dependence on the parameter $\Delta m^2_{41} L/2p$ (L is the source-detector distance and p is the neutrino momentum) as the standard two-neutrino probabilities [11]:

$$P_{\nu_\alpha\to\nu_\beta} = \frac{1}{2}\,A_{\alpha;\beta}\left(1-\cos\frac{\Delta m^2_{41}L}{2p}\right), \quad (4)$$

$$P_{\nu_\alpha\to\nu_\alpha} = 1-\frac{1}{2}\,B_{\alpha;\alpha}\left(1-\cos\frac{\Delta m^2_{41}L}{2p}\right). \quad (5)$$

The oscillation amplitudes $A_{\alpha;\beta}$ and $B_{\alpha;\alpha}$ depend on the elements on the mixing matrix U and on the form of the neutrino mass spectrum:

$$A_{\alpha;\beta} = 4\left|\sum_i U_{\beta i}\,U^*_{\alpha i}\right|^2, \quad (6)$$

$$B_{\alpha;\alpha} = 4\left(\sum_i |U_{\alpha i}|^2\right)\left(1-\sum_i |U_{\alpha i}|^2\right), \quad (7)$$

where the index i runs over the indices of the first or (because of the unitarity of U) second group of neutrino masses.

The results of SBL reactor $\bar\nu_e$ and accelerator ν_μ disappearance experiments in which no oscillations were found imply that $B_{\alpha;\alpha} \leq B^0_{\alpha;\alpha}$ for $\alpha = e,\mu$. The upper bounds $B^0_{\alpha;\alpha}$ are given by the exclusion curves obtained from the data of SBL disappearance experiments and depend on the value of Δm^2_{41}. Using Eq.(7), these upper bounds imply the following constraints for the quantities $\sum_i |U_{\alpha i}|^2$ ($\alpha = e, \mu$):

$$\sum_i |U_{\alpha i}|^2 \leq a^0_\alpha \quad \text{or} \quad \sum_i |U_{\alpha i}|^2 \geq 1 - a^0_\alpha \,, \qquad (8)$$

where

$$a^0_\alpha = \frac{1}{2}\left(1 - \sqrt{1 - B^0_{\alpha;\alpha}}\right) . \qquad (9)$$

The most stringent values of a^0_e and a^0_μ can be obtained from the results of the Bugey reactor experiment [15] and of the CDHS and CCFR accelerator experiments [16].

We have considered the range $10^{-1} \leq \Delta m^2_{41} \leq 10^3\,\text{eV}^2$. In this range $a^0_e \lesssim 4 \times 10^{-2}$ and $a^0_\mu \lesssim 2 \times 10^{-1}$ for $\Delta m^2_{41} \gtrsim 0.3\,\text{eV}^2$ (see Fig. 1 of Ref. [17]). Thus, from the results of disappearance experiments it follows that $\sum_i |U_{ei}|^2$ and $\sum_i |U_{\mu i}|^2$ can be either small or large (close to one).

From the four possible sets of values of the quantities $\sum_i |U_{ei}|^2$ and $\sum_i |U_{\mu i}|^2$ (small-small, small-large, large-small and large-large), for each neutrino mass spectrum only one set is compatible with the results of solar and atmospheric neutrino experiments [11,12]. In the case of spectra I and II we have

$$|U_{ek}|^2 \leq a^0_e \quad \text{and} \quad |U_{\mu k}|^2 \leq a^0_\mu \,, \qquad (10)$$

with $k = 4$ for spectra I and $k = 1$ for spectra II. In the case of spectrum IIIA we have

$$\sum_{i=1,2} |U_{ei}|^2 \leq a^0_e \text{ and } \sum_{i=1,2} |U_{\mu i}|^2 \geq 1 - a^0_\mu \,, \qquad (11)$$

whereas in the case of spectrum IIIB we have

$$\sum_{i=3,4} |U_{ei}|^2 \leq a^0_e \text{ and } \sum_{i=3,4} |U_{\mu i}|^2 \geq 1 - a^0_\mu \,. \qquad (12)$$

In the case of spectra I and II, $\nu_\mu \to \nu_e$ transitions in SBL experiments are strongly suppressed. Indeed, we have

$$A_{e;\mu} \leq 4\,|U_{ek}|^2\,|U_{\mu k}|^2 \leq 4\,a^0_e\,a^0_\mu \,. \qquad (13)$$

In Fig. 2 the upper bound (13) is compared with the latest LSND-allowed region (at 90% CL). Figure 2 shows that the spectra of type I and II (that include also the hierarchical spectrum) are disfavored by the result of the LSND experiment (they are compatible with the results of the LSND experiment only in the narrow region of Δm^2_{41} around $0.2 - 0.3\,\text{eV}^2$, where there is no information on $B_{\mu;\mu}$).

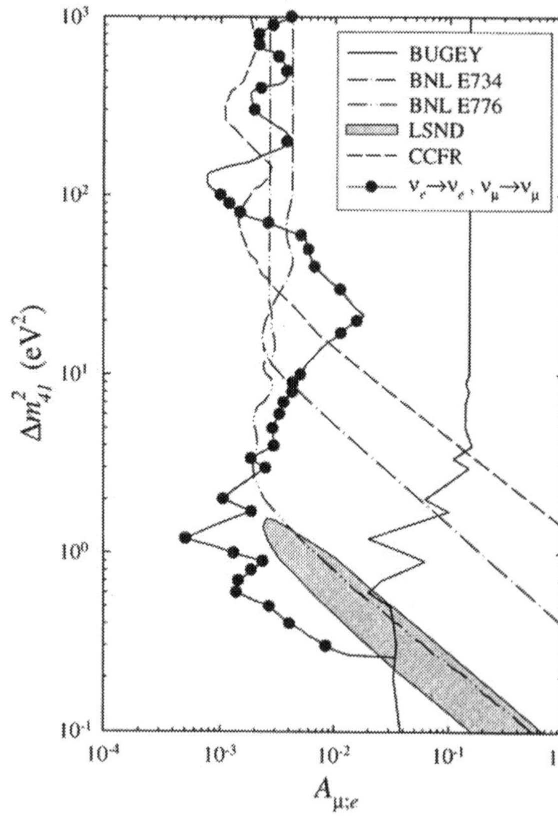

Figure 2

On the other hand, there is no incompatibility of the spectra IIIA and IIIB with the results of the LSND experiment and we conclude that these two spectra are favored by the existing data.

We discuss now some consequences for future experiments that can be inferred from the schemes IIIA and IIIB. Let us discuss first the possibilities for the effective neutrino mass $m(^3\text{H})$ measured in tritium β-decay experiments and for the effective Majorana mass $|\langle m \rangle| \equiv \left|\sum_k U^2_{ek}\, m_k\right|$ measured in neutrinoless double-β decay experiments.

In scheme IIIA we have

$$m(^3\mathrm{H}) \simeq m_4\,, \quad |\langle m\rangle| \le m_4\,, \tag{14}$$

whereas in scheme IIIB

$$\begin{aligned} m(^3\mathrm{H}) &\le a_e^0\, m_4 \ll m_4\,, \\ |\langle m\rangle| &\le a_e^0\, m_4 \ll m_4\,. \end{aligned} \tag{15}$$

Therefore, if the scheme A is realized in nature, tritium β-decay experiments experiments and neutrinoless double-β decay experiments have a possibility to see the effects of the relatively large neutrino mass $m_4 \simeq \sqrt{\Delta m^2_{\mathrm{LSND}}}$.

Let us consider now neutrino transitions in long-baseline (LBL) experiments. In the scheme IIIA the LBL transition probabilities are given by [13]

$$\begin{aligned} P^{\mathrm{LBL}}_{\nu_\alpha\to\nu_\beta} = & \left|\sum_{k=1,2} U^*_{\alpha k} e^{-i\frac{\Delta m^2_{k1} L}{2E}} U_{\beta k}\right|^2 \\ & + \left|\sum_{j=3,4} U^*_{\alpha j} U_{\beta j}\right|^2 . \end{aligned} \tag{16}$$

The transition probabilities in the scheme IIIB can be obtained from (16) with the change $1,2 \leftrightarrows 3,4$. The inequalities (11) and (12) imply strong constraints on the probabilities of $\bar\nu_e \to \bar\nu_e$ and $\nu_\mu \to \nu_e$ transitions in LBL experiments [13]. Indeed, for the probability of $\bar\nu_e \to \bar\nu_e$ transitions we have

$$P^{\mathrm{LBL}}_{\bar\nu_e\to\bar\nu_e} \ge \left(\sum_{j=3,4} |U_{ej}|^2\right)^2 \ge (1-a_e^0)^2 \tag{17}$$

in scheme IIIA and

$$P^{\mathrm{LBL}}_{\bar\nu_e\to\bar\nu_e} \ge \left(\sum_{k=1,2} |U_{ej}|^2\right)^2 \ge (1-a_e^0)^2 \tag{18}$$

in scheme IIIB. Hence, in both schemes $P^{\mathrm{LBL}}_{\bar\nu_e\to\bar\nu_e}$ is close to one and the LBL probability of transitions of $\bar\nu_e$ into any other state is small. Indeed, in both schemes we have

$$1 - P^{\mathrm{LBL}}_{\bar\nu_e\to\bar\nu_e} \le a_e^0\,(2-a_e^0)\,. \tag{19}$$

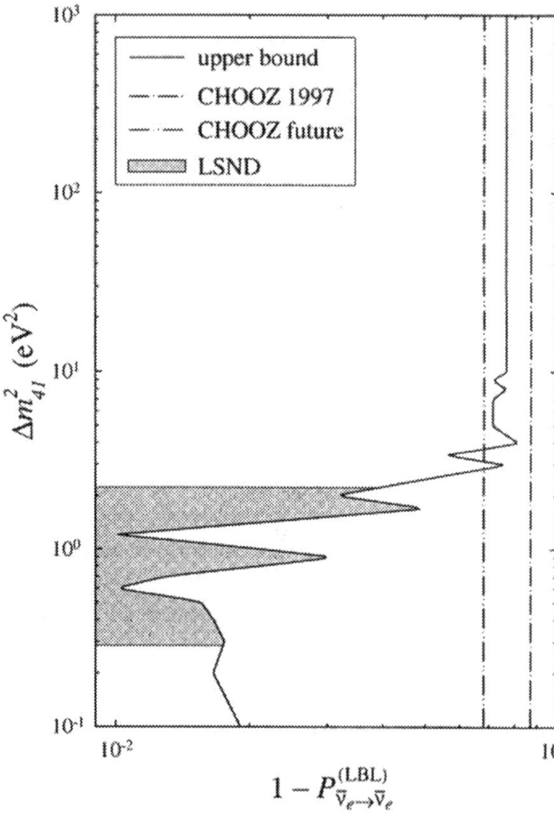

Figure 3

This limit is shown by the solid line in Fig. 3. The upper bound for the transition probability $1 - P^{\mathrm{LBL}}_{\bar\nu_e\to\bar\nu_e}$ obtained in the CHOOZ experiment [18] (dash-dotted line) and the final sensitivity of the CHOOZ experiment (dash-dot-dotted line) are also shown. It can be seen that for $\Delta m^2_{41} \lesssim 1\,\mathrm{eV}^2$ the upper bound (19) for $1 - P^{\mathrm{LBL}}_{\bar\nu_e\to\bar\nu_e}$ is much smaller than the upper bound reached in the CHOOZ experiment and than the final sensitivity of the experiment.

3. Three massive neutrinos

If the results of the LSND experiment will not be confirmed by future experiments, the most plausible scheme is the one with mixing of three massive neutrinos and a mass hierarchy [19,20,17]:

$$m_1 \ll m_2 \ll m_3\,. \tag{20}$$

The effective Majorana mass that characterize the matrix element of neutrinoless double-β decay is given in this case by [20]

$$|\langle m \rangle| \simeq |U_{e3}|^2 \sqrt{\Delta m^2_{31}} \,. \qquad (21)$$

The results of reactor neutrino experiments imply the upper bound $|U_{e3}|^2 \leq a^0_e$, with a^0_e given in Eq.(9). Therefore the effective Majorana mass is bounded by

$$|\langle m \rangle| \lesssim a^0_e \sqrt{\Delta m^2_{31}} \,. \qquad (22)$$

The value of this upper bound as a function Δm^2_{31} obtained from 90% CL exclusion plots of the Bugey [15] and CHOOZ [18] experiments is presented in Fig.4 (the solid and dashed line, respectively). The region on the right of the thick straight solid line is forbidden by the unitarity bound $|\langle m \rangle| \leq \sqrt{\Delta m^2_{31}}$.

Also the results of the Super-Kamiokande atmospheric neutrino experiment imply an upper bound for $|U_{e3}|^2$. The shadowed region in Fig.4 shows the region allowed by Super-Kamiokande results at 90% CL that we have obtained using the results of three-neutrino analysis performed by Yasuda [21].

Figure 4 shows that the results of the Super-Kamiokande and CHOOZ experiments imply that $|\langle m \rangle| \lesssim 10^{-2}\,\text{eV}$.

The observation of neutrinoless double-β decay with a probability that corresponds to a value of $|\langle m \rangle|$ significantly larger than $10^{-2}\,\text{eV}$ would mean that the masses of three neutrinos do not have a hierarchical pattern and/or exotic mechanisms (right-handed currents, supersymmetry with violation of R-parity, ..., see [22]) are responsible for the process.

Let us notice that from the results of the Heidelberg-Moscow ^{76}Ge experiment [23] it follows that $|\langle m \rangle| \lesssim 0.5 - 1.5\,\text{eV}$. The next generation of neutrinoless double-β experiments will reach $|\langle m \rangle| \simeq 10^{-1}\,\text{eV}$ [24]. Possibilities to reach $|\langle m \rangle| \simeq 10^{-2}\,\text{eV}$ are under discussion [24].

4. Conclusions

In conclusion, the neutrino mass spectrum and the structure of the neutrino mixing matrix depend crucially on the confirmation of the results of the LSND experiment. If these results will be confirmed we need (at least) four massive neutrinos with a mass spectrum of type IIIA or IIIB (see Fig. 1). If the results of the LSND experiment will not be confirmed, the most plausible scenario is the one with three massive neutrinos and a mass hierarchy. The investigation of the nature of massive neutrinos (Dirac or Majorana?) will require in this case to reach a sensitivity of $|\langle m \rangle| \lesssim 10^{-2}\,\text{eV}$ in searching for neutrinoless double-β decay.

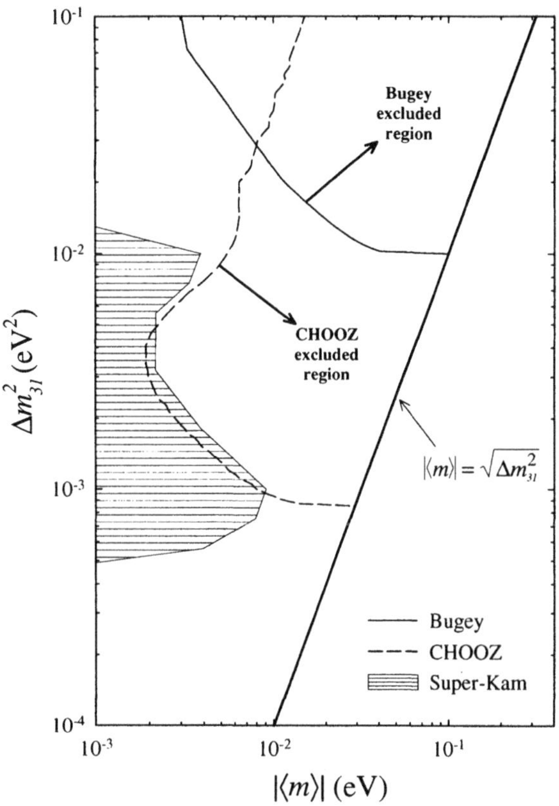

Figure 4

REFERENCES

1. T. Kajita, Talk presented at *Neutrino '98*, these Proceedings; Y. Fukuda *et al.*, hep-ex/9807003.
2. S.M. Bilenky and B. Pontecorvo, Phys. Rep. **41**, 225 (1978); S.M. Bilenky and S.T. Petcov, Rev. Mod. Phys. **59**, 671 (1987).

3. C. Caso et al, Eur. Phys. J. C **3**, 1 (1998).
4. M. Gell-Mann, P. Ramond, and R. Slansky, in *Supergravity*, ed. F. van Nieuwenhuizen and D. Freedman (North Holland, Amsterdam, 1979), p.315; T. Yanagida, Proc. of the *Workshop on Unified Theory and the Baryon Number of the Universe*, KEK, Japan, 1979; R.N. Mohapatra and G. Senjanović, Phys. Rev. Lett. **44**, 912 (1980).
5. E.J. Chun, C.W. Kim and U.W. Lee, hep-ph/9802209.
6. N. Hata and P. Langacker, Phys. Rev. D **56**, 6107 (1997).
7. G.L. Fogli, E. Lisi and D. Montanino, hep-ph/9709473.
8. S.P. Mikheyev and A.Yu. Smirnov, Yad. Fiz. **42**, 1441 (1985) [Sov. J. Nucl. Phys. **42**, 913 (1985)]; Il Nuovo Cimento C **9**, 17 (1986); L. Wolfenstein, Phys. Rev. D **17**, 2369 (1978); *ibid.* **20**, 2634 (1979).
9. C. Athanassopoulos *et al.*, Phys. Rev. Lett. **77**, 3082 (1996); D.H. White, Talk presented at *Neutrino '98*, these Proceedings.
10. J.T. Peltoniemi and J.W.F. Valle, Nucl. Phys. B **406**, 409 (1993); D.O. Caldwell and R.N. Mohapatra, Phys. Rev. D **48**, 3259 (1993); Z. Berezhiani and R.N. Mohapatra, *ibid* **52**, 6607 (1995); E. Ma and P. Roy, *ibid* **52**, R4780 (1995); R. Foot and R.R. Volkas, *ibid* **52**, 6595 (1995); S. Goswami, *ibid* **55**, 2931 (1997); A.Yu. Smirnov and M. Tanimoto, *ibid* **55**, 1665 (1997); J.R. Primack *et al.*, Phys. Rev. Lett. **74**, 2160 (1995); K. Benakli and A.Yu. Smirnov, *ibid* **79**, 4314 (1997); E.J. Chun *et al.*, Phys. Lett. B **357**, 608 (1995); J.J. Gomez-Cadenas and M.C. Gonzalez-Garcia, Z. Phys. C **71**, 443 (1996); E. Ma, Mod. Phys. Lett. A **11**, 1893 (1996); N. Okada and O. Yasuda, Int. J. Mod. Phys. A **12**, 3669 (1997); V. Barger, T.J. Weiler and K. Whisnant, hep-ph/9712495; S.C. Gibbons *et al.*,hep-ph/9803299; N. Gaur *et al.*, hep-ph/9806272; V. Barger *et al.*,hep-ph/9806328.
11. S.M. Bilenky, C. Giunti, C.W. Kim and S.T. Petcov, Phys. Rev. D **54**, 4432 (1996).
12. S.M. Bilenky, C. Giunti and W. Grimus, Proc. of *Neutrino96*, Helsinki, June 1996, edited by K. Enqvist *et al.*, p.174 (World Scientific, Singapore, 1997); Eur. Phys. J. C **1**, 247 (1998).
13. S.M. Bilenky, C. Giunti and W. Grimus, Phys. Rev. D **57** (1998) 1920.
14. S.M. Bilenky, C. Giunti and W. Grimus, Phys. Rev. D **58**, 033001 (1998); S.M. Bilenky, C. Giunti, W. Grimus and T. Schwetz, hep-ph/9804421.
15. B. Achkar *et al.*, Nucl. Phys. B **434**, 503 (1995).
16. F. Dydak *et al.*, Phys. Lett. B **134**, 281 (1984); I.E. Stockdale *et al.*, Phys. Rev. Lett. **52**, 1384 (1984).
17. S.M. Bilenky, A. Bottino, C. Giunti and C.W. Kim, Phys. Rev. D **54**, 1881 (1996).
18. M. Apollonio *et al.*, Phys. Lett. B **420**, 397 (1998).
19. A. De Rujula *et al.*, Nucl. Phys. B **168**, 54 (1980); V. Barger and K. Whisnant, Phys. Lett. B **209**, 365 (1988); S.M. Bilenky *et al.*, *ibid.* **276**, 223 (1992); S.M. Bilenky *et al.*,*ibid.* **356**, 273 (1995); K.S. Babu *et al.*, *ibid.* **359**, 351 (1995); H. Minakata, *ibid.* **356**, 61 (1995); Phys. Rev. D **52**, 6630 (1995); G.L. Fogli *et al.*, *ibid.* **52**, 5334 (1995); S. Goswami *et al.*, Int. J. Mod. Phys. A **12**, 781 (1997); S.M. Bilenky and C. Giunti, hep-ph/9802201; V. Barger *et al.*,hep-ph/9806387; A.J. Baltz *et al.*, hep-ph/9806540; V. Barger *et al.*,hep-ph/9807319; Y. Nomura and T. Yanagida, hep-ph/9807325; G. Altarelli and F. Feruglio, hep-ph/9807353; E. Ma, hep-ph/9807386; N. Haba, hep-ph/9807552.
20. S.T. Petcov and A.Yu. Smirnov, Phys. Lett. B **322**, 109 (1994); S.M. Bilenky *et al.*,Phys. Rev. D **57**, 6981 (1998).
21. O. Yasuda, hep-ph/9804400.
22. R.N. Mohapatra, hep-ph/9507234; Talk presented at *Neutrino '98*, these Proceedings.
23. M. Günther *et al.*, Phys. Rev. D **55**, 54 (1997); Phys. Lett. B **407**, 219 (1997).
24. Talks presented by A. Morales, H. Ejiri, F. Piquemal, H.V. Klapdor-Kleingrothaus and O. Cremonesi at *Neutrino '98*, these Proceedings.

Part 4

Long Baseline Experiments

Nuclear Physics B (Proc. Suppl.) 77 (1999) 159–165

NUCLEAR PHYSICS B
PROCEEDINGS
SUPPLEMENTS

Results from CHOOZ

Carlo Bemporad [a]

[a]INFN and Physics Department, University of Pisa,
Piazza Torricelli 2 , 56100 Pisa, Italy

1. Introduction

CHOOZ *, a long-baseline reactor-neutrino vacuum oscillation experiment, is named after the new nuclear power station operated by Électricité de France (EdF) near the village of Chooz in the Ardennes. Neutrino oscillation experiments probe the existence of finite neutrino mass, such experiments are therefore important for testing and possibly improving the Standard Model. A finite neutrino mass could also have an impact on many aspects of astrophysics and cosmology.

For a single neutrino energy E_ν(MeV) and a distance from the source L (meters), the survival probability can be written in terms of the mixing parameter $\sin^2 2\theta$ and the difference of the squared masses $\Delta m^2 = \left|m_2^2 - m_1^2\right|$ as follows:

$$P(\overline{\nu}_e \to \overline{\nu}_e) = 1 - \sin^2 2\theta \, \sin^2\left(\frac{1.27\,\Delta m^2(\mathrm{eV}^2)\,L(\mathrm{m})}{E_\nu(\mathrm{MeV})}\right),$$

in the context of a model with two flavour states and two mass eigenstates m_1 and m_2. When averaged over the source energy spectrum, this formula links the disappearance of neutrinos to neutrino mass.

The CHOOZ experiment [1-2] has an average value of $L/E \sim 300$ ($L \sim 1$ km, $E \sim 3$ MeV), an intense and nearly pure neutrino flavour composition ($\sim 100\%\,\overline{\nu}_e$) and an intensity known to better than 2%. It is thus ideally suited for a definitive test of $\overline{\nu}_e \to \overline{\nu}_\mu$ neutrino oscillations (or, more generally, $\overline{\nu}_e \to \overline{\nu}_x$ oscillations) down to 10^{-3} eV^2, an order of magnitude lower than previous reactor experiments [3-7]. CHOOZ overlaps the mass region indicated by the atmospheric neutrino signal [8]. It is worth mentioning that $L/E \sim 300$ is of the same order of magnitude of the ones of the new long base-line accelerator projects (KEK/SUPERKAMIOKANDE, CERN/GRAN SASSO); but these experiments need a neutrino target mass about three order of magnitude larger than the one of CHOOZ, with correspondingly larger financial investments for the beam line and for the apparatus. So, long base-line reactor experiments, although limited to $\overline{\nu}_e$ disappearance, offer a much simpler and economic answer to some of the fundamental questions about neutrino properties.

2. Description of the CHOOZ Experiment

The Chooz power station has two pressurized water reactors with a total thermal power of 8.5 GW$_{\mathrm{th}}$. The $\overline{\nu}_e$ flux and energy spectrum of similar reactors have been extensively studied. Anal-

---

* The Chooz Collaboration:
M. Apollonio[c], A. Baldini[b], C. Bemporad[b], E. Caffau[c], F. Cei[b], Y. Déclais[e], H. de Kerret[f], B. Dieterle[h], A. Etenko[d], J. George[h], G. Giannini[c], M. Grassi[b], Y. Kozlov[d], W. Kropp[g], D. Kryn[f], M. Laiman[e], C.E. Lane[a], B. Lefièvre[f], I. Machulin[d], A. Martemyanov[d], V. Martemyanov[d], L. Mikaelyan[d], D. Nicolò[b], M. Obolensky[f], R. Pazzi[b], G. Pieri[b], L. Price[g], S. Riley[g], R. Reeder[h], A. Sabelnikov[d], G. Santin[c], M. Skorokhvatov[d], H. Sobel[g], J. Steele[a], R. Steinberg[a], S. Sukhotin[d], S. Tomshaw[a], D. Veron[f], and V. Vyrodov[f]

[a] *Drexel University*
[b] *INFN and University of Pisa*
[c] *INFN and University of Trieste*
[d] *Kurchatov Institute*
[e] *LAPP-IN2P3-CNRS Annecy*
[f] *PCC-IN2P3-CNRS Collège de France*
[g] *University of California, Irvine*
[h] *University of New Mexico, Albuquerque*

PII S0920-5632(99)00412-0

ogous methods [3,9] have been used to calculate the $\overline{\nu}_e$ flux emitted by the Chooz reactors. The calculation includes

- a full description of the reactor core including the initial ^{235}U enrichment and the daily evolution of the isotopic composition of each fuel element as a function of the power produced (burn–up),
- the instantaneous fission rate derived from the thermal power of the reactors (recorded every minute),
- the $\overline{\nu}_e$ yield (as determined in [10–12]) from the four main isotopes – ^{235}U, ^{238}U, ^{239}Pu, and ^{241}Pu.

It has been shown [3] that the value of the $\overline{\nu}_e$ flux emitted by a reactor is understood to 1.4%.

The detector is located in an underground laboratory at a distance of about 1 km from the neutrino source. The 300 MWE rock overburden reduces the external cosmic ray muon flux by a factor of $\sim$ 300 to a value of 0.4 $m^{-2}s^{-1}$, significantly decreasing the most dangerous background, which is caused by fast neutrons produced by muon–induced nuclear spallation in the materials surrounding the detector. The detector is shown in fig. 1. The $\overline{\nu}_e$ are detected via the inverse beta decay reaction

$$\overline{\nu}_e + p \to e^+ + n \; ; \quad E_{e^+} = E_{\overline{\nu}_e} - 1.804\,\mathrm{MeV} \; .$$

The $\overline{\nu}_e$ signature is a delayed coincidence between the prompt e^+ signal (boosted by the two 511– keV annihilation gamma rays) and the signal from the neutron capture. The target material is a hydrogen–rich (free protons) paraffinic liquid scintillator loaded with 0.09% gadolinium. The target is contained in an acrylic vessel of precisely known volume immersed in a low energy calorimeter made of unloaded liquid scintillator. Gd has been chosen due to its large neutron capture cross section and to the high γ-ray energy released after n- capture ($\sim$ 8 MeV, well above the natural radioactivity).

The detector is simple and easily calibrated, while its behaviour can be well checked. The gadolinium loaded scintillator is unstable; it has

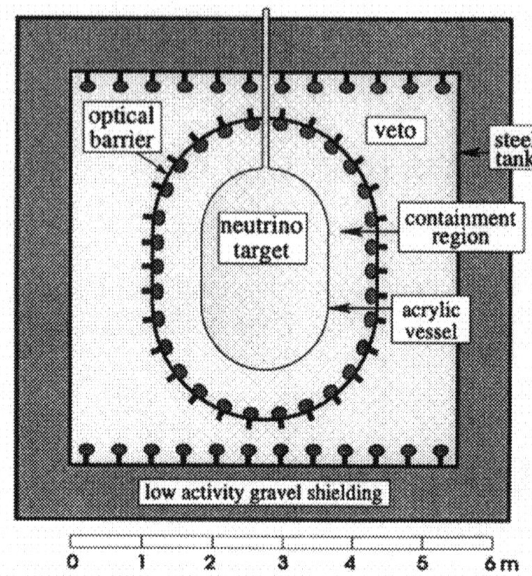

Figure 1. The CHOOZ detector.

a measured useful lifetime in the detector, due to reduced light at the PMT's, of $\tau \sim$ 750 d. We correct for these effects in the event reconstruction programs and we obtain a very good stability in the energy reconstruction. The neutron capture lines (2.2 MeV on hydrogen and 8 MeV on gadolinium) are clean and the energy resolutions are $\sigma_E/E = 9\%$ and $\sigma_E/E = 6\%$, respectively. Details about the trigger scheme, the trigger rates and the data acquisition system are given in ref. [2].

3. Data Analysis

The selection of events is based on the following requirements:

- energy cuts on the neutron candidate (6–12 MeV) and on the e^+ (from the threshold energy $E_{thr} \sim 1.3$ MeV to 8 MeV),
- a time cut on the delay between the e^+ and the neutron ($2 - 100\,\mu$s),
- spatial cuts on the e^+ and the neutron (distance from the PMT wall > 30 cm and distance $n - e^+ < 100$ cm).

A good understanding of the real nature of neutrino candidates can be obtained by representing such events in a bi-dimensional plot "secondary energy" vs. "primary energy" (see fig. 2); no event selection has been applied yet. One

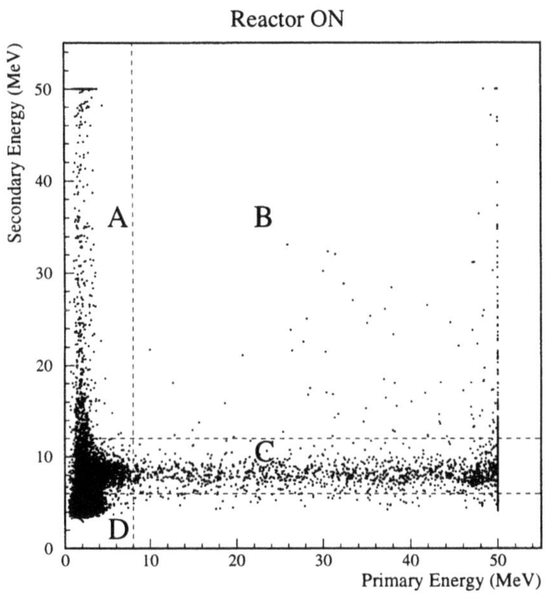

Figure 2. "Secondary energy" vs. "primary energy"; no selection applied.

observes four regions A,B,C,D and the neutrino event window at the crossing of region A and C. Region B and C are filled by primary-secondary correlated events; region B is the one of μ-stops, i.e.: cosmic μ's which entered the detector through the small dead space (detector filling pipes, support flanges, etc.) missing the anticoincidence shield; these events have large primary energy (circuit saturation present) and large secondary energies associated with the μ decay electrons. Events in region B have a secondary delay distribution in agreement with the μ lifetime. Region C events are due to fast neutrons from nuclear spallations by cosmic rays in the rock and concrete surrounding the detector; these neutrons scatter and the recoil proton is detected as "primary", while the neutron is thermalized and later captured as the "secondary" giving the characteristic 8 MeV capture energy; this fast neutron region overlaps the neutrino candidate region and is the main background source for the experiment; the secondary delay distribution is what it is expected for thermal neutron capture in the gadolinium doped scintillator (the best-fit lifetime is $\tau = 30 \pm 1\,\mu s$). Regions A and D are filled by accidental events; region D events are due to the accidental coincidence (within $100\,\mu s$) of two low energy natural radioactivity events; region A events are due to an accidental coincidence of a low energy natural radioactivity event and a large energy recoil proton from a fast neutron scattering; both delay distributions A and D are flat, as expected for accidentals. After application of all analysis cuts (apart from energy selections) the candidate event plot, in an expandent scale, is presented in fig. 3; one clearly sees that the

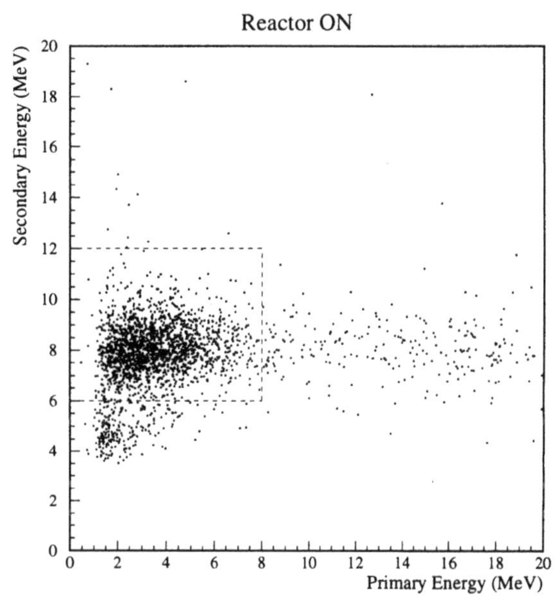

Figure 3. "Secondary energy" vs. "primary energy"; selections other than energy applied.

events spilling into the neutrino event window are mainly the ones from region C (proton recoils and neutron capture from spallation fast neutrons) and, to a lesser extent, the ones from region D (two low energy natural radioactivity events).

The neutrino signal was obtained by subtracting the small number of accidental and correlated background events which passed the neutrino selection criteria. These two components were determined separately. The accidental component was estimated by using the e^+ and n uncorrelated rates. The correlated component was estimated from reactor–off data by extrapolating the rate of high–energy neutrons followed by n-capture into the region defined by the ν event selection criteria. The background rates were found to be:

$$R_{accidental} = 0.23 \pm 0.05\,\mathrm{d}^{-1}$$
$$R_{correlated} = 0.8 \pm 0.2\,\mathrm{d}^{-1}.$$

The total background rate was measured with reactor-off and also by extrapolating the neutrino candidate rate to zero reactor power. For this purpose, neutrino candidates were collected during the period of reactor power rise. The number of events corresponding to each data run depended on the run duration ΔT and on the average reactor power W_{av}. All data were therefore fitted as a function of these two variables, thus separately determining the neutrino signal and the reactor–off background. The (grouped) data are shown in fig. 4 as a function of reactor power. The superimposed line corresponds to the fitted signal and background values. The results of the various background determinations are summarized in Table 1. It should be noted that the measurements of the background are in good agreement. From the fit we find the neutrino rate S_{fit} normalized to the full power of the two reactors ($2 \times 4.25\,\mathrm{GW_{th}}$, $2 \times 1.3 \times 10^{20}$ fissions s^{-1}) and a burn–up of 1300 MW d ton^{-1} to be:

$$S_{fit} = 25.5 \pm 1.0\,\mathrm{d}^{-1}.$$

A total of 1320 neutrino events were accumulated during 2718 hours (live time).

4. Expected Number of Events and Systematic Errors

The expected number of detected neutrino events can be written

$$N_{ev} = N_{fiss} \times \sigma_{fiss} \times \frac{1}{4\pi D^2} \times \times n_p \times \epsilon_{e^+} \times \epsilon_n \times \epsilon_{\Delta r} \times T_{live} \; ,$$

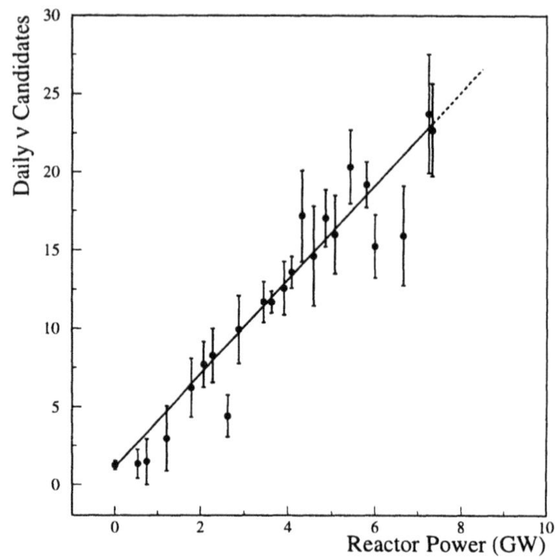

Figure 4. Number of $\overline{\nu}_e$-candidates d^{-1}, as a function of the reactor power.

where

- N_{fiss} is the number of fissions, per unit time, in the reactor cores,
- σ_{fiss} is the cross section for the reaction $\overline{\nu}_e + p \rightarrow e^+ + n$ calculated for a neutron lifetime of 887.0 ± 2.0 s, and integrated over the $\overline{\nu}_e$ energy spectrum and the fuel composition of the core,
- D is the distance of the detector from the reactors;
- n_p is the effective number of free protons in the target, including small corrections for edge effects involving the acrylic vessel and the region 2 scintillator buffer;
- ϵ_{e^+} and ϵ_n are the positron and neutron detection efficiencies averaged over the entire sensitive target; $\epsilon_{\Delta r}$ is the efficiency of the $e^+ - n$ distance cut;
- T_{live} is the live time.

Table 2 shows the various parameters together with their errors. The combined systematic error is estimated at $\sim 4\%$.

Table 1
Background Rates

estimated rate	$1.03 \pm 0.21\,\mathrm{d}^{-1}$
reactor–off rate	$1.2 \pm 0.3\,\mathrm{d}^{-1}$
rate by extrapolation to zero power	$1.1 \pm 0.25\,\mathrm{d}^{-1}$

Table 2
Normalization Parameters and Error

Parameter	Value	Error
cross section per fission	$6.327 \times 10^{-43}\,\mathrm{cm}^2$	2.7%
reactor 1 distance	$1.1146 \times 10^5\,\mathrm{cm}$	10 cm
reactor 2 distance	$0.9979 \times 10^5\,\mathrm{cm}$	10 cm
number of free protons in the Gd–loaded scintillator	3.601×10^{29}	1.5%
effective number of free protons in the target	3.637×10^{29}	1.5%
e^+ efficiency	0.968	0.7%
measured n efficiency at centre (^{252}Cf data)	0.757	1.3%
calculated n efficiency at centre (Monte Carlo)	0.759	1.3%
n efficiency averaged over the effective volume	0.739	2.0%
efficiency of the $e^+ - n$ distance cut	0.965	1.4%

5. Results

The ratio of the measured to expected neutrino signal is

$$R_{m/e} = 0.98 \pm 0.04(\text{stat}) \pm 0.04(\text{syst}).$$

One can also compare the measured positron spectrum with what is expected in the case of no oscillations (fig. 5). The 90% C.L. exclusion plot is presented in fig. 6, together with the exclusion plots of previous experiments and the atmospheric neutrino signal reported by Kamiokande [8].

After publishing the first CHOOZ results [2], we were requested by the *Particle Data Group* to repeat the analysis of our data according to the method proposed by G.Feldman and Cousins [13] to unify the treatment of neutrino oscillation experiments; in fig. 7 we present the already published 90% C.L. exclusion plot (curve A) compared with the one obtained by the new method (curve B). It is interesting to perform a different analysis based on the comparison of the positron spectra from the two, different distance, nuclear reactors. The statistical error greatly increases but the comparison is unaffected by the absolute value of the $\overline{\nu}_e$ flux, the cross section, the number of target protons and the detector efficiencies; the residual systematic errors are related to the stability of the detector performances and to the knowledge of the relative reactor $\overline{\nu}_e$ fluxes and spectral evolution. The comparison of the positron spectra due to the two different reactors was made possibile by the fact that each of the reactors was, from time to time, turned-off for several days; furthermore, when simultaneously on, the reactors worked, most of the times, at different power levels. In the calculation of the expected $\overline{\nu}_e$ flux and spectrum, we keep into account the burn-up of the reactor individual fuel elements and their distance from the detector. The comparison of the positron spectra from the two separated reactors gives a ratio $R_{far/near}$:

$$R_{far/near} = 0.96 \pm 0.06(\text{stat}) \pm 0.015(\text{syst}).$$

The correponding plot, even if excludes a smaller region, still overlaps most of the KAMIOKANDE atmospheric neutrino signal, and it does this in a way which is much less dependent on systematic effects (see fig. 8).

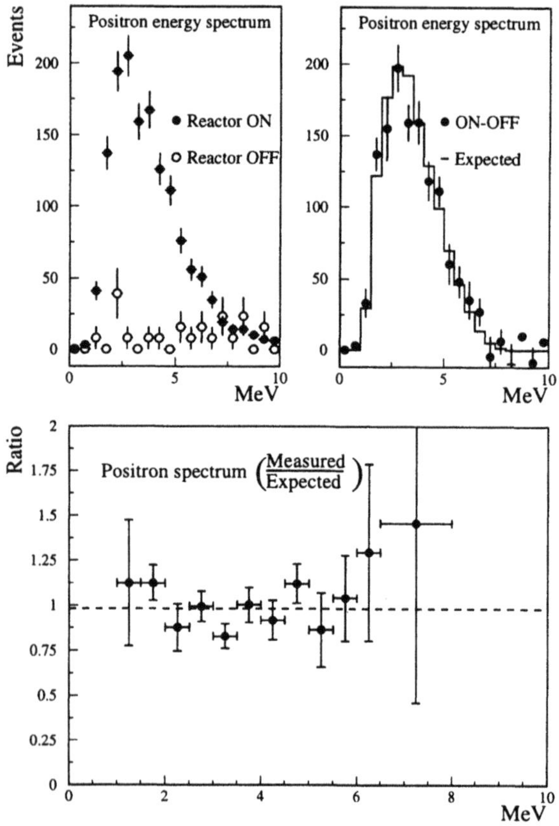

Figure 5. a) Positron energy spectrum and corresponding reactor-off background for the same live-time; the neutrino-signal expected positron spectrum is also shown. b) Ratio of the measured (background subtracted) to the expected positron spectrum.

6. A side question: do we see the neutrino "source" ?

We detect $\overline{\nu}_e$'s by a reaction for which for the e^+ emission is almost isotropic. We are nevertheless able to identify the direction of the incoming $\overline{\nu}_e$'s. This is due to the fact that the neutron produced in the reaction has a limiting angle and therefore it preserves a memory of the neutrino direction. If one observes a large number of events and reconstructs the position x_e at which the positron was generated and the

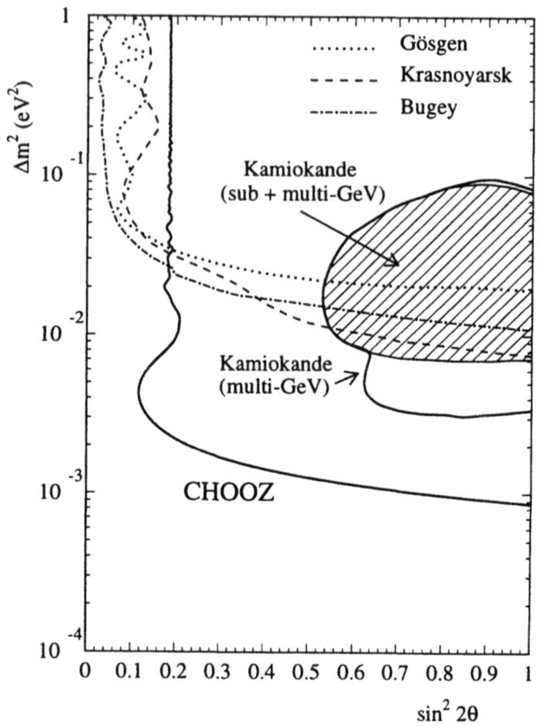

Figure 6. The 90% C.L. exclusion plot for CHOOZ, compared with previous experimental limits and with the KAMIOKANDE allowed region.

position x_n at which the neutron was captured, even with the limited resolution of the experiment ($\sigma_x \sim 20\,\text{cm}$), there should be a net effect $< x_n - x_e >\sim 1.5\,\text{cm}$ along the neutrino direction. The present statistics is small, but the effect is visible and its significance is improving. The experimental identification of the neutrino source (the CHOOZ reactor, in this case) means that a large scintillator project, like BOREXINO or KAMLAND [14], might be able to identify the incoming neutrino direction, in the case of a galactic SN explosion, in competition with water-Cherenkov experiments.

7. Conclusions

The CHOOZ experiment finds, at 90% C.L., no evidence for neutrino oscillations in the disappearance mode $\overline{\nu}_e \to \overline{\nu}_x$ for the parameter re-

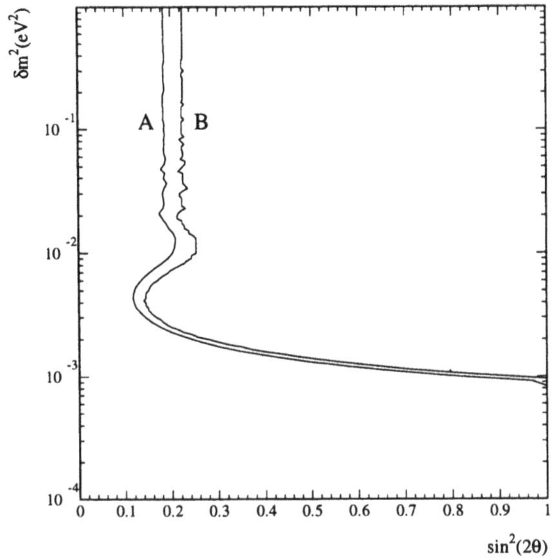

Figure 7. A): The 90% C.L. published exclusion plot [2]; B) exclusion plot by the Feldman's method [13].

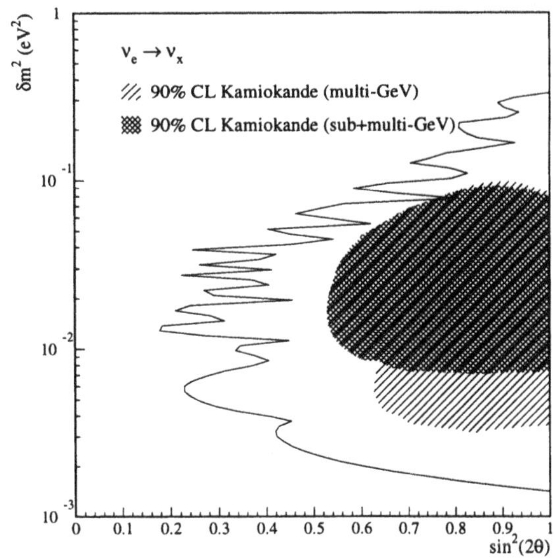

Figure 8. The 90% C.L. exclusion plot for CHOOZ, from the comparison of the two reactor data (preliminary analysis).

gion given approximately by $\Delta m^2 > 0.9\ 10^{-3}\,\mathrm{eV}^2$ for maximum mixing and $\sin^2 2\theta > 0.18$ for large Δm^2, as shown in fig. 6. The experiment is continuing to take data in order to achieve better statistics and to improve understanding of systematic effects.

8. Acknowledgements

Construction of the laboratory was funded by Électricité de France (EdF). Other work was supported by IN2P3–CNRS (France), INFN (Italy), the United States Department of Energy, and by RFBR (Russia).

REFERENCES

1. The CHOOZ Collaboration, Proposal. Available on the internet at http://duphy4.physics.drexel.edu/chooz_pub/.
2. M. Apollonio *et al.*, Phys. Lett. B**420** (1998) 397.
3. Y. Declais *et al.*, Phys. Lett. B**338** (1994) 383.
4. B. Achkar *et al.*, Nucl. Phys. B**434** (1995) 503.
5. G.S. Vidyakin *et al.*, JETP Lett. **59** (1994) 364.
6. G. Zacek *et al.*, Phys. Rev. D**34** (1986) 2621.
7. Z.D. Greenwood *et al.*, Phys. Rev. D**53** (1996) 6054.
8. Y. Fukuda *et al.*, Phys. Lett. B**335** (1994) 237.
9. B. Achkar *et al.*, Phys. Lett. B**374** (1996) 243.
10. K. Schreckenbach *et al.*, Phys. Lett. B**160** (1985) 325.
11. P. Vogel, Phys. Rev. C**24** (1981) 1543.
12. A.A. Hahn *et al.*, Phys. Lett. B**218** (1989) 365.
13. G. Feldman *et al.*, Phys. Rev. D**57** (1998) 3873.
14. BOREXINO, KAMLAND: see this Conference Proceedings.

ELSEVIER

Nuclear Physics B (Proc. Suppl.) 77 (1999) 166–170

NUCLEAR PHYSICS B
PROCEEDINGS
SUPPLEMENTS

The Palo Verde Reactor Neutrino Oscillation Experiment

Presented by Andreas Piepke on behalf of the Palo Verde Collaboration
F. Boehm[a], J. Busenitz[b], M. Dugger[c], G. Gratta[d], J. Hanson[a], H. Henrikson[a], J. Kornis[b], D. Lawrence[c], K.B. Lee[a], D. Michael[a], L. Miller[d], V.M. Novikov[a], A. Piepke[a], B. Ritchie[c], D. Tracy[d], A. Vital[b], P. Vogel[a], Y.F. Wang[d], J. Wolf[b]

[a]Division of Physics and Astronomy, Caltech, Pasadena CA 91125

[b]Department of Physics and Astronomy, University of Alabama, Tuscaloosa AL 35487

[c]Physics Department, Arizona State University, Tempe AZ 85287

[d]Physics Department, Stanford University, Stanford CA 94305

Our collaboration is operating a segmented scintillation detector, containing 11.3 tons of 0.1% Gd loaded liquid scintillator, at 750, 888 and 889 m distance from three nuclear power reactors. Goal of the experiment is to search for neutrino oscillations in disappearance mode. The anti-neutrino capture on the proton serves as detection reaction. The experiment is expected to reach a sensitivity of $\Delta m^2 > 1.3 \cdot 10^{-3}$ eV^2 and $sin^2 2\Theta > 0.1$. Our range of sensitivity is tuned to test the $\nu_\mu \leftrightarrow \nu_e$ solution of the atmospheric neutrino anomaly as reported by the Kamiokande experiment. Assembly of the detector in an underground lab, with 32 mw.e. effective overburden, is completed. Since June 1998 we are taking data at full reactor power. Detector design and performance are discussed.

1. Introduction

Motivated by the intriguing possibility to explain the apparent lack of muon type neutrinos in the atmospheric neutrino flux, as reported by the Kamiokande experiment [1], by either $\nu_\mu \leftrightarrow \nu_\tau$ or $\nu_\mu \leftrightarrow \nu_e$ neutrino oscillations, our collaboration has built a long-baseline reactor-neutrino oscillation experiment to test the latter solution. Although the recent Super Kamiokande data apparently disfavors the $\nu_\mu \leftrightarrow \nu_e$ solution [2] it seems prudent to us to perform an independent test utilizing a terrestrial, artificial source, based on a completely different systematics. Super Kamiokande makes a strong case for $\nu_\mu \leftrightarrow \nu_\tau$ oscillations by analyzing the zenith angle dependence of the flavor ratio. Our above assessment, however, receives support from the fact that Super K's allowed $\Delta m^2 - sin^2 2\theta$ parameter space for $\nu_\mu \leftrightarrow \nu_\tau$ oscillations has only marginal overlap with the original Kamiokande result.
The competing Chooz reactor neutrino experiment, aiming at the same problem, has published an early result [3] which is incompatible with $\nu_\mu \leftrightarrow \nu_e$ oscillations within the Kamiokande parameter space of [1].
We will now describe the status of our Palo Verde experiment, which commenced full operation in June 1998.

2. Neutrino Source

The Palo Verde Nuclear Generating station, the largest in the US, is located about 70 km west of Phoenix, Arizona. Three identical pressurized water reactors deliver a thermal power of 10.9 GW. The detector, installed in an underground vault with an overburden of 32 mw.e., is located 750, 888 and 889 m from each reactor core, respectively. At this distance the detector is exposed to an anti-neutrino flux of $6.9 \cdot 10^9\ \overline{\nu}_e/(cm^2 \cdot s)$ (for $E_\nu > 1.8\ MeV$). The cosmic ray muon flux is measured to be 22 $\mu/(m^2 \cdot s)$ at the above overburden. We will use the refueling outages to measure the detector background.

0920-5632/99/$ – see front matter
PII S0920-5632(99)00413-2

3. Detector

In the Palo Verde experiment the inverse beta decay, $\overline{\nu}_e + p \rightarrow e^+ + n$, serves as detection reaction. We require a correlated positron-neutron signature. 11.34 tons of 0.1% Gd loaded liquid scintillator, contained in 66 acrylic tanks, each 9 m long, are acting as detector and target, simultaneously. The measurement of the kinetic energy of the e^+, being stopped in the liquid scintillator, allows us to reconstruct the anti-neutrino energy. The energy threshold of the reaction hence is 1.8 MeV. After the positron has come to rest each annihilation quantum will totally or partially escape from the primary detector segment and deposit energy in one or more neighboring elements. The requirement of a prompt triple coincidence of adjacent modules thus identifies a positron. This scheme serves to suppress correlated background events from fast neutrons. To achieve a good detection efficiency for the annihilation radiation the large detector elements need to be operated with a low threshold of 50 keV, one of the experimental challenges of the project. The reaction neutron having around 20 keV kinetic energy is moderated by the scintillator and subsequently captured on protons or Gd. By virtue of its enormous neutron capture cross section, the 0.1% Gd loading results in an 86% capture probability on Gd. The neutron capture time, as fitted from the data, is $\tau = 32 \pm 2\ \mu s$, compared to $\tau = 180\ \mu s$ in unloaded scintillators. This tight time correlation between positron energy deposit and the 8 MeV γ ray burst following neutron capture on Gd defines the full neutrino event signature as a four fold delayed coincidence.

The fiducial volume is surrounded by a 1 m thick passive de-ionized water shield weighing about 100 tons. The whole setup is placed inside an hermetic 4π active muon veto comprised of 35 tons of liquid scintillator, contained in PVC tanks. The horizontal tanks are equipped with two 5" PMT at each end, while the side detectors have one 8" tube at either end. Figure 1 sketches the front view of the detector. The end sections of the detector are covered by veto tanks mounted on a rail system so as to allow access to the central detector. The total veto rate is 2.5 kHz.

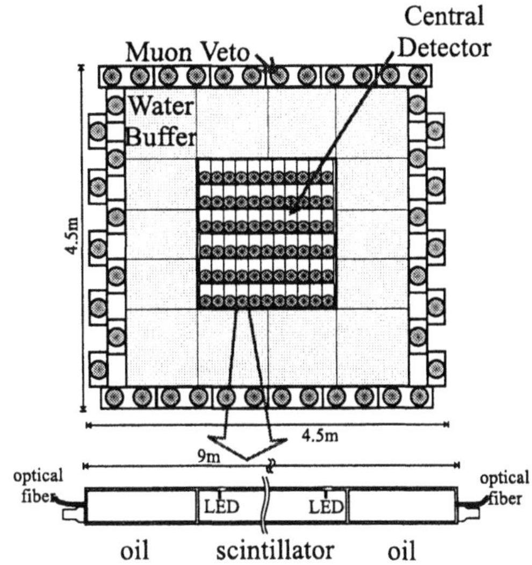

Figure 1. Detector front view. A single detector element is depicted below. The mineral oil filled buffer optically couples the PMT to the scintillator and shields the fiducial volume from γ radiation emitted by radioactive impurities contained in the PMT glass.

The underground laboratory was built from concrete based on crushed marble as to minimize the external gamma ray flux. All internal detector components were selected for their radioactivity content with the help of a low background Ge detector. Inside the shield we measure a specific Rn activity of 11 mBq/l. A Rn suppression system has been implemented. We are currently studying its effectiveness.

The Gd loaded scintillator, developed in collaboration with BICRON Inc., is based on a mixture of 60% mineral oil, 36% pseudocumene (PC) and 4% alcohol. The Gd is dissolved in form of an organo-metallic compound. The PC concentration of the mixture is limited by the requirement of its long term compatibility with the acrylic tanks. During the scintillator development great care was taken to optimize its light attenuation length (AL). At a wavelength of 440 nm (maximum of the scintillation spectrum is at 420 nm) we measure an AL of about 11 m for this scintillator. The liquid scintillator is very delicate. To avoid damage during transportation we shipped the liquid in form of a more robust Gd concentrate, containing PC and the alcohol. The scintillator was then blended with the other components on site in Palo Verde. Every batch was filtered and its mass measured. We have verified the Gd loading, AL and light yield for every production batch of Gd concentrate. For several early batches we studied the development of the transparency over time. During the first two months after pumping into the acrylic tanks we observe a decline in AL which then stabilizes.

The acrylic cells containing the Gd loaded scintillator are mounted on a roller system. Every cell can be removed for service. They are made individually light proof by wrapping them in thin Cu foil. Light collection onto the two 5" EMI PMT mounted on either end is achieved through total reflection. The corresponding light attenuation length for scintillation light is measured by scanning the detector elements with a ^{228}Th source along the detector axis. Miniature sources mounted on steel wires can be fed into the detector using Teflon guide tubes placed between each group of 4 cells. The scintillation intensity plotted as a function of distance to the PMT is a measure of the light transmission. A 14 points scan of all 66 detector elements shows that the data is well described by the sum of two exponentials corresponding to effective attenuation lengths of 4.1 ± 0.7 m and 1.2 ± 0.6 m. Such source runs also serve to define the energy scale of each detector element.

Each detector element is equipped with two blue LEDs used for regular PMT gain monitoring, by means of single photoelectron response, to calibrate the timing differences of the two PMTs in terms of distance along the detector axis, as shown in figure 2, and to define the walk correction in our timing measurements. As our PMTs are operated at a relatively high gain of $4 \cdot 10^7$ we

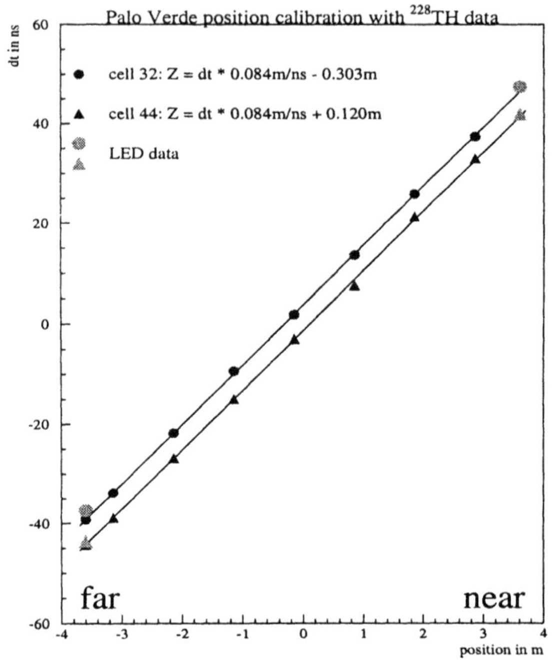

Figure 2. Arrival time differences (dt) measured for two different cells. dt-values obtained with the LED system are compared to data derived from a ^{228}Th source scan.

have to apply corrections to linearize their charge response. This is done using a fiber system capable to illuminate each PMT and simultaneously a low gain reference PMT with a linear response.

By virtue of its segmentation the detector allows a two dimensional vertex reconstruction. The resolution of 12 cm $\times$ 25 cm is given by the segment size. The timing difference of the PMTs defines the third dimension. We arrive at a longitudinal resolution of about 20 cm. To obtain the event energy we scale the linearized PMT charge, using the experimentally determined position scaling, and the event position as derived from the PMT timing.

A FASTBUS data acquisition system records the

PMT charge for every PMT. We digitize both the anode signal (high gain) and a signal derived from the 10th dynode (low gain). TDCs are used to record the timing differences of the two PMTs. We process data for a high and low threshold, for walk optimization, as well as for the leading and trailing pulse edge to have a handle on the pulse shape. To be able to measure correlated events, following in quick succession, the described electronics exists in two independent banks. An event trigger decision, based on the fast triple coincidence event topology, is obtained from a field programmable gate array [4]. All valid triple coincidences are recorded. Typical rates are of the order of 30 cps. The time correlation of the primary triple coincidence event with the delayed neutron capture is done off line. This allows us to measure random backgrounds in parallel with the signal by using time cuts beyond the positron-neutron correlation time.
For the cosmic muon veto system we have implemented a 10 μs hardware veto time. This is necessary to reject correlated events as the PMTs are very active after a direct μ-hit. On top of this we are recording the time elapsed since the last veto hit and the hit pattern for every recorded trigger. This way we can use a graded veto response off-line. A time cut on short correlation times should allow us to obtain a data set enriched in neutrons created inside the veto.

4. Results

Filling of the central detector with Gd loaded scintillator was completed on June 26, 1998. Till September 6, 1998 we have taken and analyzed 26 days of data with full reactor power. The detector-on efficiency is still low as a significant fraction of time is devoted to calibration and debugging runs in this early phase of the experiment. During two weeks of pure data taking we reached 90% on-time.
Based on Monte Carlo studies we expect a reactor induced $e^+ - n$ signal of 50 $events/day$. The same model yields a correlated background of 34^{+45}_{-25} $events/day$. It is due to fast neutron recoil imitating the e^+-signal followed by subsequent thermalization and capture of the same neutron.

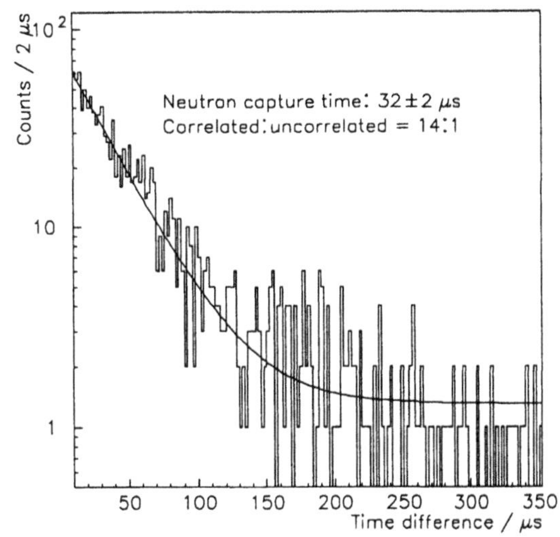

Figure 3. Distribution of time differences between prompt triple and delayed event, as derived from the data. All neutrino cuts have been applied. The signal is dominated by correlated events with the admixture of a small uncorrelated component.

Our estimate for the random coincidence background is 15 ± 10 $events/day$. It is due to γ-induced primary triples in random coincidence with a thermal background neutron. The large energy release of 8 MeV after neutron capture on Gd requires neutrons to participate in both classes of background. The dominating source of fast background neutrons is deep inelastic muon scattering outside the veto and to a lesser extent nuclear capture of stopped cosmic muons. The large error of the estimated background rates is mainly caused by our limited knowledge of the neutron production cross sections in muon reactions. Using these signal and background calculations we estimate the experimental sensitivity to be $\Delta m^2 > 1.3 \cdot 10^{-3}$ eV^2 and $sin^2 2\Theta > 0.1$, covering the allowed parameter range of [1] completely.
We will now discuss some early results obtained after 26 days of data taking. These numbers are of course still preliminary.

We estimate the resolution of the energy reconstruction for this data set by fitting the difference of the energies determined by the two PMT for each event (divided by $\sqrt{2}$) as a function of the reconstructed energy. This includes all applied corrections as we are averaging over all detector segments and vertex locations. In this way we get a resolution of 32% at 1 MeV and 20% at 5 MeV, respectively.

The raw rate of events with any delayed energy deposit within 400 μs is several thousand per day. Requiring spatial and temporal correlation between the prompt and delayed signal and within the prompt triple and tagging on Gd capture by requiring at least 3.5 MeV energy deposit in the capture signal, reduces that rate by about a factor 30. The time distribution of the triple and the delayed signal, for correlated events, as depicted in figure 3, demonstrates that the random background is only 7% of the signal, smaller than estimated.

While the Gd loading is a powerful tool to reduce the uncorrelated background we are making use of the detector segmentation to identify fast neutron recoil events. To limit the energy deposit in the two cells neighboring the positron cell to 0.6 MeV and the total energy of all "annihilation like" cell to 1.4 MeV is an effective tool for event by event recognition of neutron events. While the loss in positron efficiency due this cut is small (verified using a ^{22}Na source) the total number of events is reduced by a factor 3.7 by this cut.

We are now measuring the detector background and effectiveness of our cuts when one of the three reactors is switched off for refueling. As of this writing reactor unit 3 is down for its 35-37 days refueling. As unit 3 is with 889 m distance to the detector one of the more distant sources, the $\overline{\nu}_e$-flux is reduced to 71% of its full value. Any firm statement about detector background has to wait until this data has been analyzed.

We are planning to perform a precision calibration of positron and neutron detection efficiencies separately. To check the e^+ efficiency we are now using a sealed ^{22}Na β^+-point source which can be placed at any point inside the detector. The scattering of the 1.2 MeV gamma provides the "positron like" signal while the annihilation provides the coincidence tag. By the end of the year, after a sufficient body of reactor data has been collected, we will conduct a e^+ calibration using a ^{68}Ge-^{68}Ga β^+-source, of known activity (to $\pm 1\%$ accuracy), dissolved in the liquid scintillator of a special calibration cell [5]. The radioactive cell can replace any detector element to allow geometrical averaging and has the advantage of giving exactly the same spatial correlation as the $\overline{\nu}_e$-signal. This data will allow a precise calibration of our Monte Carlo codes.

At present we are using a ^{252}Cf spontaneous fission source to calibrate the neutron response of our detector. As this source emits an average of 3.8 neutrons and multiple gammas per fission it is not well suited for a precision calibration of the neutron efficiency. For such a calibration we will use a weak AmBe source for which we have determined the neutron emission rate to 3% precision. We are experimenting with a setup utilizing a miniaturized NaI detector to deliver an experimental tag for those neutrons resulting from a decay into the 4.4 MeV exited state of ^{12}C.

5. Conclusion

Our collaboration has build and commissioned a segmented liquid scintillation detector to search for neutrino oscillations 750-890 m from three nuclear reactors. While early findings including the low random background are encouraging a conclusion about the detector background or any physics has to await the completion of the background run in progress at the time of the writing of this article.

REFERENCES

1. Y. Fukuda et al., Phys. Lett. B 335 (1994) 237.
2. Y. Fukuda et al., Phys. Rev. Lett. 81 (1998) 1562.
3. M. Apollonio et al., Phys. Lett. B 420 (1998) 397.
4. G. Gratta et al., Nucl. Inst. Meth. A 400 (1997) 456.
5. A. Piepke and B. Cook, Nucl. Inst. Meth. A 385 (1997) 85.

NUCLEAR PHYSICS B
PROCEEDINGS
SUPPLEMENTS

ELSEVIER

Nuclear Physics B (Proc. Suppl.) 77 (1999) 171–176

Present Status of KamLAND

A. Suzuki
(for the KamLAND Collaboration)

Research Center for Neutrino Science, Tohoku University,
Aoba, Sendai 980-8578, Japan

The KamLAND project, a 1000 ton Kamioka Liquid scintillator AntiNeutrino Detector, started in 1997. Fundamental designs of the detector components have been almost finished. Civil engineering works are ready in June this year. The data-taking is expected to begin in January 2001.

1. INTRODUCTION

To research for the physics and astrophysics of neutrinos by means of detecting lower energy terrestrial and extra-terrestrial neutrinos and anti-neutrinos, the project of a 1000 ton liquid scintillator experiment, KamLAND was proposed in 1994. After the detector R&D, it was approved in 1997 [1]. Since 1998, the KamLAND project has been performed by the Japan and U.S. collaboration. Primary targets of physics in KamLAND are to search for neutrino oscillations in $\Delta m^2 > 10^{-5}$ eV^2 with reactor anti-neutrinos, to aim at the first observation of geoneutrinos, and to study neutrions from possible supernovae explosions. We are also contemplating a second phase of data-taking that, through improvements in the liquid scintillator purification will allow us to observe the ^{8}B and ^{7}Be solar neutrinos.

2. KamLAND DETECTOR

The KamLAND detector sits at the old Kamiokande site, dismantling the Kamiokande detector. The rock overburden is more than 2700 m.w.e. in any direction. The expected cosmic-ray muon rate for the detector is 0.3 Hz. The rock cavity is enlarged by 4 m in depth from the present bottom level. The site also includes the counting facilities, water and scintillator purification systems, ventilation system and electrical power station.

The detector consists of 3 layers as illustrated in Fig. 1. The cylindrical rock cavity with a 20 m diameter and a 20 m height is covered with waterproof lining materials in order to use as a water Cherenkov anti-counter. The inside of the anti-counter is segmented into small rooms by light reflection sheets so as to identify an entrance and exit points of incoming cosmic-ray muons. The old Kamiokande 20 inch-photomultiplier tubes (PMT's) are reused in this counter. A stainless steel spherical buffer tank with a 18 m diameter is solidly mounted inside the anti-counter. Mineral oil is filled into the tank to reduce buoyancy applied to the innermost liquid scintillator layer. 1280 newly developed 17 inch-PMT's are attached onto the entire inner surface of this tank, which gives 22 % photosensitive coverage. The 17 inch-PMT has the same shape and size as those of the 20 inch-PMT, but only a central part of photocathode is available. The dynode structure is a box-and-line type instead of a venecian-blind. As a result, under the conditions of single photoelectron illumination, 10^7 gain and 25 °C, a good time-resolution and a clear isolated pulse-shape are obtained: Transit Time Spread of (1-1.5) ns for 1 σ; Peak to Valley ratio of (3-5); and Dark counting rate above 0.25 photoelectron (p.e.) of 10 kHz on average. A proper operation of the PMT's is assured in the < 50 mGauss field which is realized by a set of compensating coils installed in the cavern to cancel the Earth's magnetic field.

PII S0920-5632(99)00414-4

A 1200 m^3 plastic balloon containing ~ 1000 ton liquid scintillator is deployed inside the buffer tank. A multi-layer film made of nylon + EVOH (poly vinyl alcohol) + nylon is a candidate balloon material. The EVOH film has low Radon permeability of $< 10^{-10}$ cm/s [2]. Scintillator cocktail which is made of 80% concentration of isoparaffin, 20% of pseudocumene and 2g/l of fluor (PPO), was chosen to keep the flash point greater than 60 °C. Such scintillator shows the following properties: more than 50 % of the Anthracene light output; 10 m and 20 m of the light attenuation length for 400 nm and 450 nm light, respectively; 90 % of the neutron rejection from γ-like signals; and (13.7 ± 2.1) of the α-particle quenching factor. We expect ~ 100 p.e. /MeV in KamLAND, which gives $\sigma(E)/E \sim 10\ \%/\sqrt{E}$. Contamination of U and Th in the liquid scintillator have been measured by a ICP-mass separator. Without purifying, the scintillator cocktail includes 2×10^{-13} g/g of U and $< 6 \times 10^{-12}$ g/g of Th. Our Monte Carlo study tells us that 1 order of magnitude reduction both for U and Th is required in the reactor neutrino oscillation search and more than (3-4) order of magnitude in the solar neutrino detection.

3. PHYSICS GOALS

3.1. Long baseline neutrino oscillation search

Neutrino oscillation searches in KamLAND are carried out using nuclear reactors. Anti-neutrinos produced in reactors are measured by detecting the inverse β-decay process (E_{th} = 1.8 MeV), $\bar{\nu}_e p \rightarrow e^+ n$ with the aid of a timing coincidence between a prompt e^+ signal and a delayed γ (2.2 MeV) coming from a thermal neutron capture on a proton. There are several commercial nuclear power plants around the Kamiokande site. An anti-neutrino flux of $1 \times 10^6 cm^{-2}s^{-1}$ is expected for $E_{\bar{\nu}} > 1.8$ MeV. 80 % of such flux derives from reactors at a distance between 140 km and 210 km. Thus, the flight range is limited in spite of using multi-reactors. A total thermal power flux is found to be change by at least 30 % season to season due to the high power consuming in summer and winter, and the obligatory 3 months maintenance in spring or fall .

Figure 1. Schematic view of the KamLAND detector.

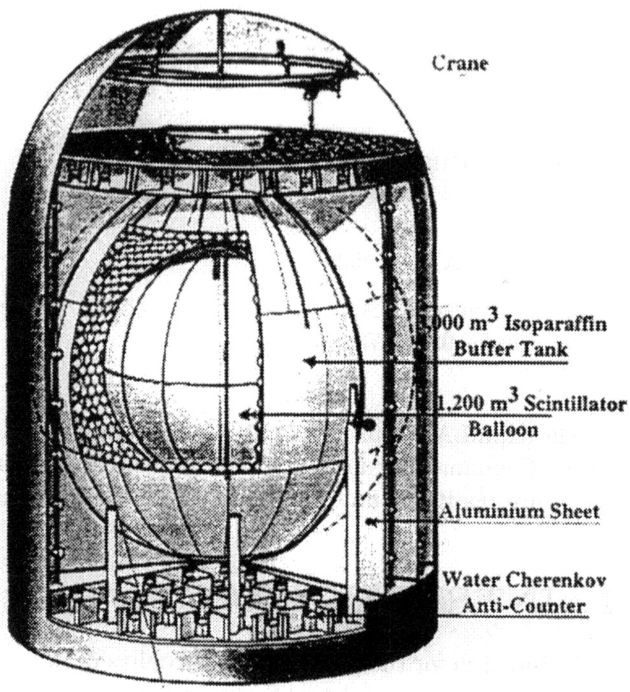

About 450 $\bar{\nu}_e p \rightarrow e^+ n$ events for one year running are expected with the 600 ton fiducial mass. On the other hand the correlated (with the timing coincidence) background event rate is estimated to be 8 in a year, assuming 10^{-14} g/g for U and Th. 10^{-12} g/g for K and 0.5 mBq/m^3 for Rn concentrations in the liquid scintillator. Fig. 2 shows the expected positron energy spectrum together with that of background events. One can see no serious effect from backgrounds.

There can be carried out three different oscillation analyses by (A) measuring the absolute flux, (B) the spectrum change and (C) the seasonal flux variation. Sensitivity of these three methods is depicted in Fig. 3, where a solid curve is obtained from (A) for 1 year data, a dotted one from (B) for 5 years data and a dashed one from (C) for 5 years data. An accessible oscillation parameter space is reached to $\Delta m^2 \sim 10^{-5}$ eV^2 and $\sin^2 2\Theta > 0.1$. This improves the present sensitivity [3] by over 2 order of magnitude and covers the MSW large angle solution to the solar neutrino deficit.

Figure 2. Expected positron energy spectra from reactor anti-neutrinos and backgrounds.

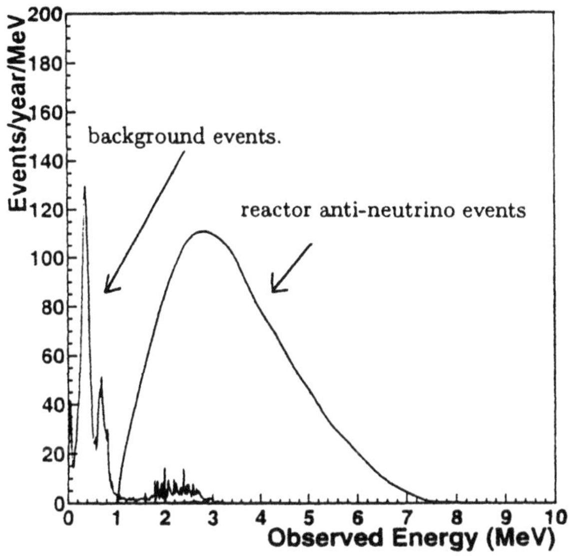

Figure 3: Parameter constraints on $\nu_e \to \nu_x$ oscillations. The solid contour represents the 95 % C.L. from the absolute flux change, the dotted contour the 90 % C.L. from the spectrum shape chage, and the dashed contour the 85 % C.L. from the seasonal flux variation. The shaded region shows the MSW large mixing solution.

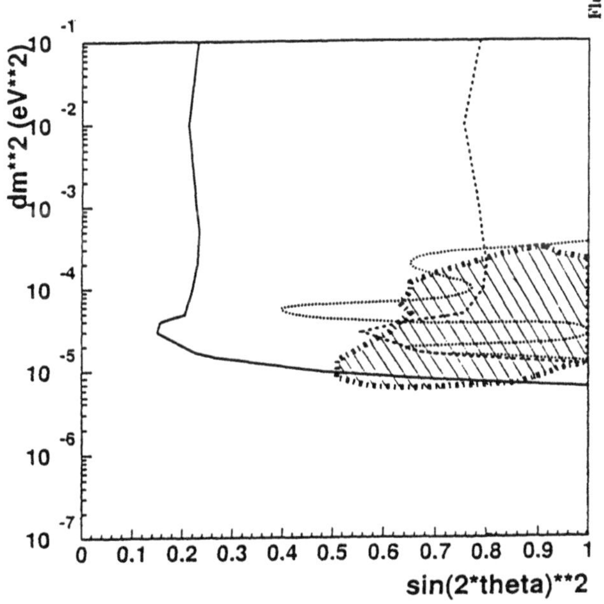

3.2. Geoneutrino detection

A first chance for terrestrial $\bar{\nu}_e$ (geoneutrino) search [4] [5] can be expected in KamLAND. A basic factor in the interior dynamics and the evolution of the present Earth is the radiogenic heat mainly from U and Th decays inside the Earth. Measuring the U and Th concentrations in the Earth interior sheds new light on geophysics. Calculation of the geoneutrino flux strongly depends on models concerning the abundance of U and Th in the continental crust, oceanic crust, upper-mantle and lower mantle. Fig. 4 is one example of the flux calculation by Krauss *et al.* [6]. Geoneutrinos are measured by the same method as reactor anti-neutrinos through a timing coincidence. For energies above the inverse β-decay threshold, 1.8 MeV only the U and U+Th components are separately detectable as seen in Fig. 4.

Figure 4: Geoneutrino flux calculation [6].

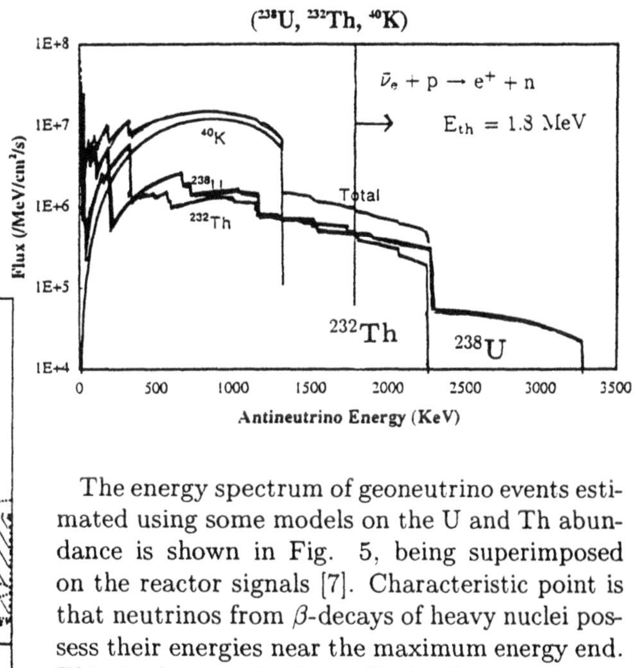

The energy spectrum of geoneutrino events estimated using some models on the U and Th abundance is shown in Fig. 5, being superimposed on the reactor signals [7]. Characteristic point is that neutrinos from β-decays of heavy nuclei possess their energies near the maximum energy end. This is due to the large Coulomb screening effect on an accompanying electron. Thus, one can see clear and sharp peaks of the U and U+Th components standing on the continuous reactor anti-neutrino signals. More than several tens of geoneutrino events can be expected for one year operation in KamLAND.

Figure 5. Positron energy spectrum induced by geoneutrinos and reactor anti-neutrinos [7].

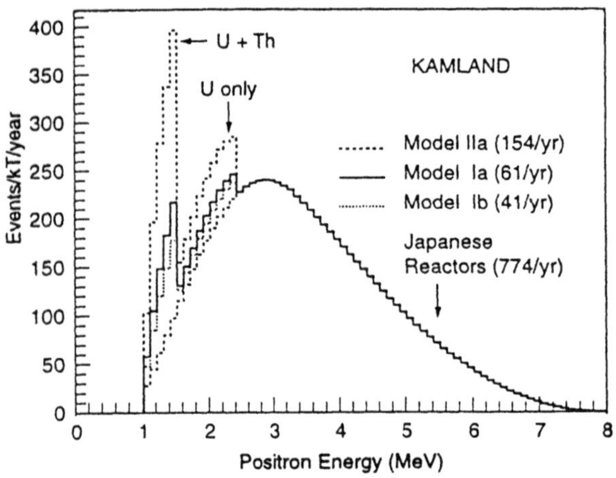

3.3. Observation of galactic supernova neutrino bursts

Through neutrino and anti-neutrino interactions with free protons and carbon nuclei, new windows in the detection of supernova neutrino bursts are opened in KamLAND. The number of events expected in a galactic supernova burst is calculated, taking the following typical parameters: the distance of 10 kpc; the released energy of 3×10^{53} erg; and the temperatures of 3.5 MeV ($< E_\nu > = 11$ MeV) for ν_e, 5 MeV ($< E_\nu > = 16$ MeV) for $\bar{\nu}_e$ and 8 MeV ($< E_\nu > = 25$ MeV) for $\nu_\mu, \bar{\nu}_\mu, \nu_\tau, \bar{\nu}_\tau$ [8]. Table 1 gives the expected number of events for different reaction channels.

The direction of the exploded star is determined by 16 single electron events from the neutrino-electron scattering. The Monte Carlo study shows that a Cherenkov ring produced by electrons with their energies above 10 MeV can be found out among scintillation light. If the distance is known, the neutrino luminosity of the supernova is obtained by 330 inverse β-decay events which are recognized by the same method as the reactor anti-neutrino detection. 58 events are expected from the neutral current reaction on ^{12}C, in which the 15.11 MeV monochromatic γ-ray is produced from the ^{12}C excited state. This gives a sensitive monitor of the supernova neutrino temperature, if the distance and the neutrino luminosity are known. Fig. 6 is the visible energy distribution of the inverse β-decay events and the neutral current events.

Table 1. Expected rates in KamLAND for a galactic supernova neutrino burst

reactions			no. of events
$\nu_x(\bar{\nu}_x)e^-$	$\rightarrow$	$\nu_x(\bar{\nu}_x)e^-$	16
$\bar{\nu}_e$ p	$\rightarrow$	e^+ n	330
$\nu_e^{12}C$	$\rightarrow$	e^- ^{12}N	2
$\bar{\nu}_e^{12}C$	$\rightarrow$	e^+ ^{12}B	7
$\nu_x(\bar{\nu}_x)^{12}C$	$\rightarrow$	$\nu_x(\bar{\nu}_x)^{12}C^*$	58

$E(^{12}C^*) = 15.11$ MeV

Figure 6. Visible energy spectrum of $\bar{\nu}_e$ p $\rightarrow e^+$n and $\nu_x(\bar{\nu}_x)^{12}C \rightarrow \nu_x(\bar{\nu}_x)^{12}C^*$ events [8].

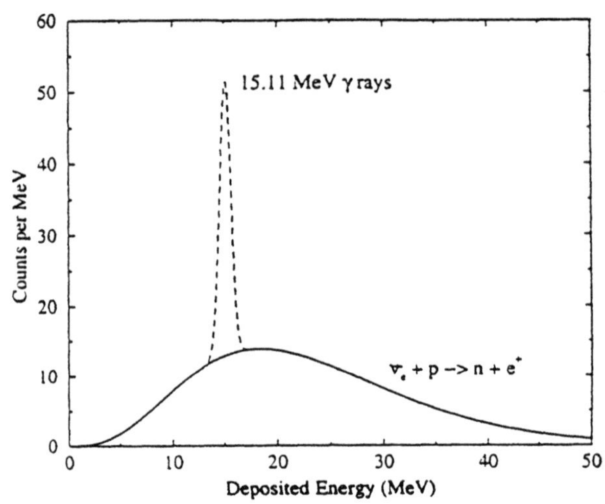

Nuclear excitation events induced by the charged current interactions, ν_e $^{12}C \rightarrow e^-$ ^{12}N and $\bar{\nu}_e$ $^{12}C \rightarrow e^+$ ^{12}B, are measured by applying the time and space correlations between the prompt e^- (e^+) and delayed decaying $e^+(e^-)$ from the β unstable ground state of ^{12}N (^{12}B). Thus they are essentially background free. These two processes provide a unique test of two neutrino oscillation solutions to solve the solar neutrino deficit. The "Just So" solution gives 14 events each for the ν_e $^{12}C \rightarrow e^-$ ^{12}N and $\bar{\nu}_e$ $^{12}C \rightarrow e^+$ ^{12}B channels. The MSW solution gives 27 events for the

ν_e $^{12}C \rightarrow e^-$ ^{12}N channel instead of 2 events, but no change in the $\bar{\nu}_e$ $^{12}C \rightarrow e^+$ ^{12}B channel. This would be a chance to determine the solution of the solar neutrino deficit.

3.4. Solar neutrino detection

The physics considered above can be achieved in the first stage KamLAND experiment with existing technology. In the next step we aim to explore the "sub-MeV" physics. Since the ν e $\rightarrow \nu$ e scatterings, involving single ionization events are dominant in this energy region, the experiment in the ultra-low background environment is required substantially. Pioneering works have been already done in Borexino and demonstrated the feasibility of the low background liquid scintillator experiment to this research field [9].

The energy sensitivity of ~ 110 p.e./MeV in KamLAND makes it possible to set a detection threshold energy of ~ 300 keV. Hence the ^{8}B and ^{7}Be solar neutrinos can be measured in principle. A skillful purification of the liquid scintillator, quite low Rn invading into the scintillator fluid, keeping the detector clean in construction and a large shielding volume for external backgrounds make it possible to detect the ^{8}B and ^{7}Be solar neutrinos in KamLAND in future.

Fig. 7 shows the expected recoil-electron energy spectrum of solar neutrino events together with the spectrum of backgrounds. 940 ^{8}B events/year with energy > 4 MeV and 110 ^{7}Be events/day with energy > 0.3 MeV are expected for the 560 ton and the 300 ton fiducial mass, respectively. The background spectrum is calculated assuming 10^{-16} g/g for U and Th, 10^{-14} g/g for K and 10 μBq/m^3 for Rn in the liquid scintillator and by quoting the data of radioactivity measurements of the detector materials and rocks. Under such a low background level the spectrum down to 3.5 MeV for the ^{8}B events is detectable, which is essential to examine whether the solar neutrino deficit is given rise to the MSW small angle or the "Just So" oscillation. The ^{7}Be signal also exceeds over the background one. This gives us variety tests on solutions to the solar neutrino problem [9]. Considering 0 SNU for the ^{7}Be neutrino observation obtained by the on-going experiments, it would be invaluable to measure the ^{7}Be neutrinos independently in Borexino and KamLAND.

Figure 7. Expected energy spectra of solar neutrino events and background events.

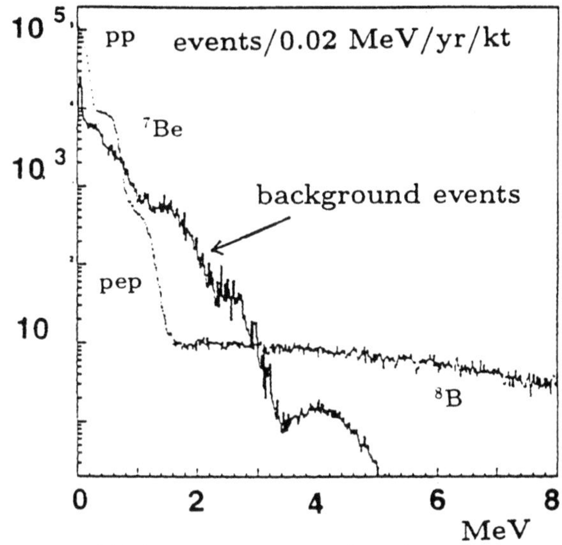

3.5. Higher energy physics

Although the first priority of the KamLAND physics is charged in "MeV" and "sub-MeV", the KamLAND detector performance allows to measure the "MeV - GeV" events. Energy response of the new 17 inch-PMT's is quite well in using at 10^7 gain as seen in Fig. 8. Here one can see not only a linear response up to 1,000 p.e., but also no saturation even at 10,000 p.e. This means the energy deposited by cosmic-ray muons passing through the detector is measurable. So recoil-protons produced by the neutral current interactions of the atmospheric neutrinos and the decays of charged pions and kaons produced in the atmospheric neutrino interactions and proton decays are identified with the aid of wave-form digitizers in the KamLAND front-end electronics.

4. Conclusions

KamLAND was approved by a 5 year project since 1997 of the JSPS program . Hence the construction schedule is so tight as shown below. All the Kamiokande detector components belonged in Tohoku Univ. in March 1998, partly from ICRR, Univ. of Tokyo and partly from KEK. Then civil engineering works at the mine side have been following since June.

Figure 8. Range of anode-current linearity of the new 17 inch-PMT at 10^7 gain as a function of light flux in units of p.e.

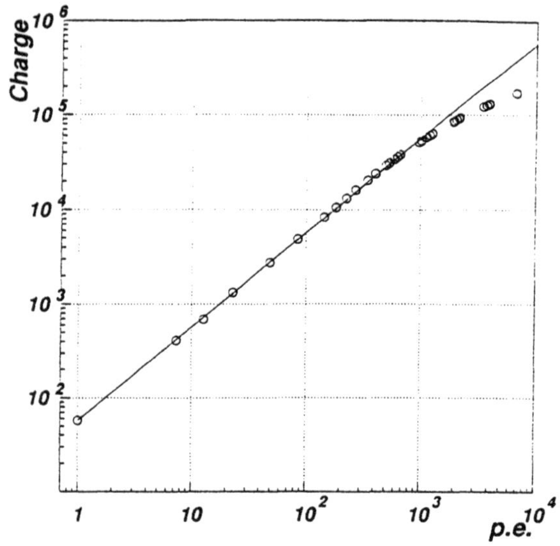

In 1998

- Remove the Kamiokande detector and expand the tunnel size and detector doom depth,
- Construct the 5,400 m^3 water pool for the water Cherenkov anti-counter;

In 1999

- Construct the 3,000 m^3 spherical buffer tank,
- Install 1280 17" PMT's into the buffer tank,
- Deploy the 1200 m^3 plastic balloon,
- Construct the water and liquid scintillator purification systems;

In 2000

- Install data taking electronics and computers,
- Fill 1000 ton liquid scintillator inside the balloon,
- Prepare and adjust all equipments; and

In 2001

- Start data taking in January 1.

We are grateful to Dr. R. Raghavan for valuable discussions and suggestions since planing this project and to Dr. M. Chen for discussions and measuring Rn permeability. We also would like to thank for discussions to the KamLAND collaborators: P. Alivisatos, S. Berridge, N. Bokor, C. Britton, W. Bryan, W. Bugg, J. Busenitz, T. Chikamatsu, H. Cohn, L. DeBraeckeleer, B. Dieterle, Yu. Efremenko, S. Enomoto, K. Furuno, S. Frank, C.R. Gould, G. Gratta, H. Hanada, E. Hart, S. Hatakeyama, G. Horton-Smith, T. Itoh, T. Iwamoto, Yu. Kamyshkov, S. Kawakami, M. Koga, J. Kornis, K.B. Lee, H.L. Liew, K. Mashiko, L. Miller, M. Nakajima, T. Nakajima, A. Nemeth, V. Novikov, H. Ogawa, K. Oki, P. Pacher, A. Piepke, S. Riley, N. Sleep, J. Shirai, F. Suekane, A. Suzuki, O. Tajima, K. Tagashira, T. Takayama, K. Tamae, H. Tanaka, D. Takagi, T. Taniguchi, W. Tornow, D. Tracy, P. Vogel, Y-F. Wang, H. Watanabe, A. Wintenberg, J. Wolf and J. Wolker. .

REFERENCES

[1] The KamLAND project is supported by Center of Excellence Grant of JSPS (Japan Society for the Promotion of Science).

[2] M. Chen, private communication (1998)

[3] M. Apollonio *et al.*, Phys. Lett., B420 (1998) 397.

[4] G. Eder, Nucl. Phys., 78 (1966) 657.

[5] G. Marx, Czech. J. of Phys., B19 (1969) 1471.

[6] L. M. Krauss, S. L. Glashow and D. N. Schramm, Nature, 310 (1984) 191.

[7] R. S. Raghavan, S. Schoenert, S. Enomoto, J. Shirai, F. Suekane and A.Suzuki, Phys. Rev. Lett., 80 (1998) 635.

[8] P. Vogel, private communication; The KamLAND proposal, Stanford-HEP-98-03, Tohoku-RCνS-98-15, (1998).

[9] L. Oberauer, Talk in this conference (1998).

N·H
ELSEVIER
Nuclear Physics B (Proc. Suppl.) 77 (1999) 177–181
NUCLEAR PHYSICS B
PROCEEDINGS
SUPPLEMENTS

A Pilot Experiment with Reactor Neutrinos in Taiwan

Henry T. Wong [a] and Jin Li[b]

[a]Institute of Physics, Academia Sinica, Taiwan.

[b]Institute of High Energy Physics, Beijing, China.

A Collaboration comprising scientists from Taiwan, mainland China and the United States has been built up since 1996 to pursue an experimental program in neutrino and astro-particle physics in Taiwan. A pilot experiment to be performed at the Nuclear Power Station II in Taiwan is now under intense preparation. It will make use of a 500 kg CsI(Tl) crystal calorimeter to study various neutrino interactions. Various possible future directions will also be explored. The conceptual design and the physics to be addressed by the pilot experiment are discussed.

1. INTRODUCTION

A Collaboration has been built up since 1996 to initiate and pursue an experimental program in neutrino and astro-particle physics in Taiwan [1]. At present, the "TEXONO" [1] Collaboration comprises more than 40 scientists with diversified expertise from Taiwan (Academia Sinica, Institute of Nuclear Energy Research, National Taiwan University, National Tsing Hua University, National Chiayi Teachers' College and Nuclear Power Plant II), China (Institute of High Energy Physics, China Institute of Atomic Energy, Nanjing University, Shandong University, University of Science and Technology at Hefei) and the United States (University of Maryland).

The goal is to conduct an international-standard particle physics experiment in Taiwan. The field of choice for the "pilot" experiment is reactor neutrino. There are operational power reactors in Taiwan. The mountainous landscape dotted with mines and tunnels makes the construction of an underground laboratory conceivable. The proximity of the reactor locations and the possible underground sites (Nuclear Power Plants I, II and IV are all about 20-30 km from Taipei city) to the city infrastructures provides an additional advantage.

The Collaboration has been intensely preparing the pilot experiment to be performed at a site of about 30 m from one of the reactor cores at Nuclear Power Station II. Meanwhile, the feasibility and conceptual studies of the "next" project will be pursued. Possible directions include long baseline reactor neutrino oscillation experiments, dark matter searches, or solar neutrino studies.

2. THE PILOT EXPERIMENT

2.1. Physics and Detector Motivations

Almost all previous reactor neutrino experiments were based on liquid scintillator techniques to study the ($\bar{\nu}_e$ p) interactions, but with different neutron-capture isotopes. An experiment focusing on gamma detection has never been attempted.

However, gamma-ray spectroscopy has been a standard technique in nuclear sciences (that is, in the investigations of physics at the MeV range). Gamma-lines of characteristic energies give unambiguous information on the presence and transitions of whichever isotopes, allowing a unique interpretation of the physical processes. The experimental difficulties of building a high-quality gamma detector for MeV neutrino physics have been the large target mass required. However, in the past few years, big electro-magnetic calorimeter systems (with mass up to 40 tons of crystals, in the case for the forthcoming B-factories detectors) have been built for high energy physics experiments, using CsI(Tl) crystals with photodiodes readout.

The properties of CsI(Tl) crystals, together with those of a few common scintillators, are

[1]**T**aiwan **EX**periment **O**n **R**eactor **N**eutrin**O**.

PII S0920-5632(99)00415-6

listed in Table 1. The CsI(Tl) crystal offers certain advantages over the other possibilities. It has relatively high light yield and high photon absorption (or short radiation length). It is mechanically stable and easy to machine, and is only weakly hygroscopic. It emission spectra well matches the response of silicon photo-diode as depicted in Figure 1, thus making a compact design with minimal passive volume and efficient shielding configuration possible.

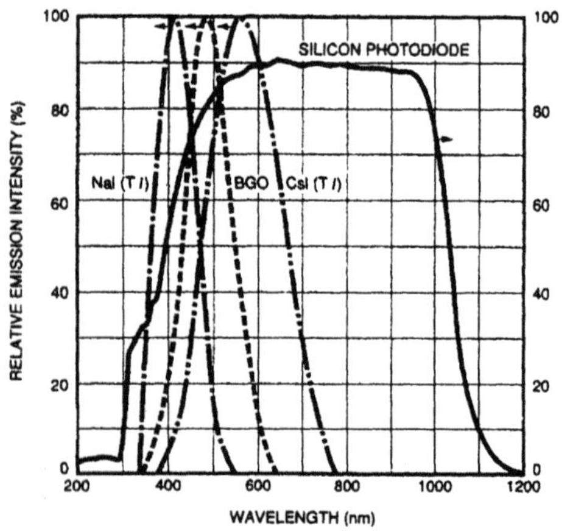

Figure 1. The sensitivities as a function of wavelength for typical silicon photo-diode, and the emission spectra of several common crystal scintillators, showing that CsI(Tl) matches best among them.

The CsI-crystal production technology is by now well matured and the cost has been reduced enormously due to the large demands. It become realistic and affordable to build a CsI detector in the range of 1-ton in target mass for a neutrino experiment. The detector mass can be further scaled up if the first experiment would yield interesting results or lead to other potential applications.

2.2. Physics Menu

Previous experiments with reactor neutrinos primarily focused on the ($\bar{\nu_e}$ p) interactions to look for neutrino oscillations. However, the use of low energy (MeV) neutrino as a probe to study particle and nuclear physics has not been well

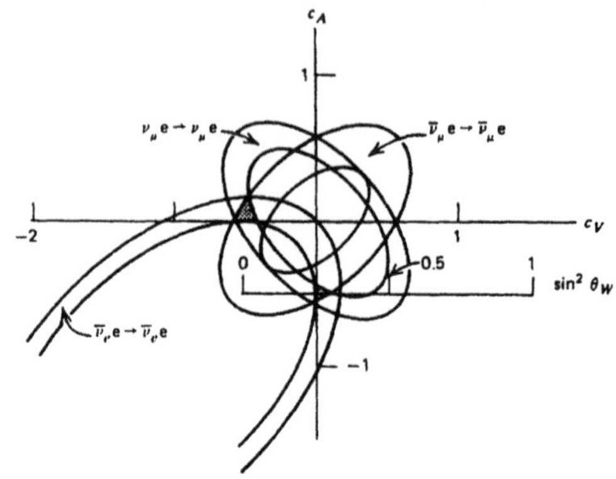

Figure 2. Sensitivities of (ν_μ e), ($\bar{\nu_\mu}$ e) and ($\bar{\nu_e}$ e) cross-sections to different regions in the g_A-g_V parameter space (axes labeled as c_A and c_V), showing their complementarity.

explored - although high energy (GeV) neutrino beams from accelerators have been very productive in investigating electroweak, QCD and structure function physics and have blossomed into a matured field. There are rooms for interesting physics with reactor neutrino experiments along this direction, some of which can be explored by a crystal calorimeter.

2.2.1. Neutrino-Electron Scattering

The cross section for the process

$$\bar{\nu_e} + e^- \rightarrow \bar{\nu_e} + e^-$$

gives information on the electro-weak parameters (g_V, g_A, and $\sin^2\theta_W$), and are sensitive to small neutrino magnetic moments (μ_ν) and the mean square charge radius ($< r^2 >$) [2]. Scatterings of the (ν_e e) and ($\bar{\nu_e}$ e) are two of the most realistic systems where the interference effects between Z and W exchanges can be studied.

The g_A Vs g_V parameter space where ($\bar{\nu_e}$ e) scatterings are sensitive to is depicted in Figure 2. The complementarity with (ν_μ e, $\bar{\nu_\mu}$ e) can be readily seen. The expected recoil energy spectrum is displayed in Figure 3, showing standard model expectations and the case with an anomalous neutrino magnetic moment at the present limit. The μ_ν term have a $\frac{1}{T}$ dependence. Accordingly, experimental searches for the neutrino magnetic moment should focus on the reduction of the threshold (usually background-limited) for

Properties	CsI(Tl)	NaI(Tl)	BGO	Liquid	Plastic	Glass
Density	4.51	3.67	7.13	0.9	1.0	~3.5
Relative Light Yield	0.45	1.00	0.15	0.4	0.35	0.15
Radiation Length (cm)	1.85	2.59	1.12	~45	~45	4
Emission Peak (nm)	565	410	480	425	425	395
Decay Time (ns)	1000	230	300	2	2	100
Refractive index	1.80	1.85	2.15	1.5	1.6	1.55
Hygroscopic	slightly	yes	no	no	no	no

Table 1
Characteristic properties of the common crystal scintillators and their comparison with typical liquid, plastic and glass scintillators.

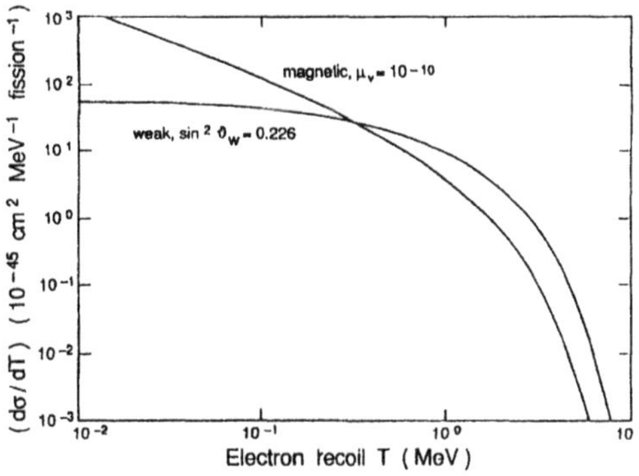

Figure 3. Differential cross section showing the electron recoil energy spectrum in $\bar{\nu}_e$-e scatterings with reactor neutrinos, for Standard Model predictions and for the case with a neutrino magnetic moment of 10^{-10} Bohr magneton, the present experimental limit.

the recoil electron energy.

Therefore, investigations of ($\bar{\nu_e}$ e) cross-sections with reactor neutrinos allow one to study electro-weak physics (measurement of $\sin^2\theta_W$) at the MeV range, to test electron-muon universality (that is, whether ($\bar{\nu_e}$ e) can be described by the same $g_{V/A}$ as (ν_μ e) and ($\bar{\nu_\mu}$ e)), and to look for an anomalous neutrino magnetic moment.

A 500 kg CsI crystal calorimeter will have more target electrons than previous experiments [3] and current projects [4], and thus can potentially improve the sensitivities of these studies. The compact detector size will also allow effective shielding design. The signature for ($\bar{\nu_e}$ e) will be a single hit out of the several hundred channels in the active target configuration. The goal is to achieve a 10% measurement on the cross-section and to probe neutrino magnetic moment down to 5×10^{-11} μ_B.

2.2.2. Neutrino Charged and Neutral Currents on Deuteron

The interactions

$$\begin{aligned} CC &: \bar{\nu_e} + d \rightarrow e^+ + n + n \\ NC &: \bar{\nu_e} + d \rightarrow \bar{\nu_e} + p + n \end{aligned}$$

have been observed [5]. Improved measurements will be of interest, especially since the NC reaction is the detection channel adopted by the forthcoming SNO experiment [6] for solar neutrino detection. Measurement of the CC/NC ratio provides a complementary method to search for neutrino oscillations, which is independent of the detailed knowledge of the neutrino source - an interesting possibility for long-baseline experiments which may receive neutrinos from many reactor cores and where the conventional "Reactor ON-OFF" subtraction may not be feasible. The SNO experiment will pursue this CC/NC ratio measurements for solar neutrino, and it would be desirable to have a laboratory experiment demonstrating the validity of the technique.

In a realistic experiment, the CsI crystal slabs will be put into a tank with 500 kg of heavy water (D_2O). Neutrons produced will mostly be captured via (n,γ) by ^{133}Cs and ^{127}I. The CC signatures will be rather spectacular: back-to-back

511 keV γs followed by two separate bursts of high energy γs from neutron capture. The NC detection will rely on a single γ-burst.

This is a complementary - and improved - technique to the previous experiments [5] which used ^{3}He proportional counters and were therefore sensitive only to neutrons. Accordingly, the CsI detector, with its γ-detection capabilities, can differentiate the signals from the other neutron-producing background channels ($\bar{\nu_e} + p \rightarrow e^+ + n$ and $\gamma + d \rightarrow p + n$) and can prevent the CC events with one undetected neutron from contaminating the NC sample. The goals are to achieve 5% and 10% measurements for $\bar{\nu_e}$d-CC and $\bar{\nu_e}$d-NC, respectively.

2.2.3. Neutral Current Excitation on ^{10}B and ^{11}B

If a compact boron-rich object like B_4C (natural boron consists of 20% ^{10}B and 80% ^{11}B) is used as the passive target, characteristic γ-lines (3.59, 5.16 MeV for ^{10}B, and 2.11, 4.45, 5.02 MeV for ^{11}B) will be emitted by the excited daughter nuclei following the NC interactions :

$$\bar{\nu_e} + {}^{10}B({}^{11}B) \rightarrow \bar{\nu_e} + {}^{10}B^*({}^{11}B^*) \ .$$

There are theoretical works [7] suggesting that these cross sections are sensitive to the axial isoscalar component of NC interactions and the strange quark content of the nucleon. Therefore, νN NC scattering may provide a complementary approach to the investigations of nucleon structure physics comparing to the eN scattering systems. The ν_e NC interaction on ^{11}B has been considered as the detection mechanism in the BOREX solar neutrino proposal [8].

A realistic experiment will consist of about 500 kg of B_4C, either in plate or powder form, inserted into a chamber with CsI crystals at optimized positions. The experimental signature will be gamma-lines of the characteristic energies which show up during reactor ON period.

If a CsI calorimeter proves itself to be optimal for studying NC excitations on nuclei [7], where the experimental signatures are the characteristic γ-lines, one can insert other passive materials to measure their cross sections, and turns the experiment into a longer-term program.

2.3. Highlights of Experimental Details

Among the various physics items mentioned above, the first to be pursued will be that of neutrino-electron scattering, using the "active target" configuration shown schematically in Figure 4. The detector will consist of about 500 kg of CsI(Tl) crystals. Individual crystal is 1 kg in mass and hexagonal in shape with 2 cm sides and 20 cm length. Two crystals are optically-coupled together and read out at both ends by two photodiodes, followed by pre-amplifiers, main amplifiers and shapers. Total energy can be derived from the sum of pulse heights from the two ends, while their difference gives the longitudinal position. The entire pulse is digitized by a FADC to be read out with a data acquisition system adopting the VME-bus.

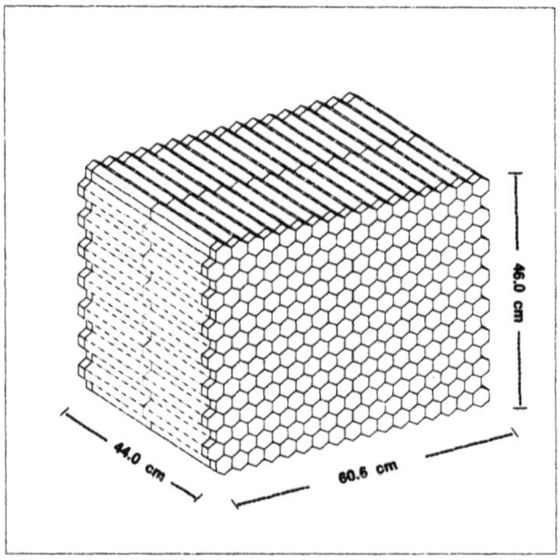

Figure 4. Schematic layout of the CsI(Tl) target, consisting of about 500 crystals, each of which is hexagonal in shape with 2 cm sides, 20 cm length and 1 kg mass. Photo-diodes and pre-amplifiers are placed at the end for readout.

The achieved energy resolution is about 16% FWHM at 660 keV. It is electronic noise-limited and hence improves linearly with energy. Pulse shape discrimination between γ/e and α events can be achieved to better than the 99% level.

The CsI target will be shielded by lead, boron-loaded polyethylene and copper, as depicted schematically in Figure 5. Cosmic rays will be vetoed by an outermost layer of plastic scintillators. The outer modules of the CsI target can be used as active veto if necessary. The whole inner target will be placed in a dry nitrogen environment to purge the radon gas, and will be kept at 5°C to reduce electronic noise.

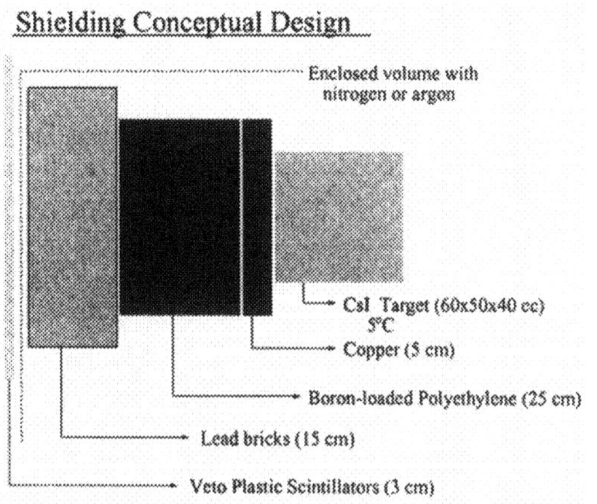

Figure 5. Schematic layout of the target and shielding. The coverage is 4π but only one face is shown.

The intrinsic radiopurity level of the CsI(Tl) crystal is very crucial to the sensitivities of this experiment, as well as to the future potential applications in low background physics. By the absence of α-peaks above 3 MeV in a measurement using a 3 kg crystal in an underground site, previous work [9] have derived that CsI crystals can be grown to a purity level where the contaminations of ^{238}U and ^{232}Th are less than the 10^{-12} g/g level. The concentration of ^{40}K is measured to be less than the ppb level, from neutron activation analysis and direct γ-counting.

3. Outlook

A Taiwan and mainland China collaboration has been built up to initiate and pursue a program in experimental neutrino physics and astroparticle physics in Taiwan. A "pilot" experiment to be performed close to the reactor core using CsI(Tl) as detector is now being prepared. Various neutrino interactions at the MeV energy range can be investigated. The feasibility and conceptual investigations of future directions are under way.

This is a pioneering "foundation" effort for Taiwan, and the importance of the outcomes of this experiment and this experience will lie besides, if not beyond, neutrino physics.

REFERENCES

1. "The Starting-Up of a Neutrino Project in Taiwan", C.Y. Chang, S.C. Lee and H.T. Wong, in *Procs. of XVI Int. Workshop on Weak Interactions and Neutrinos 1997*, to be published in Nucl. Phys. (Procs. Suppl.) **B** (1998).
2. A.V. Kyuldjiev, Nucl. Phys. **B 243**, 387 (1984); B. Kayser et al., Phys. Rev. **D 20**, 87 (1997); P.Vogel and J.Engel, Phys. Rev. **D 39**, 3378 (1989).
3. F. Reines, H.S. Gurr and H.W. Sobel, Phys. Rev. Lett. **37**, 315 (1976); G.S. Vidyakin et al, JETP Lett. **55**, 206 (1992); A.I. Derbin et al., JETP Lett. **57**, 769 (1993).
4. C. Broggin et al., Nucl. Instrum. Methods **A311**, 319 (1992); I.R. Barabanov et al., Astropart. Phys. **5**, 159 (1996); A.G. Beda et al., Preprint hep-ex/9706004 (1997).
5. T.L. Jenkins, F.E. Kinard, and F. Reines, Phys. Rev. **185**, 1599 (1969); E. Pasierb et al., Phys. Rev. Lett. **43**, 96 (1979); G.S. Vidyakin et al., JETP Lett. **49**, 151 (1988); G.S. Vidyakin et al., JETP Lett. **51**, 279 (1990).
6. G.T. Ewan et al. SNO Proposal (1987).
7. H.C. Lee, Nucl. Phys. **A 294**, 473 (1978); T.W. Donnelly and R.D. Reccei, Phys. Rep. **50**, 1 (1979); J. Bernabéu et al., Nucl. Phys. **B 378**, 131 (1992); K. Kubodera and S. Nozawa, Int. J. Mod. Phys. **E 3**, 101 (1994).
8. R.S. Raghavan and S. Pakvasa, Phys. Rev. **D 37**, 849 (1988).
9. U. Kilgus, R. Kotthaus, and E. Lange, Nucl. Instrum. Methods **A 297**, 425, (1990); R. Kotthaus, Nucl. Instrum. Methods **A 329**, 433 (1993).

Nuclear Physics B (Proc. Suppl.) 77 (1999) 182–186

NUCLEAR PHYSICS B
PROCEEDINGS
SUPPLEMENTS

Long Baseline Neutrino Oscillation Program in the United States

Stanley G. Wojcicki
Physics Department, Stanford University,
Stanford, California 94309

The long baseline neutrino oscillation program in the United States is based on a new neutrino facility at Fermilab, NuMI (Neutrinos at the Main Injector), and a new experiment with a detector underground, 730 km away. This experiment, the Main Injector Neutrino Oscillation Search (MINOS) has been designed to explore a large area in the neutrino oscillation parameter space. It has been optimized for the range of oscillation parameters which are suggested by the current and past generations of underground experiments studying atmospheric neutrinos.[1, 2, 3, 4] The main design goal of MINOS is that should be able to cover fully the suggested SuperKamiokande region.[4] Because the Δm^2 range suggested by the existing experiments is uncertain up to at least an order of magnitude, we have tried to build in the capability to vary the energy of the neutrino beam by the same factor in order to cover the full potentially interesting region.

The site of the principal MINOS detector has been chosen to be the Soudan mine in northern Minnesota, about 730km away from Fermilab. The geography of the experiment is indicated in Fig. 1. The choice of the location for the far detector was dictated by a number of different factors: the distance is well matched to the SuperKamiokande suggested oscillation parameters and the energy of neutrinos available from the Main Injector at Fermilab; there is now an established tradition of scientific cooperation with the Department of Natural Resources (DNR) of the state of Minnesota which operates the Soudan mine; the DNR will provide the infrastructure necessary to do the experiment in Soudan; the state of Minnesota has committed itself to provide financial resources to modify the existing cavern for MINOS; underground location reduces the spurious backgrounds; the existing Soudan 2 detector will provide additional measurements, complementary to those from the main detector.

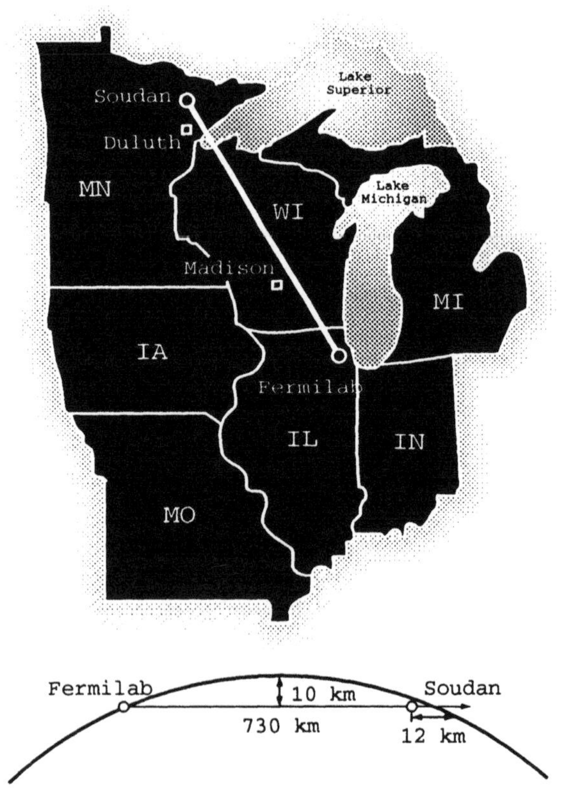

Figure 1. The trajectory of the MINOS neutrino beam between Fermilab and Soudan. The beam must be aimed into the earth at an angle of 57 mrad to reach Minnesota.

PII S0920-5632(99)00416-8

The source of the neutrinos is the new Main Injector accelerator, just completed at Fermilab, which is ideally suited for exploration of the neutrino oscillation parameters in the interesting region. The energy of the Main Injector, 120 GeV, is such as to be able to provide large flux of neutrinos in the energy range of interest: 2 - 25 GeV. This energy band, coupled with the Fermilab - Soudan distance, is not only well matched to investigation of the indicated Δm^2 range but also provides a large neutrino flux above the τ threshold.

The Main Injector is a high intensity proton synchrotron which will provide an adequate neutrino event rate even in a detector 730 km away. Furthermore, the anticipated mode of operation of the Main Injector during the next decade is ideally suited to the NuMI project. The accelerator will have to operate a significant fraction of time to produce antiprotons for $\bar{p}-p$ collisions in the Tevatron, but only one out of its six bunches is required for that purpose, the other five being available for MINOS neutrino production. Furthermore, the Main Injector is a relatively conventionally designed proton synchrotron. There is a very high likelihood that its intensity will increase with time and thus higher statistics, better sensitivity experiments will be possible in the future. We anticipate that when MINOS starts running about 5×10^{13} ppp will be available from the Main Injector every 1.9 sec.

To produce neutrinos, the 120 GeV proton beam from the Main Injector is allowed to strike a segmented carbon rod target, producing secondary pions and kaons. In order to aim the neutrino beam at the Soudan mine, the proton beam is directed downward at an angle of 57 mr when it strikes the target. Subsequently, forward going particles of interest are collected by a set of two parabolic focusing horns and allowed to propagate downstream in an evacuated beam pipe, 1 m in radius and 675 m long, placed in a specially excavated tunnel, also pointing downward towards Soudan. A beam stop is placed at the end of the decay pipe to attenuate the residual particle flux.

The desired neutrino energy selection is performed by adjusting the currents in the two horns and their locations. Thus this beam design allows modification of the energy of the accepted particles during the course of the experiment. The neutrino event rates anticipated in the far detector, for three different beam configurations, are shown in Fig. 2.

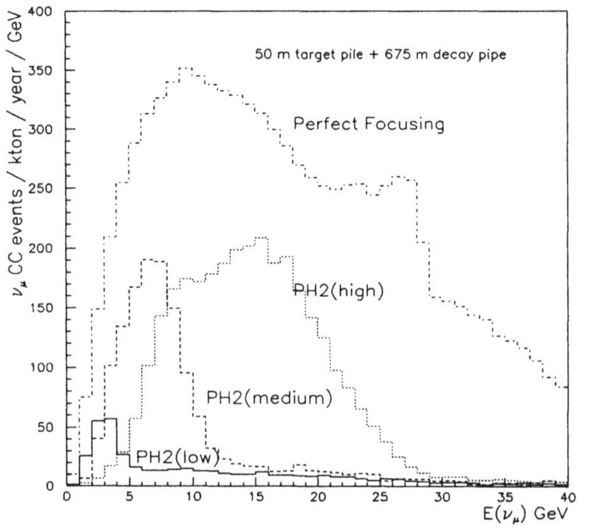

Figure 2. Neutrino interaction energy spectra predicted for different beam focusing conditions. "Perfect Focusing" assumes all secondary charged particles (with the proper sign) from the target are focused into a pencil beam with no divergence. "PH2(high)," "PH2(medium)" and "PH2(low)" are the high, medium and low energy configurations of the parabolic horn beam.

To reduce systematic errors as much as possible, MINOS will compare rates and energy spectra in two detectors: one located at Fermilab, the other one at Soudan. The Fermilab location will be about 275 m downstream of the beam stop in a newly excavated experimental hall. This location is far enough downstream so that all the muons produced in pion and kaon decays in the beam pipe will have been stopped in the intermediate earth and rock. The two detectors are made as identical as possible in their important characteristics so as to cancel out to a large extent spurious instrumental effects.

The detector in the Soudan mine will be placed in a new cavern, specially excavated for this experiment and oriented in such a way that its long dimension points toward Fermilab. The cavern will be adjacent and connected to the existing cavern housing the Soudan 2 detector, and the access to it will be provided by the currently existing shaft. The new configuration of the Soudan Laboratory is shown in Fig. 3.

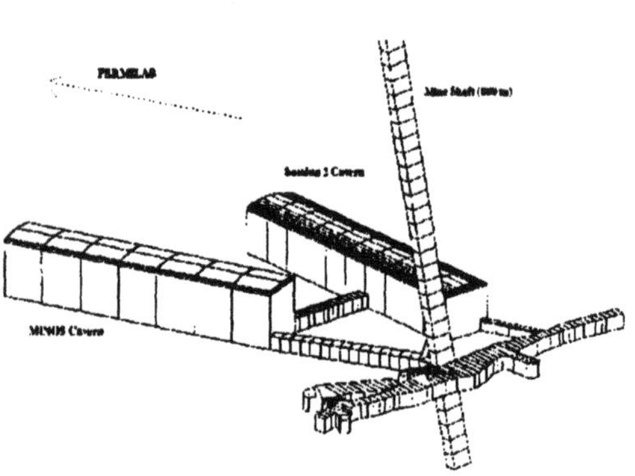

Figure 3. Sketch of the MINOS and Soudan 2 caverns in the Soudan Underground Mine State Park in northern Minnesota.

The MINOS detectors are iron/scintillator sampling calorimeters, significantly expanded in scale over the detectors used in previous neutrino experiments. The MINOS far detector has a total mass of 5.4 kt, equally divided between two supermodules. Each supermodule is composed of 243 octagonal steel plates, 8 m wide and 2.54 cm thick. A current carrying coil is provided for each supermodule to generate a toroidal magnetic field of about 1.5 T at a radius of 2 m.

Mounted on each steel plate are 192 scintillator strips, 4.1 cm wide and up to 8 m in length. The orientation of these strips differs by 90° in successive planes to provide two coordinate measurements. The scintillator is produced via extrusion process, during which a thin reflective layer is coextruded on the surface of each strip. The readout is performed via wavelength shifting fibers, glued in a groove on the surface of each strip. Some of the light produced by passage of a charged particle enters the wavelength shifting fibers and is absorbed with reemission of light in the green part of the spectrum. Some of this light in turn is trapped through internal reflection in the fiber and propagates to the two ends where it is detected by segmented position sensitive photodetectors. The photodetector is the 16 channel Hamamatsu M16 photomultiplier with 4 x 4 mm square pixels. Eight signal fibers are read by each pixel. The pixel-fiber correlation is permuted on one side of the detector to allow resolution of ambiguities by this method. Our latest results on the light output are illustrated in Fig. 4.

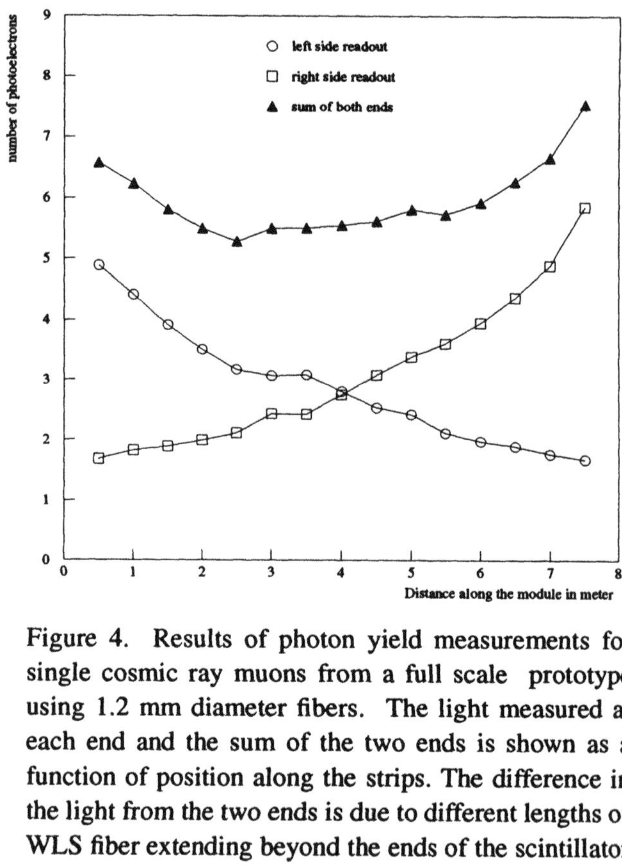

Figure 4. Results of photon yield measurements for single cosmic ray muons from a full scale prototype using 1.2 mm diameter fibers. The light measured at each end and the sum of the two ends is shown as a function of position along the strips. The difference in the light from the two ends is due to different lengths of WLS fiber extending beyond the ends of the scintillator strips.

The currently existing detector, Soudan 2, will also be used to record the neutrino data. Its mass is close to 1 kt but its fine granularity allows performance of some measurements which are difficult in a somewhat coarser main detector.

The near detector at Fermilab has a mass of about 1 kt, and to a good approximation is a scaled down version of the far detector. Only interactions in the very central part of the beam (r < 25 cm) are used for the near/far comparison, because that part of the beam has the most similar spectrum to the beam striking the far detector. This detector is offset about 1.5 m from the beam center line so that the flux of particles of interest misses the coil region.

The main measurements that MINOS will be able to perform are separation of NC and CC events, measurement of their absolute rates, and measurement of the neutrino energy spectra for the CC events. This last measurement allows determination of oscillation parameters independent of the final state. The ratio of the NC/CC rates allows identification of the oscillation mode. The detector is also capable of identifying ν_e interactions by observation of energy deposition characteristic of the electromagnetic shower. The sensitivity that can be reached in 10 kt years of exposure is shown in Fig. 5.

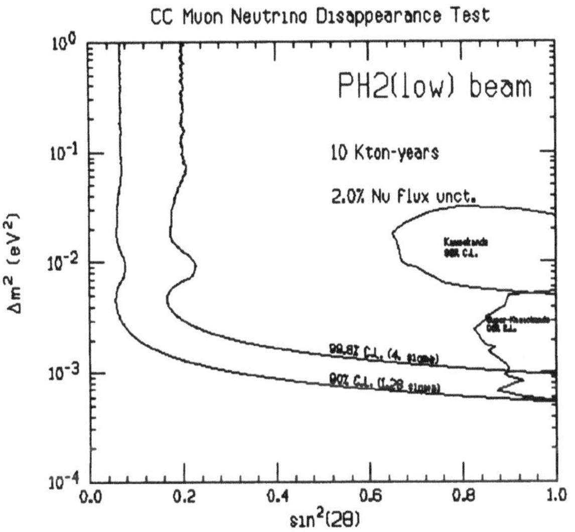

Figure 5. Sensitivity of the ν_μ disappearance test (90% CL limits and 4σ contour) for a 10 kt-yr exposure in the low energy PH2(low) neutrino beam.

Besides accelerator physics, MINOS will be able to also look at atmospheric neutrinos. Even though MINOS is considerably smaller than the SuperKamiokande, it has the capability of being able to measure muon charge and momentum, which allows it to make some complementary measurements. If Δm^2 is at the very low end of the currently allowed range, the atmospheric measurements in MINOS will be competitive in sensitivity to the accelerator measurements.

In designing the NuMI facility we have tried to maintain maximum flexibility so as to be able to react to new developments in physics and technology. The "zoom" beam design is one example of such a flexibility. The new cavern will be larger than is required for the two supermodules. The extra space will be used initially to facilitate the detector assembly but could be used subsequently to enlarge the detector by adding another supermodule should physics warrant it and financial situation allow it.

A very promising, and potentially very important, new development in neutrino experimental oscillation physics is the possibility of large hybrid emulsion detectors to see unequivocally production of taus.[5] Such observation would be the most convincing demonstration of oscillation into τ neutrinos. We are currently pursuing R&D in this general area to optimize such a detector and make it as cost effective as possible. We have allowed 10 m of free space in the upstream part of the cave as the potential future location of such a hybrid detector.

MINOS Collaboration at the present time consists of some 165 scientists and engineers from 21 institutions[6] from four countries: China, Russia, United Kingdom and Untied States. The initial funds for the NuMI/MINOS project have been already approved by the US Government. The current schedule calls for initiation of civil construction both at Fermilab and at Soudan early in 1999. The start of the data taking is planned for October, 2002.

REFERENCES

1. K.S. Hirata *et al.*, Phys. Lett. B **280**, 146 (1992); Y. Fukuda *et al.*, Phys. Lett. B **335**, 237 (1998).
2. R. Becker-Szendy *et al.*, Phys. Rev. D **46**, 3720 (1992).
3. W.M Allison *et al.*, Phys. Lett B **391**, 491 (1997).
4. Y. Fukuda *et al.*, Phys. Rev. Lett. **81**, 1562 (1998).

5. The general configuration of hybrid emulsion detector is described in the OPERA proposal submitted to CERN. Many of these ideas build on the pioneering work in automatic emulsion scanning by K. Niwa and his group at Nagoya University and further development by the CHORUS Collaboration.

6. The current institutions are: United States: Argonne National Laboratory, California Institute of Technology, Fermi National Accelerator Laboratory, Harvard University, Indiana University, Lawrence Livermore National Laboratory, University of Minnesota, University of Pittsburgh, Stanford University, Texas A&M University, University of Texas at Austin, Tufts University and Western Washington University. United Kingdom: University College London, University of Oxford, Rutherford Appleton Laboratory and University of Sussex. Russia: Joint Institute for Nuclear Research (Dubna), Institute for High Energy Physics (Protvino), Institute for Theoretical and Experimental Physics (Moscow) and Lebedev Physical Institute. China: Institute for High Energy Physics (Beijing).

ELSEVIER

Nuclear Physics B (Proc. Suppl.) 77 (1999) 187–197

NUCLEAR PHYSICS B
PROCEEDINGS
SUPPLEMENTS

Physics Projects for a Future CERN–LNGS Neutrino Programme

P. Picchi[a,b] and F. Pietropaolo[c]

[a]Dipartimento di Fisica, Universitá di Torino, via Giuria 1, I-10100 Torino, Italy

[b]Istituto di Cosmogeofisica del CNR, Corso Fiume 4, I-10100 Torino, Italy

[c]INFN, Sezione di Padova, via Marzolo 8, I-35131 Padova, Italy

We present an overview of the future projects concerning the neutrino oscillation physics in Europe. Recently a joint CERN–LNGS scientific committee has reviewed several proposals both for the study of atmospheric neutrinos and for long (LBL) and short baseline (SBL) neutrino oscillation experiments.

The committee has indicated the priority that the European high energy physics community should follows in the field of neutrino physics, namely a new massive, atmospheric neutrino detector and a ν_τ appearance campaign exploiting the new CERN–LNGS Neutrino Facility (NGS), freshly approved by CERN and INFN.

The sensitivity and the discovery potential of the whole experimental program in the Super–Kamiokande allowed region are discussed.

1. INTRODUCTION

The indication for the existence of neutrino oscillation has originally appeared in the atmospheric neutrino data of Kamiokande [1] & IMB [2] where the measurement of the ratio $R_{\mu/e}$ of μ–like and e–like events was lower than the Monte Carlo expectation.

The recent data of Super–Kamiokande (SK) [3] have strengthened the evidence for the existence of an anomaly in the flavour ratio of atmospheric neutrinos. Moreover the high statistics of SK show distortions of the angular distributions of the sub–GeV and the multi–GeV μ–like events that suggest the ν_μ oscillation hypothesis, while the angular distribution of the e–like events is consistent with the no–oscillation hypothesis.

This evidence is also supported by the SOUDAN2 [4] data and by the SK & MACRO [5] data on up-going muons.

The absence of an oscillation signal in the data of the CHOOZ [6] experiment essentially rules out the $\nu_e \leftrightarrow \nu_x$ oscillations in the interesting region of parameter space and favours the interpretation of the SK result in terms of $\nu_\mu \leftrightarrow \nu_\tau$ oscillation with Δm^2 in the range $10^{-2} - 10^{-3} eV^2$ and $sin^2(2\Theta)$ in the range $0.8 - 1.0$. More exotic interpretations, like $\nu_\mu \leftrightarrow \nu_{sterile}$, are at present not fully excluded.

A possible method to confirm these results is the development of long–baseline accelerator neutrino beams. The accelerator beams can have higher intensity and higher average energy than the atmospheric flux, and if $\nu_\mu \leftrightarrow \nu_\tau$ oscillations are indeed the cause of the atmospheric neutrino anomaly, they can produce a measurable rate of τ leptons for most of the values of the oscillation parameters that are a solution to the atmospheric data.

On the other hand measurements of atmospheric neutrinos with large statistics and/or better experimental resolutions, can also provide convincing evidence for oscillations, thanks to unambiguous detectable effects on the energy, zenith angle and L/E distributions of the events. The study of these effects can provide a precise determination of the oscillations parameters. The range of L/E available for atmospheric neutrinos ($10 - 10^4 Km/GeV$) is much larger than that of long–baseline accelerator experiments ($\simeq 100 Km/GeV$) and the sensitivity extends to lower values of Δm^2.

All these considerations call for a comprehensive physics programme, whose main goals are:
– the search for a direct neutrino oscillation signal in the full range indicated by the SK results;

PII S0920-5632(99)00417-X

– the precise test of the $\nu_\mu \leftrightarrow \nu_\tau$ oscillation hypothesis;
– the measurement of the relevant oscillation parameters: at least one squared mass difference, Δm^2, and one mixing angle, $sin^2(2\Theta)$.

The above arguments stimulated a joint CERN–INFN project for a beam towards the Gran Sasso National Laboratory (LNGS), 732 km away. At the same time several LBL and SBL experiments, based on very different techniques, as well as atmospheric neutrino experiments have been proposed and recently reviewed by a joint CERN–LNGS scientific committee.

In the following sections we will review the status of the future CERN–LNGS neutrino programme (section 2.) and of the new CERN Neutrino Beam to Gran Sasso (section 3.). In section 4. we will describe the proposed LBL experiments and discuss their sensitivity and significance in the SK allowed region of the oscillation parameter space. Finally in section 5. we will outline the characteristics and the sensitivity of a possible massive detector for atmospheric neutrino physics.

2. THE FUTURE CERN–LNGS NEUTRINO PROGRAMME

Here we faithfully report the out-come of the first meeting of the recently constituted joint CERN–LNGS scientific committee. The meeting was held at CERN on November 3-4, 1998 with the aim of reviewing the overall CERN–LNGS neutrino experimental programme and evaluating its potentiality also in view of the exsistence of other similar projects [7,8].

The committee believes that a combined experimental effort can accomplish the above programme. Elements of this programme are:

i) A large mass (larger then $20kt$) atmospheric neutrino experiment with high resolution in angle and neutrino energy, so that an explicit oscillation pattern can be put in evidence. Such a detector can be sensitive to oscillations for $\Delta m^2 = 2 \times 10^{-4} - 5 \times 10^{-3} eV^2$, covering all the relevant region also in view of the K2K experiment [7], and can measure both the mass difference and at least one of the mixing angles.

ii) A Long Base Line (LBL) beam from CERN to Gran Sasso as laid out in documents CERN 98–02 and CERN–SPSC 98–35. The feasibility of constructing a neutrino beam towards Gran Sasso has been demonstrated, being well-suited for experiments and with a built-in flexibility allowing the beam design to evolve with the field of neutrino oscillation physics.

iii) A ν_τ appearance LBL experiment, uniquely capable of precisely discriminating the $\nu_\mu \leftrightarrow \nu_\tau$ oscillation hypothesis in the range above $1 - 2 \times 10^{-3} eV^2$ with underground detectors. Ways of extending this mass range may exist, possibly in successive steps, due to extremely low experimental background and the possibility of using a detector on the surface. The search for ν_e appearance can nicely be coupled with ν_τ appearance experiments. However, due to the small number of signal events expected, a ν_τ appearance experiment may not be effective in actually determining the oscillation parameters.

iv) A ν_μ disappearance LBL experiment, with the need for a near station, again sensitive down to $1 \times 10^{-3} eV^2$ in Δm^2, provided the systematic effects can be kept under control to a sufficient level of accuracy.

The complementarity between iii) and i) or iv) is manifest. The same is not true for i) and iv), with i) having a larger reach potential at low Δm^2. The possible integration of two or more elements stated above into one combined detector deserves attention, to the extent that this can be shown to be compatible with the individual goals outlined. The committee also took note of the scientific interest expressed by:
– A short baseline experiment to search for $\nu_\mu \leftrightarrow \nu_\tau$ oscillation beyond the sensitivity reach of CHORUS and NOMAD (SPSC 98-29 & M616).
– A low energy neutrino beam derived from the PS to search for $\nu_\mu \leftrightarrow \nu_e$ oscillation in the range of parameters suggested by LSND (SPSC 98-27 & M614).
– A long-term experimental neutrino programme at CERN based on a future Neutrino Factory, offering high flux neutrino beams originating from a high intensity injector proton booster and/or muon storage ring of a $\mu^+\mu^-$ collider (SPSC 98-30 & M617, SPSC 98-31 & M618).

3. THE CERN NEUTRINO BEAM TO GRAN SASSO

A substantial part of the CERN–LNGS neutrino program will be based on a new CERN neutrino beam line (NGS) pointing to Gran Sasso, 732 Km away. The conceptual design of this facility has been studied in detail by a Technical Committee, mandated by CERN and INFN, and it feasibility has been fully demonstrated [9].

The NGS neutrino beam is produced from the decay of mesons, mostly π's and K's. The mesons are created by the interaction of a 400 GeV proton beam onto a graphite target, they are sign-selected and focused in the forward direction by two magnetic coaxial lenses, called horn and reflector and finally they are let to decay in an evacuated tunnel pointing toward Gran Sasso.

As clearly stated in the NGS report [9], the design concentrated on the civil engineering, freezing some parameters but keeping flexibility in the actual choice of the beam optics. Mainly the proton energy, the extraction from the SPS, the target room design, the geometry of the decay tunnel and the beam absorber were choosen. The main characteristics of the neutrino beam-line are listed in Table 1.

Table 1
Main parameter list of the NGS neutrino beam

Target material	graphite
Target rod length	10 cm
Target rod diameter	3 mm
Number of rods	11–13
Rod separation	1–9 cm
Horn & Reflector	parabolic
H&R length	6.65 m
H&R current	120 kA
Min horn distance from target	1.8 m
Max refl. distance from target	80 m
Decay tunnel length	992 m
Decay tunnel radius	1.22 m
Tunnel vertical slope	-50 mrad
Pressure in decay tunnel	1 Torr
Near detector pit	foreseen
Distance from target	1850 m
Proton energy	400 GeV
Expected pot/year:	
in shared SPS mode	3.95×10^{19}
in dedicated SPS mode	7.60×10^{19}

3.1. Optimization of the beam for $\nu_\mu \leftrightarrow \nu_\tau$ appearance search

As for the beam, the general strategy was to opt for a wide band neutrino beam based on the experience gathered at CERN with the design and the operation of the WANF. The beam optimization and the design of the details of the beam optics have been subject of further studies driven by the requests of the experiments. Following the indication of the CERN–LNGS committee, a first optimization of the beam has been carried out with the goal of maximizing the ν_τ CC interactions at LNGS for appearance experiments [10].

In the limit of small oscillations, where the flavour transition propability is approximated as $P(\nu_\mu \leftrightarrow \nu_x) \simeq sin^2(2\Theta) \times (1.27\Delta m^2 L/E)^2$, the ν_τ event rate a far location is given by the following formula:

$$N_\tau = K \int \phi_{\nu_\mu}(E) \times f(E) \times \epsilon(E) \times dE/E \quad (1)$$

where:
$K = N_a \times M_d \times \sigma_0 \times sin^2(2\Theta) \times (1.27\Delta m^2 L)^2$, E is the neutrino energy, ϕ_{ν_μ} is the ν_μ flux at the detector distance L, σ_0 E is the ν_μ CC interaction cross–section, f is the ratio between ν_τ and ν_μ CC interaction cross–sections, ϵ is the τ detection efficiency, N_a is the Avogadro number and M_d is the detector mass.

The integral in equation 1 is the quantity to be maximized. Note that it does not depend on the oscillation parameters. Note also that appearance experiments are only sensitive to the product $sin^2(2\Theta) \times (\Delta m^2)^2$.

A full Monte Carlo simulation of the beam has been used, based on the FLUKA97 [11] package, to evaluate neutrino fluxes and event rates. It turned out that the best optics configuration consists in a horn focusing 30 GeV mesons and a reflector tuned to 50 GeV. The predicted ν_μ event rate and the $\bar{\nu}_\mu$ and ν_e contaminations of the NGS beam at LNGS are listed in Table 2 for two modes

Table 2
Total ν_μ, $\bar{\nu}_\mu$ and ν_e CC events rate per $kt \cdot year$ at LNGS (N_μ,$N_{\bar{\mu}}$ and N_e) in the cases of shared and dedicated mode of operation of the SPS. The average energy ($< E_{\nu_\mu} >$) of the ν_μ interactions is also shown.

	shared	dedicated	$< E_{\nu_\mu} >$
N_μ	2280	4332	30.2
$N_{\bar{\mu}}$	51.3	97.5	
N_e	18.2	34.7	

Table 3
Rate (N_τ) of the ν_τ CC events per $kt \cdot year$ as a function of Δm^2 for full mixing at LNGS in the cases of shared and dedicated mode of the SPS. The average energy ($< E_{\nu_\tau} >$) of the ν_τ interactions is also shown.

$\Delta m^2 (eV^2)$	N_τ shared	N_τ dedicated	$< E_{\nu_\tau} >$
1.10^{-2}	161.	306.	19.8
8.10^{-3}	109.	206.	19.5
6.10^{-3}	64.	121.	19.1
4.10^{-3}	29.	56.	18.9
2.10^{-3}	7.5	14.2	18.7
1.10^{-3}	1.88	3.57	18.7
8.10^{-4}	1.20	2.28	18.7
6.10^{-4}	0.68	1.29	18.7

of operation of the SPS. In Table 3 we give the ν_τ event rate for values ofΔm^2 at full mixing within the range allowed by the SK atmospheric neutrino data.

Options of lower beam energy have also been considered for disappearance experiments [12]. Further studies on the optimisation of the beam are currently being done.

4. THE LBL EXPERIMENTS

In this section we give a brief description of the experiments proposed to study neutrino oscillations at LNGS with the NGS neutrino beam. The sensitivity and the discovery potentials of each experiment have been calculated for an exposure of four years and for the neutrino rates presented in the previous section.

4.1. ICARUS

ICARUS [13] is an approved experiment at LNGS, in preparation to search for proton decays in exclusive channels and to study atmospheric and solar neutrinos. Exposed at the NGS beam it will carry out $\nu_\mu \leftrightarrow \nu_\tau$ oscillation search in appearance mode.

4.1.1. The detector

The ICARUS detector is a liquid argon TPC, whose main characteristics are the following.
– It is a homogeneous tracking device, capable of dE/dx measurement. The high dE/dx resolution allows both good momentum measurement and particle identification for soft particles.
– Electromagnetic and hadronic showers are fully sampled. This allows to have a good energy resolution for both electromagnetic, $\sigma(E)/E \simeq 3\%/\sqrt{E/GeV}$, and hadronic contained showers, $\sigma(E)/E \simeq 15\%/\sqrt{E/GeV}$.
– It has good electron identification and e/π^0 discrimination thanks to the ability to distinguish single and double m.i.p. by ionization and to the bubble chamber quality space resolution.

A neutrino event detected with a small prototype (50 litres) of the ICARUS detector is shown in Figure 1 [15].

The detector has a modular structure, whose basic unit is a $0.6kt$ module. The installation of a first module at LNGS in the year 2000 has been approved. The second step of the ICARUS project should be the installation of 3 new modules (for a total mass of $2.4kt$) in 2003, when the NGS neutrino beam will be available.

Recently the ICARUS collaboration has put forward the possibility to build a Super–ICARUS [14] detector of $30kt$ to be placed just outside LNGS, with the aim of increasing the sensitivity to neutrino oscillations and cover completely the SK allowed region.

4.1.2. The $\nu_\mu \leftrightarrow \nu_\tau$ oscillation search

We report the results of the study made on the ICARUS $\nu_\mu \leftrightarrow \nu_\tau$ oscillation sensitivity assum-

Figure 1. An example of recorded neutrino interaction in a 50 liter Liquid Argon TPC prototype exposed at the CERN ν beam. The neutrino comes from the top of the picture. The horizontal axis is the time axis (drift direction) and vertically is the wire number. The visible area corresponds to $47 \times 32\ cm^2$

ing 4 modules ($2.4kt$). Because of the high resolution on measuring kinematical quantities, the ν_τ appearance search in ICARUS is based on the kinematical suppression of the background using similar techniques to those of the NOMAD experiment [16]. The basic idea consists in reconstructing, in the plane transverse to the incoming neutrino direction, the missing momentum due to the two undetected neutrinos produced in τ lepton decays. Since the missing transverse momentum is approximately Lorentz invariant, the τ detection efficiency should be constant as a function of the ν_τ energy. Nevertheless, a slight decrease with increasing energy is expected, since the cuts applied to isolate the candidate events depend on the background rate, which is an increasing function of the neutrino energy.

4.1.3. Detection efficiency and background

The ν_τ identification in ICARUS is under study for all the τ decay modes. Nevertheless very good results are already achievable with the *golden sample* of events namely the $\tau \rightarrow e$ channel whose detection efficiency has been evaluated to be about 50%.

In this channel, the main background sources are the ν_e contamination in the ν beam and the π^0's in neutral current events misidentified as electrons. The rejection power of the latter is close to 100%. A ν_e event is a background either if there are undetected neutral hadrons in the final state or because of the smearing due to nuclear effects in the target nucleon and to the detector resolution. It has been shown that a background rejection factor of about 100 is sufficient to expect less than one background event in four years [14].

4.2. OPERA

The OPERA experiment [17] is aimed to search for ν oscillation looking at the appearance of ν_τ in the NGS beam. Because of the target-detector distance, the high efficiency and the low background (less than 1 event), the experiment will be able to probe the Super-Kamiokande signal with a very high discovery potential.

4.2.1. The detector

The OPERA detector consists of a $0.75kt$ lead emulsion target. The basic element (*cell*) of the detector is composed of a 1 mm thick lead-plate followed by an emulsion sheet (ES1), a 3 mm drift space (filled with low density material) and another emulsion sheet (ES2) (see Figure reffig:ope). An ES1(ES2) is made of a pair of emulsion layers 50 micron thick, on either side of a 100(200) micron plastic base. Thirty cells are arranged together to form a *brick*, which has $15 \times 15 \times 13cm^3$ dimensions; bricks are put together to form a *module* ($2.8 \times 2.8 \times 0.15m^3$).

Since the emulsion does not have time resolution, there are electronic detectors after each module in order to correlate the neutrino interactions to the brick where they occur and to guide the scanning. Streamer tubes have been proposed as electronic detectors, but other pos-

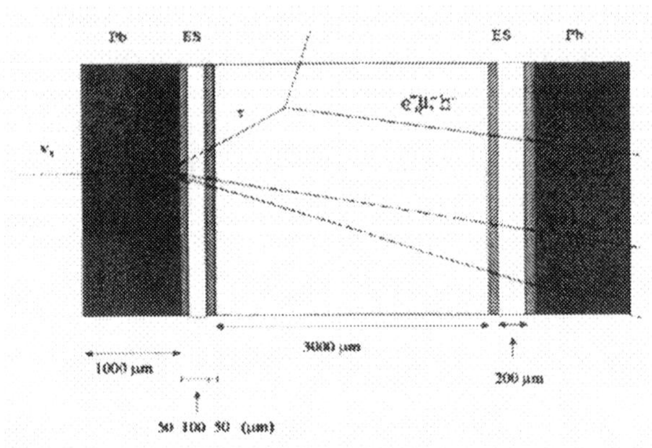

Figure 2. The basic elements of the OPERA detector

sible solutions are under study. A total of 300 modules are subdivided into 10 identical *super-modules*. The overall dimensions of the detector are $3.5 \times 3.5 \times 40m^3$.

4.2.2. The $\nu_\mu \leftrightarrow \nu_\tau$ oscillation search

The τ's produced in ν_τ CC interactions, are detected by measuring their decay kink when occurring in the drift space. The kink angle is measured by associating two high-precision 3-D track segments reconstructed in ES1 and ES2. The basic factor which, in the present design, determines the detection efficiency is the probability that the τ, before its decay, exits the lead plate (1 mm thick) where it is produced. So, the decay "kink" must occur in the drift space between consecutive emulsion layers. This drift space is filled with low density material, to eliminate the re-interaction background, otherwise relevant for the hadronic decay channels. The kink finding efficiency is related to a cut determined by the angular resolution of the emulsion trackers. Only kink angles larger then a given value ($20mrad$) are accepted. The present estimate of the OPERA τ detection efficiency is about 35%. We observe that the τ decays in the lead-target plates are not lost, but they do not offer the same golden background conditions. Studies are under way in order to use them to further increase the overall detection efficiency.

4.2.3. The background

The main source of background for the decays inside the gap is the production of charged charm particles with subsequent decay when the primary lepton is not detected. Monte Carlo simulation showed that the number of background events expected from this source is well below 1 in four years. Thus OPERA is essentially a background free experiment.

4.3. AQUA-RICH

AQUA-RICH [18] has been proposed as a long baseline experiment at LNGS. The detector, containing $125kt$ of water, uses the imaging Cerenkov technique to measure velocity, momentum and direction of almost all particles produced by neutrinos interacting in water.

Monte Carlo simulations show that hadrons are measured up to 9 GeV/c with $\Delta p/p < 7\%$ and muons up to 40 GeV/c with $\Delta p/p < 2\%$. Track direction is determined from the width of the ring image with error $\sigma(\theta) < 5mrad$, but track reconstruction (photon emission point) requires timing resolution $\sigma_t < 1ns$. The detector has to be sited outdoor, near the Gran Sasso Laboratory, and could be used also to observe atmospheric neutrinos.

4.3.1. The $\nu_\mu \leftrightarrow \nu_\tau$ oscillation search

Signal and background Monte Carlo events generated according to the NGS beam have been used to study the AQUA-RICH capability to search for $\nu_\mu \leftrightarrow \nu_\tau$ oscillations. The τ signal could be observed selecting QE events $\nu_\tau n \rightarrow \tau p$, followed by the τ muonic decay, with both the muon and the proton above threshold. A good separation between ν_τ signal and ν_μ background is possible as shown in [18] and will allow to have less than one background event in four years.

4.4. NOE

NOE [19] has been proposed as a long baseline experiment to study $\nu_\mu \leftrightarrow \nu_\tau$ and $\nu_\mu \leftrightarrow \nu_e$ oscillations.

4.4.1. The detector

The basic elements of the NOE detector are *light* transition radiation detector modules (TRD) for a total TRD mass of $2.4kt$ interleaved with modules of a massive fine grain $5.6kt$ calorimeter (CAL). A TRD and a CAL module together form the basic module of the NOE detector. The whole $8kt$ NOE detector is made of 12 subsequent basic modules.

The TRD module is built with 32 layer of marble (2 cm thick, 0.2 radiation length) interleaved with layers of polyethylene foam radiators. The marble is used as target for the ν_τ appearance search.

The CAL module is made of bars (with a cross–section of $4 \times 4cm^2$) where scintillating fibres are embedded into a distributed absorber (iron ore).

The electromagnetic and hadronic energy resolution are $\sigma(E)/E = 17\%/\sqrt{E/GeV} + 1\%$ and $\sigma(E)/E = 42\%/\sqrt{E/GeV} + 8\%$ respectively.

The muon direction and the hadronic shower axis are measured with a angular resolution $\sigma_\mu(\theta) = 0.022/\sqrt{E_\mu/GeV} + 0.040/(E_\mu/GeV)$ and $\sigma_h(\theta) = 0.175/\sqrt{E_h/GeV} + 0.351/(E_h/GeV)$ respectively.

Combining both CAL and TRD information, the rejection power to separate electrons from minimum ionising particles is $10^{-3} - 10^{-4}$. The e/π^0 discrimination is based on the fact that, because of the light TRD material, π^0's cross many TRD layers with low conversion probability.

4.4.2. The $\nu_\mu \leftrightarrow \nu_\tau$ oscillation search

The $\nu_\mu \leftrightarrow \nu_\tau$ oscillation search is performed exploiting the kinematical identification of the τ lepton decays exploiting the techniques developed by the NOMAD collaboration [16]. So far the $\tau \rightarrow e$ channel has been fully studied. The possibility to use the $\tau \rightarrow \pi$ channels is encouraging.

The τ detection efficiency in the $\tau \rightarrow e$ channel has been evaluated to be $\simeq 22\%$. As already discussed for the ICARUS experiment, a slight decrease of the efficiency with increasing neutrino energy is expected.

The corresponding background has been evaluated to be 4.6 events in four years mainly from the ν_e contamination, in the ν_μ beam. Details about the evaluation of all the background channels can be found in [19].

4.5. NICE

The NICE experiment [20] has been proposed to study the Super-Kamiokande signal using the disappearance technique in a long baseline experiment. In order to exploit the maximum potentiality of the disappearance technique, it plans to exploit a low energy version ($< E_\nu > \simeq 6-7GeV$) of the NGS neutrino beam; a close detector is also envisaged. A preliminary conceptual design of the detector is based on a large ($\simeq 10kt$) compact isotropic iron-scintillator electromagnetic/hadron calorimeter, surrounded on 4 sides by a magnetised iron spectrometer. The maximum sensitivity of the experiment on Δm^2, at full mixing, has been evaluated to be about $5 \times 10^{-4} eV^2$, provided that the systematical error is below 2%.

4.6. Sensitivity and significance of the LBL experiments

We recall that to evaluate the sensitivity and the discovery potential of the experiments searching for neutrino oscillation in appearance mode, a running time of 4 years has been considered, corresponding to 1.6×10^{20} pot operating the SPS in shared mode. The high energy NGS neutrino beam spectrum, optimized for ν_τ search, has been used.

With the these assumptions, the typical sensitivity that could be reached with an experiment at LNGS, in absence of ν_τ oscillation, is very similar for all the proposed experiments; the corresponding exclusion plot in the oscillation parameters space is shown in Figure 3.

On the other hand, when we are in presence of a claim of discovery, the relevant parameter to quote is the significance, $S = N_s/\sqrt{N_b}$ where N_s is the number of signal events and N_b is the expected background.

In Table 4 the minimum Δm^2 at full mixing satisfying the inequality $S > 4$, as well as the exclusion value at $90\% C.L.$, are shown for the proposed appearance experiments (ICARUS, Super–ICARUS, OPERA, AQUA-RICH and NOE). For most of the experiments the discovery potential extends below the SK best fit point ($\Delta m^2 = 2.2 \times 10^{-3}$ and $sin^2(2\Theta) = 1.$) in the SK allowed

Table 4
Sensitivity of the proposed ν_τ appearance experiments

Detector	Mass (kt)	Background	Signal ($\Delta m^2 = 0.005eV^2$)	Min $\Delta m^2 (eV^2)$ Exclusion $90\% C.L.$	at full mixing Discovery $S > 4$
ICARUS	2.4	2.5	21	1.1×10^{-3}	1.6×10^{-3}
Super–ICARUS	30	3.7	421	0.3×10^{-3}	0.8×10^{-3}
OPERA	0.75	0.45	37	1.2×10^{-3}	1.8×10^{-3}
AQUA-RICH	125	< 1	63	1.4×10^{-3}	2.3×10^{-3}
NOE	2.4	4.6	15	2.0×10^{-3}	3.9×10^{-3}

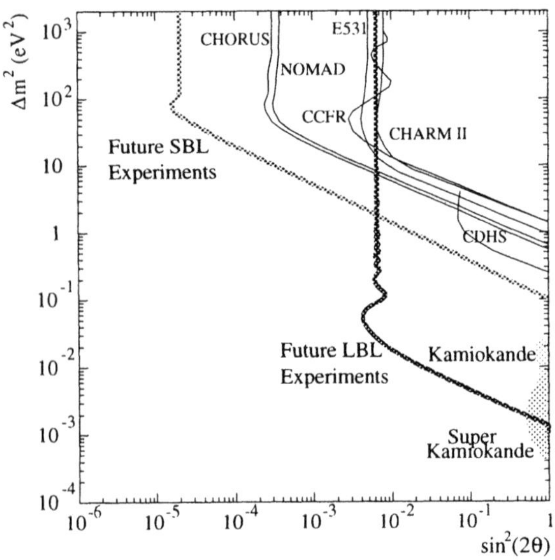

Figure 3. Oscillation parameter range that can be excluded at $90\% CL$ by the proposed LBL and SBL experiments, in the case if $\nu_\mu \leftrightarrow \nu_\tau$ appearance search.

region of the oscillation parameter space.

5. A HIGH DENSITY DETECTOR FOR ATMOSPHERIC NEUTRINOS

A new generation of massive atmospheric neutrino detectors would be particularly useful to measure precisely and separately the neutrino oscillation parameters Δm^2 and $sin^2(2\Theta)$ as explained in [21].

5.1. Experimental method

Atmospheric neutrino fluxes are not in general up/down symmetric. However, the up/down asymmetry, which is mainly due to geomagnetic effects, is reduced to the percent level for neutrino energies above 1.3 GeV. At these energies, for $\Delta m^2 < 10^{-2} eV^2$, downward muon neutrinos are not affected by oscillations. Thus, they may constitute a *near* reference source. Upward neutrinos are instead affected by oscillations, since the L/E ratio of their path length over the energy ranges up to $10^4 km/GeV$. Therefore with atmospheric neutrinos one may study oscillations with a single detector and two sources: a *near* and a *far* one.

The effects of oscillations are then searched comparing the L/E distribution for the upward neutrinos, which should be modulated by oscillations, with a reference distribution obtained from the downward neutrinos. For upward neutrinos the path length L is determined by their zenith angle as $L(\theta)$, while the reference distribution is obtained replacing the actual path length of downward neutrinos with the mirror-distance $L'(\theta) = L(\pi - \theta)$. The ratio $N_{up}(L/E)/N_{down}(L'/E)$ will then correspond to the survival probability given by

$$P(L/E) = 1 - sin^2(2\Theta)\sin^2(1.27\Delta m^2 L/E) \quad (2)$$

A smearing of the modulation is introduced by the finite L/E resolution of the detector.

We point out that results obtained by this

method are not sensitive to calculations of atmospheric fluxes.

We also remark that this method does not work with neutrinos at angles near to the horizontal, since the path lengths corresponding to a direction and its mirror-direction are of the same order.

If evidence of neutrino oscillation from the study of ν_μ disappearance is obtained, a method based on τ appearance can be used to discriminate between oscillations $\nu_\mu \leftrightarrow \nu_\tau$ and $\nu_\mu \leftrightarrow \nu_{sterile}$. Oscillations of ν_μ into ν_τ would in fact result in an excess of muon-less events produced by upward neutrinos with respect to muon-less downward. Due to threshold effects on τ production this excess would be important at high energy. Oscillations into a sterile neutrino would instead result in a depletion of upward muon-less events. Discrimination between $\nu_\mu \leftrightarrow \nu_\tau$ and $\nu_\mu \leftrightarrow \nu_{sterile}$ is thus obtained from a study of the asymmetry of upward to downward muon-less events. Because this method works with the high energy component of atmospheric neutrinos, it becomes effective for $\Delta m^2 > 3 \times 10^{-3} eV^2$.

5.2. Choice of the Detector

The outlined experimental method requires that the energy E and direction θ of the incoming neutrino be measured in each event. The latter, in the simplest experimental approach, can be estimated from the direction of the muon produced in the ν_μ charged-current interaction. The estimate of the neutrino energy E requires the measurement of the energy of the muon and of the hadrons produced in the interaction. In order to make the oscillation pattern detectable, the experimental requirement is that L/E be measured with an error smaller than half of the modulation period. This translates into requirements on the energy and angular resolutions of the detector. As a general feature the resolution on L/E improves at high energies, mostly because the muon direction gives an improved estimate of the neutrino direction. Thus the ability to measure high momentum muons (in the multi-GeV range), which is rather limited in the on-going atmospheric neutrino experiments, would be particularly rewarding.

A detector with a high efficiency on μ/π separation is required for an effective implementation of the method proposed, while, leaving aside oscillations involving electron neutrinos, no stringent requirement is put on electron identification and electromagnetic energy resolution.

5.3. A Possible Detector Structure

A large mass and high-density tracking calorimeter with horizontal sampling planes has been proposed as a suitable detector [21]. A mass of a few tens of kilotons is necessary to have enough neutrino interaction rate at high energies, while the high-density enables to operate the detector as a muon range-meter.

The detector consists in a stack of 120 horizontal iron planes 8 cm thick and 15×30 m^2 surface, interleaved by planes of sensitive elements (RPC's and/or limited streamer tubes). The sensitive elements, housed in a 2 cm gap between the iron planes, provide two coordinates with a pitch of 3 cm. The height of the detector it thus 12 metres. The total mass exceeds $34kt$. The number of read-out channels is 180,000.

5.4. Sensitivity to ν_μ oscillations

The proponents of [21] claim that with appropriate selections on μ–like events the experiment can reach the L/E resolution required to resolve the modulation periods typical of the oscillation phenomena for Δm^2 values in the range $2 \times 10^{-4} - 5 \times 10^{-3} eV^2$. As an examples, the L/E distribution obtained with the method described in section 5.1 for $\Delta m^2 = 10^{-3}$ and $sin^2(2\Theta) = 0.9$ is plotted in Figure 4. The discovery potential of the experiment, after three years of exposure, is also shown.

As indicated by the ICARUS [14], AQUA–RICH [18] and NICE [20] collaborations, similar results can be obtained with different detection techniques provided that the detector mass exceeds several tens of kt.

6. CONCLUSIONS

We believe that the neutrino oscillation search, based on the NGS facility complemented by atmospheric neutrino detection, constitutes an extremely appealing and realistic physics pro-

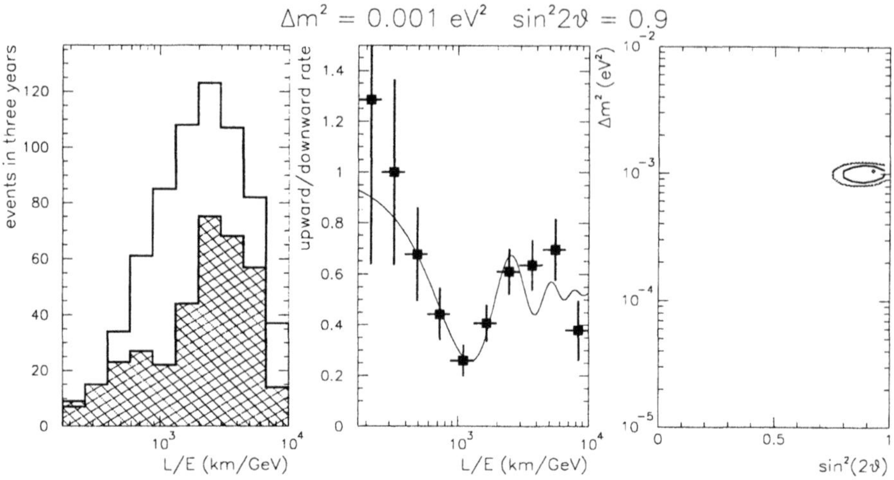

Figure 4. L/E analysis on a simulated atmospheric neutrino sample in the high density detector in presence of $\nu_\mu \leftrightarrow \nu_x$ oscillations and for an exposure of $100kt \cdot year$. From left to right: L/E spectra for upward muon events (hatched area) and downward ones (open area); their ratio with the best-fit su perimposed; the corresponding allowed regions in the oscillation parameter space at 90% and 99% C.L.

gramme for CERN and for LNGS, which will keep European neutrino physics at the frontier.

Our personal opinion, strengthened by the indications of the joint CERN–LNGS scientific committee, is that the NGS beam is extremely well suited to perform $\nu_\mu \leftrightarrow \nu_\tau$ and $\nu_\mu \leftrightarrow \nu_e$ appearance search while while ν_μ disappearance is better identified exploiting atmospheric neutrinos; to measure the oscillation parameters unambigously, a detector with very good L/E resolution is needed.

Even if the SK neutrino anomaly would turn out not to be due to neutrino oscillations, an unlikely but a priori not excluded possibility, this experimental programme would under all circumstances explore a significant region of the oscillation parameter space which is not accessible otherwise.

The joint CERN–LNGS scientific commettee has underlined the importance that the relevant decisions to establish this program, or part of it, be taken as soon as possible by the appropriate bodies in order not to undermine its effectiveness. For the same reason, it has been highly recommendable that suitable experimental proposals be presented in October 1999 along the lines given above and with appropriate strengths of the collaborations.

If promptly funded the CERN–LNGS neutrino program could start taking data by the year 2003.

ACKNOWLEDGEMENT

We gratefully acknowledge the organisers of the *XVIII International Conference on Neutrino Physics and Astrophysics* for giving us the opportunity to review the status and the perspective of the experimental neutrino oscillation programme in Europe.

REFERENCES

1. K.S. Hirata *et al*, *Phys. Lett.* B **205**, 416 (1988); *Phys. Lett.* B **280**, 146 (1992); Y. Fukuda *et al*, *Phys. Lett.* B **335**, 237 (1994).
2. R. Becker-Szendy *et al*, *Phys. Rev.* D **46**, 3720 (1992); *Phys. Rev. Lett.* **69**, 1010 (1992);
3. Y. Fukuda *et al*, *Phys. Rev. Lett.* **81**, 1562-1567 (1998).

4. H. Gallagher *et al*, *Proceedings of XIX International Conference on High Energy Physics*, Vancouver, July 23-29, 1998.
5. M. Ambrosio et al., INFN/AE–98/13, 1998.
6. M. Apollonio *et al*, *Phys. Lett.* B **420**, 397-404 (1998).
7. *Proposal for a Long Baseline Neutrino Oscillation Experiment Using KEK–PS and Super–Kamiokande*, KEK report E362, 1995. K2K collaboration, KEK-PREPRINT-97-266 & hep-ex/9803014, 1998.
8. J. Hylen *et al*, *Conceptual Design for the Technical Components of the Neutrino Beam for the Main Injector (NuMi)*, FERMILAB-TM-2018, 1997.
9. G. Acquistapace *et al*, *The CERN Neutrino Beam To Gran Sasso*, CERN 98-02 & INFN-AE-98-05, 1998.
10. A. Ereditato *et al*, ICARUS-TM-98/13 & OPERA 980722-01, 1998.
11. A. Fasso' *et al*, *Proceedings of the Third Workshop on Simulating Accelerator Radiation Environment (SARE-3)*, KEK report 97-5, 1997.
12. CERN/SPSC/98-33 & M620, 1998.
13. The ICARUS Collaboration,*Experiment Proposal*, LNGS - 94/99, 1994; ICARUS-CERN-MI Coll., CERN/SPSLC 96-58, SPSLC/P 304, 1996; J. P. Revol *et al*, ICARUS-TM-97/01, 1997.
14. The ICARUS collaboration, *ICARUS-Like Technology for Long Baseline Neutrino Oscillaitons*, CERN/SPSC/98-33 & M620, 1998.
15. F.Arneodo *et al*, *Proceedings of the Workshop on New Detectors*, Erice, 1997.
16. D. Autiero *et al*, *Proceedings of XIX International Conference on High Energy Physics*, Vancouver, 1998.
17. K. Kodama *et al*, *The OPERA ν_τ Appearance Experiment in the CERN-Gran Sasso Neutrino Beam*, CERN/SPSC/98-25 & M612, 1998.
18. P. Antonioli *et al*, *Aqua-Rich: an Atmospheric and Long Baseline Neutrino Experiment at Gran Sasso*, CERN/SPSC/98-37 & M624, 1998.
19. G. De Cataldo *et al*, *The NOE Detector for a Long Baseline Neutrino Oscillation Experiment*, CERN/SPSC/98-32 & M619, 1998.
20. M. Apollonio *et al*, *Sensitivity to Long Baseline Neutrino Oscillation of a Large Mass Calorimetric and Spectrometric Detector (NICE)*, CERN/SPSC/98-34 & M621, 1998.
21. M. Aglietta *et al*, *Measurement of Atmospheric Neutrino Oscillations with a High-Density Detector*, CERN/SPSC/98-28 & M615, 1998.

NUCLEAR PHYSICS B
PROCEEDINGS
SUPPLEMENTS

ELSEVIER

Nuclear Physics B (Proc. Suppl.) 77 (1999) 198–203

Status of K2K
(KEK to Kamioka Long Baseline Neutrino Oscillation Experiment)

K. Nishikawa[a]

[a]Institute for Particle and Nuclear Studies,
High Energy Accelerator Research organization (KEK),
1-1 Oho, Tsukuba, Japan

The purpose of K2K (KEK-PS E362) experiment [1] is to draw a definite conclusion on the neutrino oscillations with squared mass differences Δm^2 below $10^{-2}eV^2$. The experiment uses a well-defined muon neutrino (ν_μ) beam produced at the KEK-PS and two detectors, including the existing Super-Kamiokande detector. The experiment will be sensitive to the $\nu_\mu \rightarrow \nu_e$ and $\nu_\mu \rightarrow \nu_\tau$ oscillations, $\Delta m^2 \geq 3 \times 10^{-3}eV^2$ with more than the 99% confidence level for large mixing angle. The experimental methods, status, and schedule are described.

1. Goal of K2K experiment

Neutrino mass and lepton mixing are one of the key issue in investigating the physics beyond standard model. At present there is no compelling reason for neutrinos to be massless. On the contrary, most of the Grand Unified Theories(GUTs) predicts small but finite neutrino mass by introducing physics at much higher than electro-weak energy scale [2]. Although no definite theoretical guidance for the neutrino mass and their mixing angle is available, the results from Kamiokande and Super-Kamiokande[3] on neutrinos produced in the atmosphere provide a strong hint that the neutrino oscillation can be measured by accelerator neutrino experiments. At present, neutrino oscillations are reported at three Δm^2 regions, i.e., solar neutrino deficit, atmospheric neutrino anomaly, and LSND experiment [3][4]. However, if there are only three neutrinos exist in nature, those results are conflicting each other. Otherwise the results may indicate the existence of new neutrino(s). Also there is a possibility of rich mixing phenomena of three generation neutrinos. Thus the urgent task of a long base-line neutrino oscillation experiment are:
(1) to comfirm neutrino oscillation,
(2) to determine Δm^2 with better precision, and
(3) to distinguish the mode(s) of oscillation which have been observed in atmospheric neutrino observations.

Once the oscillation is observed in an accelerator neutrino oscillation experiment, Δm^2 can be determined better than atmospheric neutrino data. Since primary proton, secondary pion, and original neutrino can be measured, there is much fewer ambiguities in the measurements. Also, the distance is well defined and the contamination of ν_e is $\leq$ 1% so that ν_e appearance can be examined relatively easily.
The probability of oscillation (P) is given by the simple formula,

$$P = sin^2 2\theta \times sin^2(\frac{\Delta m^2(eV^2) \times L(km)}{E(GeV)}) \qquad (1)$$

,where L is the distance and E is the neutrino energy. The oscillation modes to be searched for are $\nu_\mu \rightarrow \nu_x$ where ν_x can be ν_τ or ν_{steril} and $\nu_\mu \rightarrow \nu_e$ appearance. The former will induces ν_μ spectrum distortion. Figure 1 shows the expected neutrino beam at Super-Kamiokande. As an example of oscillation effect, Figure 2 shows the oscillation effect for $\Delta m^2 = 5 \times 10^{-3}$ and $sin^2 2\theta = 1$.

2. Principle of Measurement

The experiment consists of six elements. The 12 GeV proton synchrotron (KEK-PS) can deliver 6×10^{12} protons every 2.2 second with a beam spill duration of 1μ second. The 12 GeV beam transport system bends extracted proton beam by about 90° to west and 1° down-ward

0920-5632/99/$ – see front matter
PII S0920-5632(99)00418-1

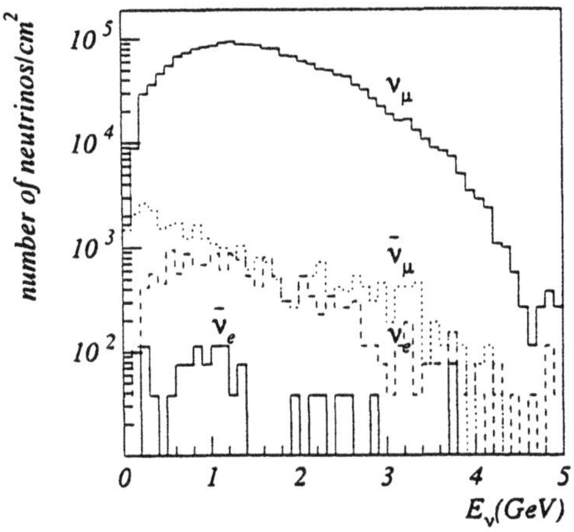

Figure 1. Expected neutrino beam at Super-Kamiokande. The contamination due to ν_e and anti-neutrino components is also shown. ν_e contamination is expected to be less than 1%

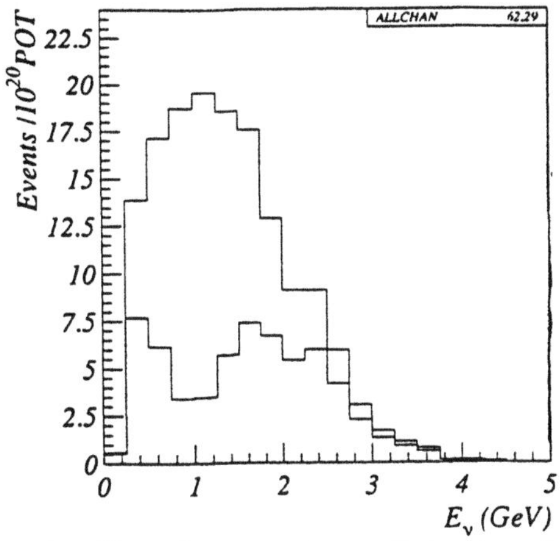

Figure 2. Neutrino spectrum distortion due to oscillation. The solid lines represent the expected neutrino spectrum without oscillation. The shaded histogram shows the spectrum distortion in the case $sin^2 2\theta = 1.0$ and $\Delta m^2 = 5 \times 10^{-3} eV^2$.

to the direction of Super-Kamiokande. Pions are produced by two interaction length aluminum target and double horn system focus pions with current 250kA. The momentum and angular distribution of pions will be periodically measured by pion monitor (ring imaging gas Cherenkov counter) before the 200m decay pipe. A muon monitor (segmented ionization chamber) measures the beam direction pulse by pulse. Finally neutrino detectors will be placed at 300m (near detector) and 250km (Super-Kamiokande) from the production target. The overall picture of the experiment is shown in Figure3.

3. Near Detector

Figure 4 shows near detector system, placed at 300m from production target. The detector consists of 1 kton water Cherenkov detector and fine grain detector system.

3.1. Fine Grain Detector

The aim is to measure neutrino flux and its spectrum. The fine grain detector consists of the water-scintillating fiber tracking detector(SFT), lead glass counters(LG), and muon range detector.

The fine grain detector was designed for the compactness to have a good geometrical acceptance and for fine granularity to have good identification of neutrino event type. Figure 5shows that the detection efficiency is more than 80% for neutrino above 0.7GeV. The SFT is a stack of the water containers and sheets of staggered scintillation fiber. The scintillation lights from the fibers are read out by image intensifier tubes and CCD chain. The charged particle tracks will be measured every 5.6 cm of water and 4 mm aluminum. To know the neutrino flux, the fiducial mass must be determined. The position resolution of the SFT was measured to be $280\mu m$ by cosmic-ray. The corresponding error of the fiducial volume determination is estimated to be about 1%. The energy of the incident neutrino can be calculated for quasi-elastic events by the formula;

$$E_\nu = \frac{(m_\mu E_\mu - m_\mu^2/2)}{(m_n - E_\mu + p_\mu cos\theta_\mu)} \quad (2)$$

The physics backgrounds($N_{BG}(E\nu)$) to the quasi-elastic events is due to $\nu_\mu + n \to \mu^- + \pi$'s. The Monte-Carlo prediction of the background is also shown in Figure 5.

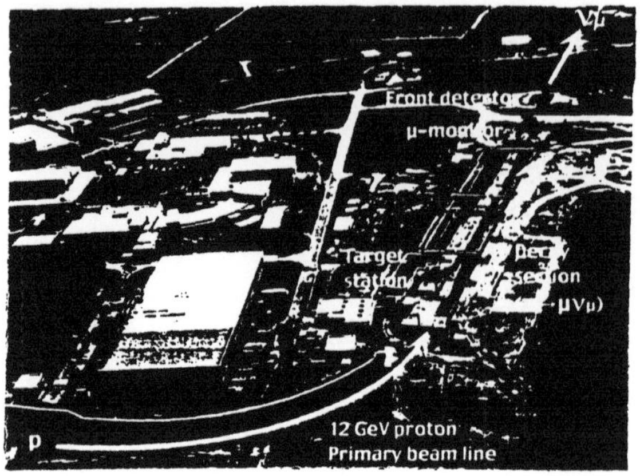

Figure 3. Overall picture of the experiment

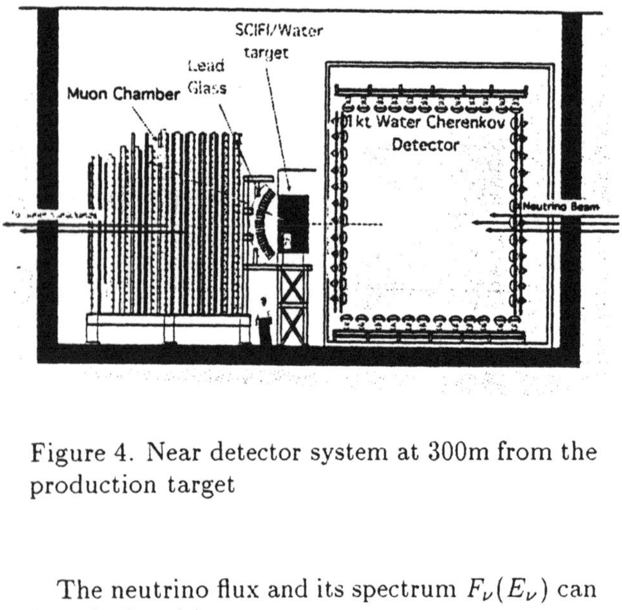

Figure 4. Near detector system at 300m from the production target

The neutrino flux and its spectrum $F_\nu(E_\nu)$ can be calculated by

$$F_\nu(E_\nu) = \frac{N(E_\nu) - N_{BG}(E_\nu)}{N_{target} \times \varepsilon \times sigma(QE(E_\nu)} \quad (3)$$

, where $N(E_\nu)$ and $N_{BG}(E\nu)$ are number of total and quasi-elastic events, respectively. $\sigma(QE(E_\nu))$ is quasi-elastic cross section, N_{target} is number of target neutrons in the fiducial volume, and ε is the detection efficiency. The errors in neutrino flux determination are estimated to be $\Delta N_{target} \sim 1\%, \Delta\varepsilon \ll 5\%$ and $\Delta\sigma(QE(E_\nu)) \sim 5\%$. Thus the accuracy of the determination of neutrino

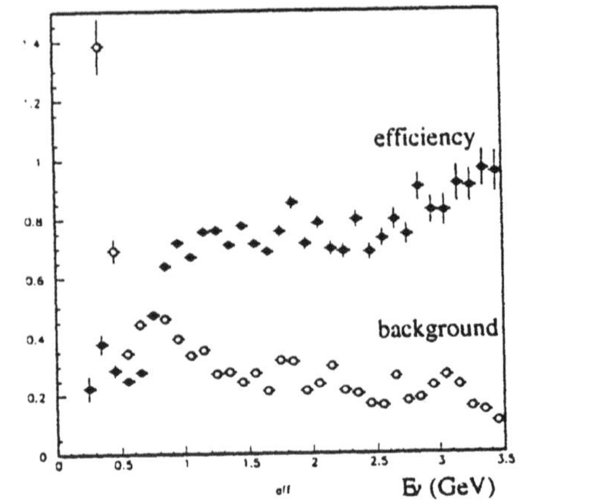

Figure 5. Detection efficiency of quasi-elastic events. The expected background is also shown in the figure.

beam is limited by the accuracy of subtraction of non-quasi-elastic events that is about 25%. Since there is about 30% ambiguity in non-quasi-elastic cross section, we estimate the error of the background subtraction to be 8% for each energy bin. This error can be improved by using data itself. The LG detector consists of 600 blocks. The contamination of ν_e in the beam will be measured with a resolution $\Delta E_e = 10\%/\sqrt{E_e(GeV)}$. Finally, muon detector will measure muon energy up to 3GeV/c. The muon detector is a stack of iron plate and drift chamber system. The muon energy can be measured with a resolution of $\Delta E_\mu/E_\mu = 8 - 10\%$. Table 1 summarizes the expected performance of fine grain detector components.

3.2. 1kton detector

The 1 kton detector is a ring imaging Cherenkov detector that works by the same principle as Super-Kamiokande. The detector volume is $10.8m\phi \times 10m$ filled with water which is viewed by 700 20-inch PMT's. The fiducial mass to be used in the oscillation analysis is 21 ton. Knowing the spectrum of neutrino at near detector site and using same technique to detect neutrino events in 1kton detector, one can predict neutrino event distribution at Super-Kamiokande for given neutrino beam. Especially, number of Cherenkov

Table 1
Expected performance of fine grain detector

	Total tonnage	Dimension HxWxD	Resolution
SFT	$4ton$	$2.4m \times 2.4m \times 1.2m$	$250\mu m$
LG	$30ton$	$3.6m \times 3.1m \times 20r.l.$	$8.2\%/\sqrt{E(GeV)}$
Muon Detector	$1200ton$	$8m \times 8m \times 2.7m$	$\Delta p_\mu/p_\mu = 10\%$

ring distribution and particle identification can be checked for given neutrino beam to predict neutrino events distribution at Super-Kamiokande.

3.3. Far Detector(Super-Kamiokande)

The Super-Kamiokande [5] is a ring imaging water Cherenkov detector located 1000m underground (2700m water equivalent) at 250 km [6] away from KEK. The total mass is 50 kton and more than 22.5 kton fiducial mass can be used for this experiment. The e/μ identification capability is important for both in the appearance and in the disappearance search. The mis-identification probability of a water Cherenkov detector was studied by using test beam of particles injected directly to a smaller detector of similar design and found to be less than 2% [7].

4. Beam monitor

The most important task of the beam monitor is to predict neutrino flux at far detector, based on the measurement at near detector. Once we know the momentum and angle of pions, neutrinos energy at the detector can be calculated by,

$$E_\nu = \frac{0.48 \times E_\pi}{1 + \theta_\pi \times \gamma_\pi} \quad (4)$$

,where $E_\pi, \theta_\pi, and \gamma_\pi$ are energy, angle, and gamma factor of pion, respectively. Since maximum proton momentum is 13 GeV/c, pions above 2 GeV/c can be separated by Cherenkov technique. A gas Cherenkov detector with phototube array read-out has been built [8]. A Monte-Calro study shows that the measurements of integrated Cherenkov light distribution at various index of reflection can be used to obtain momentum and angular distribution of pions. The relation of neutrino flux at 300m(near detector) and that of Super-Kamiokande can be calculated within 5% error as shown in Figure 6.

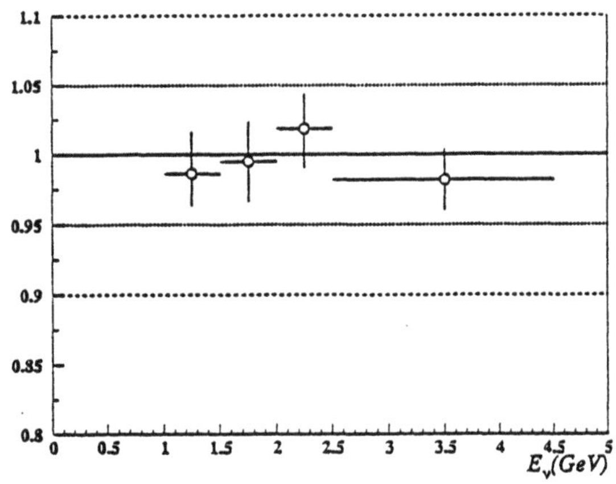

Figure 6. The calculation of ratio of neutrino spectrum at 250 km and 300 m, based on integrated Cherenkov light, is compared with neutrino beam full simulation as a function of energy. The ratio of near-far neutrino spectrum can be calculated within 5%.

This measurement has to be supplemented by a measurement of axial symmetry of the pion beam. The secondary particle profile will be measured by ionization chamber. Also the beam line direction can be measured by a muon profile monitor, placed after beam dump. The beam line and monitor system is shown in Figure 7

4.1. Extraction of Neutrino Oscillation Signal

The fine grain detector measures the neutrino flux and its spectrum to the level of 8%. The pion monitor can predict the ratio of neutrino flux at Super-Kamiokande site and near detector site with an error of 5%. From these two measurements, the neutrino beam at Super-Kamiokande can be predicted. The neutrino events at Super-Kamiokande must be compared with the pre-

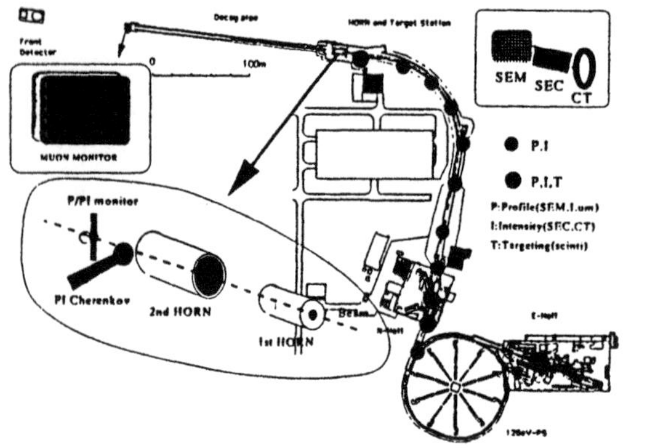

Figure 7. Neutrino beam line and monitor system for 12 GeV proton and secondary particles. The beam line consists of 300m beam transport for 12 GeV proton, target-horn system, 200m decay volume, beam dump, and muon monitor. Along the beam line, various profile and intensity monitors will be used.

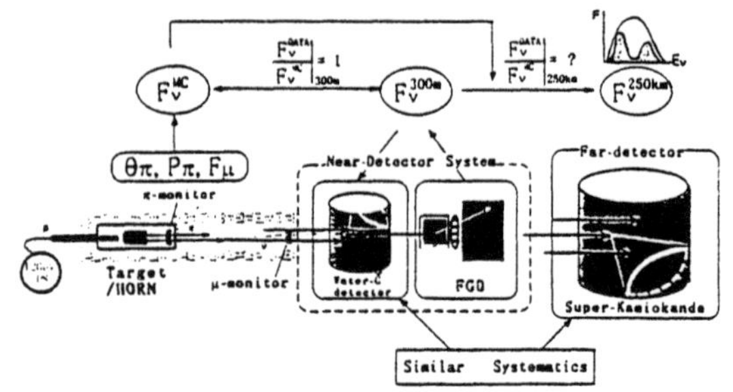

Figure 8. Principle of extraction of oscillation signal.

dicted event distributions to search for neutrino oscillation.
For the ν_μ disappearance search, events with muon type Cherenkov ring will be used. For ν_e appearance search, events with electron type Cherenkov ring will be searched. For given neutrino beam, the energy and angular distributions of the events can be predicted based on data from1 kton detector. This prediction should be compared directly with Super-Kamiokande data. The principle of the extraction of oscillation signal is summarized in Figure 8.

5. Sensitivity of Neutrino Oscillation

The sensitivity was examined, using single Cherenkov ring events for their simplicity of analysis. The expected number of events for 10^{20} protons on target(p.o.t), in the1 kton, in water target of fine grain detector(FGD), and in Super-Kamiokande(SK) are summarized in Table 2.

In Figure 9 and 10, the expected sensitivity contours of 99% C.L. for ν_e appearance and ν_μ disappearance are shown.

In making the contours, 8% bin-by bin errors in the measurement of neutrino energy distribution at near detector and 10% error in the extrapolation from 300 m to 250 km are assumed. As shown in the figure, K2K can explore the Δm^2 region down to about $3 \times 10^{-3} eV^2$ with 99% C.L..The sensitive region cover the most part of the allowed region for $\nu_\mu \to \nu_\tau$ mode which was obtained in atmospheric neutrino data in Super-Kamiokande.

6. Status and Schedule

As of this writing, all the components are ready to be installed or have been installed. We expect all the installation will be completed by the end of December 1998. The first beam is scheduled in January 1999. The first neutrino beam will be delivered in March 1999. The experiment will be running for four to six months in1999,2000,2001. and will be able to give definite answers for;
(1) Confirmation of neutrino oscillation in ν_μ disappearance for $\Delta m^2 \geq 3 \times 10^{-3} eV^2$
(2) Determination of Δm^2 with better precision
(3) Appearance measurement of $\nu_\mu \to \nu_e$ in $\Delta m^2 \geq 3 \times 10^{-3} eV^2$ region
(4) Better understanding of neutrino interactions at around $E_\nu = 1$GeV.

Table 2
Expected number of events in the fiducial volume of each detector for 10^{20}p.o.t.. The number of events in Super-Kamiokande is for the case of no oscillation.

Event type	1 kton	FDG	SK
	Fid. mass $\sim 21ton$	$\sim 4ton$	$\sim 22.5kton$
$\nu_\mu + n \to \mu^- + p$	$142k$	$44k$	120
$\nu_\mu + N \to \mu^- + N + \pi$	$130k$	$40k$	110
$\nu_\mu + N \to \mu^- + N + m\pi$	$135k$	$42k$	115
$\nu_\mu + N \to \mu^- + X$	$408k$	$127k$	350
$\nu_\mu + n \to \nu_\mu + X$	$144k$	$45k$	120
$\nu_e + n \to e^- + X$	$4k$	$12k$	4

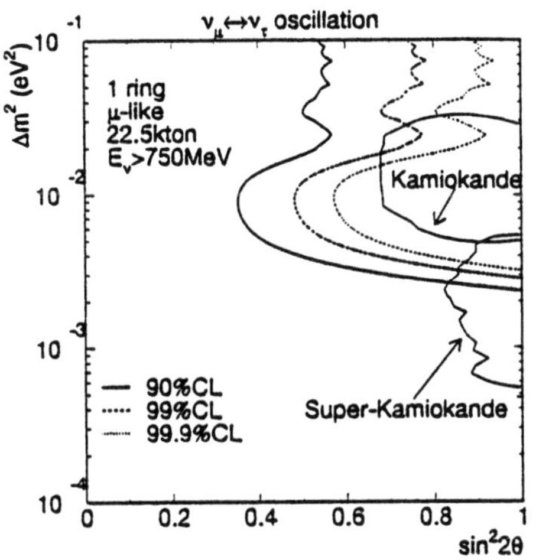

Figure 9. Expected sensitivity of K2K experiment for $\nu_\mu \to \nu_x$.

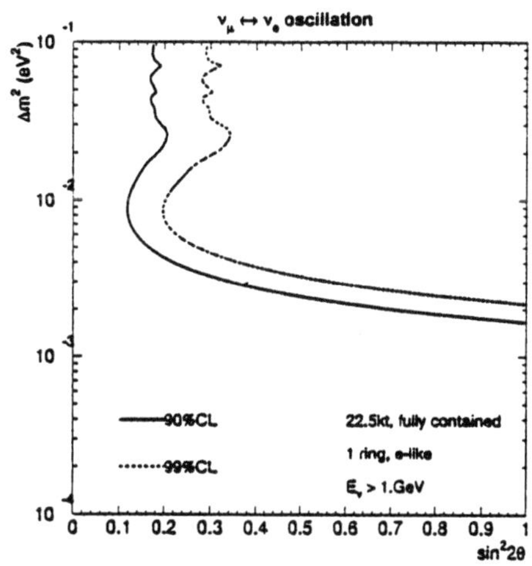

Figure 10. Expected sensitivity of K2K experiment for $\nu_\mu \to \nu_e$.

REFERENCES

1. K2K is a Japan-US-Korea collaboration. The collaborating institutions are ; KEK, ICRR,U. of Tokyo, Kobe U., Niigata U., OkayamaU., Tohoku U., Chonnam U., ,Dongshin U.,Korea U.,Seoul Nat. U.,Boston U.,Los Alamos National Lab.,State U. of New York, Stony Brook,U. of California, Irvine, U. of Hawaii, U. of Washington
2. See, for example M.Fukugita ans A.Suzuki eds.,"Physics and AstroPhysics of Neutrino"
3. T.Kajita, these proceedings
   K.S.Hirata et al. Phys.Lett. B205,416(1988)
   K.S.Hirata et al. Phys.Lett. B280,146(1992)
   E.W.Beir et al. Phys.Lett. B283,446(1992)
   Y.Fukuda et al. Phys.Lett. B335,237(1994)
   Y.Fukuda et al. Phys.Rev.Lett.81,1562(1998)
   Y.Fukuda et al. Phys. Rev. lett. 81,3319(1998)
   Y.Fukuda et al. Phys. Lett. B433,9(1998)
4. Y.Suzuki, these proceedings
   D.H.White, these proceedings
5. Y.Totsuka, in Proceedings of 18th International Symposium on Lepton Photon Interactions, Hamburg, July 1997
6. H.Noumi et.al. Nucl. Instrum. Meth. A398,399(1997)
7. S.Kasuga et al. Phys.Lett. B374,238(1996)
8. T.Inagaki "Studies of Secondary Particle Monitors in Neutrino Beam Line"(in Japanese), Master thesis, Univ. of Tokyo 1998

Part 5

Short Baseline Oscillation Experiments

NUCLEAR PHYSICS B
PROCEEDINGS
SUPPLEMENTS

ELSEVIER

Nuclear Physics B (Proc. Suppl.) 77 (1999) 207–211

Neutrino Oscillation Results from LSND

D. Hywel White, representing the LSND Collaboration [a]

[a]Mail Stop H846, Physics Division, Los Alamos National Laboratory, Los Alamos NM 87545, USA

The LSND experiment at Los Alamos has conducted a search for $\bar{\nu}_\mu \rightarrow \bar{\nu}_e$ oscillations using $\bar{\nu}_\mu$ from μ^+ decay at rest. The $\bar{\nu}_e$ are detected via the reaction $\bar{\nu}_e\, p \rightarrow e^+\, n$, correlated with the 2.2 MeV γ from $np \rightarrow d\gamma$. The use of tight cuts to identify e^+ events with correlated γ rays yielded 22 events with e^+ energy between 36 and 60 MeV and only 4.6 ± 0.6 background events. The probability that this excess is due entirely to a statistical fluctuation is 4.1×10^{-8}. A χ^2 fit to the entire e^+ sample results in a total excess of $51.8^{+18.7}_{-16.9} \pm 8.0$ events with e^+ energy between 20 and 60 MeV. If attributed to $\bar{\nu}_\mu \rightarrow \bar{\nu}_e$ oscillations, this corresponds to an oscillation probability (averaged over the experimental energy and spatial acceptance) of $0.31 \pm 0.12 \pm 0.05$.

1. INTRODUCTION

In the past several years, a number of experiments have searched for neutrino oscillations, where a neutrino of one type (say $\bar{\nu}_\mu$) spontaneously transforms into a neutrino of another type (say $\bar{\nu}_e$). For this phenomenon to occur, neutrinos must be massive and the apparent conservation law of lepton families must be violated. Data analysis proceeds under the assumption that the propbability for observing the different flavor is given by

$$P(\nu_\mu \rightarrow \nu_e) = sin^2 2\theta sin^2(1.27\Delta m^2 \ell/E_\nu) \qquad (1)$$

Δm^2 is the mass difference squared of the two neutrinos, ℓ is the propagation distance and E_ν is the neutrino energy. In 1995 the LSND experiment[1] published data showing candidate events that are consistent with $\bar{\nu}_\mu \rightarrow \bar{\nu}_e$ oscillations.[2] Additional data are reported here which provide stronger evidence for neutrino oscillations.[3]

2. DETECTOR

The Liquid Scintillator Neutrino Detector (LSND) experiment at LAMPF[4] was designed to search with high sensitivity for $\bar{\nu}_\mu \rightarrow \bar{\nu}_e$ oscillations from μ^+ decay at rest. LAMPF is a most intense source of low energy neutrinos due to its 1 mA proton intensity and 800 MeV energy. The neutrino source is well understood because almost all neutrinos arise from π^+ or μ^+ decay; π^- and μ^- are readily captured in the Fe of the shielding and Cu of the beam stop.[5] The production of kaons and heavier mesons is negligible at these energies. The $\bar{\nu}_e$ rate is calculated to be only 7×10^{-4} relative to $\bar{\nu}_\mu$ in the $20 < E_\nu < 52.8$ MeV energy range, so that the observation of a significant $\bar{\nu}_e$ rate would be evidence for $\bar{\nu}_\mu \rightarrow \bar{\nu}_e$ oscillations.

The LSND detector consists of an approximately cylindrical tank 8.3 m long by 5.7 m in diameter. The center of the detector is 30 m from the neutrino source. On the inside surface of the tank 1220 8-inch Hamamatsu phototubes provide 25% photocathode coverage. The tank is filled with 167 metric tons of liquid scintillator consisting of mineral oil and 0.031 g/l of b-PBD. This low scintillator concentration allows the detection of both Čerenkov light and scintillation light and yields a relatively long attenuation length of more than 20 m for wavelengths greater than 400 nm.[6] A typical 45 MeV electron created in the detector produces a total of $\sim$ 1500 photoelectrons, of which $\sim$ 280 photoelectrons are in the Čerenkov cone. The phototube time and pulse height signals are used to reconstruct the track with an average r.m.s. position resolution of $\sim$ 30 cm, an angular resolution of $\sim$ 12 degrees, and an energy resolution of $\sim$ 7%. A Čerenkov cone for relativistic particles and time distribution of the light, which is broader for non-relativistic particles, give excellent particle identification. Surrounding the detector is a veto shield[7] which tags cosmic ray muons going through the detector.

PII S0920-5632(99)00419-3

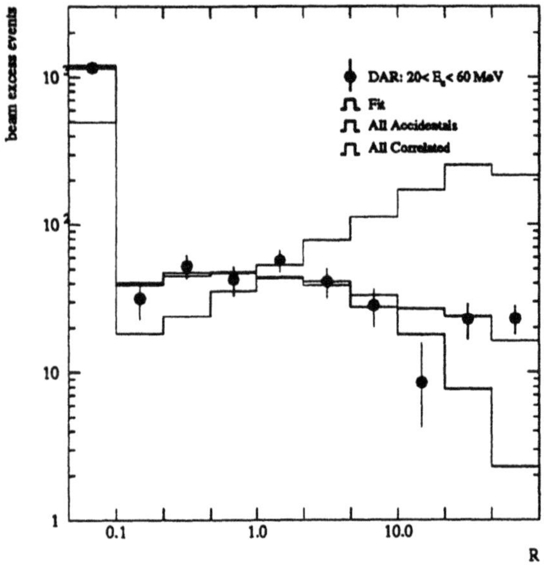

Figure 1. The R distribution, beam on minus beam off excess, for events that have energies in the range $20 < E_e < 60$ MeV. The green curve is the best fit to the data, while the blue curve is the component of the fit with an uncorrelated γ.

3. DATA

The signature for a $\bar{\nu}_e$ interaction in the detector is the reaction $\bar{\nu}_e p \rightarrow e^+ n$ followed by $np \rightarrow d\gamma$ (2.2 MeV). A likelihood ratio, R, is employed to determine whether a γ is a 2.2 MeV photon correlated with a positron or is from an accidental coincidence of a γ in the tank with an independent identified electron. R is the likelihood that the γ is correlated, divided by the likelihood that it is accidental. R depends on the number of hit phototubes for the γ, the reconstructed distance between the positron and the γ, and the relative time between the γ and positron. Figure 1 shows the R distribution, beam on minus beam off, for events with positrons in the $20 < E < 60$ MeV energy range.

The blue histogram is the result of the R fit for events without a recoil neutron, and the green histogram is the total fit, including events with a neutron. The shape of the R distribution for uncorrelated electron γ events was verified using events from the reaction $\nu_e^{12}C \rightarrow e^{-}\ ^{12}N_{gs}$. There are no neutrons in the final state and so no correlated γs. Events selected with a subsequent γ from this type of primary event are guaranteed to be accidental therefore. After subtracting the neutrino background shown in Table 1 with a recoil neutron there is a total excess of $51.8^{+18.7}_{-16.9} \pm 8.0$ events which when interpreted as due to neutrino oscillations corresponds to an oscillation probability of $(0.31 \pm 0.12 \pm 0.05)\%$.

The total number of events in different data taking conditions is shown in Table 2.

4. RESULTS

Figure 2 shows the electron energy distribution, beam on minus beam off excess, for events with an associated γ, $R > 30$.

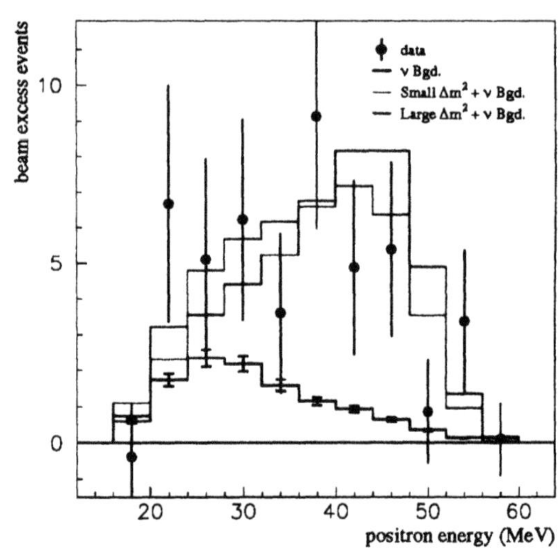

Figure 2. Electron energy distribution of events with R > 30 for data taken between 1993 and 1997

Table 1
Backgrounds: $20 < E_\nu < 60$MeV

		$R > 0$	$R > 30$
With Neutrons			
$\bar{\nu}_e p \to e^+ n$	μ^- DAR	8.6 ± 1.7	2.0 ± 0.4
$\bar{\nu}_\mu p \to \mu^+ n$	π^- DIF	3.8 ± 1.9	0.9 ± 0.4
Total	12.5 ± 2.9	2.9 ± 0.6	
Without Neutrons			
$\nu_\mu C \to \mu^- X$	π^- DIF	11.3 ± 5.6	0.1 ± 0.1
$\nu_e\ ^{12}C \to e^-\ ^{12}N$	μ^- DAR	666 ± 133	4.0 ± 0.8
$\nu_e\ ^{13}C \to e^-\ ^{13}N$	μ^- DAR	46 ± 9	0.3 ± 0.1
$\nu e \to \nu e$	μ^+ DAR	57 ± 6	0.3 ± 0.1
$\nu e \to \nu e$	π^+ DIF	8.4 ± 1.7	0.1 ± 0.1
$\nu_e C \to e^- X$	π^+ DIF	5.3 ± 1.0	
$\nu_\mu C \to \pi^0 X$	π^- DIF	0.3 ± 0.1	
$\nu_e C \to e^- X$	π^- DIF	0.9 ± 0.2	
Total	795 ± 133	4.8 ± 0.8	

Table 2
Event Totals 1993 - 1997

	Beam On	Beam Off	ν Background	Total Excess
$R > 3020 < E_\nu < 60$	61	15.6 ± 1.0	11.5 ± 1.5	33.9 ± 8.0
$R > 3036 < E_\nu < 60$	29	5.2 ± 0.6	3.0 ± 0.6	20.8 ± 5.4

For this latter requirement, the total 2.2 MeV γ detection efficiency is 23% and the probability that an event has an accidental γ in coincidence is 0.6%. The dashed histogram shows the background from expected neutrino interactions. There are 22 events beam on in the $36 < E < 60$ MeV energy range and a total estimated background (beam off plus neutrino-induced background) of 4.6 ± 0.6 events. The probability that this excess is a statistical fluctuation is $< 10^{-7}$. Table 1 gives the background estimate for events in the $36 < E_e < 60$ MeV energy range with $R \geq 0$ and $R > 30$. The R distribution is fit to extract the total number of $\bar{\nu}_e$ events and the result is shown in Table 3. The observed average value of $\cos\theta_b$, the angle between the neutrino direction and the reconstructed positron direction, is 0.20 ± 0.13, in agreement with the expected value of 0.16 for $\bar{\nu}_e p$ interactions. If the observed excess is due to neutrino oscillations, Fig. 3 shows the allowed region (90% and 99% likelihood regions) of $\sin^2 2\theta$ vs Δm^2 from a maximum likelihood fit to the L/E distribution of the 22 beam on events.

Some of the allowed region is excluded by the ongoing KARMEN experiment at ISIS,[8] the E776 experiment at BNL,[9] and the Bugey reactor experiment.[10] The favored region for decay in flight events is shown in Figure 4.

5. CONCLUSION

In summary, the LSND experiment observes an excess of events with positrons both in the $36 < E < 60$ MeV and $20 < E < 60$ MeV energy ranges that are correlated in time and space with a low energy γ. The observed excess is interpreted as $\bar{\nu}_\mu \to \bar{\nu}_e$ oscillations, corresponding to an oscillation probability of $(0.31 \pm 0.12 \pm 0.05\%)$ for the allowed regions shown in Fig. 3. More data taking is planned for the experiment, and the performance of the detector is under continuous study. If this excess is in fact due to neutrino oscillations, then the minimal standard model would need to be modified and neutrinos would have mass sufficient to influence cosmology and the evolution of the universe.

Table 3
Fitted Excess

	Fitted Excess	Total Excess	Oscillation Probability
1993 - 1995	63.5 ± 20.0	51.2 ± 20.2	$0.31 \pm 0.12 \pm 0.05$
1996 - 1997	35.1 ± 14.7	30.3 ± 14.8	$0.32 \pm 0.15 \pm 0.05$
1993 - 1997	100.1 ± 23.4	82.8 ± 23.7	$0.31 \pm 0.09 \pm 0.05$

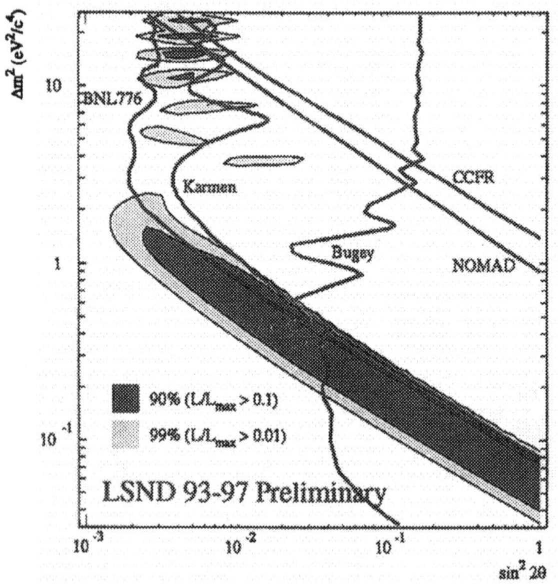

Figure 3. Plot of the LSND Δm^2 *vs* $\sin^2 2\theta$ favored regions. They correspond to 90% and 99% likelihood regions after the inclusion of the effects of systematic errors. Also shown are 90% C.L. limits from KARMEN at ISIS (dashed curve), E776 at BNL (dotted curve), and the Bugey reactor experiment (dot-dashed curve).

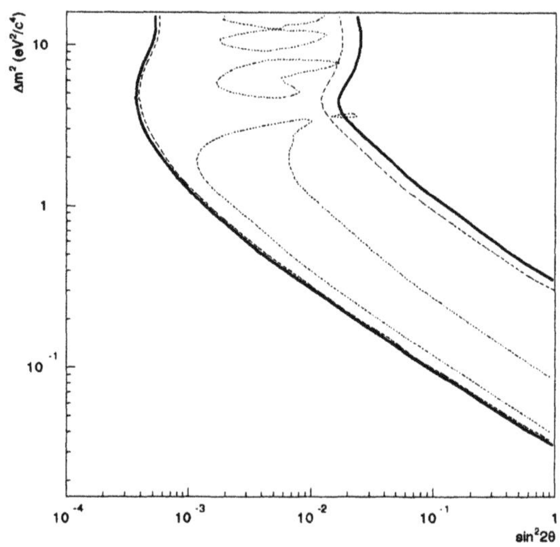

Figure 4. Plot of the LSND Δm^2 *vs* $\sin^2 2\theta$ favored regions. The solid line is for the decay in flight single electron events, the dotted line is the 90% confidence limits from the previous figure.

REFERENCES

1. The LSND Collaboration presently consists of the following people and institutions: E. Church, I. Stancu, G.J. VanDalen (Univ. of California, Riverside); W. Vernon (Univ. of California, San Diego); D.O. Caldwell, S. Yellin (Univ. of California, Santa Barbara); D. Smith, (Embry-Riddle Aeronautical Univ.); R.L. Burman, J.B. Donahue, G.T. Garvey, W.C. Louis, G.B. Mills, V. Sandberg, B. Sapp, R. Tayloe, D.H. White (Los Alamos National Laboratory); R. Imlay, H.J. Kim, A. Malik, W. Metcalf (Louisiana State Univ.): K. Johnston (Louisiana Tech Univ.); A. Fazely (Southern Univ); L.B. Auerbach, R. Majkic, (Temple Univ.).
2. C. Athanassopoulos *et al.*, *Phys. Rev. Lett.* **75**, 2650 (1995).
3. C. Athanassopoulos *et al.*, submitted to *Phys. Rev. C.*
4. C. Athanassopoulos *et al.*, submitted to *Nucl. Instrum. Methods.*.

5. R.L. Burman, M.E. Potter, and E.S. Smith, *Nucl. Instrum. Methods A* **291**, 621 (1990); R.L. Burman, A.C. Dodd, and P. Plischke, *Nucl. Instrum. Methods in Phys. Res. A* **368**, 416 (1996).
6. R.A. Reeder *et al.*, *Nucl. Instrum. Methods A* **334**, 353 (1993).
7. J.J. Napolitano *et al.*, *Nucl. Instrum. Methods A* **274**, 152 (1989).
8. B. Bodmann *et al.*, *Phys. Lett. B* **267**, 321 (1991), B. Bodmann *et al.*, *Phys. Lett. B* **280**, 198 (1992), B. Zeitnitz *et al.*, *Prog. Part. Nucl. Phys.* **32** 351 (1994).
9. L. Borodovsky *et al.*, *Phys. Rev. Lett.* **68**, 274 (1992).
10. B. Achkar *et al.*, *Nucl. Phys.* **B434**, 503 (1995).

NUCLEAR PHYSICS B
PROCEEDINGS
SUPPLEMENTS

ELSEVIER

Nuclear Physics B (Proc. Suppl.) 77 (1999) 212–219

The Search for Neutrino Oscillations $\bar{\nu}_\mu \rightarrow \bar{\nu}_e$ with KARMEN

K. Eitel[a] and B. Zeitnitz[a] for the KARMEN collaboration[1]

[a] Institut für Kernphysik I, Forschungszentrum Karlsruhe,
Postfach 3640, D-76021 Karlsruhe, Germany
e-mail: klaus@ik1.fzk.de, zeitnitz@ik1.fzk.de

The neutrino experiment KARMEN is situated at the beam stop neutrino source ISIS. It provides ν_μ's, ν_e's and $\bar{\nu}_\mu$'s in equal intensities from the π^+–μ^+–decay at rest (DAR). The oscillation channel $\bar{\nu}_\mu \rightarrow \bar{\nu}_e$ is investigated in the appearance mode with a 56 t liquid scintillation calorimeter at a mean distance of 17.7 m from the ν–source looking for p ($\bar{\nu}_e$, e^+) n reactions. The cosmic induced background for this oscillation search could be reduced by a factor of 40 due to an additional veto counter installed in 1996. In the data collected through 1997 and 1998 no potential oscillation event was observed. Using a unified approach to small signals this leads to an upper limit for the mixing angle of $\sin^2(2\Theta) < 1.3 \cdot 10^{-3}$ (90% CL) at large Δm^2 . The excluded area in ($\sin^2(2\Theta)$,Δm^2) covers almost entirely the favored region defined by the LSND $\bar{\nu}_\mu \rightarrow \bar{\nu}_e$ evidence.

1. INTRODUCTION

The search for neutrino oscillations and hence massive neutrinos is one of the most fascinating fields of modern particle physics. The **K**arlsruhe **R**utherford **M**edium **E**nergy **N**eutrino experiment KARMEN searches for neutrino oscillations in different appearance ($\nu_\mu \rightarrow \nu_e$ [2] and $\bar{\nu}_\mu \rightarrow \bar{\nu}_e$) and disappearance modes ($\nu_e \rightarrow \nu_x$ [3]). The physics program of KARMEN also includes the investigation of ν–nucleus interactions [4] as well as the search for lepton number violating decays of pions and muons and the test of the V–A structure of μ^+ decay [5].

Here, we present results of the oscillation search in the appearance channel $\bar{\nu}_\mu \rightarrow \bar{\nu}_e$ on the basis of data taken from February 1997 to April 1998 with the upgraded experimental configuration (KARMEN2). As will be shown in the following, no potential oscillation signal was observed. Therefore, special emphasis is given to the KARMEN2 capability of measuring ν induced events, the determination of the $\bar{\nu}_\mu \rightarrow \bar{\nu}_e$ evaluation cuts and the identification and measurement of the background expectation.

2. NEUTRINO PRODUCTION AND EXPERIMENT CONFIGURATION

The KARMEN experiment is performed at the neutron spallation facility ISIS of the Rutherford Appleton Laboratory, Chilton, UK. The neutrinos are produced by stopping 800 MeV protons in a beam stop target of Ta-D_2O. In addition to spallation neutrons, there is the production of charged pions. The π^- are absorbed by the target nuclei whereas the π^+ decay at rest. Muon neutrinos ν_μ therefore emerge from the decay $\pi^+ \rightarrow \mu^+ + \nu_\mu$. The produced μ^+ are also stopped within the massive target and decay via $\mu^+ \rightarrow e^+ + \nu_e + \bar{\nu}_\mu$. Because of this π^+-μ^+-decay chain at rest ISIS represents a ν-source with identical intensities for ν_μ, ν_e and $\bar{\nu}_\mu$ emitted isotropically ($\Phi_\nu = 6.37 \cdot 10^{13}\,\nu$/s per flavor for a proton beam current $I_p = 200\,\mu$A). There is a minor fraction of π^- decaying in flight (DIF) with the following μ^- DAR in the target station which again is suppressed by muon capture of the high Z material of the spallation target. This decay chain leads to a very small contamination of $\bar{\nu}_e/\nu_e < 6.2 \cdot 10^{-4}$ [6], which is even further reduced by the final evaluation cuts.

The energy spectra of the ν's are well defined due to the DAR of both the π^+ and μ^+ (Figure 1a). The ν_μ's from π^+–decay are monoen-

PII S0920-5632(99)00420-X

ergetic with $E(\nu_\mu)$=29.8 MeV, the continuous energy distributions of ν_e and $\bar{\nu}_\mu$ up to 52.8 MeV can be calculated using the V–A theory and show the typical Michel shape. Two parabolic pro-

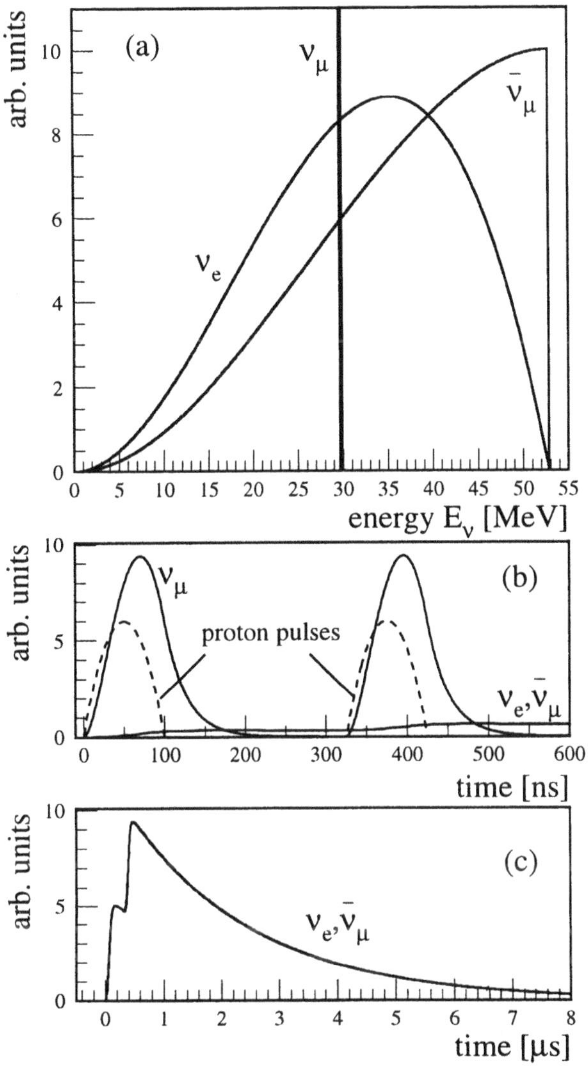

Figure 1. Neutrino energy spectra (a) and production times of ν_μ (b) and ν_e,$\bar{\nu}_\mu$ (c) at ISIS.

ton pulses of 100 ns base width and a gap of 225 ns are produced with a repetition frequency of 50 Hz. The different lifetimes of pions (τ = 26 ns) and muons (τ = 2.2 μs) allow a clear separation in time of the ν_μ-burst (Figure 1b) from the following ν_e's and $\bar{\nu}_\mu$'s (Figure 1c). Furthermore the accelerator's duty cycle of 10^{-5} allows effective suppression of any beam uncorrelated background.

The neutrinos are detected in a rectangular tank filled with 56 t of a liquid scintillator. This central scintillation calorimeter is segmented by double acrylic walls with an air gap allowing efficient light transport via total internal reflection of the scintillation light at the module walls. The event position is determined by the individual module and the time difference of the PM signals at each end of this module. Due to the optimized optical properties of the organic liquid scintillator and an active volume of 96% for the calorimeter, an energy resolution of $\sigma_E = \frac{11.5\%}{\sqrt{E[MeV]}}$ is achieved. In addition, Gd_2O_3 coated paper within the module walls provides efficient detection of thermal neutrons due to the very high capture cross section of the Gd(n,γ) reaction ($\sigma \approx$ 49000 barn). The KARMEN electronics is synchronized to the ISIS proton pulses to an accuracy of better than $\pm$2 ns, so that the time structure of the neutrinos can be exploited in full detail.

A massive blockhouse of 7000 t of steel in combination with a system of two layers of active veto counters provides shielding against beam correlated spallation neutron background, suppression of the hadronic component of cosmic radiation as well as reduction of the flux of cosmic muons. On the other hand, this shielding is a source of energetic muon induced neutrons produced by deep inelastic muon nucleon scattering and the nuclear capture of μ^-. These neutrons produced in the steel can penetrate the anti counter systems undetected and simulate a $\bar{\nu}_e$ detection sequence. The prompt signal is caused e.g. by a $n-p$ scattering followed by the delayed capture of the thermalized neutron.These neutrons were the major background source in the KARMEN1 experiment. In 1996 an additional third anti counter system with a total area of 300 m^2 was installed within the 3 m thick roof and the 2–3 m thick walls of the iron shielding [7]. By detecting the muons in the steel at a distance of 1 m from the main detector and vetoing the successive events this background has been reduced by a factor 40 compared to the KARMEN1 data.

3. SEARCH FOR $\bar{\nu}_\mu \rightarrow \bar{\nu}_e$ OSCILLATIONS

The probability for ν–oscillations $\bar{\nu}_\mu \rightarrow \bar{\nu}_e$ can be written in a simplified 2 flavor description as

$$P(\bar{\nu}_\mu \rightarrow \bar{\nu}_e) = \sin^2(2\Theta) \cdot sin^2(1.27\frac{\Delta m^2 L}{E_\nu}) \qquad (1)$$

where L and E_ν are given in meters and MeV, Δm^2 denotes the difference of the squared mass eigenvalues $\Delta m^2 = |m_1^2 - m_2^2|$ in eV^2/c^4. With $\langle L \rangle = 17.7$ m and $E_\nu < 52.8$ MeV, KARMEN is sensitive to small mixings $\sin^2(2\Theta)$ for oscillation parameters $\Delta m^2 \gtrsim 1\,eV^2/c^4$, essentially.

The signature for the detection of $\bar{\nu}_e$'s is a spatially correlated delayed coincidence of positrons from $p(\bar{\nu}_e, e^+)n$ with energies up to $E_{e^+} = E_{\bar{\nu}_e} - Q = 52.8 - 1.8 = 51.0$ MeV (Figure 2a) and γ emission of either of the two neutron capture processes $p(n,\gamma)d$ with one γ of $E(\gamma) = 2.2$ MeV or $Gd(n,\gamma)Gd$ with 3 γ–quanta in average and a sum energy of $\sum E(\gamma) = 8$ MeV (Figure 2b). The positrons are expected in a time window of several μs after beam–on–target (Figure 2c) with a 2.2 μs exponential decrease due to the μ^+ decay. The time difference between the e^+ and the capture γ is given by the thermalization, diffusion and capture of neutrons.

The raw data investigated for this oscillation search was recorded in the measuring period of February 1997 to February 1998 which corresponds to 2897 C protons on target or $8.32 \cdot 10^{20}$ $\bar{\nu}_\mu$ produced in the ISIS target. A positron candidate is accepted only if there is no activity in the central detector, the inner anti or outer shield up to 24 μs before. If only the outermost veto counter was hit, a dead time of 14 μs is applied. These conditions reduce significantly background induced by cosmic muons: penetrating μ, decay products of stopped muons and neutrons from deep inelastic muon scattering. Further cuts select sequences of events correlated in space and time. To extract a possible $\bar{\nu}_e$ induced signal these cuts were obtained from an optimization procedure to get highest sensitivity for a possible small (e^+,n) signal from $\bar{\nu}_\mu \rightarrow \bar{\nu}_e$ oscillations. This procedure described in more detail in section 5 results in the following cuts on the observed values of the prompt event: 0.6 $\mu s \leq t_p \leq 8.6$ μs, 20.0 MeV $\leq E_p \leq$ 50.0 MeV. The cuts on the delayed event are applied as follows: 5.0 $\mu s \leq t_s - t_p \leq 270$ μs, $E_s \leq 7.5$ MeV and a spatial coincidence volume of 0.97 m^3.

In the investigated data, no sequential structure fulfilled all the required properties for a (e^+,n) sequence. After all cuts, the remaining background amounts to only 2.88 ± 0.13 events caused by sequential cosmic background and ν induced sequences. These background sources are described in detail in the following section. The probability of measuring zero events with an expected number of 2.88 ± 0.13 background events is 5.6%. Applying a unified approach [8], we deduce an upper limit of $N < 1.07$ (90% CL) for a potential $\bar{\nu}_\mu \rightarrow \bar{\nu}_e$ oscillation signal. With an expectation of $N = 811 \pm 89$ for $\sin^2(2\Theta) = 1$ and large Δm^2 this corresponds to a limit of

$$\sin^2(2\Theta) < 1.3 \cdot 10^{-3} \qquad (90\%\,\mathrm{CL}) \qquad (2)$$

for $\Delta m^2 \geq 100\,eV^2/c^4$. Fig 3 shows the KARMEN2 exclusion curve in comparison with other experiments.

4. BACKGROUND SOURCES

There are in total four background sources contributing to the expected number of 2.88 ± 0.13 background events: cosmic muon induced background, ν_e induced background dominated by the charged current reaction $^{12}C(\nu_e, e^-)^{12}N_{g.s.}$ with subsequent *early* decays of the produced ^{12}N, random coincidences of *single prong* neutrino reactions (e.g. neutral current events) with low energy background events and the intrinsic contamination of the neutrino source ISIS with $\bar{\nu}_e$ from π^- and μ^- decays. All background sources except the intrinsic contamination can be measured online and with high precision parallel to the search for neutrino oscillations.

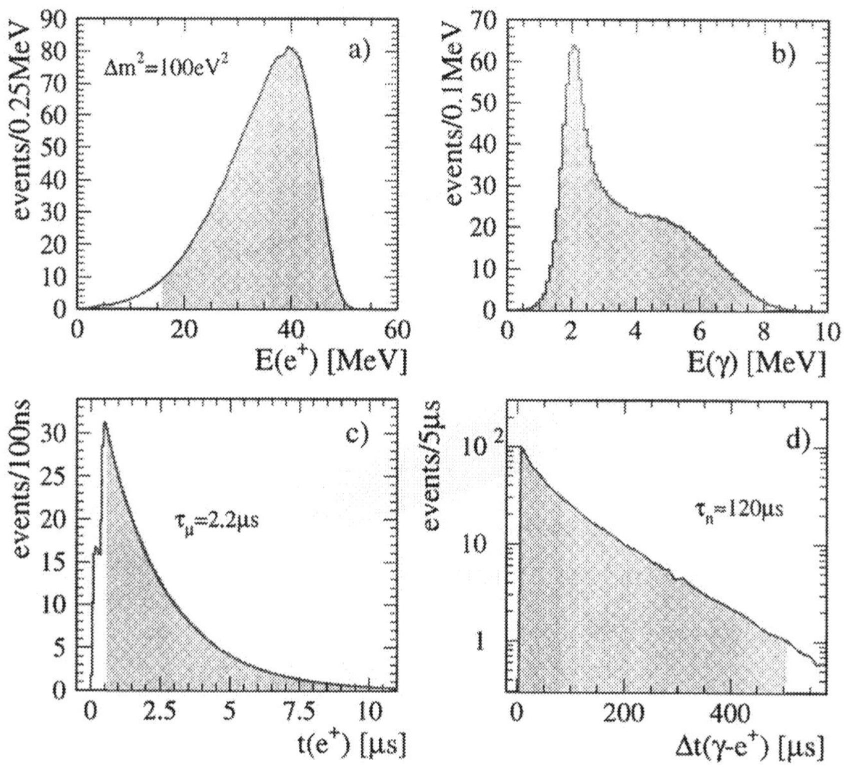

Figure 2. Expected signature for $\bar{\nu}_\mu \rightarrow \bar{\nu}_e$ full oscillation: a) MC energy of prompt positron for $\Delta m^2 = 100\,\mathrm{eV}^2/\mathrm{c}^4$; b) energy of delayed γ's; c) time of prompt event relative to ISIS beam–on–target; d) time difference between prompt e^+ and delayed γ's

4.1. Cosmic muon induced background

The KARMEN experiment is surrounded by a massive 7000 t iron shielding. Energetic muon induced neutrons produced in the steel can penetrate the anti counter systems of the detector undetected. By detecting the muons in the steel with the new veto counter at a distance of 1 m from the main detector and vetoing the successive events the background can be strongly suppressed. This background is well described by a detailed three–dimensional Monte Carlo simulation and therefore well understood. Moreover this background can be precisely measured online due to the ISIS duty factor of 5×10^{-4} for ν's from μ^+ decay. The extrapolation of the number of events recorded in a large time window outside the beam pulses to the actual measuring time interval of $0.6 \leq t_p \leq 8.6\ \mu$s is straight forward resulting in a precise cosmic background expectation of 0.64 ± 0.06 events.

4.2. ν_e induced background

The $\bar{\nu}_\mu$ for the search for $\bar{\nu}_\mu \rightarrow \bar{\nu}_e$ are produced by the DAR $\mu^+ \rightarrow e^+ + \nu_e + \bar{\nu}_\mu$. Therefore there is an equal amount of ν_e produced in the very same decay reaction. The ν_e can be detected via a sequence of a prompt electron from the inverse beta decay $^{12}\mathrm{C}\,(\nu_e\,, e^-)\,^{12}\mathrm{N}_{\mathrm{g.s.}}$ and the subsequent detection of a delayed positron from the decay $^{12}\mathrm{N}_{\mathrm{g.s.}} \rightarrow {}^{12}\mathrm{C} + e^+ + \nu_e$. The lifetime of ^{12}N is 15.9 ms and the β–decay endpoint is 16.3 MeV. About 500 sequences of this type have been clearly identified with a signal to noise ratio of 35 during the KARMEN1 data taking allowing a rich analysis of the nuclear physics

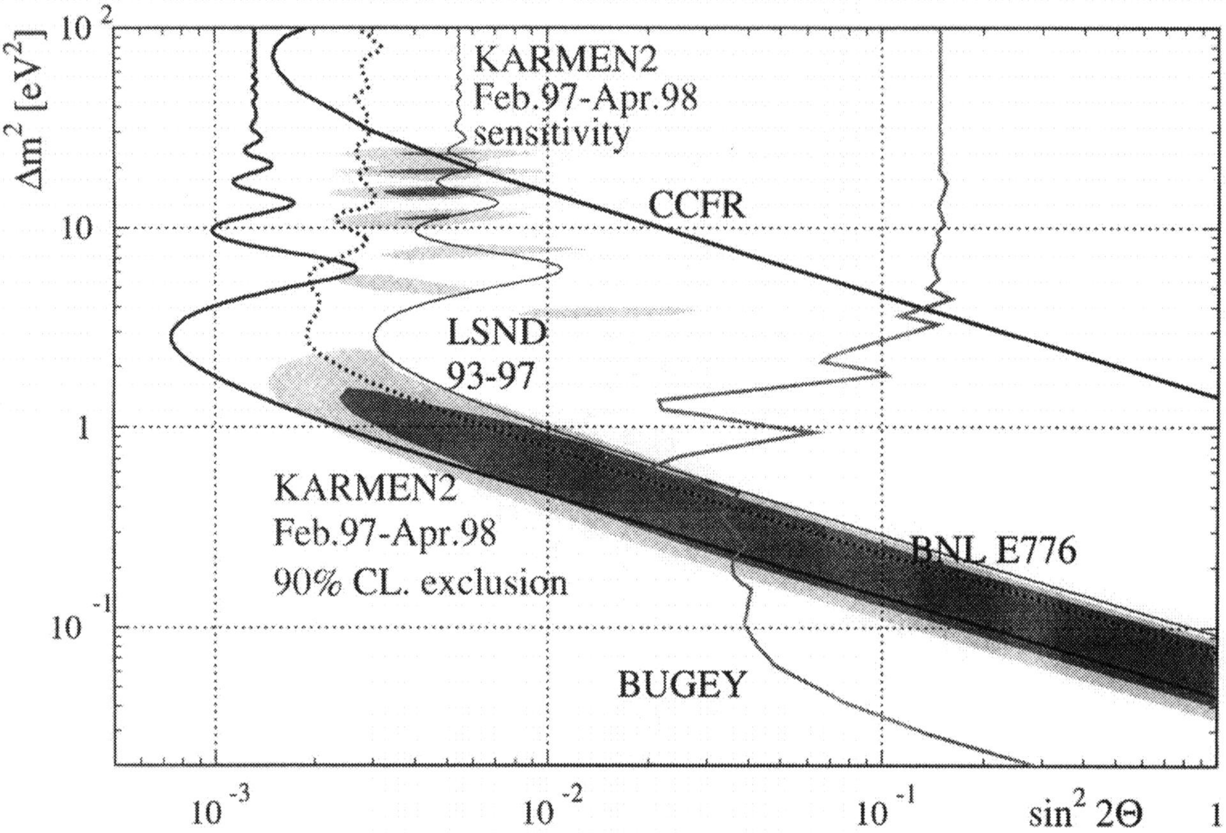

Figure 3. KARMEN2 90% CL exclusion limit and sensitivity compared to other experiments: BNL [9], CCFR [10], BUGEY [11] and the evidence for $\bar{\nu}_\mu \rightarrow \bar{\nu}_e$ oscillations reported by LSND [12].

involved [4].

The 1.7% fraction of ^{12}N decaying within the first 270 μs contributes to the expected background level. Once again this background can be recorded online and with high statistics. The extrapolation of the measured number of charged current sequences to the smaller time differences and the lower energies of delayed γ's of the oscillation search is straight forward. Looking for events with larger time differences ($t_s - t_p > 500\,\mu$s) and energies appropriate for (e^+, e^-) sequences from $^{12}C(\nu_e, e^-)^{12}N_{g.s.}$ and $^{12}N_{g.s.} \rightarrow {}^{12}C + e^+ + \nu_e$ one obtains a number of charged current events compatible with the KARMEN1 data.

4.3. Neutrino random coincidences

Due to the relative high rate of low energy radioactive background events in the KARMEN detector there is a small probability that such a low background event occurs randomly correlated in space and time to a prompt neutrino event with an energy above 20 MeV. The rate of such random coincidences is strongly suppressed by the tight spatial and time coincidence cuts. The rate of the radioactive background events is constant in time and thus the energy and position distributions of these events can be recorded with high precision and extrapolated to the actual measuring time window. The probability of finding such a low energy background event in the vicinity of a prompt neutrino event can be obtained by generating pseudo neutrino events with a Monte Carlo method and looking for correlated delayed events. The absolute number of random coincidences is obtained by multiplying this probability with the measured number of single prong neutrino reactions. Hence the amount of this background can also be monitored online and is determined from the very same dataset scanned for neutrino oscillations.

4.4. Intrinsic contamination

The only background source which can not be directly extracted from the data is the contamination of the neutrino source with $\bar{\nu}_e$ produced in the π^-–μ^- decay chain. Detailed Monte Carlo simulations [6] including a three–dimensional model of the ISIS target are used to obtain the fraction of π^- and μ^- decaying before they are captured by the nuclei of the target materials. The lifetimes of the μ^- depend on the target materials and are generally shorter than the μ^+ decay time. This effect as well as the shape of the $\bar{\nu}_e$ energy spectrum has been included into the calculation of the expected number of e^+ events generated by the contamination.

In Table 1 the individual contributions of the above described background sources are summarized. Note that the absolute number of expected

Table 1
Expected sequences from different background components within the evaluation cuts specified in the text. The last row shows the expectation of (e^+,n) sequences assuming maximal mixing, i.e. $\sin^2(2\Theta) = 1$.

background contribution	events
cosmic induced sequences	0.64±0.06
ISIS $\bar{\nu}_e$ contamination	0.56±0.09
ν induced random coincidences	0.72±0.04
(e^-,e^+) from $^{12}C(\nu_e, e^-)^{12}N_{g.s.}$	0.96±0.05
total background	2.88±0.13
$\bar{\nu}_e$ signal for $\sin^2(2\Theta) = 1$	811±89

background events is very precise. With these reliable background contributions including the detailed knowledge of the spectral distributions it is possible to optimize the cuts applied to the data without using any information about the actually measured result.

5. OPTIMIZATION OF EVALUATION CUTS

This section shows how we obtained optimal cuts for the search for $\bar{\nu}_\mu \rightarrow \bar{\nu}_e$ oscillations independently from the measured result. Second, we show that if we ignore the information of the new veto counter, which corresponds to the KARMEN1 experimental situation, and loose the optimized evaluation cuts we find background events in the neutrino window in good agreement with the measured background expectation.

As the true values of the oscillation parameters $\sin^2(2\Theta)$ and Δm^2 are not known we chose to maximize the *sensitivity* of the experiment, i.e. we optimized the experiment to deliver the most stringent upper limit on $\sin^2(2\Theta)$ assuming that there are no neutrino oscillations in the sensitive Δm^2 range of KARMEN. Therefore the maximum sensitivity is equivalent to a minimal upper limit on $\sin^2(2\Theta)$ for a fixed Δm^2. Thus we calculated for every possible evaluation interval I the ratio $S(I, \Delta m^2)$ of the expected number N_{expected} of oscillation events for maximal mixing ($\sin^2(2\Theta) = 1$) and the upper limit with α confidence level on the number of oscillation events $M_\alpha(I)$ that one would get for the measuring interval I:

$$S(I, \Delta m^2) = \frac{N_{\text{expected}}}{M_\alpha(I)} \tag{3}$$

The optimal measuring interval I is the one with maximal $S(I, \Delta m^2)$. In order to make this procedure independent of the result of the measurement the upper limit $M_\alpha(I)$ was chosen to be the *mean* expected upper limit obtained by summing up all possible upper limits weighted with the poisson probability of such a result:

$$M_\alpha(I) = \sum_{n=0}^{\infty} l_\alpha[n, \mu_b(I)] \cdot \frac{\mu_b^n}{n!}\, e^{-\mu_b} \tag{4}$$

Here $l_\alpha[n, \mu_b(I)]$ is the upper limit for n measured events with $\mu_b(I)$ expected background events (a zero oscillation signal was assumed). The mean upper limit $M_\alpha(I)$ does not only depend on Δm^2 but also on the absolute number of expected oscillation events and therefore on the measuring time. We varied the cuts with respect to the following observables: the energy of the prompt event E_p, the time of the prompt event t_p, the energy of the delayed event E_s and the time difference between delayed and prompt event $t_s - t_p$.

All cuts show a rather strong dependence on Δm^2 and change with the expected number of oscillation events, i.e. with increasing measuring time. Motivated by the result of the LSND experiment [12] we chose to be most sensitive to Δm^2 value of ≤ 0.3 eV^2 resulting in the cuts given in section 3.

To test the ability of KARMEN2 to measure events with a signature similar to that of the expected $\bar{\nu}_e$ induced events within the appropriate ν time window we ignored the information provided by the additional third layer of veto counters. If one accepts events with an additional veto hit one obtains a background situation similar to that of KARMEN1 dominated by cosmic ray induced neutron background. Moreover somewhat looser cuts on the prompt energy $11 \leq E_p \leq 50$ MeV, prompt time $0.6 \leq t_p \leq 10.6$ μs, energy of the delayed event $0 \leq E_s \leq 8$ MeV and difference between time of delayed and prompt event 0.5 $\mu s \leq t_s - t_p \leq 500$ μs were used. These cuts provide a good efficiency for muon induced neutron background. The number of measured events in the very same dataset used for the neutrino oscillation evaluation is 39. The expected background of 40.1 ± 1.4 events is dominated by 33.2 ± 1.4 cosmic induced background sequences. Fig. 4 shows energy and time distributions of the measured events compared to the expected background. The very good agreement demonstrates once again the precise knowledge of the background sources and the ability of KARMEN2 to detect events with a signature similar to that of the expected oscillation events.

6. CONCLUSION AND OUTLOOK

We have detected no $\bar{\nu}_\mu \rightarrow \bar{\nu}_e$-like sequence in the KARMEN2 data so far. The knowledge of the significantly reduced background situation after the upgrade in KARMEN2 is precise and reliable. This is important because in the unified approach the upper limit for an oscillation signal depends on the number of expected background events even in the case of a zero result.

Our result is obtained by a frequentist approach providing full coverage which was recently adopted by the Partice Data Group [13]. Of course one has to keep in mind the true meaning of 90% confidence intervals and that the obtained upper limit is due to change (it can even become less stringent) during the ongoing measuring time of KARMEN2. This is an unavoidable feature of an analysis based on such small event samples. However, the result obtained can indeed be used to infer physical implications concerning the result of the LSND collaboration which in fact shows only a 'favoured region' but not a detailed 90% confidence level area.

REFERENCES

1. KARMEN collaboration:
B. Armbruster, M. Becker, G. Drexlin, V. Eberhard, K. Eitel, H. Gemmeke, T. Jannakos, M. Kleifges, J. Kleinfeller, C. Oehler, P. Plischke, J. Rapp, M. Steidl, J. Wolf, B. Zeitnitz: *Institut für Kernphysik I, Forschungszentrum Karlsruhe, Institut für experimentelle Kernphysik, Universität Karlsruhe, Postfach 3640, D-76021 Karlsruhe, Germany*;
B.A. Bodmann, E. Finckh, S. Haug, J. Hößl, P. Jünger, W. Kretschmer, I. Stucken: *Physikalisches Institut, Universität Erlangen-Nürnberg, Erwin Rommel Straße 1, D-91058 Erlangen, Germany*;
C. Eichner, R. Maschuw, C. Ruf: *Institut für Strahlen- und Kernphysik, Universität Bonn, Nußallee 14-16, D-53115 Bonn, Germany*;
I.M. Blair, J.A. Edgington: *Physics Department, Queen Mary and Westfield College, Mile End Road, London E1 4NS, United Kingdom*;
N.E. Booth: *Department of Physics, University of Oxford, Keble Road, Oxford OX1 3RH, United Kingdom*

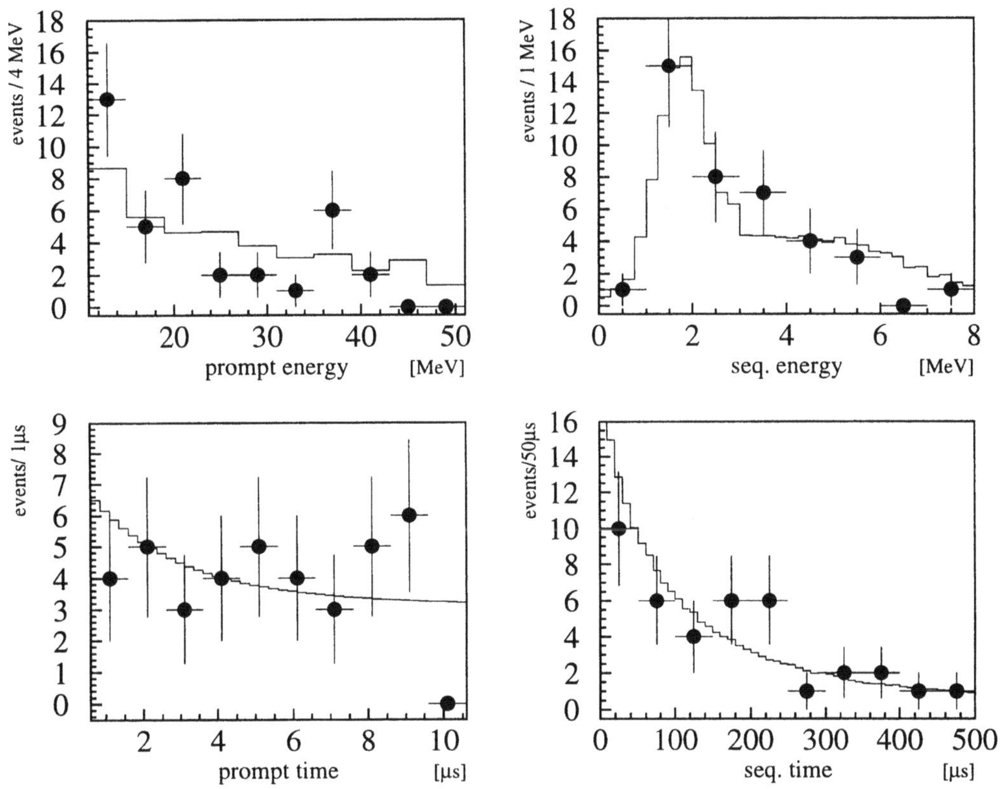

Figure 4. Event sequences with an oscillation signature found in the KARMEN2 data if one *allows* hits in the third layer veto counter. The number of 39 measured events agrees well with the expected number 40.1 ± 1.4 events. The background is dominated by cosmic muon induced energetic neutrons produced in the 7000 t steel shielding.

2. B. Zeitnitz *et al.*, Prog. Part. Nucl. Physics **40**, 169 (1998).
3. B. Armbruster *et al.*, Phys. Rev. C **57**, 3414 (1998).
4. R. Maschuw *et al.*, Prog. Part. Nucl. Physics **40**, 183 (1998).
5. B. Armbruster *et al.*, Phys. Rev. Lett. **81**, 520 (1998).
6. R.L. Burman et al., Nucl. Instr. Meth. A **368**, 416 (1996).
7. G. Drexlin *et al.*, Prog. Part. Nucl. Physics **40**, 193 (1998).
8. G.J. Feldman and R.D. Cousins, Phys. Rev. D **57**, 3873 (1998).
9. L. Borodovsky *et al.*, Phys. Rev. Lett. **68**, 274 (1992).
10. A. Romosan *et al.*, Phys. Rev. Lett. **78**, 2912 (1997)
11. B. Achkar *et al.*, Nucl. Phys. B **434**, 503 (1995).
12. C. Athanassopoulos *et al.*, Phys. Rev. C **54**, 2685 (1996); D.H. White, these proceedings
13. C. Caso *et al.* (PDG), The European Physical Journal **C3**, 1 (1998)

ELSEVIER

Nuclear Physics B (Proc. Suppl.) 77 (1999) 220–224

NUCLEAR PHYSICS B
PROCEEDINGS
SUPPLEMENTS

CHORUS results

Osamu Sato [a]
CHORUS Collaboration

[a]Department of Physics, Nagoya University,
464-8602, Nagoya, JAPAN

The CHORUS experiment aims to detect neutrino oscillation $\nu_\mu \to \nu_\tau$, especially at the dark matter mass scale ~ 10 eV.

The CHORUS apparatus has been installed in the CERN West Area and exposed to the SPS Wide Band neutrino beam for about 150 days per year from May '94 to November '97.

The emulsion stacks were replaced every 2-year period and developed. A limit on $\nu_\mu \to \nu_\tau$ oscillation has recently been published from the analysis of a subsample of neutrino interactions, taken in '94 and '95 [1] [2]. This paper contains an update of the result using the statistics taken in '96.

At present the ν_τ interaction search has been performed with 66,304 "1μ" events and 7,081 "0μ" events in nuclear emulsion. No neutrino oscillation signal was found, which translates into a mixing angle limit $sin^2(2\theta) \leq 1.3 \times 10^{-3}$ (90%CL at large Δm^2, of the order of 100 eV^2).

1. The CHORUS experiment

The neutrino oscillation signal is searched as tau neutrino events produced by an originally ν_τ-free beam. The CERN wide band neutrino beam contains mainly ν_μ (94%) and the background coming from the prompt ν_τ in the beam is estimated as low as 3.3×10^{-6} ν_τ CC interaction per ν_μ CC interaction.

The average beam energy of ν_μ is 27 GeV, the distance from neutrino beam source to the CHORUS apparatus is about 600m.

770 kg of nuclear emulsion are used as neutrino target. Thanks to the very sharp space resolution in nuclear emulsion (sub-μm), the τ decay followed by the ν_τ interaction can be detected directly by the short distance decay topology ($\leq$ 3mm).

The target setup is a sandwich structure of nuclear emulsion and of scintillating fiber trackers (Target Tracker). The Target Tracker reconstructs the neutrino events and gives angle and position information about the tracks, to be used for emulsion scanning.

An air core magnet behind the target setup measures the momentum of the tracks. An electromagnetic and hadronic calorimeter measures the energy of the event. The most downstream part is a muon spectrometer.

A background free identification of ν_τ interactions can be achieved by combining the data from the electronic detector to the topological information in the emulsion.

Detailed performance on the apparatus are described elsewhere [3].

2. The event selection

In 85% of the cases the tau decays into a single charged particle ("kink" topology), while the remaining 14% show a charged-3-prong topology.

Therefore the current analysis focuses on the following channels :

$$\tau^- \to \mu^- + \bar{\nu_\mu} + \nu_\tau \qquad (Br \sim 18\%)$$
$$\tau^- \to h^- + n\pi^0 + \nu_\tau \qquad (Br \sim 50\%)$$

The first channel will be searched among the events with a negative muon identified by the electronic detectors (1-μ sample) while the search for the second channel will use the events with no muon associated to the vertex (0-μ sample).

In the first class, only the μ^- of momentum lower than 30 GeV/c have been analyzed. In the 0μ class, all negative particles with a reconstructed momentum between 1 and 20 GeV/c have been considered as possible tau decay prod-

PII S0920-5632(99)00421-1

Table 1
Neutrino events numbers

Run (Year)	1994	1995	1996	1997	All
1μ Prediction	66,911	110,916	129,669	151,105	458,601
1μ Scanned frac	63%	34%	56%	0%	33%
1μ Located	18,286	17,890	30,128	0	66,304
0μ Prediction	17,731	27,841	32,548	37,929	116,049
0μ Scanned frac	50%	29%	0%	0%	15%
0μ Located	3,401	3,680	0	0	7,081

uct.

3. The scanning system

The tracks passing the previous selection are scanned by automatic microscope systems from the downstream emulsion in the stack where the vertex is predicted.

An emulsion plate, the size of which is 36cm × 72cm and the thickness is 790μm (350μm of emulsion on both side of a 90 μm thick acetylcellulose-base) is set on the microscope stage.

The movement of the 3 axes(x,y,z) is controlled by computer and a high refresh rate CCD takes microscope images around the predicted track location.

The objective lens magnification ×50 makes the size of one CCD view as 150 μm × 120μm of emulsion. For one track, 16 frames of CCD pixel data are taken changing the depth of the focal plane (every 6 μm). A device called Track Selector, developed at Nagoya University, recognizes a track from the CCD 16 frames pixel information.

The principle of the Track Selector is to add linear offsets to the CCD pixel positions (calculated from the given angle information) and overlay the pulse height of the 16 frames.

If the track predicted by the scintillating fiber tracker is there, a big pulse height peak appears. The track finding efficiency of the Track Selector is above 98% for track angles up to 400 mrad. The speed of the track recognition is 0.3 second per microscope view(150 μm × 120μm).

4. Event location and Decay search

One stack of emulsion module consists of 36 plates with a total thickness of 3cm, corresponding to one cascade length.

All tracks are scanned from downstream up to the vertex (scan-back). Scanning is done at the upstream surface of every plate. Only 100μm depth CCD pixel information, represented by white spots in figure 1, are used to judge whether the track element exists. The plate in which the track disappears (i.e. the track element is not found in two successive plates, represented by grey spots in figure 1) is called the Vertex Plate. The definition of the Vertex Plate is the starting point of the decay search.

For most of the events, one or more predicted tracks exist in addition to the scan-back track. They are searched in the plate downstream of the Vertex Plate. The presence of the neutrino interaction vertex in the Vertex Plate can then be confirmed if at least one track is found which has a small impact parameter with the scan-back track.

The procedure used for the decay search is detailed below. It takes into account the different situations illustrated in figure 1. If a decay topology is found, it will have to satisfy the cuts given in Table 2 to be accepted as a ν_τ signal.

4.1. Long flight with Small angle decay search

If the τ decay angle is small ($\leq$25mrad), the primary vertex will be found by the automatic procedure at the Vertex Plate (bottom type in Figure 1). In this case, the candidate ν_τ interactions are in the small impact parameter sample.

Table 2
Definition of candidate ν_τ interactions in CHORUS

Applied cut	Main suppressed event type
No other lepton (μ or e) at primary vertex	ν_μ CC+charm decay
Kink decay is found in emulsion	hadron interaction
Charge negative (decayed particle)	ν_μ CC+charm decay
Flight length < 3 or 5 plates downstream of Vertex plate	hadron interaction/decay
Decay transverse momentum > 250 MeV/c	hadron interaction/decay

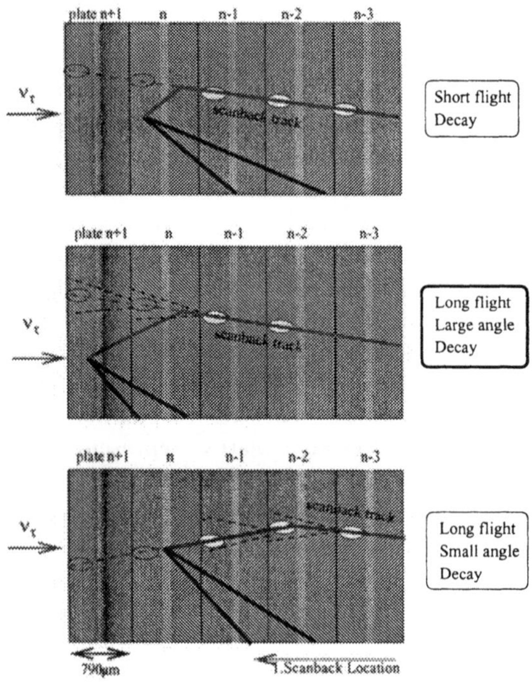

Figure 1. Decay types searched for : The plate n is the Vertex Plate

The decay possibility is checked by transverse momentum (pt), which is calculated from the track momentum and the difference between the angles at the most upstream plate angle and at the most downstream plate angle.

If pt is bigger than 250 MeV/c, the events are scanned by eye and re-measured carefully. No signal event has been found by this manual scanning and all the large pt values could be attributed to the measurement inaccuracy of the automatic scanning system.

4.2. Short flight decay , Long flight with Large angle decay search

The sample of events with large impact parameter contains τ decays of the type shown in the top and middle part of Figure 1.

In these cases the τ decay is in the Vertex Plate and the primary vertex can be found in the same plate or in a more upstream plate. To hunt a decay in the Vertex Plate, full depth information of the Vertex Plate is taken by means of 48 frames of CCD pixel information (Image Data). Tracks are followed in the Image Data and if an angle difference is detected in the emulsion (pt>250MeV/c), the event is checked by eye.

The "decay " position and the topology are carefully checked. Almost all hadron interactions make fragment of nuclei or Auger electron blob at the "decay " position. They can be easily found because of the sub-micron space resolution of nuclear emulsion.

If a τ decay candidate is found in the Vertex Plate, the primary vertex is searched by human eye check and the topological and kinematical criteria on Table 2 are applied. Up to now, no signal event was found.

4.3. Long flight with Large angle decay search

The image analysis doesn't work near the base and near the emulsion surface. To recover decay finding efficiency, we developed another type of decay analysis. In the case of Long flight with Large angle decay, we can detect the τ track at the upstream surface of Vertex Plate and detect the τ daughter track in the downstream plate(middle type in Figure 1).

A general search for all angle tracks is performed on the upstream surface of the Vertex Plate to find the parent(could be τ) and to calculate the impact parameter with the decay daughter (measured in downstream plate).

When the impact parameter is small enough (I.P $< 15\mu m$), they are considered as REAL parent and daughter, and scanned manually. The manual check procedure is the same as in the previous section. As a result, no signal event was found in this procedure.

5. Efficiency check

The kink finding efficiency has been evaluated by Monte Carlo simulation. The validity of this calculation can be checked by looking at the observed samples of hadron interactions and charm decays.

From the 0μ event sample, 21 hadron interactions have been observed by the decay search procedure. The result is in good agreement with the Monte Carlo expectation 24 ± 2.

In a partial sample of 2μ events, 25 charged charm muonic decays were found. This statistics is also in good agreement with the Monte Carlo expectation 22.8 ± 3.9.

6. Current limit for neutrino oscillation

66,304 "1μ" events and 7081 "0μ" events have been found in emulsion and the τ decay search was performed.

The sensitivity to $\nu_\mu \rightarrow \nu_\tau$ oscillation can be derived from the numbers listed in Table 3 and the following relations :

Number of ν_τ events in the case of full mixing :

$$N_\tau = N_{cc}\left(\frac{\sigma_\tau}{\sigma_\mu}\right)\left(\frac{A_\tau}{A_\mu}\right) \cdot Br \cdot \eta_{kink}$$

$\frac{\sigma_\tau}{\sigma_\mu}$: CC interaction cross section ratio
$\frac{A_\tau}{A_\mu}$: CHORUS Acceptance ratio
Br : Branching ratio for search
η_{kink} : kink finding efficiency

Number of observed events :

$$N_{\tau obs} = N_\tau \cdot P(\nu_\mu \rightarrow \nu_\tau)$$
$$P(\nu_\mu \rightarrow \nu_\tau) = sin^2(2\theta) \cdot sin^2(1.27\Delta m^2 \tfrac{L}{E})$$

θ :mixing angle
Δm^2 :squared mass difference (eV^2)
L :neutrino flight length (km)
E :neutrino energy (GeV)

No neutrino oscillation signal event has been observed. This yields a limit on the mixing parameter $sin^2(2\theta) < 1.3 \cdot 10^{-3}$ at large squared mass difference (of the order of 100 eV^2), where the second factor in the above expression for the neutrino oscillation probability averages to 1/2.

The neutrino oscillation parameters are shown in Figure 2.

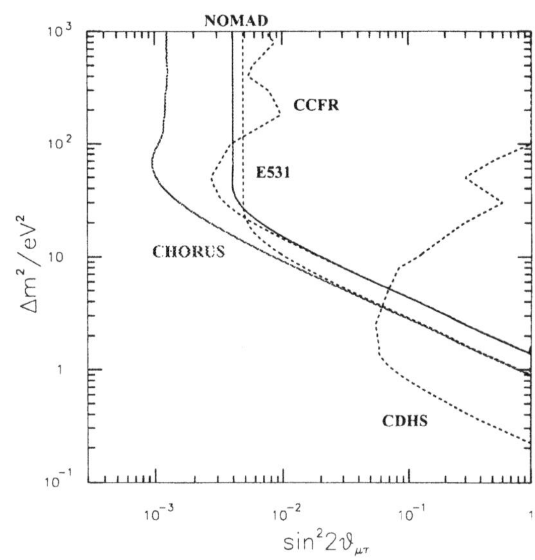

Figure 2. CHORUS exclusion plot ($\nu_\mu \rightarrow \nu_\tau$)

7. Conclusions

The CHORUS detector recorded 600,000 neutrino events in nuclear emulsion target after 4 years exposure to neutrino beam (CERN SPS).

Table 3
Summary of the events presently analyzed by CHORUS

...	...	1μ event	0μ event
Number of analyzed event		66,304	7,081
Number of ν_μ CC event	N_{cc}	66,304	31,333
Cross section ratio	$\frac{\sigma_\tau}{\sigma_\mu}$	0.530	0.530
Acceptance ratio	$\frac{A_\tau}{A_\mu}$	1.08	0.43
Branching ratio	Br	0.18	0.50
kink finding efficiency	η_{kink}	0.426	0.234
Number of tested ν_τ CC event	N_τ	2910	835
Number of observed ν_τ CC event	$N_{\tau obs}$	0	0

Up to now about 30 % of the events which were reconstructed by the Target Tracker have been scanned by automatized scanning systems and found in nuclear emulsion.

From the detailed decay search at the Vertex plate, no ν_τ event was found.

With the same procedure, several charm decays and hadron interactions were found. Since their topology in emulsion is very similar to tau decay, these events were used to check the Monte Carlo evaluation of kink finding efficiency.

From analyzed sample of 66,304 "1μ" events and 7,081 "0μ" , a new limit on the neutrino oscillation parameters was derived. We excluded the possibility of $sin^2(2\theta) > 1.3 \cdot 10^{-3}$ area at large Δm^2 (90%CL).

About 70% of the events are still to be analyzed. Thanks to the continuous improvements of automatic scanning devices and reconstruction algorithms, this analysis will be completed with an improved efficiency.

If no signal events will be found at the end of the analysis, the sensitivity will reach $sin^2(2\theta) = 2 \cdot 10^{-4}$ at large Δm^2.

REFERENCES

1. E.Eskut et al.,CHORUS Collaboration, Phys.Lett. *B424* 202 (1998)
2. E.Eskut et al.,CHORUS Collaboration, Phys.Lett. *B434* 205 (1998)
3. E.Eskut et al.,CHORUS Collaboration, Nucl.Inst.Meth *A401* 7 (1997)
4. Van de Vyver,B. et al., Nucl.Inst.Meth *A385* 91-99(1997)
5. Ya.Zel'dovic and I.D.Novikov,Relativistic Astrophysics,Nauka,Moscow 1967
6. Harari,H. Phys.lett.B *216* 413-418(1989)
7. Ushida,N. et al.,E531 Collaboration ,Phys.Rev.Lett.*57* 2897-2900(1997)

NORTH-HOLLAND
ELSEVIER
Nuclear Physics B (Proc. Suppl.) 77 (1999) 225–231
NUCLEAR PHYSICS B
PROCEEDINGS
SUPPLEMENTS

A Search for $\nu_\mu \leftrightarrow \nu_\tau$ Oscillations Using the NOMAD Detector

J. J. Gómez-Cadenas * [a]

[a] CERN, EP Division Genève, Switzerland and
IFIC, Valencia, Spain

This talk presents the current status of the search for $\nu_\mu \leftrightarrow \nu_\tau$ oscillations with the NOMAD detector. Significant progress has been made with respect to our published results. The analysis techniques have been improved and a larger data sample, corresponding to the 1995-1997 runs has been used. The search yields no evidence of the presence of oscillations. The updated limit on the neutrino oscillation mixing angle is $\sin^2 2\theta_{\mu\tau} < 1.2 \times 10^{-3}$, in the limit of large Δm^2.

1. Introduction

NOMAD is an experiment searching for $\nu_\mu \leftrightarrow \nu_\tau$ oscillations in the CERN-SPS wide band neutrino beam. The experiment is sensitive to the so-called cosmologically relevant mass range [1] ($\Delta m^2 > 1\ eV^2$) and to small mixing angles. Potential ν_τ candidates are identified through their charged current (CC) interactions in which a τ lepton is produced. NOMAD uses a purely kinematical technique to identify those τ's. The detector is sensitive to most τ decay modes. In this talk emphasis will be placed on the search for oscillations using the decay modes $\tau^- \to e^-\bar{\nu}_e\nu_\tau, \tau^- \to h^- + X$ and $\tau^- \to \pi^-\pi^+\pi^-\nu_\tau$, which have been recently updated. Our published results [2] were based on the analysis of the 1995 data, while the updated results presented here are based on the analysis of 1995-1997 data (1995-1996 for the $\tau^- \to \pi^-\pi^+\pi^-\nu_\tau$ analysis). For a full discussion about the various NOMAD analyses see [2].

2. Apparatus and Experimental Technique

2.1. The detector

The detector (Fig. 1) has been described elsewhere [3]. It consists of an active target of 2.7 tons fiducial mass made of 44 drift chambers, located perpendicular to the beam axis. The chambers also provide measurement of the momenta of charged tracks emanating from neutrino interactions. The target is followed by a transition radiation detector providing electron identification, a preshower and an electromagnetic calorimeter. All the above systems are located in a dipole magnet which creates a magnetic field of 0.4 T. External to the magnet there is a hadronic calorimeter and two muon detection stations.

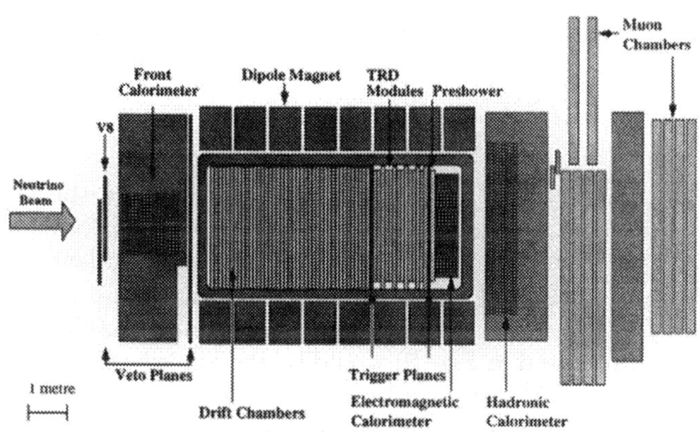

Figure 1. Side view of the NOMAD detector.

2.2. The kinematical method

The kinematical technique used by NOMAD is illustrated with two examples. First consider the identification of τ candidates which decay electronically, for which the main background are ν_e CC events. The difference between the two types of events is illustrated in Fig. 2. The relevant quantities are $\vec{p}_T^{\,l}$, the transverse momentum of

*on behalf of the NOMAD collaboration

0920-5632/99/$ – see front matter
PII S0920-5632(99)00422-3

the observed electron, $\vec{p}_T^H$, the transverse momentum of the hadronic jet and $\vec{p}_T^{\nu}$, the missing transverse momentum. In the case of a ν_τ CC this missing transverse momentum is essentially due to the undetected neutrinos emitted in the τ decay, while in the case of a ν_e CC it is due to undetected neutrals, badly reconstructed tracks, and the Fermi motion.

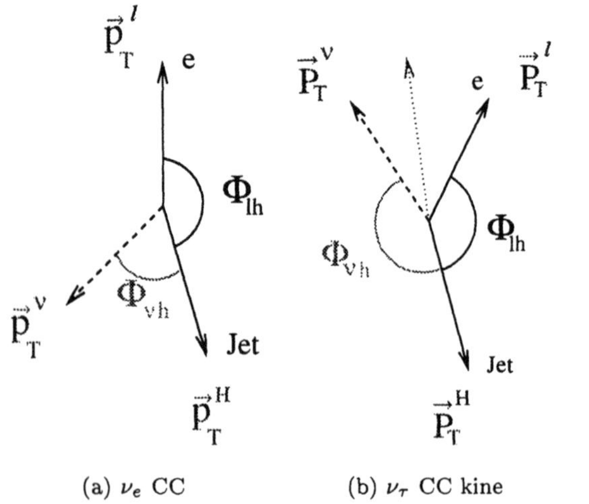

Figure 2. Kinematical Triangle in the transverse plane.

The above three vectors form a triangle in the transverse plane since $\vec{p}_T^{\,l} + \vec{p}_T^{\,H} + \vec{p}_T^{\,\nu} = 0$. Both the triangle shape (specified by two independent variables) and its size are different for ν_e and ν_τ CC events. A possible choice for these variables (illustrated in Fig. 2) would be the angle between $\vec{p}_T^{\,l}$ and $\vec{p}_T^{\,H}$, Φ_{lh}, and the angle between $\vec{p}_T^{\,\nu}$ and $\vec{p}_T^{\,H}$, $\Phi_{\nu h}$, for the shape of the triangle and the transverse mass $m_\perp = \sqrt{(|\vec{p}_T^{\,\nu}| + |\vec{p}_T^{\,l}|)^2 - (\vec{p}_T^{\,H})^2}$, which specifies the size.

If the τ decays hadronically (consider for example the decay $\tau^- \to \pi^-(K^-)\nu_\tau$) the main backgrounds are ν_μ induced neutral currents (NC) in which one pion from the hadronic jet is mistaken as the pion emanating from the τ decay. The most relevant kinematical difference between the τ signal and the NC background is the isolation of the τ daughter candidate w.r.t. the hadronic jet (see Fig. 3), which can be described, for example, with the variable $Q_T = \sqrt{(\vec{p}_T^{\,l})^2 - (\vec{p}_T^{\,l} \cdot \vec{P_{tot}})^2 / P_{tot}^2}$.

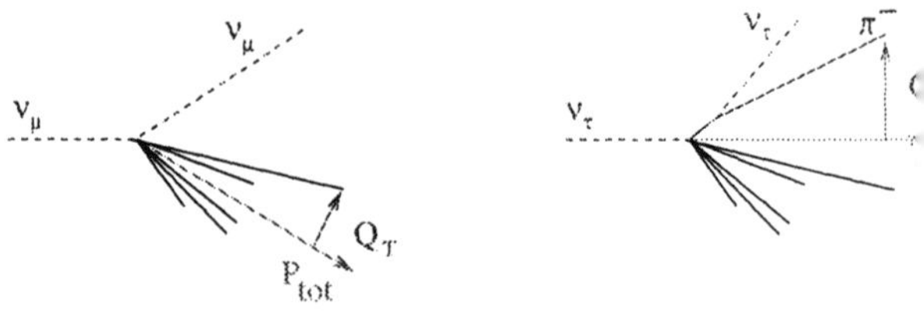

Figure 3. Isolation of the τ daughter candidate for ν_μ NC and ν_τ CC events.

2.3. Use of Likelihood Techniques

In NOMAD, no single variable unambiguously signals the presence of a τ. Instead one exploits the fact that the distributions of several kinematical variables (eg, $\Phi_{lh}, \Phi_{\nu h}, Q_T, m_\perp$) are different for the τ signal and the various backgrounds. Therefore, information is optimized by combining these variable into a likelihood ratio function. In the next section several examples will be discussed of the application of this technique.

2.4. The data simulator

For the oscillation search a background rejection at the level of 10^{-5} must be achieved. On the other hand, neither the generator of the neutrino interactions nor the modelling of the detector can be trusted to predict reliably the behavior of the data over five orders of magnitude. In order to *correct* the Monte Carlo predictions for both the background estimation and the signal efficiency, extensive use is made of various *data simulators* obtained from the data themselves. The technique is as follows. One starts from two samples of ν_μ CC, which are identified by the presence of a well reconstructed negative muon. One is a data sample (DS), and the other is a Monte Carlo sample (referred as Monte Carlo Simulator, MCS). In both cases, the negative muon is removed and substituted by either a τ decay, an electron or nothing. Thus one obtains samples of "fake" signal, ν_e CC and NC. The Monte Carlo prediction of the efficiency for every sample is corrected using the ratio of DS to MCS which corrects for all the effects in the hadronic jet which may have not been correctly described by the Monte Carlo, i.e,

$$\epsilon = \epsilon_{MC} \times \frac{\epsilon_{DS}}{\epsilon_{MCS}}. \quad (1)$$

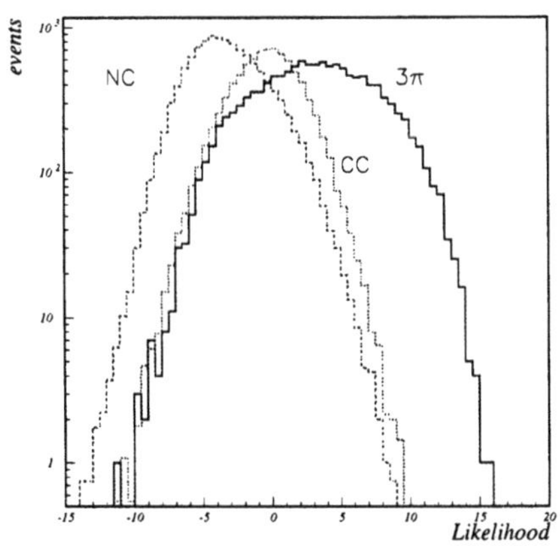

Figure 4. The predicted values of the likelihood function for signal and backgrounds, for the $\tau^- \to \pi^-\pi^+\pi^-\nu_\tau$ analysis.

2.5. Blind Analysis

A "Blind Analysis" technique has been used for the updated NOMAD analysis presented here, in order to avoid the biases which are introduced if the data is examined in the "signal region" before the analysis is demonstrated to be robust.

As an example, consider the oscillation search in the $\tau^- \to \pi^-\pi^+\pi^-\nu_\tau$ mode. Fig. 4 shows the likelihood function for the signal and the major backgrounds (ν_μ CC and NC). The "signal region" can be defined as the one with likelihood values larger than, say, 7.5, for which the signal to background ratio is very high. The technique consists in defining a "box" around this signal region inside which the data cannot be analyzed until the analysis can predict the backgrounds *outside* the box, as well as in the τ^+ search (Since no τ^+ signal is expected, the positive data must be explained by the data simulator corrected Monte Carlo prediction of the backgrounds).

Fig. 5 shows the likelihood function for the data (points with error bars) and the calculated backgrounds (solid line). One can see than *outside* the box the likelihood is consistent with the background prediction, and that the likelihood for the "positive" τ's is also consistent with the expected background. At this point the analysis is "allowed" to "open the box", and examine the data in the region which is expected to be populated by the signal.

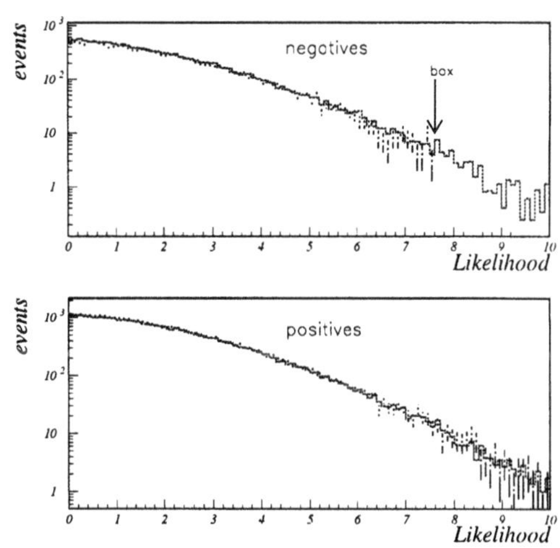

Figure 5. Data points and predicted likelihood for the sum of all the backgrounds, for the $\tau^- \to \pi^-\pi^+\pi^-\nu_\tau$ analysis. Shown is the negative data (with the box around the signal region) and the positive data. Both show good agreement between the data and the predicted background.

3. Search for $\nu_\mu \leftrightarrow \nu_\tau$ oscillations

3.1. $\tau^- \to e^- \bar{\nu}_e \nu_\tau$ channel

The search of τ candidates in the electronic channel is the most sensitive to the possible existence of oscillations, due to the reduced backgrounds (the contamination of ν_e's in the SPS beam made primarily of ν_μ's is of the order of 1 %) and the good capabilities of the NOMAD detector to identify electrons.

The first step of the analysis is a topological selection. It is required that one and only one prompt electron is identified in the event (for a discussion about the identification of electrons with the NOMAD detector see [2] and [3]). The electron must emerge from the primary vertex and must be inconsistent with being the result of a gamma conversion. About 34 % of the signal events (and of the ν_e CC backgrounds) satisfy the above conditions while the background due to NC (in which, for example, a missed conversion simulates the primary electron) is reduced by a factor 2500. The residual NC backgrounds are eliminated by requiring the electron candidate to be isolated from the hadronic jet (see [2] for further discussion). This condition leaves about 12 % of the signal and of the ν_e CC background, and essentially no NC.

The ν_e CC background is eliminated by a cut on the transverse mass $0.2 < m_\perp < 1.8$ *GeV* and a cut on a likelihood ratio which exploits the difference between the signal and the ν_e CC background in several variables, such as the primary electron momentum, the transverse plane angles Φ_{lh} and $\Phi_{\nu h}$, etc (for a detailed discussion see [2]). The data simulator corrections to this function for both ν_e CC and the signal are shown in Fig. 6. Fig. 7 shows the data simulator corrected likelihood, for simulated ν_e and ν_τ CC events, compared with the data. One can see that the data is consistent with the predicted ν_e CC background, showing no evidence of oscillations.

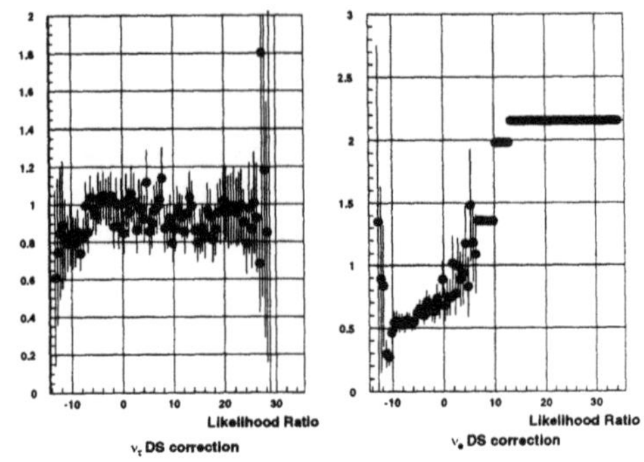

Figure 6. Data simulator corrections to the signal efficiency and ν_e CC backgrounds predictions

The sensitivity of the analysis (defined as the efficiency divided by the square root of the expected background) is shown in Fig. 8. One could choose to "open the box" (i.e, compare the events found in the data with the background predictions and establish whether there is a significant excess) on the maximum of this figure of merit (where the signal efficiency is optimized w.r.t. the background). However, slightly more sensitive results are obtained choosing three bins near the maximum, as this also exploits the difference in the likelihood shape between signal and background. The results of the analysis are shown in Table 1. Notice that N_τ corresponds to the number of τ's that would be observed in the case of full oscillation, and is a measure of the analysis efficiency. As can be seen, the data is consistent with the expected background in all three bins chosen.

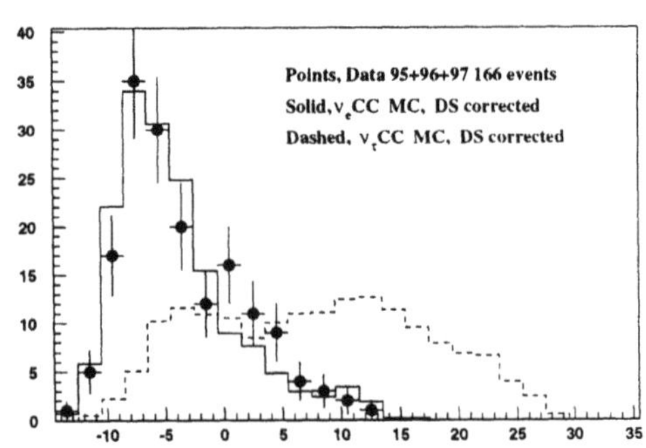

Figure 7. Data simulator corrected likelihood ratios for the signal and for the ν_e CC background compared with the data.

Table 1
Results for the $\tau^- \to e^- \bar{\nu}_e \nu_\tau$ analysis

Likelihood	N_τ	Expected Bkgnd	Data
6.5-10	693	3.4 ± 0.7	4
10-13	623	4.1 ± 0.9	2
> 13	1594	0.6 ± 0.4	1

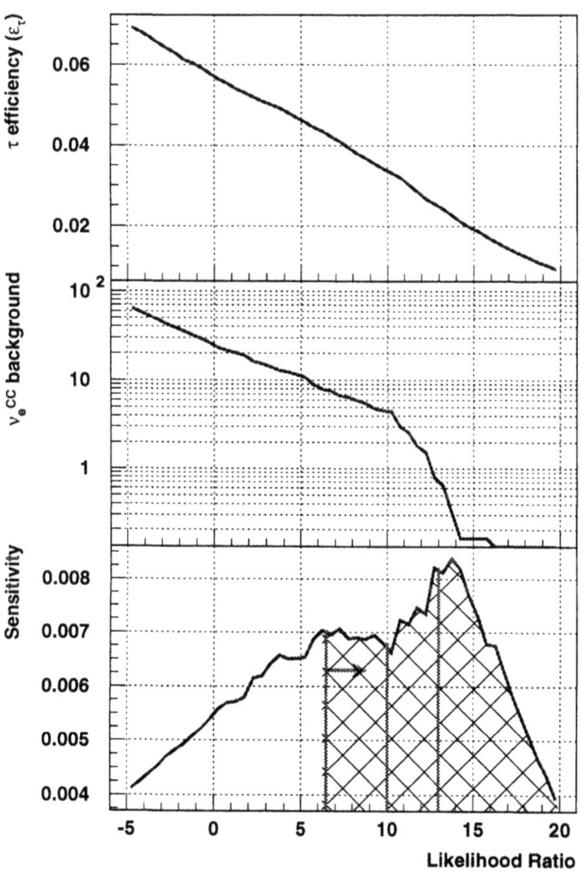

Figure 8. Sensitivity of the $\tau^- \to e^- \bar{\nu}_e \nu_\tau$ analysis.

3.2. $\tau^- \to h^- + X$ channel

The first step of the analysis is the rejection of ν_μ and ν_e CC events. For that purpose, all events in which either a primary electron or muon is identified are rejected. Events in which a high $p_\perp$ track ($p_\perp > 0.8$ GeV) escapes detector acceptance are also rejected, to avoid CC events in which the primary lepton escapes identification. Next, the τ daughter candidate (in this case a negative pion) is chosen as the most isolated negative track in the event. These "topological" requirements keep 25 % of the signal, while rejecting ν_μ (ν_e) CC by a factor 3.4×10^{-3} (6×10^{-3}) and NC by a factor 8.5×10^{-2}.

The rejection of the dominant NC background, as well as the remaining CC background (due to events in which the lepton has not been identified) is achieved by building a three-dimensional likelihood. The variables used are $m_\perp, Q_T$ and $\rho_m = (p_T^\nu)/(p_T^l + p_T^\nu + p_T^H)$. The likelihood exploits the difference in the transverse plane triangle between signal and NC (the triangle shape is given by ρ_m while the triangle size is given by $m_\perp$) and the isolation of the τ daughter (given by Q_T). The variables used to build the likelihood functions are shown in Fig. 9. In Fig. 10 the likelihood ratio is shown for CC, NC and the signal. The procedure to open the analysis box is identical to the one explained above. Three bins are chosen near the maximum of the relevant figure of merit. The results are shown in Table 2, and are consistent with the absence of oscillations.

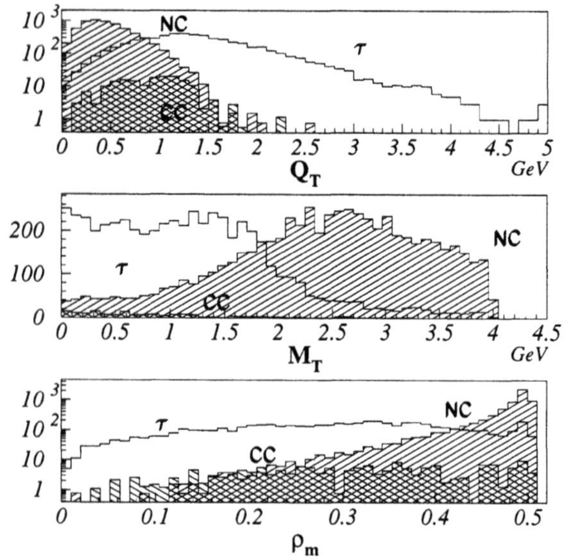

Figure 9. The variables used to form the likelihood function for the $\tau^- \to h^- + X$ analysis.

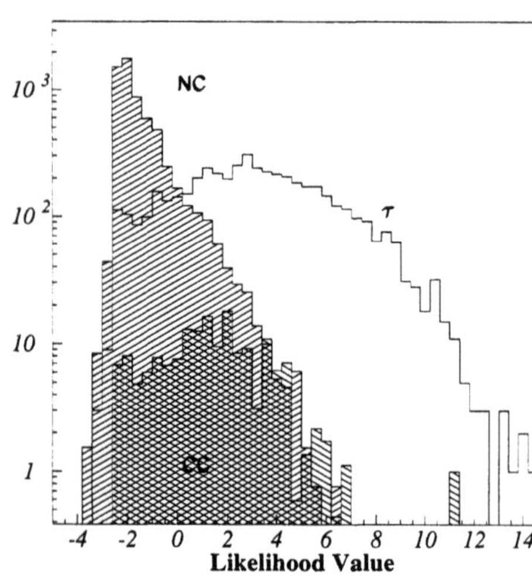

Figure 10. The likelihood function for the $\tau^- \to h^- + X$ analysis.

Table 2
Results for the $\tau^- \to h^- + X$ analysis

Likelihood	N_τ	Expected Bkgnd	Data
7-9	664	2.3 ± 0.8	3
9-11	234	$1.1^{+0.8}_{-0.6}$	2
> 11	1133	$1.1^{+0.7}_{-0.5}$	0

3.3. $\tau^- \to \pi^-\pi^+\pi^-\nu_\tau$ channel

The analysis for this channel is conceptually very similar to the one for the $\tau^- \to h^- + X$ channel. It vetoes events in which a track consistent with being a primary muon or electron is identified, as well as events in which a large transverse momentum track escapes the detector. It also builds likelihood functions based on kinematical variables which are equivalent to the ones described above (shown in Fig. 4). However, the analysis adds one interesting feature which is the selection of the τ daughter candidates. In this case one has to decide among all the possible $h^-h^+h^-$ combinations in the event. Use is made here of the fact that the decay $\tau^- \to \pi^-\pi^+\pi^-\nu_\tau$ is mediated by the A_1 and ρ resonances, $\tau \to A_1\nu \to \rho\pi\nu \to \pi\pi\pi\nu$. To choose the "right combination" (i.e, precisely the three pions that arise from the τ decay) from the "wrong' combinations (all the random combinations in which one or more tracks do not arise from the τ decay but from the hadronic jet) a likelihood is formed which exploits the existence of an internal structure in this decay, as shown in Fig. 11 (s_1 and s_2 are the two invariant masses which arise from the two possible h^+h^- combinations). The τ daughter candidates are chosen by selecting the combination with maximum likelihood value. The rest of the analysis proceeds in a similar way to the $\tau^- \to h^- + X$ analysis and it is not described here for brevity. This analysis "opens the box" in a single bin, for which they obtain $N_{tau} = 1011$, an expected background of 7 ± 2.7 events and 5 events observed, consistent with the absence of oscillations.

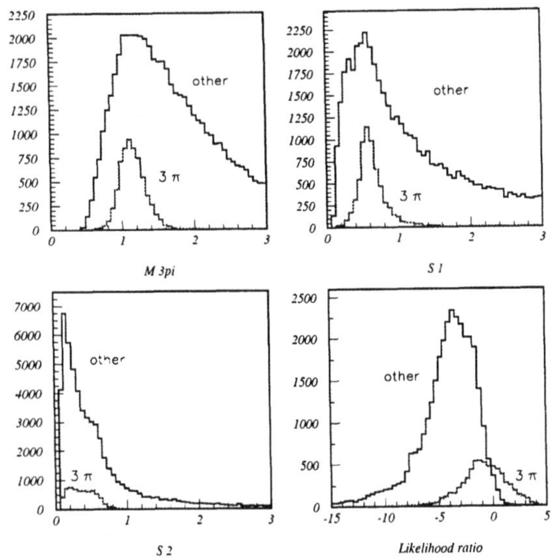

Figure 11. The likelihood function for the selection of the right combination in the $\tau^- \rightarrow \pi^-\pi^+\pi^-\nu_\tau$ analysis.

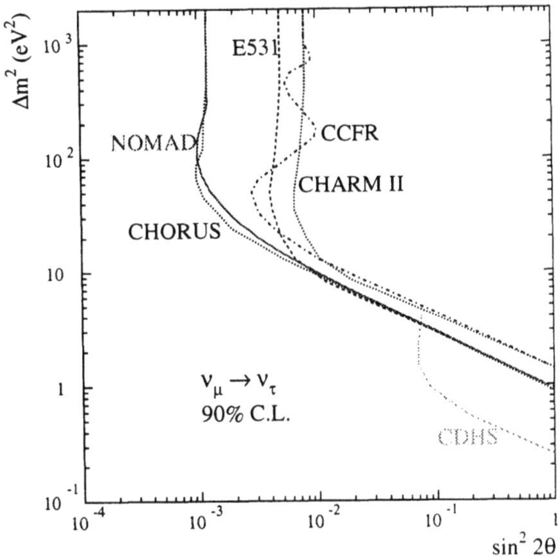

Figure 12. NOMAD exclusion plot calculated using the two neutrino families oscillation formalism and the statistical method described in [4].

4. Results

The results of the updated analysis (which has been improved since the presentation of this talk, see [5]) combined with our previous results [2] show no evidence for oscillations and allow us to set a new limit, at the 90 % C.L. on the probability of $\nu_\mu \leftrightarrow \nu_\tau$ of

$$P(\nu_\mu \rightarrow \nu_\tau) < 6 \times 10^{-4},$$

which improves our previous limit ([2]) by a factor 3.5 and the limit from the E531 [6] experiment by a factor of four. Fig. 12 shows the conventional $\Delta m^2 - \sin^2 2\theta_{\mu\tau}$ exclusion plot, assuming two family oscillations.

REFERENCES

1. H. Harari, Phys. Lett B 216(1989)413; J. Ellis, J.L. Lopez, D.V. Nanopoulos, Phys. Lett. B 292 (1992) 189.
2. NOMAD Collaboration, J. Altegoer et al., Phys. Lett. B 431 (1998) 219.
3. NOMAD Collaboration, J. Altegoer et al., Nucl. Instr. and Meth. A 404(1998)96.
4. G.J. Feldman and R.D. Cousins, Phys. Rev. D57(1998)3873.
5. D. Autiero, Proceedings of the ICHEP-98 conference, Vancouver, Canada, July 23-29, 1998.
6. E531 Collaboration, N. Ushida et al., Phys. Rev. Lett. 57(1986)2897.

Nuclear Physics B (Proc. Suppl.) 77 (1999) 232–238

NUCLEAR PHYSICS B
PROCEEDINGS
SUPPLEMENTS

ELSEVIER

Future Short Baseline Neutrino Oscillation Experiments

Leslie Camilleri[a]

[a]CERN, EP Division, Geneva, Switzerland

1. Introduction

The previous talks have delineated the neutrino oscillation panorama (Figure 1). The atmospheric neutrino experiments [1] indicate neutrino oscillations with $\Delta m^2 \sim 10^{-2} - 10^{-3} eV^2$ and maximal mixing, the solar neutrino experiments [2] indicate either $\Delta m^2 \sim 10^{-10} eV^2$ and maximal mixing for vacuum oscillations or $\Delta m^2 \sim 10^{-5} eV^2$ and $\sin^2 2\theta \sim 5 \times 10^{-3}$ or $\sin^2 2\theta$ close to 1 for matter oscillations, the favoured region of LSND [3] not ruled out by other experiments extends from $\Delta m^2 \sim 0.1 eV^2$ to $\Delta m^2 \sim$ a few eV^2, CHORUS [4] and NOMAD [5] do not see oscillations down to $\sin^2 2\theta \sim 10^{-3}$ and cosmology [6] would welcome a neutrino mass up to a few eV. This talk will describe future experiments addressing the cosmological region and experiments aimed at the LSND region.

2. The Cosmological Region

A neutrino mass that would make a significant contribution to the hidden mass of the universe and thus contribute to the solving of the dark matter puzzle is still the most valuable prize in neutrino physics. This would presumably be through a mixed dark matter scenario and would involve a neutrino mass of 1-2 eV. Assuming the Δm^2 observed in neutrino oscillations is the difference between this mass and a negligible mass of a second neutrino, CHORUS and NOMAD would only have a sensitivity of $\sin^2 2\theta \sim 10^{-3}$ in this domain. The aim of future $\nu_\mu - \nu_\tau$ oscillation searches is therefore to improve the sensitivity of the search by about an order of magnitude. NOMAD has a number of events looking exactly like a ν_τ interaction should but, in spite of the good kinematical capabilities of the experiment,

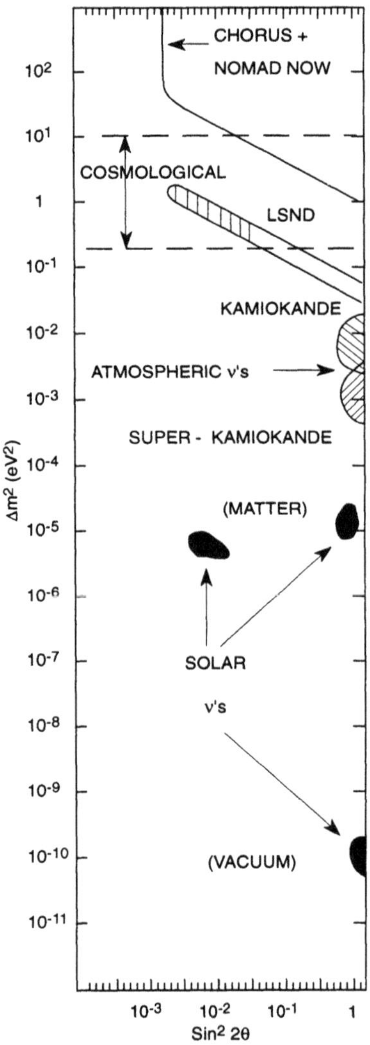

Figure 1. A summary of neutrino oscillation searches.

the number of such events is consistent with the number of expected background events. Therefore to improve on this situation it is imperative to be able to determine whether the τ decay

0920-5632/99/$ – see front matter
PII S0920-5632(99)00423-5

daughter candidate originates from a secondary vertex located at a distance $\sim$ 1 mm from the primary vertex. In short the topology capabilities of CHORUS must be combined with the kinematical capabilities of NOMAD. This is the strategy of TOSCA [7] and, until it was withdrawn earlier this year, of COSMOS [8].

2.1. TOSCA

In order to achieve this increase in sensitivity TOSCA is intending to work on three factors.

- Double the number of protons on target by using the cycles within the SPS super cycle of 14.4 s currently used to accelerate leptons for LEP to accelerate more protons. Note that LEP will stop running in the year 2000.

- Increase the emulsion target mass from the 800 kg currently used in CHORUS to 2400 kg. This however introduces Challenge #1, which is to achieve this increase while maintaining a good energy-momentum resolution.

- Improve the τ detection efficiency by a factor of 2-3. Challenge # 2 is to accomplish this while maintaining the background at the 1 event level.

All together this yields an increase in sensitivity of a factor of $\sim$ 15. However Challenge # 3 is to reduce the emulsion scanning time to cope with the 8 million events expected in such an exposure.

2.1.1. Challenge # 1

In order to avoid the degradation of the energy-momentum resolution caused by the 12 radiation lengths of 2400 kg of emulsion, the target is split into 6 stacks each followed by its own spectrometer. All 6 stacks are housed in the NOMAD magnet and share a common muon detection. By instrumenting the return yoke of the magnet with RPC's the muon coverage of NOMAD is extended to larger angles thus reducing the number of muons from ν_μ CC events that escape identification. Each TOSCA module consists of a 400 kg emulsion stack followed by a special emulsion sheet, 2 silicon planes each providing two orthogonal coordinates, a gaseous tracker (TPC, TRT or DC) and an optional calorimeter. The silicon planes are assembled from 72 cm long ladders consisting of 12 detectors bonded together. Fifty such ladders have been operating in NOMAD for a year [9]. The strategy is to observe a charged particle in the tracker then in the silicon, extrapolate it to the special plate, scan for it there and finally find it in the bulk emulsion. The advantage of this setup is that the silicon detectors provide an extrapolation accuracy of 20 μm. Therefore the special plate need only be scanned over a 100 $\times$ 100 μm^2 area which is 100 times smaller than currently being used in CHORUS and less than one microscope view. This results in much faster scanning at the special plate and in a smaller probability to pick up the wrong track.

2.1.2. Challenge # 2

In the $\tau^- \to \mu^- \bar{\nu}_\mu \nu_\tau$ channel the background is mostly due to the reaction
$\bar{\nu}_\mu + N \to D^- + \mu^+, D^- \to \mu^- \bar{\nu}_\mu X$
in which the μ^+ fails to be recognized as a muon and the D^- decay simulates exactly a τ decay, including the displaced vertex. This is minimized with the increased muon identification coverage described earlier. In the $\tau^- \to \pi^- \nu_\tau$ channel, the main background comes from white kinks, neutral current events in which a pion undergoes, very close to the primary vertex, a secondary interaction producing a single outgoing track. Here the good kinematics capabilities of the detector helps. For a genuine τ interaction the angle ϕ in the transverse plane between the parent τ direction as measured in the emulsion and the hadronic jet direction as reconstructed with the spectrometer peaks at π rads (Figure 2a). For a background event the parent π is usually close to the rest of the hadronic jet and this angle has a smaller value (Figure 2b). The third background, the intrinsic ν_τ component of the beam coming from the decay of D_s mesons originating at the neutrino production target, is decreased by using 350 GeV/c instead of 450 GeV/c protons to produce neutrinos. The background in three τ decay channels is summarized in Table 1 and adds up to 1.3 events for a 3 year run amounting to 1.3 $\times$ 10^{20} protons on target.

Table 1
Expected contributions to the background in TOSCA for the muon, electron and hadron τ decay channels.

τ Decay Mode	Charm	White Kink	Intrinsic ν_τ	Total
μ	0.08	-	0.29	0.37
e	0.02	-	0.07	0.09
$h^-(n\pi^0)$	0.11	0.38	0.36	0.85
Total	0.21	0.38	0.72	1.31

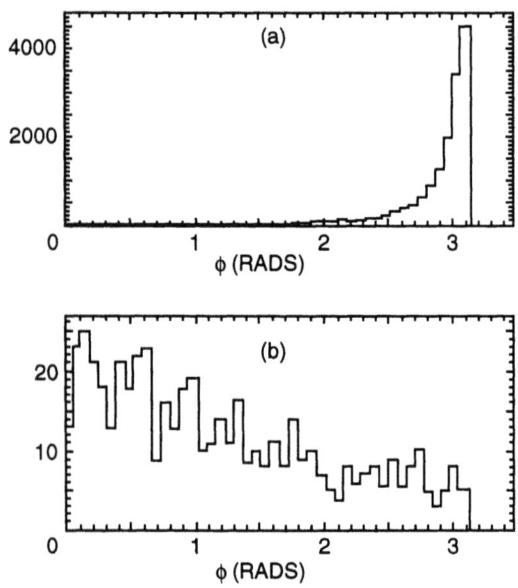

Figure 2. The distribution of the transverse angle ϕ between the measured tau direction and the reconstructed hadronic vector direction (a) for ν_τ events, (b) for background events.

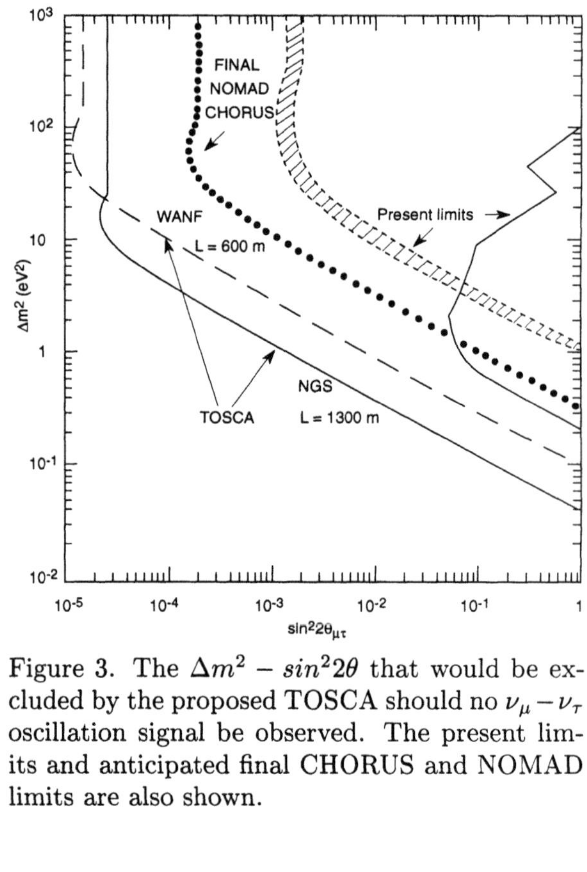

Figure 3. The $\Delta m^2 - sin^2 2\theta$ that would be excluded by the proposed TOSCA should no $\nu_\mu - \nu_\tau$ oscillation signal be observed. The present limits and anticipated final CHORUS and NOMAD limits are also shown.

The improvement in the muon channel efficiency over CHORUS comes from a better muon identification coverage (0.90 instead of 0.65), a better scanning efficiency (0.75 instead of 0.45) due to a better prediction from the silicon and less edge effects because of the use of bigger microscope stages and bigger emulsion plates, a better kink finding efficiency (0.75 instead of 0.61) due to less edge effects and recognizing kinks closer to the primary vertex. However kinematics cuts, not used in CHORUS result in a loss of efficiency of 0.7 All together the efficiency goes from 0.18 in CHORUS to 0.35 in TOSCA.

2.1.3. Challenge # 3

The emulsion scanning rate is currently 12000 events/microscope/year. Because of the better position prediction at the special plate and of the explosive growth in automatic scanning techniques one can safely assume an order of magnitude increase in this rate. Thus 20 automatic microscopes can scan the 8 million events in 2-3 years.

TOSCA could either be run in the present West

Area Neutrino Facility (WANF) with a mean neutrino flight path of 600 m or in the planned Neutrino to the Gran Sasso (NGS) beam [10] at a location resulting in a mean flight path of 1300 m. In the WANF a $\sin^2 2\theta$ limit at high Δm^2 of 1.5×10^{-5} and a Δm^2 limit of $0.1 eV^2$ at maximal mixing could be achieved. The corresponding numbers for the NGS would be 2.5×10^{-5} and $0.04 eV^2$. The exclusion plots are shown in Figure 3, together with the foreseen NOMAD and CHORUS final limit and the present ones.

3. The LSND region

There is only one indication of ν oscillations in experiments using man-made beams: the LSND signal in $\bar{\nu}_\mu - \bar{\nu}_e$. They have now increased their statistics and the excess of positron events is confirmed [3]. In addition the experiment also sees evidence of oscillations in the $\nu_\mu \to \nu_e$ channel. It is therefore of prime importance to confirm or infirm this result. KARMEN II [11] with their improved veto shield is in the process of doing so but with limited ultimate sensitivity. Two experiments, MiniBoone [12], approved to run at Fermilab, and I216 [13], a letter of intent submitted to CERN, can check LSND decisively.

3.1. MiniBoone

This is an experiment that will use a low energy neutrino beam produced using the 8 GeV protons from the Fermilab booster. The average neutrino energy is 1 GeV and the predominantly ν_μ beam has a 0.3% ν_e component. The experiment, located 500 m from the source, intends to detect an excess of ν_e interactions coming from $\nu_\mu \to \nu_e$ oscillations. The detector consists of a 6 m radius spherical tank containing 769 tons of mineral oil. The main detector consists of the inner 5.5 m radius volume and is viewed by 1220 photomultipliers. The outer 0.5 m, which is optically isolated, is viewed by an additional 269 photomultipliers and is used as a veto. The products of neutrino interactions are detected through the emission of both scintillation light and Čerenkov light. The ring pattern formed by the latter is used to distinguish between electrons, muons and pizero's.

Muons, used to identify ν_μ CC interactions, produce a sharp ring when they stop in the detector or a filled ring (or disk) when they exit the detector. Electrons, used to identify ν_e CC interactions, produce a fuzzy ring because of the width of the shower they generate in the oil. Pizero's tend to produce two fuzzy rings, one from each of their decay photons.

The major background is due to the intrinsic ν_e component of the beam. Unlike in high energy beams for which this component comes from K decay, the ν_e's in this low energy beam are due predominantly to muon decay. However these muons are produced in the decay of the pions that, at the same time produce the ν_μ component of the beam. Thus measuring this ν_μ component (which is more than 100 times more intense than the ν_e component) will provide a good estimate of this background.

The other backgrounds consist of muons produced in ν_μ CC interactions and π^0's produced in neutral current interactions simulating a single fuzzy ring. These backgrounds will be computed from the large number of recognized muons and of reconstructed two-ring π^0's and a Monte Carlo simulation.

A one year exposure amounting to 5×10^{20} protons on target will produce 1800 electrons from intrinsic ν_e interactions, 600 muons and 600 π^0's simulating electrons. The corresponding signal will amount to 1200 events if $\Delta m^2 = 0.4 eV^2$ and $sin^2 2\theta = 0.02$.

The oscillation parameters can be determined as the energy distribution of the excess events depends on them (Figure 4). The error on these parameters and the significance of the measurement are given in Table II for two sets of parameters in the LSND region. As can be seen the significance is greater than 15 σ's.

Table 2
Expected precision with which MiniBoone could measure the oscillation parameters, for two examples of these parameters.

Δm^2	$sin^2 2\theta$	$\delta(\Delta m^2)$	$\delta(sin^2 2\theta)$	$\chi^2_{NoOsc}/51df$	Signal Signif.
0.3 (eV^2)	0.03	0.10 (eV^2)	0.02	441	44 σ
2.0 (eV^2)	0.002	0.10 (eV^2)	0.0002	152	15 σ

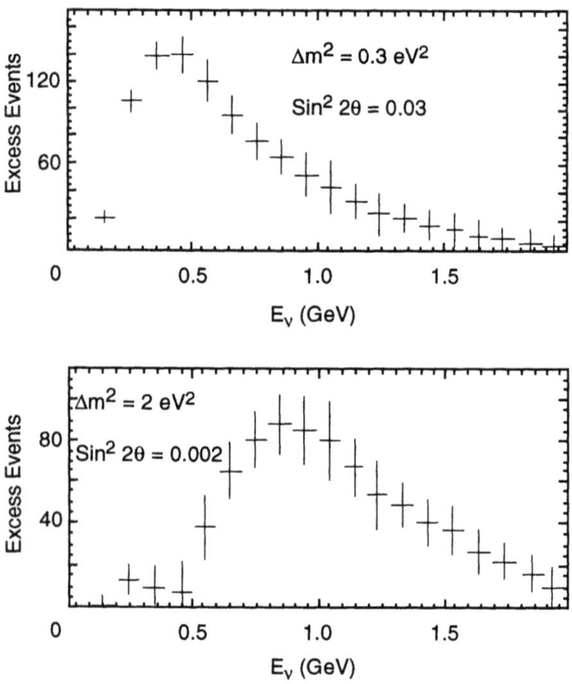

Figure 4. The expected energy distribution of excess events in the MiniBoone experiment for $\Delta m^2 = 0.3eV^2, sin^2 = 0.03$ and for $\Delta m^2 = 2eV^2$, $sin^2 2\theta = 0.002$.

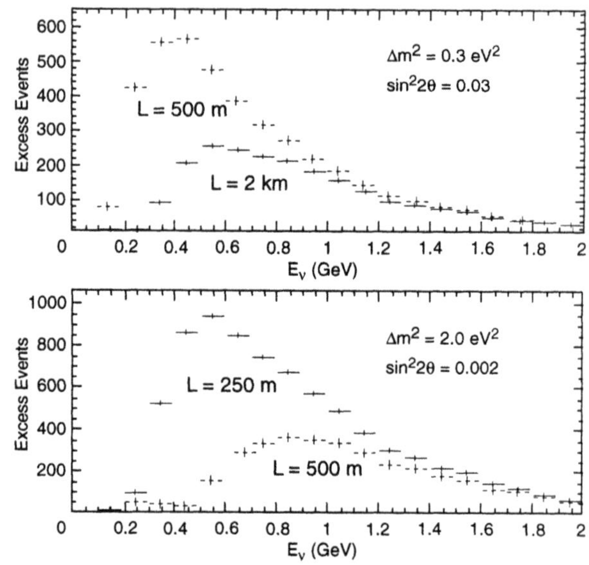

Figure 5. The expected distribution of excess events in the Boone experiment for a source-detector distance, L, of 500 m and 2000 m for $\Delta m^2 = 0.3eV^2$, $sin^2 = 0.03$ and for L = 500 m and 250 m for $\Delta m^2 = 2.0eV^2$, $sin^2 2\theta = 0.002$.

If LSND is confirmed through the observation of a positive signal, the parameters can be determined even more accurately through the use of a second detector. From Figure 5, it can be seen that for low Δm^2 a second detector at 2 km or for high Δm^2 a second detector at 250 m, would produce very different energy distributions for the excess events. It is estimated that Δm^2 then could be determined to $\pm 0.014eV^2$ and $sin^2 2\theta$ to ± 0.002 in one year.

3.2. I216

This experiment would use a ν_μ beam with a 0.6% ν_e contamination from the CERN PS. The average beam energy is also about 1 GeV. It is also a ν_e appearance experiment. The difference with MiniBoone is that it would use two detectors, one at 130 m and one at 885 m from the source, in order to measure the difference in the ratio of ν_e CC to ν_μ CC, $R_{e\mu}$, at the two locations. The experiment uses three identical modules, one in the near and two in the far location. Each module consists of a tracking calorimeter (300 streamer planes interleaved with 300 scintillator planes) followed by a tail catcher (20 scintillator planes interleaved with 1 cm thick iron plates) and a muon catcher (10 streamer tube planes interleaved with 10 cm iron plates). A module is 17 m long, 3.67 m × 3.67 m in cross-section, weighs 128 tons and includes 75 basic units of tracking calorimeter. A basic unit is built of four 2cm-thick scintillator planes assembled from 24.5 cm wide slabs and four streamer tube planes. There are two planes each of vertical scintillator slabs and vertical streamer tubes and similarly for the horizontal direction. Thus each unit, which is 0.36 radiation lengths thick, provides two points in each of the two coordinates transverse to the beam and four energy measurements. The basic discrimination between electrons and muons comes, of course, from track length. However additional discrimination between electrons, muons and π^0's is provided by the different shower characteristics of these particles resulting in different energy deposition patterns near the beginning of the track (Figure 6). Further discrimination against the background is obtained by limiting the oscillation search to electron shower energies between 0.8 and 2.2 GeV. The lower cut removes π^0 background and the upper cut intrinsic ν_e CC events (Figure 7). For a two year run of 2.5×10^{20} protons on target 130000 events of the type $\nu_\mu n \to \mu^- p$ are expected. The 0.6% ν_e component of the beam is reduced by a factor of 0.4 coming from the energy cut and a factor of 0.8 from the energy deposition pattern resulting in 248 events. The π^0's in neutral current events are reduced by factors of 0.12 and 0.02 by the two cuts resulting in 61 events.

For $\Delta m^2 = 2eV^2$ and $\sin^2 2\theta = 6 \times 10^{-3}$, the values of $R_{e\mu}$ measured at the two sites would be $R_{e\mu}\,|_{FAR} = (7.10 \pm 0.29) \times 10^{-3}$ and $R_{e\mu}\,|_{NEAR} = (4.00 \pm 0.05) \times 10^{-3}$ yielding a difference of $(3.10 \pm 0.29(stat) \pm 0.10(syst)) \times 10^{-3}$, a significance of 10σ. The systematic error is due primarily to the differences in the intrinsic ν_e component between the two locations.

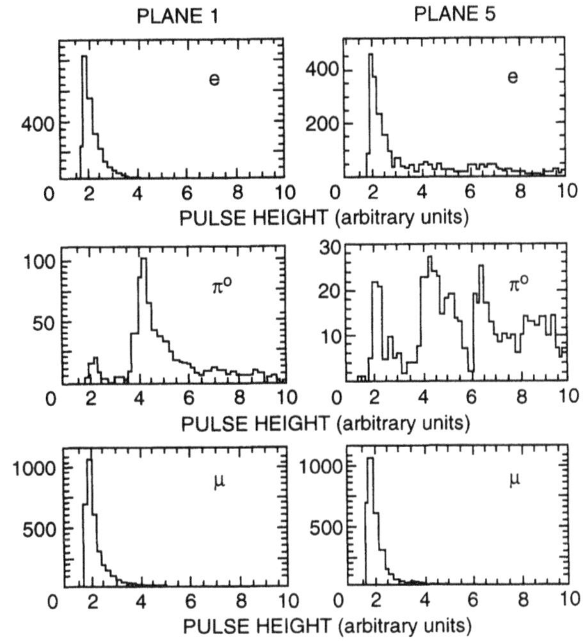

Figure 6. The pulse height distribution for electrons, π^0's and muons in planes 1 and 5 of the I216 detector.

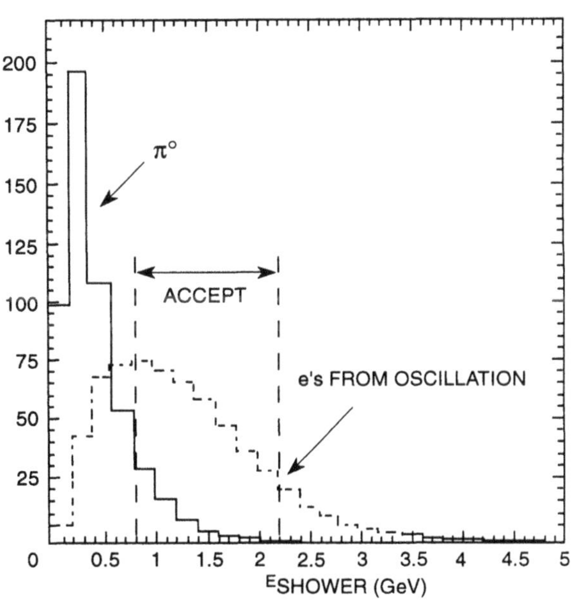

Figure 7. The energy distribution of π^0's and expected electrons from $\nu_\mu \to \nu_e$ oscillations in the I216 detector.

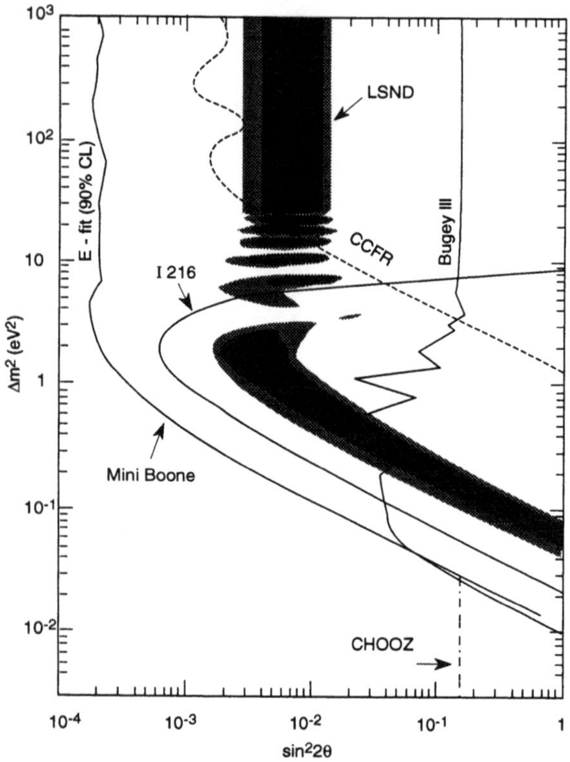

Figure 8. The $\Delta m^2 - sin^2 2\theta$ to be excluded by the MiniBoone and I216 should no $\nu_\mu \to \nu_e$ oscillation signal be observed by these two detectors. The LSND signal region is also shown.

If no signals are found in the two experiments the exclusion plot of Figure 8 would be obtained. As can be seen, the LSND region is well within the capabilities of the two experiments and a definitive answer to this question should be forthcoming.

REFERENCES

1. Results from Super-Kamiokande and Kamiokande, T. Kajita, these proceedings.
   Upwards going muons and MACRO, F. Ronga, these proceedings.
   Contained events and Soudan-2, E. Peterson, these proceedings.
2. Results from Super-Kamiokande and Kamiokande, Y. Suzuki, these proceedings.
   Results from Gallex and GNO, T. Kirsten, these proceedings.
   Results from SAGE, V. N. Gavrin, these proceedings.
3. Results from LSND, D. H. White, these proceedings.
4. Results from CHORUS, O. Sato, these proceedings.
5. Results from NOMAD, J.-J. Gomez-Cadenas, these proceedings.
6. H. Harari, Phys. Lett. **B216** (1989) 413; J. Ellis, J. L. Lopez and D. V. Nanopoulos, Phys. Lett. **B292** (1992) 189.
7. TOSCA, A high sensitivity short baseline experiment to search for $\nu_\mu \to \nu_\tau$ oscillation, CERN-SPSC/97-5 SPSC/ I 213.
8. COSMOS, Muon neutrino to tau neutrino oscillations, Fermilab proposal P803, October 1993.
9. A B_4C–Silicon Target for the Detection of Neutrino Interactions, G. Barichello et al. CERN-EP-98-21, 9 February 1998. To be published in Nucl. Inst. and Meth.
   Performance of Long Modules of Silicon Microstrip Detectors, G. Barichello et al. CERN-PPE 97-162, 11 December 1997. To be published in Nucl. Inst. and Meth.
10. The CERN Neutrino beam to Gran Sasso CERN 98-02, INFN/AE-98/05.
11. Results from KARMEN, B. Zeitnitz, these proceedings.
12. BOONE, A proposal for an experiment to measure $\nu_\mu \to \nu_e$ oscillations and ν_μ disappearance at the Fermilab Booster, December 7, 1997. Experiment 898.
13. Search for $\nu_\mu \to \nu_e$ oscillations at the CERN PS CERN-SPSC/97-21, SPSC/I216.

Part 6

Implications of the Solar and Atmospheric Neutrino Data

Nuclear Physics B (Proc. Suppl.) 77 (1999) 241–256
NUCLEAR PHYSICS B
PROCEEDINGS
SUPPLEMENTS

Implications of Solar and Atmospheric Neutrinos

Paul Langacker[a]

[a]Department of Physics and Astronomy
University of Pennsylvania, Philadelphia PA 19104-6396, USA

The importance of non-zero neutrino mass as a probe of particle physics, astrophysics, and cosmology is emphasized. The present status and future prospects for the solar and atmospheric neutrinos are reviewed, and the implications for neutrino mass and mixing in 2, 3, and 4-neutrino schemes are discussed. The possibilities for significant mixing between ordinary and light sterile neutrinos are described.

1. NEUTRINO MASS

Neutrino mass and properties are superb simultaneous probes of particle and astrophysics:

- Decays and scattering processes involving neutrinos have been powerful probes of the existence and properties of quarks, tests of QCD, of the standard electroweak model and its parameters, and of possible TeV-scale physics.
- Fermion masses in general are one of the major mysteries/problems of the standard model. Observation or nonobservation of the neutrino masses introduces a useful new perspective on the subject.
- Nonzero ν masses are predicted in most extensions of the standard model. They therefore constitute a powerful window on new physics at the TeV scale, intermediate scales (e.g., 10^{12} GeV), or the Planck scale.
- There may be a hot dark matter component to the universe. If so, neutrinos would be (one of) the most important things in the universe.
- The neutrino masses must be understood to fully exploit neutrinos as a probe of the Solar core, of supernova dynamics, and of nucleosynthesis in the big bang, in stars, and in supernovae.

2. THEORY OF NEUTRINO MASS

There are a confusing variety of models of neutrino mass. Here, I give a brief survey of the principle classes and of some of the terminology. For more detail, see [1].

A Weyl two-component spinor is a left (L)-handed[1] particle state, ψ_L, which is necessarily associated by CPT with a right (R)-handed antiparticle state[2] ψ^c_R. One refers to active (or ordinary) neutrinos as left-handed neutrinos which transform as $SU(2)$ doublets with a charged lepton partner. They therefore have normal weak interactions, as do their right-handed anti-lepton partners,

$$\begin{pmatrix} \nu_e \\ e^- \end{pmatrix}_L \overset{\text{CPT}}{\longleftrightarrow} \begin{pmatrix} e^+ \\ \nu^c_e \end{pmatrix}_R . \tag{1}$$

Sterile[3] neutrinos are $SU(2)$-singlet neutrinos, which can be added to the standard model and are predicted in most extensions. They have no ordinary weak interactions except those induced by mixing with active neutrinos. It is usually convenient to define the R state as the particle and the related L anti-state as the antiparticle.

$$N_R \overset{\text{CPT}}{\longleftrightarrow} N^c_L . \tag{2}$$

(Sterile neutrinos will sometimes also be denoted ν_s.)

Mass terms describe transitions between right (R) and left (L)-handed states. A Dirac mass

[1] The subscripts L and R really refer to the left and right chiral projections. In the limit of zero mass these correspond to left and right helicity states.
[2] Which is referred to as the particle or the antiparticle is a matter of convenience.
[3] Sterile neutrinos are often referred to as "right-handed" neutrinos, but that terminology is confusing and inappropriate when Majorana masses are present.

PII S0920-5632(99)00424-7

term, which conserves lepton number, involves transitions between two distinct Weyl neutrinos ν_L and N_R:

$$-L_{\rm Dirac} = m_D(\bar{\nu}_L N_R + \bar{N}_R \nu_L) = m_D \bar{\nu}\nu, \qquad (3)$$

where the Dirac field is defined as $\nu \equiv \nu_L + N_R$. Thus a Dirac neutrino has four components ν_L, ν_R^c, N_R, N_L^c, and the mass term allows a conserved lepton number $L = L_\nu + L_N$. This and other types of mass terms can easily be generalized to three or more families, in which case the masses become matrices. The charged current transitions then involve a leptonic mixing matrix (analogous to the Cabibbo-Kobayashi-Maskawa (CKM) quark mixing matrix), which can lead to neutrino oscillations between the light neutrinos.

For an ordinary Dirac neutrino the ν_L is active and the N_R is sterile. The transition is $\Delta I = \frac{1}{2}$, where I is the weak isospin. The mass requires $SU(2)$ breaking and is generated by a Yukawa coupling

$$-L_{\rm Yukawa} = h_\nu(\bar{\nu}_e \bar{e})_L \begin{pmatrix} \varphi^0 \\ \varphi^- \end{pmatrix} N_R + H.C. \qquad (4)$$

One has $m_D = h_\nu v/\sqrt{2}$, where the vacuum expectation value (VEV) of the Higgs doublet is $v = \sqrt{2}\langle\varphi^o\rangle = (\sqrt{2}G_F)^{-1/2} = 246$ GeV, and h_ν is the Yukawa coupling. A Dirac mass is just like the quark and charged lepton masses, but that leads to the question of why it is so small: one requires $h_{\nu_e} < 10^{-10}$ to have $m_{\nu_e} < 10$ eV.

A Majorana mass, which violates lepton number by two units ($\Delta L = \pm 2$), makes use of the right-handed antineutrino, ν_R^c, rather than a separate Weyl neutrino. It is a transition from an antineutrino into a neutrino. Equivalently, it can be viewed as the creation or annihilation of two neutrinos, and if present it can therefore lead to neutrinoless double beta decay. The form of a Majorana mass term is

$$\begin{aligned} -L_{\rm Majorana} &= \frac{1}{2} m_T(\bar{\nu}_L \nu_R^c + \bar{\nu}_R^c \nu_L) = \frac{1}{2} m_T \bar{\nu}\nu \\ &= \frac{1}{2} m_T(\bar{\nu}_L C \bar{\nu}_L^T + H.C.), \end{aligned} \qquad (5)$$

where $\nu = \nu_L + \nu_R^c$ is a self-conjugate two-component state satisfying $\nu = \nu^c = C\bar{\nu}^T$, where C is the charge conjugation matrix. If ν_L is active then $\Delta I = 1$ and m_T must be generated by either an elementary Higgs triplet or by an effective operator involving two Higgs doublets arranged to transform as a triplet.

One can also have a Majorana mass term

$$-L_{\rm Majorana} = \frac{1}{2} m_N(\bar{N}_L^c N_R + \bar{N}_R N_L^c) \qquad (6)$$

for a sterile neutrino. This has $\Delta I = 0$ and thus can be generated by the VEV of a Higgs singlet[4].

Some of the principle classes of models for neutrino mass are:

- A triplet majorana mass m_T can be generated by the VEV v_T of a Higgs triplet field. Then, $m_T = h_T v_T$, where h_T is the relevant Yukawa coupling. Small values of m_T could be due to a small scale v_T, although that introduces a new hierarchy problem. The simplest implementation is the Gelmini-Roncadelli (GR) model [2], in which lepton number is spontaneously broken by v_T. The original GR model is now excluded by the LEP data on the Z width.

- A very different class of models are those in which the neutrino masses are zero at the tree level (typically because no sterile neutrino or elementary Higgs triplets are introduced), but only generated by loops [3], *i.e.*, radiative generation. Such models generally require the *ad hoc* introduction of new scalar particles at the TeV scale with nonstandard electroweak quantum numbers and lepton number-violating couplings. They have also been introduced in an attempt to generate large electric or magnetic dipole moments. They also occur in some supersymmetric models with cubic R parity violating terms in the superpotential [4].

- In the seesaw models [5], a small Majorana mass is induced by mixing between an active neutrino and a very heavy Majorana sterile neutrino M_N. The light (essentially active) state has a naturally small mass

$$m_\nu \sim \frac{m_D^2}{M_N} \ll m_D. \qquad (7)$$

[4]In principle this could also be generated by a bare mass, but this is usually forbidden by higher symmetries in extensions of the standard model.

There are literally hundreds of seesaw models, which differ in the scale M_N for the heavy neutrino (ranging from the TeV scale to grand unification scale), the Dirac mass m_D which connects the ordinary and sterile states and induces the mixing (e.g., $m_D \sim m_u$ in most grand unified theory (GUT) models, or $\sim m_e$ in left-right symmetric models), the patterns of m_D and M_N in three family generalizations, etc. One can also have mixings with heavy neutralinos in supersymmetric models with R parity breaking [4], induced either by bilinears connecting Higgs and lepton doublets in the superpotential or by the expectation values of scalar neutrinos.

- Superstring models often predict the existence of higher-dimensional (nonrenormalizable) operators (NRO) such as

$$-L_{\text{eff}} = \bar{\psi}_L H \left(\frac{S}{M_{\text{str}}}\right)^P \psi_R + H.C., \tag{8}$$

where H is the ordinary Higgs doublet, S is a new scalar field which is a singlet under the standard model gauge group, and $M_{\text{str}} \sim 10^{18}$ GeV is the string scale. In many cases S will acquire an intermediate scale VEV (e.g., 10^{12} GeV), leading to an effective Yukawa coupling

$$h_{\text{eff}} \sim v \left(\frac{\langle S\rangle}{M_{\text{str}}}\right)^P \ll v. \tag{9}$$

Depending on the dimensions P of the various operators and on the scale $\langle S\rangle$, it may be possible to generate an interesting hierarchy for the quark and charged lepton masses and to obtain naturally small Dirac neutrino masses [6]. Similarly, one may obtain triplet and singlet Majorana neutrino masses, m_T and m_N by analogous higher-dimensional operators. The former are small. Depending on the operators [6] the latter may be either small, leading to the possibility of significant mixing between ordinary and sterile neutrinos [7], or large, allowing a conventional seesaw.

- Mixed models, in which both Majorana and Dirac mass terms are present, will be further discussed in the section on sterile neutrinos.

3. SOLAR NEUTRINOS

Tremendous progress has been made recently in solar neutrinos [8]. For many years there was only one experiment, while now there are a number that are running or finished, and more are coming on line soon. The original goal of using the solar neutrinos to study the properties of the solar core underwent a 30 year digression on the study of the properties of the neutrino itself. The quality of the experiments themselves and of related efforts on helioseismology, nuclear cross sections, and solar modeling is such that the revised goal of simultaneously studying the properties of the Sun and of the neutrinos is feasible.

3.1. Experiments

The experimental situation is very promising. We now have available the results of five experiments, Homestake (chlorine) [9], Kamiokande [10], GALLEX [11], SAGE [12], and Superkamiokande [10]. Especially impressive are the successful ^{51}Cr source experiments for SAGE and GALLEX (which probe a combination of the extraction efficiencies and the neutrino absorption cross section, yielding $0.95 \pm 0.07^{+0.04}_{-0.03}$ of the expected rate), and the successful ^{71}As spiking experiment completed at the end of the GALLEX run to test the extraction efficiency (yielding $R = 1.00 \pm 0.01$ for the ratio of actual to expected extractions).

Coming soon, there should be results from SNO, Borexino, The Gallium Neutrino Observatory (GNO), and the next phase of SAGE, which will yield much more detailed, precise, or model independent information on the ^8B (SNO [13]), ^7Be (Borexino [14]), and *pp* (GNO, SAGE) neutrinos. Future generations of even more precise experiments should especially be sensitive to the ^7Be and *pp* neutrinos [15]. The overall goal of the program should be very ambitious, *i.e.*, to measure the arriving flux of $\nu_e, \nu_{\mu+\tau}$, and ν_s (sterile neutrinos), and even possible antineutrinos, for each of the initial flux components, as well as to measure or constrain possible spectral distortions, day-night (earth) effects, seasonal and solar cycle variations, and mixed (*e.g.*, simultaneous spectral and day-night) effects.

Table 1
Results of Solar neutrino experiments, compared with the predictions of BP 98. The chlorine and gallium results are in units of SNU ($10^{-36}s^{-1}$ captures per target atom), and the water Cerenkov results are in units of $10^6/cm^2s$.

	experiment	BP-98
Homestake (chlorine)	2.56 ± 0.23	$7.7^{+1.2}_{-1.0}$
GALLEX, SAGE (gallium)	72.2 ± 5.6	129^{+8}_{-6}
Kamiokande, SuperK ($\nu e \to \nu e$)	2.44 ± 0.10	$5.15 \times (1^{+0.19}_{-0.14})$

3.2. Interpretation

The observed fluxes are in strong disagreement with the predictions of the standard solar model (SSM). The overall rates are compared with the predictions of the new Bahcall-Pinsonneault 1998 (BP 98) model [16] in Table 1, where it is seen that all of the fluxes are much lower than the expectations. BP 98 contains a number of refinements compared to earlier theoretical calculations, but the most important changes are a 20% (1.3 σ) lower ^8B flux, as described below, and 1.1 σ decreases in the ^{37}Cl and ^{71}Ga capture rates.

Recent results in helioseismology [16–20] leave little room for deviations from the standard solar model. The eigen-frequencies effectively measure the sound speed T/μ, where T and μ are respectively the temperature and density, as a function of radial position, down to 5% of the solar radius. The results agree with the predictions of BP 98 to $\sim 10^{-3}$, even though T and μ individually vary by large values over the radius of the Sun. This leaves very little room for non-standard solar models (NSSM), which would typically have to deviate by several percent to have much impact on the neutrino flux predictions. The only aspect of the SSM relevant to the neutrino fluxes that is not severely constrained are nuclear cross sections, especially S_{17} and S_{34}, which are respectively proportional to the cross sections for ^8B and ^7Be production, and to the absorption cross sections for the radiochemical experiments.

The experimental and theoretical status of the nuclear cross sections were critically examined at a workshop at the Institute for Nuclear Theory in 1997 (INT 97) [20,21]. The participants recommended a lower S_{17}, by relying on the best documented individual measurements rather than an average, and also a larger uncertainty in S_{34}, both of which were incorporated in BP 98. Haxton has recently argued [22] that there are still considerable uncertainties in the Ga absorption cross sections, but this possibility is strongly disfavored by the ^{51}Cr source and ^{71}As spiking experiments.

Even the relatively large shift in S_{17} advocated by INT 97 and used by BP 98 does little to change the basic disagreement between the observations and the standard solar model. Even if a particular NSSM could be consistent with helioseimology, it would be difficult to account for the observations. The Kamiokande and Superkamiokande results can be regarded (in the absence of neutrino oscillations) as a measurement of the ^8B flux. Subtracting this "experimental" ^8B flux from either the gallium or chlorine predictions, the observed fluxes are still inconsistent with the observed solar luminosity.

This line of reasoning is developed in the "model-independent" analyses of the neutrino flux components [23–25], which can be viewed as a measurement of "global" spectral distortions. The idea is that all plausible astrophysical or nuclear physics modifications of the standard solar model do not significantly distort the spectral shape of the pp or ^8B neutrinos: all that they can do is modify the overall magnitude of the pp, ^7Be, ^8B, and minor flux components. Furthermore, the observed solar luminosity places a linear constraint on the pp, ^7Be, and CNO fluxes (provided that the time scale for changes in the solar core is long compared to the 10^4 yr required for a photon to diffuse to the surface).

By combining the different experiments, each class of which has a different spectral sensitivity, one concludes that

$$\frac{\phi(^7Be)}{\phi(^7Be)_{\rm SSM}} \ll \frac{\phi(^8B)}{\phi(^8B)_{\rm SSM}}, \tag{10}$$

where SSM refers to the standard solar model predictions. The same result holds even if one

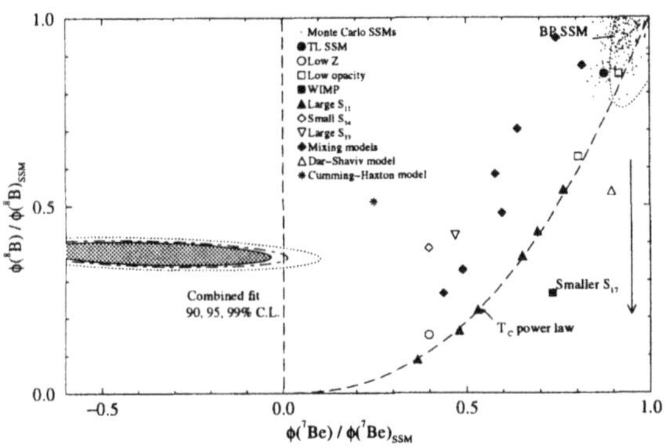

Figure 1. Allowed regions for the ^7Be and ^8B fluxes (normalized by BP 98), compared with the predictions and uncertainties in the SSM and various non-standard solar models. Courtesy of N. Hata.

discards any one of the three types of experiment (chlorine, gallium, water), *or* ignores the luminosity constraint. No plausible astrophysical model has succeeded in suppressing ^7Be neutrinos significantly more than ^8B neutrinos, mainly because ^8B is made from ^7Be. Models with a lower core temperature or with a lower S_{17} do not come anywhere near the data. The Cumming-Haxton model [26] with large ^3He diffusion comes closest, but even that is far from the data. That model is probably also excluded by helioseismology, but Haxton has argued [22] that final judgment should wait until a self-consistent model with ^3He diffusion is constructed to be compared with the helioseismology data.

3.3. Possible Solutions

As discussed in the previous section, an astrophysical/nuclear explanation of the solar neutrinos experiments is unlikely. The most likely particle physics explanations include:

- A matter enhanced (MSW) transition of ν_e into ν_μ or ν_τ. There are the familiar small (SMA) and large (LMA) mixing angle solutions [23] with $\Delta m^2 \sim 10^{-5}$ eV^2, as well as the low mass (LOW) solution with $\Delta m^2 \sim 10^{-7}$ eV^2 and near maximal mixing. The latter is a very poor fit, but sometimes shows up in fits at the 99% cl.

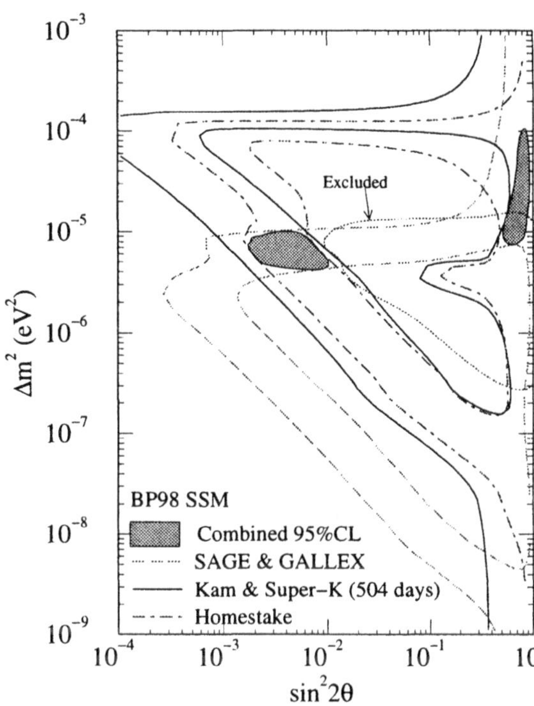

Figure 2. Allowed MSW solutions, not including Superkamiokande spectral data. Courtesy of N. Hata.

- There is also a small mixing angle MSW solution for ν_e into a sterile neutrino ν_s. The major difference between $\nu_{\mu,\tau}$ and ν_s, and the reason there is no LMA solution, is that in the first case the $\nu_{\mu,\tau}$ can scattering elastically from electrons in the water Cerenkov experiments, with about $1/6^{th}$ the ν_e cross section, leading to a lower survival probability for ν_e than for astrophysical or sterile neutrino solutions. There is also a small difference for the MSW conversion rate for sterile neutrinos in the Sun, but that is proportional to the neutron density, and is much less important.

- The vacuum ("just so" [27]) oscillation solutions [23], with near maximal mixing and $\Delta m^2 \sim 10^{-10}$ eV^2 are another possibility. These are somewhat fine-tuned, with Δm^2 such that the Earth-Sun distance is at roughly half an oscillation length, $L_{\rm osc}$, or an odd multiple. Since $L_{\rm osc} = 4\pi E/\Delta m^2$, one expects a significant variation of the ν_e survival probability with neutrino energy.

- The above solutions are such that only two neutrinos are important for the Solar neutrinos. However, it is possible that transitions between all three neutrinos are important. There could be generalized MSW solutions involving more than one value of Δm^2, or mixed MSW and vacuum solutions [28]. In both cases, there could be considerably different spectral distortions than in the two-neutrino case.

Other possibilities include:

- Maximal mixing [29] (i.e., vacuum oscillations with $\Delta m^2 \gg 10^{-10}$ eV^2), combined with a low S_{17}. Such solutions lead to an energy independent suppression of the ν_e survival probability. Even allowing a suppressed ^8B production rate, this possibility is viable only if one ignores (or greatly expands the uncertainties in) the Homestake Chlorine experiment.

- RSFP [30] (resonant spin flavor precession), involving rotations of left handed neutrinos into sterile right handed neutrinos, combined with MSW flavor transitions. These were motivated by possible hints (not confirmed by other experiments) of time dependence correlated with the Sunspot activity in the chlorine experiment. This could only occur if there are extremely large neutrino electric or magnetic dipole moments or transition moments, which would present a considerable challenge to the model builder. Although such effects have not been reported by other groups, there is still a somewhat surprising difference in rates observed by the GALLEX collaboration in their third and fourth data taking intervals. However, this could also be a statistical fluctuation. In any case, such RSFP effects could be probed experimentally by studying the $\bar{\nu}_e$ and $\bar{\nu}_\mu$ spectra [31].

- Flavor changing neutral current effects [32], possibly generated by R-parity violating terms in supersymmetry, could be an alternative means of generating enhanced neutrino flavor conversions in the Sun.

- The possible violation of Lorentz invariance [33] could affect not only the Solar neutrinos, but could also be relevant to the observed ultra high energy cosmic rays.

- There could be a lepton flavor dependent violation of the the equivalence principle [34].

Perhaps the most important possibility or complication is that more than one thing could be going on simultaneously. There could be any of the above effects in conjunction with non-standard properties of the Sun or nuclear cross sections. Many but not all such NSSM possibilities are excluded by helioseismology and neutrino source experiments. While it is very unlikely than such effects could by themselves account for the data, their combination with new neutrino properties could considerably confuse the interpretation of future experimental results. This is one or the reasons that it is important to have as many independent precise experimental results as possible.

3.4. Needs

To distinguish the many possibilities we need as much precise data as possible. Especially useful are observables that are independent of or insensitive to the initial ν_e fluxes, and therefore to the astrophysical and nuclear cross section uncertainties. Such observables include:

- The neutral to charged current interaction ratio (NC/CC), which will be measured by SNO for deuteron dissociation. Since the NC cross section is the same for all active neutrinos, the NC rate measures the sum of the ν_e, ν_μ, and ν_τ fluxes, while the CC only measures ν_e. An anomalous NC/CC ratio would provide definitive evidence for transitions of ν_e into ν_μ or ν_τ, either by MSW or vacuum oscillations. Although the NC measurement is difficult, SNO should have the requisite sensitivity. A confirmation could be obtained by comparing the SNO CC rate with the fluxes determined in $\nu e \to \nu e$ measurements, since $\nu_{\mu,\tau}$ also contribute to the latter, with about $1/6^{\rm th}$ the ν_e cross section. (The Borexino experiment will similarly allow an indirect determination of the transitions of ^7Be neutrinos into $\nu_{\mu,\tau}$ by comparing with the ν_e flux inferred from radiochemical experiments.)

  Transitions of ν_e into a sterile neutrino ν_s would not lead to an anomalous NC/CC ratio. This would make it much harder to verify ν_s transitions, but would serve as evidence for sterile neutrinos if MSW or vacuum oscillations are established by other means.

- There is no known astrophysical mechanism that can significantly distort the ^8B neutrino spectrum from the expected β decay shape. Not only would a spectral distortion establish a non-astrophysical solution to the solar neutrinos, but it would be a powerful probe of the mechanism. Study of the ^8B spectrum can be viewed as a cleaner extension (by individual experiments) of the "global" spectral distortion inferred from the combined experiments.

  One expects significant spectral distortions for the MSW SMA solution, for vacuum oscillations, and for hybrid solutions, but not for the LMA solution. The ratio of observed to expected spectrum can be conveniently parametrized by the first two moments [35], i.e., a linear approximation, for the SMA case, while the other cases can exhibit more complicated shapes. Measurement of the spectral distortion is very difficult, and requires excellent energy calibrations and extending the measurement to as low an energy as possible. Both SuperKamiokande and SNO have the capability to measure a spectral distortion. SuperK has the advantage of higher statistics. However, the ν energy is shared between the final electron and neutrino, so any spectral distortion is partially washed out in the observed e^- spectrum. SNO, on the other hand, has the advantage that the electron in the CC reaction carries all of the neutrino energy (plus the known binding energy), leading to a harder electron spectrum and an essentially direct measurement of the ν spectrum.

  One of the highlights of this conference was the preliminary new statistics-limited SuperKamiokande spectrum, from around 6.7 to 14.5 MeV, obtained after a series of careful calibrations of their detector using an electron Linac [10]. The lower energy data are consistent with no distortions, but there is evidence for a significant excess of events in the three energy bins above 13 MeV. These data, for the first time, give a statistically significant indication of a spectral distortion: the no oscillation hypothesis (and also LMA solution) is disfavored at the 95-99% CL level. The SMA MSW solution is also a very poor fit, although it is allowed at 95% CL. The best fit favors vacuum oscillations. The favored $\Delta m^2 \sim 4\times10^{-10}$ eV^2 gives a much better fit to the data than for the lower range Δm^2 around 10^{-10} eV^2 found in recent global analyses of the total event rates. However, new studies based on BP 98 with its larger S_{17} allow a larger Δm^2, consistent with the spectral distortions.

  An alternate interpretation of the high energy excess is that the flux of hep neutrinos ($^3He + p \to {}^4He + e^+ + \nu_e$) has been seriously underestimated. Their flux would have to be larger by a factor of twenty or so from the usual estimates for them to contribute significantly to the excess, but it has been emphasized that there is no direct experimental measure of or rigorous theoretical bound on the cross section [36]. The issue can be resolved by a careful study of the energy range 14-18.8 MeV, above the endpoint of the ^8B spectrum. (The highest energy SuperK bin is centered above this endpoint, but there is a significant energy uncertainty.)

The SuperK spectrum has important implications, but it is still preliminary. In additional to finalizing the analysis, additional lower energy points are expected that should help clarify the situation.

- For some regions of MSW parameters, one expects an asymmetry between day and night event rates due to regeneration of ν_e at night as the converted neutrinos travel through the Earth [37]. Superkamionde has binned their data for daytime and for a number of different nighttime zenith angles (i.e., different paths through the earth). They see no evidence for a zenith angle dependence, and their overall day-night asymmetry is

$$\frac{D-N}{D+N} = -0.023 \pm 0.020 \pm 0.014, \qquad (11)$$

where D (N) refers to day (night) rates and the first (second) error is statistical (systematic). The absence of an effect excludes a significant region of MSW parameter space independent of the details of the solar model (and with only a small uncertainly from the Earth's density profile). This excludes the lower Δm^2 part of the LMA solution, but has little impact on the SMA solution. (The part of the SMA solution with the largest $\sin^2 2\theta$ was expected to have a barely observable day-night asymmetry, but the effect is predicted to be smaller with the new BP 98 fluxes, which shift the SMA region to slightly smaller $\sin^2 2\theta$.)

Several authors have emphasized recently that the Earth effect is signficantly enhanced for neutrinos passing through the core of the Earth [38]. (There is an analogous effect for atmospheric neutrinos.) This parametric (or oscillation length) resonance, in which the oscillation length is comparable to the diameter of the core, was included automatically in previous numerical studies, but not explicitly commented on. It is larger for transitions into $\nu_{\mu,\tau}$ than for ν_s. Since relatively few of the solar neutrinos pass through the core for the existing high latitude detectors, it has been suggested that there should be a dedicated experiment at low latitude [39].

- For vacuum oscillations [23], the Earth-Sun distance is typically at a node of the oscillations. This is somewhat fine-tuned, leading to the name "just so". Since the oscillation length is $4\pi E/\Delta m^2$ there is a strong energy dependence to the survival probability. One also expects a strong seasonal variation, due to the eccentricity of the Earth's orbit. However, the seasonal variation can be partially washed out as one averages over energies, so one should ideally measure the spectral shape binned with respect to the time of the year [40].

- RSFP could lead to long term variations in the neutrino flux, e.g., correlated with Sunspots or Solar magnetic fields. Other changing magnetic effects could conceivably alter the solar neutrinos in other ways, e.g., by changing the local density. Only the Homestake experiment has seen any significant hint of a time variation, and that hint has been considerably weakened by more recent Homestake data. Nevertheless, it is conceivable that time dependent effects are energy dependent, and therefore different experiments have different sensitivity. They could also have been somewhat hidden in the water Cerenkov experiments because of the neutral current. It would be useful to run all of the experiments simultaneously through a solar cycle.

- RSFP [30] could also lead to the production of $\bar{\nu}_e$ which can be observed in the SNO detector by the delayed coincidence of the γ ray emitted by the capture of the neutron from $\bar{\nu}_e p \to e^+ n$.

3.5. Outlook

The model independent observables that can be measured by SuperK and SNO for the ^8B neutrinos should go far towards distinguishing the different possibilities. However, it will be especially difficult to establish transitions into sterile neutrinos. It will also be very difficult to sort out what is happening in a three-flavor or hybrid scenario, such as MSW transitions combined with non-standard solar physics. For these reasons, we would like to have accurate information on spectral distortions, day-night effects (especially for neutrinos passing through the Earth core), NC/CC ratios, and absolute fluxes arriving at the Earth for the ^7Be and pp neutrinos as well. There is a strong need for the next generations of experiments.

A challenging but realistic goal is to simultaneously establish the neutrino mechanism(s) (e.g., MSW SMA solution), determine the neutrino parameters, *and* study the Sun [41]. Even with existing data, if one assumed two-flavor MSW but allowed an arbitrary solar core temperature T_C, it was possible to simultaneously determine the MSW parameters (with larger uncertainties than when the SSM is assumed) and T_C, with the result that $T_C = 0.99^{+0.02}_{-0.03}$ with respect to the SSM prediction of 1.57 $\times 10^7$ K [24]. In the future, it should be possible to determine the neutrino parameters and simultanously the ^8B and ^7Be fluxes, for comparison with the SSM predictions. It will also be possible to constrain density fluctuations in the Sun [42], which can smear out the MSW affects. However, recent estimates suggest that such effects are negligible [43].

To fully exploit the future data, it will be important to carry out global analyses of all of the observables in all of the experiments (possibly incorporating helioseismology data as well). Global analyses are difficult because of difficulties with systematic errors. However, they often contain more information than the individual experiments, and allow uniform treatment of theoretical uncertainties. For this purpose, it is important that each experiment publish all of their data, such as double binning the data with respect to energy and zenith angle, including full systematics and correlations.

4. ATMOSPHERIC NEUTRINOS

Although the prediction for the absolute number of μ or e produced by the interactions of neutrinos produced in cosmic ray interactions in the atmosphere has a theoretical uncertainty of around 20%, it is believe that the ratio $N(\mu)/N(e)$ can be predicted to within 5% [44]. To zeroth approximation, the ratio is just two, independent of the details of the cosmic ray flux or interactions, because each produced pion decays into two ν_μ and one ν_e (I am not distinguishing ν from $\bar{\nu}$), and for energies large compared to m_μ the interaction cross sections are the same. Of course, the actual ratio depends on the neutrino energies, and therefore on the details of the hadronic energies, polarization of the intermediate muons from π decay, etc.

For years the ratio R of observed $N(\mu)/N(e)$, normalized by the predicted value, found in the water Cerenkov experiments (Kamiokande, IMB, SuperKamiokande) has been around 0.6 [45]. This has recently been confirmed by the higher SuperK [46] statistics (R = 0.63(3)(5) for sub-GeV events and 0.65(5)(8) for multi-GeV), and independently by the iron calorimeter experiment at Soudan [47] (0.58(11)(5)) and by Macro [48] (0.53(15) for upward events and 0.71(21) for stopping or downgoing events). This depletion of μ events suggests the possibility of ν_μ oscillations into ν_e, ν_τ, or ν_s, with near-maximal mixing ($\sin^2 2\theta > 0.8$) and $\Delta m^2 \sim 10^{-3} - 10^{-2}$ eV^2.

To confirm oscillations, more detailed information is needed. Already, the CHOOZ [49] (France) reactor $\bar{\nu}_e$ disappearance experiment excludes the $\nu_\mu \to \nu_e$ interpretation of the atmospheric neutrino anomaly for $\Delta m^2 > 10^{-3}$ eV^2. This should be extended by the coming Palo Verde experiment [50], and the planned KamLand [51] experiment at Kamiokande (sensitive to many nearby reactors) should extend the senstivity down to the MSW LA solar neutrino range.

In the future [52], there will also be accelerator long baseline experiments for $\nu_\mu \to \nu_{e,\tau}$ appearance, or ν_μ disappearance (into $\nu_{e,\tau,s}$). The KEK to Kamiokande (K2K) experiment will be sensitive to ν_μ disappearance down to $\Delta m^2 \sim 5\times 10^{-3}$ eV^2, while the Fermilab to Soudan (MINOS) experiment will probe both appearance and disappearance down to 10^{-3} eV^2. There are also proposals for a CERN to Gran Sasso experiment (ICARUS, OPERA), which would be sensitive to most of the parameter range suggested by Superkamiokande. These experiments should be able to confirm or refute the atmospheric neutrino oscillations, except[5] possibly for the smallest $\Delta m^2 \sim 10^{-3}$ eV^2.

Much more detailed information can be derived from the atmospheric neutrino data itself, by searching for indications of the $\sin^2(1.27\Delta m^2 L/E)$ dependence of the transition probability characteristic of neutrino oscillations. (L is the distance traveled and E is the neutrino energy.) This can be studied by considering the zenith angle distribution for fixed neutrino energy (in practice, the data is divided into sub-GeV and mutli-GeV bins), or by up-down asymmetries $(U-D)/(U+D)$, where U and D are respectively the number of up and downgoing muons or electrons [53]. The data can also be plotted as a function of L/E, but that is less direct since the full neutrino energy is not measured on an event by event basis in the water Cerenkov experiments.

The Kamiokande collaboration observed an indication of oscillations in their zenith angle distribution for contained events [54]. However, the new Superkamiokande zenith angle distributions for contained events have much better statistics. They strongly indicate a zenith angle distribution in muon events consistent with oscillations, with an enhanced effect in the multi-GeV sample, consistent with expectations. There is no anomaly or excess in the electron events. This implies that ν_μ is oscillating into ν_τ or possibly a sterile neutrino ν_s, and not into ν_e. The latter result confirms the conclusions of CHOOZ. (Subdominant oscillations into ν_e in three-neutrino schemes are still possible.) The SuperK events virtually establish neutrino oscillations. Independent evidence is obtained by the zenith angle distributions for upward through-going muon events from SuperK, MACRO[6], and very preliminary results from SOUDAN.

Future atmospheric neutrino observations could possibly shed further light on the question of whether ν_μ is oscillating into ν_τ or into ν_s, although they are all very difficult. These include (a) subtle (e.g., parametric resonance) effects on neutrinos propagating through the Earth's core [55], which would affect $\nu_\mu \to \nu_s$, but not $\nu_\mu \to \nu_\tau$ (because ν_μ and ν_τ have the same neutral current interactions). In either scenario, secondary $\nu_\mu \to \nu_e$ oscillations would also be modified by Earth core effects. (b) The NC/CC ratio, including its zenith angle distribution and up-down asymmetry [56]. The NC rate could in principle be measured in $\nu N \to \nu \pi^0 X$, although this is a very difficult measurement. The preliminary SuperK result [10] $R(\pi^0/e) = 0.93(7)(19)$ on the ratio or π^0 to e events compared to expectations slightly favors $\nu_\mu \to \nu_\tau$ but does not exclude $\nu_\mu \to \nu_s$. (c) Direct observation of events in which ν_τ produces a τ would establish $\nu_\mu \to \nu_\tau$ oscillations [57]. However, this is extremely difficult.

There may also be significant three neutrino effects. For example, even if the dominant transition for the atmospheric neutrinos involves $\nu_\mu \to \nu_\tau$, there could be important subdominant ν_e effects.

There have been several careful phenomenological analyses of the atmospheric neutrino data in two neutrino and three neutrino mixing schemes [58]. One important theoretical issue posed by the atmospheric neutrinos, is why is there nearly maximal mixing (i.e., $\sin^2 2\theta \sim 1$), when most theoretical schemes involving hierarchies of neutrino masses, as well as the analogs in the quark mixing sector, yield small mixings.

5. IMPLICATIONS FOR NEUTRINO MIXING

5.1. The Global Picture

Various scenarios for the neutrino spectrum are possible, depending on which of the experimental indications one accepts. The simplest scheme, which accounts for the Solar (S) and Atmospheric (A) neutrino results, is that there are just three light neutrinos, all active, and that the mass eigenstates ν_i have masses in a hierarchy, analogous to the quarks and charged leptons. In that case, the atmospheric and solar neutrino mass-squared differences are measures of the mass-squares of the two heavier states, so that $m_3 \sim (\Delta m^2_{\rm atm})^{1/2} \sim 0.03 - 0.1$ eV;

[5]The long baseline experiments were proposed when the earlier Kamiokande results suggested a somewhat larger $\Delta m^2 \sim 10^{-2}$ eV^2.

[6]The MACRO results [48] are not in very good agreement with oscillations, but the oscillation hypothesis nevertheless fits much better than the no-oscillation case.

$m_2 \sim (\Delta m^2_{solar})^{1/2} \sim 0.003$ eV (for MSW) or $\sim 10^{-5}$ eV (vacuum oscillations), and $m_1 \ll m_2$. The weak eigenstate neutrinos $\nu_a = (\nu_e, \nu_\mu, \nu_\tau)$ are related to the mass eigenstates ν_i by a unitary transformation $\nu_a = U_{ai}\nu_i$. If one makes the simplest assumption (from the Superkamiokande and CHOOZ data), that the ν_e decouples entirely from the atmospheric neutrino oscillations, $U_{e3} = 0$, (of course, one can relax this assumption somewhat) and ignores possible CP-violating phases [59], then

$$\begin{pmatrix} \nu_e \\ \nu_\mu \\ \nu_\tau \end{pmatrix} = \begin{pmatrix} 1 & 0 & 0 \\ 0 & c_\alpha & -s_\alpha \\ 0 & s_\alpha & c_\alpha \end{pmatrix} \times \begin{pmatrix} c_\theta & -s_\theta & 0 \\ s_\theta & c_\theta & 0 \\ 0 & 0 & 1 \end{pmatrix} \begin{pmatrix} \nu_1 \\ \nu_2 \\ \nu_3 \end{pmatrix}, \quad (12)$$

where α and θ are mixing angles associated with the atmospheric and solar neutrino oscillations, respectively, and where $c_\alpha \equiv \cos\alpha$, $s_\alpha \equiv \sin\alpha$, and similarly for c_θ, s_θ.

For maximal atmospheric neutrino mixing, $\sin^2 2\alpha \sim 1$, this implies $c_\alpha = s_\alpha = 1/\sqrt{2}$, so that

$$U = \begin{pmatrix} c_\theta & -s_\theta & 0 \\ \frac{s_\theta}{\sqrt{2}} & \frac{c_\theta}{\sqrt{2}} & -\frac{1}{\sqrt{2}} \\ \frac{s_\theta}{\sqrt{2}} & \frac{c_\theta}{\sqrt{2}} & \frac{1}{\sqrt{2}} \end{pmatrix}. \quad (13)$$

For small θ, this implies that $\nu_{3,2} \sim \nu_{+,-} \equiv (\nu_\tau \pm \nu_\mu)/\sqrt{2}$ participate in atmospheric oscillations, while the solar neutrinos are associated with a small additional mixing between ν_e and ν_-. Another limit, suggested by the possibility of vacuum oscillations for the solar neutrinos, is $\sin^2 2\theta \sim 1$, or $c_\theta = s_\theta = 1/\sqrt{2}$, yielding

$$U = \begin{pmatrix} \frac{1}{\sqrt{2}} & -\frac{1}{\sqrt{2}} & 0 \\ \frac{1}{2} & \frac{1}{2} & -\frac{1}{\sqrt{2}} \\ \frac{1}{2} & \frac{1}{2} & \frac{1}{\sqrt{2}} \end{pmatrix}, \quad (14)$$

which is referred to as bi-maximal mixing [60]. A number of authors have discussed this pattern and how it might be obtained from models, as well as how much freedom there is to relax the assumptions of maximal atmospheric and solar mixing (the data actually allow $\sin^2 2\alpha \gtrsim 0.8$ and $\sin^2 2\theta \gtrsim 0.6$) or the complete decoupling of ν_e from the atmospheric neutrinos. Another popular pattern,

$$U = \begin{pmatrix} \frac{1}{\sqrt{2}} & -\frac{1}{\sqrt{2}} & 0 \\ \frac{1}{\sqrt{6}} & \frac{1}{\sqrt{6}} & -\frac{2}{\sqrt{6}} \\ \frac{1}{\sqrt{3}} & \frac{1}{\sqrt{3}} & \frac{1}{\sqrt{3}} \end{pmatrix}, \quad (15)$$

known as democratic mixing [61], yields maximal solar oscillations and near-maximal (8/9) atmospheric oscillations.

In this hierarchical pattern, the masses are all too small to be relevant to mixed dark matter (in which one of the components of the dark matter is hot, i.e., massive neutrinos) or to neutrinoless double beta decay ($\beta\beta_{0\nu}$). However, the solar and atmospheric oscillations only determine the differences in mass squares, so a variant on this scenario is that the three mass eigenstates are nearly degenerate rather than hierarchical [62], with small splittings associated with Δm^2_{atm} and Δm^2_{solar}. For the common mass m_{av} in the 1-several eV range, the hot dark matter could account for the dark matter on large scales (with another, larger, component of cold dark matter accounting for smaller structures) [63]. If the neutrinos are Majorana they could also lead to $\beta\beta_{0\nu}$ [64]. Current limits imply an upper limit of

$$\langle m_{\nu_e} \rangle = \sum_i \eta_i U^2_{ei} |m_i| < 0.46 - 1 \text{ eV}, \quad (16)$$

on the effective mass for a mixture of light Majorana mass eigenstates, where η_i is the CP-parity of ν_i and the uncertainty on the right is due to the nuclear matrix elements. (There is no constraint on Dirac neutrinos.) The combination of small $\langle m_{\nu_e} \rangle \ll m_{av}$, maximal atmospheric mixing, and $U_{e3} = 0$ would imply cancellations, so that $\eta_1 \eta_2 = -1$ and $c_\theta = s_\theta = 1/\sqrt{2}$, i.e., maximal solar mixing. Even the more stringent limit in (16) is large enough that there is room to relax all of these assumptions considerably. Nevertheless, there is strong motivation to try to improve the $\beta\beta_{0\nu}$ limits.

The LSND experiment [65] has reported evidence for $\nu_\mu \to \nu_e$ and $\bar{\nu}_\mu \to \bar{\nu}_e$ oscillations with $\Delta m^2_{LSND} \sim 1$ eV^2 and small mixing $\sim 10^{-3} - 10^{-2}$, while the KARMEN experiment sees no candidates. KARMEN [66] is sensitive to most of the same parameter range as LSND, although there is a small window of oscillation parameters for which both experiments are consistent. A resolution of the situation may have to wait for the mini-BOONE experiment at Fermilab. However, it is interesting to consider the implications if the LSND result is confirmed. In that case, there are three distinct mass-squared differences, $\Delta m^2_{\rm LSND} \sim 1$ eV^2, $\Delta m^2_{\rm atm} \sim 10^{-3} - 10^{-2}$ eV^2, and $\Delta m^2_{\rm solar} \sim 10^{-5}$ eV^2 (MSW) or 10^{-10} eV^2 (vacuum), implying the need for a fourth neutrino[7]. Since the Z lineshape measurements at LEP only allow 2.992 ± 0.011 light, active neutrinos [70], any light fourth neutrino would have to be sterile, ν_s.

Several mass patterns for the four neutrinos have been suggested [71]. (There course also be more than four light neutrinos [72].) To be consistent with both LSND and CHOOZ, states containing ν_μ and ν_e must be separated by about 1 eV. Assuming the atmospheric neutrinos involve $\nu_\mu \to \nu_\tau$, one could have nearly degenerate $\nu_{+,-} \equiv (\nu_\tau \pm \nu_\mu)/\sqrt{2}$ at around 1 eV, with the solar neutrinos described by a dominantly ν_s state at ~ 0.003 eV or $\sim 10^{-5}$ eV and a much lighter (dominantly) ν_e. (Solar neutrinos can be accounted for by a SMA MSW solution or possibly by vacuum oscillations, but not by a LMA MSW.) Alternatively, one could reverse the pairing, with a nearly degenerate ν_s and ν_e at ~ 1 eV, and $\nu_{+,-}$ around 0.03-0.1 eV. The other models involve $\nu_\mu \to \nu_s$ with near-maximal mixing for the atmospheric neutrinos, and $\nu_e \to \nu_\tau$ for the solar neutrinos. Again, there are two possibilities, with the nearly degenerate $\nu_s - \nu_\tau$ pair around 1 eV and a lighter $\nu_\tau - \nu_e$, or the other way around.

All of these patterns involve two neutrinos in the eV range, and therefore the possibility of a significant hot dark matter component. The two which have the (dominantly) ν_e state around 1 eV could contribute to $\beta\beta_{0\nu}$ if the neutrinos are Majorana. A very small $\langle m_{\nu_e} \rangle$ due to cancellations would suggest near maximal mixing for the solar neutrinos, but this could again be relaxed significantly given all of the uncertainties.

6. PARTICLE PHYSICS IMPLICATIONS: FROM THE TOP DOWN

Almost all extensions of the standard model predict non-zero neutrino mass at some level, often in the observable $10^{-5} - 10$ eV range. It is therefore difficult to infer the underlying physics from the observed neutrino masses. However, the neutrino mass spectrum should be extremely useful for top-down physics; i.e., the predicted neutrino masses and mixings should provide an important test, complementary to, e.g., the sparticle, Higgs, and ordinary fermion spectrum, of any concrete fundamental theory with serious predictive power.

Prior to the precision Z-pole measurements at LEP and SLC there were two promising paths for physics beyond the standard model: compositeness at the TeV scale (e.g., dynamical symmetry breaking, composite Higgs, or composite fermions), or unification, which most likely would have led to deviations from the standard model prediction at the few % or few tenths of a % level, respectively. The absence of large deviations [73] strongly supports the unification route, which is the domain of supersymmetry, grand unification, and superstring theory. The implication is that non-zero neutrino masses are most likely not the result of unexpected new physics at the TeV scale, such as by loop effects associated with new ad hoc scalar fields. (They could, however, be due to neutrino-neutralino mixing or loop effects in supersymmetric models with R parity breaking.) Alternatively, they could be associated with new physics at very high energy scales, most likely either seesaw models or higher dimensional operators.

[7]There have been several attempts to get by with only three neutrinos. However, attempts to take $\Delta m^2_{\rm solar} = \Delta m^2_{\rm atm}$ [67] fail because they lead to an unacceptable energy-independent suppression of the solar neutrinos. Similarly, $\Delta m^2_{\rm atm} = \Delta m^2_{\rm LSND}$ were marginally compatible with the earlier Kamiokande atmospheric data [68], but do not describe the zenith angle distortions (and lower $\Delta m^2_{\rm atm}$) observed by Superkamiokande. There is still a possibility of combining a three neutrino scheme with anomalous interactions [69], which could, e.g., affect the zenith distribution and allow a larger $\Delta m^2_{\rm atm}$, or affect the LSND results and allow a lower $\Delta m^2_{\rm LSND}$.

7. ORDINARY-STERILE NEUTRINO MIXING

As discussed in Section 5.1 the combination of solar neutrinos, atmospheric neutrino oscillations, and the LSND results, if confirmed, would most likely imply the mixing of ordinary active neutrinos with one (or more) light sterile neutrinos. One difficulty is that the sterile neutrinos could have been produced in the early universe by the mixings. For the range of mass differences and mixings relevant to LSND and the atmospheric neutrinos, the sterile neutrino would have been produced prior to nucleosynthesis, changing the freezeout temperature for $\nu_e n \leftrightarrow e^- p$ and leading to too much ^4He [74]. However, Foot and Volkas have recently [75] argued that MSW effects involving sterile neutrinos could amplify a small lepton asymmetry, leading to an excess of ν_e compared to $\bar{\nu}_e$, reducing the ^4He. It has also been argued that ordinary-sterile neutrino mixing could facilitate heavy element synthesis by r-processes in the ejecta of neutrino-heated supernova explosions [76,74].

Most extensions of the standard model predict the existence of sterile neutrinos. For example, simple $SO(10)$ and E_6 grand unified theories predict one or two sterile neutrinos per family, respectively. The only real questions are whether the ordinary and sterile neutrinos of the same chirality mix significantly with each other, and whether the mass eigenstate neutrinos are sufficiently light. When there are only Dirac masses, the ordinary and sterile states do not mix because of the conserved lepton number. Pure Majorana masses do not mix the ordinary and sterile sectors either. In the seesaw model the mixing is negligibly small, and the (mainly) sterile eigenstates are too heavy to be relevant to oscillations. The only way to have significant mixing and small mass eigenstates is for the Dirac and Majorana neutrino mass terms to be extremely small and to also be comparable to each other. This appears to require two miracles in conventional models of neutrino mass.

One promising possibility involves the generation of neutrino masses from higher-dimensional operators in theories involving an intermediate scale [6], as described in Section 2. Depending on the intermediate scale and the dimensions of the operators naturally small Dirac and Majorana masses are possible, and in some cases they are automatically of the same order of magnitude [7]. Another interesting possibility [77] involves sterile neutrinos associated with a parallel hidden sector of nature as suggested in some superstring and supergravity theories. Other mechanisms in which one can obtain ordinary-sterile neutrino mixing are described in [78].

8. CONCLUSIONS

- Neutrino mass is an important probe of particle physics, astrophysics, and cosmology.

- There are several experimental indications or suggestions: (a) The Superkamiokande and other results on atmospheric neutrinos provide strong evidence for ν_μ oscillations. (b) The combination of solar neutrino experiments implies a global spectral distortion, strongly supporting neutrino transitions or oscillations. The preliminary SuperK results on the ^8B spectrum suggests a spectral distortion, most consistent with vacuum oscillations but possibly with small angle MSW. (c) LSND has candidate events in both decay at rest and decay in flight. The non-observation of candidates by KARMEN is close to being an experimental contradiction, also there is still a small parameter space consistent with both. (d) Mixed dark matter is an interesting hint for eV scale masses, but is not established.

- In the future many solar neutrino experiments and (model independent) observables will be needed to identify the mechanism, determine the neutrino parameters, and simultaneously study the Sun. This program is complicated by possible three neutrino effects, possible sterile neutrinos, and the possibility that there are both neutrino mass effects and nonstandard solar physics (although the latter is constrained by helioseismology). Experiments that are sensitive to the pp and ^7Be neutrinos are needed. Important observables include neutral to charged current ratios, spectral distortions, day-night effects (possibly involving parametric core enhancement), and seasonal variations (especially for vacuum oscillations).

- For the atmospheric neutrinos, we need more detailed spectral and zenith angle information, and the neutral to charged current ratio as a function of the zenith angle. Independent information, including possible ν_τ appearance, for the same parameter range should be forthcoming from long baseline experiments.
- The planned Mini-BOONE experiment at Fermilab should clarify the LSND-KARMEN situation.
- Future cosmic microwave anisotropy experiments and large scale sky surveys should be able to determine whether neutrinos contribute significantly to the dark matter.
- Significant improvements in $\beta\beta_{0\nu}$ would be very powerful probes of the Majorana nature of neutrinos in the mass ranges suggested by the LSND and atmospheric neutrino results.
- Most extensions of the standard model predict nonzero neutrino masses, so it is difficult to determine their origin in a "bottom-up" matter. However, the neutrino spectrum will be a powerful constraint on "top-down" calculations of fundamental models.
- The possibility of mixing between ordinary and light sterile neutrinos should be taken seriously.

ACKNOWLEDGMENTS

This work was supported by U.S. Department of Energy Grant No. DOE-EY-76-02-3071. I am grateful to Naoya Hata for collaborations on the implications of Solar neutrinos.

REFERENCES

1. For detailed reviews, see G. Gelmini and E. Roulet, Rept. Prog. Phys.**58**,1207 (1995); P. Langacker in *Testing The Standard Model*, ed. M. Cvetic and P. Langacker (World, Singapore, 1991) p. 863; B. Kayser, F. Gibrat-Debu, and F. Perrier, *The Physics of Massive Neutrinos*, (World Scientific, Singapore, 1989); Report of the *DPF Long Range Study: Neutrino Mass Working Group*, P. Langacker, R. Rameika, and H. Robertson, conveners, Feb. 1995.
2. G. B. Gelmini and M. Roncadelli, Phys. Lett. **99B**, 411 (1981); H. Georgi *et al.*, Nucl. Phys. **B193**, 297 (1983).
3. A. Zee, Phys. Lett. **93B**, 389 (1980). For later references, see [1].
4. L. J. Hall and M. Suzuki, Nucl. Phys. **B231**, 419 (1984); T. Banks et al., Phys. Rev. **D52**, 5319 (1995); R. Hempfling, Nucl. Phys. **B478**, 3 (1996); B. de Carlos and P. L. White, Phys. Rev. **D54**, 3427 (1996); A. Yu Smirnov and F. Vissani, Nucl. Phys. **B460**, 37 (1996); R, M. Borzumati et al., Phys. Lett. **B384**, 123 (1996); H. P. Nilles and N. Polonsky, Nucl. Phys. **B499**, 33 (1997); E. Nardi, Phys. Rev. **D55**, 5772 (1997).
5. M. Gell-Mann, P. Ramond, and R. Slansky, in *Supergravity*, ed. F. van Nieuwenhuizen and D. Freedman, (North Holland, Amsterdam, 1979) p. 315; T. Yanagida, *Proc. of the Workshop on Unified Theory and the Baryon Number of the Universe*, KEK, Japan, 1979; S. Weinberg, Phys. Rev. Lett. **43**, 1566 (1979).
6. G. Cleaver, M. Cvetic, J. R. Espinosa, L. Everett, P. Langacker, Phys. Rev. **D57**, 2701 (1998).
7. P. Langacker, Phys. Rev. **D58** (1998), hep-ph/9805281.
8. For a general background, see J. N. Bahcall, *Neutrino Astrophysics*, (Cambridge Univ. Press, Cambridge, 1989).
9. K. Lande, these proceedings.
10. Y. Suzuki, these proceedings.
11. T. Kirsten, these proceedings.
12. V. N. Gavrin, these proceedings.
13. A. McDonald, these proceedings.
14. L. Oberauer, these proceedings.
15. R. E. Lanou, hep-ex/9808033, these proceedings; R. Raghavan, these proceedings.
16. J. N. Bahcall, S. Basu, M. H. Pinsonneault, Phys. Lett. **B433**, 1 (1998).
17. V. Castellani et al., Nucl. Phys. Proc. Suppl. **70**, 301 (1999).
18. S. Degl'Innocenti and B. Ricci, Astropart.

Phys. **8**, 293 (1998).
19. D. Gough, these proceedings.
20. J. N. Bahcall, astro-ph/9808162, these proceedings.
21. E. Adelberger et al., astro-ph/9805121.
22. W. Haxton, nucl-th/9810024, these proceedings.
23. See [24,25] and references theirin.
24. N. Hata and P. Langacker, Phys. Rev. **D56**, 6107 (1997).
25. J. N. Bahcall, PI. Krastev, and A. Yu. Smirnov, Phys. Rev. **D58** (1998), hep-ph/9807216.
26. A. Cumming and W. C. Haxton, Phys. Rev. Lett. **77**, 4286 (1996).
27. S. L. Glashow and L. M. Krauss, Phys. Lett. **B190**, 199 (1987).
28. Q. Y. Liu and S. T. Petcov, Phys. Rev. **D56**, 7392 (1997); K. S. Babu, Q. Y. Liu, and A. Yu. Smirnov, Phys. Rev. **D57**, 5825 (1998).
29. A. J. Baltz, A. S. Goldhaber, and M. Goldhaber, hep-ph/9806540.
30. For a recent review, see E. Kh. Akhmedov, hep-ph/9705451.
31. S. Pastor, V. B. Semikoz, and J. W. F. Valle, Phys. Lett. **B423**, 118 (1998).
32. S. Bergmann and A. Kagan, hep-ph/9803305; G. Brooijmans, hep-ph/9808498.
33. S. L. Glashow et al., Phys. Rev. **D56**, 2433 (1997) and these proceedings.
34. A. Halprin and C. N. Leung, Phys. Lett. **B416**, 361 (1998).
35. J. N. Bahcall, P. I. Krastev, and E. Lisi, Phys. Rev. **C55**, 494 (1997); G. L. Fogli, E. Lisi, and D. Montanino, Phys. Lett. **B434**, 333 (1998); B. Faid et al., hep-ph/9805293.
36. J. N. Bahcall and P. I. Krastev, Phys. Lett. **B436**, 243 (1998); G. Fiorentini et al., astro-ph/9810083.
37. For recent discussions, see J. N. Bahcall and P. I. Krastev, Phys. Rev. **C56**, 2839 (1997) and [23].
38. E. Akhmedov, P. Lipari, and M. Lusignoli, Phys. Lett. **B300**, 128 (1993); S. T. Petcov, Phys. Lett. **B434**, 321 (1998); M. Maris and S. T. Petcov, hep-ph/9803244, Phys. Rev. **D58** (1998); M. Chizhov, M. Maris, and S. T. Petcov, hep-ph/9810501; E. K. Akhmedov, hep-ph/9805272; A. Smirnov, hep-ph/9809481, these proceedings; S. T. Petcov, hep-ph/9809587, these proceedings.
39. J. M. Gelb, W.-k. Kwong, and S. P. Rosen, Phys. Rev. Lett. **78**, 2296 (1997).
40. S. L. Glashow, P. J. Kernan, and L. M. Krauss, hep-ph/9808470.
41. N. Hata and P. Langacker, Phys. Rev. **D52**, 420 (1995).
42. A. B. Balantekin, J. M. Fetter, and F. N. Loreti, Phys. Rev. **D54**, 3941 (1996); H. Nunokawa et al., Nucl. Phys. **B472**, 495 (1996); C. P. Burgess and D. Michaud, Ann. Phys. (NY) **256**, 1 (1997).
43. P. Bamert, C. P. Burgess, and D. Michaud, Nucl. Phys. **B513**, 319 (1998).
44. P. Lipari, T. Stanev, and T. K. Gaisser, Phys. Rev. **D58** (1998); T. K. Gaisser, hep-ph/9809481, these proceedings.
45. References to earlier work may be found in [46].
46. T. Kajita, hep-ex/9810001, these proceedings; Y. Fukuda et al., Phys. Rev. Lett. **81**, 1562 (1998).
47. E. Peterson, these proceedings.
48. F. Ronga, hep-ex/9810008, these proceedings. M. Ambrosio et al., Phys. Lett. **B434**, 451 (1998).
49. C. Bemporad, these proceedings; M. Apollonio et al., Phys. Lett. **B420**, 397 (1998).
50. A. Piepke, these proceedings.
51. A. Suzuki, these proceedings.
52. Articles by S. Wojcicki, F. Pietropaolo, and K. Nishikawa in these proceedings.
53. J. W. Flanagan, J. G. Learned, and S. Pakvasa, Phys. Rev. **D57**, 2649 (1998).
54. Y. Fukuda et al., Phys. Lett. **B335**, 237 (1994).
55. E. K. Akhmedov et al., hep-ph/9808270; P. Lipari and M. Lusignoli, Phys. Rev. **D58** (1998), hep-ph/9803440; Q. Y. Liu, S. P. Mikheyev, and A. Yu. Smirnov, hep-ph/9803415; Q. Y. Liu and A. Yu. Smirnov, Nucl. Phys. **B524**, 505 (1998); and [38].
56. F. Vissani and A. Yu. Smirnov, Phys. Lett. **B432**, 376 (1998); J. G. Learned, S. Pakvasa, and J. L. Stone, Phys. Lett. **B435**, 131 (1998); M. Goldhaber and M. Diwan, unpub-

lished.
57. L. J. Hall and H. Murayama, hep-ph/9810468.
58. M. C. Gonzalez-Garcia, H. Nunokawa, O. L. G. Peres, and J. W. F. Valle, hep-ph/9807305; O. Yasuda, hep-ph/9809206, these proceedings, and hep-ph/9809205; V. Barger, T. J. Weiler, and K. Whisnant, hep-ph/9807319; G. L. Fogli, E. Lisi, A. Marrone, and G. Scioscia, hep-ph/9808205.
59. See G. C. Branco, M. N. Rebelo, and J. I. Silva-Marcos, hep-ph/9810328, and references theirin.
60. V. Barger, S. Pakvasa, T. J. Weiler, and K. Whisnant, Phys. Lett. **B437**, 107 (1998); M. Jezabek and Y. Sumino, hep-ph/9807310; R. N. Mohapatra and S. Nussinov, hep-ph/9809415; and [29,59].
61. H. Fritzsch and Z.-z. Xing, hep-ph/9808272 and 9807234.
62. D. O. Caldwell and R. N. Mohapatra, Phys. Rev. D **48**, 3259 (1993), **50**, 3477 (1994); J. T. Peltoniemi and J. W. F. Valle, Nucl. Phys. B **406**, 409 (1993). For recent discussions, see D. Caldwell, hep-ph/9804367; H. Georgi and S. L. Glashow, hep-ph/9808293.
63. For recent discussions, see J. Primack, Science **280**, 1398 (1998); E. Gawiser and J. Silk, Science **280**, 1405 (1998).
64. H. V. Klapdor-Kleingrothaus, these proceedings; R. N. Mohapatra, hep-ph/9808284, these proceedings.
65. D. H. White, these proceedings; C. Athanassopoulos et al., Phys. Rev. Lett. **81**, 1774 (1998).
66. K. Eitel et al., hep-ex/9809007, these proceedings.
67. A. Acker and S. Pakvasa, Phys. Lett. **B397**, 209 (1997).
68. C. Cardall and G. Fuller, Phys. Rev. D **53**, 4421 (1996); G. L. Fogli et al., Phys. Rev. D **54**, 3667 (1996), **56** 4365 (1997); K.S. Babu, J. C. Pati, and F. Wilczek, Phys. Lett. B **359**, 351 (1995), erratum: **364**, 251 (1995).
69. E. Ma and P. Roy, Phys. Rev. Lett. **80**, 4637 (1998); L. M. Johnson and D. W. McKay, Phys. Lett. **B433**, 355 (1998).
70. C. Caso et al., Eur. Phys. J. **C3**,1 (1998).
71. D. O. Caldwell and R. N. Mohapatra [62]; J. T. Peltoniemi and J. W. F. Valle [62]; S. C. Gibbons et al., Phys. Lett. **B430**, 296 (1998); Q. Y. Liu and A. Yu. Smirnov, Nucl. Phys. **B524**, 505 (1998); V. Barger, T. J. Weiler, and K. Whisnant, Phys. Lett. **B427**, 97 (1998); S. M. Bilenkii, C. Giunti, and W. Grimus, hep-ph/9809368, these proceedings; E. J. Chun, C. W. Kim, and U. W. Lee, Phys. Rev. **D58** (1998), hep-ph/9802209; N. Okada and O. Yasuda, Int. J. Mod. Phys. **A12**,3669 (1997).
72. A. Geiser, CERN-EP-98-056; W. Krolikowski, hep-ph/9808307 and 9808207; D. Suematsu, hep-ph/9808409.
73. For a recent study, see J. Erler and P. Langacker, hep-ph/9809352.
74. For a complete set of references, see [7].
75. R. Foot and R. R. Volkas, Phys. Rev. **D55**, 5147 (1997), **D56**, 6653 (1997); N. F. Bell, R. Foot, and R. R. Volkas, Phys. Rev. **D58** (1998), hep-ph/9805259.
76. D. O. Caldwell, G. M. Fuller, and Y.-Z. Qian, in preparation; H. Nunokawa et al., Phys. Rev. **D56**, 1704 (1997); J. Peltoniemi, hep-ph/9511323.
77. B. Brahmachari and R. N. Mohapatra, Phys. Lett. **B437**, 100 (1998); R. N. Mohapatra, hep-ph/9808236.
78. U. Sarkar, hep-ph/9808277 and 9807466; N. Arkani-Hamed and Y. Grossman, hep-ph/9806223; K. Benakli and A. Yu Smirnov, Phys. Rev. Lett. **79**, 4314 (1997); E. Ma, Phys. Lett. **B380**, 286 (1996); R. Foot and R. R. Volkas, Phys. Rev. **D52**, 6595 (1995); J. T. Peltoniemi and J. W. F. Valle [62]; J. T. Peltoniemi, D. Tommasini, and J. W. F. Valle, Phys. Lett. **B298**, 383 (1993).

Part 7

Accelerator and Reactor Neutrino Experiments

NUCLEAR PHYSICS B
PROCEEDINGS
SUPPLEMENTS

ELSEVIER

Nuclear Physics B (Proc. Suppl.) 77 (1999) 259–264

Result from DONUT -Direct Observation of ν_τ interaction-

NAKAMURA Mituhiro[a] on behalf of DONUT collabolation

[a]Department of Physics, NAGOYA University,
Furo-cho, Chikusa-ku, Nagoya, 464-8601, JAPAN

Current status of DONUT(FNAL/E872) is reported. We have exposed emuslion target in '97 fixed target run and accumulated neutrino events corresponding to the total proton exposure of $\sim 4.55 \times 10^{17}$ 800GeV protons and the target weight of $\sim 250kg$.

In the preliminary analysis, we have located 34 neutrino interactions in Emulsion target and found one event which can be explained as $\nu_\tau(\overline{\nu_\tau}) + N \rightarrow \tau^-(\tau^+) + X$ followed by $\tau^- \rightarrow e^- + \nu_\tau + \overline{\nu_e}$ or $\tau^+ \rightarrow e^+ + \nu_e + \overline{\nu_\tau}$.

1. Introduction

DONUT(Fermilab E872) is a beam dump experiment which intend to observe tau neutrino charge current interactions for the first time in the world.

In this experiment, a prompt neutrino beam containing ν_τ was created by dumping 800GeV protons on tungsten beam dump. The source of ν_τ and $\overline{\nu_\tau}$ is the cascade decay of $Ds^-(Ds^+) \rightarrow \tau^-(\tau^+) + \overline{\nu_\tau}(\nu_\tau)$ followed by $\tau^-(\tau^+) \rightarrow \nu_\tau(\overline{\nu_\tau}) + X$. The ratio of ν_τ charge current interactions is safely expected to be $\sim 5\%$ due to reliable measurements of the branching ratio of $Ds^- \rightarrow \mu^- + \overline{\nu_\mu}$ in the previous experiments using hybrid emulsion detector [1,2].

The method to detect ν_τ CC interaction is the topological detection of the decay of the emerging τ from the CC interaction. For this purpose, nuclear emulsion was utilized as a target and a tracking device. Nuclear emulsion has three dimensional resolution of $\sim 1\mu m$ and is especially suited to the detection of short decay path of the τ. Same kind targets were used in CHORUS of which purpose is to detect the oscillation from ν_μ to ν_τ [3].

In the next section, DONUT experimental set up including the prompt neutrino beam line will be described. In the following section, the status of '97 exposure and the current status of the emulsion analysis will be reported.

2. Prompt Neutrino Beam Line

The preparation of prompt neutrino beam line which is suited to the emulsion exposure was one of the key of this experiment.

The expected most serious problem was the high density penetrative muon flux from the beam dump target (μ from charm semileptonic decay etc.) In order to sweep out them from the beam line, two magnets are installed at the downstream of the dump target (tungsten: 40inch long). The maximum Pt kick of the first magnet is 8 GeV. Adding to them, passive shields made of steel and lead were set in front of the emulsion target in order to stop low energy components, like scattered muon and electromagnetic components from muon bremstralung. At the position of 36m downstream from the dump, background free zone of $\sim 1m$ width was kept and at where the emulsion target was installed. Fig.1 shows a schematic view of the beam line.

When we received the first beam on the dump target at Oct.'96, we observed a lot of low energy γ ray backgrounds from nuclear captures of thermalized neutron and activated gas components which we did not expect. We modified the shields around beam dump, added extra γ shields around the emulsion target and suceeded to supress the background level below allowed level for emulsion exposure at the end May '97.

PII S0920-5632(99)00425-9

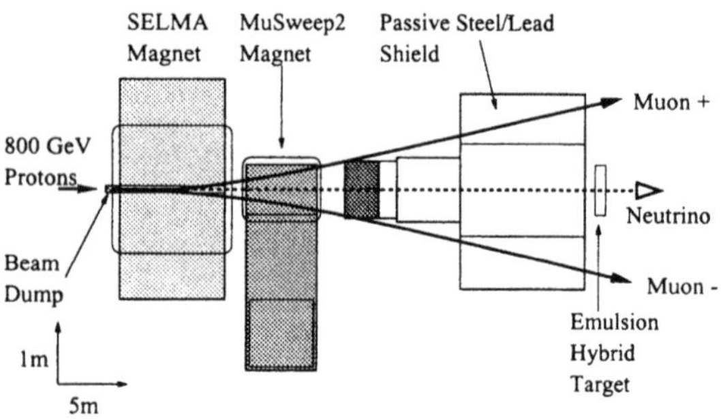

Figure 1. DONUT Prompt Neutrino Beam Line.

3. DONUT Detector

Fig.2 shows a schematic view of the DONUT detector installed downstream of the prompt neutrino beam line. It contains an emulsion hybrid target, a charged particle spectrometer, an electromagnetic calorimeter and a muon identifier.

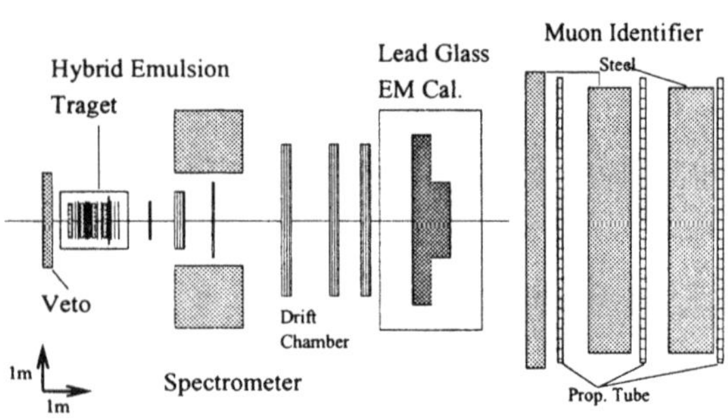

Figure 2. Schematic view of the DONUT detector.

3.1. Emulsion Hybrid Target

The concept of the target design is almost same as that of CHORUS[3]. The schematic figure of the hybrid target is shown in Fig.3. There are four emulsion target modules and four scintillating fiber tracker(SFT) modules. Each module was installed on the support individually. The emulsion target act as neutrino target and micro vertex detector. The SFT module give macroscopic tracking infromation of the emerging tracks from emulsion target.

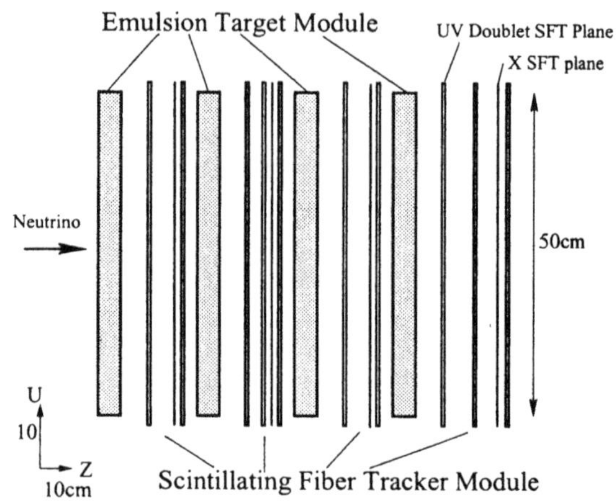

Figure 3. Schematic view of the Emulsion Hybrid Target.

3.1.1. Emulsion target module

The size of emulsion target is $50cm \times 50cm$ and the thickness is 7cm in maximum. There are two types of emulsion target module, i.e. Bulk type and ECC type.

In bulk type, emulsion target is constructed from bulk emulsion sheets as shown in Fig.4(a). One bulk emulsion sheet has thick emulsion layer(thickness: $350\mu m$) on the both sides of a thin plastic film (thickness: $90\mu m$). About 90 sheets are stacked in one bulk module. Because the ratio of nuclear emulsion is 96% of total mass and 89% of total length, almost all of the neutrino interaction vertex and the decay vertex will be observed in the emulsion.

The other type, i.e. ECC, is a sandwitch of emulsion films and iron plates as shown in Fig.4(b). One emulsion film has thin emulsion layer of $100\mu m$ on both side of a plastic film of medium thickness(200μm or 800μm.) The thickness of one iron plate (precisely stainless steel plate of SUS316) is 1mm. The emulsion occupation ratio is $\sim$ 8% in mass and 10 $\sim$ 14% in length. Therefore, emulsion sheets act as three dimensional tracking devices of 1 micron precision placed between passive targets. The corresponding angle resolution obtained from one emulsion film is $\sim$ 1$mrad$. The decay of τ and charm is detected by utilizing an impact parameter analysis and an detection angle reflection.

ECC has a long history in cosmic ray experiment. The discovery of charmed particle was done by using this type emulsion chamber in 1971[4].

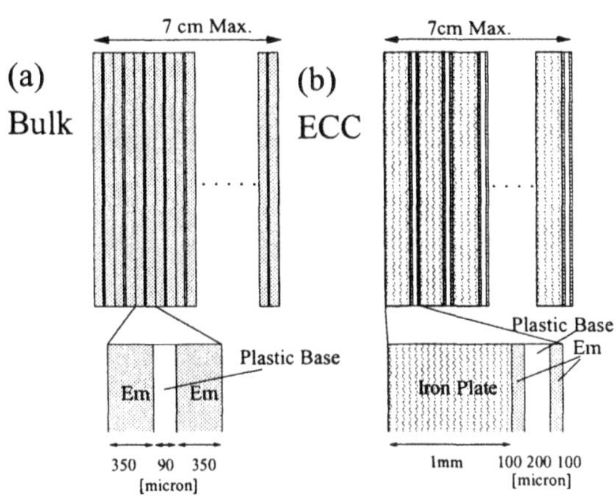

Figure 4. Structure of the emulsion target. (a) Bulk. (b) ECC.

3.1.2. Scintillating Fiber Module

In order to support the event location of neutrino interaction in emulsion target, scintillating fiber tracker was utilized. The construction method of the SFT is almostly same as that of CHORUS fiber tracker[5]. The achieved detection efficiency of single plane is $\sim$ 95% and the position resolution is $\sim$ 170μm.

3.2. Downstream Detector

Charged charm production in ν CC interaction of any kinds neutrino becomes a background, if the lepton is mis-identified or not identified.

For Muon identification, a muon detector made of steel and position detectors (proportional tubes and scintillating pad detectors) was installed at the most downstream part. Because of the thick target, 2 $\sim$ 3 Xo per module, electron identification by spectrometer plus EM calorimeter can be partially applied to the event occured in most down stream emulsion target. For other module, emulsion technique must be utilized to identify electrons.

Hadronic secondary interaction becomes a background, if it has a decay like topology (the number of daughter track is odd) and the daughter track has reasonable longitudinal and transvers momentum compatible to the tau decay. About 95% of hadronic secondary interaction show nuclear activity like nuclear evapolation tracks or auger electrons, therefore, if the "decay point" is in an emulsion plate, rejection of this kind background is quite easy. Even if the "decay point" is not in emulsion plates, once the daughter is identified as a lepton, it becomes free from this background.

4. Status of '97 Emuslion exposure

Emulsion exposure was started at April '97. Until the beam stop at Oct. '97, seven emulsion modules, one full bulk module, two full ECC modules and 4 hybrid modules of bulk and ECC, have been exposed. The total proton on target shared to our experiment was 4.55 × 10^{17}. Table 1 shows the summary of the exposure. The number of expected accumulated ν charged current interactions is around one thousand. Which correspond to the τ charged current interaction of $\sim$ 50.

5. Event Analysis

The emulsion development was finished at the end of January of '98. The event analysis was started at the summer of '97 when the first module ECC3 was extracted from the beam line.

Table 1
Summary of Exposed Emulsion targets in '97

Module	Type	Weight[kg]	POT	Expected ν^{cc}	Expected ν_τ^{cc}
ECC1	ECC	105	2.6×10^{17}	210	10
E/B1	Hybrid	69	2.0×10^{17}	110	5
E/B2	Hybrid	68	4.2×10^{17}	230	12
ECC3	ECC	105	1.2×10^{17}	100	5
E/B3	Hybrid	67	4.2×10^{17}	220	11
E/B4	Hybrid	65	1.9×10^{17}	100	5
B4	Bulk	56	2.0×10^{17}	90	4
Total			4.55×10^{17}	~ 1000	~ 50

5.1. Event Selection

The total number of triggers recorded is 10^7, on the other hand the expected ν interaction is $\sim 10^3$. Almost all of the recorded events were triggerd by remnant soft component like low energy γ rays or electrons from muon interactions occured in the shield steel or the return york of the spectrometer magnet. These events show a few hits in the SFT, so that elimination is not so difficult. The first pass applied to the raw data is the requirement of (1)$\geq$ 1 track in the drift chamber pointing within 50cm of the emulsion target. OR (2) a vertex in an emulsion volume made from the U view of the SFT. OR (3)$\geq$ 30GeV deposited in the lead glass calorimeter. The events passed this step were further decoded and investigated by human eye.

5.2. Event Location in Emulsion Target

In order to search for τ decays, recorded ν interaction vertexes must be investigated under microscope.

At first, the interaction vertex must be located in the emulsion. Though the scanning speed of emulsion using automatic scanning system becomes faster and faster [6], current speed is not enough to do the event location without restricting the scanning position by extra tracking devices. In DONUT, SFT was utilized for this purpose and gave informations to two different location method, "Scan Back" and "Net scan".

"Scan back" was utilzed in the analysis of E531 and CHORUS. In this method, at least one three dimensionaly reconstructed track is required per event. The track is traced back from SFT to the most downstream emulsion plate and traced to the interaction vertex plate by plate. Because nuclear emulsion has no dead time, all trajectories of ionizing particles are recorded. When the background density is higher corresponding to the position resolution, a removable emulsion sheet, "Changeable Sheet" is installed. In DONUT, eight changeable sheets were installed at upstream and downstream of the emulsion targets and replaced once per week during the exposure.

"Net Scan" is a newly developed location method for DONUT. In this method, only rough vertex position predictions are required. Scanning is started from the plate placed downstream of the predicted vertex. In this plate, almost all of the tracks which have emission angle smaller than 250mrad is picked up using automatic scanning system. The average number of picked up tracks are around several thousand. All tracks are traced back to the upstream plates which cover the expected vertex position. In the next step, penetrating tracks from the start plate to the last upstream plate are rejected and only stopped tracks in the volume are remained. The remained tracks are investigated whether they construct a vertex or not requiring small impact parameter like $\leq 10\mu m$. In principle, if there is a vertex accompanied with at least two tracks, we can locate it.

5.3. Category of the located events

Until this conference, preliminary event location in ECC1/3 and a part of E/B 4 and B4 was tried. 34 events were located in emulsion targets,

28 events in ECC and 6 events in Bulk.

Table 2 shows the event category identified by downstream detector or emulsion information + SFT information. Especially the electron identification was carried out using emulsion information. Electron is identified by detecting an initiation of electromagnetic shower. The existence of narrowly accompanied electron pairs to the track and the increase of multiple scattering of the track (means energy loss by bremstralung) are the key of the electron idenfication in nuclear emulsion. The efficiency of electron identification is depend on its energy and the followed length. According to a Monte Carlo simulation, the efficiency is greater than 95% in the case that the energy is $\geq 2GeV$ and the followed length is $\geq$ 2 Xo.

The number of identified electron event is smaller than the identified muon events. This may indicate a existence of location biases which must be studied.

5.4. Decay search

All charged tracks emitted from the located vertex within the angle of 400mrad are followed to the downstream side and investigate the exsitence of decay topology, especially one prong decay topology(kink).

Among the 34 events, we found one kink candidate in "NC like sample" and one lost track candidate in "with μ" sample. Lost track means the track disappeared in the iron plate. Which is considered to be a charm decay or a hadronic secondary interaction, of which daughter track has a large angle out of our acceptance ($\leq$ 400 mrad). Because of the existence of μ from the primary vertex, it is not so interesting.

5.5. ν_τ CC Candidate Event

The ν_τ CC interaction candidate is RUN 3024 - EVENT 30175 located in ECC1 module. Schematic figures of this event are shown in Fig.5.

The primary vertex was in the iron plate just upstream of the emulsion film number 12. Three tracks were emitted from the primary vertex. One of them which was emitted to the most forward direction shows kink decay topology after 4.5mm from the primary vertex. The kink vertex was in the plastic base of emulsion film number 9. The distance between the base upstream surface to the kink point is $65\mu m$. No other tracks including nuclear evapolation tracks from this vertex are recognized. The kink angle was measured to be 95mrad.

One electron pair was found at the emulsion film number 2 which is $\sim 1cm$ (0.4 Xo) downstream of the kink vertex. The conversion point is in the iron plate of plate number 2. There were hits in SFT corresponding to the one track of the electron pair. The other track of the electron pair is apparantly low momentum because multiple scattering was easily recognized within one emulsion layer of $100\mu m$ ($\sim$ 0.3% Xo) and shows low dE/dX compatible to minimum ionizing particle, so it was uniqely identified as an electron.

The distance between the electron pair and the kink daughter track becomes zero within the measurement error($\sim 10\mu m$) at about 2 mm upstream from the surface of emulsion film number 2. The direction of the electron pair is incompatible to the case that it comes from the primary vertex.

The existence of such kind associated electron pair indicates that the daughter track is an electron, we intend to extend the identification process to the next emulsion target(ECC3) of which thickness is $\sim$ 3 Xo. Using the emulsion information and SFT information, the duagher track was connected to ECC3 and followed from upstream plate to the bottom plate. The daughter track shows typical feature of an electron, i.e. bremstralung and conversion of the γray.

The energy was estimated from two different method, measurement of multiple scattering of the daughter particle and shower track number counting after 54mm Iron plate passing. Precision of the estimation is under study but from these two method, the energy is estimated to be $2 \sim 3GeV$.

The background was limited, because the daughter is identified as an electron. The only possible one is a charged charm semi-electronic decay of which accompanied lepton escapes from the lepton identification. In the current event, the remained two tracks emitted from the primary vertex are out of muon identifier acceptance. The

Table 2
Summary of Exposed Emulsion targets in '97

Type of the event	Total	Sign +	Sign -	Sign Unknown
with μ	16	5	6	5
with e	7	1	0	6
NC like	11			
Total	34			

expected number of charm background in the current sample is (number of charm in 34 events $\sim$ 2.2) $\times$ (ratio of charged charm $\sim$ 0.5) $\times$ (semi-electronic decay of charged charm $\leq$ 0.17) $\times$ (possibility of lepton out of acceptance $\sim$ 0.05) $\leq$ 0.01event in the current sample. Futhermore, we can reject the case that the escaped lepton is an electron, because the two tracks passed the downstream target of 3 Xo thickness and show no electron feature. Therefore, the expected number is much smaller than the value.

On the other hand, the expected number of tau signal is (number of tau in 34 events $\sim$ 1.5) $\times$ (electronic decay of tau 0.18) $\sim$ 0.3 event in the current sample.

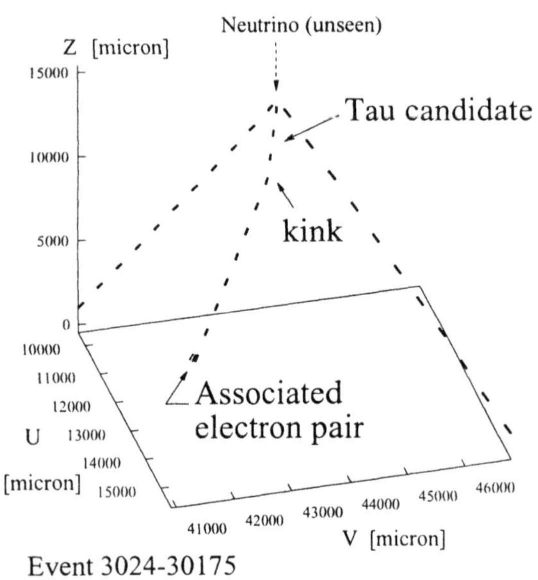

Figure 5. ν_τ CC interaction candidate event.

6. Summary

DONUT analysis of '97 run is in progress. We have a convince to show further evidence of ν_τ CC interactions, though there are several points must be impoved. Within one year, first analysis of '97 run will be finished, using current location efficiency, that is one of the point must be improved, around 10 τ events will be identified and analyzed.

REFERENCES

1. S. Aoki et al., WA75 collaboration, Prog. Theor. Phys. 89(1993)131-138.
2. K. Kodama et al., E653 collaboration, Phys. Lett. B382(1996)299-304.
3. E.Eskut et al., CHORUS collaboration, Nucl. Instr. and Meth. A401(1997)7.
4. K.Niu et al., Prog. Theor. Phys. 46(1971)1644.
5. Nakamura M., Proceedings of SCIFI93, (1995)194-201. Nakano T. et al., Proceedings of SCIFI93, (1995)525-533. Nakano T. et al., IEEE on NS, NS39(1992)680-684.
6. Nakano T., will be appered in Proceedings of Internatinal workshop of Nuclear Emuslion Techniques, 1998 NAGOYA. S. Aoki et al., Nucl. Instr. and Meth. B51(1990)466-472.

NUCLEAR PHYSICS B
PROCEEDINGS
SUPPLEMENTS

ELSEVIER

Nuclear Physics B (Proc. Suppl.) 77 (1999) 265–270

DETERMINATION OF $\sin^2\theta_W$ FROM NEUTRINO-NUCLEON SCATTERING AT NuTeV

R. H. Bernstein[3], T. Adams[4], A. Alton[4], S. Avvakumov[7], L. de Barbaro[5], P. de Barbaro[7], A. Bodek[7], T. Bolton[4], J. Brau[6], D. Buchholz[5], H. Budd[7], L. Bugel[3], J. Conrad[2], R. B. Drucker[6], R. Frey[6], J. Goldman[4], M. Goncharov[4], D. A. Harris[7], R. A. Johnson[1], S. Koutsoliotas[2,*], J. H. Kim[2], M. J. Lamm[3], W. Marsh[3], D. Mason[6], K. S. McFarland[3,†], C. McNulty[2], D. Naples[4], P. Nienaber[3], A. Romosan[2,‡], W. K. Sakumoto[7], H. Schellman[5], M. H. Shaevitz[2], P. Spentzouris[3], E. G. Stern[2], M. Vakili[1,‖], A. Vaitaitis[2], V. Wu[1], U. K. Yang[7], J. Yu[3] and G. P. Zeller[5]

[1]University of Cincinnati, Cincinnati, OH 45221 USA
[2]Columbia University, New York, NY 10027 USA
[3]Fermi National Accelerator Laboratory, Batavia, IL 60510 USA
[4]Kansas State University, Manhattan, KS 66506 USA
[5]Northwestern University, Evanston, IL 60208 USA
[6]University of Oregon, Eugene, OR 97403 USA
[7]University of Rochester, Rochester, NY 14627 USA

† Current address: University of Rochester, Rochester, NY 14627 USA
* Current address: Bucknell University, Lewisburg, PA 17837 USA
‡ Current address: Lawrence Berkeley National Laboratory, Berkeley, CA 94720 USA
‖ Current address: Texas A&M University, College Station, TX 77801 USA

Submitted to the Proceedings of Neutrino '98
June 1998, Takayama, Japan

Abstract

We report the determination of $\sin^2\theta_W$ in $\nu - N$ deep inelastic scattering from the NuTeV experiment. Using separate neutrino and anti-neutrino beams, NuTeV is able to extract $\sin^2\theta_W$ with low systematic errors from the Paschos-Wolfenstein variable R^-, a ratio of differences of neutrino and anti-neutrino neutral-current and charged-current cross-sections. NuTeV measures $\sin^2\theta_W{}^{(\text{on-shell})} = 0.2253 \pm 0.0019(\text{stat}) \pm 0.0010(\text{syst})$, which implies $M_W = 80.26 \pm 0.11$ GeV.

0920-5632/99/$ – see front matter
PII S0920-5632(99)00426-0

1 Introduction

In the past, neutrino scattering experiments have played a key role in establishing the validity of the electroweak Standard Model. Today, even with the large samples of on-shell W and Z bosons at e^+e^- and $p\overline{p}$ colliders, precision measurements in neutrino-nucleon scattering still play an important role. The measurement reported herein is competitive in precision with direct probes of weak boson parameters and tests the validity of the electroweak theory by determining $\sin^2\theta_W$ in a different process and at small q^2. In this respect, if neutrino scattering observed deviations from expectations based on direct measurements from W and Z bosons, this would be an exciting hint of new physics entering in tree-level processes or in radiative corrections. In particular, neutrino scattering would be sensitive to non-Standard Model effects ranging from leptoquark exchange to neutrino oscillations[1, 2].

Experimental quantities sensitive to electroweak physics that are most precisely measured in neutrino scattering are the ratios of charged-current (W exchange) to neutral-current (Z exchange) scattering cross-sections from quarks in heavy nuclei. The ratio of these cross-sections for either neutrino or anti-neutrino scattering from isoscalar targets of u and d quarks can be written as[3]

$$R^{\nu(\overline{\nu})} \equiv \frac{\sigma(\overset{(-)}{\nu}_\mu N \rightarrow \overset{(-)}{\nu}_\mu X)}{\sigma(\overset{(-)}{\nu}_\mu N \rightarrow \mu^{-(+)} X)} = (g_L^2 + \frac{g_R^2}{r}), \quad (1)$$

where

$$r \equiv \frac{\sigma(\overline{\nu}_\mu N \rightarrow \mu^+ X)}{\sigma(\nu_\mu N \rightarrow \mu^- X)} \sim \frac{1}{2}, \quad (2)$$

and $g_{L,R}^2 = u_{L,R}^2 + d_{L,R}^2$, the isoscalar sums of the squared left or right-handed quark couplings to the Z. At tree level in the Standard Model, $q_L = I_{\text{weak}}^{(3)} - Q_{\text{EM}}\sin^2\theta_W$ and $q_R = -Q_{\text{EM}}\sin^2\theta_W$; therefore, R^ν is particularly sensitive to $\sin^2\theta_W$.

In a real target, there are corrections to Eqn. 1 resulting from the presence of heavy quarks in the sea, the production of heavy quarks in the target, non leading-order quark-parton model terms in the cross-section, electromagnetic radiative corrections and any isovector component of the light quarks in the target. In particular, in the case where a charm-quark is produced from scattering off of low-x sea quarks, the uncertainties resulting from the effective mass suppression of the heavy final-state charm quark are large. The uncertainty in this suppression ultimately limited the precision of previous νN scattering experiments which measured electroweak parameters[4, 5, 6].

To eliminate the effect of uncertainties resulting from scattering from sea quarks, one can instead form a quantity suggested by Paschos and Wolfenstein[7],

$$R^- \equiv \frac{\sigma(\nu_\mu N \to \nu_\mu X) - \sigma(\overline{\nu}_\mu N \to \overline{\nu}_\mu X)}{\sigma(\nu_\mu N \to \mu^- X) - \sigma(\overline{\nu}_\mu N \to \mu^+ X)} \quad (3)$$

$$= \frac{R^\nu - rR^{\overline{\nu}}}{1-r} = (g_L^2 - g_R^2). \quad (4)$$

Since $\sigma^{\nu q} = \sigma^{\overline{\nu}\overline{q}}$ and $\sigma^{\overline{\nu} q} = \sigma^{\nu \overline{q}}$, the effect of scattering from sea quarks, which is symmetric under $q \leftrightarrow \overline{q}$, cancels in the difference of neutrino and anti-neutrino cross-sections. The remaining contribution from d_V yields a factor of roughly five smaller error from this process. R^- is a more difficult quantity to measure than R^ν, primarily because neutral current neutrino and anti-neutrino scattering have identical observed final states and can only be separated by *a priori* knowledge of the initial state neutrino.

2 The NuTeV Experiment and Neutrino Beam

The NuTeV detector consists of an 18 m long, 690 ton target calorimeter with a mean density of 4.2 g/cm^3, followed by an iron toroid spectrometer. The target calorimeter consists of 168 iron plates, 3m × 3m × 5.1cm each. The active elements are liquid scintillation counters spaced every two plates and drift chambers spaced every four plates. There are a total of 84 scintillation counters and 42 drift chambers in the target. The toroid spectrometer is not directly used in this analysis. NuTeV used a continuous test beam of hadrons, muons and electrons to calibrate the calorimeter and toroid response. The testbeam illuminated the front of the calorimeter in-between extractions of the fast-spill neutrino beam ($\approx$ 4 msec) and the testbeam was pointed to study transverse variations in detector response.

In this detector $\nu_\mu/\overline{\nu}_\mu$ charged-current events are identified by the presence of an energetic muon in the final state which travels a long distance in the target calorimeter. Quantitatively, a length is measured for each event based on the number of neighboring scintillation counters above a low threshold. Charged-current candidates are those events with a length of greater than 20 counters (2.1 m of steel-equivalent) and all other events are neutral-current candidates.

NuTeV's target calorimeter sits in the Sign Selected Quadrupole Train (SSQT) neutrino beam at the FNAL TeVatron. The observed neutrinos result from decays of pions and kaons produced from the interactions of 800 GeV protons in a production target. Immediately downstream of the target, a dipole magnet with $\int Bdl = 5.2$ T-m bends pions and kaons of one charge in the direction of the NuTeV detector, while oppositely charged and neutral mesons are stopped in dumps. Focusing magnets then direct the sign-selected mesons into a 0.5 km decay region which ends 0.9 km upstream of the NuTeV detector. The resulting beam is either almost purely neutrino or anti-neutrino, depending of the selected sign of mesons. Anti-particle backgrounds are observed at a level of less than 1–2 parts in 10^3. The beam is almost entirely muon neutrinos, with electron neutrinos creating 1.3% and 1.1% of the observed interactions from the neutrino and anti-neutrino beams, respectively.

Because charged-current electron neutrino interactions usually lack an energetic muon in the final state they are almost always identified as neutral-current interactions in the NuTeV detector. Therefore, the electron neutrino content of the beam must be very precisely known. Most (93% in the neutrino beam and 70% in the anti-neutrino beam) observed $\nu_e/\overline{\nu}_e$ result from $K^\pm_{e3}$ decays. The remainder are products of prompt decays of charmed particles or neutral kaons, or decays of secondary muons. Prediction of the former component comes from a beam Monte Carlo, tuned to reproduce the observed $\nu_\mu/\overline{\nu}_\mu$ flux (Figure 1). Because of the precise alignment of the magnetic optics in the SSQT (checked by observing deflections of the primary proton beam in a special

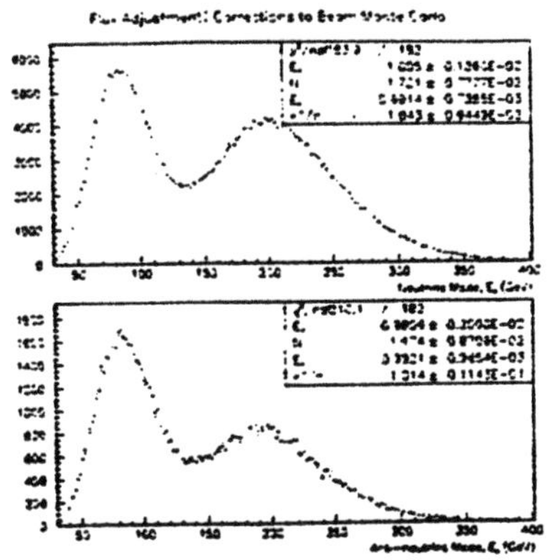

Figure 1: The ν_μ and $\overline{\nu}_\mu$ energy spectra from the data and the tuned beam Monte Carlo.

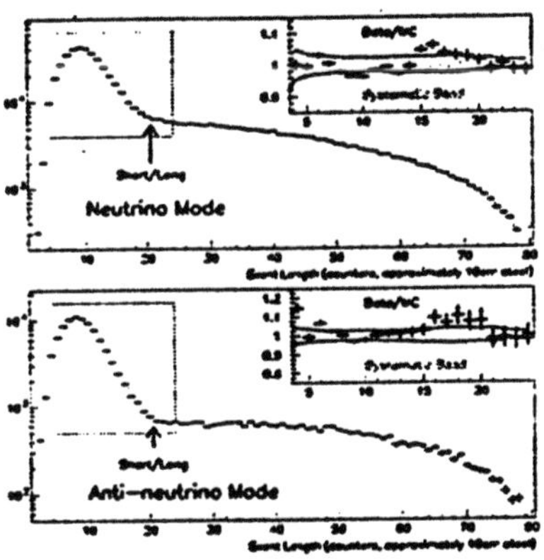

Figure 2: Length distributions in the data from the neutrino and anti-neutrino beams. Neutral-current/charged-current separation is made at a length of 20 counters, approximately 2.1 m of steel.

low-energy Tevatron run), this procedure results in a fractional uncertainty on the prediction of $\nu_e/\overline{\nu}_e$ from $K^\pm_{e3}$ of $\approx 1.5\%$, dominated by the $K^\pm_{e3}$ branching ratio uncertainty. Small detector calibration uncertainties, 0.5% on the calorimeter and muon toroid energy scale, affect the measured $\nu_\mu/\overline{\nu}_\mu$ flux and also contribute substantial uncertainties to both the muon and electron neutrino fluxes. Sources of $\nu_e/\overline{\nu}_e$ other than $K^\pm$ decay have larger uncertainties, at the 10–20% level, because of the lack of a direct constraint from the data.

3 Extraction of $\sin^2\theta_W$

Events selected for this analysis are required to deposit at least 20 GeV in the target calorimeter to ensure efficient triggering and vertex identification. The location of the neutrino interaction must be within the central $2/3^{\text{rds}}$ of the calorimeter's transverse dimensions, at least 0.4 m of steel-equivalent from the upstream end of the calorimeter, and at least 2.4 m from the downstream end. The first requirement reduces the misidentification of $\nu_\mu/\overline{\nu}_\mu$ events with muons exiting the side of the calorimeter; the second reduces non-neutrino backgrounds, and the third ensures sufficient calorimeter downstream of the interaction to measure the event length. Small backgrounds from cosmic-ray and muon induced events are subtracted from the sample. After all cuts, 1.3 million and 0.30 million events are observed in the neutrino and anti-neutrino beam, respectively. The ratios of neutral-current candidates (short events) to charged-current candidates (long events), R_{meas}, are 0.4198 ± 0.0008 in the neutrino beam and 0.4215 ± 0.0017 in the anti-neutrino beam.

R_{meas} is related to the ratios of cross-sections and $\sin^2\theta_W$ using a detailed detector and cross-section Monte Carlo simulation with the tuned flux (Figure 1) as input. This Monte Carlo must predict the substantial cross-talk between the samples. In the neutral-current sample, the backgrounds in the neutrino and anti-neutrino beam from $\nu_\mu/\overline{\nu}_\mu$ charged-current events are 19.3% and 7.4%, and the backgrounds from $\nu_e/\overline{\nu}_e$ charged-currents are 5.3% and 5.8%. The charged-current sample has only a 0.3% background from neutral-current events for each beam.

The important details of the detector for this analysis are the calorimeter response to muons, the measurement of the neutrino interaction vertex, and the range of hadronic showers in the calorimeter. The efficiency, noise and active areas of the scintillation counters are all measured using neutrino data or muons from the testbeam. Longitudinal and transverse vertex resolutions and biases are studied using a GEANT-based detector Monte Carlo. The longitudinal bias arises from splashback in the hadronic interaction and is measured from the data using track-based vertices in events with two energetic final state muons. Hadronic shower length in the calorimeter is measured using hadrons from the testbeam. To study possible effects from the difference in strange-quark content between neutrino-induced and π^--induced showers, hadronic showers from K^-s are used as a cross-check. No significant differences are observed. Measured detector parameters are varied within their uncertainties in the Monte Carlo to study systematic errors associated with this simulation.

The cross-section model is of paramount importance to this analysis. Neutrino-quark deep-inelastic scattering processes are simulated using a leading-order cross-section model. Neutrino-electron scattering and quasi-elastic scattering are also included. Leading-order parton momentum distributions come from a modified Buras-Gaemers parameterization[8] of structure function data from the CCFR experiment[9] which used the same target-calorimeter and cross-section model as NuTeV. The parton distributions are modified to produce u and d valence and sea quark asymmetries consistent with muon scattering[10] and Drell-Yan[11] data. The shape and magnitude of the strange sea come from an analysis of events in CCFR with two oppositely charged muons (e.g., $\nu q \rightarrow \mu^- c$, $c \rightarrow \mu^+ X$)[12]. Mass suppression from heavy quark production is generated in a slow-rescaling model whose parameters are measured from the same dimuon data. The charm sea is taken from the CTEQ4L parton distribution functions[13]. The magnitude of the charm sea is assigned a 100% uncertainty and the slow-rescaling mass for $(\nu/\overline{\nu})c \rightarrow (\nu/\overline{\nu})c$ is varied from m_c to $2m_c$. Our parameterization of $R_{\rm long} = \sigma_L/\sigma_T$ is based on QCD predictions and data[14] and is varied by 15% of itself in order to estimate uncertainties. Electroweak and pure QED radiative corrections to the scattering cross-sections are applied using computer code supplied by Bardin[15], and uncertainties are estimated by varying parameters of these corrections. Possible higher-twist corrections are considered with a 100% uncertainty using a VMD-based model which is constrained by lepto-production data[16].

The key test of the Monte Carlo is its ability to predict the length distribution of events in the detector. Figure 2 shows good agreement between the data and Monte Carlo within the systematic uncertainties.

SOURCE OF UNCERTAINTY		$\delta \sin^2\theta_W$
Statistics:	Data	**0.00188**
	Monte Carlo	0.00028
TOTAL STATISTICS		0.00190
	$\nu_e/\overline{\nu}_e$	0.00045
	Energy Measurement	0.00051
	Event Length	0.00036
TOTAL EXP. SYST.		0.00078
	Radiative Corrections	0.00051
	Strange/Charm Sea	0.00036
	Charm Mass	0.00009
	u/d, $\overline{u}/\overline{d}$	0.00027
	Longitudinal Structure Function	0.00004
	Higher Twist	0.00011
TOTAL PHYSICS MODEL		0.00070
TOTAL UNCERTAINTY		0.0022

Table 1: Uncertainties in $\sin^2\theta_W$

To compute $\sin^2\theta_W$, a linear combination of $R^{\nu}_{\rm mea}$ and $R^{\overline{\nu}}_{\rm meas}$ was formed,

$$R^-_{\rm meas} \equiv R^{\nu}_{\rm meas} - \alpha R^{\overline{\nu}}_{\rm meas}, \quad (5$$

where α is calculated using the Monte Carlo suc that $R^-_{\rm meas}$ is insensitive to small changes in the slow rescaling parameters for charm production. α = 0.5136 for this measurement. This technique is similar to an explicit calculation of R^-, but here the background subtractions, the cross-section corrections t Eqn. 4, and the dependence on $\sin^2\theta_W$ are calculate by Monte Carlo. This approach explicitly minimize uncertainties related to the suppression of charm production, largely eliminates uncertainties related t scattering from sea quarks, and reduces many of th detector uncertainties common to both the ν an $\overline{\nu}$ samples. Uncertainties in this measurement o $\sin^2\theta_W$ are shown in Table 1.

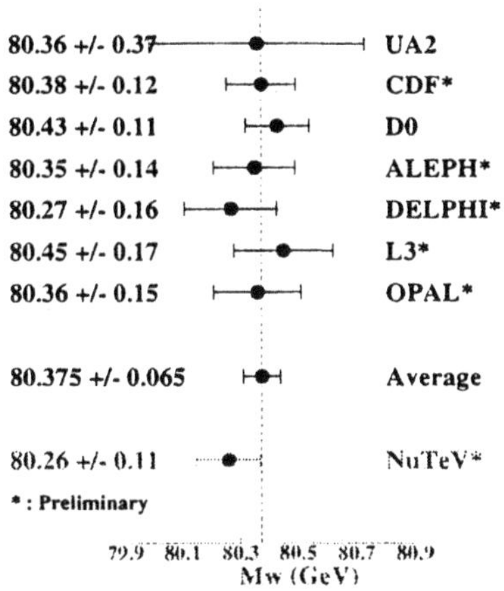

Figure 3: Current direct M_W measurements compared with this result

The preliminary result from the NuTeV data is *

$$\begin{aligned}\sin^2\theta_W{}^{(\text{on-shell})} &= 0.2253 \pm 0.0019(\text{stat}) \quad (6)\\ &\pm 0.0010(\text{syst})\\ &-0.00142 \times \left(\frac{M_{\text{top}}{}^2 - (175\ \text{GeV})^2)}{(100\ \text{GeV})^2}\right)\\ &+0.00048 \times \log_e\left(\frac{M_{\text{Higgs}}}{150\ \text{GeV}}\right). \quad (7)\end{aligned}$$

The small residual dependence of our result on M_{top} and M_{Higgs} comes from the leading terms in the electroweak radiative corrections[15]. Since $\sin^2\theta_W{}^{(\text{on-shell})} \equiv 1 - M_W^2/M_Z^2$, this result is equivalent to

$$\begin{aligned}M_W &= 80.26 \pm 0.10(\text{stat}) \quad (8)\\ &\pm 0.05(\text{syst})\\ &+0.073 \times \left(\frac{M_{\text{top}}{}^2 - (175\ \text{GeV})^2)}{(100\ \text{GeV})^2}\right)\\ &-0.025 \times \log_e\left(\frac{M_{\text{Higgs}}}{150\ \text{GeV}}\right). \quad (9)\end{aligned}$$

A comparison of this result with direct measurements of M_W is shown in Figure 3.

It is possible to extract the NuTeV result in a model-independent framework, where the result is expressed in terms of combinations of the left and right-handed quark couplings. The linearized constraint (expanded around one-loop couplings at an average $\log_{10}\left(\frac{-q^2}{1\ \text{GeV}^2}\right) \approx 1$ for NuTeV's central value of $\sin^2\theta_W$) is

$$\begin{aligned}0.4530 - \sin^2\theta_W &= 0.2277 \pm 0.0022\\ &= 0.8587u_L^2 + 0.8828d_L^2\\ &\quad -1.1657u_R^2 - 1.2288d_R^2. \quad (10)\end{aligned}$$

Note the similarity of this result to $1/2 - \sin^2\theta_W = g_L^2 - g_R^2$, the definition of the Paschos-Wolfenstein R^- in Eqn. 4.

(It is also possible to combine the NuTeV result with data from NuTeV's predecessor, the CCFR experiment. Adding the CCFR data[4] in the R^-_{meas}-based method described above, we obtain a slight improvement in precision, $\sin^2\theta_W = 0.2255 \pm 0.0018(\text{stat}) \pm 0.0010(\text{syst})$.)

4 Conclusions

The NuTeV experiment has completed its data-taking and has extracted a preliminary result for $\sin^2\theta_W{}^{(\text{on-shell})}$, which is equivalent to M_W in the Standard Model. The precision of this result is approximately a factor of two improvement over previous measurements in νN scattering because of the reduced systematics associated with measuring the Paschos-Wolfenstein ratio, R^-. This result is consistent with the average of direct M_W data.

We would like to gratefully acknowledge the substantial contributions in the construction and operation of the NuTeV beamlines and the refurbishment

*The weak radiative correction applied to extract $\sin^2\theta_W{}^{\text{on-shell}}$ from the measured quantities has changed since the presentation at Moriond due to an error in the implementation of the Bardin code for radiative corrections. Two other small experimental corrections, for muon energy deposition and for charm semi-leptonic decays, were improved as well. The net shift in the result, 0.0054, is dominated by the fix in the implementation of the radiative corrections.

of the NuTeV detector from the staff of the Fermilab Beams and Particle Physics Divisions.

References

[1] P. Langacker, *et al.*, *Rev. Mod. Phys.* **64**, 87 (1991) .

[2] K.S. McFarland, D. Naples, *et al.*, *Phys. Rev. Lett.* **75**, 3993 (1995) .

[3] C.H. Llewellyn Smith, *Nucl. Phys.* **B228**, 205 (1983) .

[4] K.S. McFarland, *et al.*, *Eur. Phys. Jour.* **C1**, 509 (198) .

[5] A, Blondel, *et al.*, *Zeit. Phys.* **C45**, 361 (1990) .

[6] J. Allaby, *et al.*, *Zeit. Phys.* **C36**, 611 (1985) .

[7] E.A. Paschos and L. Wolfenstein, *Phys. Rev.* **D7**, 91 (1973) .

[8] A.J. Buras and K.J.F. Gaemers, *Nucl. Phys.* **B132**, 249 (1978) .

[9] W.G. Seligman, *et al*, *Phys. Rev. Lett.* **79**, 1213 (1997) .

[10] M. Arneodo, *et al.*, *Nucl. Phys.* **B487**, 3 (1997) .

[11] E.A. Hawker, *et al.*, *Phys. Rev. Lett.* **80**, 3715 (1998) .

[12] S.A. Rabinowitz, *et al.*, *Phys. Rev. Lett.* **70**, 134 (1993) .

[13] CTEQ Collaboration, *Phys. Rev.* **D55**, 1280 (1997) .

[14] L.W. Whitlow, *SLAC-Report*-357, 109 (1990) .

[15] D.Yu. Bardin, V.A. Dokuchaeva, JINR-E2-86-260 (1986); and private communication.

[16] J. Pumplin, *Phys. Rev. Lett.* **64**, 2751 (1990) . $S_0 \leq 2$ GeV^2 is allowed by data summarized in M. Virchaux and A. Milsztajn, *Phys. Lett.* **B274**, 221 (1992) .

NUCLEAR PHYSICS B
PROCEEDINGS
SUPPLEMENTS

Nuclear Physics B (Proc. Suppl.) 77 (1999) 271–275

Events with Isolated Charged Leptons and Missing Momentum observed at the e^+p Collider HERA

D. Haidt[a]

[a]DESY
D-22603 Hamburg, Germany

Ten events observed by H1 and ZEUS are presented and discussed

1. Introduction

In the history of high energy physics *leptons* and *missing momentum* have proved to be powerful tools in searches for new phenomena.

The e^+p Collider HERA is operated with positrons of 27.5 GeV and protons of 820 GeV providing final states with invariant masses up to 300 GeV. Due to the large transverse momentum of the final state lepton the value of the quantity $\sum \vec{p}_T$, the vectorial sum over the transverse momenta of all observed final state particles, is a simple and efficient discriminator between the two inclusive processes $e^+p \to e^+ + anything$ (NC) and $e^+p \to \bar{\nu} + anything$ (CC), being small in the first and large in the latter case. During the year 1994 the H1 collaboration [1] observed in their study of events with *large missing momentum*, namely $|\sum \vec{p}_T| > 25$ GeV, an outstanding event (see figure 1) consisting of an isolated μ^+ with large transverse momentum and a hadron jet. In the meantime the available luminosity has increased by an order of magnitude and the two collaborations H1 and ZEUS have performed searches for events with the two signatures *missing momentum* and *isolated lepton* following complementary approaches :

- H1 selecting events with missing momentum and searching for isolated charged particles
- ZEUS selecting events with isolated charged leptons and looking for large transverse momentum imbalance in the final state.

The H1 analysis [2] based on 36.5 pb^{-1} is final, while the ZEUS analysis [3] based on 47 pb^{-1} is preliminary.

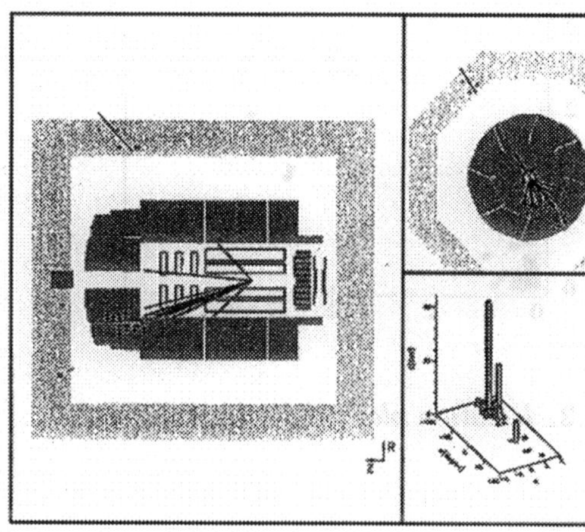

Figure 1. *The 1994 H1-event*

2. The two searches

The major steps in the analysis by H1 [2] are :

- Two independent requirements :
  1. Missing calorimetric momentum : $|\sum \vec{p}_T| > 25$ GeV
  2. Charged particle : $p_T > 10$ GeV and polar angle : $\Theta > 10^0$
- Reject $e^+p \to e^+ + anything$, if event balanced in azimuth or $E - p_L$

PII S0920-5632(99)00427-2

The resulting sample contains 124 events and is dominated by $e^+p \rightarrow \overline{\nu} + anything$. Are there events with *isolated* high-p_T charged particles ? This question is answered by introducing a distance measure between two points $(\eta, \phi)_i$ on the LEGO plot :

$$D_{12} = \sqrt{(\eta_1 - \eta_2)^2 + (\phi_1 - \phi_2)^2} \tag{1}$$

For any chosen high-p_T track the distance of closest approach is determined with respect to a jet (D_{jet}) and any other track (D_{track}). Jets are reconstructed by a cone algorithm with $R = 1$ and $E_T > 5$ GeV. As a matter of fact, every event has a jet. Tracks are accepted in the polar range $5^0 < \theta < 153^0$.

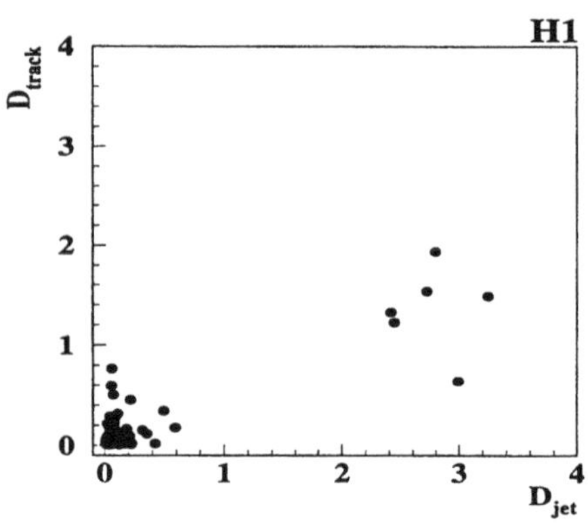

Figure 2. *Isolation plot*

Figure 2 shows, as expected, the majority of the events in the left lower corner in contrast to 6 completely isolated charged particles, which are uniquely identified as leptons. None of the events has an isolated charged hadron, nor are there events with 2 high-p_T particles. The lepton flavor in each of the 6 events (1 e^-, 2 μ^+, 2 μ^-, 1 $\mu^\pm$) differs from the one of the initial state e^+.

The major steps in the analysis by ZEUS [3] are :

- Isolated charged lepton:
  High p_T-track within $15^0 < \theta < 164^0$
  $E_e > 15$ GeV resp. $p_T^\mu > 5$ GeV
  Matching track with distance of closest approach < 10 (e) and 20 cm(μ)
  Isolation : no other track within cone of R=0.5 and accompanying energy in cone of R=1 smaller than 5 GeV
  Lepton identification
- Missing transverse momentum :
  $p_T(\text{calo}) > 19$ GeV (e), > 18 GeV (μ)
  $p_T(\text{calo}+\mu) > 18$ GeV
- Suppress NC events : candidate with isolated e must have $\theta_e < 1.3$ rad and $\theta_{aco} > 0.3$ rad (if $p_T^{had} > 4$ GeV).

The final sample consists of 4 events, each with a e^+. Figure 3 displays their second event.

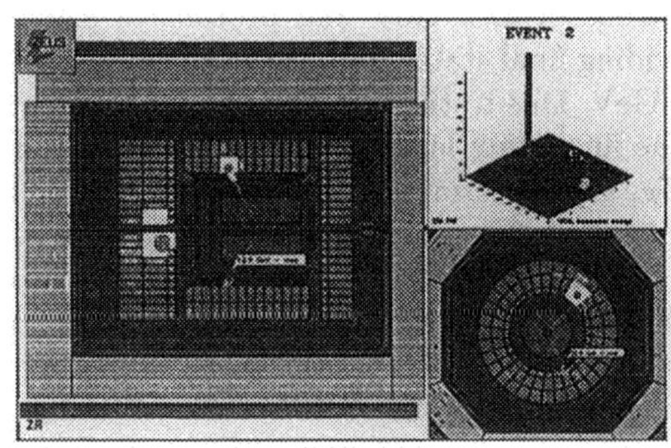

Figure 3. *The second ZEUS event*

3. Discussion

The kinematic properties of the two event samples are summarized in table 1. An examination of the relevant observables exhibits similar characteristics.

For the understanding of the signal various Standard Model processes have been considered as listed in the 2 tables below.

	ELECTRON	MUON-1	MUON-2	MUON-3 *)	MUON-4	MUON-5
The isolated high-P_T lepton						
Charge	Neg.(5σ)	Pos.(4σ)	Pos.(4σ)	Neg.(4σ)	Neg.(2σ)	unmeasured
P_T^l	$37.6^{+1.3}_{-1.3}$	$23.4^{+7.5}_{-5.5}$	$28.0^{+8.7}_{-5.4}$	$38.6^{+12.0}_{-7.4}$	$81.5^{+75.2}_{-26.4}$	> 44
θ^l	27.3 ± 0.2	46.2 ± 0.1	28.9 ± 0.1	35.5 ± 0.1	28.5 ± 0.1	31.0 ± 0.1
The hadronic system						
P_T^X	8.0 ± 0.8	42.2 ± 3.8	67.4 ± 5.4	27.4 ± 2.7	59.3 ± 5.9	30.0 ± 3.0
$P_{\parallel}^X$	-7.2 ± 0.8	-42.1 ± 3.8	-61.9 ± 4.9	-12.5 ± 2.1	-57.0 ± 5.5	-28.6 ± 3.1
$P_{\perp}^X$	-3.4 ± 0.9	-2.7 ± 1.8	26.8 ± 2.7	-24.3 ± 2.5	-16.3 ± 3.2	-9.1 ± 2.3
P_z^X	79.9 ± 4.4	153.1 ± 9.1	247.0 ± 18.9	183.7 ± 13.6	118.9 ± 12.1	145.4 ± 8.2
E^X	81.1 ± 4.5	162.0 ± 10.0	256.9 ± 19.5	186.8 ± 14.0	141.7 ± 13.7	154.8 ± 9.1
Global properties						
P_T^{miss}	30.6 ± 1.5	$18.9^{+6.6}_{-8.3}$	$43.2^{+6.1}_{-7.7}$	$42.1^{+10.1}_{-5.9}$	$29.4^{+71.8}_{-13.9}$	> 18
δ	10.4 ± 0.7	$18.9^{+3.9}_{-3.2}$	$17.1^{+2.5}_{-1.7}$	$26.9^{+4.2}_{-2.9}$	$43.5^{+19.3}_{-7.2}$	> 22
$M_T^{l\nu}$	67.7 ± 2.7	$3.0^{+1.5}_{-0.9}$	$22.8^{+6.7}_{-4.2}$	$75.8^{+23.0}_{-14.0}$	94^{+157}_{-54}	> 54
*) Positron in MUON-3 : $P_T^e = 6.7 \pm 0.4$, $P_{\parallel}^e = 6.1 \pm 0.4$, $P_{\perp}^e = -2.8 \pm 0.2$, $P_z^e = -3.7 \pm 0.2$						

Candidate	1	2	3	4
Year	95	97	97	97
Corrected Electron p_T (GeV)	24.7 ± 1.2	47.64 ± 1.9	44.4 ± 1.9	36.8 ± 1.6
Electron Polar Angle	37°	58°	70°	54°
Corrected Hadronic p_T (GeV)	0.4 ± 0.2	18.9 ± 2.7	24.6 ± 1.5	18.6 ± 2.4
Corrected Missing p_T (GeV)	24.3 ± 1.2	33.8 ± 2.4	22.7 ± 2.1	32.6 ± 1.4
Corrected Transverse Mass (GeV)	49.0 ± 1.6	79.2 ± 3.3	62.3 ± 3.3	67.8 ± 2.2
Matching Track p_T (GeV/c)	$22.3^{+6.8}_{-4.2}$	$44.1^{+33.0}_{-13.2}$	$42.9^{+17.8}_{-9.7}$	$35.9^{+13.8}_{-7.8}$
Track Charge	+1	+1	+1	+1

Table 1
Kinematic properties of the 6 H1 and the 4 ZEUS events

H1	Electron Channel	Muon Channel
Data	0 e^+,1 e^-	5
W prod.	1.65 ± 0.47	0.53 ± 0.11
Z prod.	0.01 ± 0.01	0.01 ± 0.01
CC	0.02 ± 0.01	0.01 ± 0.01
NC : e^+	0.51 ± 0.10	0.09 ± 0.06
NC : e^-	0.02 ± 0.01	
γ prod.	< 0.02	< 0.02
Heavy q	< 0.04	< 0.04
$\gamma\gamma$ e^+	0.09 ± 0.03	$0.14^{+0.14}_{-0.07}$
$\gamma\gamma$ e^-	0.04 ± 0.01	

ZEUS	Electron Channel	Muon Channel
Data	4 e^+,0 e^-	0
W prod.	2.22 ± 0.02	0.46 ± 0.02
CC	0.65 ± 0.17	0.37 ± 0.13
NC	0.32 ± 0.14	< 0.03
$\gamma\gamma(e)$	0.27 ± 0.27	
$\gamma\gamma(\mu)$		0.41 ± 0.18
$\gamma\gamma(\tau)$	< 0.06	0.06 ± 0.06

The most prominent contribution is W-production with subsequent leptonic decay. This hypothesis explains naturally the flavor properties of the final state lepton and the large imbalance in transverse momentum. Both collaborations have used the leading order calculation of Baur, Vermaseren and Zeppenfeld [4] to predict the distributions of various observables such as transverse mass (Jacobian peak), missing momentum, transverse momentum of the charged lepton and the hadron system (see figure 4; equivalent figures exist also for the ZEUS analysis). The most prominent event is displayed in figure 5.

Other processes do not contribute significantly to the signal. The smallness of several contributions reflects directly the severe selection criteria. The events with μ observed only by H1 attract some attention, as they occur more frequently than expected with the Standard Model.

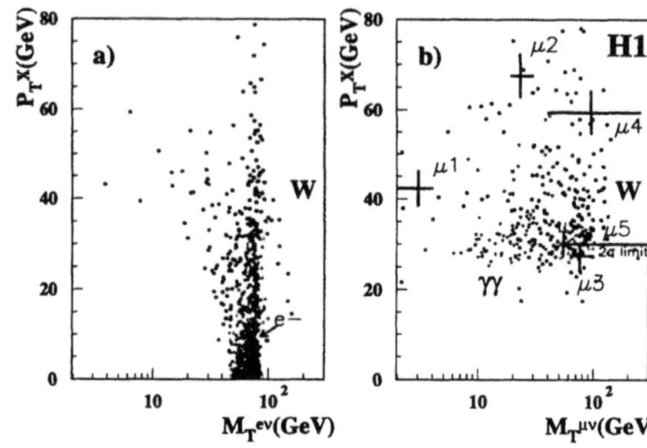

Figure 4. *Comparison of the 6 H1 events (crosses) with the W-hypothesis (dots) and the $\gamma\gamma$ (circles) background ; the simulation is based on a luminosity of 500 pb^{-1}; the asymmetry between the e-channel (a) and the μ-channel (b) originates from the fact that the e contributes to the calorimetric p_T, while the μ does not.*

4. Conclusions

The two collaborations H1 and ZEUS have reported 10 outstanding events with isolated high-p_T lepton and missing transverse momentum corresponding to a visible cross section of about 0.1 pb.

A definitive comparison of the two analyses is not yet possible, since the ZEUS results are preliminary and in addition the selection criteria are different. On the other hand, the two analyses are sufficiently similar to allow for a qualitative comparison, which is the author's assessment. Within the presently small statistics the event samples are compatible in size with each other and so are the background estimates. The flavor composition may be compared as follows : (a) given the 4 ZEUS e^+ events H1 should expect 1.6, while 0 observed (but 1 e^-); (b) given the 5 H1 μ-events ZEUS should expect 3.2, while none is observed. Future HERA running will ascertain the nature of the events with μ.

There is evidence for $e^+p \to W + anything$. It is rewarding to observe W-production finally

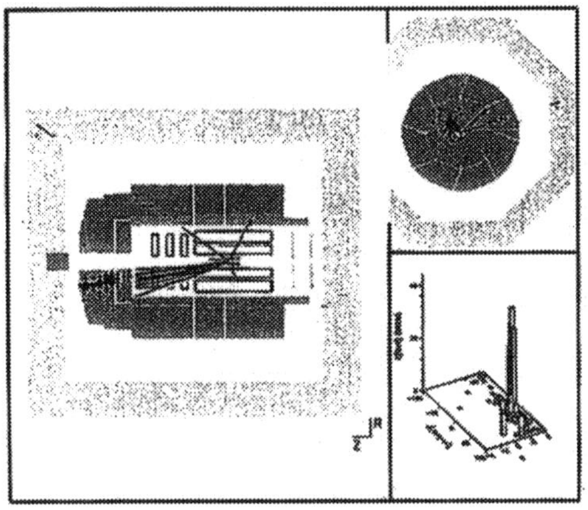

Figure 5. *Display of H1-event MUON-3 as candidate for the process* $e^+p \rightarrow e^+W^- + jet$ *: the final state* e^+ *taken as the scattered* e^+ *fixes the kinematics of the event; the* μ^- *together with the missing momentum combine to an invariant mass compatible with the* W *mass.*

in lepton-nucleon scattering, where the search started - 35 years ago - in the first neutrino experiments [5]. All events satisfy the W-kinematics, though in some cases the probablity is quite low. For the time being, the program used to simulate W-production, both for the total rate and the differential distributions, is based on lowest-order only, since no higher order calculation is yet available.

5. Acknowledgement

It is a pleasure to thank Drs. Yoji Totsuka and Yoichiro Suzuki for the invitation to this inspiring and well organised conference. In preparing the talk I appreciated discussions with J. Meyer, M. Kuze and S. Schlenstedt.

REFERENCES

1. H1-Collaboration : DESY 94-248 (1994)
2. H1-Collaboration : C. Adloff et al., Eur.Phys.J. **C**; DOI 10.1007/s100529800973 and DESY-98-063 (May 1998).
3. ZEUS-Collaboration : I. Negri : Talk given at *Lake Louise Winter Institute*,February 1998, see http://www-zeus.desy.de/conferences98/ U. Schneekloth : Talk given at *Moriond Electroweak*, March 1998, see http://www-zeus.desy.de/conferences98/ and DESY report 98-060
4. U. Baur, J.A.M Vermaseren and D. Zeppenfeld, Nucl. Phys. **B375** (1992) 3
5. D. Perkins : contribution to *Proceedings of the Siena International Conference on Elementary Particles*, 1963, p.555; F. Ferrero : ibidem, p.571.

NUCLEAR PHYSICS B PROCEEDINGS SUPPLEMENTS

ELSEVIER

Nuclear Physics B (Proc. Suppl.) 77 (1999) 276–284

Neutrino Physics with a Muon Collider

Panagiotis Spentzouris[a]

[a]Fermi National Accelerator Laboratory,
P.O. Box 500, Batavia IL 60510, USA

In the last few years, the idea of a muon accelerator complex has been actively pursued as an option for the next generation lepton collider. The very intense muon source ($\sim 10^{21}$/ per year) required for the collider, could provide very intense neutrino beams from muon decay in flight. The neutrino sources could either be the straight sections of the accelerator and the collider ring or a dedicated muon storage ring. In this note I discuss the physics potential of such neutrino sources for neutrino oscillations and Deep Inelastic Scattering measurements.

1. Introduction

A muon collider is a very appealing option for the next generation lepton machine. Compared to more conventional accelerators (hadron colliders or e^+e^- machines), a muon collider has some clear advantages both on the technical and the physics output aspects of the machine [1]. Since the mass of the muon is 207 times the electron mass, the radiative losses are much smaller for a muon circular machine compared to an electron circular machine (the radiative losses are proportional to m^4, where m is the mass of the lepton). Because of this simple fact, a muon collider could have a much smaller ring, and it could reach much higher energies than an electron machine. The small radiative losses also lead to a very small beam energy spread, allowing for precise measurements of the masses and widths of any new resonant state produced at the collider. In addition, a muon collider compared to any electron machine, has the advantage of s-channel Higgs production (the cross-section is proportional to m^2). Since in order to understand Electroweak symmetry breaking in the Standard model, and to explore potential new physics we need to probe elementary particle interactions at the TeV scale and beyond, the muon collider option seems very appropriate. It is widely accepted that hadron and lepton colliders offer complementary advantages in exploring the high energy frontier; hadron machines have access to a broader spectrum of states, while lepton machines are more suitable for precision measurements. A future muon collider should be viewed as a complementary machine to the LHC at CERN. Of course, the realization of a muon collider accelerator facility poses significant technical challenges, both in cooling and accelerating the very diffuse initial muon beam to a beam suitable for a collider, without significant losses. The effort to solve these technical problems is well underway, and the results of the design studies are very encouraging [1,2].

An additional important aspect of a muon collider accelerator complex is that it provides the ability to do very interesting physics measurements using the various stages of the accelerator complex. This simply means that the machine could be operational for physics measurements even before the collider is completed. In this note we will concentrate on the physics potential of a cold muon source for neutrino physics, in comparison with conventional neutrino sources.

2. Neutrino beams from a muon collider

In order to understand the scaling of the output of a muon collider neutrino source, we will start with a brief description of the current design scheme for a muon collider complex, since the intensity of the useful (cooled) muon beam determines the average neutrino beam intensity. A detailed description of this type of muon source can be found in Ref. [1–3]. In this design, the muon source consists of a proton accelerator, a

PII S0920-5632(99)00428-4

charged pion production target and collection system, a pion decay channel, and a muon cooling channel. The muon source receives protons from an accelerator complex which accelerates bunches of 5×10^{13} particles to energies of 16 GeV. The protons subsequently interact in a target to produce approximately 3×10^{13} charged pions of each sign per proton bunch. The pions are then collected using a high field co-axial solenoid with a typical field value of 20 Tesla, and an inner radius of 7.5 cm. Muons are produced by allowing the pions to decay in a 20 m long decay channel, consisting of a 7 Tesla solenoid with a radius of 25 cm. The muon yield of such a channel is on average 0.2 muons of each charge per incident proton. Since for the collider operation two proton bunches are used for every accelerator cycle (to produce both positive and negative muons), there would be 1×10^{13} muons of each charge available at the end of the decay channel per cycle. For a proton accelerator of 15 Hz, this results in about 1.5×10^{21} positive and negative muons in an operational year (10^7 secs). The decay channel is followed by a beam cooling section, where the six-dimensional phase-space of the muon beam is reduced to values suitable for the next stage of the machine. The proposed method of cooling is ionization cooling [4]. In this scheme, the muons lose both longitudinal and transverse momentum by ionization losses in absorbers placed within strong focusing magnetic fields. The longitudinal momentum is replaced using rf accelerating cavities, and the energy spread is reduced by using wedge shaped absorbers in a region of dispersion, by forcing higher energy particles to pass through more material than lower energy particles. At the end of the ionization cooling channel each bunch is expected to contain about 5×10^{12} muons, with a momentum of order 100 MeV/c. The muons are then accelerated using rapid recirculating linacs (RLA), and are injected in a storage ring. The amount of acceleration depends on the desired properties of the final stage of the machine. The final destination is the collider ring, but an intermediate stage of a dedicated muon storage ring for neutrino production could be considered. Neglecting acceleration losses, the number of cooled muons at the final stage is on the order of 7.5×10^{20} muons of each charge per operational year.

Neutrinos result from decays in flight of muons, and collimated neutrino beams could be produced in any straight section of the muon-collider complex. These neutrino sources are divided in two categories: neutrinos from the muon-accelerator stages (RLA) or the straight sections of the collider ring, and neutrinos from a dedicated storage ring. The first type of source is parasitic to the operation of the collider, and ideal for neutrino experiments which are located close to the source (the location of the source defines the location of the experiment). The parameters for this type of source are discussed in detail in Ref. [5]; the neutrino yields relevant to this discussion are tabulated in Table 1, where the beams from the last phase of the muon-accelerator (RLA3) and from a 10 m straight section of the collider are considered (see also Ref. [6–8]).

The second type of source is a dedicated muon storage ring, ideal for neutrino oscillation experiments, as we will discuss in the following section. For a ring with straight sections equal in length to the arcs of the ring, 25% of the muons would decay in each straight section, yielding 2×10^{20} neutrinos and 2×10^{20} anti-neutrinos per straight section per year, pointing at the direction of the straight section. Detailed designs for various storage ring configurations can be found in Ref. [11]. A summary of neutrino interaction yields for various experimental configurations is tabulated in Table 2; for a detailed discussion on the various storage ring possibilities see Ref. [9]. An important design issue is to keep the beam angular divergence small [9], so that the angular divergence of the produced beam is dominated by the muon decay kinematics which scales as $1/\gamma_\mu$.

2.1. Neutrinos from Muon Decays, comparison with conventional Neutrino Sources

Neutrinos produced from muon decay have a precisely known mixture of neutrino types: 50% muon neutrinos (anti-neutrinos) and 50% electron anti-neutrinos (neutrinos) when a μ^- (μ^+) beam is used. In the muon rest-frame the distribution of muon anti-neutrinos (neutrinos) and

Table 1
Neutrino fluxes and event rates for two parasitic neutrino beams, numbers are for muon neutrinos only. The RLA3 rates assume that the machine is ramping through 20 turns.

Source	RLA3	250 GeV Muon Collider
E_μ	150 – 250 GeV	250 GeV
turns/pulse	12	1560
decay length	533 m	10 m
$< E_{\nu_\mu} >$	135 GeV	178 GeV
beam radious at 600 m (50% of ν's)	25 cm	15 cm
Rate per 40 tons/year	5×10^9	5×10^9

Table 2
Number of ν_e charged current interactions per year and mean energies of the interacting neutrinos for a detector of mass m_{DET} at a distance L from a storage ring which circulates 7.5×10^{20} unpolarised positive muons per year with momenta p_μ. 25% of the muons are assumed to decay in the straight section pointing at the experiment.

p_μ (GeV/c)	m_{DET}	L (km)	$< E_\nu >$ (GeV)	L/$< E_\nu >$ (km/GeV)	ν_e CC interactions/yr
20	10 kT	10000	13	744	1×10^3
10	10 kT	732	6.6	111	3×10^4
20	10 kT	732	13	57	2×10^5
1.5	20 T	1	1	1	1×10^5

electron neutrinos (anti-neutrinos) from the decay $\mu^\pm \to e^\pm + \nu_e\ (\overline{\nu}_e) + \overline{\nu}_\mu\ (\nu_\mu)$ is predicted by the V-A theory:

$$\frac{d^2 N_{\nu_\mu}}{dx d\Omega} \propto \frac{2x^2}{4\pi}\left[(3-2x) \mp (1-2x)P_\mu \cos\theta\right], \quad (1)$$

and

$$\frac{d^2 N_{\nu_e}}{dx d\Omega} \propto \frac{12x^2}{4\pi}\left[(1-x) \mp (1-x)P_\mu \cos\theta\right], \quad (2)$$

where $x \equiv 2E_\nu/m_\mu$, θ is the angle between the neutrino momentum vector and the average muon spin direction, P_μ is the average muon polarisation along the beam direction, and m_μ is the muon rest mass. Thus, the neutrino and anti-neutrino differential flux depends on the parent muon energy, decay angle, and polarisation. Since the muon beam energy and intensity are very well constrained by the accelerator requirements, and the beam polarisation can be measured from the muon decay electron spectrum, the spectra of all of the components of the neutrino beam could be known with great accuracy.

In contrast, neutrino production from charged pion and kaon decays in conventional neutrino beamlines introduces much larger uncertainties in the flux determination. In such beamlines, the exact momentum distributions of the decaying mesons are difficult to measure in situ. The additional contributions from neutral kaon and charm decays, the latter produced in the target and dumps, further complicate the determination of the flux, requiring accurate simulations of the beamline and the underlying physical processes. (For a review of various configurations of conventional neutrino beamlines and their performances see Ref. [10]). The resulting beam composition is mostly muon neutrinos and anti-neutrinos, with a small admixture of electron and possibly tau neutrino and anti-neutrino components, which are not well constrained and constitute backgrounds to high precision measurements. Since the accuracy with which the absolute neutrino flux is known is one of the most important issues in precision neutrino experiments, it appears that a muon accelerator neutrino source has a great advantage compared to conventional meson decay sources.

3. Neutrino Physics with a Muon Collider Neutrino Source

Physics measurements with high energy neutrino beams could be divided into oscillation and non-oscillation neutrino physics; the latter includes precision tests of the Standard Model, and structure function and QCD measurements. In this section we will summarize some of the very interesting possibilities provided by a muon accelerator complex neutrino source.

3.1. Oscillation neutrino physics

In order to comment on the neutrino oscillation possibilities we will use the two-flavor vacuum oscillations framework, where the probability $P(\nu_1 \rightarrow \nu_2)$ that a neutrino of type 1 oscillates into a neutrino of type 2, after traveling a distance L is given by:

$$P(\nu_1 \rightarrow \nu_2) = \sin^2(2\theta)\ \sin^2(1.27\Delta m^2\ \frac{L}{E_\nu})\ , \quad (3)$$

where θ is the mixing angle, $\Delta m^2 \equiv m_2^2 - m_1^2$, and m_1 and m_2 are the masses of the two species; Δm^2 is measured in eV^2/c^4, L in km, and the neutrino energy E_ν in GeV. Experiments measure this probability, either by determining a finite value ϵ for the probability or by quoting a limit $P(\nu_1 \rightarrow \nu_2) < \epsilon$. There are two kinds of high energy accelerator oscillation experiments, flavor appearance and flavor disappearance. In the first case, the experiment attempts to detect interactions of a neutrino type not present in the generated beam; the sensitivity mainly depends on the accurate knowledge of the flavor content of the initial beam. In the second case, the experiment attempts to measure a deficit of the initial flux, by comparing it to calculations or to a measurement from a second detector very close to the neutrino source. With the second detector, the sensitivity depends on how well the "near" flux can be projected to the far detector, which has a different solid angle coverage. From equation 3 we can see that searches for small mixing are in practice constrained to $\Delta m^2 \ll \frac{E_\nu}{1.27\,L}$, while for large mixing $P(\nu_1 \rightarrow \nu_2) \approx (1.27\Delta m^2\ \frac{L}{E_\nu})^2$. In the $\sin^2(2\theta)$, Δm^2 parameter space there are two regions of great interest to future accelerator neutrino experiments: the atmospheric neutrino region [12], at maximal mixing and with Δm^2 which could be as low as $6 \times 10^{-4} eV^2$, and the LSND region [13] where $\Delta m^2 \sim 1 eV^2$ and $\sin^2(2\theta) \sim 10^{-3} - 10^{-2}$. The most satisfactory explanation for the oscillations observed in the first region is $\nu_\mu \rightarrow \nu_\tau$ or $\nu_\mu \rightarrow \nu_s$. In the second region $\nu_\mu \rightarrow \nu_e$ has been observed, and must be verified. We will examine the capabilities of a muon collider source in these two regions. Since there are conventional neutrino source experiments which have been proposed to explore these regions ([14–18] for the atmospheric, [19,20] for the LSND), it is useful not only to examine the sensitivity but also to stress the complementarity of the experimental technique.

The beam from a muon collider source is a mixture of electron and muon-like neutrinos, with one of the two flavors being an anti-neutrino. The key point here is that the charge of the leptons produced from the charged current interactions of these neutrinos are opposite. For example, if the initial beam is generated by positive muons, then the charged current interactions from the produced muon–anti-neutrino and electron–neutrino will produce μ^+ and e^- at the detector. Assuming that the detector is magnetic allowing charge identification, and can identify electrons with good efficiency, then both appearance and disappearance experiments could be performed, exploiting all the possibilities from both available beams. This is very important, since in a three (or more...) flavor oscillation scenario, the presence of the kurrently observed oscillation signals implies a rich mixing matrix for $\nu_x \rightarrow \nu_y$. This is a unique feature of a muon collider neutrino source, since conventional beams do not have a usable electron neutrino component. Furthermore, the presence of two beams provides a very reliable in situ flux calibration source for most cases of disappearance experiments, removing in principle the requirement for a near detector. For example, for a $\nu_\mu \rightarrow \nu_x$ study, the ν_e flux and Equations 1 and 2) could be used to determine

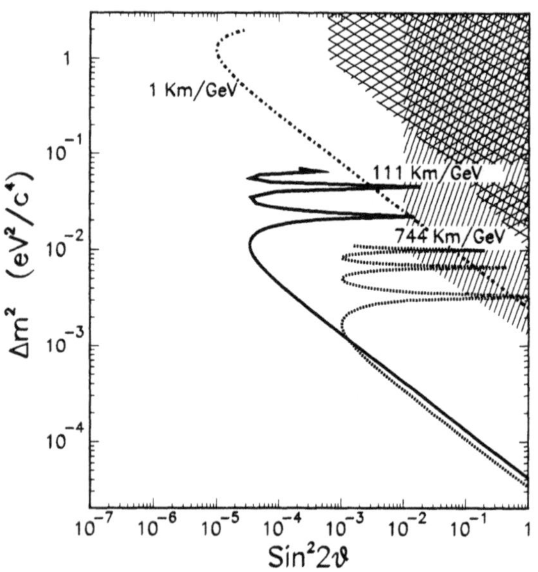

Figure 1. Contours of single–event sensitivity for a "wrong charge" ν_e–ν_μ oscillation search and for 1 year of running. The values of L/E are specified on the figure, and they correspond to the detector configurations summarized in Table 2. The hatched and cross-hatched areas show the regions which are expected to be explored by the MINOS experiment [14] after 2 years of running and the MiniBooNe experiment [19] after 1 year of running.

the initial ν_μ flux. For appearance experiments, the big advantage is the well defined flavor composition of the initial neutrino beam.

To illustrate the physics potential of the muon storage ring neutrino sources discussed in the previous section, we will consider the sensitivity of an experiment searching for ν_e–ν_μ or ν_e–ν_τ oscillations performed by searching for the appearance of charged current interactions producing "wrong-sign" muons [9]. Based on the discussion above, if there is a $\nu_e \rightarrow \nu_\mu$ oscillation, the ν_μ charged current interactions will produce the wrong sign muons (μ^-'s) for an initial $\overline{\nu_\mu}$, ν_e beam. Similarly, if the $\nu_e \rightarrow \nu_\tau$ and the neutrino energy is sufficiently large, the charged current interaction will produce a τ^- which, with a branching ratio of 17%, will decay to produce a μ^-. Since in the absence of backgrounds or systematic uncertainties, neutrino oscillation searches could be characterized by the total number of neutrino interactions observed, we use the parameters tabulated in Table 2 to define single event sensitivities for these configurations. The single event sensitivity contours are shown in Figure 1 for ν_e–ν_μ oscillations. The distances listed in the first 2 rows of Table 2 roughly correspond to a storage ring sited at the Fermi National Accelerator Laboratory with far sites located in Japan and at the Soudan mine, the last row corresponds to a typical short-baseline distance. The first option will require a storage ring tilted at a large angle ($\sim 50°$) with respect to the horizon. Note that the next generation of long baseline neutrino oscillation experiments currently under design [14–18] are expected to probe down to $\Delta m^2 \sim 1 \times 10^{-3}\ eV^2/c^4$ for $sin^2(2\theta) = 1$. Hence, the long baseline cases considered in this example would provide a significant improvement beyond the next generation of experiments. At small mixing–angles and large Δm^2 the sensitivity could go down to $\sin^2(2\theta) \sim 10^{-5}$, which is a significant improvement on the expected reach of the next generation of neutrino oscillation experiments [19,20]. Figure 3 shows the calculated single–event contours for ν_e–ν_τ oscillations, based on a search for wrong–sign muons [9].

At large mixing angles the long baseline experiments would be sensitive to values of Δm^2 ap-

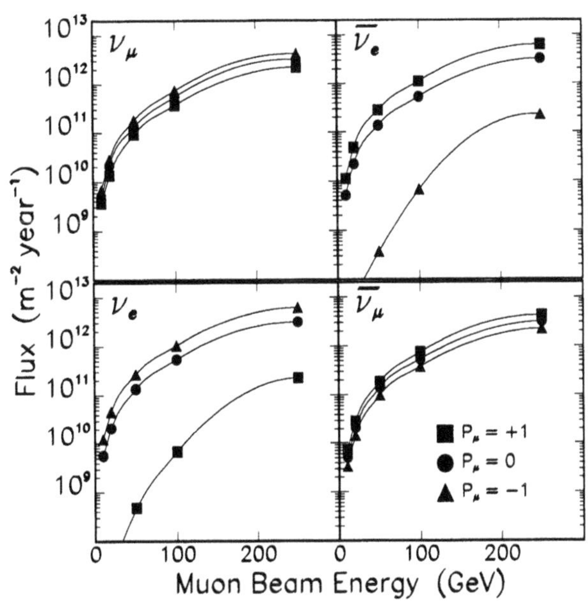

Figure 2. Neutrino and anti-neutrino fluxes at a far site located 100000 km from a muon storage ring neutrino source. The fluxes are shown as a function of the energy of the stored muons for negative muons (top plots) and positive muons (bottom plots), and for three muon polarisations as indicated.

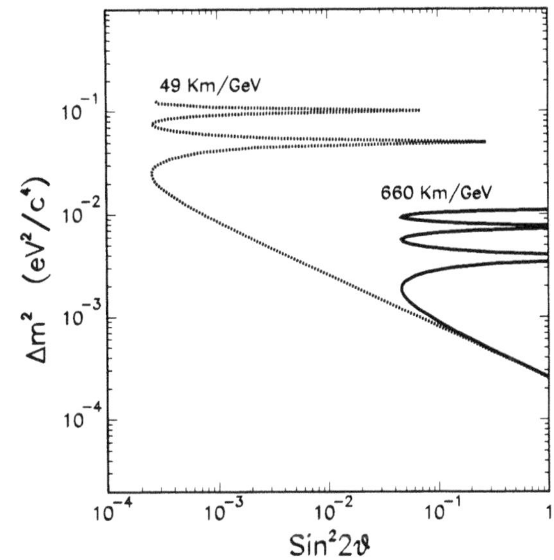

Figure 3. Contours of single–event sensitivity for ν_e–ν_τ oscillations based on searching for "wrong-sign" muons with the configurations for L/E of 660 km/GeV (solid contour) and 49 km/GeV (dotted contour). The contours correspond to a 10 kT-year exposure with 20 GeV/c unpolarised muons stored in a muon ring ring, with the straight section pointing at detectors 10000 km and 732 km from the ring.

proaching 10^{-4} eV/c^2, an improvement of several orders of magnitude beyond the sensitivities of past ν_e–ν_τ oscillation searches [21,22]. It is worth noticing that for a ν_τ appearance experiment in the absence of backgrounds and for maximal mixing, higher muon beam energies are better and L is not a very important parameter. The oscillation probability 3 goes as $\sim (\frac{L}{E_\nu})^2$, so the $1/L^2$ dependence due to flux loss from beam divergence cancels. The dependence of the flux on the energy of the stored muons is faster than quadratic (see Figure 2), thus the E_ν dependence of the flux will approximately cancel the probability dependence. The net effect is from the energy dependence of the $\sigma_{\nu_\tau}/\sigma_{\nu_\mu}$ cross-section ratio which rises with E_ν.

3.2. Non-oscillation neutrino physics

The main constraint for experiments with conventional neutrino beams is statistics. In order to gain reasonable statistical power, massive targets of high A materials are used. The main problems of this approach are lack of full event reconstruction, poor resolution, and possible nuclear effects. Despite these substantial experimental limitations, these kind of neutrino experiments have provided competitive measurements of Kobayashi-Maskawa matrix elements, the strong coupling constant and the Weinberg Angle, and are still our only source of information on the anti-quark content in the proton [10]. The flux numbers listed in Table 1 indicate a factor of a thousand over present neutrino experiments. An experiment with a small hydrogen target would see interaction rates/year on the order of 1 million events, comparable to the highest statistics present neutrino experiments which use a 680 ton iron target [10,6]. These neutrino beams, available at the muon-collider complex open up new and exciting possibilities. They allow the use of light thin targets, and detectors with good spatial resolution and final state reconstruction ability [23,6]. The improved experimental resolution will allow high precision neutrino structure function measurements in the very interesting kinematic region of low Bjorken-x, with direct impact on the determination of the strong-coupling constant [8], the flavor dependence of structure functions [6,7], and the nuclear dependence of the structure functions [7]. Specifically, since neutrino scattering from protons is sensitive to d, s and $\overline{u}$ quarks, and anti-neutrino scattering is sensitive to $\overline{d}$, $\overline{s}$ and u quarks, experiments using H_2 and D_2 targets could distinguish the contributions from each light quark species (the d and s quarks can be untangled via charm production, if there is good detector resolution and charm tagging). In addition, the nuclear effects on structure functions from neutrino Deep Inelastic Scattering could be measured, since both thin light and thin nuclear targets can be used in this type of detector. There is very little known experimentally about nuclear effects in neutrino DIS, while there is an abundance of very precise data from charged-lepton scattering. This measurent would be very interesting, since the origin of structure function nuclear effects is not theoretically understood [26], and current discrepancies between structure functions extracted in neutrino versus muon DIS [24,25] could be explained by different nuclear effects. If nuclear effects is not the reason for the discrepancy, (excluding the possibility for experimental error on the current measurements), then the conventional picture of DIS would have to be modified. In addition, the measurement of nuclear effects in F_3 would allow the differentiation between the nuclear effects in the valence and sea quarks. The structure function possibilities extend to spin physics too, neutrino beams are 100% polarised, so neutrinos only scatter from left-handed quarks or right-handed anti-quarks, while anti-neutrinos do the opposite. Matching the above property of the beam with a polarised target, clean measurements of the spin asymmetries in parton densities could be obtained [8]. Another exciting possibility is that of charm physics. Around 5% of charged current neutrino interactions involve charm production. With the beam parameters given in Table 1 we could expect $\approx$ 100000 reconstructed charm events from a 1 m hydrogen target/year. The final state reconstruction ability allows for inclusive charm production measurements with an extended kinematic reach and higher statistics, compared to the conventional measurements which use only the dimuon events [10] (produced from charm semileptonic decays). The dimuon rates alone are an order of magnitude higher with a muon collider source, even with a 1m liquid H_2 target (see Fig. 4). These measurements can provide information on a wide spectrum of physics topics which include the direct measurement of $|V_{cd}|$, the study of charmed hadron fragmentation properties, the measurement of branching fractions for D^0, D^+, D_s^+, Λ_c^+, and the measurement of the total charm production cross-section. Assuming good vertexing capabilities, there is also a possibility for a flavor changing neutral current search, using events with a single charm vertex and no primary muon. The sensitivity for this kind of search is high, being proportional to the total number of neutral current interactions. For all of the above topics the existing measurements are not high precision, since the only existing neutrino experiments with final state reconstruction are either emulsion or bubble chamber experiments.

There are two important issues which should be taken into account for the design of neutrino fixed target experiments at a muon collider. First, since most of the structure function related measurements require both neutrino and anti-neutrino DIS measurements (from the same neutrino flavor), the ability to change the beam polarity is necessary. Second, the backgrounds produced by muon halo and neutrino interactions upstream at the various elements of the beamlines will have to be seriously studied. The neutrino detectors must be located far enough away from the accelerator for the primary muons to range out in earth. For 250 GeV muons going through concrete this distance is on the order of 600 meters. However, a detector with a large amount of concrete directly proceeding it will be overwhelmed by muons produced in neutrino interactions in the shielding. A possible solution could be magnetized iron which would range out primary muons and sweep away most of the products of neutrino interactions in the shield [6].

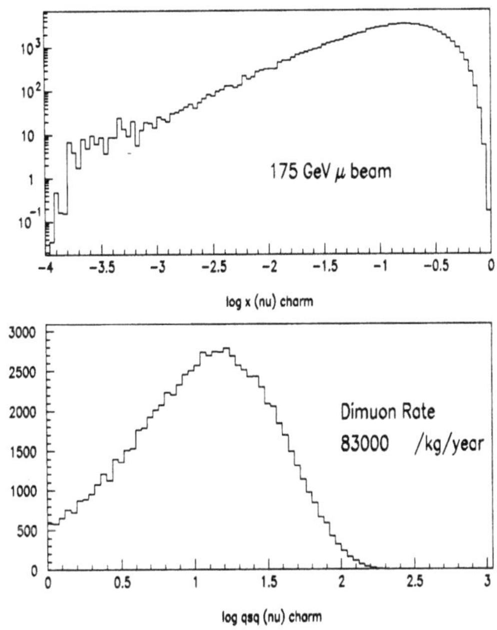

Figure 4. Charm induced dimuon production rates as a function of Bjorken-x and 4-momentum transfered squered (Q^2).

4. Conclusions

The very intense muon source which is currently being developed for a future high-luminosity muon collider would provide sufficient muons to make very intense neutrino and anti-neutrino beams. These intense neutrino beams could be produced from any straight sections of the acceleration components and the collider ring or from a dedicated muon storage ring. The beams could be used for high sensitivity neutrino oscillation experiments and for high precision neutrino Deep Inelastic Scattering experiments. If O(10^{20}) muons per year were allowed to decay within a 20 GeV/c storage ring with a straight section pointing in the desired direction, the resulting beams would produce hundreds of charged current neutrino interactions per year in a 10 kT detector on the other side of the Earth, which would be suitable for neutrino oscillation experiments. In addition to the high intensities, the advantages of these beams over conventional neutrino beams for oscillation experiments include high initial flavor purity, absolute flux determination and the possibility for in situ flux measurements by using the two flavor content of the beam. The physics program that can be pursued with the muon storage ring neutrino sources could provide significant sensitivity improvements for both short– and long–baseline neutrino experiments and for various beam energies. On the non-oscillation physics front, the muon-collider complex will allow factor of 1000 improvements (based on statistics) over present neutrino experiments, even running parasitically simultaneously with the collider. The great intensity improvement allows the use of light targets and high resolution experiments, which then provide the opportunity for physics measurements never done before in neutrino physics with high precision.

REFERENCES

1. "$\mu^+\mu^-$ Collider Feasibility Study", The Muon Collider Collaboration, Fermilab-Conf-96/092, July 1996, unpublished.
2. R. Palmer, A. Tollestrup, and A. Sessler, "Status Report of a High Luminosity Muon Collider and Future Research and Development Plans", Proc. of the 1996 DPF / DPB Summer Study on New Directions for High-energy Physics (Snowmass 96), Snowmass, CO, 25 June - 12 July 1996.
3. S. Holmes et al., "A Development Plan for the Fermilab Proton Source", FERMILAB-TM-2021, September 1997, unpublished.
4. A.N. Skrinsky and V.V. Parkhomchuk, Sov. J. Part. Nucl. **12**, 223 (1981).
5. C. Ankenbrandt amd S. Geer, "Accelerator Scenario and Parameters for the First Muon Collider and Front-End of a Muon Collider", Proc. Workshop on Physics at the First Muon Collider and Front-End of a Muon Collider, Fermilab 6-9th November, 1997.
6. H. Schellman, "Deep Inelastic Scattering at a Muon Collider-Neutrino Physics", Proc. Workshop on Physics at the First Muon Collider and Front-End of a Muon Collider, Fermilab 6-9th November, 1997.
7. P. Spentzouris, "Deep Inelastic Scattering and Neutrino Physics", Proc. Workshop on Physics at the First Muon Collider and Front-End of a Muon Collider, Fermilab 6-9th November, 1997.
8. D.A. Harris and K.S. McFarland, "A small target Neutrino Deep Inelastic Scattering Experiment" Proc. Workshop on Physics at the First Muon Collider and Front-End of a Muon Collider, Fermilab 6-9th November, 1997.
9. S. Geer, Phys. Rev. **D57**, 6989 (1998).
10. Janet M. Conrad, Michael H. Shaevitz, and Tim Bolton "Precision measurements with high-energy neutrino beams", preprint: hep-ex/9707015, July 1997, Submitted to Rev.Mod.Phys.
11. C. Johnstone, "High Intensity Muon Storage Rings for Neutrino Production: Lattice Design", Proc. Workshop on Physics at the First Muon Collider and Front-End of a Muon Collider, Fermilab 6-9th November, 1997.
12. Y.Fukuda et al. preprint hep-ex/9803006
13. C.Athanassopoulos et al. Phys. Rev. **C54**, 2685 (1996)
14. E. Ables et al; "P-875: A Long-baseline Neutrino Oscillation Experiment at Fermilab, NuMI-L-63 Minos Proposal", Feb. 1995, unpublished.
15. K.Nishikawa et al. (KEK-PS E362 Collab.), "Proposal for a Long Baseline Neutrino Oscillation Experiment, using KEK-PS and Super-Kamiokande", 1995, unpublished; INS-924, April 1992, Submitted to J.Phys.Soc.Jap.
16. M. Ambrosio et al. (NOE Collab.), Nucl. Instr. Meth. **A363**, 604 (1995).
17. P. Ce Cennini et al., "ICARUS II: A Second Generation Proton Decay Experiment and Neutrino Observatory at the Gran Sasso Laboratory: Proposal", LAGS-94/99-I (1994), unpublished.
18. T. Ypsilantis, Nucl. Instr. Meth. **A371**, 330 (1996).
19. E. Church et al. (BooNE Collab.), "A letter of intent for an experiment to measure $\nu_\mu \to \nu_e$ oscillations and ν_μ disappearance at the Fermilab Booster", May 16, 1997, unpublished.
20. ICARUS-CERN-Milano Coll. CERN/SPSLC 96-58, SPSLC/P 304, December 1996
21. N. Ushida et al. (The Fermilab E531 Collab.), Phys. Rev. Lett. **57**, 2897 (1986).
22. M. Talebzadeh et al. (BEBC WA66 Collab.), Nucl. Phys. **B291**, 503 (1987).
23. B.J. King "Neutrino Physics at a Muon Collider", Proc. Workshop on Physics at the First Muon Collider and Front-End of a Muon Collider, Fermilab 6-9th November, 1997.
24. M. Arneodo et al., Nucl. Phys. **B483**, 3 (1997)
25. W.G. Seligman et al., Phys. Rev. Lett. **79**, 1213 (1997).
26. D.F. Geesaman, K. Saito, A.W. Thomas, Ann. Rev. Nucl. Part. S ci. **45**, 337 (1995).

N·H
ELSEVIER
Nuclear Physics B (Proc. Suppl.) 77 (1999) 285–289
NUCLEAR PHYSICS B
PROCEEDINGS
SUPPLEMENTS

Status of the MUNU experiment

G. Jonkmans[a]*

[a]Institut de Physique, Université de Neuchâtel,
CH-2000 Neuchâtel, Switzerland

The MUNU experiment, a low background 1 m^3 time projection chamber surrounded by active anti-Compton shielding, is now under way at the Bugey nuclear reactor. It is dedicated to the experimental study of $\overline{\nu}_e e^-$ scattering down to 500 keV. The experiment is sensitive to a $\overline{\nu}_e$ magnetic moment down to 3×10^{-11} Bohr magneton.

1. INTRODUCTION

The MUNU experiment[1] is designed to measure $\overline{\nu}_e e^- \rightarrow \overline{\nu}_e e^-$ scattering at low energies (down to 500 keV). A precise investigation of this fundamental process provides informations on basic features of the weak interaction and on neutrino properties. The measurement of the differential cross section for $\overline{\nu}_e e^-$ scattering probes the destructive interference that take place between the charged weak current (CC) and the neutral weak current (NC). By doing so, a measurement of Weinberg's angle, $\sin^2\theta_W$, is in principle possible down to unprecedent energies.

Perhaps, more interestingly, is the sensitivity at low energies of possible electromagnetic interactions of the neutrino through a hypothetical neutrino magnetic moment. Moreover, MUNU will attempt to measure for the first time the scattering angle of the recoil electron. This additional information improves the sensitivity to possible electromagnetic interactions of the neutrino as the weak scattering cross section vanishes at zero angle for $E_\nu = m_e c^2$: the so-called "zero-dynamic" regime[2].

Only a few attemps have been made to measure the electron neutrino magnetic moment[3,4]. The best limit so far is $\mu_\nu < 2 \times 10^{-10}\mu_B$.

2. DETECTOR DESCRIPTION

The different energy dependence of the weak and (possible) electromagnetic cross sections in

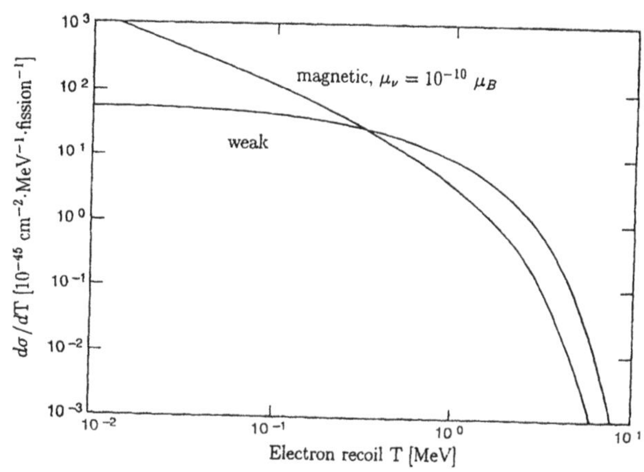

Figure 1. Expected spectrum of recoil electrons for $\overline{\nu}_e e^-$ scattering averaged over the reactor spectrum. The contribution from weak interaction alone and from a magnetic moment alone are showned seperately[5].

$\overline{\nu}_e e^-$ scattering (see figure 1) places stringents constraints on any experiment willing to have a high sensitivity on a neutrino magnetic moment: first, a low neutrino energy beam and secondly a low electron detection threshold.

The MUNU experiment is a tracking detector where, both, the electron energy and direction are determined. This has several advantages. Since

*On behalf of the MUNU collaboration.

0920-5632/99/$ – see front matter
PII S0920-5632(99)00429-6

the $\overline{\nu}_e e^-$ differential scattering cross section is forward peaked, the signal and the background can be measured online by looking at the forward-backward asymetry of events. Also, it becomes possible to define a fiducial volume thereby contributing to the background rejection.

Since the event rate is small (~10 ev./day) and the energy threshold low (500 keV), the detector components must be constructed from selected material of low natural radio-activity.

The general layout of the MUNU detector is displayed in figure 2. The main component is a 1 m^3 time projection chamber (TPC) filled with CF_4 gas which acts both as target material and drift gas. CF_4 was choosen for its high density (3.68 g/l at 1 bar) which maximizes the event rate and for its relatively low Z which reduces the multiple scattering of the observed electrons. The TPC is immersed in 10 m^3 of liquid scintillator viewed by 48 hemispherical photomultipliers (PMTs) and contained in a stainless-stell tank. This arrangement serves as anti-Compton to eliminate events that originate from gamma's scattering in the TPC. The anti-Compton also provides low-activity shielding. Moreover, an additional 8 cm of Boron loaded polyethylene absorbs neutron entering the outer lead shielding (15 cm).

2.1. The central detector

Aside from the advantages previously noted, the CF_4 offers low cosmogenic activation of C and F. Moreover, CF_4 does not contain free protons and the background from the reaction $\overline{\nu}_e p \rightarrow e^+ n$ is absent. The CF_4 is contained in a cylindrical acrylic vessel of diameter 90 cm and lenght 160 cm. The acrylic was choosen because of its low natural radioactivity. The contamination in ^{232}Th ^{238}U is of order 10^{-12} g/g and that of ^{40}K of order 10^{-13} g/g. The TPC is designed to operate at up to a pressure of 5 bar with a differential pressure between inside and outside of up to ±100 mbar. The drift field is delimited by a cathode plane at one end and a grid plane a the other end. The homogeneity of the electric field is reinforced by field shaping rings made from copper strips wrapped around the acrylic vessel and connected by a chain of high voltage resistors from the cathode plane to the grid. In this way, a field of 120 V cm^{-1} bar^{-1} is obtained and a drift velocity of about 3 cm μs^{-1} is achieved. The drift length is measured at 5 bar to be 22^{+14}_{-6} m[1]. The CF_4 gas is circulated continuously through an Oxysorb filter to remove oxygen and through a cold trap which removes freon contaminents and water vapors.

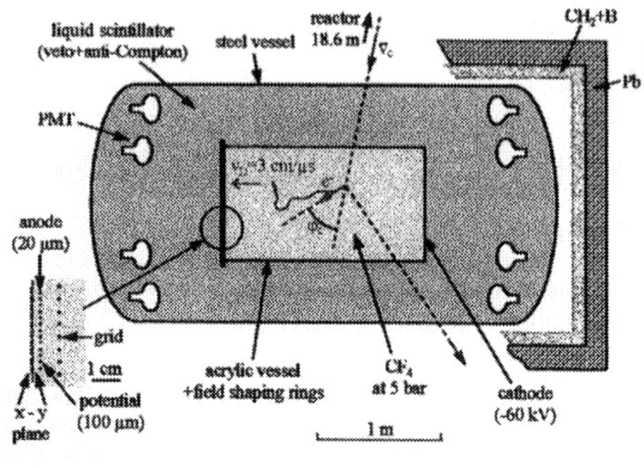

Figure 2. Cross-sectional view of the MUNU detector. The central TPC is showned surrounded by the anti-Compton scintillator and the various shielding layers.

The drifting electrons are collected about 1 cm behind the grid by an anode plane. The anode wires are connected together and give the total charge collected ("the energy signal"). Behind the anode plane is the $x-y$ plane which provides the spatial information. The spatial resolution is of order 1 mm in x, y and z. The z coordinate being determined by the time evolution of the signal. The energy resolution, measured with a ^{113}Sn source, is 20% FWHM at 370 keV and at 5 bar.

2.2. The active veto

The 48 PMTs viewing the liquid scintillator are placed in groups of 24 on each lid of the stainless steel tank. The hemispherical photomultipliers are 20 cm diameter EMI 9354 made from low activity glass. They are completely immersed in the scintillator and held in place by a polyethylene structure. The stainless steel tank is lined with TiO_2 based paint to increase the light collection efficiency. The dimension of the tank always ensure a minimum of 50 cm of liquid scintillator shielding of the TPC. Given that the mineral oil based liquid scintillator (NE235) has a measured attenuation length of 8 m at 430 nm and that individual PMTs will trigger at the 1 photo-electron level, the anti-Compton efficiency is evaluated to be 98% for gamma's above 100 keV.

3. SIGNALS AND BACKGROUNDS

The MUNU detector is situated at 18 m from the core of the 2800 MWth Bugey reactor ($5 \times 10^{20} \overline{\nu}_e s^{-1}$). The expected events rates from $\overline{\nu}_e e^-$ have been calculated and are given in table 1 for two ranges of energy and with and without a neutrino magnetic moment μ_ν. The total event rate above 500 keV in the absence of a neutrino magnetic moment is 9.5 events/day to be compared with 13.4 events/day for $\mu_\nu = 10^{-10} \mu_B$.

Table 1
Event rates predicted in the MUNU TPC for two energy ranges and for two values of μ_ν.

T (MeV)	Events/day	
	$\mu_\nu = 0$	$\mu_\nu = 10^{-10}$
0.5 - 1.0	5.3	8.1
>1.0	4.2	5.3

The main background comes from natural radiactivity who can yield compton electrons in the TPC. To minimize this background, all components entering in the assembly of the detector have been carefully choosen from radiochemically clean materials. Gamma activities where measured by various means and for various specific contaminents (see [1] and reference therein for a complete list). Based on those measured activities and assuming a threshold of 100 keV for the anti-Compton plus a forward angle selection, the total background rate from natural radio-activity is estimated to be 4 events per day above 500 keV.

A second background of importance is that induced by cosmic rays. Although the Bugey lab has a large overburden of steel and concrete corresponding to about 20 m water equivalent which almost completely eliminate cosmic neutrons, there still remains a muon flux of 32 $m^{-2}s^{-1}$. Muons can stop in the gas and produce β activity through muon capture or spallation products. But this background is small. However, muon capture in the surrounding materials yield neutrons which in turn capture to produce gammas. The estimated Compton background from these is about 2 events per day.

The background from the reaction $\overline{\nu}_e p \rightarrow e^+ n$ is deemed negligible. The total backgound rate is thus of 6 events per day.

With the rates given above, we can than estimate that after one year of data taking we can measure the energy spectrum above 500 keV with a statistical error of about 3%. Combined with a systematic error of 5% mainly from the detection efficiency, the reactor spectrum, the reactor power and burn-up, this yields to a sensitivity of $3 \times 10^{-11} \mu_B$ (90% C.L.) for the neutrino magnetic moment. This is an order of magnitude better than previous experiments. As can be seen in table 1, the ratio of rates in the different energy range differs markedly in the absence or presence of a neutrino magnetic moment. This feature can be used to cross-check the result with reduced systematics. Moreover, the study of the angular distribution of the recoil electron, precisely in the "zero-dynamic" regime would be further evidence of a neutrino magnetic moment.

Assuming the standard form of the weak interaction for $\overline{\nu}_e e^-$ scattering *i.e.* in the absence of a neutrino magnetic moment, we can measure Weinberg's angle, $\sin^2\theta_W$, with an accuracy of 5%. It is comparable to that achieved by the CHARM II collaboration for the $\nu_\mu e^-$ scattering[6].

4. STATUS AND PROSPECTS

The status of the experiment is as follow. Commissioning of the detector is well under way. We have been running at a pressure of 3 bar for several months now. Tracks of acceptable quality are obtained. See for example figure 3 which shows a cosmic muon inducing a delta electron where the blob at the end of the track is clearly visible. This demonstrate that we are able to observe minimum ionizing tracks and that we can determine the track direction. The anti-Compton threshold is measured to be around 100 keV and the count rate is consistent with expectations based on the background estimates above. We have taken background runs during the recent reactor shut-down and have found no events yet above 1 MeV.

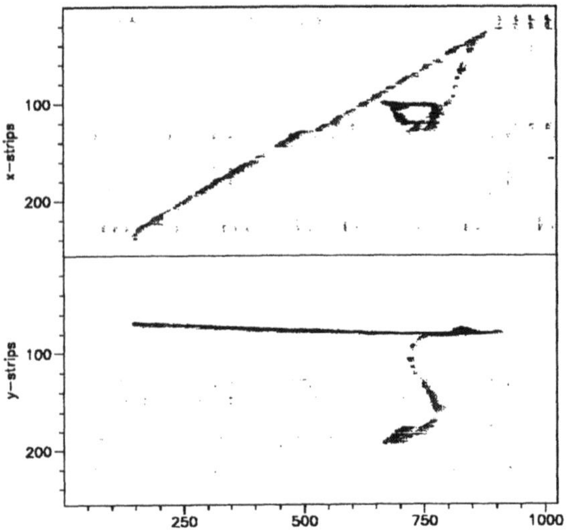

Figure 3. The x and y projection, versus drift time, of a muon event recorded in the TPC. The delta electron produced is easely identifiable.

The MUNU detector also serves as a low background prototype for a much larger TPC for the detection of low energy solar neutrinos from the Sun. A 2000 m^3 TPC filled with CF_4 at 1 bar would observe about 10 solar neutrinos per day above 100 keV. The energy and incident direction of the neutrino can be reconstructed from the electron recoil energy and scattering angle. In this way, spectroscopy of low energy solar neutrinos from the p-p, ^{7}Be and pep branch becomes feasible as show a simulation of rates expected after one year of data taking (see figure 4).

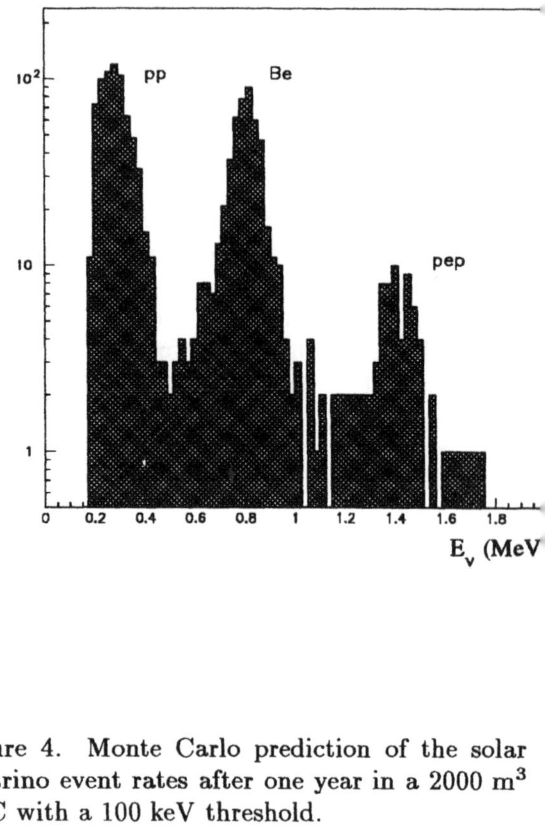

Figure 4. Monte Carlo prediction of the solar neutrino event rates after one year in a 2000 m^3 TPC with a 100 keV threshold.

REFERENCES

1. C. Amsler *et al.*, NIM **A396**, 115 (1997).
2. J. Segura *et al.*, Phys. Rev. **D49**, 1633 (1994).
3. F. Reines, H.S. Gurr, H.W. Sobel, Phys. Rev. Lett. **37**, 315 (1976).
4. A.I. Derbin *et al.*, JETP Lett. **57**, 768 (1993).

5. P. Vogel, J. Engel, Phys. Rev. **D39**, 3378 (1989).
6. D. Geiregat *et al.*, Phys. Lett. **B259**, 499 (1991).

Part 8

Neutrino and Particle Physics

ELSEVIER

Nuclear Physics B (Proc. Suppl.) 77 (1999) 293–298

NUCLEAR PHYSICS B
PROCEEDINGS
SUPPLEMENTS

Large Lepton Mixing in Seesaw Models*
- Coset-space Family Unification -

J. Sato and T. Yanagida
Department of Physics and RESCUE, University of Tokyo,
Hongo, Bunkyo-ku, Tokyo, 113, Japan

We show that the large mixing between ν_μ and ν_τ observed by the SuperKamiokande collaboration is a quite natural prediction in a large class of seesaw models. This large mixing is basically due to the unparallel family structure suggested from the observed mass hierarchies in quark and lepton mass matrices. We show that the unparallel family structure is automatically realized in "coset-space family unification" model based on $E_7/SU(5)\times U(1)^3$. This model also suggests the small angle MSW solution to the solar neutrino problem.

1. Introduction

T. Kajita from the SuperKamiokande collaboration has reported, in this conference, very convincing evidence of neutrino oscillation in their atmospheric neutrino data[1]. It is now clear that the long-standing puzzle of muon neutrino deficit in underground detectors[2] is due to the neutrino oscillation. A remarkable feature of the oscillation is almost maximal mixing between ν_μ and ν_τ ($\sin^2\theta_{23} \geq 0.8$), in sharp contrast to the quark sector for which mixing angles among different generations are all small. At first glance the rule governs the lepton mass matrices seems significantly different from the one relevant for the quark sector. We first show, in this talk, that the large mixing between ν_μ and ν_τ is quite naturally understood in a large class of seesaw models[3].

2. General Consideration in Seesaw Models

We adopt the SU(5) grand unification (GUT) as an example to make our point clearer, in which the lepton doublets belong to $\mathbf{5}^*$ of SU(5) GUT. We also assume supersymmetry(SUSY).

Let us discuss first the up-type quark mass matrix that is given by the following superpotential:

$$W = h_{ij}\mathbf{10}_i\mathbf{10}_j < H(\mathbf{5}) > . \qquad (1)$$

The most natural explanation of the mass hierarchy is given by the Froggatt-Nielsen mechanism[4]. We here assume a U(1) symmetry which is broken by a condensation of a superfield ϕ. The observed mass hierarchy,

$$m_t : m_c : m_u \simeq 1 : \epsilon^2 : \epsilon^4, \qquad (2)$$

suggests that $\epsilon =< \phi > /M_G \sim 1/20$ and the U(1) charges are 0, 1, 2 and -1 for the third, second, first families of $\mathbf{10}$'s and the ϕ. Here M_G is the gravitational scale $M_G \simeq 2.4\times 10^{18}$GeV.

The down-type quark/charged lepton mass matrix is given by

$$W = f_{ij}\mathbf{10}_i\mathbf{5}^*_j < \overline{H}(\mathbf{5}^*) > . \qquad (3)$$

The observed mass hierarchy,

$$m_b : m_s : m_d = m_\tau : m_\mu : m_e \simeq 1 : \epsilon : \epsilon^3, \qquad (4)$$

suggests that the third, second and first families of $\mathbf{5}^*$ have the U(1) charges A, A, and $A+1$, respectively. A crucial point is that the third and the second families of $\mathbf{5}^*$ have the same U(1) charge A. A could be 0 or 1. We take $A = 0$ for simplicity. We should stress here that the observed mass hierarchies in quark and lepton

*Talk is given by T. Yanagida

0920-5632/99/$ – see front matter
PII S0920-5632(99)00431-4

			U(1) charge
$\mathbf{10}_3$	$\mathbf{5}^*_3$	$\mathbf{5}^*_2$	[0]
$\mathbf{10}_2$	$\mathbf{5}^*_1$		[1]
$\mathbf{10}_1$			[2]

Table 1
Unparallel Family Structure

mass matrices already suggest an unparallel family structure in Table 1.

Now, let us discuss the neutrino mass matrix. In a generic seesaw model it is given by the following effective superpotential:

$$W_{eff} = \frac{\kappa_{ij}}{M_{\nu_R}} \mathbf{5}^*_i \mathbf{5}^*_j < H(\mathbf{5}) H(\mathbf{5}) > \quad (5)$$

The U(1) charge assignment for $\mathbf{5}^*_i$ leads to

$$\kappa_{ij} \sim \begin{pmatrix} 1 & 1 & \epsilon \\ 1 & 1 & \epsilon \\ \epsilon & \epsilon & \epsilon^2 \end{pmatrix}. \quad (6)$$

Notice that the U(1) charges for the superheavy right-handed neutrino ν_R are canceled out in the effective neutrino mass matrix in eq.(5). From eq.(6) we easily see a large mixing close to the maximal between ν_μ and ν_τ. The appearance of the large mixing is originated from the unparallel family structure discussed above.[2] On the contrary to the ν_μ-ν_τ mixing, we have small mixing between ν_e and ν_μ or ν_τ. Thus, the small angle MSW solution [5] to the solar neutrino problem [6] is also a quite natural expectation in a large class of seesaw models.

3. Coset-space Family Unification on $E_7/SU(5)\times U(1)^3$

In this section we show that the unparallel family structure discussed in the previous section is naturally obtained in the coset-space family unification[7] based on E_7.

The $E_7/SU(5)\times U(1)^3$ model[8,9] contains three families of $\mathbf{10}_i + \mathbf{5}^*_i + \mathbf{1}_i$ $(i = 1-3)$ and one

[2]This crucial point is emphasized by T. Yanagida and P. Ramond in this conference.

SU(5)	$U(1)_1$	$U(1)_2$	$U(1)_3$
$\mathbf{10}_1$	0	0	4
$\mathbf{10}_2$	0	3	-1
$\mathbf{10}_3$	2	-1	-1
$\mathbf{5}^*_1$	0	3	3
$\mathbf{5}^*_2$	2	-1	3
$\mathbf{5}^*_3$	2	2	-2
$\mathbf{1}_1$	0	3	-5
$\mathbf{1}_2$	2	-1	-5
$\mathbf{1}_3$	2	-4	0
$\mathbf{5}$	2	2	2

Table 2
U(1) charges of the NG multiplets. The $U(1)_1$, $U(1)_2$ and $U(1)_3$ are the unbroken U(1)'s of coset-subspaces $E_7/E_6\times U(1)$, $E_6/SO(10)\times U(1)$ and $SO(10)/SU(5)\times U(1)$, respectively.

$\mathbf{5}$ as NG multiplets. Here, the SU(5) is the usual GUT gauge group. Their quantum numbers under the unbroken subgroup are given in Table 2. Notice that the first family $\mathbf{10_1}$ has non-vanishing charge only for the $U(1)_3$ which means that the $\mathbf{10_1}$ is the NG multiplet for $SO(10)/SU(5)\times U(1)$. Similarly, we find that $\mathbf{10_2}$, $\mathbf{5}^*_\mathbf{1}$ and $\mathbf{1_1}$ are NG multiplets for $E_6/SO(10)\times U(1)$ and the remaining fields are NG multiplets for $E_7/E_6\times U(1)$. Thus, it is now clear that the unparallel family structure is an automatic prediction of this coset-space family unification [10].

This model can not be quantized in the original form, since there is a nonlinear-sigma model anomaly[11,9]. However, this global obstruction is easily removed[9] by introducing a matter multiplet $\mathbf{5}^*$ which is also needed for an SU(5) gauge-anomaly cancellation[8]. We assume that some explicit breaking induces an invariant mass for the NG $\mathbf{5}$ and this matter $\mathbf{5}^*$ and we neglect them in our discussion.

In addition to the NG multiplets we introduce a pair of Higgs multiplets $\mathbf{5}_H$ and $\mathbf{5}^*_H$. As long as the global E_7 is exact these Higgs multiplets never have Yukawa couplings to the NG quarks and leptons. Thus, the observed hierarchy in quark-lepton mass matrices is regarded as a consequence of a hierarchy in the explicit breaking of the global E_7. This situation is very similar to that in the QCD, where the mass hierarchy between NG pions and kaons ($m_K^2 \gg m_\pi^2$) is originated from the hierarchy in quark masses ($m_s \gg m_{u,d}$) which are explicit breaking parameters of the chiral $SU(3)_L \times SU(3)_R$.

We consider three steps for the explicit breaking:

$$E_7 \underset{\epsilon_0}{\longrightarrow} E_6 \underset{\epsilon_1}{\longrightarrow} SO(10) \underset{\epsilon_2}{\longrightarrow} SU(5), \quad (7)$$

which leads to the mass hierarchy

$$\begin{aligned} &m_t \gg m_c \gg m_u \\ &m_b \gg m_s \gg m_d \\ &m_\tau \gg m_\mu \gg m_e. \end{aligned} \quad (8)$$

To realize this hierarchy we assume that the global E_7 is broken explicitly by the fundamental representation of E_7, $\mathbf{56}$, which contains six breaking parameters, $\epsilon_0, \bar\epsilon_0, \epsilon_1, \bar\epsilon_1, \epsilon_2, \bar\epsilon_2$ that are all singlets of SU(5). They carry U(1) charges as

$$\begin{aligned} &\epsilon_0(-3,0,0), && \bar\epsilon_0(3,0,0) \\ &\epsilon_1(-1,-4,0), && \bar\epsilon_1(1,4,0) \\ &\epsilon_2(-1,-1,-5), && \bar\epsilon_2(1,1,5) \end{aligned} \quad (9)$$

where the numbers in each parenthesis denote charges of $U(1)_1 \times U(1)_2 \times U(1)_3$. The desired hierarchy in eq.(7) is represented by

$$\epsilon_0 \gg \epsilon_1 \gg \epsilon_2. \quad (10)$$

The structure of Yukawa couplings for the NG quarks and leptons depends on U(1) charges of the Higgs $\mathbf{5}_H$ and $\mathbf{5}_H^*$. To determine them, we consider that the Higgs multiplets $\mathbf{5}_H$ and $\mathbf{5}_H^*$ belong to $\mathbf{27}$ of E_6 in $\mathbf{133}$ of E_7. Then, U(1) charges for the $\mathbf{5}_H$ are given by

$$\mathbf{5}_H \ (2,2,2). \quad (11)$$

The Higgs $\mathbf{5}_H^*$ is a linear combination of two $\mathbf{5}^*$'s in $\mathbf{27}$ of E_6[3] as

$$\mathbf{5}_H^* = \sin\theta \mathbf{5}_{16}^* + \cos\theta \mathbf{5}_{10}^* \quad (12)$$

where U(1) charges for $\mathbf{5}_{16}^*$ and $\mathbf{5}_{10}^*$ are given by[4]

$$\mathbf{5}_{16}^*(2,-1,3) \ \text{and} \ \mathbf{5}_{10}^*(2,2,-2). \quad (13)$$

We now discuss Yukawa couplings for the quark and lepton multiplets. In general, Yukawa couplings are given in a form $a_n \epsilon^n \psi\psi H$ where ϵ, ψ and H stand for the explicit breaking parameters, the NG multiplets and the Higgs multiplets, respectively. By our choice of the U(1) charges for the explicit breaking parameters and Higgs multiplets, Yukawa couplings take the following form in the leading order of the explicit breaking parameters, ϵ's;

$$W = W_U + W_D + W_E + W_\nu, \quad (14)$$

$$W_U = \sum_{ij} a_{ij} Y_{Uij} \mathbf{10}_i \mathbf{10}_j \mathbf{5}_H, \quad (15)$$

$$W_D = W_E = \sum_{ij} b_{ij} Y_{D/Eij} \mathbf{5}_i^* \mathbf{10}_j \mathbf{5}_H^*, \quad (16)$$

$$W_\nu = \sum_{ij} c_{ij} Y_{\nu ij} \mathbf{5}_i^* \mathbf{1}_j \mathbf{5}_H, \quad (17)$$

where W_U, W_D, W_E and W_ν represent superpotentials of Yukawa couplings for up-type quarks, down-type quarks, charged leptons and neutrinos.

[3] $\mathbf{27}$ of E_6 is decomposed to $\mathbf{16} + \mathbf{10} + \mathbf{1}$ of SO(10). The $\mathbf{16}$ and $\mathbf{10}$ contain one $\mathbf{5}$ and two $\mathbf{5}^*$ of SU(5).

[4] The orthogonal combination of the $\mathbf{5}_{16}^*$ and $\mathbf{5}_{10}^*$ is assumed to have a GUT scale mass. We also assume that color triplets in $\mathbf{5}_H$ and $\mathbf{5}_H^*$ receive a GUT scale mass after the spontaneous breakdown of the SU(5) GUT. This requires a fine tuning. We do not, however, discuss this fine tuning problem here, since it is beyond the scope of this talk.

In these expressions Y's are given by,[5, 6]

$$Y_U \simeq \begin{pmatrix} \epsilon_2^2 & \epsilon_1\epsilon_2 & \epsilon_0\epsilon_2 \\ \epsilon_1\epsilon_2 & \epsilon_1^2 & \epsilon_0\epsilon_1 \\ \epsilon_0\epsilon_2 & \epsilon_0\epsilon_1 & \epsilon_0^2 \end{pmatrix}, \qquad (18)$$

$$Y_{D/E} \simeq \begin{pmatrix} \epsilon_1\epsilon_2\cos\theta & \epsilon_1^2\cos\theta & \epsilon_0\epsilon_1\cos\theta \\ \epsilon_0\epsilon_2\cos\theta & \epsilon_0\epsilon_1\cos\theta & \epsilon_0^2\cos\theta \\ \epsilon_0\epsilon_2\sin\theta & \epsilon_0\epsilon_1\sin\theta & \epsilon_0^2\sin\theta \end{pmatrix}, \qquad (19)$$

$$Y_\nu \simeq \begin{pmatrix} \epsilon_1^2 & \epsilon_0\epsilon_1 & \epsilon_0\epsilon_2 \\ \epsilon_0\epsilon_1 & \epsilon_0^2 & 0 \\ 0 & 0 & \epsilon_0^2 \end{pmatrix} \qquad (20)$$

We have assumed the E_7 representations for ϵ_i, $\mathbf{5}_H$ and $\mathbf{5}_H^*$ to determine their U(1) charges. However, we consider that this assumption is over statement since the E_7 is already spontaneously broken. What is relevant to our analysis is only their charges of the unbroken subgroup $SU(5)\times U(1)^3$. With this general consideration it is impossible to estimate the coefficients a_{ij}, b_{ij} and c_{ij} in eqs.(15), (16) and (17) and hence we assume that they are of O(1).

From the above Yukawa couplings in eqs.(18) and (19) we easily derive the following mass relations;

$$\frac{m_u}{m_c} \sim \frac{\epsilon_2^2}{\epsilon_1^2},$$
$$\frac{m_c}{m_t} \sim \frac{\epsilon_1^2}{\epsilon_0^2},$$
$$\frac{m_e}{m_\mu} = \frac{m_d}{m_s} \sim \frac{\epsilon_2}{\epsilon_0}\sin^{-1}\theta, \qquad (21)$$
$$\frac{m_\mu}{m_\tau} = \frac{m_s}{m_b} \sim \frac{\epsilon_1}{\epsilon_0}\sin\theta\cos\theta.$$

These relations describe very well the observed mass relations provided that

$$\frac{\epsilon_1}{\epsilon_0} \sim 0.05, \quad \frac{\epsilon_2}{\epsilon_1} \sim 0.05 \quad \text{and} \quad \tan\theta \sim 1. \qquad (22)$$

We see that the Cabibbo-Kobayashi-Maskawa mixing angles for quarks between the 1st and the 2nd, the 2nd and the 3rd, and the 3rd and the 1st family are of the order ϵ_2/ϵ_1, ϵ_1/ϵ_0, and ϵ_2/ϵ_0, respectively. It also describes the observed mixing angles very well provided that the relations in eq.(22) are satisfied.

We do not further mention details of the mass relations since there should be corrections to the mass matrices in eqs.(18) and (19) from some higher dimensional operators which may affect masses for lighter particles significantly. Otherwise, we have a SU(5) GUT relation, $m_d = m_e$, which seems unrealistic[12].

So far, we have discussed the mass matrices for quarks and charged leptons and found that the qualitative global structure of the obtained matrices fits very well the observed mass spectrum for quarks and charged leptons (except for $m_d = m_e$) and mixing angles for quarks if the relations in eq.(22) are satisfied[7].

We are now at the point to discuss neutrino masses and lepton mixings. We assume that Mayorana masses for right-handed neutrinos N_i are induced by SU(5) singlet Higgs multiplets $\bar{s}_i(\mathbf{1})$. We introduce two singlets $\bar{s}_1(\mathbf{1})$ and $\bar{s}_2(\mathbf{1})$ whose U(1) charges[8] are given by

$$\bar{s}_1(1,4,0) \quad \text{and} \quad \bar{s}_2(1,1,5). \qquad (23)$$

Their vacuum expectation values, $\langle\bar{s}_1\rangle$ and $\langle\bar{s}_2\rangle$ are expected to be of order of the SU(5) GUT scale $\sim 10^{16}$ GeV.

Majorana masses for N_i are induced from non-

[5] One may wonder that in eq.(19) the (3,1) element of $Y_{D/E}$, has a term of $\epsilon_0\epsilon_1$. We do not think that such a term appears there, since in the limit $\epsilon_2 \to 0$, the global SO(10) symmetry becomes exact and the $\mathbf{10}_1$ is the true NG multiplet which has no Yukawa interaction in the superpotential.

[6] Precisely speaking, our coset-space $E_7/SU(5)\times U(1)^3$ contains three dimensional parameters f_0, f_1 and f_2. We assume $f_0 \sim f_1 \sim f_2$ here, for simplicity. However, even if it is not the case, one obtains the same form of Yukawa couplings as in eqs.(18), (19) and (20) by redefining ϵ's as $\epsilon_i = \tilde{\epsilon}_i/f_i$ (i =0,1,2) where $\tilde{\epsilon}_i$ are original dimensional parameters for the explicit E_7 breakings.

[7] The observed mass for the strange quark seems somewhat smaller than the SU(5) GUT value[12].

[8] These $\bar{s}_i(\mathbf{1})$ are regarded as SU(5) singlet components of $\mathbf{56}$ of E_7.

renormalizable interactions of a form;[9]

$$W_N = \frac{\epsilon^2}{M_G} N_i N_j \bar{s}_k \bar{s}_l. \quad (24)$$

Here, M_G is the gravitational scale $M_G \simeq 2.4 \times 10^{18}$ GeV. Then, the matrix of the Majorana masses takes the following form;[10]

$$M_{\nu_R} = \frac{1}{M_G} \begin{pmatrix} \epsilon_1^2 \bar{s}_2{}^2 & \epsilon_0 \epsilon_1 \bar{s}_2{}^2 & \epsilon_0 \epsilon_1 \bar{s}_1 \bar{s}_2 \\ \epsilon_0 \epsilon_1 \bar{s}_2{}^2 & \epsilon_0^2 \bar{s}_2{}^2 & \epsilon_0^2 \bar{s}_1 \bar{s}_2 \\ \epsilon_0 \epsilon_1 \bar{s}_1 \bar{s}_2 & \epsilon_0^2 \bar{s}_1 \bar{s}_2 & \epsilon_0^2 \bar{s}_1{}^2 \end{pmatrix}, \quad (25)$$

where all elements are multiplied by undetermined factors of O(1) like in the case for quarks and leptons.

The neutrino masses are given by[3]

$$m_\nu \simeq m_{\nu_D} M_{\nu_R}^{-1} m_{\nu_D}^T, \quad (26)$$

where

$$(m_{\nu_D})_{ij} = c_{ij} Y_{\nu ij} \langle \mathbf{5}_H \rangle. \quad (27)$$

Three eigenvalues of the matrix in eq.(26) are of order, $m_{\nu_1} \sim \epsilon_1^2 M_G \langle \mathbf{5}_H \rangle^2 / \langle \bar{s}_2 \rangle^2$, $m_{\nu_2} \sim \epsilon_0^2 M_G \langle \mathbf{5}_H \rangle^2 / \langle \bar{s}_2 \rangle^2$ and $m_{\nu_3} \sim \epsilon_0^2 M_G \langle \mathbf{5}_H \rangle^2 / \langle \bar{s}_1 \rangle^2$. It is remarkable that for $\langle \mathbf{5}_H \rangle \sim 100$GeV, $\epsilon_0 \sim 1$ and $\langle \bar{s}_i \rangle \sim 10^{16}$GeV we get the desired mass for neutrino $m_{\nu_i} \sim 0.1$ eV.

From the Mikheev-Smirnov-Wolfenstein solution(MSW)[5] to the solar neutrino problem, we have[13,14]

$$\delta m^2_{\nu_e \nu_\mu} \simeq 10^{-6} - 10^{-5} \mathrm{eV}^2. \quad (28)$$

We see that there are two choices

$$\left(\frac{\langle \bar{s}_1 \rangle}{\langle \bar{s}_2 \rangle} \right)^2 \sim 10^{-2} - 10^{-1} \text{ or } \left(\frac{\langle \bar{s}_2 \rangle}{\langle \bar{s}_1 \rangle} \right)^2 \sim 10^{-2} - 10^{-1} \quad (29)$$

to account for atmospheric and solar neutrino anomalies, simultaneously. Thus, all off-diagonal elements of the diagonalization matrix for the neutrino mass matrix in eq.(26) are of O(0.1) in either cases.

[9]Other mass terms such as $\epsilon^2 N_i N_j$ can be forbidden by some chiral symmetry.

[10]The mass term of the form $\epsilon^4 N_i N_j$ may produce a similar form to eq.(25) if $\bar{\epsilon}_0 = 0$ and $\bar{\epsilon}_1, \bar{\epsilon}_2 \neq 0$.

However, it is very interesting that the mixing angle for lepton doublets which mixes charged leptons in the second and the third family is of order $\tan\theta$ (see eq.(19)) and hence of the order 1. This means, together with the above result, that the weak mixing angle relevant for ν_μ-ν_τ oscillation can be so large, $\sin^2 2\theta_{\nu_\mu \nu_\tau} \simeq 1$, as required for explaining the observed atmospheric neutrino anomaly. On the other hand, the mixing angle for $\nu_\mu - \nu_e$ oscillation is very small, $\theta_{\nu_\mu \nu_e} \sim$ O(0.1), which may fit the small angle MSW solution[13,14] to the solar neutrino problem.

4. Conclusion and Discussion

In this talk we have shown that the coset-space family unification on $E_7/SU(5) \times U(1)^3$ naturally accommodates the large lepton mixing, $\sin^2 2\theta_{\nu_\mu \nu_\tau} \simeq 1$, necessary for explaining the atmospheric neutrino anomaly reported by the SuperKamiokande collaboration[1]. The main reason why we have a large mixing of the SU(2) lepton doublets in the second and the third family is the twisted structure of family. Namely, the $\mathbf{5}^*$'s in the second and the third family both live on the same coset-subspace $E_7/E_6 \times U(1)$. On the other hand the $\mathbf{10}$'s in the third, the second and the first family live on the separate coset-subspaces, $E_7/E_6 \times U(1)$, $E_6/SO(10) \times U(1)$ and $SO(10)/SU(5) \times U(1)$, respectively. This unparallel family structure is an unique feature of the present coset-space family unification.

It is quite natural that the NG multiplets carry no $U(1)_R$ charge. Thus, the dangerous lower $(d = 4, 5)$ dimensional operators contributing to proton decays are forbidden by imposing the R-invariance $U(1)_R$. However, the R invariance is broken at the gravitino scale at least and hence we may expect small R-violating $d = 4$ operators.

The existence of approximate global E_7 symmetry is the most crucial assumption in our coset-space family unification. We hope that it is understood by some underlying physics at the gravitational scale. The Horava and Witten M theory[15] will be a hopeful example, since it is known[16] that there appear enhanced global symmetries on the 10 dimensional boundary of 11 dimensional space-time .

REFERENCES

1. SuperKamiokande Collaboration, T. Kajita, this proceeding; Y. Fukuda et al, hep-ex/9807003.
2. Kamiokande Collaboration, K.S. Hirata et al., Phys. Lett. B 205 (1988) 416; Phys. Lett. B 280 (1992) 146; Y. Fukuda *et al.*, Phys. Lett. B 335 (1994) 237;
   IMB Collaboration, D. Casper *et al.*, Phys. Rev. Lett. 66 (1991) 2561; R. Becker-Szendy *et al.*, Phys. Rev. D 46 (1992) 3720;
   SOUDAN2 Collaboration, T. Kafka, Nucl. Phys. B (Proc. Suppl.) 35 (1994) 427; M. C. Goodman, Nucl. Phys. B (Proc. Suppl.) 38 (1995) 337; W. W. M. Allison et al., Phys. Lett. B 391 (1997) 491.
3. T. Yanagida, in *Proceedings of the Workshop on Unified Theory and Baryon Number in the Universe*, edited by A. Sawada and H. Sugawara, (KEK, Thukuba, Japan, 1979);
   M. Gell-Mann, P. Ramond and R. Slansky, in *Supergravity*, edited by F. van Nieuwenhuizen and D. Freedman, (North Holland, 1979).
4. C.D. Froggatt and H.B. Nielsen, Nucl. Phys. B 147 (1979) 277.
5. L. Wolfenstein, Phys. Rev. D 17 (1978) 2369; S. P. Mikheev and A. Yu. Smirnov, Sov. J. Nucl. Phys. 42 (1985) 913.
6. Homestake Collaboration, K. Lande. this proceeding;
   SuperKamiokande & Kamiokande Collaboration, Y. Suzuki, this proceeding;
   GALLEX Collaboration, T. Kirstenys, this proceeding;
   SAGE Collaboration, V. N. Gavri, this proceeding.
7. W. Buchmuller, R. D. Peccei and T. Yanagida, Nucl. Phys. B 227 (1983) 503.
8. T. Kugo and T. Yanagida, Phys. Lett. B 134 (1984) 313.
9. T. Yanagida and Y. Yasui, Nucl. Phys. B 269 (1986) 576.
10. J. Sato and T. Yanagida, Phys. Lett. B430 (1998) 127.
11. G. Moore and P. Nelson, Phys. Rev. Lett. 53 (1984) 1519; Commun. Math Phys. 100 (1985) 83;
    P. di Vecchia, S. Ferrara and L. Girardello, Phys. Lett. B 151 (1985) 199;
    E. Cohen and C. Gomez, Nucl. Phys. B 254 (1985) 235.
12. Particle Data Group, Phys. Rev. D 54 (1996) 1.
13. J. N. Bahcall and P. I. Krastev, Phys. Rev. D 53 (1996) 4211;
    G. L. Fogli, E. Lisi and D. Montanino, Astropart.Phys.9 (1998) 119.
14. See, for a review, M. Fukugita and T. Yanagida, in *Physics and Astrophysics of Neutrinos*, edited by M. Fukugita and A. Suzuki (Springer-Verlag, Tokyo, 1994).
15. P. Horava and E. Witten, Nucl. Phys. B460 (1996) 506; Nucl. Phys. B475 (1996) 94.
16. E. Sharpe, Nucl. Phys. B523 (1998) 211.

NUCLEAR PHYSICS B
PROCEEDINGS
SUPPLEMENTS

ELSEVIER

Nuclear Physics B (Proc. Suppl.) 77 (1999) 299–307

IMPLICATIONS OF THE SUPERKAMIOKANDE RESULT ON THE NATURE OF NEW PHYSICS*

Jogesh. C. Pati[a†]

[a]Department of Physics, University of Maryland, College Park, MD-20742, U. S. A.

It is remarked that the SuperKamiokande (SK) discovery of ν_μ to ν_τ (or ν_X)-oscillation, with a $\delta m^2 \approx 10^{-2} - 10^{-3} eV^2$ and $\sin^2 2\theta > 0.8$, provides a clear need for the right-handed (RH) neutrinos. This in turn reinforces the ideas of the left-right symmetric gauge structure $SU(2)_L \times SU(2)_R$ as well as SU(4)-color, for which the RH neutrinos are a compelling feature. It is noted that by assuming (a) that B-L and I_{3R}, contained in a string-derived $G(224) = SU(2)_L \times SU(2)_R \times SU(4)^c$ or SO(10), break near the GUT-scale, as opposed to an intermediate scale, (b) the see-saw mechanism, and (c) the SU(4)-color relation between the Dirac mass of the tau neutrino and m_{top}, one obtains a mass for ν_L^τ which is just about what is observed. This is assuming that the SK group is actually seeing $\nu_L^\mu - \nu_L^\tau$ (rather than $\nu_L^\mu - \nu_X$)-oscillation. Following a very recent work by Babu, Wilczek and myself, it is furthermore noted that by adopting familiar ideas of understanding Cabibbo-like mixing angles in the quark-sector, one can quite plausibly obtain a large $\nu_L^\mu - \nu_L^\tau$ oscillation angle, as observed, in spite of highly non-degenerate masses of the light neutrinos: e.g. with $m(\nu_L^\mu)/m(\nu_L^\tau) \approx 1/10 - 1/20$. In this case, $\nu_L^e - \nu_L^\mu$ oscillation can be relevant to the small angle MSW explanation of the solar neutrino-puzzle. Implications of the mass of ν_L^τ and the large oscillation angle suggested by the SK result, on proton decay are noted.

1. Introduction

The SuperKamiokande (SK) result, convincingly showing the oscillation of ν_μ to ν_τ (or ν_X), with a value of $\delta m^2 \approx 10^{-2}$ to 10^{-3} eV^2 and $\sin^2 2\theta > 0.8$[1], appears to be the *first clear evidence* for the existence of new physics beyond the standard model. The purpose of this talk is to make two points regarding the implications of the SK result, which though simple, seem to be far-reaching. The first is the argument as to why one needs new physics beyond the standard model. The second is the remark that the SK result already tells us much about the nature of the new physics. In particular, it seems to suggest clearly the existence of right-handed neutrinos, a new form of matter, accompanying the observed left-handed ones. This in turn reinforces the twin ideas of the left-right symmetric gauge structure $SU(2)_L \times SU(2)_R$ and of SU(4)-color, which were proposed some time ago as a step towards higher unification [2]. Either one of these symmetries require the existence of the right-hand neutrinos. I note that by assuming (a) that B-L and I_{3R}, contained in a string or a GUT-derived $G(224) = SU(2)_L \times SU(2)_R \times SU(4)^c$, break near the GUT-scale as opposed to an intermediate or a low-energy scale, (b) the see-saw mechanism [3], and (c) the SU(4)-color relation between the Dirac mass of ν_τ and m_{top}, one obtains a mass for ν_L^τ which is just about what is observed. This is presuming that the SK group is actually observing $\nu_L^\mu - \nu_L^\tau$, (rather than $\nu_L^\mu - \nu_X$), oscillation and that the neutrino masses are hierarchical ($m(\nu_L^e) << m(\nu_L^\mu) << m(\nu_L^\tau)$), so that the observed value of δm^2 in fact represents the $(mass)^2$ of ν_L^τ. Such a hierarchical pattern, as opposed to near degeneracy of two or three neutrino flavors, is of course naturally expected within the see-saw formula. Following a very recent work by Babu, Wilczek and myself [4], I furthermore note that by combining contributions to the oscillation angle from the neutrino and the charged lepton-sectors, and by following familiar ideas on the understanding of Cabibbo-like mixing angles in the quark-sector, one can quite plausibly obtain a large $\nu_L^\mu - \nu_L^\tau$-oscillation angle, as observed, in spite of hierarchical masses of the light neutrinos: e.g. with $m(\nu_L^\mu)/m(\nu_L^\tau) \approx 1/10 - 1/20$.

†

0920-5632/99/$ – see front matter
PII S0920-5632(99)00432-6

In this case, $\nu_L^e - \nu_L^\mu$ oscillation can be relevant to the small angle MSW explanation of the solar neutrino puzzle. The results on δm^2 and mixing obtained in the context of G(224) can of course be obtained within any extension of G(224), such as SO(10) [5], together with supersymmetry. At the end, implications of the neutrino mass-scale observed at SuperKamiokande on proton decay are noted. Comments are made on how the SK result supplements that of LEP in selecting out the route to higher unification.

2. The Need for New Physics:

First, as we know, the standard model (SM), based on the gauge symetry $SU(2)_L \times U(1)_Y \times SU(3)^C$, contains 15 two-component objects in each family – e.g. for the electron-family they are: $[Q = (u_L, d_L), L = (\nu_L^e, e_L^-), u_R, d_R$ and $e_R]$ - and the Higgs doublet $H = (H^+, H^o)$. Notice that in the standard model, the left-handed neutrino ν_L is an odd ball in that it is the only member in each family which does not have a right-handed counterpart ν_R. This feature in fact carries over to its grand unifying extension SU(5) as well [6]. In other words, the standard model (as also SU(5)) provides a clear distinction between left and right, in the spectrum as well as in the gauge interactions, and thus explicitly violates parity and charge conjugation.

Without a right-handed counterpart, a left-handed neutrino ν_L cannot acquire a Dirac mass. But it may still acquire a Majorana mass (like $m_L \nu_L^T C^{-1} \nu_L$), by utilizing the effects of quantum gravity, which may induce a lepton-number violating non-renormalizable operator (written schematically) in the form[7]

$$\lambda_L \, LLHH/M_{Pl} + hc. \tag{1}$$

Here, M_{Pl} denotes the reduced Planck mass $= 2\times 10^{18}$ GeV and λ_L is the effective dimensionless coupling. Using the VEV of $< H > \approx 250$GeV, such an operator would then give:

$$m(\nu_L) \approx \lambda_L \frac{(250\text{GeV})^2}{2 \times 10^{18}\text{GeV}} \approx (\lambda_L)(3 \times 10^{-5}\text{eV}) \tag{2}$$

Such a mass would lead to values of δm^2 (for any two light neutrino-species) $\leq \lambda_L^2(10^{-9}\text{eV}^2)$. This is far too small (even for ridiculously large $\lambda_L \sim 10^2$, say) compared to the observed value of $\delta m^2 \approx 10^{-2} - 10^{-3}\text{eV}^2$.[1] It thus follows rather conclusively that the specific range of values of δm^2 reported by SuperKamiokande cannot reasonably be accommodated within the standard model, even with the inclusion of quantum gravity, and thus there must exist new physics beyond the standard model.

3. The Nature of New Physics:

We now go further and turn to the second point about the nature of the new physics, suggested by the SK result. The only reasonable way to understand a mass for the neutrino or δm^2, as observed, it seems to me, is to introduce a right-handed (RH) neutrino (ν_R) and utilize the see-saw mechanism (described below).[2] This in turn has far-reaching implications. The existence of a RH neutrino becomes compelling by extending the SM symmetry to include either SU(4)-color or the left-right symmetric gauge-structure $SU(2)_L \times SU(2)_R$, [2]. Thus the SK result motivates, on observational ground, the route to higher unification via the gauge-structure:

$$G(224) = SU(2)_L \times SU(2)_R \times SU(4)^C. \tag{3}$$

This is the minimal extension of the SM that specifies all quantum numbers (given a representation), quantizes electric charge and introduces ν_R. With respect to G(224), quarks and leptons

[1] One may ask whether the mass-scale in the denominator of eq. (1) could plausibly be the GUT scale ($\approx 2 \times 10^{16}$GeV), instead of the reduced Planck mass. That would give $m(\nu_L) \approx \lambda_L(3 \times 10^{-3}\text{eV})$, which is closer but still a bit low compared to the SuperKamiokande value of (10^{-1} to 3×10^{-2}eV), unless $\lambda_L \approx 30$ to 10. But, more to the point, in the context of the standard model, supplemented by just gravity, while Planck mass seems to have every reason to appear in eq. (1), there does not seem to be any simple reason for the relevance of the GUT scale. I thank S. Weinberg, who had considered operators like eq. (1) long ago [7] for raising this point and for discussions.

[2] The alternative of giving a Majorana mass to ν_L through renormalizable interaction by introducing a $SU(2)_L$ Higgs-triplet ξ and choosing the corresponding (Yukawa coupling) $\times$ (VEV of ξ) to be nearly (1/10 - 1/30)eV seems to be rather arbitrary.

of a given family fall into the neat pattern [2]:

$$F^{e}_{L,R} = \begin{bmatrix} u_r & u_y & u_b & \nu_e \\ d_r & d_y & d_b & e^- \end{bmatrix}_{L,R} \quad (4)$$

with the transformation properties $F^{e}_{L} = (2,1,4)$, and $F^{e}_{R} = (1,2,4)$; likewise for the μ and the τ-families. We see that the RH neutrino (ν_R) arises as the fourth color partner of the RH up-quarks and, also, as the left-right conjugate partner of the LH neutrino (ν_L). It is worth noting that the symmetry G(224), subject to left-right discrete symmetry [2,8], possesses some additional advantages, even without being embedded into a simple group like SO(10) [5] or E_6 [9]. These include: (i) inclusion of all members of a family into one multiplet, (ii) quark-lepton unification through SU(4)-color, (iii) quantization of electric charge, mentioned above, (iv) spontaneous violations of parity [2,8] and of CP [10], (v) (B-L), as a local symmetry whose spontaneous violation may be needed to implement baryogenesis [11], (vi) a promising solution to the strong CP problem in the context of supersymmetry [12], and (vii) a possible resolution of the μ-problem in the same context [13]. Embedding G(224) into SO(10), for which ($F^{e}_{L} + \overline{F^{e}_{R}}$) yield the 16 of SO(10), would of course retain most of these advantages, except possibly (vii). Last, but not least, the symmetry G(224) can emerge from strings with three chiral families (see e.g. Refs. 14 and 15). In this case, the gauge coupling unification [16] at string scale would still hold [17] even without having the covering SO(10), below the string scale.[3] It is worth noting that in the string context there is a distinct advantage if the preferred string solution would contain G(224) rather than SO(10), because it appears rather difficult to implement doublet-triplet splitting for string-derived SO(10) so as to avoid rapid proton decay.[20] For string-derived G(224) [14], on the other hand, the dangerous color triplets are either projected out or naturally become superheavy.

[3]Possible resolutions of a mismatch between MSSM (M_X) and string-unification scales by about a factor of 20 have been proposed, including one that suggests two vector-like families ($16 + \overline{16}$) at the TeV-scale, that leads to semi-perturbative unification and raises M_X to a few $\times 10^{17}$ GeV[18]; and also one that makes use of string duality[19] and allows for a re-evaluation of M_{string} compared to that of Ref. [17]. In general, both ideas may play a role.

4. The Mass of ν^{τ}_{L}

I now turn to an estimate of the masses of the light neutrinos, that are observed in the laboratory, especially the ν^{τ}_{L}, allowing for the existence of the RH neutrinos (ν^{i}_{R}s). For this purpose, I will work with either G(224) or its natural extension SO(10). With a string or a GUT-origin, one can motivate the symmetry-breaking scale for either G(224) or SO(10), to be around $M_{string}/10$, which is nearly the (empirical) GUT-scale $\approx 2 \times 10^{16}$GeV.

The amusing thing is that, in contrast to the case of the SM (eq.(1)), now the mass of ν^{τ}_{L} comes out to be just in the right range, so as to be relevant to the SK result.

The simplest reason for the known neutrinos to be so light (< 30eV (say)) is provided by the so-called see-saw mechanism [3]. It attributes Dirac masses $m(\nu^{i}_{D})$ which would be related to the up-flavor quark-mass (m_u, m_c or m_t), depending upon the Higgs representation (see below), by SU(4)-color. Simultaneously it assigns superheavy Majorana masses (M_{iR}) to the RH neutrinos, preserving the SM symmetry; by utilizing the VEV of a suitable Higgs multiplet (call it Σ), which would be involved in breaking SO(10) or G(224) to the SM symmetry G(213). Before discussing the choice of Σ and its coupling, let us recall that a mass-matrix involving Dirac and superheavy Majorana masses, as mentioned above, would diagonalize to yield three superheavy RH neutrinos with masses M_{iR} and three light LH neutrinos with masses [3]:

$$m(\nu^{i}_{L}) \approx m(\nu^{i})^2_D / M_{iR} \quad (5)$$

In writing this, we have neglected (for simplicity) possible off diagonal mixings between different flavors. (For a more general analysis, see e.g. Ref. 4 and 21). Since the Dirac masses enter quadratically into (5), and are highly hierarchical (e.g. $m_u : m_c : m_t \approx 1 : 300 : 10^5$), we expect, even allowing for a rather large hierarchy (by successive factors of order 100, say) in M_{iR}, that the masses of the left-handed neutrinos will be light but hierarchical ($m(\nu_L^e) << m(\nu_L^\mu) << m(\nu_L^\tau)$).

The Higgs multiplet Σ, mentioned above, and its conjugate $\bar{\Sigma}$ (needed for supersymmetry), can either be in a symmetric tensorial representation[3] - i.e. (126_H, $\overline{126}_H$) of SO(10) or equivalently $[(1,3,10), (1,3,\overline{10})]$ of G(224) - or in the spinorial representation - i.e. (16_H, $\overline{16}_H$) [22] of SO(10) - or equivalently in $[(1,2,\bar{4})_H, (1,2,4_H)]$ [2] of G(224).

We first remark that, in string theory, the tensorial representations 126_H and $\overline{126}_H$, and likewise $(1,3,10)_H$ and $(1,3,\overline{10})_H$, which can have renormalizable Yukawa interactions with quarks and leptons, are hard, perhaps impossible, to realize [23], and have not been realized in any solution yet. By contrast, the spinorial 16_H and $\overline{16}_H$, as also $(1,2,4)_H$ and $(1,2,\bar{4})_H$, do emerge quite simply in string-solutions (see e.g. Ref. 14 for G(224) and Ref. 20 for SO(10)). Taking this as a good guide, we will work only with the spinorial 16_H and $\overline{16}_H$, or equivalently with $(1,2,4)_H$ and $(1,2,\bar{4})_H$.

The effective non-renormalizable interaction, involving these multiplets, which we expect might be induced by Planck-scale physics, and would give Majorana masses to the RH neutrinos, are then[4]

$$\mathcal{L}_M(SO(10)) = \lambda_R^{ij} 16_i \cdot 16_j \overline{16}_H \cdot \overline{16}_H / M_{P\ell} + hc \quad (6)$$

$$\mathcal{L}_M(G(224)) = \lambda_R^{ij}(1,2,4)_i(1,2,4)_j \times (1,2,\bar{4})_H(1,2,\bar{4}_H)/M_{p\ell} + hc \quad (7)$$

Here, $i, j = 1, 2, 3$, correspond respectively to e, μ and τ-families. Such effective non-renormalizable interactions may well arise – in part or dominantly – by renormalizable interactions through tree-level exchange of superheavy states, such as those in the string-tower.

Judging from the string-side, one naturally expects the VEVs of fields which break GUT-like symmetries – i.e. SO(10) or G(224) – to the standard model symmetry to be of order $M_{string}/(5 \text{ to } 20) \approx 2-8\times 10^{16}$GeV [see, e.g. Ref. 24 and 14], where $M_{string} \approx 4 \times 10^{17}$ GeV.[17]. This is also nearly the GUT-scale ($M_{GUT} \approx 2 \times 10^{16}$GeV), judged from the MSSM extrapolation of the three gauge-couplings,[3] Thus, both from the viewpoint of connection with string theory, as well as comparison with the MSSM unification-scale, we expect the VEV's of the respective fields to be given by:

$$\text{For SO(10)}: <\overline{16}_H> \approx 3 \times 10^{16} \text{ GeV}.\eta \quad (8)$$

$$\text{For G(224)}: <(1,2,\bar{4})_H> \simeq 3 \times 10^{16} \text{ GeV}.\eta \quad (9)$$

with $\eta \approx 1/2$ to 2, being the most plausible range. Thus, using (6) – (7) and (8) – (9), for either SO(10) or G(224), the Majorana masses of the RH neutrinos are given by:

$$M_{iR} \approx \lambda_{ii} \frac{(3\times10^{16}\text{GeV})^2}{2\times10^{18}\text{GeV}}\eta^2 \approx \lambda_{ii}(4.5 \times 10^{14} GeV)\eta^2 \quad (10)$$

In writing (10), we have ignored the effects of off-diagonal mixing.

Now using SU(4)-color and the Higgs multiplet $(2,2,1)_H$ for G(224) or equivalently 10_H for SO(10), one obtains the relation $m_\tau(M_X) = m_b(M_X)$, which is known to be successful. Thus, there is a good reason to believe that the third family gets its masses primarily from the 10_H or equivalently $(2,2,1)_H$, which automatically gives the same Dirac mass to the quark and the lepton of a given flavor. In turn this implies:

$$m(\nu_D^\tau) \approx m_{top}(M_X) \approx (100 - 120)\text{GeV} \quad (11)$$

combining (10) and (11) via the see-saw relation (5), we obtain:

$$m(\nu_L^\tau) \approx \frac{(100\text{GeV})^2(1 \text{ to } 1.44)}{\lambda_{33}(4.5\times10^{14}\text{GeV})\eta^2} \approx (1/45)eV(1 \; to \; 1.44)/\lambda_{33}\eta^2 \quad (12)$$

Now, considering that we expect $m(\nu_L^\mu) << m(\nu_L^\tau)$ (by using eq. (5)), and assuming that SuperKamiokande observation represents $\nu_L^\mu \to \nu_L^\tau$-oscillation, so that the observed $\delta m^2 \approx 10^{-2}to10^{-3}$eV^2 corresponds to $m(\nu_L^\tau)_{obs} \approx 1/10$ to $1/30$eV, it seems *truly remarkable* that the expected magnitude of $m(\nu_L^\tau)$, given by eq. (12), is just about what is observed, if $\lambda_{33}\eta^2 \approx$

[4]We are not exhibiting the interactions of $(2,1,\bar{4})_H$ because, either it is absent (as in Ref. 14) or has zero VEV.

3

1 to 1/4. Such a range for $\lambda_{33}\eta^2$ seems most plausible and natural (see remarks below). This observation regarding the agreement between the expected and the observed value of δm^2 (in this case $m(\nu_L^\tau)$), in the context of the ideas mentioned above, is the main point of this note.

We remark that this agreement has come about without making any parameter unnaturally small or large. In particular, the effective Majorana coupling of the third family (λ_{33}) is needed to be of order one, barring the effect of mixing, for this agreement to hold. One is tempted to compare with the top-Yukawa coupling (h_{top}) which is also nearly one. This common feature regarding maximality of the dimensionless couplings associated with the third family (i.e. $\lambda_{33} \sim h_{top} \sim 1$) may well find its explanation in the context of string solutions for which such couplings may be given just by the gauge coupling [e.g. $h_{top} = \sqrt{2}g \approx 1$, [see e.g. Ref. [24]] and are thus of order one[5], while those associated with the second and the first families are progressively smaller, because, subject to string symmetries and selection rules, they are induced only at the level of higher dimensional operators utilizing VEV's of fields which are small (by nearly factor of 10) compared to the string-scale. In addition to λ_{33}, the value of $m(\nu_L^\tau)$ depends on the Dirac mass $m(\nu_D^\tau)$ (see eq. (5)) and on the VEV of $< \overline{16}_H >$ or $< (1,2,4)_H >$, and thus on η^2 The use of SU(4)-color plays a crucial role in that it enables one to determine $m(\nu_D^\tau)$ fairly reliably from m_{top}. As regards the VEVs of fields, the use of string as well as GUT-related ideas suggest most plausibly nearly the same value for the VEV of $< 16_H >$ (or $< (1,2,\bar{4})_H >$), within a factor of 2 to 4, which is reflected in the uncertainty in $\eta(\approx 1/2$ to $2)$ (see eqs. (8)/(9)). It is for these reasons that the value of $m(\nu_L^\tau)$ obtained in eq. (12), with $\lambda_{33}\eta^2 \approx 1$ to 1/4, seems most plausible.

Together with the result $\delta m^2 \simeq 10^{-2} - 10^{-3} eV^2$, the SuperKamiokande group reports another puzzling feature that $\nu_\mu \to \nu_\tau$ (or ν_X) - oscillation angle is nearly maximal - i.e. $\sin^2 2\theta > 0.8$. Ordinarily, such large oscillation angle is attributed to nearly degenerate masses of the $(\nu_\mu - \nu_\tau)$ or $(\nu_\mu - \nu_X)$ systems, as many authors in fact have. However, considering that nearly degenerate masses for the light neutrinos seem to be rather unnatural in the context of the see-saw formula, Babu, Wilczek and I have very recently observed [4] that such degeneracy is not even needed to obtain large oscillation angle. By *combining* the contributions from the mixing angle of the neutrinos (i.e. $\nu_L^\mu - \nu_L^\tau$) with that from the charged leptons $(\mu - \tau)$, and by following familiar ideas on the understanding of Cabibbo-like quark-mixing angles, one can in fact obtain, quite simply and naturally, large $(\nu_L^\mu - \nu_L^\tau)$ oscillation angle, as observed, *in spite of a highly non-degenerate $\nu^\mu - \nu^\tau$ system*, e.g. with $m(\nu_L^\mu)/m(\nu_L^\tau) \approx 1/10 - 1/20$. Briefly, a simple and plausible origin of the large mixing angle is as follows. If one assumes that the lighter eigenvalue for a hierarchical 2×2-system arises entirely or primarily by the off-diagonal mixing of the (would-be) light with the heavier state (as in a symmetrical see-saw type mass matrix), one obtains the familiar square root-formula[25] for the mixing angle, like $\theta_{d,u} \approx (\sqrt{m_d/m_s}, \sqrt{m_u/m_c})$, and the Cabibbo angle is obtained by combining θ_d with θ_u, allowing for a relative phase between them. Regardless of the phase, such an expression for the Cabibbo angle is known to be fairly successful (to better than 30 %). Assuming analogous mass-matrices for the $\nu_\mu - \nu_\tau$ system (Dirac and Majorana) as well as for the charged leptons $(\mu - \tau)$, one obtains, ignoring CP violation (and assuming the exact see-saw form for each of the three matrices): $\theta_{osc}(\nu_\mu - \nu_\tau) = \theta(\nu_L^\mu - \nu_L^\tau) \pm \theta(\mu - \tau) \approx [m(\nu_L^\mu)/m(\nu_L^\tau)]^{1/2} \pm [m_\mu/m_\tau]^{1/2} \approx 0.31 \pm 0.25 \approx 0.56 \,\text{or}\, 0.06$, where we have put $m(\nu_L^\mu)/m(\nu_L^\tau) \approx 1/10$. This yields, choosing a positive relative sign between the two mixing angles, $\sin^2 2\theta_{osc} \approx 0.8$. *In short, a large oscillation angle can arise quite plausibly, without near degeneracy and without large mixing in the mass*

[5]Although λ_{ij} are associated with effective non-renormalizable couplings, as mentioned before, they may well arise, in part or dominantly, through the exchange of superheavy states $\{\phi_\alpha\}$ (such as those in the string-tower or just below string-scale), if these possess Yukawa couplings of the form $h_{i\phi} 16_i \overline{16}_H \phi$, together with invariant mass-term $(M_\phi \phi\phi + hc)$. If $h_{(i\phi)}$ are family-hierarchical with $h_{3\phi}$ being maximal (i.e. O(1) like h_{top}) and leading, λ'_{ij}s would also be hierarchical, with $\lambda_{33} (= h_{3\phi}^2 (M_{Pl}/M_\phi))$ being maximal ($\mathcal{O}(1)$) and leading.

eigenstates of the neutral and the charged leptons. Various sources of departures from the simple square root formula for the mixing angle corresponding for example to departures from exactly symmetrical see-saw mass matrices are needed to account for quark masses and mixings such as V_{bc}. These can lead to even larger oscillation angles (for $m(\nu_L^\mu)/m(\nu_L^\tau) \approx 1/10 - 1/20$), as discussed in Ref[4]: In this case, $\nu_e - \nu_\mu$ - oscillation can become relevant to the small angle MSW explanation[26] of the solar neutrino-puzzle. I refer the reader to Ref. 4 for a full discussion of this explanation of the large oscillation angle for the $\nu_\mu - \nu_\tau$ system, with hierarchical masses for the neutrinos. The purpose of the present note has primarily been to emphasize the implications of the observed *magnitude* of δm^2- or equivalently, in our case of $m(\nu_L^\tau)$, on the nature of new physics.

5. Link Between Neutrino Masses and Proton Decay

Proton decay is one of the hallmarks of grand unification [[2],[6]]. In a recent paper, Babu, Wilczek and I noted that, contrary to common impression, in a class of supersymmetric unified theories such as SUSY SO(10) or SUSY G(224), there is likely to be an *intimate link* between the neutrino masses and proton decay[21]. This is because, in the process of generating light neutrino masses via the see-saw mechanism, one inevitably introduces a new set of color-triplets (unrelated to electroweak doublets), with effective couplings to quarks and leptons, which are related to the superheavy Majorana masses of the RH neutrinos (see eqs. (6) and (7)). Exchange of these new color-triplets give rise to a new set of d=5 proton decay operators, which are thus directly related to the neutrino-masses. Assuming that $\nu_e - \nu_\mu$ oscillation is relevant to the MSW explanation of the solar neutrino puzzle, so that $m(\nu_L^\mu) \approx 3 \times 10^{-3}$eV, which corresponds to $M(\nu_R^\mu) \approx 2 \times 10^{12}$ GeV, the strength of the new d=5 operators turns out to be just about right ($\tau_P \approx 10^{31.5\pm2.5}$ yrs), for proton decay to be observable at SuperKamiokande. This is the case even when one makes allowance (within reason) for uncertainties in proton decay rate due to those in the SUSY spectrum, hadronic matrix element and color triplet mass etc. to extend proton lifetime.

The flavor-structure of the new d=5 operators are, however, expected to be distinct from those of the standard d=5 operators, which are related to the highly hierarchical Dirac masses of quarks and leptons. In contrast to the standard d=5 operators, the new ones can lead to prominent charged lepton decay modes, such as $\ell^+\pi^o, \ell^+K^o$ and $\ell^+\eta$, especially for $\ell = \mu$ even for low or moderate values of $\tan\beta \leq 10$. The intriguing feature thus is that owing to the underlying SO(10) or just SU(4)-color symmetry, *proton decay operator knows about neutrino masses and vice versa.*

With a *maximal* effective Majorana-coupling for the third family (i.e. $\lambda_{33} \sim \mathcal{O}(1)$), as suggested here, that corresponds to $M_{3R} \approx$ (few $\times\ 10^{14}$GeV) for the case of no mixing (see eq. (10)), one might however worry that proton may decay too fast, because of an enhancement in the new d=5 operators, relative to that considered in Ref. 21. It turns out, however, that because τ^+ is heavier than the proton and because $\bar{\nu}_\tau K^+$ mode receives a strong suppression-factor from the small mixing angle associated with the third family ($V_{ub} \approx 0.002 - 0.005$), a maximal Majorana-coupling of the third family ($\lambda_{33} \sim \mathcal{O}(1)$), and thus $m(\nu_L^\tau) \approx (1/10 - 1/30)$eV, is perfectly compatible with present limit on proton lifetime[4]. With a family-hierarchical Majorana coupling - i.e. $\lambda_{33} \sim \mathcal{O}(10)\lambda_{23} \approx \mathcal{O}(10^2)\lambda_{22}$ etc. - ν_L^τ and ν_L^μ - masses can be relevant respectively to the atmospheric and the solar-neutrino-problems, yet the new neutrino-mass related d=5 operator does not conflict with limits on proton lifetime. The mass of ν_τ and the large oscillation angle, suggested by the SuperK result, it turns out, enhance our expectations significantly for observation of proton decay in the near future at SuperK and the proposed ICARUS [4].

6. Concluding Remarks and a Summary

As noted in the introduction and the subsequent sections, the impressive result of SuperKamiokande clearly has far-reaching implications on the nature of new physics. These are summarized below and some remarks are added:

6.1. The Right-Handed Neutrino: A New Form of Matter

As noted in the introduction, the most reasonable explanation for the neutrino mass-scale observed at SuperKamiokande needs a RH neutrino (ν_R). Many in the past, motivated by the possible masslessness of neutrinos, have preferred to view the neutrino as an "odd ball," believing that it is the messenger that nature is intrinsically left-right asymmetric (parity-violating). This is reflected by the two-component neutrino hypothesis of Lee, Yang, Landau and Salam, as well as by the hypothesis of the grand unification-symmetry SU(5). The SuperKamiokande result (especially its value for δm^2) clearly suggests, however, that that is in fact not the case. *Neutrino is "elusive" but not an odd ball after all.* It has its RH counterpart (one for each flavor) just like all the other fermions.

Nevertheless, the neutrino has a *unique character.* It is the only fundamental fermion, among the members of a quark-lepton family, that is electrically neutral (not counting possible SUSY gauge matter such as photino or gluino). Therefore, it is the *only fermion* that can acquire both a Dirac mass ($\Delta F = \Delta L = 0$), combining ν_R and ν_L, and a Majorana mass for either ν_R or ν_L ($\Delta F = \Delta L = \Delta(B-L) = 2$), conserving electric charge. The Majorana masses of the RH neutrinos can be superheavy, because they do not break the Standard model symmetry. The lightness of ν_L is in fact a reflection of the heaviness of ν_R. By the same token, the light neutrinos know about both mass-scales – the Dirac and the Superheavy Majorana – and thereby simultaneously of the physics at the electroweak and the string/GUT-scales. In short, neutrino masses carry a *gold mine of information* about the nature of new physics.

6.2. Minimal Extension Needed of the Standard Model

: In suggesting the need for the RH neutrino, the SuperKamiokande result in turn suggests, following discussions presented here, that the standard model symmetry must be extended minimally to the symmetry-structure $G(224) = SU(2)_L \times SU(2)_R \times SU(4)^c$. The need for $SU(4)^c$ has been noted above and is summarized below. Strictly speaking, for an understanding of $(\delta m)^2$, as presented here, the extension of the SM symmetry to just $G(214) = SU(2)_L \times I_{3R} \times SU(4)^c$ would suffice.[6] The further extension of G(214) to G(224) (that also quantizes electric charge by replacing I_{3R} by $SU(2)_R$) may however be needed by some of the other considerations, listed in Sec. 3, as well as those of fermion masses and mixings.

6.3. The Three Necessary Ingredients

: Understanding the neutrino mass-scale observed at SuperKamiokande, as discussed here, utilizes three concepts in an essential manner. They are: (a) SU(4)-color that not only enforces ν_R, but more importantly gives the Dirac mass of ν^τ, fairly reliably, by relating it to the mass of the top quark (eq. (11));[7] (b) String/GUT-scale physics that determines the Majorana mass of the RH tau-neutrino (subject to maximality of the effective coupling) (eqs. (8)-(10)); and (c) the see-saw relation (eq. (5)).

6.4. Selecting the Route to Higher Unification

: Unlike proton decay, which can probe directly into the full grand unification symmetry (including gauge transformations of $q \to \overline{q}$ and $q \to \overline{\ell}$), neutrino physics probes directly into $SU(2)_L \times SU(2)_R \times SU(4)^c$, but not necessarily beyond. For example, the results discussed here, such as determinations of the Dirac and the Majorana mass of the τ neutrino utilize only G(224), but not the full SO(10). They have also utilized supersymmetry, at least indirectly, because without it, there would be no rationale for the use of string-GUT-related scale for the VEV of $\overline{16}_H$ or

[6] For a string-origin of G(214), see Ref. [27].

[7] It is, of course, possible that a string-derived solution containing, for example, only $G(2213) = SU(2)_L \times SU(2)_R \times (B-L) \times SU(3)^C$ or $G(2113) = SU(2)_L \times I_{3R} \times (B-L) \times SU(3)^C$ [24], or flipped SU(5) × U(1)[28], all of which yield RH neutrinos, may still relate $m(\nu_D^\tau)$ to m_{top} at string-scale. This comes about because such a solution still remembers its origin through SU(4)-color or SO(10). Here, I am only discussing the minimal *underlying symmetry* needed to remove arbitrariness in the choice of $m(\nu_D^\tau)$, which appears to be SU(4)-color.

$(1, 2, \overline{4})_H$.

At the same time, by providing clear support for G(224), the SK result selects out SO(10) or E_6 as the underlying grand unification symmetry, rather than SU(5). Either SO(10) or E_6 or both of these symmetries ought to be relevant at some scale, and in the string context, that may, of course, well be in higher dimensions, above the compactification-scale, below which there need be no more than just the G(224)-symmetry. If, on the other hand, SU(5) were regarded as a fundamental symmetry, first, there would be no compelling reason, based on symmetry alone, to introduce a ν_R, because it is a singlet of SU(5). Second, even if one did introduce ν_R^i by hand, the Dirac masses, arising from the coupling $h^i \bar{5}_i < 5_H > \nu_R^i$, would be unrelated to the up-flavor masses and thus rather arbitrary (contrast with eq. (11)). So also would be the Majorana masses of the ν_R^i's, which are SU(5)-invariant and thus can even be of order Planck scale (contrast with Eq. (10)). This would give $m(\nu_L^\tau)$ in gross conflict with the observed value. We thus see that the SK result clearly disfavors SU(5) as a fundamental symmetry, with or without supersymmetry.

In summary, it seems that the single discovery of atmospheric neutrino-oscillation has brought to light the existence of the right-handed neutrino and has reinforced the ideas of SU(4)-color, left-right symmetry and see-saw. The agreement between the simplest estimate of the mass of the tau-neutrino, presented here, and the"observed value" suggests the correctness of these three ideas. Simultaneously, it suggests the relevance of the string/GUT-scale-symmetry-breaking, as opposed to intermediate or TeV-scale breaking of (B-L). *Any symmetry* containing G(224) = $SU(2)_L \times SU(2)_R \times SU(4)^c$, such as SO(10) or E_6, would of course possess the same desirable features as regards neutrino physics, as G(224). Given the wealth of insight already provided by the SuperKamiokande result, one looks forward eagerly to further revelations of deeper physics in the coming years from the neutrino-system through the many existing and the forthcoming facilities, involving atmospheric, solar and accelerator neutrinos. In particular, one would like a clarification of whether the SK result is observing $\nu_L^\mu - \nu_L^\tau$ (as assumed here) as opposed to $\nu_L^\mu - \nu_X$ oscillation, and whether the resolution to the solar neutrino-problem would favor the small angle MSW solution (supported here) as opposed to $\nu_e - \nu_X$ or vacuum oscillation [8]. One of course also looks forward to learning much about further aspects of unification from searches for proton decay, which, as we saw [21] [4], is intimately related to neutrino masses, because of SU(4)-color and supersymmetry.

7. Acknowledgement:

I wish to thank the Organizers of the Neutrino-98 Conference, especially Y. Totuska, Y. Suzuki and T. Kajita, for inviting me to speak at the Conference and for their kind hospitality. I have greatly benefitted from discussions and communications with Schmuel Nussinov, Qaisar Shafi, Steven Weinberg, and Edward Witten, and especially from collaborative discussions with Kaladi S. Babu and Frank Wilczek. The research is supported in part by NSF Grant No. Phy-9119745, and in part by the Distinguished Research Fellowship awarded by the University of Maryland.

REFERENCES

1. Results of the SuperKamiokande and Kamiokande Collaboration, on Atmospheric Neutrino Oscillations, reported by T. Kajita at the Neutrino-98 Conference, Takayama, Japan, June 4-9, 1998.
2. J. C. Pati and A. Salam, Phys. Rev. Lett. **31**, 661 (1973); Phys. Rev. D**10**, 275 (1974).
3. M. Gell-Mann, P. Ramond and R. Slansky, in: *Supergravity*, eds. F. van Nieuwenhuizen and D. Freedman (Amsterdam, North Holland, 1979) p. 315; T. Yanagida, in: *Workshop on*

[8] Clarification is of course needed also about the LSND result. Should this result get confirmed, a sterile neutrino with a mass of a few eV could play the role of an intermediary so as to induce $\nu_e \to \nu_s \to \nu_\mu$ indirect oscillation (as in K.S. Babu, J.C. Pati and F. Wilczek, Phys. Lett. B**359**, 351 (1995). This would preserve the success of the interpretation of atmospheric and solar neutrino oscillations in terms of $\nu_\mu - \nu_\tau$ and direct $\nu_e - \nu_\mu$ oscillation, presented here.

the Unified Theory and Baryon Number in the Universe, eds. O. Sawada and A. Sugamoto (KEK, Tsukuba) 95 (1979); R. N. Mohapatra and G. Senjanovic, Phys. Rev. Lett. **44**, 912 (1980).

4. K. S. Babu, J. C. Pati and F. Wilczek - "Link Between Neutrino masses, SuperKamiokande Result and Proton Decay" - to appear.
5. H. Georgi, in Particles and Fields, ed. by C. Carlson (AIP, NY, 1975), p. 575; H. Fritzsch and P. Minkowski, Ann. Phys. **93**, 193 (1975).
6. H. Georgi and S. L. Glashow, Phys. Rev. Lett. **32**, 438 (1974).
7. Operators of this form with a varying or a GUT scale characteristic mass were considered by S. Weinberg, Phys. Rev. Lett. **43**, 1566 (1979); Proc. XXVI Int'l Conf. on High Energy Physics, Dallas, TX, 1992; For a recent work based on Planck scale effects, see E. Akhmedov, Z. Berezhiani and G. Senjanovic, Phys. Rev. D **47**, 3245 (1993).
8. R. N. Mohapatra and J. C. Pati, Phys. Rev. **D11**, 566 (1974); G. Senjanovic and R. N. Mohapatra, Phys. Rev. D**12**, 1502 (1975).
9. F. Gürsey, P. Ramond and P. Sikivie, Phys. Lett. **B 60**, 177 (1976).
10. R. N. Mohapatra and J. C. Pati, Phys. Rev. **D11**, 2558 (1975).
11. V. Kuzmin, Va. Rubakov and M. Shaposhnikov, Phys. Lett **BM155**, 36 (1985); M. Fukugita and T. Yanagida, Phys. Lett. **B 174**, 45 (1986); M. A. Luty, Phys. Rev. **D45**, 455 (1992); W. Buchmuller and M. Plumacher, hep-ph/9608308.
12. R. N. Mohapatra and A. Raisin, Phys. Rev. D**54**, 5385 (1996).
13. S. F. King and Q. Shafi, hep-ph/9711288. For certain desirable features of G(224) as regards fermion mass-matrices, involving nonrenormalizable operators, see e.g. B. Allanach, S. F. King, G. Leontaris and S. Lola, Phys. Lett. B**407**, 275 (1997).
14. I. Antoniadis, G. Leontaris and J. Rizos, Phys. Lett **B245**, 161 (1990); G. Leontaris, Phys. Lett. **B372**, 212 (1996).
15. G. Shiu and S. H. Henry Tye, hep-th/9805157.
16. H. Georgi, H. Quinn and S. Weinberg, Phys. Rev. Lett. **33**, 451 (1974).
17. P. Ginsparg, Phys. Rev. D**197**, 139 (1987); V. Kaplunovsky, Nucl. Phys. B**307**, 145 (1988).
18. K. S. Babu and J. C. Pati, Phys. Lett. **B384**, 140 (1996).
19. E. Witten, hep-th/9602070
20. Z. Kakushadze and S. H. Tye, Phys. Rev. lett. **77**, 2612 (1996); Z. Kakushadze et al. hep-ph/9705202 and references therein.
21. K. S. Babu, J. C. Pati and F. Wilczek, Phys. Lett. **B423**, 337 (1998), hep-ph/9712307.
22. See, e.g., E. Witten, Phys. Lett. **91 B**, 81 (1980).
23. K. R. Dienes and J. March-Russell, hep-th/9604112; K. R. Dienes, hep-ph/9606467.
24. A. Faraggi, Phys. Lett. **B278**, 131 (1992); Phys. Lett. **B274**, 47 (1992); Nucl. Phys. **B403**, 101 (1993); A. Faraggi and E. Halyo, Nucl. Phys. **B416**, 63 (1994).
25. S. Weinberg, I. I. Rabi Festschrift (1977); F. Wilczek and A. Zee, Phys. Lett. **70B**, 418 (1977); H. Fritzsch, Phys. Lett **70B**, 436 (1977).
26. S. Mikheyev and a. Smirnov, Nuov. Cim. **9C**, 17 (1986), and L. Wolfenstein, Phys. Rev. **D17**, 2369 (1978).
27. Z. Kakashudze, hep-th/9806044 V4.
28. I. Antoniadis, J. Ellis, J. Hagelin and D. Nanopoulos, Phys. Lett B**231**, 65 (1989).

ELSEVIER

Nuclear Physics B (Proc. Suppl.) 77 (1999) 308–312

NUCLEAR PHYSICS B
PROCEEDINGS
SUPPLEMENTS

Implications of a Minimal $SO(10)$ Higgs Structure

Carl H. Albright,[a] K.S. Babu[b*] and S.M. Barr[c†]

[a] Department of Physics, Northern Illinois University, DeKalb, IL 60115
and Fermi National Accelerator Laboratory, P.O. Box 500, Batavia, IL 60510

[b] School of Natural Sciences, Institute for Advanced Study, Princeton, NJ 08540

[c] Bartol Research Institute, University of Delaware, Newark, DE 19716

A minimal $SO(10)$ Higgs structure involving a single adjoint field along with spinors, vectors and singlets has been shown to break the $SO(10)$ gauge symmetry to the standard model while stabilizing the F-flat directions and solving the doublet-triplet splitting problem naturally. With this minimal set of Higgs fields, we show how to construct quark and lepton mass matrices which explain well the many features of the observed spectrum, including the Georgi-Jarlskog mass relations. A large $\nu_\mu - \nu_\tau$ mixing angle results naturally as observed in the atmospheric neutrino data. A particular model relying on a family symmetry has been constructed which realizes the desired mass matrices.

A brief discussion is given of the implications of a minimal $SO(10)$ Higgs structure that have been developed in a recent series of papers. Barr and Raby [1] have shown how this minimal set of Higgs fields breaks the $SO(10)$ gauge symmetry to the standard model while stabilizing the F-flat directions and thus solves the double-triplet splitting problem. Following this lead, the authors [2] have used this Higgs structure to construct quark and lepton mass matrices which are fairly tightly constrained with some interesting features emerging. Of special interest to this Conference is the large $\nu_\mu - \nu_\tau$ mixing angle resulting from the special textures of the Dirac matrices, as opposed to the more conventional large hierarchical structure for the Majorana neutrino matrix [3].

1. MINIMAL HIGGS STRUCTURE

We begin with a summary of the minimal $SO(10)$ Higgs structure [1] which solves the doublet-triplet splitting problem naturally rather than by fine-tuning. The Higgs fields which are involved consist of a pair of **10**'s, one **45**, two pairs of $\mathbf{16} + \overline{\mathbf{16}}$'s and four singlets. The Higgs superpotential is written

$$\begin{aligned} W &= T_1 A T_2 + M_T T_2^2 + W_A + W_C \\ &\quad + W_{CA} + W_{TC} \\ W_A &= \mathrm{tr} A^4/M + M_A \mathrm{tr} A^2 \\ W_C &= X(\overline{C}C)^2/M_C^2 + f(X) \\ W_{CA} &= \overline{C}'(PA/M_1 + Z_1)C \\ &\quad + \overline{C}(PA/M_2 + Z_2)C' \\ W_{TC} &= \lambda T_1 \overline{C}\,\overline{C} \end{aligned} \qquad (1)$$

Here T_1 and T_2 label the two **10**'s, A labels the **45**, C, $\overline{C}$, C', $\overline{C}'$ label the two pairs of $\mathbf{16}+\overline{\mathbf{16}}$'s, while P, X, Z_1, Z_2 label the four singlets.

The W_A terms produce the Dimopoulos - Wilczek mechanism [4] by generating a VEV for the single **45** in the $B-L$ direction. The $T_1 A T_2$ term gives superheavy masses to the color triplets in T_1 and T_2. The mass term $M_T T_2^2$ gives superheavy masses to the T_2 doublets as well. As a result of the presence of W_C, the $F_X = 0$ condition forces the C and $\overline{C}$ pair to get VEVs in the $SU(5)$-singlet direction. The VEVs of A and C then break $SO(10)$ to the standard model. The term W_{CA} couples C and $\overline{C}$ to A and prevents the production of colored pseudo-goldstone bosons in the breaking of $SO(10)$. Since no GUT-

*Work supported in part by the Department of Energy Grant No. DE-FG02-90ER-40542.
†Work supported in part by the Department of Energy Grant No. DE-FG02-91ER-40626.

PII S0920-5632(99)00433-8

scale VEVs are generated for C' and $\overline{C}'$, the Dimopoulos - Wilczek hierarchical form of $\langle A\rangle$ is not destabilized by the presence of W_{CA}, thus solving the doublet-triplet splitting problem. Finally, the presence of the term W_{TC} induces an electroweak breaking VEV for C' which mixes with that in T_1. Hence the two Higgs doublets appear in the combinations

$$H = \mathbf{5}(T_1), \quad H' = \overline{\mathbf{5}}(C')\cos\theta - \overline{\mathbf{5}}(T_1)\sin\theta \quad (2)$$

in terms of the $SU(5)$ representations present in T_1 and C'. The combination orthogonal to H' gets massive and drops out of the picture.

An important point to be made is that the above form of the Higgs superpotential can be uniquely obtained by the introduction of a $U(1)\times Z_2\times Z_2$ family symmetry [1] with the appropriate assignment for the charges of the Higgs fields as follows:

$$\begin{aligned} &A(0)^{+-}, T_1(1)^{++}, T_2(-1)^{+-} \\ &C(\tfrac{1}{2})^{-+}, \overline{C}(-\tfrac{1}{2})^{++}, C'(\tfrac{1}{2}-p)^{++}, \overline{C}'(-\tfrac{1}{2}-p)^{-+} \\ &X(0)^{++}, P(p)^{+-}, Z_1(p)^{++}, Z_2(p)^{++} \qquad (3) \end{aligned}$$

2. FERMION MASS MATRICES FROM MINIMAL SET OF HIGGS FIELDS

We can then attempt to construct fermion mass matrices from the VEVs appearing in the minimal set of Higgs fields. The VEVs in question appear at the GUT scale and at the electroweak scale as follows:

$$\begin{aligned} \Lambda_G: &\quad \langle A\rangle,\ \langle C\rangle,\ \langle\overline{C}\rangle,\ \langle P\rangle,\ \langle X\rangle,\ \langle Z_1\rangle,\ \langle Z_2\rangle \\ \Lambda_{ew}: &\quad \langle T_1\rangle,\ \langle C'\rangle \qquad (4) \end{aligned}$$

Note that since the VEVs of the doublets of the T_1 $SO(10)$ $\mathbf{10}$ appear in the $SU(5)$ $\mathbf{5}+\overline{\mathbf{5}}$ pair, $\langle T_1\rangle$ couples symmetrically in family space to all members of a pair of $\mathbf{16}$ fermions, whether up or down quarks, neutrinos or charged leptons. On the other hand, since the C' VEV of the doublet appears only in the $SU(5)$ $\overline{\mathbf{5}}$ of the $\mathbf{16}$, this VEV couples only to the down quarks and charged leptons in a $\mathbf{16}$ and $\mathbf{10}$ fermion pair and asymmetrically at that. All the GUT scale VEVs except $\langle A\rangle$ are $SU(5)$ singlets, with $\langle A\rangle$ of the single $SO(10)$ $\mathbf{45}$ assigning an antisymmetric $B-L$ quantum number of magnitude 1/3 or 1 to the quarks and leptons, respectively.

Yukawa coupling unification at the GUT scale suggests as usual the coupling of $\langle T_1\rangle$ to the third generation quarks and leptons according to $\mathbf{16}_3\mathbf{16}_3T_1$. Now, however, because of the linear combination appearing in (2), the top-to-bottom quark mass ratio at the GUT scale assumes the form:

$$m_t^o/m_b^o = \tan\beta/\sin\theta \qquad (5)$$

in terms of the $\langle\overline{\mathbf{5}}(T_1)\rangle - \langle\overline{\mathbf{5}}(C')\rangle$ mixing angle θ. Hence $\tan\beta$ can assume any value in the range 2 - 55.

The Georgi-Jarlskog relations [5], $m_s^0 \cong m_\mu^0/3$ and $m_d^0 \cong 3m_e^0$, together with the minimal Higgs structure then suggest the following textures for the Dirac mass matrices [2]:

$$\begin{aligned} U^0 &= \begin{pmatrix} 0 & \sigma & 0 \\ \sigma & 0 & \epsilon/3 \\ 0 & -\epsilon/3 & 1 \end{pmatrix} m \\ N^0 &= \begin{pmatrix} 0 & \sigma & 0 \\ \sigma & 0 & -\epsilon \\ 0 & \epsilon & 1 \end{pmatrix} m \\ D^0 &= \begin{pmatrix} 0 & \sigma+\sigma' & 0 \\ \sigma+\sigma' & 0 & \rho+\epsilon/3 \\ 0 & -\epsilon/3 & 1 \end{pmatrix} \tilde{m} \\ L^0 &= \begin{pmatrix} 0 & \sigma+\sigma' & 0 \\ \sigma+\sigma' & 0 & -\epsilon \\ 0 & \rho+\epsilon & 1 \end{pmatrix} \tilde{m} \qquad (6) \end{aligned}$$

where the matrices are written so that the left-handed antifermions multiply them from the left and the left-handed fermions from the right. The 2 - 3 sector of the above matrices is essentially uniquely determined. Here the ϵ terms arise from the $B-L$ VEVs, $\langle A\rangle$, of the antisymmetric $\mathbf{45}$, while the ρ terms arise from the $\langle C'\rangle$ VEV. The 1 - 2 sector has more uncertainty. We have made the simplest choices here; for example, the σ terms may arise from $\langle T_1\rangle$ Higgs VEVs after integrating out superheavy $\mathbf{16}$ fermions, while the σ' terms appear after integrating out superheavy $\mathbf{10}$ fermions.

If we assume that $\rho \gg \epsilon \gg \sigma' \gg \sigma$, by diagonalizing the matrices we find:

$$\frac{m_b^0}{m_\tau^0} \cong 1 - \frac{2}{3}\frac{\rho}{\rho^2+1}\epsilon\cos\alpha$$

$$\frac{m_\mu^0}{m_\tau^0} \cong \epsilon\frac{\rho}{\rho^2+1}\left(1 - \frac{\rho^2-1}{\rho(\rho^2+1)}\epsilon\cos\alpha\right)$$

$$\frac{m_s^0}{m_b^0} \cong \frac{1}{3}\epsilon\frac{\rho}{\rho^2+1}\left(1 - \frac{1}{3}\frac{\rho^2-1}{\rho(\rho^2+1)}\epsilon\cos\alpha\right)$$

$$\frac{m_c^0}{m_t^0} \cong \frac{1}{9}\epsilon^2$$

$$\frac{m_d^0}{m_e^0} \cong 3\left(1 + \frac{2}{3\rho}\epsilon\cos\alpha\right)$$

$$V_{cb}^0 \cong \frac{1}{3}\epsilon\frac{\rho^2}{\rho^2+1}\left(1 + \frac{2}{3}\frac{1}{\rho(\rho^2+1)}\epsilon\cos\alpha\right)$$

$$|V_{us}^0| \cong \left|\sqrt{\frac{m_d^0}{m_s^0}}\frac{1}{(\rho^2+1)^{1/4}} - \sqrt{\frac{m_u^0}{m_c^0}}e^{i\phi}\right|$$

$$|V_{ub}^0| \cong \left|\sqrt{\frac{m_d^0 m_s^0}{m_s^0 m_b^0}}\frac{\rho}{(\rho^2+1)^{1/4}} - \sqrt{\frac{m_u^0}{m_c^0}}e^{i\phi}\left(\sqrt{\frac{m_c^0}{m_t^0}} - \frac{m_s^0}{m_b^0}\frac{1}{\rho}\right)\right| \quad (7)$$

Here α is the relative phase between ϵ and ρ, while ϕ is the relative phase between σ and σ'. In addition to the Georgi-Jarlskog relations [5], we observe that $m_b^0 \simeq m_\tau^0$; V_{cb}^0, m_μ^0/m_τ^0 and $m_s^0/m_b^0 \sim O(\epsilon)$; while $m_c^0/m_t^0 \sim O(\epsilon^2)$.

Of special interest is the issue of neutrino masses and mixings. The light neutrino mass matrix is given by $M_\nu = -N^T M_R^{-1} N$, in terms of the Dirac neutrino matrix and the superheavy right-handed Majorana neutrino mass matrix. If we simply take M_R diagonal and similar to the identity matrix, a large mixing emerges by virtue of the form of the Dirac matrices N^0 and L^0 in Eq. (6) as indicated below. In fact, the mixing will generally be very large, unless the form of M_R is fine-tuned. As a result of the asymmetrical ρ contributions appearing in D^0 and L^0, we can then understand why V_{cb} mixing is small in the quark sector while the $\nu_\mu - \nu_\tau$ mixing is large in the neutrino sector. The atmospheric anomaly [6] can thus be understood without resorting to a very hierarchical form for the Majorana matrix.

3. NUMERICAL RESULTS

In order to obtain numerical comparisons with experiment, the fermion masses and mixings have been evolved [2] from the unification scale, M_G, to the supersymmetry scale $M_{SUSY} \sim m_t$, by making use of 2-loop MSSM β functions and from M_{SUSY} to the running mass scales with the use of 3-loop QCD and 1-loop QED or EW beta functions. We find the known quark mass and mixing data is best fitted with $\tan\beta \simeq 30$. For this value, and the known m_μ, m_τ and V_{cb}, the two parameters ρ and ϵ are found to be

$$\rho = 1.73(1-\Delta_{cb}), \quad \epsilon = 0.136(1-0.5\Delta_{cb}) \quad (8)$$

in terms of the chargino loop correction $\Delta_{cb} \simeq -0.05$ for V_{cb}.

The following predictions then emerge with $\cos\alpha = 1$:

- Good agreement with the experimental value for $m_b(m_b) = 5.0(1+\Delta_b)$ GeV is reached with the combined gluino and chargino loop correction $\Delta_b \cong -0.15$.

- With $\Delta_s \simeq \Delta_b \cong -0.15$, $m_s(1GeV) = 176(1+\Delta_s) = 150$ MeV compared with 180 ± 50 MeV.

- We find $m_c(m_c) = (1.05 \pm 0.11)(1-\Delta_{cb}) \sim (1.10 \pm 0.11)$ GeV, in reasonable agreement with the experimental value of (1.27 ± 0.1) GeV.

- For a non-hierarchical diagonal form for M_R, we find $\sin^2 2\theta_{\mu\tau} \simeq 0.7$. This large neutrino mixing occurs not because of a hierarchy in the right-handed Majorana neutrino mass matrix but rather because of the asymmetrical form appearing in the charged lepton mass matrix as a result of the minimal Higgs structure assumed.

- For the form of the first generation contributions to the mass matrices given in (6), acceptable results for $|V_{us}|$ and $|V_{ub}|$ emerge with the phase $\phi \sim 180^o$. The leptonic mixings $|(U_\nu)_{e\nu_2}|$ and $|(U_\nu)_{e\nu_3}|$ are small and consistent with the small angle MSW solution for the solar neutrinos, but their precise values are sensitive to the assumed structure of M_R.

In [2], detailed results have been obtained for a broader range of the input parameters ρ, ϵ, $\cos\alpha$ and ϕ.

4. SPECIFIC $SO(10)$ SUPERSYMMETRIC GRAND UNIFIED MODEL

It is of interest to construct a specific $SO(10)$ supersymmetric grand unified model which leads to the textures for the mass matrices postulated in Eq. (6). This has been accomplished in [2] for the second and third generation contributions which are essentially uniquely determined. The first generation contributions, being higher order, are less well determined and are subject to further study as are the contributions to the right-handed Majorana matrix.

Considering only the second and third generations, we are led to the following Yukawa superpotential,

$$\begin{aligned} W_{Yukawa} &= \mathbf{16}_3\mathbf{16}_3 T_1 + \mathbf{16}_2\mathbf{16} T_1 + \mathbf{16}_3\overline{\mathbf{16}}A \\ &\quad + \mathbf{16}\overline{\mathbf{16}}P + \mathbf{10}\mathbf{10}'\overline{C}C/M_P \\ &\quad + \mathbf{16}_2\mathbf{10}C + \mathbf{16}_3\mathbf{10}'C' \end{aligned} \quad (9)$$

In addition to the two light fermion families, one pair of $\mathbf{16}+\overline{\mathbf{16}}$ and one pair of $\mathbf{10}+\mathbf{10}'$ fermions have been introduced which get superheavy as a result of the interactions present in Eq. (9). By making use of the previous $U(1)\times Z_2\times Z_2$ family assignments for the Higgs fields given in Eq. (3), the above terms for the Yukawa superpotential are uniquely obtained if we extend the following $U(1)\times Z_2\times Z_2$ assignments to the fermions:

$$\begin{aligned} &\mathbf{16}_3(-\tfrac{1}{2})^{++}, && \mathbf{16}_2(-\tfrac{1}{2}+p)^{++} \\ &\mathbf{16}(-\tfrac{1}{2}-p)^{++}, && \overline{\mathbf{16}}(\tfrac{1}{2})^{+-} \\ &\mathbf{10}(-p)^{-+}, && \mathbf{10}'(p)^{++} \end{aligned} \quad (10)$$

The desired 22, 23, 32 and 33 entries in the Dirac matrices of Eq. (6) are then obtained with the Yukawa interactions in Eq. (9) by integrating out the superheavy fermions introduced above. The relevant diagrams are pictured in Fig. 1 where the asymmetrical nature of the contributions is readily apparent.

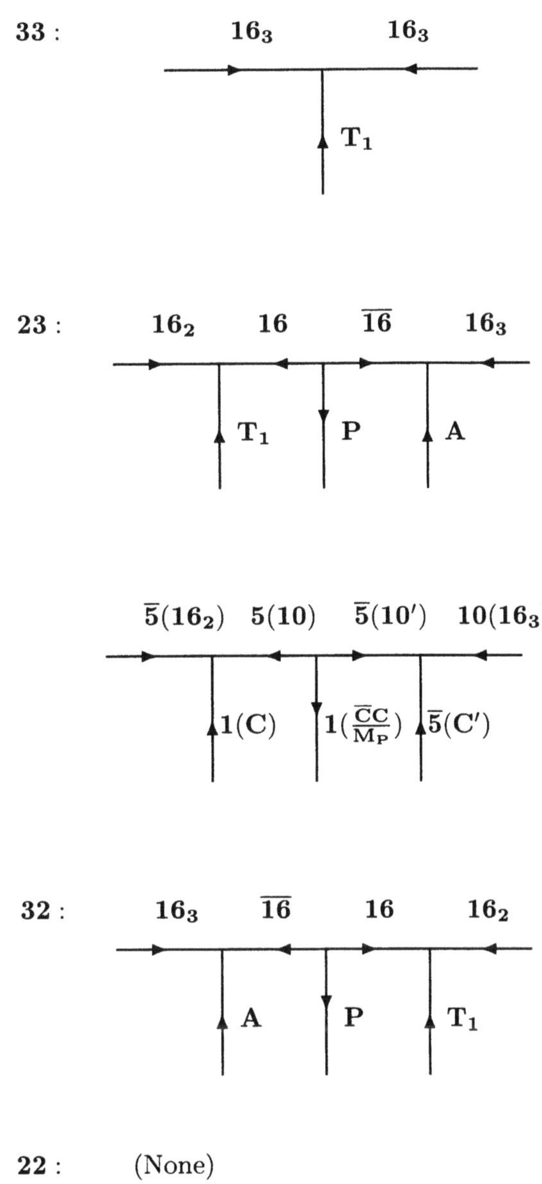

Figure 1. Diagrams that generate the 33, 23, 32 and 22 entries in the quark and lepton mass matrices of Eq. (6). The second diagram of the 23 entry appears only for the down quark mass matrix. A similar diagram in reverse order would appear for the 32 entry of the charged lepton mass matrix.

5. SUMMARY

In summary, we have shown that with the minimal set of $SO(10)$ Higgs fields introduced in Eq. (1) to solve the doublet-triplet splitting problem, fermion mass matrices can be constructed which explain well the known quark mass and mixing data and lead to the suggestion of large $\nu_\mu - \nu_\tau$ mixing responsible for the atmospheric neutrino anomaly. Unlike previous studies, this large neutrino mixing arises not from a large hierarchy in the right-handed Majorana matrix but rather as a result of the skewed spinor $\mathbf{16}'$ Higgs and antisymmetrical $B - L$ adjoint **45** contributions to the Dirac matrices.

REFERENCES

1. S.M. Barr and S. Raby, Phys. Rev. Lett. **79**, 4748 (1997).
2. C.H. Albright and S.M. Barr, Phys. Rev. D **58**, 013002 (1998); C.H. Albright, K.S. Babu and S.M. Barr, Phys. Rev. Lett. **81**, 1167 (1998).
3. B. Brahmachari and R.N. Mohapatra, Phys. Rev. D **58**, 015003 (1998); J. Sato and T. Yanagida, hep-ph/9710516; M. Bando, T. Kugo, and K. Yoshioka, Phys. Rev. Lett. **80**, 3004 (1998).
4. S. Dimopoulos and F. Wilczek, Report No. NSF-ITP-82-07 (1981), in *The unity of fundamental interactions*, Proceedings of the 19th Course of the International School of Subnuclear Physics, Erice, Italy, 1981, ed. A. Zichichi (Plenum Press, New York, 1983).
5. H. Georgi and C. Jarlskog, Phys. Lett. **B86**, 297 (1979).
6. K.S. Hirata *et al.*, Phys. Lett. B **205**, 416 (1988); K.S. Hirata *et al.*, Phys. Lett. B **280**, 146 (1992); Y. Fukuda *et al.*, Phys. Lett. B **335**, 237 (1994); D. Caspar *et al.*, Phys. Rev. Lett. **66**, 2561 (1991); R. Becker-Szendy *et al.*, Phys. Rev. D **46**, 3720 (1992); Nucl. Phys. B (Proc. Suppl.) **38**, 331 (1995); T. Kafka, Nucl. Phys. B (Proc. Suppl.) **35**, 427 (1994); M. Goodman, *ibid.* **38**, 337 (1995); W.W.M. Allison *et al.*, Phys. Lett. B **391**, 491 (1997).

ELSEVIER

Nuclear Physics B (Proc. Suppl.) 77 (1999) 313–318

NUCLEAR PHYSICS B
PROCEEDINGS
SUPPLEMENTS

Cosmic Ray and Neutrino Tests of Special Relativity*

Sheldon L. Glashow
Harvard University

Sidney Coleman and I developed a formalism with which to describe tiny departures from exact Lorentz invariance, and from which to explore their potentially observable consequences[1][2]. To the Lagrangian of the standard model—which is invariant under Lorentz transformations and $SU(3)\times SU(2)\times U(1)$ gauge symmetries—we add all possible small perturbations that are:

1. Renormalizable, *i.e.*, of dimension no greater than four,
2. Invariant under the gauge symmetries of the standard model, and
3. Rotationally invariant in a preferred frame.

Except for the latter restriction, our approach is similar to that of Kostalecký and Colladay[3]. However, we address different questions, namely tests of Lorentz invariance involving highly relativistic particles. The preferred frame, plausibly the rest frame of the cosmic background radiation (CBR), is assumed to have a non-relativistic velocity relative to Earth. This relative velocity may be neglected for our considerations, although it plays an essential role in other tests of Lorentz invariance.

Lorentz violation alters the dispersion relations for freely moving elementary particles. In particular, each particle species a can have its own maximal attainable velocity (MAV) in the preferred frame. We denote these MAV by c_a (with $c_\gamma \equiv 1$) and define the Lorentz-violating parameters $\delta_{ab} \equiv c_a^2 - c_b^2$. A strong limit on this type of Lorentz violation, obtained by Lamoreaux *et al.*[4] from a search for anisotropies of nuclear transitions due to Earth's motion relative to the preferred frame, sets the scale for our subsequent discussion. They find:

$$|\delta_{\gamma m}| \equiv |c_\gamma^2 - c_m^2| < 6 \times 10^{-22}\,, \qquad (1)$$

where they assume c_m, the MAV of material matter, to be the same for all massive particles. A stronger (but one-sided) constraint, $\delta_{p\gamma} < 10^{-22}$, that is independent of Earth's motion follows from the reported observation of primary cosmic-ray protons with energies $\geq 10^{20}$eV[1]. These observations[5]lead us to the first of the three issues to be discussed herein.

Evading the GZK cosmic-ray cutoff: Soon after the discovery of the cosmic background radiation (CBR), Greisen[6] and Zatsepin and Kuz'min[7] saw how it limits the propagation of ultra-high energy (UHE) nucleons. Primary nucleons with sufficient energy will suffer inelastic impacts with CBR photons. This results in what is known as the GZK cutoff, saying that nucleons with energies $> 5 \times 10^{19}$ eV cannot reach us from further than a few dozen Mpc. However, the cosmic-ray energy spectrum seems to extend well beyond this energy.

The mechanism producing UHE cosmic rays is unknown. Exotic origins have been proposed, among them: topological defects, active galactic nuclei, and gamma-ray bursts[8]. These schemes are constrained, if not ruled out, by the GZK cutoff. Another explanation, that the most energetic cosmic rays are decay products of hypothetical super-heavy relic particles[9], is consistent with observations of six events by AGASA with energies $> 10^{20}$ eV that are widely separated in position and not coincident with any known astrophysical source[5].

We have little to say about the origin of UHE

*Research supported in part by the National Science Foundation under grant number NSF-PHY/98-02709.
†We thank M. Wise for relieving the tedium by providing this result and others.

PII S0920-5632(99)00435-1

cosmic rays *per se*. Rather, we point out that there may not be a GZK cutoff after all. Tiny departures from Lorentz invariance, too small to have been detected otherwise, have effects that increase rapidly with energy and can kinematically prevent cosmic-ray nucleons from undergoing inelastic collisions with CBR photons. The cutoff thereby undone, a deeply cosmological origin of UHE cosmic rays becomes tenable.

We do not anticipate a failure of special relativity nor are we aware of any reasonable theory predicting one. Nonetheless, strict Lorentz invariance should not be accepted on faith but rather as a plausible hypothesis subject to experimental test. It follows that we cannot accept the GZK cutoff as an indisputable fact. Further observations of UHE cosmic rays could confirm a predicted 'bump' just below the cutoff[10] resulting from products of inelastic collisions of primary protons with CBR photons. This would prove the GZK cutoff to be at least partially effective. Or, they could indicate sources at cosmological distances that would belie the cutoff.

To see how the GZK cutoff is affected by Lorentz violation, we begin by considering the formation of the first pion-nucleon resonance:

$$p + \gamma\ (\mathrm{CBR}) \longrightarrow \Delta(1232), \tag{2}$$

by a proton of energy E colliding with a CBR photon of energy ω. The target photon energies have a thermal distribution with temperature $T = 2.73$ K, corresponding to $\omega_0 = 2.35 \times 10^{-4}$ eV. Energy conservation provides the condition under which reaction (2) can proceed:

$$4\omega \geq \delta_{\Delta p} E + \frac{M_\Delta^2 - M_p^2}{E}. \tag{3}$$

If Lorentz symmetry is unbroken, $\delta_{\Delta p} = 0$ and Eq. (3) yields the conventional threshold for a head-on impact: $E_f = (M_\Delta^2 - M_p^2)/4\omega$. Otherwise inequality (3) is a quadratic form in E, satisfiable if and only if $\delta_{\Delta p} \leq \omega/E_f$. As $\delta_{\Delta p}$ increases from zero, the threshold grows toward $2E_f$ as $\delta_{\Delta p}$ approaches a critical value:

$$\hat{\delta}(\omega) \equiv \frac{4\omega^2}{M_\Delta^2 - M_p^2} \simeq 3.5 \times 10^{-25}\ [\omega/\omega_0]^2. \tag{4}$$

For $\delta_{\Delta p} > \hat{\delta}$, reaction (2) is kinematically forbidden for all E. Recalling that the photon spectrum is thermal, we see that if $\delta_{\Delta p}$ is comparable to $\hat{\delta}(\omega_0)$, the GZK cutoff due to resonant $\Delta(1232)$ formation would be relaxed. Should it much exceed this value, formation would be precluded off virtually all CBR photons

Reaction (2) is the dominant process leading to the GZK cutoff as originally formulated. However, if $\Delta(1232)$ formation is not possible, a weakened version of the cutoff may result from nonresonant photo-production:

$$p + \gamma\ (\mathrm{CBR}) \longrightarrow p + \pi. \tag{5}$$

Ordinarily (for $\delta_{\pi p} = 0$), the threshold energy is $E_p = M_\pi(2M_p + M_\pi)/4\omega$. If Lorentz invariance is violated and $\delta_{\pi p}$ is imagined to increase from zero, the threshold grows. For a fixed photon energy ω, the threshold diverges as $\delta_{\pi p}$ approaches a critical value:

$$\tilde{\delta}(\omega) \equiv \frac{4\omega^2}{M_\pi^2} \simeq 1.1 \times 10^{-23}\ [\omega/\omega_0]^2. \tag{6}$$

For larger values of $\delta_{\pi p}$, reaction (5) (as well as multiple pion production) is kinematically forbidden *at all proton energies.* For the actual case of a thermal distribution of photon energies, values of $\delta_{\pi p}$ comparable to or greater than $\tilde{\delta}(\omega_0)$ would suppress photo-pion production, or even eliminate it entirely so that no vestige of the cutoff survives.

We have shown how tiny values of the *a priori* unknown Lorentz-violating parameters $\delta_{\Delta p}$ or $\delta_{\pi p}$ can suppress or forbid the processes underlying the GZK cutoff. Note that much larger (and experimentally intolerable) violations of Lorentz invariance would be needed to affect significantly the interactions of UHE cosmic rays with nuclei in the atmosphere.

Existing bounds on departures from special relativity are insufficient to disfavor those required to mitigate the GZK cutoff. Fortunately, several bounds can be considerably strengthened. Laboratory tests of Lorentz invariance far more precise than any done before are now feasible[11]. Dedicated searches for velocity oscillations of solar neutrinos, or of accelerator-produced ~TeV neutrinos at baselines of ~ 1000 km, can reveal neutrino velocity differences as small as 10^{-25}. At present, lacking detailed observations of the highest energy cosmic rays and more precise tests of special relativity, we must regard as intriguingly open questions both the existence of the GZK cut-off and a cosmologically remote origin of UHE cosmic radiation.

Lorentz-violating neutrino oscillations: We assume there are three chiral neutrinos with Majorana masses given by the complex symmetric matrix m in a flavor-diagonal basis. Conventional neutrino oscillations are described in terms of the hermitean squared-mass matrix $m^2 = mm^\dagger$. In particular, observable oscillation effects depend on two differences of squared masses and four parameters akin to the Kobayashi-Maskawa angles and phase in the quark sector.

If Lorentz symmetry is violated, the description of neutrino oscillations is much more complicated. The neutrino dispersion relation is affected both by TCP-conserving terms (the matrix c_ν of maximal attainable velocities) and by TCP-violating terms (an energy-shifting matrix a_ν). Consider a neutrino with definite momentum p. Under virually all experimentally relevant circumstances, we may take the neutrino to be ultra-relativistic. Its energy eigenstates are determined by the following matrix equation:

$$E \simeq p + m^2/2E + a_\nu + c_\nu E \qquad (7)$$

where m^2, a_ν and c_ν are each 3×3 hermitean matrices. The matrix c_ν defines the energy eigenstates as velocity eigenstates in the limit of high energy, just as the matrix m^2 defines them at low energy as mass eigenstates.

To proceed, we limit ourselves to a discussion of two-flavor neutrino oscillations. Imagine neutrinos produced with a definite momentum and flavor (ν_ℓ, where $\ell = e$ or μ) and detected after travelling a distance R through empty space. Their oscillations satisfy a seemingly conventional formula:

$$P(\nu_\ell \to \nu_\ell) = 1 - \sin^2 2\Theta \sin^2 \{\Delta R/4\}. \qquad (8)$$

However, the mixing angle Θ and phase factor Δ appearing in Eq. (8) are given implicitly in terms of eight parameters:

$$\begin{aligned}\Delta \sin 2\Theta &= \delta m^2 \sin 2\theta_m/E \\ &+ 2\delta a e^{i\eta} \sin 2\theta_a + 2\delta v\, e^{i\eta'} \sin 2\theta_v, \\ \Delta \cos 2\Theta &= \delta m^2 \cos 2\theta_m/E \\ &+ 2\delta a \cos 2\theta_a + 2\delta v \cos 2\theta_v. \qquad (9)\end{aligned}$$

The observable parameters are three mixing angles, two complex phases, and the differences between the eigenvalues of the matrices m^2, a_ν, and c_ν (denoted respectively by δm^2, δa and δv). To illustrate the possibilities inherent in Eq.(9), we describe a few special cases of Lorentz-violating two-flavor neutrino oscillations:

$$\begin{aligned}&P(\nu_\ell \to \nu_\ell) \\ &= 1 - \frac{\sin^2\left\{(\delta m^2 R/4E)\sqrt{1+(E/E_0)^2}\right\}}{1+(E/E_0)^4} \qquad (10) \\ &= 1 - \frac{\sin^2\left\{(\delta v R E/2)\sqrt{1+(E_0/E)^2}\right\}}{1+(E_0/E)^4} \qquad (11) \\ &= 1 - \sin^2 2\theta \sin^2 \\ &\quad \left\{R(\delta m^2/4E + \delta a/2 + \delta v E/2)\right\} \qquad (12)\end{aligned}$$

where $E_0 \equiv \delta m^2/(2\delta v)$. Eq. (10) corresponds to $\theta_m = \pi/4$ with $\delta a = \sin 2\theta_v = 0$. It yields maximal oscillations for $E \ll E_0$, but essentially none for $E \gg E_0$. Eq. (11) corresponds to a converse case with $\theta_v = \pi/4$ and $\delta a = \sin 2\theta_m = 0$: maximal oscillations at high energy, none at low energy.

To obtain Eq.(12), we set all three mixing angles equal and put $\eta = \eta' = 0$. This example encompasses the three scenarios discussed by Foot, Leung and Yasuda [12] for atmospheric neutrino oscillations — each of which they found to be compatible with current atmospheric neutrino data. In principle, the study of neutrino oscillations can provide exceedingly strong constraints on both TCP-conserving and TCP-violating departures from Lorentz invariance. But it will be a challenging problem for the experimenter to extract them from the data.

Testing lorentz invariance with muon colliders: That radiative lepton decays ($\ell \to \ell' + \gamma$), are often searched for but never seen is not surprising. They are forbidden in the minimal standard model. They are induced by radiative corrections in models with neutrino masses, but the expected branching ratios are far too small to be detected. However, an accidental symmetry ordinarily preventing radiative decay is lifted by Lorentz-violating perturbations, provided that the velocity eigenstates of leptons differ from their mass eigenstates—and there is no reason to expect these terms, if present, to respect flavor conservation.

Indeed, tiny Lorentz-violating effects too small to have been detected otherwise may cause $\mu \to e + \gamma$ to become the dominant decay mode of muons with sufficiently high energies. Flavor-violating departures from special relativity (characterized by ϵ, a parameter soon to be defined) lead to branching ratios for this 'forbidden' process $\sim \alpha\epsilon^2\gamma^4 M\tau_0$, where M and τ_0 are the mass and rest-lifetime of the decaying muon and γ is its Lorentz factor in the preferred frame. The branching ratio increases with the third power of the muon energy. It is controlled by the magnitude of the departure from Lorentz invariance but is otherwise a first-order radiative decay, not a weak decay.

Departures from Lorentz invariance can affect the $\pi^\pm$ lifetime, and as well the rate for the allowed process $\mu \to e + \nu + \bar{\nu}$, but these cases are radically different. Lorentz-violating or not, these decay modes are intrinsically weak. Lorentz violation may yield corrections of the form $\Gamma_0(1 + \epsilon^2\gamma^4)/\gamma$, but without the enormous enhancement factor $\alpha M\tau_0 \simeq 2.6 \times 10^{15}$ present for radiative muon decay. Thus the most sensitive tests of Lorentz invariance in this context are obtained from the study of muons, and in particular, from the search for a lifetime anomaly of muons at ultra-high energies. The relevant Lorentz-violating addition to the standard model Lagrangian (where mixing to the tau lepton is ignored) takes the following form in the preferred frame:

$$\sum \left(\overline{\mu_a} \quad \overline{e_a}\right) \vec{\gamma}\cdot(\vec{p} - e\vec{A}) \cdot \begin{pmatrix} 2\bar{c}_a + \delta c_a \cos 2\theta_a & \delta c_a \sin 2\theta_a \\ \delta c_a \sin 2\theta_a & 2\bar{c}_a - \delta c_a \cos 2\theta_a \end{pmatrix} \begin{pmatrix} \mu_a \\ e_a \end{pmatrix} \quad (13)$$

with μ and e denoting fields corresponding to the mass eigenstates and the summation extending over the helicity states $a = L, R$. The parameter θ_L determines the velocity eigenstates of left-handed leptons (or right-handed antileptons) whose MAVs are $\bar{c}_L \pm \frac{1}{2}\delta c_L$. Similarly, θ_R determines the velocity eigenstates of right-handed leptons (or left-handed antileptons), whose MAVs differ by δc_R.

Electroweak gauge invariance requires the parameters $\bar{c}_L$, δc_L and θ_L to pertain to neutrinos as well as charged leptons. Consequently they are constrained by various neutrino experiments. If θ_L is large, the severe bound $|\delta c_L| < 10^{-21}$ has been deduced [1]. However, there is no analogous constraint on δc_R (nor any significant constraint on δc_L if θ_L is small). Constraints from observations of neutrinos from supernova 1987a [13] are too weak to be relevant to our subsequent discussion. Neutrino signals from gamma-ray bursts could reveal [14] Lorentz-violating velocity differences as small as $|\delta c_L| = 10^{-18}$. Of course, no such signals have been detected and no such constraint would apply to δc_R.

It is convenient to define the small parameters:

$$\epsilon_a \equiv |\delta c_a \sin 2\theta_a|^2\,, \qquad a = L,\, R\,.$$

A straightforward but tedious computation[†] yields the $\mu \to e + \gamma$ branching ratio *for muons at rest:*

$$\begin{aligned} B &= \frac{\alpha M\tau_0}{4}(\epsilon_L + \epsilon_R) \\ &\simeq 6.4 \times 10^{14}\,(\epsilon_L + \epsilon_R). \end{aligned} \quad (14)$$

The current experimental limit [15] is $B < 4.9 \times 10^{-11}$, and yields the following upper limit on the relevant Lorentz violating parameter:

$$\epsilon_L + \epsilon_R < 8 \times 10^{-26}$$
from muon decay at rest. (15)

The branching ratio for Lorentz-violating radiative muon decay is a rapidly increasing function of the energy. While direct searches for this decay mode of ultra-relativistic muons do not seem to be feasible, one might hope to detect the onset of this mode through its effect on the muon lifetime The lifetime of a ultrarelativistic left-handed μ^- (or right-handed μ^+) is:

$$\tau_L(\gamma) = \frac{\gamma\tau_0}{1+b_L\gamma^4}\,, \quad \text{where} \quad b_L \equiv \frac{\alpha M \tau_0}{30}\,(68\,\epsilon_R + \epsilon_L)\,. \tag{16}$$

A similar result pertains to the lifetime of ultra-relativistic right-handed μ^- (or left-handed μ^+):

$$\tau_R(\gamma) = \frac{\gamma\tau_0}{1+b_R\gamma^4}\,, \quad \text{where} \quad b_R \equiv \frac{\alpha M \tau_0}{30}\,(68\,\epsilon_L + \epsilon_R)\,. \tag{17}$$

At sufficiently high energy, the lifetime of muons with either helicity decreases with γ^{-3}, rather than increasing with γ.

The CERN $g-2$ experiment, aside from measuring the muon's anomalous magnetic moment, offers a precise test of the energy dependence of its lifetime. At $\gamma = 29.3$ (corresponding to the 'magic energy' at which the experiment was performed) the results confirm the expected muon lifetime to an accuracy of one part in a thousand [16]. Because the muons in the ring are racemic, we obtain the limit $b_L + b_R < 2.7 \times 10^{-9}$, or

$$\epsilon_L + \epsilon_R < 5 \times 10^{-25} \quad \text{from muon } g-2\,. \tag{18}$$

which is inferior to that obtained from the direct search, Eq.(15),but not by much! The agreement between theoretical and experimental values of $g-2$ provides a much weaker test of Lorentz invariance.

What does all this have to do with the much-discussed and possibly-feasible muon collider? The design and proper operation of such a machine is contingent on the relativistic extension of the muon lifetime, which as we have seen, is affected by flavor-dependent violations of Lorentz symmetry. As working hypotheses, we assume (1) that the muons being accelerated are unpolarized, and (2) that the muon decay rate cannot exceed twice its normal value at the design energy $E = \gamma M$ lest the machine be compromised. This criterion says that the mean decay rate of the muons at the collision energy satisfies: $\frac{1}{2}(\Gamma_L + \Gamma_R) < 2/(\gamma\tau_0)$, whence $\frac{1}{2}(b_L + b_R)\gamma^4 < 1$.

For a beam energy of 0.5 TeV (the 'lesser muon collider' or lmc) our criterion becomes $b_L + b_R < 3 \times 10^{-15}$. For a beam energy three times larger (the 'greater muon collider' or gmc), we would need $b_L + b_R < 4 \times 10^{-17}$. Lorentz violating effects of this magnitude are not excluded by the preceding arguments. It would be particularly unfortunate if the muon collider would detect a breakdown of Lorentz invariance by not working! We believe that this potential disaster can be averted through judicious analyses of present and future data concerning underground muons.

The underground muon flux has been, and can be, measured at a wide variety of rock depths and inclinations. In particular, available data about muon fluxes measured at a given slant depth but different inclinations (and hence, different flight times) seem to yield concordant results. A careful analyses of these data can reveal or constrain a lifetime anomaly for ultra-relativistic muons. Recall however, that cosmic-ray muons arise primarily from forward decays of pions and are longitudinally polarised μ_R^- and μ_L^+. So searches for a lifetime anomaly can bound b_R, but not b_L. b_R is proportional to the linear combination $68\,\epsilon_L + \epsilon_R$. Although ϵ_L involves parameters that may be constrained by observations of neutrino oscillations, it may be that $\epsilon_R \gg \epsilon_L$. Absent any *a priori* knowledge of the ratio ϵ_L/ϵ_R, a bound on b_R yields a 69-fold weaker bound on $b_R + b_L$. Nonetheless, it is not implausible to presume that cosmic-ray physicists can establish the bound $b_R < 5 \times 10^{-17}$, which would be sufficient to safeguard the operation of the muon collider.

Note added in proof: The challenge of the previous paragraph has been met! Cowsik and Sreekantan [17] set the bound $b_R < 10^{-25}$ from their analysis of horizontal air showers. This bound is more than sufficient to avoid any detrimental effect of Lorentz violation on the functioning of proposed muon colliders.

REFERENCES

1. S. Coleman and S.L. Glashow, Phys. Lett. B405 (1997) 249;'Evading the GZK Cutoff,' **hep-ph/9808446**; and ms in preparation.
2. S.Glashow, Nucl. Phys. B (Proc. Suppl.) 70 (1999) 180; AIP Conf. Proc. 444 (1998) 119.
3. D. Colladay and V.A. Kostalecký, Phs. Rev. D55 (1997) 6760 and **hep-ph/9809521**.
4. S.K. Lamoreaux, J.P. Jacobs, B.R. Heckel, F.J. Raab, and E.N. Fortson, Phys. Rev. Lett. 57 (1986) 3125.
5. M. Takeda *et al.*, Phys. Rev. Lett. 81 (1998) 1163
6. K. Greisen, Phys. Rev. Lett. 16 (1966) 748.
7. G.T. Zatsepin and V.A. Kuz'min, JETP Lett. 41 (1966) 78.
8. See 5 and V. Berezinsky, **hep-th/9802351** for further references in these connections.
9. V. Berezinsky, M. Kachelrieß, and V. Vilenkin, Phys. Rev. Lett. 79 (1997) 4302.
10. C.T, Hill and D.N. Schramm, Phys. Rev. D31 (1985) 564.
11. E.N. Fortson, private communication.
12. R. Foot, C.N. Leung, and O. Yasuda, **hep-ph/9809458**.
13. *E.g.*, M.J. Longo, Phys. Rev. D36 (1987) 3276; L. Stodolsky, Phys. Lett. B201 (1988).
14. E. Waxman and J. Bahcall, Phys. Rev. Lett. 78 (1997) 2292.
15. Bolton, Phys. Rev. D38 (1988) 2077.
16. *E.g.*, F. Combley, F.J.M. Farley and E. Picasso, Phys. Rep.68 (1981) 93.
17. R. Cowsik and B.V. Sreekantan, **hep-ph/9811241**.

Part 9

Direct Search for Neutrino Mass

ELSEVIER

Nuclear Physics B (Proc. Suppl.) 77 (1999) 321–326

NUCLEAR PHYSICS B
PROCEEDINGS
SUPPLEMENTS

New results from the Mainz Neutrino Mass Experiment

H. Barth[a], A. Bleile[a], J. Bonn[a], L. Bornschein[a], B. Degen[a], L. Fleischmann[a], O. Kazachenko[b], A. Kovalik[c], E.W. Otten[a], M. Przyrembel[a], Ch. Weinheimer[a]

presented by Christian Weinheimer

[a]Institute of Physics, Joh. Gutenberg University, 55099 Mainz, Germany

[b]on leave from Institute for Nuclear Research of the Russian Academy of Sciences, Troitsk/Russia

[c]on leave from Joint Institute for Nuclear Research, Dubna/Russia

The present status of the Mainz tritium β decay experiments is given. The very recent improvement of the Mainz setup and the first results from tritium data are presented. The former trend towards negative values of m_ν^2 for increasing data intervals was identified to be a roughening transition of the quench-condensed T_2 film, which now can be avoided.

1. Introduction

Our knowledge on the fundamental questions whether neutrinos have mass and whether they mix has improved recently: At the Neutrino 98 conference the Super-Kamiokande experiment reported "evidence for neutrino oscillation" for the atmospheric neutrinos [1]. In addition the deficit of solar neutrinos comes out more and more clearly from different experiments and the hypothesis of neutrino oscillation seems to be the only possible explanation [2]. These neutrino oscillation experiments measure flavour mixing angles and differences of squares of the neutrino mass eigenstates (Δm^2), but they are not able to determine the absolute neutrino mass values.

For this purpose a direct determination of the neutrino mass eigenstates is needed. Although the search for neutrinoless double β decay seems to be the most sensitive method, it has the limitation to require Majorana-type neutrinos and some assumptions about the mixing matrix. Therefore the determination of the mass of the electron antineutrino[1] from tritium β decay remains very important and adds crucial information to the understanding of the neutrino masses and mixing. In spite of facing some problems in understanding the recently measured spectra fully, the sensitivity of this method is currently reaching a few eV/c^2.

The eV/c^2 neutrino mass range might be very interesting since in both oscillation scenarios mentioned above the differences between the square values of the neutrino mass eigenstates are very small ($2 \cdot 10^{-3}$eV^2/c^4 or less than 10^{-4}eV^2/c^4, respectively). If the neutrino mass eigenstates would obey a similar generation hierarchy as the charged leptons or the quarks the neutrinos would not contribute to the dark matter in the universe significantly. Therefore it is also very attractive to assume a degenerate neutrino mass scheme with masses of about 1-2.5 eV/c^2 [3,4] to fulfil the requirements of the missing hot dark matter component in the universe [5]).

This paper is structured as follows: The problem of negative values of m_ν^2 is discussed in section 2. The improved Mainz setup and the first results are presented in section 3. The conclusion and an outlook are given in section 4.

[1] Just for correctness, it should be mentioned that speaking of the mass of the electron antineutrino is not right in the case of neutrino flavour mixing. In this case each neutrino mass eigenstate m_i contributes with an own β spectrum of relative amplitude $|U_{ei}|^2$ according to its mixing to the electron antineutrino to the β electron energy spectrum. If the different neutrino masses are not resolved the β spectrum is determined by an average electron neutrino mass $\overline{m_\nu} = \sum |U_{ei}|^2 \cdot m_i$, which may differ from the so-called effective electron neutrino mass $\langle m_\nu \rangle = |\sum U_{ei}^2 \cdot m_i|$ measured in neutrinoless double beta decay.

0920-5632/99/$ – see front matter
PII S0920-5632(99)00436-3

2. Problem of $m_\nu^2 < 0$

After our first experiments in 1991 [8] the tritium β spectrum has been investigated with the Mainz setup in 1994 for the second time [9]. As tritium source a film of about 37 monolayers of T_2 molecules quench-condensed on a graphite substrate at a temperatures of about 3 K was used. The square of the neutrino mass m_ν^2, which is the true observable, was determined from a reasonably small interval of the last 140 eV below the endpoint, obtaining

$$m_\nu^2 = -22 \pm 17_{\rm stat} \pm 14_{\rm sys} \ \ {\rm eV}^2/{\rm c}^4 \ , \qquad (1)$$

for which a limit of $m_\nu < 5.6$ eV/c^2 (95% C.L.) can be deduced [9].

Looking to the experimental spectrum further below this interval a clear excess was observed compared to the extrapolation of the fit over the last 140 eV similar to the 1991 run [8]. This excess drives the fit result on m_ν^2 towards significantly negative values, as observed by many other experiments before (see references in [10]). A more detailed analysis of our data showed that the excess is most likely due to an unknown or underestimated energy loss process.

After checking all sources of systematic uncertainties one question was still remaining: Has the T_2 film undergone a transition from the quench-condensed homogeneous film into a rough inhomogeneous one [11] resulting in a harder energy loss spectrum, which could have produced the measured β spectrum?

In cooperation with the group of P. Leiderer at Konstanz/Germany detailed investigations for different hydrogen isotopes were performed using conversion electron spectroscopy and scattered light techniques obtaining the following results: The roughening transition cannot be avoided but its speed can drastically be slowed down by using lower temperatures. Extrapolating from the stable hydrogen isotopes a T_2 film at 2 K should have a time constant of tens of years if the β decay does not matter. This gave a clear recipe for our new experiments described in section 3, but no valid statement could be given how much a roughening transition had disturbed the experimental spectra of the Mainz 1991 and 1994 measurements.

At the same time the Troitsk experiment, which uses a similar spectrometer but a gaseous T_2 source, reported in their first measurements also an excess count rate further below the endpoint similar to the one observed at Mainz [12]. Meanwhile the Troitsk group is also able to explain their excess by electrons trapped in the source which escape into the spectrometer by large angle scattering [13].

The Troitsk group reported on a second excess rate near the endpoint E_0, known now as the "Troitsk anomaly" [12,14]. They claim that the spectral shape of this anomaly can be described by a sharp step in the experimental electron spectrum at a few eV below E_0. Since their spectrometer is integrating, this sharp step in the measured data corresponds to a monoenergetic line in the primary β spectrum with an amplitude of about $6 \cdot 10^{-11}$ of the total β spectrum, too small to be checked by the Mainz 1994 data. At the Neutrino 98 conference the Troitsk group reported that the position oscillates with a frequency of 0.5 years between 5 eV and 15 eV below E_0 [14]. If not considered in their analysis, the fits give slightly but significantly negative values for m_ν^2 in the range of -10 to -20 eV^2/c^4. There exist some speculative ideas about the origin of this anomaly, but there is no explanation within standard physics for the existence of a monoenergetic line within the continuous β spectrum. It is clear that an independent experimental check is mandatory.

3. The improved setup and first data

As pointed out in sections 1 and 2 the motivation to improve the Mainz setup significantly is twofold, namely to increase the sensitivity on m_ν to a few eV/c^2 and to check the anomaly in the tritium β spectrum, which was reported by the Troitsk group. To fulfil these expectations the Mainz experiment had to solve the following problems:

1. to avoid the T_2 film roughening transition
2. to increase the signal rate
3. to decrease the background rate
4. to allow long-term runs

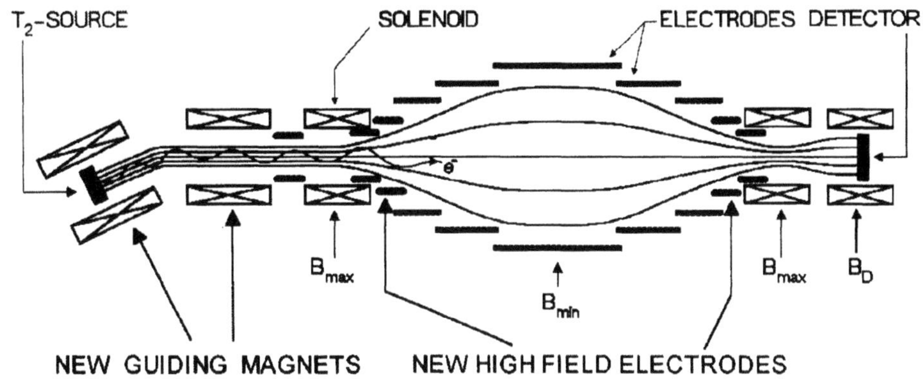

Figure 1. The improved and enhanced Mainz setup schematically, not in realistic scale. The outer diameter amounts to 1 m, the distance from source to detector is 6 m.

For these reasons the Mainz experiment has improved its setup substantially (see figure 1). In two first long-term measurements of the tritium β spectrum (4 weeks in summer 1997 and 3 weeks in winter 1998), it was demonstrated that the improved setup is working as nicely as expected:

- The new, automatically controlled source cryostat, which was installed to slow down the T_2 film roughening transition to a negligible speed by working at temperatures below 2 K, was running during the two long-term measurements at a stable temperature of 1.84 K within a range of $\pm$ 0.03 K.
- A new doublet of superconducting solenoids, tilted by 20° to each other, was installed. β decay electrons from the source in the most left solenoid are guided into the spectrometer without losses as before, whereas tritium molecules evaporating from the source are prohibited from contaminating the spectrometer, which was the biggest source of background for the 94 run. Although the thickness of T_2 films used for the measurements (1997: 973 ± 55 Å, 1998: 493 ± 5 Å) were much larger than the one of the measurement in 1994 (126 Å) the background was even significantly lower (see fig. 2) and close to the background rate without tritium source of 0.010 s^{-1}.
- The high field electrodes were redesigned to lower the background contribution from the spectrometer itself. Due to a better alignment of the whole system the spectrometer could run at a higher energy resolution of 4.4 eV compared to 6.3 eV in 1994.
- An experiment control system was installed in order to run the experiment automatically. Human intervention was needed only for filling of LHe and LN_2. It is worth mentioning that the spatial separation of the source and the spectrometer allows a valve to be closed by a control system in the case of any problems, which is an essential feature for automatic long-term running.

Figure 2 shows the endpoint region of the β spectra taken with the improved Mainz setup compared to the 1994 Mainz data. The statistical improvement of the 1997 and 1998 data due to longer data taking periods, but moreover due to a larger signal rate and a lower background rate, thus resulting in an about 10 times higher signal-to-background-ratio, is clearly visible.

Three main sources contribute to the systematic uncertainties of the measurements:

1. By recent intensive studies on the thickness determination of our T_2 films and on the cross section for inelastic scattering of β electrons within the T_2 film we could reduce the principle uncertainties by a factor two. Although the film thickness was increased significantly compared to 1994 the increase of inelastic scattering was partly compensated by changing the maximum starting angle for electrons accepted

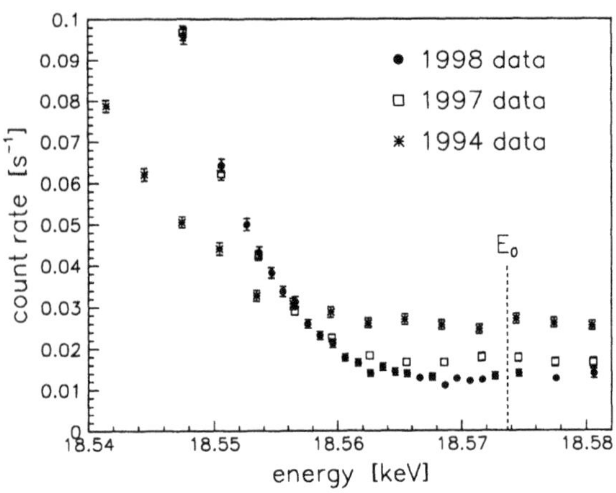

Figure 2. Mainz tritium β spectra close to the endpoint

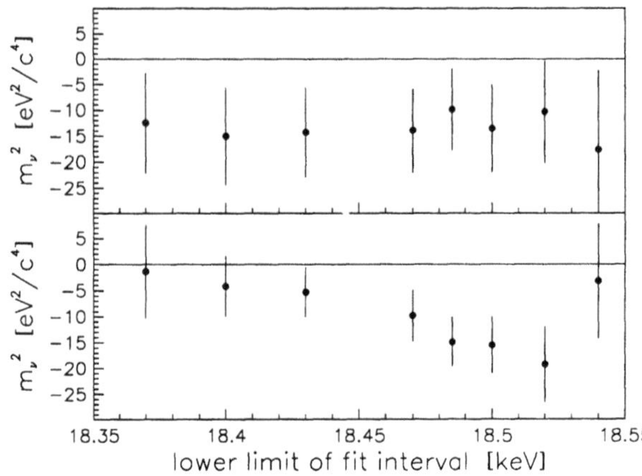

Figure 3. Fit results of m_ν^2 with total uncertainties (1 σ) in dependence on the lower limit of the data interval used for the analysis for the Mainz measurements of 1997 (top) and of 1998 (bottom).

by the spectrometer from $\vartheta_{\max} = 78.5°$ in 1994 to $\vartheta_{\max} = 45°$. Additionally, the much higher signal-to-background-ratio allows to use much smaller energy intervals below the endpoint for the final analysis, thus reducing the fraction of inelastically scattered electrons within the analysed data.

2. Due to our new better knowledge of the properties of quench-condensed hydrogen films by our energy loss measurements we reconsidered the situation for the excitation of neighbour molecules due to the fast change of nuclear charge during the β decay. A small energy shift of 1 eV and an amplitude reduction of 33 % was applied and also fully taken into account as systematic uncertainty.
3. For the thicker T_2 films used in 1997 and 1998 we observed a charging up of the film by several volts due to the emitted β electrons. By our measurements on line shifts of the ^{83m}Kr K conversion line on top and in between different thick T_2 films we could prove, that the potential within the film increases linearly with the distance to the substrate. A more detailed description and a simple model explaining quantitatively this effect can be found in reference [15]. The charging effect can be considered in the data analysis, but decreases slightly the effective energy resolution with which the β spectrum is investigated.

Fig. 3 shows the results on m_ν^2 including the total uncertainties for the measurements of 1997 and 1998, as function of the lower limit of the data interval used for the analysis. Two things can be observed:

- There is no trend anymore towards negative values of m_ν^2 for larger fit intervals as it was the case for the Mainz 1991 and 1994 data. This shows that a roughening transition of the T_2 film was the reason for this behaviour in the former measurements, which now does not appear anymore due the much lower temperature of the T_2 film.
- Still slightly but significantly negative values for m_ν^2 are obtained which are incompatible with $m_\nu^2 \geq 0$ at the 1–3 σ level.

Although the problem for the former trend towards negative values of m_ν^2 seems now to be solved the much higher sensitivity of the new data shows, that the description of the spectra still does not match the data completely. There is a small, but significant anomaly in the Mainz data. However, the data are statistically not yet good enough to extract the shape of this residual

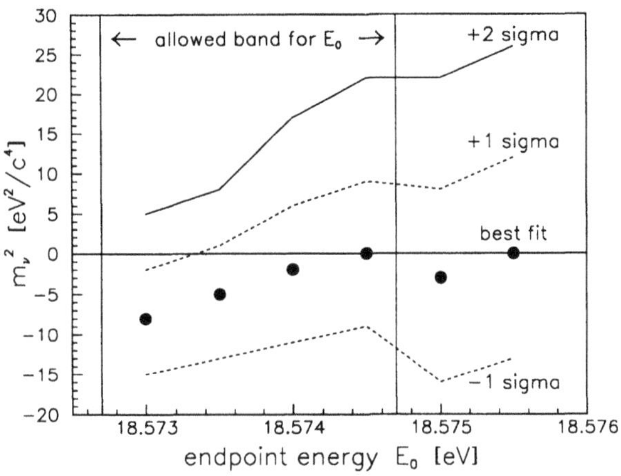

Figure 4. 1998 fit results on m_ν^2 for the last 15 eV of the β spectrum as function of the endpoint energy E_0, dashed lines: $\pm 1\ \sigma$ statistical uncertainty on m_ν^2, solid line: $+2\ \sigma$ statistical uncertainty on m_ν^2, vertical line: the allowed band for E_0 obtained from independent fits over medium large fit intervals.

anomaly clearly. Therefore the existence of the "Troitsk anomaly" can neither be proven nor disproven by the present Mainz data. Although the Mainz data are now competitive to the Troitsk data on a single run comparison, more data taking is necessary to check the special features of the "Troitsk anomaly" like its time dependence.

Due to the presence of an unresolved anomaly in the Mainz data it is not straightforward to give an upper limit on the neutrino mass, therefore only the following alternatives have been presented at this conference: two orthogonal cases will be considered to get information on the neutrino mass:

1. Let us assume first that the description of the anomaly by the Troitsk group as a sharp step would be also right for the Mainz data. Then we can fit this sharp step together with m_ν^2, as the Troitsk group does, to our 1998 data resulting for the last 70 eV of the spectrum in $m_\nu^2 = -9 \pm 8 \pm 2\mathrm{eV}^2/\mathrm{c}^4$, which would correspond to an upper limit of $m_\nu < 3.4$ eV/c^2 (95 % C.L.).
2. Let us assume now, that the anomaly is not caused by a Troitsk-like sharp step, but still by some ordinary but underestimated systematic effect. Then it is very likely that the last 15 eV of the β spectrum below E_0 are free of such disturbances, since all atomic physics effects in hydrogen have about that threshold. Fig. 4 shows the fit result on m_ν^2 for the last 15 eV of the 1998 β spectrum in dependence on the endpoint energy E_0 as input parameter. E_0 is determined independently by fits of the β spectrum over medium large fit intervals (≈ 100 eV). These fits give E_0 within a conservative error band of ± 1 eV (see fig. 4, since E_0 does not correlate much to systematic effects for these medium large fit intervals. When this restriction on E_0 is applied to the fits over the last 15 eV of the β spectrum, upper limits at the 2 σ level for m_ν^2 and m_ν of ≤ 22 eV^2/c^4 and ≤ 5 eV/c^2, respectively are obtained as seen from fig. 4.

4. Conclusion and outlook

To improve the sensitivity on m_ν down to a few eV/c^2 and to check the anomaly close to the endpoint, which was reported by the Troitsk group, we have improved our setup significantly in 1995–1997. By two runs in 1997 and 1998 we have demonstrated that the sensitivity of our data has become competitive to the one of the Troitsk experiment due to the improvement of our signal-to-background-ratio by a factor of 10. By using a much lower temperature of the T_2 source the roughening transition of the quench-condensed T_2 film was avoided, which formerly caused a trend towards negative values of m_ν^2 for increasing fit intervals. But still there is a small unresolved anomaly in our data. Whether it corresponds to the "Troitsk anomaly" or not can neither be proven nor disproven at the moment. More data taking is needed to extract the shape of our anomaly and to check special features of the "Troitsk anomaly" like the time dependence. During this conference already another 5 weeks data taking period has been started. More data taking is planned for the near future.

Because of these anomalies and moreover due to the importance to check directly neutrino masses in the cosmological relevant range down to 1 eV/c^2 the present experiments at Mainz and Troitsk should fully explore their sensitivity limit. But it seems that the decisive answer to these questions could come only from a new experiment with superior sensitivity. For this purpose the Mainz group has suggested to build a large version of the present spectrometer, up-scaled by a factor of 5 in outer dimensions. An additional time-of-flight method could turn such an integrating spectrometer into a narrow band filter to clearly check the "Troitsk anomaly"[16].

Acknowledgement

This work was supported by the Deutsche Forschungsgemeinschaft under contract number Ot33/13.

REFERENCES

1. T. Kajita, *these proceedings*, and hep-ex/9807003, *submitted to* Phys. Rev. Lett.
2. V.N. Gavrin, T. Kirsten, Y. Suzuki (*these proceedings*)
3. G. Barenboim and F. Scheck, *accepted for publication by* Phys. Lett. B, and hep-ph/9808327
4. H. Georgi and S.L. Glashow, hep-ph/9808293
5. D. Caldwell, *these proceedings*
6. A. Picard *et al.*, Nucl. Inst. Meth. **B63** (1992) 345
7. V.M. Lobashev *et al.*, Nucl. Instr. Meth. **A240** (1985) 305
8. Ch. Weinheimer *et al.*, Phys. Lett. **B300** (1993) 210
9. H. Backe *et al.*, Proc. XVII Conference on Neutrino Physics and Astrophysics, Neutrino 96, Helsinki/Finnland, June 1996, World Scientific/Singapure
10. C. Caso *et al.* (compilation of Particle Data Group), Eur. Phys. J. **C3** (1998) 1
11. P. Leiderer *et al.*, Jour. Low Temp. Phys **89** (1992) 219
12. A.I. Belesev *et al.*, Phys. Lett. **B350** (1995) 263
13. V.M. Lobasehv, Proc. XVII Conference on Neutrino Physics and Astrophysics, Neutrino 96, Helsinki/Finnland, June 1996, World Scientific/Singapure
14. V.M. Lobashev, *these proceedings*
15. H. Barth *et al.*, Prog. Part. Nucl. Phys. **40** (1998) 353
16. J. Bonn *et al.*, *submitted to* nucl. inst. and meth.

ELSEVIER

Nuclear Physics B (Proc. Suppl.) 77 (1999) 327–332

NUCLEAR PHYSICS B
PROCEEDINGS
SUPPLEMENTS

NEUTRINO MASS AND ANOMALY IN THE TRITIUM BETA-SPECTRUM. RESULTS OF THE "TROITSK ν-MASS" EXPERIMENT.

V. M. Lobashev, V.N.Aseev, A.I.Belesev, A.I.Berlev, E.V.Geraskin, A.A.Golubev, N.A.Golubev, O.V.Kazachenko, Yu.E.Kuznetsov, R.P.Ostroumov, L.A.Ryvkis, B.E.Stern, N.A.Titov, S.V.Zadorozhny, Yu.I.Zakharov

Institute for Nuclear Research; Academy of Sciences of Russia; 60-th October Anniversary Prospect 7a; 117312 Moscow, Russia

Abstract

Results of the "Troitsk ν-mass" experiment on the search for the neutrino rest mass in the tritium beta-decay are presented. Study of time dependence of anomalious, bump-like structure at the end of beta spectrum reported earlier gives indication of periodic shift of the position of the bump with respect to end-point energy with period of 0.5 year. New upper limit for electron antineutrino rest mass $m_\nu < 2.5eV/c^2$ is derived after accounting for the bump. Possible variants of more sensitive facility are discussed.

1. Introduction.

The direct or kinematical approach to the search for the neutrino rest mass is based on the study of neutrino momentum-energy balance in weak semileptonic decays. In this case any dependence on the leptonic or flavor quantum numbers is excluded. Straightforward variant is to prove such balance in two-body decay by measuring masses and momenta of charged particles as it is made in the $\pi \to \mu\nu$ decay. In practice this way is limited by precision of knowledge of the particles masses.

Maximal sensitivity to mass effect may be attained when neutrino energy is minimal. Such situation usually can be obtained in three-body or multibody decay. Total energy spectrum of visible particles in the vicinity of maximal energy is dominated by the neutrino phase space volume which is proportional to pE where p is momentum and E total energy of the neutrino. Deviation of this product from p^2 allows one to deduce the mass of neutrino. Smallness of this product defines fast decreasing of the measured spectrum intensity by approaching end point energy and makes main difficulty of the experiment. At the moment the most dramatical situation arised in experiments on the search for the electron antineutrino mass in tritium beta spectrum. Putting in operation of new spectrometric facilities in Troitsk (Moscow) [1] and in Mainz [2] allowed to observe details of

0920-5632/99/\$ – see front matter
PII S0920-5632(99)00438-7

beta-spectrum at about $5 - 15\,eV$ below the end point. Besides significant improvement of upper limit for the neutrino mass the experiment in Troitsk revealed existence of anomalious structure of bump-like shape (for differential spectrum mode) in the region of $5 - 15\,eV$ below end point with integral intensity about 10^{-10} of total decay rate. Very enigmatic feature of this structure appeared to be shift of its position with time.

This structure in the condition of absence of understanding of its nature plays role of systematics for the search of neutrino mass, strongly increasing possible error.

2. Search for electron antineutrino rest mass

The shape of the beta-spectrum with nonzero neutrino mass is

$$W(E,Z) = A\,F(E,Z)\;E\,p\,\Sigma W_i(E_{oi} - E) \sqrt{(E_{oi}-E)^2 - m_\nu^2 c^4} \qquad (1)$$

where E is the total energy and p is the momentum of electron; W_i is the probability, and E_{oi} is the end-point energy of the partial decay into i-th final state. The effect of nonzero neutrino mass emerges as a cut-off of the spectrum at $E_{oi} - E = m_\nu c^2$, and intensity deficiency smoothly declining to lower energy. At present lowest limit for electron neutrino mass was achieved by the study of the shape of tritium beta spectrum near its end point. The decay of tritium provides a unique opportunity for such experiments due to low end-point energy, high specific activity, the lowest Z, and possibility to calculate most of the corrections for its super allowed spectrum.

3. Integral electrostatic spectrometer with adiabatic magnetic collimation

The development of a new approach to spectroscopy of tritium started at the end of 1982 at the Institute for Nuclear Research of the Russian Academy of Sciences (Troitsk). The main ideas were published in [3, 4, 5]. Independently similar ideas emerged at the Institute for Physics of Mainz University [6]. The main feature of this approach is a new type of an integral electrostatic spectrometer with strong inhomogeneous magnetic field providing guiding and collimation of the electrons. Main advantage of such spectrometer is large improvement in energy resolution, amounting to $3,5 - 4\,eV$ (FW) and luminosity.

The strong guiding magnetic field in the spectrometer permitted to couple it in a natural way with the gaseous windowless tritium source also with strong magnetic field. This approach was developed in Troitsk set-up. An essential part of the spectrometer and of the tritium source is a set of superconducting solenoids which produce a continuous longitudinal magnetic field through the whole setup. The cylindrical electrode in the central part of the spectrometer is an integral electrostatic analyzer. The details of the set-up design and of the measurement procedure may be found in [1],[7], [8] and [9].

The tritium spectrum was measured by changing the spectrometer high voltage in steps. Direction of high voltage scanning was reversed each cycle ($1-2$ hours). Altogether, in the period of 1994-1998 years the time of measurement amounted to about 180 days. The measurements were made in the range of the spectrometer potential from 18000 to 18770 V. The differential theoretical spectrum with $m_\nu^2 \neq 0$ has been calculated according to Eq. 1.

As a basic set of variable parameters in χ^2 fit procedure we used 4 parameters : normalization factor, end point energy, background and m_ν^2. The final state spectrum of decay product (FSS) was taken from most resent theoretical calculations [10].

Corrections for inelastic interactions of electrons in tritium gas are also very important. These corrections as well as the FSS spectrum are strongly correlated with mass of neutrino and some other parameters of the spectrum. Special system with electron gun and magnetic transportation to the rear part of the source allowed to measure integral spectrum of inelastic losses of electrons in tritium as well as density of the source.

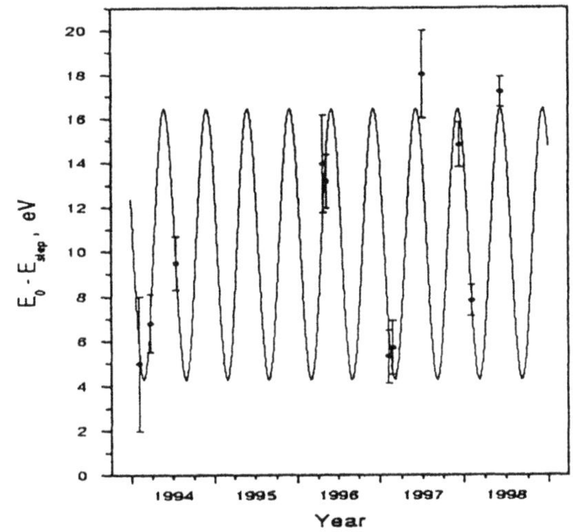

Figure 1. The step position dependence on the calendar time of measurements. Parameters of the fitted sinusoid are: Period $0,5036 \pm 0,0023$ year, mean value $10,3$ $\pm 0,4\,eV$, amplitude $6,1 \pm 0,55\,eV$, phase $2,8 \pm 0,16\,rad$.

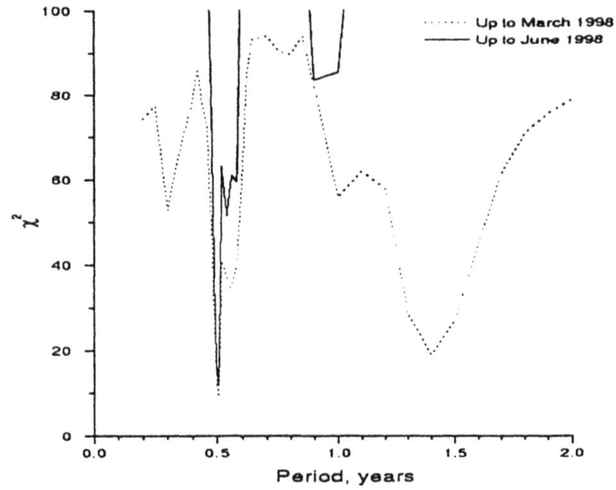

Figure 2. The χ^2 dependence on the period of sinusoid fitted to step position plot versus calendar time of measurement. Period value was scanned and 3 other parameters were left variable. Solid line corresponds to all the run fit and dotted line with the last run (June 1998) being omitted.

4. Anomalious structures in the spectrum

Fitting of the first data of 1994 run with 4 basic variable parameters resulted in the value for m_ν^2 equal to $-22 \pm 5\,eV^2$ for the truncated spectrum with the truncation energy (further refered as E_{low}) more than $18300\,eV$. At lower truncation energy (down to $18000\,eV$) the m_ν^2 value increased to $-58\,eV^2$ and was accompanied by a strong increase in χ^2. The negative values for m_ν^2 obviously indicated that there exist some systematic effects not taken into account in the calculation of the theoretical spectrum [1].

Low-energy increase of the negative value of m_ν^2 proved to be the result of a rise in the counting rate at lower energies. It was found that this effect was due to underestimation of the trapping of electrons in the tritium source. After correction for this effect the low energy anomaly in the spectrum disappeared within error. In order to take into account the step-like structure the fit was made with the theoretical spectrum summed up with a spectrum in the form of a θ-function, which depended on two variable parameters: ΔN_{step} and E_{step}. It was taken that $\Delta N_{step} \neq 0$ for $E \leq E_{step}$ and $\Delta N_{step} \equiv 0$ for $E > E_{step}$. The fit with these additional variables resulted in values for ΔN_{step} in average $6 \cdot 10^{-11}$ of total decay intensity and $E_0 - E_{step}$ varying within $5 - 16\,eV$ In all the runs fit with step function made m_ν^2 value about zero within fit errors.

Positions of the step with respect to end point energy for all the runs and subruns versus calendar time of measurement are plotted in Fig. 1. Most surprising feature of this plot revealed when a sinusoidal curve was fitted to all the points.

The period of oscillation of step position proved to be equal to $0,504 \pm 0,003$ years, mean value of the position $10,3\,eV$ and amplitude $6,1\,eV$. Size of the steps also undergo synchronous variation so that the maximal size correspond to maximal shift. Unfortunately the relative error of this parameter is significally larger than the position value error. Dependence of χ^2 on the value of the period is shown in Fig. 2. It demonstrates that half a year period gives unique description of the data.

Combining data of all the 4 years in one year plot confirm that the variation have biseasonal character with maximal shift of the step position in beginning of June and December (see Fig. 3). It is worth while to point out that examining of dependence of the m_ν^2 fitted with only basic set of variables versus E_{low} reveals correlation of the m_ν^2 value with $E_0 - E_{step}$. Unlike homogeneous

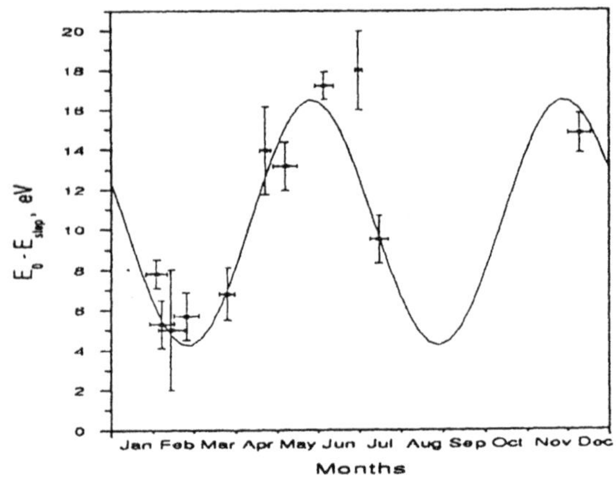

Figure 3. The plot of step positions versus time of the year. Fitted sinusoid is the same as in Fig. 1, but with the period being 0.500 year. Horizontal bars are lengh of the run.

negative $m_\nu^2 \approx -(10-15)eV^2$ for 18000 $eV < E_{low} <$ 18450 eV, the m_ν^2 for $E_{low} >$ 18500 eV in different runs changes the sign to positive ($5 - 10\ eV^2$) at the maximal $E_0 - E_{step}$ and to negative ($20 - 30\ eV^2$) for minimal $E_0 - E_{step}$. Such correlation is naturally explained by enhancement (bump) at the end of the spectrum and supports previous observation. Of course present set of data needs to be sufficiently extended but main features of this phenomenon make excusable considering of exotic explanation for them. One of such explanations stems from long-standing discussion of the effect produced by capture of relic very low energy neutrino by tritium atoms with emission of monochromatic electrons [11]. For free neutrino the energy of such electron must exceed or be the same as end-point energy. In order to produce the bump intensity, corresponding to 10^{-10} of total decay rate it is necessary to suppose existence of neutrino cloud with density as high as $0,5 \cdot 10^{15}\nu/cm^3$, that is 10^{13} times more than generally accepted average density of relic neutrino. Such density corresponds to degenerated neutrino spectrum with Fermi-energy of about $5\,eV$.

In order to observe bump below end point of beta spectrum, that corresponds to capture of neutrino with negative energy, it is naturally to assume binding of neutrino in the cloud. The binding energy from the above data may be estimated as $15 - 20\,eV$ and could vary over the cloud. Binding of neutrino seems to be necessary for stability of such cloud provided for example by the neutrino long range (about the size of cloud) self interaction. In order to explain the half year modulation period one may suppose that the neutrino cloud has shape of flattered disc with axis of symmetry inclined with respect to normal direction to Ecliptic plane. In case of neutrino density gradually decreasing to the periphery of the cloud, the Earth in its movement twice a year crosses most dense and less dense area of the cloud. It is interesting to point out that time of the maximal shift in Fig. 3 corresponds to position of the Earth on the orbit when axis of the Sun rotation is perpendicular to Earth-Sun direction.

The size of neutrino cloud is of the order $10^{14}\,cm$ and it does not contradict to average density of relic neutrino in the Universe. The possibility of existence of such cloud from the point of view of contradiction to astrophysical and elementary particle data was ones considered in [12] with conclusion that existing data do not contradict to such picture if to abstract oneself from the problem of trapping of relativistic neutrino in potential well. Of course this scenario is extremely speculative and need totally convincing experimental confirmation as well as certain theoretical background.

Experimental data up to now does not exclude, that shape of the end-point region is more complicated than one-bump structure. Nevertheless it appears to be well established that centrum of gravity of the enhancement is below end-point of the tritium beta-spectrum, and it undergoes periodical shift with respect to end-point.

The bump-like structure in the tritium beta spectrum with the $B.R. \sim 10^{-10}$ could not have been seen by other experimental groups due to insufficient statistical accuracy and energy resolution. A bump like structure with relative intensity $3,0 \pm 0,6 \cdot 10^{-9}$ which was reported in [13] obviously cannot be identified as the same effect owing to much larger value.

5. Neutrino mass upper limit

Deduction of the neutrino mass from the data in presence of unexplained anomaly requires a special approach. As it was mentioned earlier the procedure accepted for this purpose consisted in addition to theoretical spectrum of the step function with two variable parameters supposing that such addition may describe in the first approximation local enhancement in the beta-spectrum near to end-point. Distortion of beta-spectrum imitating the m_ν^2 effect should also be concentrated near end point, otherwise the effect relatively rapidly sinks in growing statistical errors at increasing $E_o - E$, but unlike the local enhancement it appears as an addition to (for negative m_ν^2) or deficiency (positive m_ν^2) of the spectrum that is linearly increasing with $E - E_0$.This difference allows to separate both effects in fit procedure. Experimental spectra near the end point indicate that accounting for the step-like bump distortion is necessary for adequate description of the experiment data. Of course the size and position of the step being introduced as a free parameter, correlates with m_ν^2 and increases the final error of neutrino mass thus acting as a kind of systematic error. This increase sufficiently compensates the uncertainty of substitution of an a priory unknown anomaly shape by the step-like function. Other systematical errors come mostly from the uncertainties of parameters of the correction factors which are introduced in the spectrum before the fit. These factors are: trapping effect, source density, possible variation of excitation and ionization parts of the inelastic cross section, dead time, and influence of high exited FSS part. A remarkable property of total systematic error is its decreasing with increasing of E_{low}, Taking into account that fit error of m_ν^2 increases with increasing of E_{low} one may select the optimal E_{low}, when the total error, including both the fit and systematic error taken in quadrature, is minimal. The results for m_ν^2 for all the runs are given below:

$$1994 \ m_\nu^2 = -2,7 \pm 10,1_{fit} \pm 4,9_{syst}\, eV^2/c^4 \quad (2)$$

$$1996 \ m_\nu^2 = +0,5 \pm 7,1_{fit} \pm 2,5_{syst}\, eV^2/c^4 \quad (3)$$

$$1997(1) \ m_\nu^2 = -8,6 \pm 7,6_{fit} \pm 2,5_{syst}\, eV^2/c^4 \quad (4)$$

$$1997(2) \ m_\nu^2 = -3,2 \pm 4,8_{fit} \pm 1,5_{syst}\, eV^2/c^4 \quad (5)$$

$$1998 \ m_\nu^2 = -0,6 \pm 8,1_{fit} \pm 2.0_{syst}\, eV^2/c^4 \quad (6)$$

The combined value in quadrature:

$$m_\nu^2 = -2.0 \pm 3,4_{fit} \pm 2,3_{syst}\, eV^2/c^4 \quad (7)$$

Combined systematic error is obtained by averaging with weights of fit errors. From here one may obtain the 95% C.L. Bayesian upper limit for m_ν:

$$m_\nu < 2,5\, eV/c^2; \quad (8)$$

6. Further study of the effects in tritium beta-spectrum.

Further investigation of the bump-like anomaly and the neutrino mass search at the level about $1\, eV/c^2$ require major improvement of tritium beta-spectrometry. One of the obvious way is enlargement of existing set-up by a few times. The other way could be the development of a differential spectrometer with the resolution and luminosity on a par with the integral one. Differential spectrometer allows better study of local anomalies in continuous spectra and will serve both for search of the kinks from heavy neutrino and above mentioned tasks. In the report [9] a new type of differential spectrometer designed on the principles of adiabatic motion of electrons in electric and magnetic fields was proposed. The spectrometer consists of the integral electrostatic spectrometer with adiabatic magnetic collimation with the central part which is lengthed and bent on 180 or 360 degree. The input part of the spectrometer cuts electrons with the energy below the potential of the central electrode with relative spread less than $E_0 \cdot H_{min}/H_0$. At the central region of the spectrometer the electrons fly in the weak magnetic field with their momenta being lined up along the magnetic lines and the energy being $E_{in} - e \cdot V_0$, where E_{in} is the initial electron energy and V_0 is the potential of the analyzing electrode. Magnetic field in the central part has toroid like shape. Electrons moving adiabatically inside the toroidal electrode are in zero electrostatic field and drift perpendicularly to the toroidal plane owing to well known transverse drift. The magnitude of the drift with respect to the magnetic force lines depends on the velocity of the electron and rapidly increases with the electron energy. Although the drift is not big, mounting inside the toroidal electrode a multislot collimator with the slots parallel to the toroid plane allows to cut electrons with the drift more than the width of a slot. One can expect some increase of the background due to bombardment of the collimator ends by ions which are accelerated in the detector part of the spectrometer. To avoid this a slot mask will be mounted on the detector so that the adiabatic images of the end plates of the collimator are projected onto the covered regions of the detector. Thus only electrons with the energy not more than several eV inside the toroidal electrode can reach the detector. Electrons with higher or with very low energy will die on the collimator plates.

Luminosity of such spectrometer will depend on the cross section of the central electrode and on the dimension of the tubes in the tritium source. The optimal parameters of the spectrometer should be studied in details but it seems quite possible to design a device with resolution about $2\,eV$ and luminosity $1\,cm^2$. This could give substantial improvement of all the tritium spectrometry.

7. Acknowledgments

This work was partially supported by the Russian Foundation for Basic Research (grants 3903 and 18633a), by Program for Fundamental Nuclear Physics and INTAS-RFBR grant 95-819. One of the author (V.M.L.)is very thankful to Alexander-von-Humboldt Foundation for a grant for Scientific Research.

References

[1] A. I. Belesev et al., Phys. Lett., B 350(1995) 263.
[2] Ch. Weinheimer et. al., Phys. Lett., B 300(1993) 210.
[3] V. M. Lobashev and P. E. Spivak, Preprint INR P-0291, Moscow (1983)
[4] V. M. Lobashev and P. E. Spivak, Nucl. Instr. Methods A 240 (1985) 305.
[5] V. M. Lobashev et al., Nucl. Instr. Methods A 238(1985) 496.
[6] A. Picard et al., Nucl. Instr. Meth., B63(1992)345.
[7] V. M. Lobashev et al., Proceedings of the International Conference NEUTRINO-96, Helsinki, Finland; June 13-19, 1996, World Scientific p.264-277.
[8] V. M. Lobashev et al., Proceedings of WIN-97, Capri, June 22-28, 1997.
[9] V. M. Lobashev, Progress in Particle and Nuclear Physics 40, 337-351,(1998).
[10] S. Jonsell and H. Monkhorst, Phys. Rev. Lett. 76 (1996) 4476.
[11] T. Goldman and G. J. Stephenson Jr., hep-ph/9309308.
[12] R. N. Mohapatra and S. Nussinov, Phys.Lett.B 395 (1997) 63-68.
[13] W. Stoeffl and D. J. Decman, Phys. Rev. Lett. 75 (1995) 3237.

Part 10

Double Beta Decay

NUCLEAR PHYSICS B
PROCEEDINGS
SUPPLEMENTS

ELSEVIER

Nuclear Physics B (Proc. Suppl.) 77 (1999) 335–345

Review on Double Beta Decay Experiments and Comparison with Theory

Angel Morales[a]

[a]Laboratory of Nuclear Physics and High Energy Physics. Faculty of Science, University of Zaragoza, Pedro Cerbuna 12, 50009 Zaragoza, Spain

1. INTRODUCTION & MOTIVATION

In the Standard Model of Particle Physics neutrinos are strictly massless, although there is no theoretical reason for such prejudice. On the experimental side, there is not yet conclusive evidence that the neutrino has a non-zero mass, although the results of several experiments (widely reported to this Conference) with solar, atmospheric and terrestrial neutrinos lead to inconsistencies in the standard theory, unless it is assumed that neutrinos have indeed masses. Moreover, galaxy formation requires hot (as well as cold) non-baryonic dark matter to match properly the observed spectral power at all scales of the universe. A light neutrino of a few eV could make the hot dark matter, and help to solve the neutrino oscillation problem.

In the Standard Model, neutrinos and antineutrinos are supposed to be different particles, but no experimental proof has been provided so far. The nuclear double beta decay addresses both questions: whether the neutrino is self-conjugated and whether it has a Majorana mass. In fact, the lepton number violating neutrinoless double beta decay (A, Z) $\rightarrow$ (A, Z+2) + $2e^-$ is the most direct way to determine if neutrinos are Majorana particles. For this non-standard $2\beta0\nu$ process to happen, the emitted neutrino in the first neutron decay must be equal to its antineutrino and match the helicity of the neutrino absorbed by the second neutron. Phenomenologically that implies the presence of a mass term or a right-handed coupling. A well-known argument of Schechter and Valle [1] shows that in the context of any gauge theory, whatever mechanism be responsible for the neutrinoless decay, a Majorana neutrino mass is required. Moreover [23], the observation of a $2\beta0\nu$ decay implies a lower bound for the neutrino mass, i.e. at least one neutrino eigenstate has a non-zero mass.

Another form of neutrinoless decay, (A, Z) $\rightarrow$ (A, Z+2) + $2e^-$ + χ may reveal also the existence of the Majoron (χ), the Goldstone boson emerging from the spontaneous symmetry breaking of B–L, of most relevance in the generation of Majorana neutrino masses and of far-reaching implications in Astrophysics and Cosmology.

These and other issues, like the verification of SUSY models, compositeness, leptoquarks, etc. make the search for the neutrinoless double beta decay an invaluable exploration of non-standard model physics, probing mass scales well above those reached with accelerators. In this overview we will refer basically to the question of the neutrino mass in connection with the current results of the double beta decay searches.

The two-neutrino decay mode (A, Z) $\rightarrow$ (A, Z+2) + $2e^-$ + $2\overline{\nu}_e$ is a conventional [2], although rare, second order weak process ($2\beta2\nu$), allowed within the Standard Model. The half-lives are customary expressed as $[T^{2\nu}_{1/2}\ (0^+ \rightarrow 0^+)^{-1} = G_{2\nu} \mid M^{2\nu}_{GT} \mid^2$, where $G_{2\nu}$ is an integrated kinematical factor [3] and $M^{2\nu}_{GT}$ the nuclear double Gamow Teller matrix element.

The neutrinoless decay half-life (as far as the mass term contribution is concerned) is expressed as $(T^{0\nu}_{1/2})^{-1} = F_N < m_\nu >^2 / m_e^2$, where $F_N \equiv G_{0\nu} \mid M^{0\nu} \mid^2$ is a nuclear factor-of-merit and $M^{0\nu}$ is the neutrinoless nuclear matrix-element, $M^{0\nu} = M^{0\nu}_{GT} - (g_V/g_A)^2\ M^{0\nu}_F$, with $M^{0\nu}_{GT,F}$ the corresponding Gamow-Teller and Fermi contributions. $G_{0\nu}$ is an integrated kinematic factor [3]. The quantity $< m_\nu >= \Sigma\lambda_j m_j U^2_{ej}$ is the so-

0920-5632/99/$ – see front matter
PII S0920-5632(99)00440-5

called effective neutrino mass parameter, where U_{ej} is a unitary matrix describing the mixing of neutrino mass eigenstates to electron neutrinos, λ_j a CP phase factor, and m_j the neutrino mass eigenvalue.

As far as the neutrinoless decay with the emission of Goldstone bosons is concerned, various Majoron models have been invented to circumvent the Z^0 width constrain on the number of neutrino species—which ruled out the original Majoron models—and to allow, at an observable rate, double beta neutrinoless decays with Majoron (or other massless or light bosons) emission ($2\beta 0\nu\chi$). The ($0\nu\chi$) half-life is expressed as $T^{-1}_{0\nu\chi} =< g >| M^{0\nu} |^2 G_{0\nu\chi}$, where $M^{0\nu}$ is the same matrix element as in the $2\beta 0\nu$ and g the Majoron coupling to neutrinos ($g\chi\overline{\nu}_e\gamma_5\nu_e$).

Concerning the neutrino mass question, the discovery of a $2\beta 0\nu$ decay will tell that the Majorana neutrino has a mass equal or larger than $< m_\nu >= m_e/(F_N T^{0\nu}_{1/2})^{1/2}$ eV, where $T^{0\nu}_{1/2}$ is the neutrinoless half life. On the contrary, when only a lower limit of the half-life is obtained (as it is the case up to now), one gets only an upper bound on $< m_\nu >$, but not an upper bound on the masses of any neutrino. In fact, $< m_\nu >_{exp}$ can be much smaller than the actual neutrino masses. The $< m_\nu >$ bounds depend on the nuclear model used to compute the $2\beta 0\nu$ matrix element. The $2\beta 2\nu$ decay half-lives measured till now constitute bench-tests to verify the reliability of the nuclear matrix element calculations which, obviously, are of paramount importance to derive the Majorana neutrino mass upper limit.

2. STRATEGIES FOR DOUBLE BETA DECAY SEARCHES

The experimental signatures of the nuclear double beta decays are in principle very clear: In the case of the neutrinoless decay, one should expect a peak (at the $Q_{2\beta}$ value) in the two-electron summed energy spectrum, whereas two continuous spectra (each one of well-defined shape) will feature the two-neutrino and the Majoron-neutrinoless decay modes (the first having a maximum at about one third of the Q value, and the latter shifted towards higher energies). In spite of such characteristic imprints, the rarity of the processes under consideration make very difficult their identification. In fact, double beta decays are very rare phenomena, with two-neutrino half-lives as large as 10^{19} y to 10^{24} y and with neutrinoless half-lives as long as 10^{25} y (and above), as the best lower limit stands by now. Such remotely probable signals have to be disentangled from a (much bigger) background due to natural radioactive decay chains, cosmogenic-induced activity, and man-made radioactivity, which deposit energy on the same region where the 2β decays do it but at a faster rate. Consequently, the main task in 2β-decay searches is to diminish the background as much as possible by going underground and using state-of-the-art ultralow background techniques to supress it or to identify it and subtract it. All the experiments follow this general strategy because the experimental sensitivity in 2β decay searches is limited by the level of background achieved.

To measure 2β decays, three general approaches have been followed: The geochemical experiments, where isotopic anomalies in noble gases daughter of 2β decaying nucleus over geological time scales are looked for. Some examples are the decays of ^{82}Se, ^{96}Zr, ^{128}Te, ^{130}Te. Another method is that of the radiochemical experiments, which are based on the fact that when the daughter nuclei of a double beta emitter are themselves radioactive, they can be accumulated, extracted and counted. Examples are ^{238}U, ^{244}Pu.

Most of the recent activity, however, refers to direct counting experiments, which measure the energy of the 2β emitted electrons and so the spectral shapes of the 2ν, 0ν, and $0\nu\chi$ modes of double beta decay. Some experimental devices track also the electrons (and other charged particles), measuring the energy, angular distribution, and topology of events. The tracking capabilities are essential to discriminate the 2β signal from the background. The types of detectors currently used are:

- Calorimeters where the detector is also the 2β source (Ge diodes, scintillators—CaF_2, $CdWO_4$—,thermal detectors, ionization chambers). They are calorimeters which measure the two-electron sum energy and discriminate partially signal from background by pulse shape analysis (PSD).

- Tracking detectors of source≠detector type (Time Projection Chambers TPC, drift chambers, electronic detectors). In this case, the 2β source plane(s) is placed within the detector tracking volume, defining two—or more—detector sectors.

- Tracking calorimeters: They are tracking devices where the tracking volume is also the 2β source. Only one of this type of device is operating (a Xenon TPC), but there are others in project.

Well-known examples of 2β emitters measured in direct counting experiments are ^{48}Ca, ^{76}Ge, ^{96}Zr, ^{82}Se, ^{100}Mo, ^{116}Cd, ^{130}Te, ^{136}Xe, ^{150}Nd.

The strategies followed in the 2β searches are varied. Calorimeters of good energy resolution and almost 100% efficiency (Ge-detectors, Bolometers) are well suited for 0ν searches. However, they lack the tracking capabilities to identify the background on an event-by-event basis. Pulse Shape Discrimination (PSD) will help. Simultaneous measure of heat and ionization would do it. The Monte Carlo (MC) modeling of the background spectrum to be subtracted from the data is approximate. So, one should first reduce the radioimpurities as much as possible and then trace back and MC-model the remaining contaminations and subtract them. On the contrary, the identification capabilities of the various types of chambers make them very well suited for 2ν and $0\nu\chi$ searches. However, their energy resolution is rather modest and the efficiency is only of a few percent. Furthermore, the ultimate major background source in these devices when looking for $2\beta0\nu$ decay will be that due to the standard $2\beta2\nu$ decay. The rejection of background provided by the tracking compensates, however, the figure of merit in 0ν searches. Modular calorimeters can have large amounts of 2β emitters (Heidelberg/Moscow, IGEX, CUORE and GENIUS project). However, current operating chambers—except the Xe/TPC—cannot accommodate large amounts of 2β emitters in the source plate. Future tracking devices will have 10 kg and more (NEMO3, MUNU).

As a general rule, the detector must optimize the so-called detector factor-of-merit or neutrinoless sensitivity (introduced by the pioneer work of E. Fiorini), which for source=detector devices reads $F_D = 4.17\times10^{26}(f/A)(Mt/B\Gamma)^{1/2}\varepsilon_\Gamma$ years where B is the background rate (c/keV kg y), M the mass of 2β emitter (kg), ε_Γ the detector efficiency in the energy bin Γ around $Q_{2\beta}$ (Γ = FWHM) and t the time measurement in years (f is the isotopic abundance and A the mass number). The other guideline of the experimental strategy is to choose a 2β emitter of large nuclear factor of merit $F_N = G_{0\nu} \mid M^{0\nu} \mid^2$, where the kinematical factor qualifies the goodness of the $Q_{2\beta}$ value and $M^{0\nu}$ the likeliness of the transition. Notice that the upper limit on $< m_\nu >$ is given by $< m_\nu >< m_e/(F_D F_N)^{1/2}$.

3. OVERVIEW OF EXPERIMENTAL SEARCHES

In the following we will overview some of the direct counting experiments, reporting only on 2β transitions to the ground state. A considerable activity has been done recently on transitions to excited states but we will omit them for lack of space.

There exist two experiments in operation looking for the double beta decay of ^{76}Ge. They both employ large amounts of enriched ^{76}Ge in sets of detectors. The Heidelberg/Moscow Collaboration experiment (a set of five large Ge detectors amounting to 10.2 kg) running in Gran Sasso [4] (exposed by H.V. Klapdor-Kleingrothaus in these Proceedings), and the IGEX Collaboration in Canfranc (Spain), which is described below.

The International Germanium Experiment (IGEX) has three large enriched (up to 86%) detectors ($\sim$ 2 kg) and three smaller ones ($\sim$ 1 kg). The FWHM energy resolutions of the large detectors at 1333-keV are 2.16, 2.37, and 2.13 keV, and the energy resolution of the summed data is 4 keV (at the $Q_{2\beta}$ value of 2038 keV). They feature a unique electroformed copper technology in the cryostat and use ultralow background mate-

Table 1
Theoretical half-lives $T^{2\nu}_{1/2}$ in some representative nuclear models versus direct experiments.

	Theory								*Experiment*
	SM			QRPA		$1^{+}D$	OEM	MCM	
	[2]	[12]	[13]	[14]	[16]	[19]	[18]	[17]	
^{48}Ca(10^{19}y)	2.9	7.2	3.7						$4.3^{+2.4}_{-1.1} \pm 1.4$ UCI
^{76}Ge(10^{21}y)	0.42	1.16	2.2	1.3	3.0		0.28	1.9	$1.77^{+0.14}_{-0.12}$ H/M
									1.45 ± 0.15 IGEX
^{82}Se(10^{20}y)	0.26	0.84	0.5	1.2	1.1	2.0	0.88	1.1	$1.08^{+0.26}_{-0.06}$ UCI
									$0.83 \pm 0.09 \pm 0.06$ NEMO
^{96}Zr(10^{19}y)				0.85	1.1			.14-.96	$2.1^{+0.8}_{-0.4} \pm 0.2$ NEMO
^{100}Mo(10^{19}y)				0.6	0.11	1.05	3.4	0.72	$1.15^{+0.30}_{-0.20}$ Osaka
									$1.16^{+0.34}_{-0.08}$ UCI
									$0.95 \pm 0.04 \pm 0.09$ NEMO
^{116}Cd(10^{19}y)					6.3	0.52		0.76	$2.6^{+0.9}_{-0.5}$ Osaka
									$2.7^{+0.5}_{-0.4}{}^{+0.9}_{-0.6}$ Kiev
									$3.75 \pm 0.35 \pm 0.21$ NEMO
^{136}Xe(10^{21}y)			2.0	0.85	4.6				≥ 0.55 Gothard
^{150}Nd(10^{19}y)					0.74				$1.88^{+0.66}_{-0.39} \pm 0.19$ ITEP
									$0.675^{+0.037}_{-0.042} \pm 0.068$ UCI

rials. The first stage FET (mounted on a Teflon block a few centimetres apart from the centre contact of the crystal) is shielded by 2.6 cm of 500 y old lead to reduce the background. Also the protective cover of the FET and the glass shell of the feedback resistor were removed for such purpose. Further stages of amplification are located 70 cm away from the crystal. All the detectors have preamplifiers modified for pulse shape analysis (PSD) for background identification.

The Canfranc IGEX setup consists in an innermost shield of 2.5 tons ($\sim$ 60 cm cube) of archaeological lead (2000 yr old)—having a ^{210}Pb(^{210}Bi) content of < 0.01 Bq/kg—, where the 3 large detectors are fitted into precision-machined holes to minimize the empty space around the detectors available to radon. Nitrogen gas evaporated from liquid nitrogen, is forced into the remaining free space to minimize radon intrusion. Surrounding the archaeological lead block there is a 20-cm thick layer of low activity lead ($\sim$ 10 tons), sealed with plastic and cadmium sheets. A cosmic muon veto and a neutron shield close the assembly.

The background recorded in the energy region between 2.0 and 2.5 MeV is about 0.2 c/keV kg y prior to PSD. Background reduction through pulse shape discrimination is in progress to eliminate multisite events, characteristic of non-2β events. This technique is currently capable of rejecting about one third of the background events, so the current IGEX background is ≤ 0.07 c/keV kg y. Further preamplifier development and pulse shape simulations are expected to improve the background rejection efficiency, pursuing the goal of probing Majorana neutrino masses corresponding to half-lives of 10^{26} years. The current results of IGEX, both for the $2\beta 2\nu$ and $2\beta 0\nu$ decay modes, are given in Tables 1 and 2. The two-electron summed energy spectrum around $Q_{2\beta} =$ 2038 keV region is shown in Figure 1 for an exposure of 92.68 mole years. Data from one of the large detectors—which went underground in Canfranc more than three years ago—corresponding to 291 days, were used to set a value for the 2ν-decay mode half-life by simply subtracting MC-simulated background. Figure 2a shows the best fit to the stripped data corresponding to a half-life $T^{2\nu}_{1/2} = (1.45 \pm 0.20) \times 10^{21}$ y, whereas Figure 2b shows how the experimental points fit the double beta Kurie plot.

Table 2
Limits on Neutrinoless Decay Modes

Emitter	*Experiment*	$T^{0\nu}_{1/2}$	*C.L.*
^{48}Ca	HEP Beijing	$> 1.1 \times 10^{22}$ y	68%
^{76}Ge	MPIH/KIAE	$> 1.2 \times 10^{25}$ y	90%
	IGEX	$> 0.8 \times 10^{25}$ y	90%
^{82}Se	UCI	$> 2.7 \times 10^{22}$ y	68%
	NEMO 2	$> 9.5 \times 10^{21}$ y	90%
^{96}Zr	NEMO 2	$> 1.3 \times 10^{21}$ y	90%
^{100}Mo	LBL/MHC/ UNM	$> 2.2 \times 10^{22}$ y	68%
	UCI	$> 2.6 \times 10^{21}$ y	90%
	Osaka	$> 2.8 \times 10^{22}$ y	90%
	NEMO 2	$> 6.4 \times 10^{21}$ y	90%
^{116}Cd	Kiev	$> 3.2 \times 10^{22}$ y	90%
	Osaka	$> 2.9 \times 10^{21}$ y	90%
	NEMO 2	$> 5 \times 10^{21}$ y	90%
^{130}Te	Milano	$> 7.7 \times 10^{22}$ y	90%
^{136}Xe	Caltech/UN/ PSI	$> 4.4 \times 10^{23}$ y	90%
^{150}Nd	UCI	$> 1.2 \times 10^{21}$ y	90%

The Time Projection Chamber TPC of the UC Irvine group is a rectangular box filled with helium and located underground at 290 m.w.e. (Hoover Dam). A central 2β source plane divides the volume into two halves. A magnetic field of 1200 Gauss is placed perpendicular to the source plane. Electrons emitted from the source follow helical trajectories from where the momentum and the angles of the β-particles are determined. The 2β signal is recognized as two electron emitted from a common point in the source with no other associated activity during some time before and after the event. The 2β source is thin enough (few mg/cm^2) to allow α-particles to escape and be detected for tagging the background. The UCI TPC has measured the two-neutrino double beta decay of ^{82}Se, ^{100}Mo, ^{150}Nd and ^{48}Ca (this last case in a collaboration with Caltech and the Kurchatov Institute), with efficiencies of about $\sim 11\%$ and energy resolution of $\sim 10\%$ at the Q value. Figures 3.1, 3.2 and 3.3 show respectively [5] the UCI $2\beta 2\nu$ decay spectra of ^{100}Mo, ^{150}Nd and ^{48}Ca, depicting in each case the measured spectra and their background components as well as the corresponding 2β-decay best fits. Results are quoted in Tables 1 and 2.

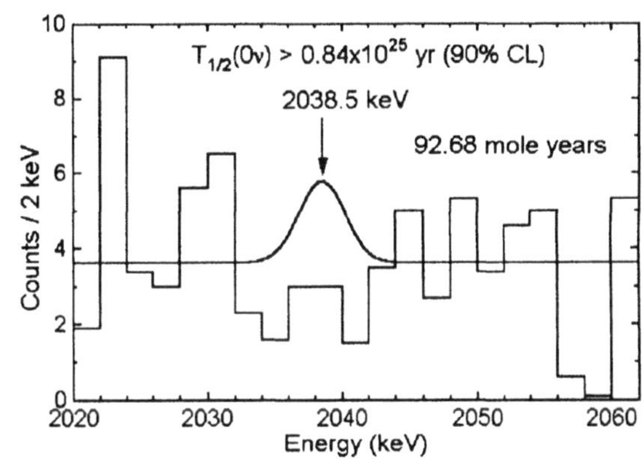

Figure 1.

The NEMO 2 apparatus [6] is an electron tracking detector (with open Geiger cells) filled with helium gas. An external calorimeter (plastic scintillator) covers the tracking volume and measures the β energies and time of flight. The 2β source is placed in a central vertical plane and is divided in two halves, one enriched and another of natural abundance (of about 150 grams each), to monitor and subtract the background. To identify a 2β signal, one should have a 2e-track with a

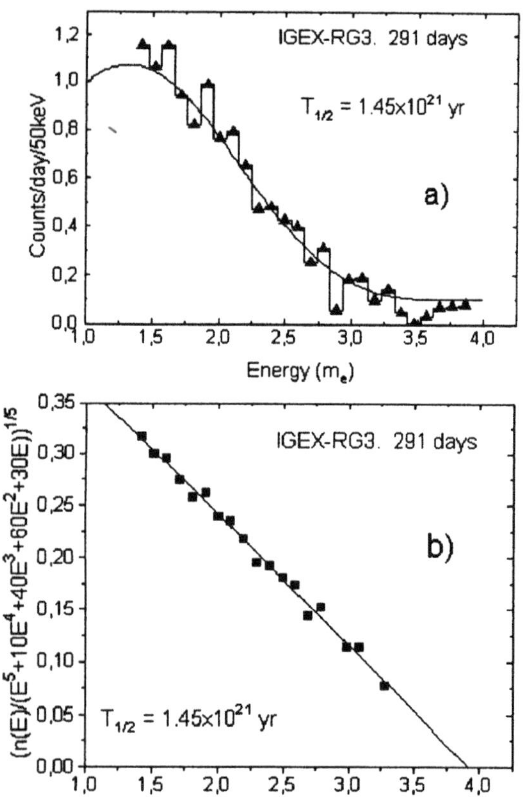

Figure 2.

common vertex ($cos\alpha < 0.6$) in the source plus two fired plastic scintillators (E deposition>200 keV each). The two-electron events are selected by time of flight analysis (in the energy range of 2β). NEMO 2 has been operating for several years at the Modane Underground Laboratory (Frejus Tunnel) at 4800 m.w.e and has measured the $2\beta 2\nu$ decays of ^{100}Mo, ^{116}Cd, ^{82}Se and ^{96}Zr (see Figures 4a,b,c,d) with an efficiency of about $\varepsilon_{2\nu} \sim 2\%$ and an energy resolution Γ (1 MeV)= 18% (for results refer to Table 1 and Table 2). A new, bigger detector of the NEMO series, NEMO 3, is ready to start running next year, with 10 kg of ^{100}Mo.

The ELEGANTS V detector of the University of Osaka (placed successively in Kamioka and Otho) is an electron tracking detector which consists of two drift chambers for β-trajectories, sixteen modules of plastic scintillators for β energies and timing measurement, and twenty modules of NaI for X- and γ-rays identification. The 2β signals should appear as two tracks in the drift

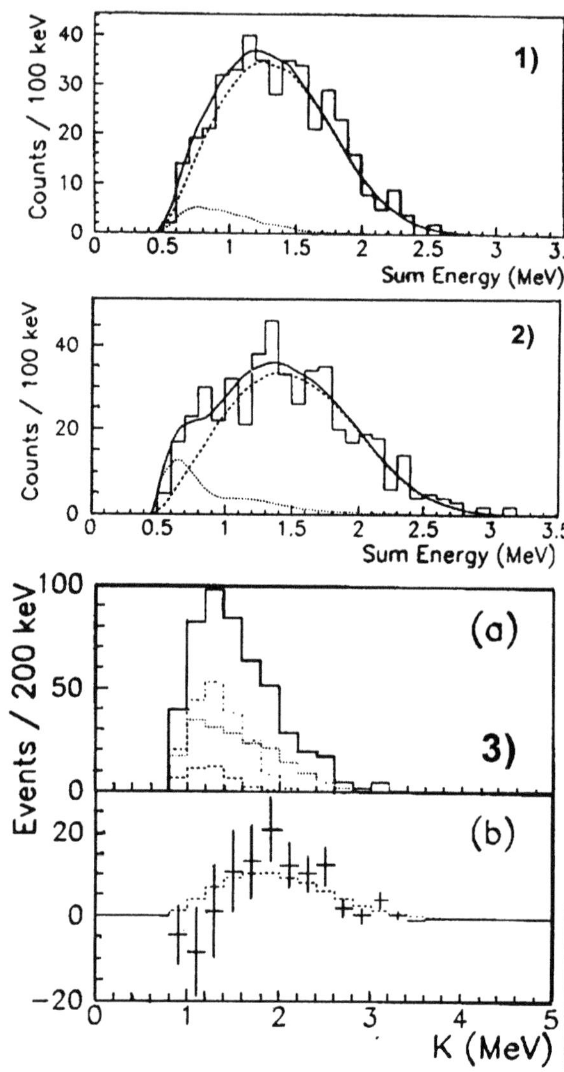

Figure 3.

chamber with the vertex in the source plus two signals from two plastic scintillators segments. Both enriched and natural sources (of about 100 grams) are employed in the detector for background monitoring and subtraction. This detector has measured [7] the $2\beta 2\nu$ decay of ^{116}Cd, ^{100}Mo (see Figure 5a,b) with efficiencies of $\varepsilon_{2\nu} \sim 7\% \sim 10\%$ and $\varepsilon_{0\nu} \sim 20\%$, and energy resolution of 150 keV at 1 MeV (the results of ELEGANTS V are quoted in Tables 1 and 2). A new variant of ELEGANTS is searching for the double beta decay of ^{48}Ca.

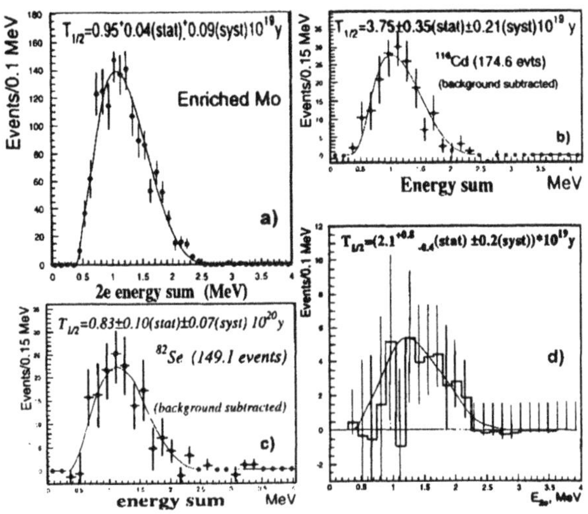

Figure 4.

The Caltech/PSI/Neuchatel Collaboration [8] investigates the double beta decay of ^{136}Xe in the Gothard Tunnel (3000 m.w.e.) by using a time projection chamber where the Xenon is at the same time the source and the detector medium, i.e. a calorimeter plus a tracking device. It has a cylindrical drift volume of 180 fiducial litres at a pressure of 5 atm. The Xenon is enriched up to 62.5% in ^{136}Xe, with a total mass of m=3.3 kg. The energy resolution is 6.6% at 2.48 MeV and the $2\beta 0\nu$ efficiency $\varepsilon_{0\nu} \sim 30\%$. The 2β signal appears as a continuous trajectory with distinctive end features: a large angle multiple scattering and increase charge deposition (charge "blobs") at both ends. As usual, the 2β topology gives powerful background rejection, leading to a figure of B $\sim 10^{-1} - 10^{-2}$ c/keV kg y (at 2480 keV). In the neutrinoless decay mode search, the experimental set up has already reached its limit (Table 2 and Figure 6). In the two-neutrino decay mode, the comparison of the single-electron and two-electron background spectra before and after a recent upgrading [8] of the readout plane (a factor 4 reduction in single e^- background above 1800 keV) shows that the two-electron spectrum is not reduced as much as the single-electron one. That implies that a significant 2β signal is contained in the 2e data, and so a new run (at low pressure) is in progress in a search for the $2\beta 2\nu$ mode.

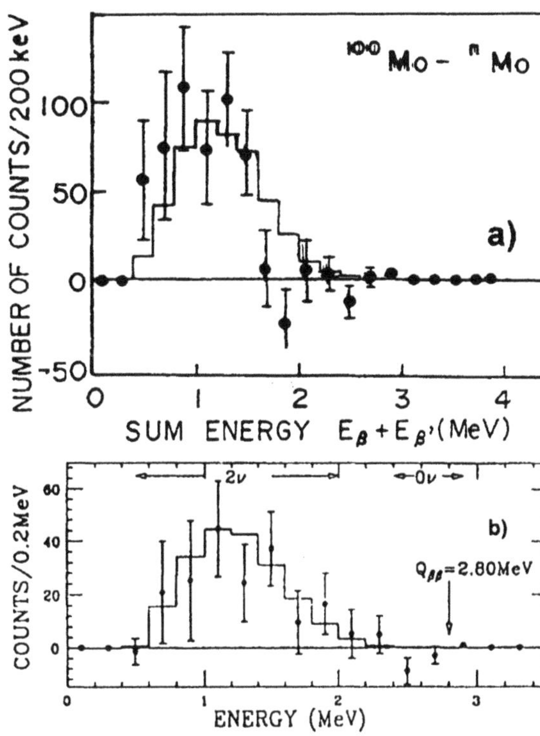

Figure 5.

The ITEP group has measured [9]the double beta decay of ^{150}Nd (40 g) with a TPC of $\sim$ 300 litres filled with CH_4 at atmospheric pressure, in a 700 gauss magnetic field. The detection efficiency is $\varepsilon_{2\nu} \sim 3\%$ (see results in Table 1). A large ($13m^3$) TPC is underway for Xe (7.5 kg) and Nd (5 kg).

The group of INR at Kiev [10]is investigating the double beta decay of ^{116}Cd with cadmium tungstate ($^{116}CdWO_4$) scintillator crystals of 12 to 15 cm^3 which feature an energy resolution of $\Gamma = 7\%$ at 2614 keV. A series of test experiments to reduce the background has lead to a figure of B$\sim$ 0.6 c/keV kg y. Results are quoted in Tables 1 and 2.

A series of bolometer experiments have been

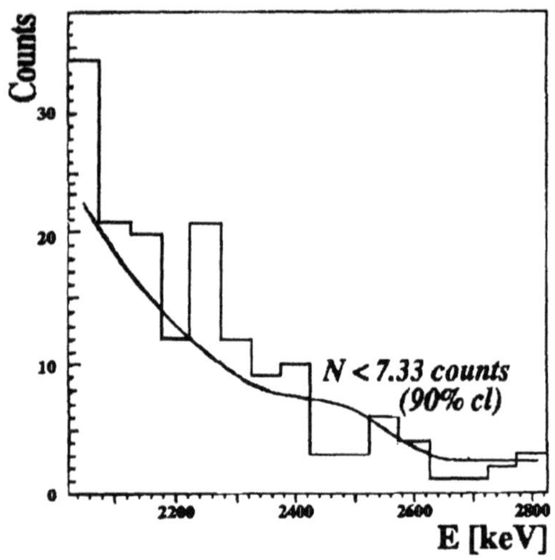

Figure 6.

carried out by the Milan group since 1989 in the Gran Sasso Laboratory, searching for the double beta decay of ^{130}Te [11]. The increase of the temperature produced by the energy released in the crystal due to a nuclear event (i.e. 2β), is measured by means of a sensor in thermal contact with the absorber. The Milan group uses Tellurium oxide crystals as absorbers, and glued NTD Ge thermistors as sensors. Notice that natural Tellurium contains 34% of ^{130}Te. After using successively TeO_2 crystals of 73 g and 334 g, as well as a set of four of these large crystals, a tower-like array of 20 crystals of 340 g in a copper frame is currently taking data at a temperature of $\sim$ 10 mK. In a recent run, featuring an energy resolution (summed over the twenty energy spectra) of $\sim$ 10 keV at 2615 keV, and a background of about 0.5 c/keV kg day in that region, they got in only a few days a better neutrinoless half-life limit than in all their previous experiments (See Table 2). The calibration spectrum of the summed twenty-crystal spectra and the background around the $Q_{2\beta}$ region corresponding to a short running have been presented to this Conference [11] and are shown in Figures 7a,b. An enlarged version of this experiment, CUORE (a Cryogenic Underground Observatory for Rare Events) consisting of an array of 1000 crystals of TeO_2 of 750 g each (with NTD Ge sensors), operating at 7 $\sim$ 10 mK, is planned to be installed at Gran Sasso [11].

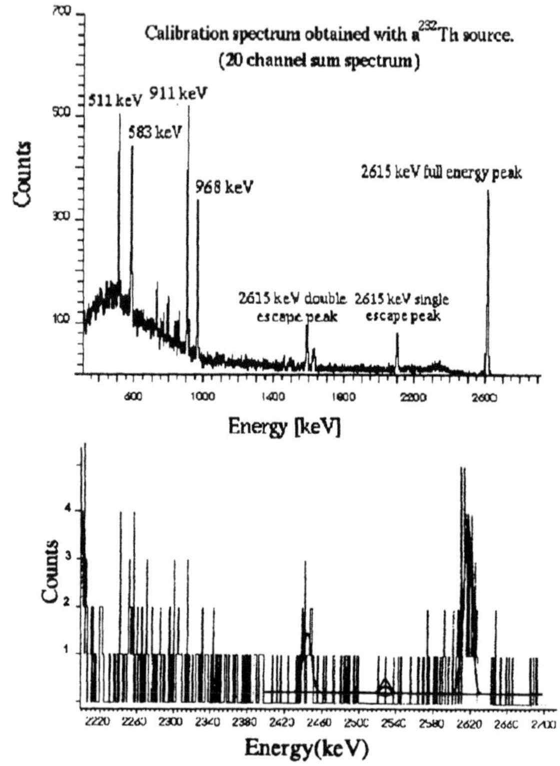

Figure 7.

4. EXPERIMENTAL RESULTS CONFRONT THEORY

Two main lines have been followed in computing the 2β-decay nuclear matrix elements: Shell Model (SM) and Quasiparticle Random Phase Approximation (QRPA). Both approaches have been widely applied with various degrees of success. The current theoretical predictions of the 2ν decay modes have provided a general framework of concordance with the experiment (within a factor 2–5). That gives confidence in the reasonable reliability of the $2\beta0\nu$ decay matrix elements used to extract $< m_\nu >$ bounds.

The first attempts to calculate 2β nuclear matrix element were made by using the nuclear shell model, but as most 2β emitters are heavy or medium heavy nuclei, it was necessary to use a weak coupling limit shell model [2,12] and/or truncation of the model space to cope with the calculation. Such truncations excluded configurations relevant for the final results. Predictions of such former calculations are given in Tables 1 and 3. Until recently, large SM calculations were possible only in the case of ^{48}Ca [13]a. New progress in SM codes have allowed to perform large model space SM calculations [13]b in heavy and medium heavy nuclei using realistic single particle basis. Still there are important truncations because of the large valence space. For ^{48}Ca, ^{76}Ge and ^{82}Se, the results are in good agreement with the experiment, whereas for ^{136}Xe there exists some discrepancy (See Table 1). Estimates of the neutrinoless decays in this large model space SM calculation give longer neutrinoless decay half-lives (for equal $< m_\nu >$ values) than the QRPA results.

QRPA is simple from a computing point of view; it includes many features of the two-body interaction which plays a relevant role in 2β decays; it is very sensitive to the $J = 1^+$, $T = 0$ particle-particle interaction and have contributed significantly to understand the large suppression of the experimental rates which failed to be explained by the earlier theoretical approaches. QRPA was first applied to compute the $2\beta 2\nu$ matrix elements by the Caltech group [14] using a zero range force. Results in agreement with experiment were obtained for various 2β measured decays, when the value of the strength g_{pp} of the particle-particle interaction used was the one fitting the β^+ decay of nuclei with magic number of neutrons. Subsequent works of the groups of Tubingen [15] and of Heidelberg [16] (both in 2ν and 0ν decays) confirmed and refined the results with more realistic NN interactions. The suppression of the $2\beta 2\nu$ matrix elements is extremely sensitive to the strength g_{pp} of the particle-particle interaction, which in fact may lead to almost null matrix elements for values of g_{pp} in its physical range. The great sensitivity of $M_{GT}^{2\nu}$ on g_{pp} makes difficult to make definite rate prediction, contrary to the Shell Model case. The value of g_{pp} has to be adjusted, otherwise the QRPA rates span a wide range of values. On the contrary, the $(2\beta)0\nu$ rates are not so sensitive. The neutrino potential makes the difference with the $(2\beta)2\nu$ case. The various multipolarities (besides $J^\pi = 1^+$) arising because of its radial dependence, wash out much of the suppression. The nuclear sensitivity of the 0ν rates is rather smooth and the predictions are much more reliable. The QRPA has been applied to most of the 2β emitters.

Several QRPA variants (like the Multiple Commutator Method, MCM [17]) or extensions have been also applied, as well as some alternative methods, like the Operator Expansion Method (OEM) [18], the SU(4) symmetry, the 1^+ intermediate state dominance model (1^+D) [19], the pseudo SU(3), and quite a few more (see Ref. [17] for a recent theoretical review). The OEM, for instance, which avoids summation over intermediate states, predicted results much less sensitive to g_{pp}, but has also several drawbacks. The alterntative 1^+D model of Zaragoza/Osaka [19,7], suggested a long time ago [19], relies on the fact that in a double beta transition, the intermediate state (odd-odd nucleus) having 1^+ ground state (gs) can decay by EC to the initial gs, and by β^- to the gs of the final nucleus and so the feeding of pertinent ft-values provide the 2β decay nuclear matrix elements. An archetypical example is provided by the transition ^{100}Mo–^{100}Tc–^{100}Ru, which in most of the calculations is predicted to decay faster than observed ($\sim 10^{19}$ y). The QRPA did not work either for ^{100}Mo, nor did some of their cures like the OEM (almost insensitive to g_{pp}), which fall a factor three apart from the experimental value. However, by assuming a dominant contribution of the lowest state of the intermediate nuclei, the correct value of $M_{2\nu}$ could be reproduced [20], as already noted quite a few years ago [19] in this and other transitions. Working out this model (i.e. feeding the single GT transition matrix element as given by experiment [say from β^- and EC decays and/or (pn),(np), (^{3}He,t) reactions, presently being carried out at RCNP (Osaka)], Ejiri et al. obtained $2\beta 2\nu$ half-life values in fair agreement with the experiment [6,21].

Results of 2ν and 0ν theoretical half-lives are given in Tables 1 and 3 according to various nuclear models. The reader can derive by himself from Tables 3 and 2 the $< m_\nu >$ upper bounds

Table 3
Neutrinoless half-lives in various Theoretical Models (for the $< m_\nu >$ Term) $T^{0\nu}_{1/2} < m_\nu >^2$ values are given in 10^{24} (eV)^2 y.

	^{76}Ge	^{82}Se	^{100}Mo	^{128}Te	^{130}Te	^{136}Xe	^{150}Nd	^{116}Cd	^{48}Ca
Weak Coupl. SM [2,12]	1.67	0.58		4.01	0.16				
$g_A = 1.25 (g_A = 1)$	(3.3)	(1.2)		(7.8)	(0.31)				
Large Space SM [13]	17.5	2.39				12.1			6.25
QRPA [14]	14	5.6	1.9	15	0.66	3.3			
QRPA [16]	2.3	0.6	1.3	7.8	0.49	2.2	0.034	0.49	
QRPA [15]	2.15	0.6	0.255	12.7	0.52	1.51	0.045		
OEM [18]	2.75	0.704		12.6	0.723	4.29	0.056	0.583	
QRPA with	18.4	2.8	350	150	2.1	2.8		4.8	28
(without) np pair. [22]	(3.6)	(1.5)	(3.9)	(19.2)	(0.86)		(4.7)	(2.4)	

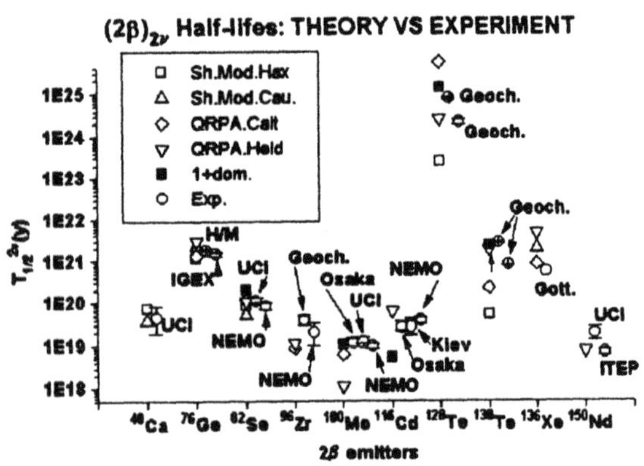

Figure 8.

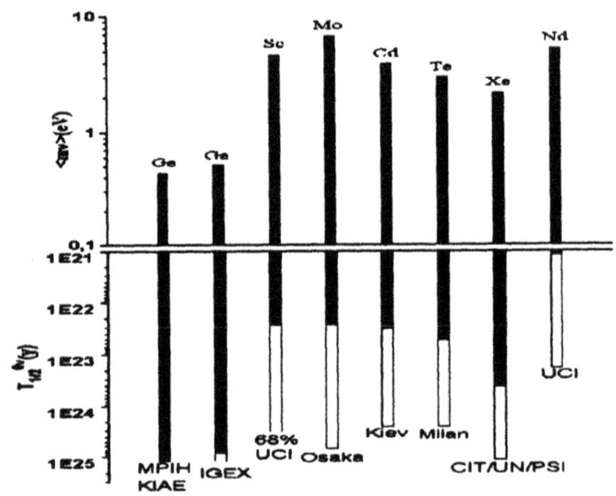

Figure 9.

according to his preferred nuclear model. Figure 8 sumarizes the confrontation theory vs. experiment in the two-neutrino decay modes, whereas Figure 9 gives a comparison of the neutrinoless half-live results of the various major experiments, and the corresponding neutrino mass bounds. The white hystograms represent the half-life limit each experiment must reach to match the current bounds for the neutrino mass obtained in germanium experiments.

5. CONCLUSIONS AND OUTLOOK

The standard $2\beta 2\nu$ decay has been directly observed in several nuclei: ^{48}Ca, ^{76}Ge, ^{82}Se, ^{96}Zr, ^{100}Mo, ^{116}Cd and ^{150}Nd and others are under investigation (^{130}Te, ^{136}Xe). QRPA reproduces reasonably well the measured half lives with some fine tuning of the nuclear parameter. Recent (large model space) shell model calculations give also good predictions in the cases where they have been applied so far, reinforcing the confidence on the matrix elements needed to extract the experimental limits on $2\beta 0\nu$ decay. Data from the most sensitive experiments on $2\beta 0\nu$ lead to the limit $< m_\nu >< 0.4 - 1.5$ eV for the effective neutrino mass, according to the nuclear model.

The Ge experiments provide the stringest bound to the neutrino mass parameter and they seem to offer, for the next future, the best prospectives to reach the lowest values of $< m_\nu >$. The Heidelberg-Moscow experiment and IGEX on ^{76}Ge will continue the data taking with a background reduced by pulse shape discrimination. These experiments will achieve sensitivities of order $T_{1/2}^{0\nu} \approx 5 \times 10^{25}$ y or close to 10^{26} in ^{76}Ge, corresponding to $< m_\nu > \approx 0.2 - 0.6$ eV (according to the nuclear matrix element used). As proved by the 20-crystal array bolometers of the Milan group, the low temperature thermal detection of 2β decays is now mastered. The cryogenic detectors are supposed to provide better energy resolution and more effective absorption of the particle energy (thermal vs ionization) and so the CUORE project is a promising (and feasible) undertaking.

Summarizing, currently running or planned experiments (H/M, IGEX, NEMO 3, MUNU, Bolometer Arrays), will explore effective neutrino masses down to about 0.1—0.3 eV. To increase the sensitivity it is necessary to go to larger source masses and reduce proportionally the background. That would bring the sensitivity to neutrino mass bounds below the tenth of electronvolt. Projects like CUORE [11] or GENIUS [4] go in that direction. To pursue the goal even further, huge detector masses, and still better event identification are needed.

In spite of the progress, a long way is still ahead of us. What is at stake is worth the effort.

6. ACKNOWLEDGEMENTS

I am indebted to my colleagues of the IGEX Collaboration, in particular to J. Morales for discussion and comments, and to CICYT (Spain) and the Commission for Cultural, Education and Scientific Exchange between the United States of America and Spain for financial support.

REFERENCES

1. J. Schechter, J.W.F. Valle, Phys. Rev. D 25 (1982) 2951.
2. W.C. Haxton, G.J. Stephenson Jr., Prog. Part. Nucl. Phys. 12 (1984) 409.
3. M. Doi, T. Kotani, E. Takasugi, Progr. Theor. Phys. Suppl. 83 (1985) 1.
4. H.V. Klapdor, these Proc. and Refs. therein.
5. A. De Silva et al., Phys. Rev. C56 (1997) 2451, and A. Balysh et al., Phys. Rev. Lett. 77 (1996) 5186.
6. F. Piquemal, these Proc. and Refs. therein.
7. H. Ejiri, these Proc. and Refs. therein.
8. J-C. Vuilleumier et al., Phys. Rev. D48 (1993) 1009, and J. Farine, Proc. Neutrino 96, Helsinki, June 1996. Ed. K. Enqvist et al., World Scientific, p. 347.
9. V. Artemiev et al., Phys. Lett. B345 (1995) 564.
10. F.A. Danevich et al., Phys. Lett. B344 (1995) 72.
11. O. Cremonesi, these Proc. and Refs. therein.
12. W.C. Haxton, Nucl. Phys. B (Proc. Suppl.) 31 (1993) 82 and Refs. therein.
13. a) E. Caurier et al., Phys. Let.. B 252 (1990) 13 and Erratum, Phys. Rev. C 50 (1994) 223. b) E. Caurier et al., Phys. Rev Lett. 77 (1996) 1954; J. Retamosa et al., Phys. Rev. C 51 (1995) 371; A. Poves et al., Phys. Lett. B 361 (1995).
14. P. Vogel and coll., Phys. Rev. Lett. 57 (1986) 3148; Phys. Rev. C 37 (1988) 731. See also M. Moe and P. vogel, Ann. Rev. Nucl. Part. Sci. 44 (1994) 247.
15. A. Faessler and coll., Phys. Lett. B 194 (1987) 11; B 199 (1987) 473. See also T. Tomoda, Rep. Prog. Phys. 54 (1991) 53.
16. A. Staudt et al., Europhys. Lett. 13 (1990) 31 and Refs. therein.
17. J. Suhonen, O. Civitarese, Phys. Rep. 300 (1998) 123.
18. X.R. Wu et al., Phys. Lett. B 272 (1991) 169 and B 276 (1992) 274 and J. G. Hirsch et al., Nucl. Phys. A 589 (1995) 445.
19. J. Abad, A. Morales, R. Núñez-Lagos, A.F. Pacheco, Ann. Fis. A 80 (1984) 9; J. Phys. C3 Suppl. 45 (1984) 147.
20. A. Griffiths and P. Vogel, Phys. Rev. C 47 (1993) 2910.
21. H. Ejiri, Int. J. of Mod. Phys. E 6 (1997) 1.
22. G. Pantis et al., Phys. Rev. 53 C (1996)695.
23. B. Kayser et al., "New and Exotic Phenomena", Ed. O. Facker, Editions Frontieres 1987, p. 349.

ELSEVIER

Nuclear Physics B (Proc. Suppl.) 77 (1999) 346–351

NUCLEAR PHYSICS B
PROCEEDINGS
SUPPLEMENTS

Double Beta Decays and Neutrino Nuclear Responses

H. Ejiri [a]

[a]Research Center for Nuclear Physics, Osaka University.
Ibaraki, Osaka 567-0043, Japan
ejiri@rcnp.osaka-u.ac.jp

Neutrinos (ν) beyond the standard theory are studied by investigating double beta decays ($\beta\beta$). The present status of $\beta\beta$ studies at RCNP is briefly reported. The $\beta\beta$ decays on ^{100}Mo and ^{48}Ca are studied at the Oto Cosmo Observatory. The Oto observatory is a new underground laboratory with low Rn and cosmic-ray backgrounds. The sensitivities expected there are 0.5$\sim$1eV for the Majorana ν-mass, 10^{-6} $\sim 10^{-8}$ for the right-handed weak currents, 2$\sim$4.10^{-5} for the ν- Majoron coupling, and so on. Nuclear axial weak responses for $\beta\beta$-ν are investigated by charge-exchange spin-flip nuclear reactions.

1. INTRODUCTION

Neutrinos, which are current nuclear and particle physics interests, have extensively been studied at RCNP (Research Center for Nuclear Physics), Osaka University. RCNP is a laboratory complex. It consists of the cyclotron laboratory, the Oto underground laboratory, and of the laser electron photon laboratory. The cyclotron laboratory aims at studies of nucleon meson nuclear physics by means of the sub-GeV protons and light ions from the 0.4 GeV cyclotron, while the Oto laboratory aims at studies of lepton nuclear physics, including studies of neutrinos and dark matters, by means of high sensitive detectors. The laser electron photon laboratory aims at studies of quark nuclear physics by means of multi-GeV polarized photons. The photons are produced by Compton backscattering of laser photons from 8 GeV electrons at Spring-8. Important subjects of nuclear and astroparticle physics interests can be studied by various probes and detectors at these laboratories of RCNP.

Neutrino studies at RCNP are being carried out by investigating double beta decays at the underground laboratory (Oto Cosmo Observatory), and by investigating neutrino nuclear responses at the cyclotron laboratory.

Neutrino-less double beta decays ($0\nu\beta\beta$) are very sensitive to the Majorana neutrino mass, the right-handed weak currents, the Majoron neutrino coupling, the SUSY particle coupling, and to other weak processes [1, 2]. These are all beyond the minimal electroweak standard theory. They are studied by means of the high sensitive ELEGANTs detector at the Oto Cosmo Observatory.

Neutrino nuclear responses in $0\nu\beta\beta$ are crucial for getting physics quantities beyond the standard theory. The two neutrino double beta decay ($2\nu\beta\beta$) followed by the two neutrinos is the process within the standard theory. Nuclear responses for $2\nu\beta\beta$ are derived from observed $2\nu\beta\beta$ transition rates.

The nuclear responses for $0\nu\beta\beta$ and $2\nu\beta\beta$ processes are associated with charged weak nuclear currents. They are obtained from corresponding charge-exchange nuclear reactions [3]. Charge exchange (^{3}He, t) and (d, ^{2}He) reactions have extensively been studied by using medium energy ^{3}He and d beams at the cyclotron laboratory.

2. NEUTRINO STUDIES BY DOUBLE BETA DECAYS

2.1. Double Beta Decays

Neutrinos (ν) have extensively been studied by measuring double beta decays in nuclei. Two neutrino and neutrino-less double beta decays are written as

$$2\nu\beta\beta \quad A \rightarrow B + \beta + \beta + \nu + \nu \qquad (1)$$

0920-5632/99/$ – see front matter
PII S0920-5632(99)00442-9

$$0\nu\beta\beta \quad A \rightarrow B + \beta + \beta \qquad (2)$$

In eq. (1), ν is an anti-neutrino and a neutrino is case of double β^- and β^+ decays, respectively. Then, $2\nu\beta\beta$ conserves the lepton number L, and is a process within the standard $SU(2)_L \times U(1)$ theory. On the other hand, $0\nu\beta\beta$ violates the lepton number conservation law by ΔL=2, and thus is a process beyond the standard theory.

The $0\nu\beta\beta$ process is used to study neutrinos beyond the standard theory, while $2\nu\beta\beta$ is used to study neutrino nuclear responses associated with $\beta\beta$ decays. The $2\nu\beta\beta$ and $0\nu\beta\beta$ diagrams are shown in Fig. 1.

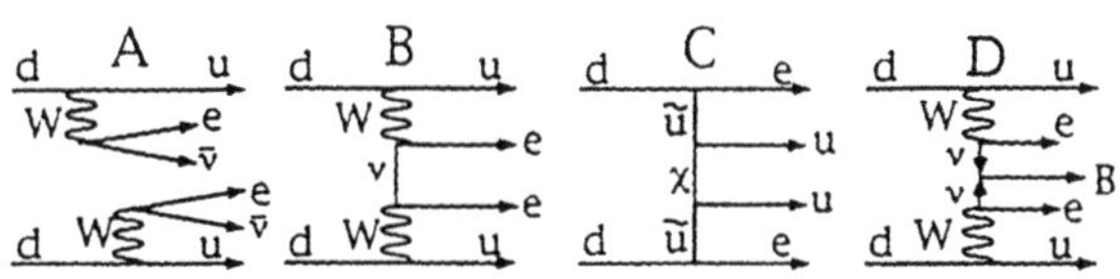

Figure 1. Schematic diagrams of $\beta\beta$ process. A: $2\nu\beta\beta$, B: $0\nu\beta\beta$, C: $0\nu\beta\beta$ with SUSY exchange, D: $0\nu\beta\beta$ followed by Majoron (B).

The transition rates for $2\nu\beta\beta$ and $0\nu\beta\beta$ are written as [1, 2]

$$T^{2\nu} = G^{2\nu} \mid M^{2\nu} \mid^2, \qquad (3)$$

$$T^{0\nu} = G^{0\nu} \mid M^{0\nu} \mid^2 [< m_\nu >^2 + < \lambda >^2 + < \eta >^2 + < m_\nu >< \lambda > + < \lambda >< \eta > + < m_\nu >< \eta >]^2, \qquad (4)$$

where $G^{2\nu}(G^{0\nu})$ and $M^{2\nu}(M^{0\nu})$ are the phase space factor and the nuclear matrix element for $2\nu\beta\beta$ ($0\nu\beta\beta$) process, respectively. $T^{0\nu}$ given in eq. (4) is the $0\nu\beta\beta$ process due to the Majorana neutrino mass terms of $< m_\nu >$ and right-handed weak current terms of $< \lambda >$ and $< \eta >$. The rates for other processes are written by using relevant terms such as the ν-Majoron coupling of $< g_B >$, the SUSY coupling of $< \lambda_{ijk} >$, and so on.

The nuclear responses for $2\nu\beta\beta$ and $0\nu\beta\beta$ are given by the matrix elements of $M^{2\nu}$ and $M^{0\nu}$, respectively. $M^{2\nu}$ is derived from the observed $2\nu\beta\beta$ transition rate as given in eq. (3). The nuclear response $M^{0\nu}$ is crucial for getting values for (limits on) the neutrino mass and other quantities of particle physics interest from the observed values for (limits on) the $0\nu\beta\beta$ transition rate, as given in eq. (4).

2.2. Oto Underground Laboratory

Precision measurements at low-background underground laboratories are necessary for studying very rare decays such as $2\nu\beta\beta$ and $0\nu\beta\beta$.

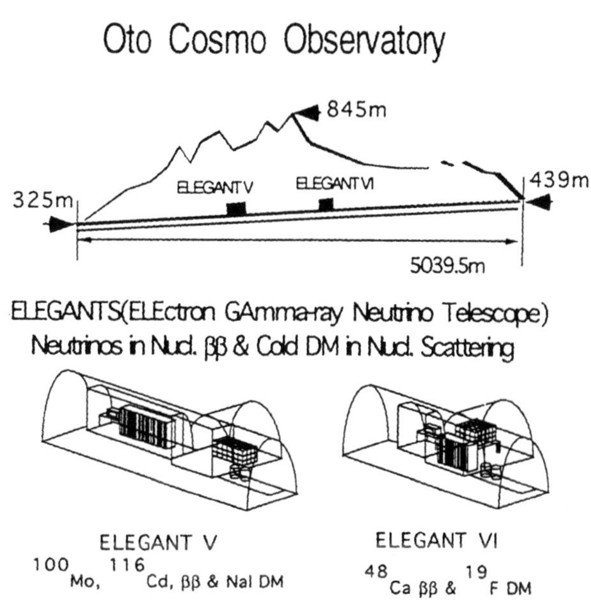

Figure 2. Oto Cosmo Observatory and ELEGANT V and VI

The Oto Cosmo Observatory has recently been opened for neutrino and dark-matter studies as an underground laboratory of RCNP. It is located at the middle of the unused rail-way tunnel with

5km in length and 470m in depth. Presently, two experimental rooms are set, as shown in Fig. 2.

The background rate at the middle of the tunnel are $4{\cdot}10^{-3}/\mathrm{m}^2/\mathrm{sec}$ for cosmic muons, $4{\cdot}10^{-1}/\mathrm{m}^2/\mathrm{sec}$ for neutrons and about 10 $\mathrm{Bq/m}^3$ for Rn. The Rn content is further reduced down to 0.1~0.05 $\mathrm{Bq/m}^3$ inside the detector container by circulating parified N_2 gas. These background rates are satisfactory. In particular, the low Rn content is crucial for low-energy studies of rare nuclear decays such as nuclear $\beta\beta$ decays, DM nuclear scatterings, and others. Here DM candidates to be studied in nuclei are weakly interacting massive particles (WIMPS).

High sensitive detectors of ELEGANT (ELEctron GAmma-ray Neutrino Telescope: EL.) V and VI are set at the experimental rooms for studying $\beta\beta$ decyas and WIMPS.

2.3. Double Beta Decays By ELEGANT V

ELEGANT V (EL V) is now being used to study $0\nu\beta\beta$ of ^{100}Mo. The $2\nu\beta\beta$ and $0\nu\beta\beta$ processes for ^{100}Mo and ^{116}Cd were first studied by using EL V at the Kamioka underground laboratory [4, 5].

EL V consists of multi-layer drift chambers for $\beta\beta$ trajectories, plastic scintillator arrays for $\beta\beta$ energies and times, NaI scintillator arrays for X and γ rays [4]. EL V is shown in Fig. 3.

The previous studies of ^{100}Mo and ^{116}Cd showed for the first time finite halflives for the $2\nu\beta\beta$ rate and stringent limits on the $0\nu\beta\beta$ rates [5]. Here, ^{100}Mo and ^{116}Cd were chosen because of the large phase space factors of $G^{0\nu}$ and $G^{2\nu}$ due to the large $\beta\beta$ Q-values.

The $2\nu\beta\beta$ matrix elements are derived from the observed $2\nu\beta\beta$ rates as $M^{2\nu}$=0.096 and 0.07 for ^{100}Mo and ^{116}Cd, respectively.

The $2\nu\beta\beta$ matrix elements for these nuclei and others are very small because of the destructive interference of the single particle (SP) process and the GT giant resonance one. They are analyzed in terms of the single particle-hole (|S>), the spin isospin giant resonance (|G>), and their coupling [6]. The $2\nu\beta\beta$ processes between the 0^+ ground states of even-even nuclei are found to proceed through low-lying intermediate nuclei, as shown in Fig. 4.

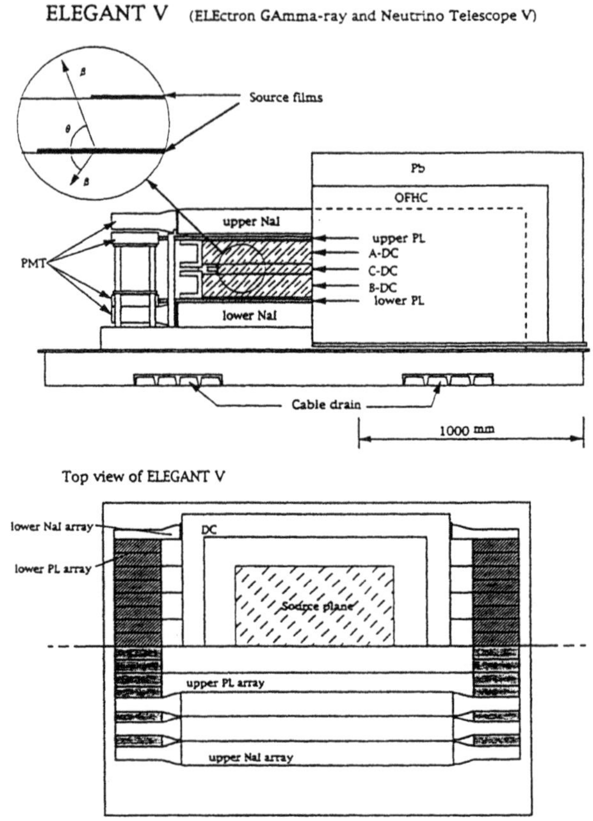

Figure 3. ELEGANT V used for studying $\beta\beta$ of ^{100}Mo and ^{116}Cd. A-DC and B-DC are upper and lower drift chambers for tracking β rays and C is the central one for tracking β and α rays. PL's are plastic scintillators. [ref. 4]

No finite $0\nu\beta\beta$ yields were observed for ^{100}Mo and ^{116}Cd. Quite stringent limits with 68% CL on $< m_\nu >$, $< \lambda >$, $< \eta >$, and $< g_B >$ are derived from the observed limits on the $0\nu\beta\beta$ for ^{100}Mo and matrix elements [7], as follows.

$T^{0\nu}(m_\nu)$	$>5.2{\cdot}10^{22}$y	$\langle m_\nu \rangle$	<2.2eV
$T^{0\nu}(\lambda)$	$>3.9{\cdot}10^{22}$y	$\langle \gamma \rangle$	$<3.7{\cdot}10^{-6}$
$T^{0\nu}(\eta)$	$>5.1{\cdot}10^{22}$y	$\langle \eta \rangle$	$<2.5{\cdot}10^{-6}$
$T^{0\nu}(g_B)$	$>5.4{\cdot}10^{21}$y	$\langle g_B \rangle$	$<7.3{\cdot}10^{-5}$

EL V is now in operation for studying the $0\nu\beta\beta$ of ^{100}Mo at the experimental room II of the Oto

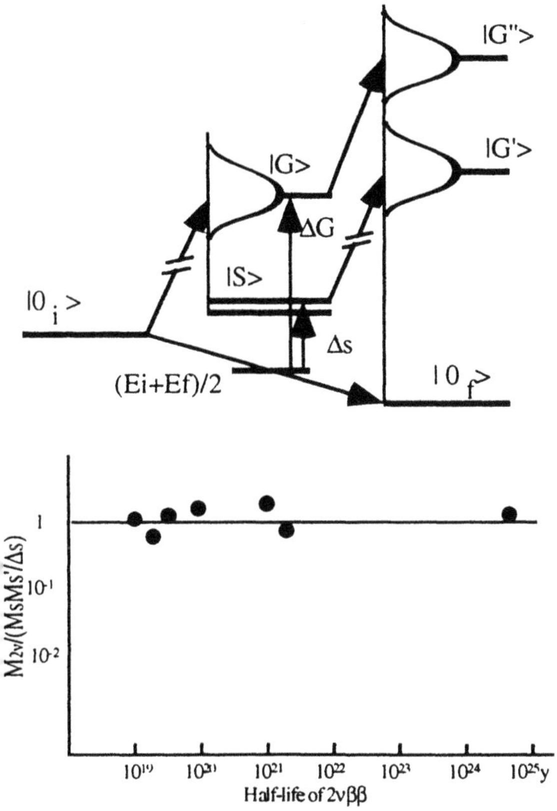

Figure 4. Top: Level and transition scheme for $2\nu\beta\beta$. Bottom: Rations of $2\nu\beta\beta$ matrix element $M^{2\nu}$ to the single particle-hole matrix elements $M_S^{2\nu}$ (ref. 6).

Cosmo Observatory. Improvements were made on the detector components. The whole detector system is set inside an air-tight container. The Rn content inside the detector container was reduced to the level of around 0.1~0.05 Bq/m^3 by introducing purified Rn-free N_2 gas into the detector container.

The sensitivities of the improved EL V at the Oto Cosmo Observatory are evaluated by using the observed and evaluated background rates. They are around 1~1.5·10^{23}y for $T^{0\nu}$ and 1.7~1.3 eV for the Majorana neutrino mass of $< m_\nu >$ and 5~3·10^{-5}for the Majoron neutrino coupling of $< g_B >$.

2.4. Double Beta Decays By ELEGANT VI

ELEGANT VI (EL VI) is a newly developed detector for studying neutrinos by $\beta\beta$ of ^{48}Ca and WIMPS by scatterings from ^{19}F. It consists of CaF_2 scintillator array surrounded by CsI scintillators, as shown in Fig. 5 [8]. The CaF_2 scintillator array consists of 25 modules of detectors, and one detector module is made of 3 CaF_2 crystals, each with 7.5×4.5×4.5 cm^3. The total CaF_2 weight is 35.2kg. The CaF_2 array contains 0.034 kg of ^{48}Ca with the largest $Q_{\beta\beta}$value of 4.3 MeV. A preliminal test run was made at Oto underground laboratories. The goal is to get high sensitivities of sub eV for the Majorana neutrino mass and of ~2·10^{-5} for the Majoron neutrino coupling.

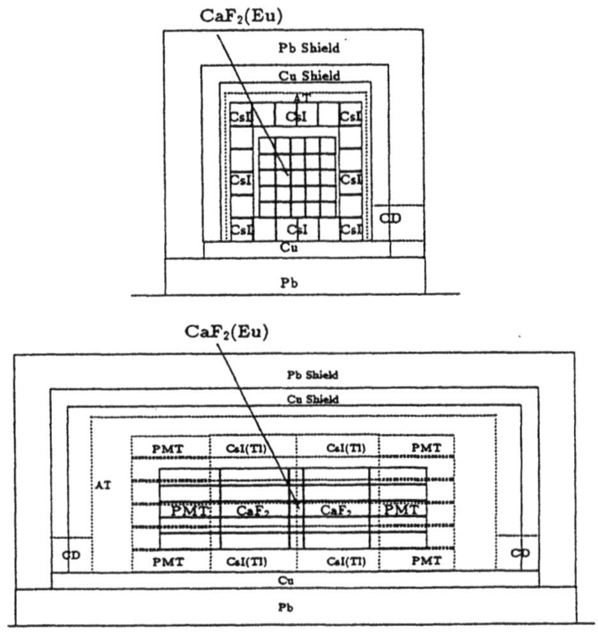

Figure 5. ELEGANT VI is used to study WIMP's and ^{48}Ca $\beta\beta$-decays. The central CaF_2 (Eu) detectors, each with 4.5×4.5×4.5cm^3 are used for WIMP's studies, and they are replaced by the CaF_2 (pure) ones, each with 7.5×4.5×4.5cm^3 for studying neutrinos by $\beta\beta$ of ^{48}Ca .

Precision measurements of $0\nu\beta\beta$ and $2\nu\beta\beta$ decays on several nuclei are necessary because of uncertainty in nuclear responses.

3. NEUTRINO NUCLEAR RESPONSES FOR $2\nu\beta\beta$ AND $0\nu\beta\beta$

3.1. Neutrino Nuclear Respsnses Studied By Nuclear Reactions

Fundamental properties of neutrinos and weak interactions are studied by measuring single and double beta decays in nuclei. Here nuclei, which consist of nucleons in good quantum states, are used as excellent microlaboratories for studying neutrinos and weak interactions. Here neutrino nuclear responses for β and $\beta\beta$ decays are crucial for getting physics variables of the neutrinos and the weak interactions.

Neutrino nuclear responses are given by β and $\beta\beta$ nuclear matrix elements. Matrix elements of charged-current axial-vector (GT) type and charged-current vector (F) type are studied by investigating charge-exchange spin-flip and charge-exchange spin-nonflip nuclear reactions, as shown in Fig. 6.

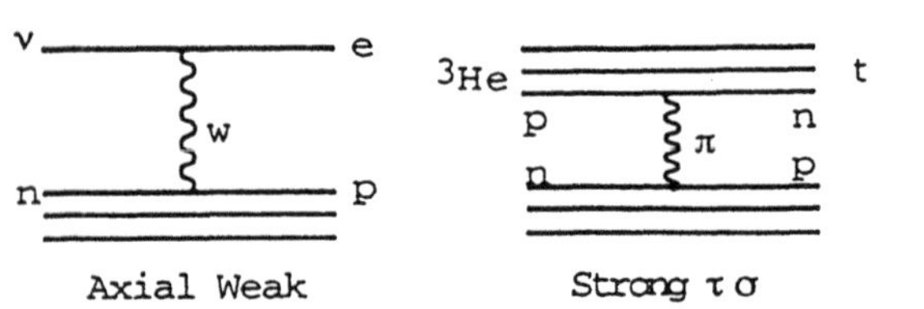

Figure 6. Nuclear spin isospin responses for the charged-current axial-vector weak process and for the charge-exchange spin-flip strong process (ref. 3).

Charge-exchange reactions for neutrino nuclear responses have extensively been investigated by using medium energy light ions from the RCNP ring cyclotron with K=0.4 GeV. The ring cyclotron provides p, d and ^{3}He beams with 0.2~0.5 GeV, where the isospin spin interaction becomes relatively large. Thus they are very useful for investigating isospin-flip (charge-exchange) spin-flip (axial-vector) responses.

3.2. Nuclear Responses for $2\nu\beta\beta$

Nuclear responses for $2\nu\beta\beta$ of ^{100}Mo are studied by the (^{3}He, t) at E (^{3}He)=0.45 GeV [9]. The GT strengths for single β decays from ^{100}Mo to the intermediate nucleus of ^{100} Tc were obtained, as shown in Fig. 7.

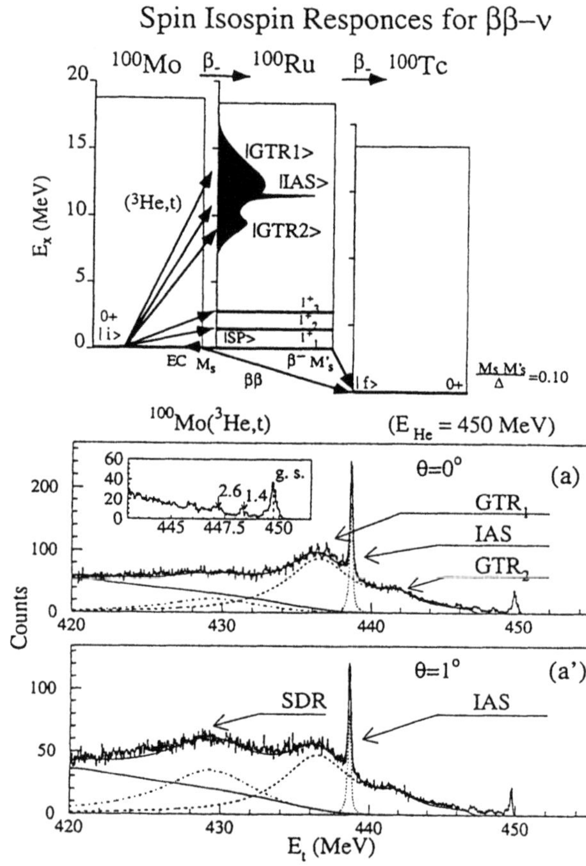

Figure 7. Nuclear responses for $\beta\beta$-ν on ^{100}Mo studied by the (^{3}He, t) reaction on ^{100}Mo. g.s.: single particle-hole ground state. IAS: isobar analogue state. GTR: GT giant resonances [ref.9].

The nuclear matrix element $M_S(\beta)$ for the β^- decay from the 0^+ ground state of ^{100}Mo to the single particle-hole 1^+ state (SP) of ^{100}Tc was obtained from the GT strength derived from the (^{3}He, t) cross-section. The nuclear matrix element $M_S(\beta^-)$ for the β^- transition from the SP 1^+ state in ^{100}Tc to the 0^+ ground state of ^{100}Ru was evaluated from the known ft value of the β^- decay. Using these matrix element, the $2\nu\beta\beta$ matrix element $M_S^{2\nu}(\beta\beta)$ through the SP 1^+ intermediate state was obtained. The obtained value of $M_S^{2\nu}(\beta\beta)$=0.10 agrees with the measured value of 0.096 [5]. Thus, we get

$$M^{2\nu}(\beta\beta) \sim M_S^{2\nu}(\beta\beta) = \frac{M_S(\beta)M_S(\beta')}{\Delta s}, \quad (5)$$

where Δs is the energy denominator. The $2\nu\beta\beta$ process proceeds mainly through the SP 1^+ state in the intermediate nucleus. The GT giant resonance, which absorbs most of the axial vector β^- strengths does not contribute to $2\nu\beta\beta$. This supports the experimental and theoretical analyses [6].

3.3. Nuclear Responses for $0\nu\beta\beta$

The $0\nu\beta\beta$ process is mainly the ν-exchange process between two nucleons in the nucleus. Thus the $0\nu\beta\beta$ matrix element, M^ν, involves the neutrino potential term, $h_+(r, E_K)$, with r and E_K being the distance between the two nucleons and the energy of the Kth intermediate state. Since the neutrino energy ω is much larger than E_K, the neutrino potential is given mainly by the Coulomb potential as $h_+(r, E_K) \sim R/r$. The neutrino potential for the two nucleons in the nucleus is confined in the region of r=r_0 and R, where r_0 and R are the nucleon and nuclear radii, respectively.

The confined potential is expressed approximately by the Beteman-type separable representation as $h_+(r, E) \sim R/r = f(r_1)f(r_2)$ [10]. Then $M^{0\nu}$ is written, as in case of $M^{2\nu}$, by a product of the successive single β matrix elements. Consequently one gets

$$M^{0\nu} = \sum_K M_K^*(lsJ)M_K(lsJ), \quad (6)$$

where J, s, and l are angular momentum, spin and orbital angular momentum, respectively. Since the ν exchange involves a large momentum of q ~0.2~0.5GeV/c, the angular momentum of $J = 0$ ~4 are to be considered.

The $0\nu\beta\beta$ matrix element expressed in the separable form is related to the single β responses of the multi polarities of J. Then, the $0\nu\beta\beta$ process with J can be evaluated as the successive single β processes with J through low-lying single particle-hole states, in analogy to the $2\nu\beta\beta$ case.

Experimental studies of $M^{0\nu}$ and relevant single beta ones of $M_s(lsJ)$ are now under progress by using single charge exchange reactions of (^{3}He, t), and (d, ^{2}He) and double charge exchange reactions of light ions.

REFERENCES

1. W. C. Haxton and G. J. Stephenson, Jr. Prog. Part. Nucl. Phys. 12 (1984) 409
2. M. Doi, T. Kotani and E. Takasugi, Prog. Theor. Phys. 83 (Supp 1) (1985) 1
3. H. Ejiri, Int. J. Modern Phys. E Vol. 6. No1 (March 1997) 1
4. H. Ejiri et al., Nucle. Instr. Methods A 302 (1991) 304
5. H. Ejiri et al., Phys. Lett. 258B; Nucl. Phys. A 611 (1996) 85
6. H. Ejiri and H. Toki, J. Phys. Soc. Japan Lett. 65 (1996) 7
7. T. Tomoda and A. Faessler. Phys. Lett. B 199 (1987) 475
8. R. Hazama et al., Proc. Int. Symp. WEIN '95, Osaka June 1995. ed.
   H. Ejiri et al., World Scientific Pub. (1995) 635
9. H. Akimune, H. Ejiri et al., Phys. Lett. 394B (1997) 23
10. H. Ejiri, V. B. Belyaev, and H. Toki, to be published (1998)

ELSEVIER

Nuclear Physics B (Proc. Suppl.) 77 (1999) 352–356

NUCLEAR PHYSICS B
PROCEEDINGS
SUPPLEMENTS

Results from the NEMO experiment

F. Piquemal [a] for the NEMO collaboration

[a]Centre d'Etudes Nucléaires de Bordeaux-Gradignan,
CNRS-IN2P3 et Université de Bordeaux I, F-33175 Gradignan Cedex, FRANCE

The NEMO collaboration is building a detector to search for neutrinoless double beta decay. The main results of the second prototype NEMO 2 are presented as well as the expected performance of the final detector NEMO 3.

1. Introduction

In 1989 the NEMO (Neutrinoless Experiment with Molybdenum) collaboration[1] started a R&D program to build a detector able to lower the sensitivity to the effective neutrino mass down to about 0.1 eV by looking for the neutrinoless double beta decay process ($\beta\beta0\nu$). The observation of such a process will prove the existence of a massive Majorana neutrino.

The experimental setup has been designed with a removable source in order to study different $\beta\beta$ emitters. It must also be able to reconstruct trajectories and to measure the energy and the time of flight of the emitted electrons. Two prototypes, NEMO 1 and NEMO 2, have been built to test the techniques of detection and to study the background. This paper reports the main results obtained in 6 years of running with the NEMO 2 prototype detector installed in the Fréjus Underground Laboratory (4800 m.w.e.). A brief description and the expected performance of the final detector NEMO 3 currently under construction are also given.

2. The NEMO 2 detector

A more detailed description can be found in Ref. [1]. Only the main characteristics of the detector will be presented here. The NEMO 2 detector (Figure 1) is made of a 1 m^2, 50 μm

[1]CEN-Bordeaux-Gradignan, France; CFR-Gif/Yvette, France; CRN-Strasbourg, France; Department of Physics-Jyvaskyla, Finland; FNSPE-Prague, Czesch. Republik ;INR-Kiev, Ukraine; ITEP-Moscow, Russia; JINR-Dubna, Russia; LAL-Orsay, France; LPC-Caen, France, MHC-South Hadley, USA

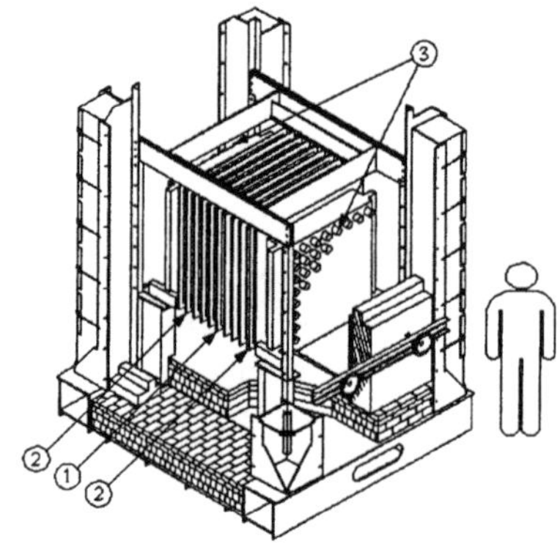

Figure 1. The NEMO 2 prototype without shielding. (1) Central frame with the source plane capable of supporting plural source foils. (2) tracking device of 10 frames, each consisting of two perpendicular planes of 32 geiger cells. (3) Two scintillator arrays of 8 by 8 counters each.

source foil sandwiched by two tracking volumes composed of Geiger cells and two plastic scintillator arrays. The tracking volumes are filled with a mixture of helium gas and 4% ethyl alcohol in order to miminize multiple scattering effects. The detector is able to track electrons with energy as low as 100 keV. A delayed trigger allows the possibility of detecting alpha particles.

The calorimeter part is made of two arrays of plastic scintillators coupled to photomultiplier tubes (PMT). These counters allow energy and time of flight measurements. The resolution is 17.4 % (FWHM) in energy and 250 ps (σ) in time at 1 MeV. The detection threshold is 50 keV. Time and energy calibrations were checked daily.

The shielding is composed of an internal 5 cm

0920-5632/99/$ – see front matter
PII S0920-5632(99)00443-0

Table 1
Comparison between measured and expected number of events in the $\beta\beta$ 0ν energy region for ^{100}Mo, ^{116}Cd, ^{82}Se.

	^{100}Mo	^{116}Cd	^{82}Se
Windows (MeV)	[2.6-3.0]	[2.3-2.8]	[2.4-3.1]
Running time (h)	6140	5960	10357
Mass (g)	172	152	157
Measured events	1	0	1
Expected background events	2	0.2	1

lead layer surrounded by 20 cm of iron. All materials used in the construction were selected using low background Ge detectors.

The signature of any $\beta\beta$ decay event is given by the pure electron-electron channel. However, other channels like electron-gamma or electron-delayed alpha channels have shown to be very important for background studies. For example, delayed alpha emission is a good signature of a ^{214}Bi contamination in the foil [2].

3. NEMO 2 results

3.1. Background studies near $Q_{\beta\beta}$ energy

The main goal of the NEMO 2 prototype was to study and to understand the components of the background in the $\beta\beta 0\nu$ energy region for several nuclei: ^{100}Mo ($Q_{\beta\beta}$=3.034 MeV) [3], ^{116}Cd ($Q_{\beta\beta}$=2.805 MeV) [4] and ^{82}Se ($Q_{\beta\beta}$=2.995 MeV) [5]. Two origins of background called *internal* and *external* components are identified.

The *internal* background corresponds to the events coming from radioactive contaminations in the foil. In the energy region of interest for $\beta\beta 0\nu$, the only activities are from ^{214}Bi and ^{208}Tl. These nuclei decay by β emission and a secondary electron can be produced by internal conversion, by the Compton effect from photons of the cascades or by Möller scattering. The tail of the $\beta\beta 2\nu$ spectrum contributes also because of the energy resolution.

The *external* background is due to high energy gamma rays (> 2.6 MeV)crossing the source foil. Their origin is from neutron capture occuring inside the detector. The interactions of these photons in the foil can lead to the production of 2 electrons by e^+e^- pair creation, double Compton effect or Compton + Möller effect. To understand this background component, several tests with different types of shielding have been performed . The low energy photon flux coming from photomuplier tubes and other surrounding materials don't contribute to the background at the $\beta\beta 0\nu$ energy.

The results of the background measurements are summarized in Table 1 . The number of expected events is calculated by simulation taking into account all known sources of background. The good agreement between measured and expected number of events in the energy window of the $\beta\beta 0\nu$ decay makes us confident in our ability to control the background in the final NEMO 3 detector.

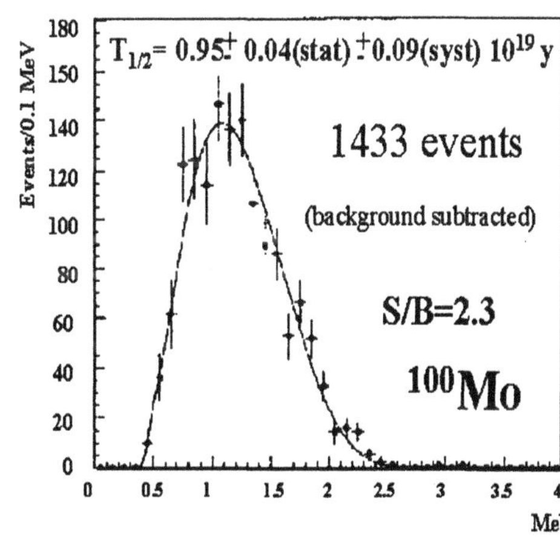

Figure 2. Experimental energy spectrum of 2e events for ^{100}Mo (6140 h), background subtracted.

Table 2
Measured $\beta\beta2\nu$ half-lifes with NEMO 2 detector and Signal/Background (Sig/Back) ratio.

	$T_{1/2}^{\beta\beta2\nu}$ (y)	Sig/Back
^{100}Mo	0.95±0.04(stat)±0.09(syst) 10^{19}	2.3
^{116}Cd	3.75±0.35(stat)±0.21(syst) 10^{19}	4
^{82}Se	0.83±0.10(stat)±0.07(syst) 10^{20}	3
^{76}Zr (preliminary)	$2.1^{+0.8}_{-0.4}$(stat)±0.1(syst) 10^{19}y	

3.2. $\beta\beta2\nu$ results

Even if the NEMO 2 prototype was first of all dedicated for the $\beta\beta0\nu$ background studies, its performance also allowed measurement of the half-life of the $\beta\beta2\nu$ process. As an exemple, Figure 2 shows for ^{100}Mo the measured 2e energy sum spectrum after background subtraction. The removed background below 2.6 MeV is essentially due to the photon flux coming from the photomultiplier tubes. The contamination of the source (measured with Germanium detector and NEMO 2 itself) in ^{214}Bi, ^{208}Tl and ^{234}Pa gives a negligible contribution.

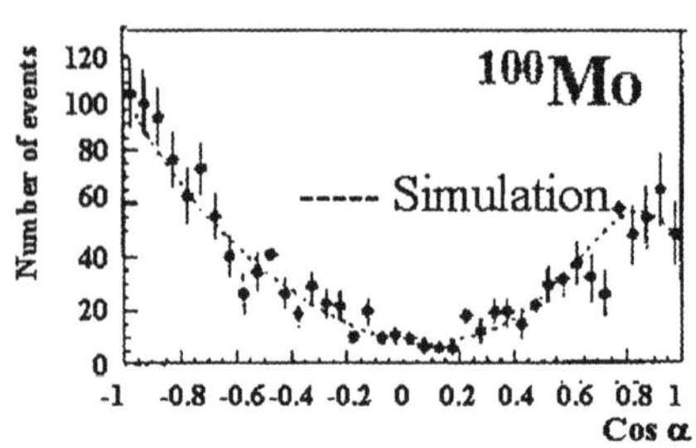

Figure 3. Angular distribution of 2e events for ^{100}Mo (6140 h, background subtracted). The broken line corresponds to the simulation.

The experimental value of the $\beta\beta2\nu$ half-life, reported on the figure 2, is obtained by a fit of the data with the expected $\beta\beta2\nu$ shape (solid line).The results for several isotopes are reported in Table 2 together with the corresponding signal/background ratio.

As shown in Figure 3, the detector also allows measurement of the angular distribution between the 2 emitted electrons. This distribution is significantly distorted by the detector geometry and the multiple scattering in the foil. The simulation (dashed line) is in good agreement with the data.

Detailed analysis can be found in Ref [3],[4] and [5].

4. The NEMO 3 detector

4.1. Structural design of NEMO 3

The NEMO 3 detector, Figure 4, will be similar in function to the earlier detector, NEMO 2. More specifically, the NEMO 3 detector will also operate in the Fréjus Underground Laboratory and will house up to 10 kg of double beta decay isotopes. To date, much attention has been focused on 10 kilograms of enriched Mo samples (97% ^{100}Mo). Also currently available are the following isotopes: 1 kg of ^{82}Se; 1 kg of ^{116}Cd; and 1 k

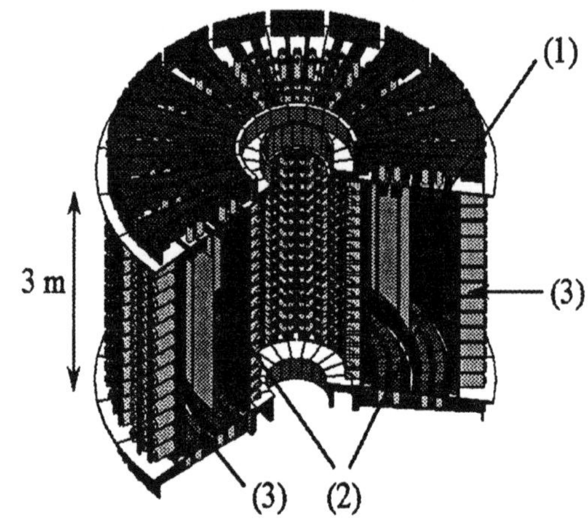

Figure 4. NEMO 3 detector. (1) Source foil, (2) tracking volumes consisting of 3 m vertical Geiger cells, (3) calorimeter made of plastic scintillators coupled to photomultiplier tubes.

The detector is cylindrical in design and divided into 20 equal sectors. A thin (40-50 μm) cylindrical source foil will be constructed from either a metal film or powder bound by an organic

Table 3
NEMO 3 Expected background rate and maximum acceptable activities (mBq/kg) in ^{214}Bi and ^{208}Tl.

Isotope	Events/year			mBq/kg	
	^{214}Bi	^{208}Tl	$\beta\beta 2\nu$	^{214}Bi	^{208}Tl
^{100}Mo	0.4	0.4	1.1	0.3	0.02
^{82}Se	0.1	0.1	0.1	0.07	0.005
^{150}Nd	none	0.4	1.1	none	0.02

glue to mylar strips. The source will hang between two concentric cylindrical tracking volumes consisting of open octagonal drift cells operating in Geiger mode. These cells run vertically and are staged in a 4, 2, and 3 row pattern to optimize track reconstruction. The design of the drift cells calls for 50 μm anode and cathode wires to prevent rapid aging.

The external walls of these tracking volumes are covered by calorimeters made of large blocks of plastic scintillator coupled to very low radioactivity 3" and 5" Hammamatsu PMTs . The energy resolution depends on the scintillator shape and the associated PMT. It ranges from 11% to 14.5% (FWHM) for 1 MeV electrons.

The complete detector contains 6180 Geiger cells and 1940 scintillators.

Additionally, a solenoid capable of producing a 30 Gauss field will surround the detector to reject pair production events. Finally, external shielding in the form of 20 cm of low activity iron will reduce gamma ray fluxes and thermal neutrons. If needed, additional shielding will be introduced to suppress the contribution of fast neutrons. More details on this are given below.

4.2. Neutrons and radioactivity requirements

At the depth of the experimental hall in the Frejus Underground Laboratory any effect of cosmic rays has been found negligible. Vigorous flushing of the air in the hall reduces the radon levels to 10-20 Bq/m^3. The presence of ^{214}Bi decays in the detector from this level of radon contamination is below that introduced by the PMTs.

Thermal and fast neutrons fluxes in the hall have been measured at levels of 1.6×10^{-6} neutrons/s·cm^2 and 4×10^{-6} neutrons/s·cm^2, respectively [6]. From NEMO 2 studies, it appears that the effects of photons coming from neutron capture are expected to be negligible. The magnetic field will be used to study the pair production and confirm the prediction of a negligible contribution.

Radioactivity of the materials which have gone into the construction of the detector have been measured with HP Ge detectors at the Fréjus Underground Laboratory or at the CENBG laboratory in Bordeaux. The activity in the mechanical pieces which frame the detector are required to be less than 1 Bq/kg.

As expected, the radioactive contamination in the experiment is dominated by the low radioactivity glass in the PMTs. The total activity of all of the 0.6 tons of PMTs is 800 Bq, 300 Bq and 18 Bq for ^{40}K, ^{214}Bi, and ^{208}Tl, respectively. These levels are three orders of magnitude below standard PMT levels. In the energy region of interest for $\beta\beta 0\nu$ decays i.e., around 3 MeV, the above external background component doesn't give any contribution.

However, the internal component from ^{214}Bi, ^{208}Tl contaminations in the source foil and tail of the $\beta\beta 2\nu$ decays have to be seriously minimized. The $\beta\beta 2\nu$ decays ultimately define the half-life limits to which the $\beta\beta 0\nu$ decays can be studied. To insure that $\beta\beta 2\nu$ defines these limits, maximum acceptable activities of ^{214}Bi and ^{208}Tl in the source foil were calculated (Table 3). For ^{100}Mo it is believed that these limits can be reached, whereas for Se with a longer $\beta\beta 2\nu$ decay half-life, more stringent levels are sought and will require some additional research. Note that the energetic decay of ^{150}Nd removes concerns of contamination by ^{214}Bi, but new techniques to enrich Nd will have to be developed for this to be realized.

4.3. Expected performance of NEMO 3

In Figure 5, the projected performance of the NEMO 3 detector with 10 kg of isotopes (^{100}Mo, ^{82}Se, ^{150}Nd) and 5 years of data are compared to the other double beta decay experiments or projects in terms of the effective neutrino mass limit.

For the running experiments with ^{76}Ge (Heidelberg-Moscow, IGEX) or ^{136}Xe (Neuchatel-Caltech), the neutrino mass has been deduced from the already published $\beta\beta0\nu$ half-life limits.

In case of the recent proposal GENIUS (see this proceedings) with 1 ton of ^{76}Ge, the two limits have been calculated using the published background level of the Heidelberg-Moscow experiment and assuming 2 or 3 orders of magnitude background reduction.

The broad range of results for the effective neutrino mass results from the use of various nuclear matrix elements (nme's) calculations. In Figure 5 calculations for the above experiments have been performed with QRPA, Shell model, and SU(3) nme's. It is worth noting that there seems to be a movement towards greater acceptance of shell model nme's.

4.4. Status of the NEMO 3 detector

12 sectors over 20 are already built. The placement of completed sectors on the frame in Frejus laboratory will start in December 1998. This final stage of the construction is expected to continue until January 1, 2000. Presently, it is planned to start operating with 7 kg of ^{100}Mo and 1 kg of ^{82}Se, with some sectors filled with foils especially designed to check background.

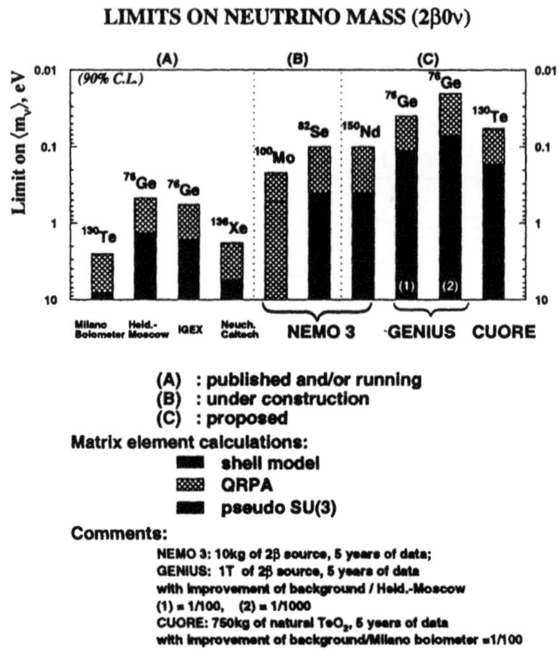

Figure 5. Projected performance of NEMO 3 detector in terms of limit on neutrino effective mass $< m_\nu >$ for 10 kg of 2β source and 5 years of data, compared to published and/or running experiment: Heidelberg-Moscow, IGEX, Neuchatel-Caltech and to the proposed GENIUS and CUORE experiments. GENIUS numbers are given for 1 ton of 2β source, 5 years of data with the hypothesis of 2 or 3 orders of magnitude improvement of their present background (respectively (1) and (2) on the figure) and CUORE numbers for 750 kg of natural TeO_2, 5 years of data with an improvement of the background of 2 order of magnitude compared to Milano bolometer.

REFERENCES

1. R. Arnold et al.,*Nucl. Instr. Meth.* **A354** (1995)338.
2. R. Arnold et al.,*Nucl. Instr. Meth.* **A401** (1997)144.
3. D. Dassie et al.,*Phys. Rev.* **D 51** (1995)2090.
4. R. Arnold et al.,*Z. Phys.* **C 72** (1996)239.
5. R. Arnold et al.,*Nucl. Phys.* **A636** (1998)209.
6. V. Chazal, Thesis, Université de Lyon (1996).

NUCLEAR PHYSICS B
PROCEEDINGS
SUPPLEMENTS

ELSEVIER Nuclear Physics B (Proc. Suppl.) 77 (1999) 357–368

Double Beta Decay with Ge-detectors – and the future of Double Beta and Dark Matter Search (GENIUS)

H.V. Klapdor–Kleingrothaus [a]

[a] Max–Planck–Institut für Kernphysik,
P.O. Box 103980, D–69029 Heidelberg, Germany

Nuclear double beta decay provides an extraordinarily broad potential to search for beyond Standard Model physics, probing already now the TeV scale, on which new physics should manifest itself. These possibilities are reviewed here. First, the results of present generation experiments are presented. The most sensitive one of them – the Heidelberg–Moscow experiment in the Gran Sasso, using enriched ^{76}Ge – probes the electron neutrino mass now in the sub eV region and will reach a limit of $\sim$ 0.1 eV in a few years. Basing to a large extent on the theoretical work of the Heidelberg Double Beta Group in the last two years, results are obtained also for SUSY models (R–parity breaking, sneutrino mass), leptoquarks (leptoquark-Higgs coupling), compositeness, right–handed W boson mass and others. These results are comfortably competitive to corresponding results from high–energy accelerators like TEVATRON, HERA, etc. Second, future perspectives of $\beta\beta$ research are discussed. A new Heidelberg experimental proposal (GENIUS) is presented which would allow to increase the sensitivity for Majorana neutrino masses from the present level of at best 0.1 eV down to 0.01 or even 0.001 eV. Its physical potential would be a breakthrough into the multi-TeV range for many beyond standard models. Its sensitivity for neutrino oscillation parameters would be larger than of all present terrestrial neutrino oscillation experiments and of those planned for the future. It would further, already in a first step, cover almost the full MSSM parameter space for prediction of neutralinos as cold dark matter, making the experiment competitive to LHC in the search for supersymmetry.

1. Introduction – Motivation for the search for double beta decay – and a future perspective: GENIUS

Double beta decay yields – besides proton decay – the most promising possibilities to probe beyond standard model physics beyond accelerator energy scales [61,62,66,67]. Propagator physics has to replace direct observations. That this method is very effective, is obvious from important earlier research work and has been stressed, e.g. by [85], etc.. Examples are the properties of W and Z bosons derived from neutral weak currents and β–decay, and the top mass deduced from LEP electroweak radiative corrections.

The potential of double beta decay includes information on the neutrino and sneutrino mass, SUSY models, compositeness, leptoquarks, right–handed W bosons, test of special relativity and eaquivalence principle in the neutrino sector and others (see [62,71] and [70]). The recent results of the Heidelberg–Moscow experiment using enriched ^{76}Ge detectors, which will be reported here, demonstrate that $0\nu\beta\beta$ decay probes already now the TeV scale on which new physics should manifest itself according to present theoretical expectations. To give just one example, inverse double beta decay $e^-e^- \rightarrow W^-W^-$ requires an energy of at least 4 TeV for observability, according to present constraints from double beta decay [19]. Similar energies are required to study, e.g. leptoquarks [33,40,13,74,26,22].

To increase by a major step the present sensitivity for double beta decay and dark matter search, a new project has been proposed recently [62,71] which would operate one ton of 'naked' enriched **GE**rmanium detectors in liquid **NI**trogen as shielding in an **U**nderground **S**etup (GENIUS). It would improve the sensitivity from the present potential of at best $\sim$ 0.1 eV to neutrino masses down to 0.01 eV, a ten ton version even to 0.001 eV. The first version would allow to test a $\nu_e \rightarrow \nu_\mu$ explanation of the atmospheric

PII S0920-5632(99)00444-2

neutrino problem, the second directly the large angle solution of the solar neutrino problem, and, for degenerate ν mass models even the small angle solution. The sensitivity for neutrino oscillation parameters would be larger than for all present accelerator neutrino oscillation experiments, or those planned for the future. GENIUS would, in its extended version, further allow to test the recent hypothesis of a sterile neutrino and the underlying idea of a shadow world (see section 2). Both versions of GENIUS would definitely be a breakthrough into the multi-TeV range for many beyond standard models currently discussed in the literature, and the sensitivity would be comparable or even superior to LHC for various quantities such as right-handed W-bosons, R-parity violation, leptoquark or compositeness searches.

Another issue of GENIUS is the search for Dark Matter in the universe. Almost the full MSSM parameter space for predictions of neutralinos as cold dark matter could be covered already in a first step of the full experiment using only 100 kg of ^{76}Ge or even natural Ge, making the experiment competitive to LHC in the search for supersymmetry.

We restrict ourselves in this paper to the most recent results obtained for various particle physics parameters from double beta decay (skipping theoretical background, for which we refer to [61,62,67] and references given there), and on its future possibilities. For a more detailed review of the particle physics potential of neutrinoless double beta decay we refer to [62].

2. Double Beta Decay Experiments: Present Status and Results

2.1. Present Experimental Status

Fig. 1 shows an overview over measured $0\nu\beta\beta$ half-life limits and deduced mass limits. The largest sensitivity for $0\nu\beta\beta$ decay is obtained at present by Germanium experiments using enriched ^{76}Ge [50,52,55,57,61].

Only a few of the present most sensitive experiments may probe the neutrino mass in the next years into the sub-eV region, the Heidelberg-Moscow ^{76}Ge experiment being the by far most advanced and most sensitive one, see Fig. 1 .

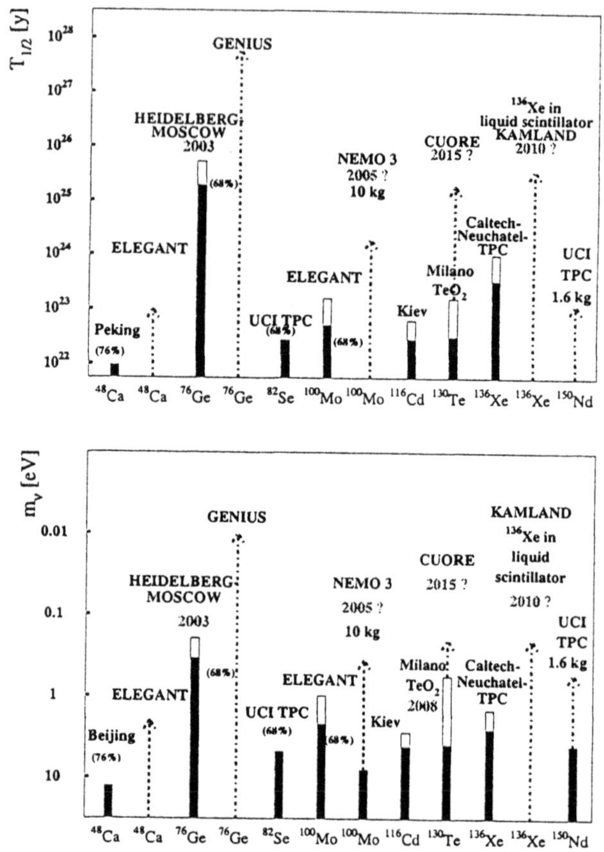

Fig. 1 *Present situation, 1998, and expectation for the near future and beyond, of the most promising $\beta\beta$-experiments concerning accessible half life (a) and neutrino mass limits (b). The filled bars correspond to the present status, open bars to expectations for running experiments, dashed lines to experiments under construction and dash-dotted lines to proposed experiments.*

No one of them will pass below $\sim 0.1 - 0.2$ eV. A detailed discussion of the various experimental possibilities can be found in [62]. A useful listing of existing data from the various $\beta\beta$ emitters is given in [92].

2.2. Present limits on beyond standard model parameters

The sharpest limits from $0\nu\beta\beta$ decay are presently coming from the Heidelberg-Moscow experiment [55,50,57,52,61]. They will be given in the following. With five enriched (86% of ^{76}Ge)

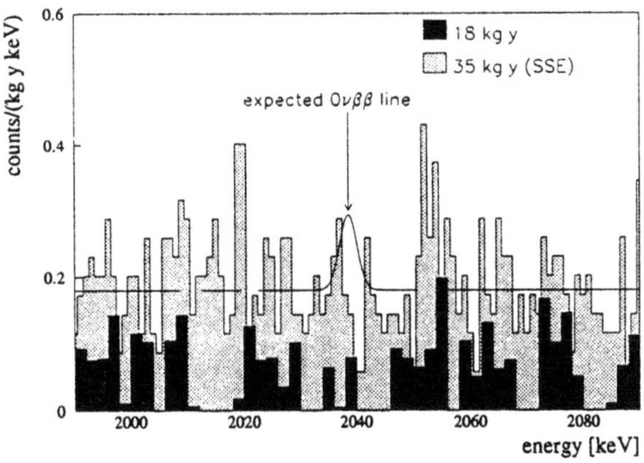

Fig. 2 *Integral spectrum in the region of interest after subtraction of the first 200 days of measurement of each detector, leaving 35 kg y of measuring time. The solid curve corresponds to the signal excluded with 90%C.L. It corresponds to* $T_{1/2}^{0\nu} > 1.2 \cdot 10^{25}$ *y. The darkened histogram corresponds to data accumulated meanwhile using a new pulse shape analysis method [35] in a measuring time of 18 kg y.*

detectors of a total mass of 11.5 kg taking data in the Gran Sasso underground laboratory, and with a background of at present 0.07 counts/kg year keV, the experiment has reached its final setup and is now exploring the sub–eV range for the mass of the electron neutrino. Fig. 2 shows the spectrum taken in a measuring time of 35 kg y. The experiment will allow to test the half life region up to $\sim 6 \cdot 10^{25} y$, corresponding to a neutrino mass limit of $\sim 0.1 - 0.2$ eV, during the next five years.

Half–life of neutrinoless double beta decay

The deduced half–life limit for $0\nu\beta\beta$ decay is

$$T_{1/2}^{0\nu} > 1.2 \cdot 10^{25} y \;\; (90\% C.L.) \qquad (1)$$

$$> 1.9 \cdot 10^{25} y \;\; (68\% C.L.) \qquad (2)$$

Neutrino mass

Light neutrinos: The deduced upper limit of an (effective) electron neutrino

Majorana mass is, with the matrix element from [90]

$$\langle m_\nu \rangle < 0.44 eV \;\; (90\% C.L.) \qquad (3)$$

$$< 0.34 eV \;\; (68\% C.L.) \qquad (4)$$

This is the sharpest limit for a Majorana mass of the electron neutrino so far. (In contrast to this value all limits 'deduced' sometimes from solar and atmospheric neutrino experiments are essentially speculations, since they are highly model-dependent: from the assumptions on the neutrino mass hierarchy (type of see-saw mechanism), of special models, like MSSM, or others). Within the typical uncertainty of the matrix elements of a factor of 2 [90,87] the present value is not yet sufficient to rule out degenerate neutrino mass scenarios [75,89,77], popular at present for the description of experimental hints on non–vanishing neutrino masses from solar and atmospheric neutrino and dark matter experiments.

Superheavy neutrinos: For a superheavy *left*–handed neutrino [50] exploiting the mass dependence of the matrix element (for the latter see [78]) a lower limit of

$$\langle m_H \rangle \geq 100 TeV \qquad (5)$$

has been deduced [19,50].

Right–handed W boson

For the right–handed W boson a lower limit of (Fig. 3)

$$m_{W_R} \geq 1.2 TeV \qquad (6)$$

is obtained [43].

SUSY parameters – R–parity breaking and sneutrino mass

The constraints on the parameters of the minimal supersymmetric standard model with explicit R–parity violation deduced [38,42,39] from the $0\nu\beta\beta$ half–life limit are more stringent than those from other low–energy processes and from the largest high energy accelerators (Fig. 4). The limits are

$$\lambda_{111}^{'} \leq 3.9 \cdot 10^{-4} \left(\frac{m_{\tilde{q}}}{100GeV}\right)^2 \left(\frac{m_{\tilde{g}}}{100GeV}\right)^{\frac{1}{2}} \qquad (7)$$

with $m_{\tilde{q}}$ and $m_{\tilde{g}}$ denoting squark and gluino masses, respectively, and with the assumption

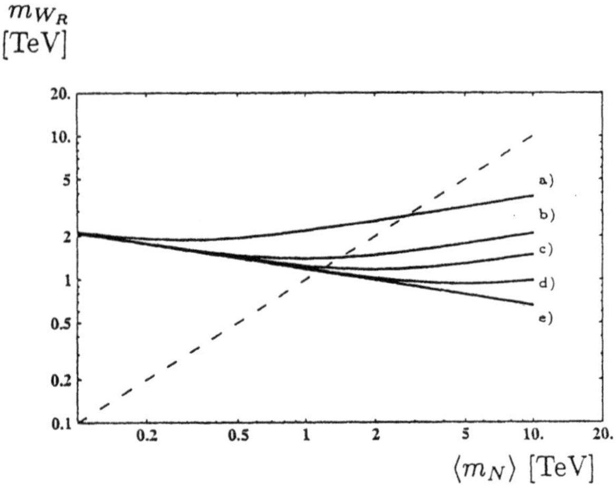

Fig. 3 *Limits on the mass of the right-handed W-boson from neutrinoless double beta decay (full lines) and vacuum stability (dashed line). The five full lines correspond to the following masses of the doubly charged Higgs, $m_{\Delta^{--}}$: a) 0.3, b) 1.0, c) 2.0, d) 5.0 and e) ∞ [TeV] (from [43]).*

$m_{\tilde{d}_R} \simeq m_{\tilde{u}_L}$. This result is important for the discussion of new physics in the connection with the high–Q^2 events seen at HERA. It excludes the possibility of squarks of first generation (of R–parity violating SUSY) being produced in the high–Q^2 events [27,2,46].

We find further [39]

$$\lambda'_{113}\lambda'_{131} \leq 1.1 \cdot 10^{-7} \tag{8}$$

$$\lambda'_{112}\lambda'_{121} \leq 3.2 \cdot 10^{-6}. \tag{9}$$

For the $(B-L)$ violating sneutrino mass $\tilde{m}_M$ the following limits are obtained [45]

$$\tilde{m}_M \leq 2\left(\frac{m_{SUSY}}{100GeV}\right)^{\frac{3}{2}} GeV, \quad \chi \simeq \tilde{B} \tag{10}$$

$$\tilde{m}_M \leq 11\left(\frac{m_{SUSY}}{100GeV}\right)^{\frac{7}{2}} GeV, \quad \chi \simeq \tilde{H} \tag{11}$$

for the limiting cases that the lightest neutralino is a pure Bino $\tilde{B}$, as suggested by the SUSY solution of the dark matter problem [53], or a pure Higgsino. Actual values for $\tilde{m}_M$ for other choices of the neutralino composition should lie in between these two values.

Another way to deduce a limit on the 'Majorana' sneutrino mass $\tilde{m}_M$ is to start from the experimental neutrino mass limit, since the sneutrino contributes to the Majorana neutrino mass m^{ν}_M at the 1–loop level proportional to $\tilde{m}^2_M$. This yields under some assumptions [45]

$$\tilde{m}_{M_{(i)}} \leq (60-125)\left(\frac{m^{exp}_{\nu(i)}}{1eV}\right)^{1/2} MeV \tag{12}$$

Starting from the mass limit determined for the electron neutrino by $0\nu\beta\beta$ decay this leads to

$$\tilde{m}_{M_{(e)}} \leq 22MeV \tag{13}$$

This result is somewhat dependent on neutralino masses and mixings. A non–vanishing 'Majorana' sneutrino mass would result in new processes at future colliders, like sneutrino–antisneutrino oscillations. Reactions at the Next Linear Collider (NLC) like the SUSY analog to inverse neutrinoless double beta decay $e^-e^- \to \chi^-\chi^-$ (where χ^- denote charginos) or single sneutrino production, e.g. by $e^-\gamma \to \tilde{\nu}_e\chi^-$ could give information on the Majorana sneutrino mass, also. This is discussed by [44,45] and by [46,72] in [66] and by [48]. A conclusion is that future accelerators can give information on second and third generation sneutrino Majorana masses, but for first generation sneutrinos cannot compete with $0\nu\beta\beta$–decay.

Compositeness

Evaluation of the $0\nu\beta\beta$ half–life limit assuming exchange of excited Majorana neutrinos ν^* yields for the mass of the excited neutrino a lower bound of [82,91].

$$m_N \geq 3.4 m_W \tag{14}$$

for a coupling of order O(1) and $\Lambda_c \simeq m_N$. Here, m_W is the W–boson mass.

Leptoquarks

Assuming that either scalar or vector leptoquarks contribute to $0\nu\beta\beta$ decay, the following constraints on the effective LQ parameters can be derived [40]:

$$\epsilon_I \leq 2.8 \times 10^{-9}\left(\frac{M_I}{100\text{GeV}}\right)^2, \tag{15}$$

$$\alpha^{(L)}_I \leq 3.5 \times 10^{-10}\left(\frac{M_I}{100\text{GeV}}\right)^2, \tag{16}$$

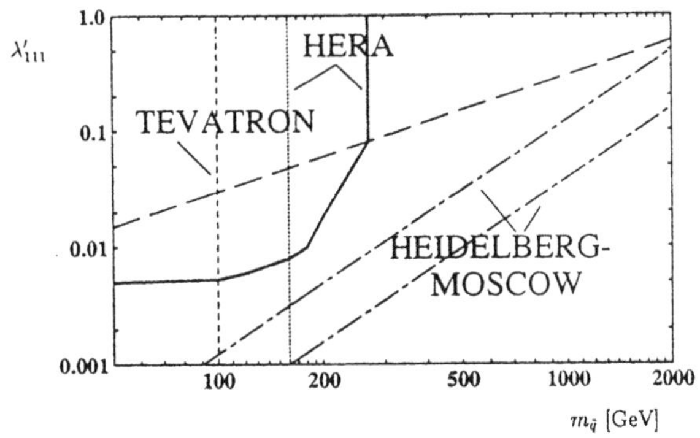

Fig. 4 *Comparison of limits on the R–parity violating MSSM parameters from different experiments in the λ'_{111}–$m_{\tilde{q}}$ plane. The dashed line is the limit from charged current universality according to [6]. The vertical dashed line is the limit from the data of Tevatron [84]. The thick full line is the region which might be explored by HERA in case of studying resonant squark production while the dotted vertical line is the mass reach of HERA probing R_p violating decays [25]. . The two dash–dotted lines to the right are the limits obtained from the half-life limit for $0\nu\beta\beta$ decay of ^{76}Ge, for gluino masses of (from left to right) $m_{\tilde{g}}$ =1 TeV and 100 GeV, respectively. The regions to the upper left of the lines are forbidden. (from [38])*

$$\alpha_I^{(R)} \leq 7.9 \times 10^{-8} \left(\frac{M_I}{100\mathrm{GeV}}\right)^2 . \qquad (17)$$

Since the LQ mass matrices appearing in $0\nu\beta\beta$ decay are (4×4) matrices [40], it is difficult to solve their diagonalization in full generality algebraically. However, if one assumes that only one LQ-Higgs coupling is present at a time, the (mathematical) problem is simplified greatly and one can deduce from, for example, eq. (15) that either the LQ-Higgs coupling must be smaller than $\sim 10^{-(4-5)}$ or there can not be any LQ with e.g. couplings of electromagnetic strength with masses below $\sim$ 250 GeV. These bounds from $\beta\beta$ decay are of interest in connection with recently discussed evidence for new physics from HERA [37,3,54,27]. Assuming that actually leptoquarks have been produced at HERA, double beta decay (the Heidelberg–Moscow experiment) would allow to fix the leptoquark–Higgs coupling to a few 10^{-6} [46]. It may be noted, that after the first consideration of leptoquark–Higgs coupling in [40] recently Babu et al. [4] noted that taking into account leptoquark–Higgs coupling reduces the leptoquark mass lower bound deduced by TEVATRON – making it more consistent with the value of 200 GeV required by HERA.

Half–life of $2\nu\beta\beta$ decay

The Heidelberg–Moscow experiment produced for the first time a high statistics $2\nu\beta\beta$ spectrum ($\gg$ 20000 counts, to be compared with the 40 counts on which the first detector observation of $2\nu\beta\beta$ decay by [30] (for the decay of ^{82}Se) had to rely). The deduced half–life is [52]

$$T_{1/2}^{2\nu} = (1.77^{+0.01}_{-0.01}(stat.)^{+0.13}_{-0.11}(syst.)) \cdot 10^{21} y \qquad (18)$$

This result brings $\beta\beta$ research for the first time into the region of 'normal' nuclear spectroscopy and allows for the first time statistically reliable investigation of Majoron–accompanied decay modes.

Majoron–accompanied decay

From simultaneous fits of the 2ν spectrum and one selected Majoron mode, experimental limits for the half–lives of the decay modes of the newly introduced Majoron models [24] are given for the first time [81,51].

The small matrix elements and phase spaces for these modes [81,41] already determined that these modes by far cannot be seen in experiments of the present sensivity if we assume typical values for the neutrino–Majoron coupling constants around $\langle g \rangle = 10^{-4}$.

3. GENIUS – A Future Large Scale Double Beta and Dark Matter Experiment, and its Physical Potential

It is obvious from Fig. 1 that *none* of the present experimental approaches, or plans or even vague ideas has a chance to surpass the border of 0.1 eV for the neutrino mass to lower values (see also [79]). At present there is only one way visible to reach the domain of lower neutrino masses,

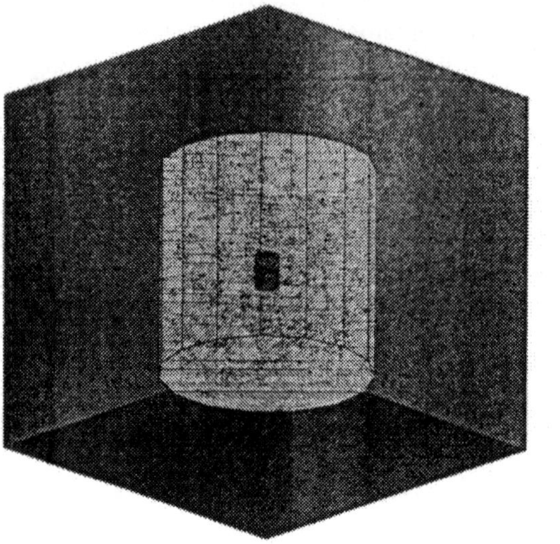

Fig. 5 *Simplified model of the GENIUS experiment: 288 enriched* ^{76}Ge *detectors with a total of one ton mass in the center of a 9 m high liquid nitrogen tank with 9 m diameter; GEANT Monte Carlo simulation of 1000 2.6 MeV photons randomly distributed in the nitrogen is also shown.*

suggested by the author of this report [62] and meanwhile investigated in some detail concerning its experimental realization and and physics potential in [64,36,69,68,11].

A simplified model of GENIUS is shown in Fig. 5 consisting of about 300 enriched ^{76}Ge detectors with a total of one ton mass in the center of a 12 m high liquid nitrogen tank with 12 m diameter. Monte Carlo simulations, using the CERN GEANT code, of the background [36,11], starting from purity levels of the nitrogen being in general an order of magnitude less stringent than those already achieved in the CTF for the BOREXINO experiment show, that a count rate in the region of interest for neutrinoless double beta decay of 0.04 counts/ keV · y · ton can be reached. Below 100 keV the background count rate is about 10 counts/keV · y · ton. Two neutrino double beta decay would dominate the spectrum with $4 \cdot 10^6$ events per year.

Starting from these numbers, a lower half-life limit of

$$T^{0\nu}_{1/2} \geq 5.8 \cdot 10^{27} \qquad (68\% C.L.) \qquad (19)$$

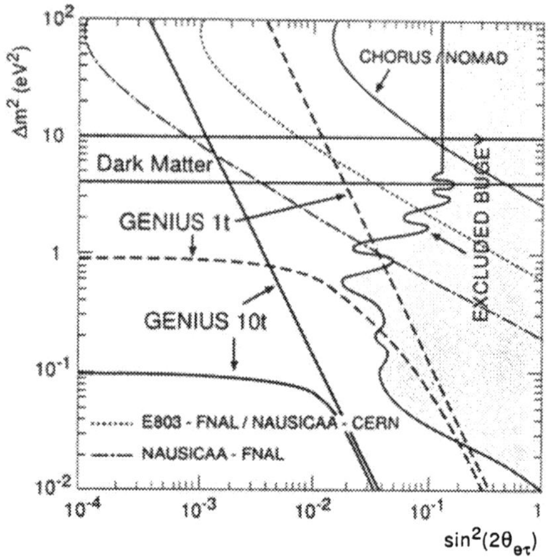

Fig. 6 *Current limits and future experimental sensitivity on* $\nu_e - \nu_\tau$ *oscillations. The shaded area is currently excluded from reactor experiments. The thin line is the estimated sensitivity of the CHORUS/NOMAD experiments. The dotted and dash-dotted thin lines are sensitivity limits of proposed accelerator experiments, NAUSICAA and E803-FNAL [32]. The thick lines show the sensitivity of GENIUS (broken line: 1 t, full line: 10 t), if neutrino masses are strongly hierarchical, the lines bending to the left correspond to R=0.01 (from [64])*

can be reached within one year of measurement (following the highly conservative procedure for analysis recommended by [80]). This corresponds – with the matrix elements of [90] – to an upper limit on the neutrino mass of

$$\langle m_\nu \rangle \leq 0.02 eV \qquad (68\% C.L.) \qquad (20)$$

In the case of zero background (this assumption might be justified since our assumed impurity concentrations are still more conservative than proved already now for example by Borexino) the final sensitivity of the experiment can be defined by the limit, which would be obtained after 10 years of measurement. For the one ton experiment this would be:

$$T^{0\nu}_{1/2} \quad \geq \quad 6.4 \cdot 10^{28}\, y \quad \text{(with 68\% C.L.)} \qquad (21)$$

and

$$\langle m_\nu \rangle \quad \leq 0.006 eV \quad \text{(with 68\% C.L.)} \qquad (22)$$

The ultimate experiment could test the $0\nu\beta\beta$ half life of ^{76}Ge up to a limit of $5.7 \cdot 10^{29}$y and the neutrino mass down to $2 \cdot 10^{-3}$eV using 10 tons of enriched Germanium. We just present a few further examples for the physics potential of GENIUS, three for particle physics, one for astrophysics. For an extensive discussion and more examples see [62,64].

GENIUS – neutrino mass textures and neutrino oscillations

GENIUS will allow to probe the neutrino mass down to $10^{-(2-3)}$ eV, and thus surpass the existing neutrino mass experiments by a factor of 50-500. GENIUS will test the structure of the neutrino mass matrix and thereby also neutrino oscillation parameters * superior in sensitivity to the best proposed dedicated terrestrial neutrino oscillation experiments. For a detailed discussion see [64,62]. It would test the LSND result in the case of quasi–degenerate neutrino masses [64,62]. This could be of interest since KARMEN will not cover the full range of LSND in future [29]. In its 10 ton version it will also allow to test the hypothesis of a shadow world underlying introduction of a sterile neutrino mentioned in section 1. Figure 6 shows part of this potential. Fig. 6 compares the potential of GENIUS with the sensitivity of CHORUS/NOMAD and with the proposed future experiments NAUSIKAA–CERN and NAUSIKAA–FNAL, looking for $\nu_e \leftrightarrow \nu_\tau$ oscillations, for different assumptions on m_1/m_2. Already in the worst case for double beta decay of $m_1/m_2 = 0$ GENIUS 1 ton is more sensitive than the running CERN experiments. In the case of quasi–degenerate neutrino masses (R=0.01), the sensitivity of GENIUS 1 ton is larger than that of all accelerator neutrino oscillation experiments planned for the future. Fig. 7 shows a summary of currently known constraints on neutrino oscillation parameters (original taken from [34]), but including the $0\nu\beta\beta$ decay sensitivities of GENIUS 1 ton and GENIUS 10 tons, for different assumptions on m_1/m_2 (and for $\eta^{CP} = +1$). It is seen that already GENIUS 1 ton tests all degenerate or quasi–degenerate ($m_1/m_2 \gtrsim 0.01$) neutrino mass models in any range where neutrinos are interesting for cosmology, and also the atmospheric neutrino problem, if it is due to $\nu_e \leftrightarrow \nu_\mu$ oscillations. GENIUS in its 10 ton version would directly test the large angle MSW solution of the solar neutrino problem, and for almost degenerate models even the small angle solution.

GENIUS and left–right symmetry:

If GENIUS is able to reach down to $\langle m_\nu \rangle \leq$ 0.01 eV, it would at the same time be sensitive to right-handed W-boson masses up to $m_{W_R} \geq 8$ TeV (for a heavy right-handed neutrino mass of 1 TeV) or $m_{W_R} \geq 5.3$ TeV (at $\langle m_N \rangle = m_{W_R}$). Such a limit would be comparable to the one expected for LHC, see for example [83], which quotes a final sensitivity of something like 5 – 6 TeV. Note, however that in order to obtain such a limit the experiments at LHC need to accumulate about $100 fb^{-1}$ of statistics. A 10 ton version of GENIUS could even reach a sensitivity of $m_{W_R} \geq 18$ TeV (for a heavy right-handed neutrino mass of 1 TeV) or $m_{W_R} \geq 10.1$ TeV (at $\langle m_N \rangle = m_{W_R}$). This means that already GENIUS 1 ton could be sufficient to definitely test recent supersymmetric left–right symmetric models having the nice features of solving the strong CP problem without the need for an axion and having automatic R–parity conservation [73,76].

---

*The double beta observable, the effective neutrino mass, can be expressed in terms of the usual neutrino oscillation parameters, once an assumption on the ratio of m_1/m_2 is made. E.g., in the simplest two–generation case

$$\langle m_\nu \rangle = |c_{12}^2 m_1 + s_{12}^2 m_2 e^{2i\beta}|, \qquad (23)$$

assuming CP conservation, i.e. $e^{2i\beta} = \eta = \pm 1$, and $c_{12}^2 m_1 << \eta s_{12}^2 m_2$,

$$\Delta^2_{m12} \simeq m_2^2 = \frac{4\langle m_\nu \rangle^2}{1 - \sqrt{1 - sin^2 2\theta}} \qquad (24)$$

A little bit more general, keeping corrections of the order (m_1/m_2) one obtains

$$m_2 = \frac{\langle m_\nu \rangle}{|(\frac{m_1}{m_2}) + \frac{1}{2}(1 - \sqrt{1 - sin^2 2\theta})(\pm 1 - (\frac{m_1}{m_2}))|}. \qquad (25)$$

For the general case see [64].

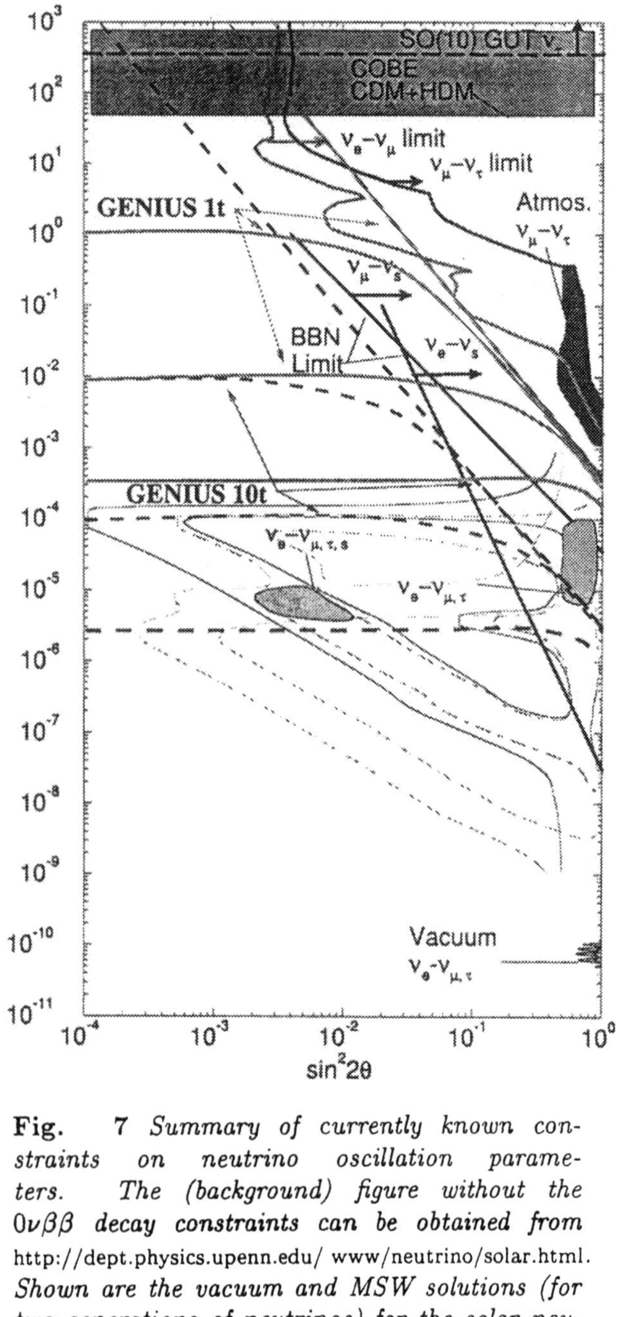

Fig. 7 *Summary of currently known constraints on neutrino oscillation parameters. The (background) figure without the* **$0\nu\beta\beta$ *decay constraints can be obtained from*** http://dept.physics.upenn.edu/ www/neutrino/solar.html. *Shown are the vacuum and MSW solutions (for two generations of neutrinos) for the solar neutrino problem, the parameter range which would solve the atmospheric neutrino problem and various reactor and accelerator limits on neutrino oscillations. In addition, the mass range in which neutrinos are good hot dark matter candidates is indicated, as well as limits on neutrino oscillations into sterile states from considerations of big bang nucleosynthesis. Finally the thick lines indicate the sensitivity of GENIUS (full lines 1 ton, broken lines 10 ton) to neutrino oscillation parameters for three values of neutrino mass ratios $R = 0, 0.01$ and 0.1 (from top to bottom). For GENIUS 10 ton also the contour line for $R = 0.5$ is shown. The region beyond the lines would be excluded. While already the 1 ton GENIUS would be sufficient to constrain degenerate and quasi-degenerate neutrino mass models, and also would solve the atmospheric neutrino problem if it is due to $\nu_e \leftrightarrow \nu_\mu$ oscillations, the 10 ton version of GENIUS could cover a significant new part of the parameter space, including the large angle MSW solution to the solar neutrino problem, even in the worst case of $R = 0$ (from [64]), and for the almost degenerate case($R = 0.5$), even the small angle MSW solution.*

GENIUS and R_p–violating SUSY:

The improvement on the R–parity breaking Yukawa coupling λ'_{111} (see section 2.2) is shown in Fig. 8, which updates Fig. 4. The full line to the right is the expected sensitivity of the LHC – in the limit of large statistics. The three dashed–dotted lines denote (from top to bottom) the current constraint from the Heidelberg–Moscow experiment and the sensitivity of GENIUS 1 ton and GENIUS 10 tons, all for the conservative case of a gluino mass of 1 TeV. If squarks would be heavier than 1 TeV, LHC could not compete with GENIUS. However, for typical squark masses below 1 TeV, LHC could probe smaller couplings. However, one should keep in mind, that LHC can probe squark masses up to 1 TeV only with several years of data taking.

The potential of GENIUS for Cold Dark Matter Search:

Weakly interacting massive particles (WIMPs) are candidates for the cold dark matter in the universe. The favorite WIMP candidate is the lightest supersymmetric particle, presumably the neutralino. The expected detection rates for neutralinos of typically less than one event per day and kg of detector mass [14–16,53,17,18], however, make direct searches for WIMP scattering

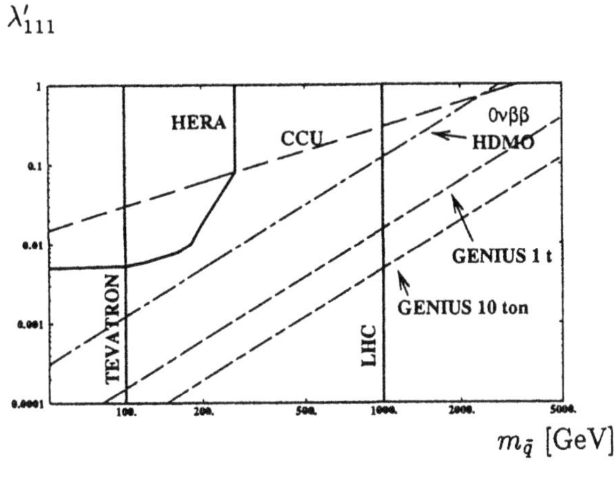

Fig. 8 *Comparison of sensitivities of existing and future experiments on $\not R_p$ SUSY models in the plane $\lambda'_{111} - m_{\tilde{q}}$. Note the double logarithmic scale! Shown are the areas currently excluded by the experiments at the TEVATRON, the limit from charged-current universality, denoted by CCU, and the limit from absence of $0\nu\beta\beta$ decay from the Heidelberg-Moscow collaboration ($0\nu\beta\beta$ HDMO). In addition, the estimated sensitivity of HERA and the LHC is compared to the one expected for GENIUS in the 1 ton and the 10 ton version. The figure is essentially an update of Fig. 4 (from [64]).*

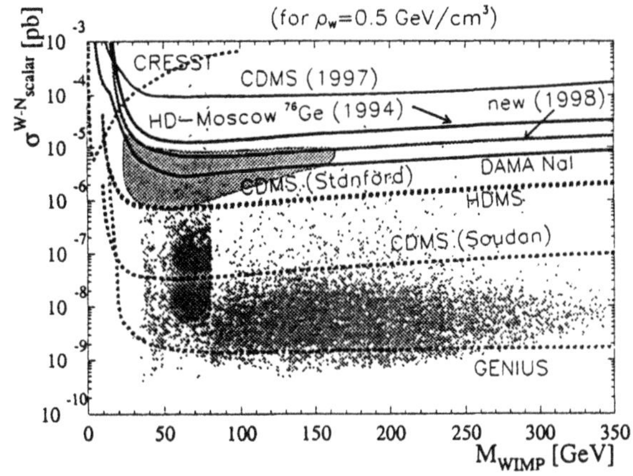

Fig. 9 *WIMP–nucleon cross section limits in pb for scalar interactions as function of the WIMP–mass in GeV. Solid lines denote the already achieved limits: HEIDELBERG–MOSCOW experiment 1994 [49] and 1998 [12] (the UKDMC NaI experiment [88] is similar to the 1994 HEIDELBERG–MOSCOW limits); the 1997 CDMS nat. Ge [1] and the new DAMA NaI results [20]. Dashed lines denote sensitivities of experiments under construction (for HDMS [10,65], CDMS [9,1,86], CRESST [28] and for GENIUS). These limits are compared to theoretical expectations (scatter plot) for WIMP–neutralino cross sections calculated in the MSSM framework with non–universal scalar mass unification [16]. The 90 % allowed region claimed by [21] (light filled area), which is further restricted by indirect dark matter searches [23] (dark filled area), could be easily tested with the GENIUS experiment. The GENIUS limit given would be reached within three years using only 100 kg of natural Ge.*

experimentally a formidable task.

Fig. 9 shows a comparison of existing constraints and future sensitivities of cold dark matter experiments, together with theoretical expectations for neutralino scattering rates [16]. Obviously, GENIUS could easily cover the range of positive evidence for dark matter recently claimed by DAMA [21,23]. It would also be by far more sensitive than all other dark matter experiments at present under construction or proposed, like the cryogenic experiment CDMS. Furthermore, obviously GENIUS will be the only experiment, which could seriously test the MSSM predictions almost over the whole SUSY parameter space. In this way, GENIUS could compete even with LHC in the search for SUSY, see for example the discussion in [5,17,18].

It is interesting to note, that if WIMP scattering is found by GENIUS it could be used to constrain the amount of R-parity violation within supersymmetric models. Due to the fact that neutralinos then are bound in the galaxy even today, neutralino decays via R-parity violating operators would have to be highly suppressed. The details depend, of course, on the neutralino mass and composition. However, finding the neutralino with GENIUS would imply typical limits on R-parity violating couplings of the order of $10^{-(16-20)}$ for any of the λ_{ijk}, λ'_{ijk} or λ''_{ijk} in the superpotential [47].

4. Conclusions

Double beta decay has a broad potential for providing important information on modern particle physics beyond present and future high energy accelerator energies which will be competitive for the next decade and more. This includes SUSY models, compositeness, left–right symmetric models, leptoquarks, and the neutrino and sneutrino mass, and even tests of special relativity and equivalence principle (see [70]). Based to a large extent on the theoretical work of the Heidelberg Double Beta group, results have been deduced from the HEIDELBERG–MOSCOW ^{76}Ge experiment for these topics and have been presented here. For the neutrino mass double beta decay now is particularly pushed into a key position by the recent possible indications of beyond standard model physics from the side of solar and atmospheric neutrinos, dark matter COBE results and others. New classes of GUTs basing on degenerate neutrino mass scenarios which could explain these observations, can be checked by double beta decay in near future. The HEIDELBERG–MOSCOW ^{76}Ge experiment has reached a leading position among present $\beta\beta$ experiments and as the first of them now yields results in the sub–eV range. We have presented a new idea and proposal of a future double beta experiment (GENIUS) with highly increased sensitivity based on use of 1 ton or more of enriched 'naked' ^{76}Ge detectors in liquid nitrogen. This new experiment would be a breakthrough into the multi-TeV range for many beyond standard models. The sensitivity for the neutrino mass would reach down to 0.01 or even 0.001 eV. The experiment would be competitive to LHC with respect to the mass of a right –handed W boson , in search for R–parity violation and others, and would improve the leptoquark and compositeness searches by considerable factors. It would probe the Majorana electron sneutrino mass more sensitive than NLC (Next Linear Collider). It would yield constraints on neutrino oscillation parameters far beyond all present terrestrial $\nu_e - \nu_X$ neutrino oscillation experiments and could test directly the atmospheric neutrino problem and the large angle solution of the solar neutrino problem. GENIUS would cover almost the full SUSY parameter space for prediction of neutralinos as cold dark matter and compete in this way with LHC in the search for supersymmetry. Even if SUSY would be first observed by LHC, it would still be fascinating to verify the existence and properties of neutralino dark matter, which could be achieved by GENIUS. Concluding GENIUS has the ability to provide a major tool for future particle– and astrophysics.

Finally it may be stressed that the technology of producing and using enriched high purity germanium detectors, which have been produced for the first time for the Heidelberg–Moscow experiment, has found meanwhile applications also in pre-GENIUS dark matter search [49,31,65,10] and in high–resolution γ-ray astrophysics, using balloons and satellites [56,57,7,8,63].

REFERENCES

1. D.S. Akerib et al., preprint astro–ph/9712343
2. G. Altarelli, J. Ellis, G.F. Guidice, S. Lola, M.L. Mangano, preprint hep-ph/9703276
3. K.S. Babu et al., preprint hep-ph/9703299 (March 1997)
4. K.S. Babu *et al.*, preprint hep-ph/9705414v2 (1997)
5. H. Baer. M. Bhrlik, hep-ph/9706509
6. V. Barger, G.F. Guidice, T. Han. Phys.Rev. **D 40** (1989) 2987
7. S.D. Barthelmy *et al.*, AIP Conf. Proc. **280** (1993) 1166
8. S. D. Barthelmy et al., Astrophys. J. **427** (1994) 519
9. P.B. Barnes *et al.*, Nucl. Instr. Meth. **A 370** (1996) 233
10. L. Baudis, J. Hellmig, H.V. Klapdor–Kleingrothaus, A. Müller, F. Petry, Y. Ramachers, H. Strecker, Nucl. Inst. Meth. **A 385** (1997) 265
11. L. Baudis, H.V. Klapdor–Kleingrothaus, B. Majorovits, Y. Ramachers, subm. for publ.
12. L. Baudis et al., subm . to Phys. Rev. Lett., 1998
13. E. Baver, M. Leurer, Phys. Rev. **D 51** (1995) 260
14. V. Bednyakov, H.V. Klapdor–Kleingrothaus,

S.G. Kovalenko, Phys. Lett. **B 329** (1994) 5 and Phys. Rev. **D 50** (1994) 7128
15. V. Bednyakov, H.V. Klapdor–Kleingrothaus, S.G. Kovalenko, Phys. Rev. **D 55** (1997) 503
16. V. Bednyakov, H.V. Klapdor–Kleingrothaus, S.G. Kovalenko, Y. Ramachers, Z. Phys. **A 357** (1997) 339
17. V.A. Bednyakov, H.V. Klapdor–Kleingrothaus, subm. to Phys. Rev. Lett.
18. V.A. Bednyakov, H.V. Klapdor–Kleingrothaus, hep-ph/9802344 vers. 3, subm. to Phys. Rev. Lett.
19. G. Belanger, F. Boudjema, D.London, H. Nadeau, Phys. Rev. **D 53** (1996) 6292
20. R. Bernabei *et al.*, Phys. Lett. **B 389** (1997) 757
21. R. Bernabei *et al.*, ROM2F/97/33; P. Belli at TAUP, Gran Sasso, Sept. 7 (1997)
22. J. Blümlein, E. Boos, A. Pukhov, preprint DESY 94-072 (April 1994)
23. A. Bottino *et al.*, hep-ph/9709292
24. C.P.Burgess, in [60]
25. J. Butterworth, H. Dreiner, Nucl. Phys. **B 397** (1993) 3 and H. Dreiner, P. Morawitz, Nucl. Phys. **B 428** (1994) 31
26. D. Choudhury, hep-ph/9408250 (1994)
27. D. Choudhury, S. Raychaudhuri, preprint hep-ph/9702392
28. S. Cooper in ICHEP96 (Singapore, World Scientific); W. Seidel et al., Phys. Lett. **B 323** (1994) 95; M. Bühler et al., Nucl. Inst. Meth. **A 370** (1996) 237
29. G. Drexlin, Proc. Internat. School on Neutrino Physics, Erice, Italy, Sept. 1997, to be publ. in Plenum Press
30. S.R. Elliot, A.A. Hahn, M.K. Moe, Phys. Rev. Lett. **59** (1987) 1649
31. T. Falk, A. Olive, M. Srednicki, Phys. Lett. **B 339** (1994) 248
32. M. Gonzalez–Garcia, hep-ph/9510419
33. H1 Collab., Phys. Lett. **B 353** (1995) 578, Z. Phys. **C 64** (1994) 545
34. N. Hata, P. Langacker, Phys. Rev. **D 50** (1994) 632
35. J. Hellmig, PhD Thesis, Univ. of Heidelberg, 1996
36. J. Hellmig, H.V. Klapdor–Kleingrothaus, Z. Phys. **A 359** (1997) 351
37. J.L. Hewett, T.G. Rizzo, preprint hep-ph/9703337v3 (May1997)
38. M. Hirsch, H.V. Klapdor–Kleingrothaus, S.G. Kovalenko, Phys. Rev. Lett. **75** (1995) 17
39. M. Hirsch, H.V. Klapdor–Kleingrothaus, S.G. Kovalenko, Phys. Lett. **B 372** (1996) 181, Erratum: Phys. Lett. **B381** (1996) 488
40. M. Hirsch, H.V. Klapdor–Kleingrothaus, S.G. Kovalenko, Phys. Lett. **B 378** (1996) 17 and Phys. Rev. **D 54** (1996) R4207
41. M. Hirsch, H. V. Klapdor–Kleingrothaus, S. G. Kovalenko, H. Päs, Phys. Lett. **B 372** (1996) 8
42. M. Hirsch, H.V. Klapdor–Kleingrothaus, S. Kovalenko, Phys. Rev. **D 53** (1996) 1329
43. M. Hirsch, H.V. Klapdor–Kleingrothaus, in [60]; M. Hirsch, H.V. Klapdor–Kleingrothaus, O. Panella, Phys. Lett. **B 374** (1996) 7
44. M. Hirsch, H.V. Klapdor–Kleingrothaus, S.G. Kovalenko, Phys. Lett. **B 398** (1997) 311 and **B 403** (1997) 291
45. M. Hirsch, H.V. Klapdor–Kleingrothaus, S.G. Kovalenko, Phys. Rev. **D 57** (1998) 1947
46. M. Hirsch, H.V. Klapdor–Kleingrothaus, S. Kovalenko, in [66]
47. M. Hirsch, H.V. Klapdor–Kleingrothaus, Proc. Int. Workshop on Dark Matter in Astro– and Particle Physics (DARK96), Heidelberg, Sept. 1996, Eds. H.V. Klapdor–Kleingrothaus and Y. Ramachers (World Scientific, Singapore) 1997, p. 640
48. M. Hirsch, H.V. Klapdor–Kleingrothaus, S. Kolb, S.G. Kovalenko, Phys. Rev. **D 57** (1998) 2020
49. HEIDELBERG–MOSCOW collab., Phys. Lett. **B 336** (1994) 141
50. HEIDELBERG–MOSCOW collab., Phys. Lett. **B 356** (1995) 450
51. HEIDELBERG–MOSCOW collab., Phys. Rev. **D 54** (1996) 3641,
52. HEIDELBERG–MOSCOW collab., Phys. Rev. **D 55** (1997) 54 and Phys. Lett. **B 407** (1997) 219
53. G. Jungmann, M. Kamionkowski, K. Griest, Phys. Rep. **267** (1996) 195
54. J. Kalinowski et al., preprint hep-ph/9703288v2 (March 1997)
55. H.V. Klapdor–Kleingrothaus, Internal Re-

port, MPI–H 1987–V17, proposal
56. H.V. Klapdor–Kleingrothaus, Proc. Int. Symposium on γ-Ray Astrophysics, Paris 1990, AIP Conf. Proc. 232 (1991) 464
57. H.V. Klapdor–Kleingrothaus, Progr. Part. Nucl. Phys. **32** (1994) 261
58. H. V. Klapdor–Kleingrothaus, A. Staudt, Non–Accelerator Particle Physics, IOP Publ., Bristol, Philadelphia, 1995; and Teilchenphysik ohne Beschleuniger, Teubner Verlag, Stuttgart, 1995
59. H.V. Klapdor–Kleingrothaus, in [60]
60. H.V. Klapdor–Kleingrothaus and S. Stoica (Eds.), Proc. *Int. Workshop on Double Beta Decay and Related Topics*, Trento, 24.4.–5.5.'95, World Scientific Singapore (1996)
61. H.V. Klapdor–Kleingrothaus, Invited talk at NEUTRINO 96, Helsinki, June 1996, World Scientific Singapore 1997, p. 317
62. H.V. Klapdor–Kleingrothaus, hep-ex/9802007, in [66], and in CERN–Courier 37 vol. **9**, p. 16 (1997)
63. H.V. Klapdor–Kleingrothaus, M.I. Kudravtsev, V.G. Stolpovski, S.I. Svertilov, V.F. Melnikov, I. Krivosheina, J. Moscow. Phys. Soc. **7** (1997) 41
64. H.V. Klapdor–Kleingrothaus, M. Hirsch, Z. Phys. **A 359** (1997) 361
65. H.V. Klapdor–Kleingrothaus, Y. Ramachers, in: Proc. Int. Workshop on Dark Matter in Astro– and Particle Physics (DARK96) Sept. 1996, Eds. H.V. Klapdor–Kleingrothaus and Y. Ramachers, Heidelberg, (World Scientific Singapore) 1997, p. 459
66. H.V. Klapdor–Kleingrothaus and H. Päs (Eds.), Proc. Beyond the Desert - Accelerator- and Non–Accelerator Approaches, Castle Ringberg, June 1997, IOP, Bristol, 1998
67. H.V. Klapdor–Kleingrothaus, in Neutrino Physics, ed. K. Winter, 2nd edition, Cambridge University Press, 1998
68. H.V. Klapdor–Kleingrothaus, Y. Ramachers, Eur. Phys. J. **A 3** (1998) 85
69. H.V. Klapdor–Kleingrothaus, Yu. Zdesenko, Eur. Phys. J. **A 3** (1998)
70. H.V. Klapdor–Kleingrothaus, H. Päs, U. Sarkar, hep-ph/9809396
71. H.V. Klapdor–Kleingrothaus, J. Hellmig, M. Hirsch, J. Phys. **G 24** (1998) 483
72. S. Kolb, M. Hirsch, H.V. Klapdor–Kleingrothaus, S. Kovalenko, in [66]
73. R. Kuchimanchi, R.N. Mohapatra, Phys. Rev. Lett. **75** (1995) 3939
74. M. Leurer, Phys. Rev. **D 49** (1994) 333
75. R.N. Mohapatra, Progr. Part. Nucl. Phys. **32** (1994) 187
76. R.N. Mohapatra, A. Rasin, Phys. Rev. Lett. **76** (1996) 3490 and Phys. Rev. **D 54** (1996) 5835
77. R. N. Mohapatra, Proc. *Neutrino 96*, Helsinki, 1996, World Scientific Singapore, 1997, p. 290
78. K. Muto, E. Bender, H.V. Klapdor, Z. Phys. **A 334** (1989) 177, 187;
79. D. Normile, Science 276 (1997) 1795
80. Particle Data Group, Phys. Rev. **D 50** (1994)
81. H. Päs *et al.*, in [60];
82. O. Panella, in [66]
83. T.G. Rizzo, hep/ph/9612440
84. D.P. Roy, Phys.Lett. **B 283** (1992) 270
85. C. Rubbia, Proc. TAUP 95, Toledo, Sept. 17–21 (1995), Nucl. Phys. **B** (Proc. Supl.) 48 (1996) 172
86. B. Sadoulet, Proc. NEUTRINO 98, World Scientific, Singapore
87. F. Simkovic *et al.*, Phys. Lett. **B 393** (1997) 267
88. P.F. Smith *et al.*, Phys. Lett. **B 379** (1996) 299
89. A. Yu. Smirnov, Proc. Int. Conf. on High Energy Physics, Warsaw 1996, hep-ph/9611465v2 (Dec 1996), (Singapore, World Scientific 1997)
90. A. Staudt, K.Muto, H.V. Klapdor–Kleingrothaus, Europhys. Lett. **13** (1990) 31
91. E. Takasugi, in [66]
92. V.I. Tretyak, Yu. Zdesenko, At. Data Nucl. Data Tables **61** (1995) 43

ELSEVIER

Nuclear Physics B (Proc. Suppl.) 77 (1999) 369–375

NUCLEAR PHYSICS B
PROCEEDINGS
SUPPLEMENTS

Present and future of low temperature detectors

O. Cremonesi
INFN Sez. Milano, Via Celoria 16, 20133 Milano, Italy
Oliviero.Cremonesi@mi.infn.it

Low temperature particle detectors are briefly presented and discussed. Their role in a few relevant aspects of neutrino physics, such as neutrinoless double beta decay (0ν-DBD), neutrino mass and solar neutrinos, is reviewed. The possible use of large arrays of bolometers to realize high sensitivity experiments is considered.

1. Introduction

Proposed more than ten years ago [1,2] as possible detectors for single particle interactions through the measurement of temperature rises in proper materials, low temperature calorimeters are going to be used in many fields of scientific research, from Biology to X-ray Astrophysics [3]. For what concerns Non-Accelerator physics, many experiments exploiting their peculiar features have been proposed and some of them are already producing the first relevant physics results. Before discussing the most important applications of thermal detectors (or *bolometers* as they are often defined) to neutrino physics, a short introduction to their properties and concepts will be given. Special relevance will be given to their applications in experiments on Solar Neutrinos, Double Beta Decay and direct neutrino mass measurements. No mention will be given to experiment on WIMPS interactions which were discussed by B.Sadoulet at this conference [4].

2. Low temperature detectors (LTD)

A very simple idea is at the base of the phonon-mediated particle detection: the specific heat of a dieletric diamagnetic material cooled down at temperatures in the few mK ranges can be so low that appreciable temperature increases can be induced in macroscopic amounts of material even by the tiny energy released by a single particle interaction. The specific heat of a diamagnetic dielectric crystal at low temperatures (lattice contribution) is in fact ruled by the Debye law

$$C = 1944 \cdot n \cdot N_{at} \cdot \left(\frac{T}{\Theta_D}\right)^3 J/K \qquad (1)$$

where n is the number of moles, N_{at} is the number of atoms per molecule, T the crystal temperature and Θ_D the material Debye temperature. According to (1), C decreases very rapidly with temperature. Materials with high Θ_D are therefore preferred. However, lattice vibrations are not the only contribution to the specific heat of materials. Metals, for example, are ruled out because of the electrons contribution which, at low temperatures, is proportional to T [5]. Materials with atomic or nuclear magnetism are also dangerous, while superconductive absorbers deserve a longer discussion. Well below the critical temperature T_c, in fact, their specific heat should be dominated by the only lattice contribution (1). However, because of the possible trapping of a large fraction of the deposited energy in quasiparticle states, the thermalization times could be very long, leading to a partial integration of the deposited energy and to a deterioration of the detector performances. Actually, contradictory results were obtained in real tests giving a general suggestion that superconductors with low Debye temperatures are preferable, being better and faster thermalizers [6].

In a very naive approach, a low temperature calorimeter can be schematized as a device consisting of a particle absorber (characterized by the lowest heat capacity C) and a sensitive "thermometer" in good thermal contact with it, in order to measure the absorber temperature variations. A weak thermal link (characterized by

0920-5632/99/$ – see front matter
PII S0920-5632(99)00445-4

a thermal conductance G) between the detector and a heat bath at constant temperature must also be considered; thus, after a temperature pulse, the original temperature will be restored in a characteristic time τ given by C/G. It can be shown[2] that the fluctuations of the internal energy of such a system are given by

$$\Delta E_{rms} \simeq \sqrt{k_b T^2 C} \qquad (2)$$

This expression represents also the intrinsic energy resolution of thermal detectors and can be as low as a few eV.

Generally, the immediate consequence of a particle interaction in the detector absorber is the fast production ($< 1\mu s$) of *high energy* non-equilibrium phonons (tens of meV) which, on much slower time scales (ms or tens of ms), convert to low energy and thermal phonons (thermalization) producing a temperature rise. The actual goal of "thermometers" is the detection of such phonons and they can therefore be divided in two main categories: fast devices, such as superconductive films or superconductive tunnel junction (STJ), which are sensitive to the fast phonon component, and thermometers (or actual temperature sensors) which measure the thermal phonon component.

Generally, the performance of a low temperature detector (and therefore its possible applications) is determined by the peculiar choice of the absorber (mass, composition, etc.) and phonon sensor properties. In particular, most of the experiments of relevance for neutrino physics are based on the use of thermometers and belong therefore to the cathegory of thermal detectors (or low temperature calorimeters).

The typical thermometers used in these applications are semiconductor thermistors doped at a critical concentration near the metal-insulator transition or superconducting films operated just across the superconducting phase transition (Transition Edge Sensors: TES). Both devices exhibit a very steep dependance of the resistance on the temperature with logarithmic sensitivities ($A = -d\log R(T)/d\log T$) of the order of 10 and 1000 respectively. They are usually embedded in a proper bias circuit and the detector temperature rise is measured as a voltage pulse whose amplitude is given by

$$(\frac{\Delta V}{V}) = A(\frac{\Delta T}{T}) \qquad (3)$$

where T is the working temperature and V the bias across the thermometer. Since many informations on the initial particle interaction are washed out by the thermalization process (position, momentum and interaction type), true thermal detectors are sensitive only to the total deposited energy which, however, can be measured very accurately.

Besides the already mentioned energy resolution (and consequently energy threshold), the main advantages of thermal detectors with respect to conventional ones are represented by their sensitivity to low- or non-ionizing events [7] and material choice flexibility. The first characteristic, joined to the low achievable thresholds, makes thermal detectors ideal devices for WIMP's search, while the second (the only constraint on the absorber material is to have a proper Θ_D) is of crucial importance in all applications in which the experiment sensitivity can be enormously increased if one has the freedom to choose the detector composition (e.g. 0ν-DBD).

Hybrid detectors, consisting of both conventional and low temperature devices, are also possible. They can be of crucial importance in all applications in which more informations besides the total deposited energy are needed and have been proposed and tested, in particular, for Dark Matter searches [4] and 0ν-DBD [8].

3. Thermal detectors for neutrino physics

Thermal detectors are no more at an R&D stage and their characteristics have already been exploited in many fields of physics (Dark Matter searches, X-ray Astrophysics, measurement of bulk material radioactive contaminations) [9]. Only experiments of relevance for neutrino physics and their future perspectives will be however discussed here.

3.1. Solar neutrinos

Both direct and indirect applications of thermal detectors have been so far suggested for solar neutrino spectroscopy. The only direct application of low temperature calorimeters was proposed some year ago by the Milano group and consisted in the realization of a large array of low temperature NaBr calorimeters (100 tons of total mass) to measure the flux of ^{7}Be solar neutrinos [10]. Solar neutrinos should be detected through the CC reaction

$$^{81}Br(\nu_e, e^-)^{81}Kr^* \quad (E_{thr} = 471.2 keV) \qquad (4)$$

to the first excited state of ^{81}Kr (the transition to the ground state of ^{81}Kr is forbidden). The experimental signature should consist in the *fast* coincidence (~13 sec) between the prompt electron (4) and the delayed 190 keV de-excitacion photon (or internal conversion electron) and give rise to a background rejection at the level of 10^{-3} events/day to be compared with an estimated signal of 0.3 events/day. A statistics of $\simeq$ 100 events was thus expected for the higher energy ^{7}Be line in one year measurement. By exploiting the expected good energy resolution (few keV FWHM), an accurate measurement of the ^{7}Be line shape and therefore a direct measurement of the interior solar measurement was suggested [11]. Unfortunately, the first tests on NaBr crystals as thermal detectors were not encouraging [10], and more systematic investigation is required.

Two indirect applications of thermal detectors for solar neutrino experiments have been so far proposed. In both cases the peculiar properties of low temperature calorimeters are exploited to realize efficient counting system for radiochemical detectors. The first detector has been proposed by the Genova-Moscow [12] collaboration to study the reaction

$$^7Li(\nu_e, e)^7Be \quad (E_{thr} = 860 keV) \qquad (5)$$

followed, with a lifetime of 53.38 days by the EC decay of ^{7}Be. Originally it was suggested to detect this decay (thus counting the number of neutrino induced reactions) by observing the emission of the 478 γ-ray following the reaction (5) to an excited state of ^{7}Li, which occur with a branching ratio of 10 %. The estimated counting efficiency was however only 8 %. Recently, the Genoa group has demonstrated the possibility to detect the EC decay of ^{7}Be by directly observing the X-rays and Auger electrons following it, by means of a μ-bolometer [9]. An impressive increase of the experiment sensitivity is thus achievable. The necessity to use low temperature μ-calorimeters is related to the fact that the total energy released in the EC decay of ^{7}Be to the g.s. of ^{7}Li is only 112 eV. The developed μ-calorimeter consisted of a NTD#12 (Neutron Transmutation Doped) Ge thermistor glued to an absorber containing ^{7}Be nuclei (Be metal and BeO_2 samples irradiated with protons at the Moscow Meson Facility). The LTD absorber material choice seems compatible with the successful chemical extraction methods under test in Moscow. No analysis of the counting background conditions has however been so far carried out. Problems could also rise from the stocking of large amounts of Li deep underground. A statistical accuracy of 12 % on the total solar ν flux is expected in one year of exposure of 10 tons of Lithium.

The second application of an LTD to the counting system of a solar ν radiochemical experiment has been recently suggested by the GNO collaboration [13] to improve the counting efficiency of the experiment, by substituting the low activity μ-proportional counters developed and successfully operated for GALLEX with a proper thermal μ-calorimeter. A prototype detector, consisting of a Ir/Au (proximity effect) TES deposited on a sapphire substrate, was developed by the Munich group. The spectrum obtained after depositing a spot of ^{71}Ge on the sapphire substrate shows an excellent energy resolution but background measurements and the development of a *fast* method to prepare the actual detectors for the experiment (synthesis and deposition of the ^{71}Ge extracted from the Ga target) still need a more systematic investigation.

3.2. Neutrinoless Double Beta Decay

Neutrinoless double beta decay is a rare nuclear process in which two electrons are simultaneously emitted by an even-even nucleus (A,Z). It violates lepton number conservation and its observation would have direct implications in the field of Gran Unified or Supersymmetric theories [14]. The most sensitive method to investigate 0ν-DBD is to build a high energy resolution detector containing the candidate nuclei and acting therefore also as the decay source [15]. So far, the highest sensitivity experiment has been realized using this technique to study the candidate nucleus ^{76}Ge, taking advantage of the good energy resolution of large germanium semiconductor detectors [16]. The experimental signature is a sharp peak in the background energy spectrum at the DBD transition energy. The use of thermal detectors to search for such a decay has been suggested in 1984 [1] essentially to allow an extension of the "source=detector" technique to candidate nuclei other than ^{76}Ge. Up to now, only the Milano group has been involved in this search. Calorimeters of TeO_2, $CdWO_4$, $PbMoO_4$ and CaF_2 have already been developed and tested to investigate the isotopes ^{130}Te, ^{116}Cd, ^{100}Mo and ^{48}Ca. However, most efforts have been devoted during the last years to the realization of a sensitive experiment on ^{130}Te $0\nu\beta\beta$, and the first results have been already produced. Because of its large natural isotopic abundance (33.87 %), its reasonably high transition energy (2528 keV) and the favourable predicted 0ν decay rate [14], ^{130}Te is an interesting candidate to 0ν-DBD. Moreover, large single crystals of TeO_2 with excellent features as thermal detectors can be grown. Preliminary experiments with single TeO_2 crystals of 73 g and 334 g and with a preliminary array of four detectors [17] were performed in a low activity cryostat located in the Gran Sasso underground laboratories (LNGS, Italy), at a depth of about 3500 m.w.e.

Since heat capacity imply a strong constraint on the dimensions of the single detector, the only way to improve the experimental sensitivity is to use arrays consisting of a large number of proper mass single crystals. This technique is very powerful and could represent a valid alternative to the recently proposed future experiments with conventional detectors. The present experiment consists in a tower of five floors of 4 detectors mounted inside a low radioactivity copper frame and operating in Hall A of LNGS under a heavy shield of radiopure materials (Cu and Pb). The twenty absorbers are crystals of natural TeO_2 of $3 \times 3 \times 6$ cm^3 with a total active mass of about 6.8 kg while the temperature sensors are NTD Ge thermistors glued to each of them. A resistor of 10 to 100 kΩ, realized with a heavily doped meander on a 1 mm^3 silicon chip, has been attached to each absorber and acted as a heater to calibrate and stabilize the gain of the bolometer. A preliminary test run was carried out in May 1998 [18]. The array base and working temperatures were about 8 mK and 13 mK respectively. The 20 detectors were calibrated using a ^{232}Th radioactive source positioned just outside the cryostat vessel. The corresponding spectrum, obtained adding the spectra of the 20 detectors, is presented in Fig.1 and shows the excellent reproducibility of the array. The FWHM energy resolutions at the 2615 keV ^{208}Tl line is about 10 keV. A preliminary search on $0\nu\beta\beta$ decay was carried out by combining the pulses from the 20 detectors collected in 0.145 y·kg of effective running time background measurement.

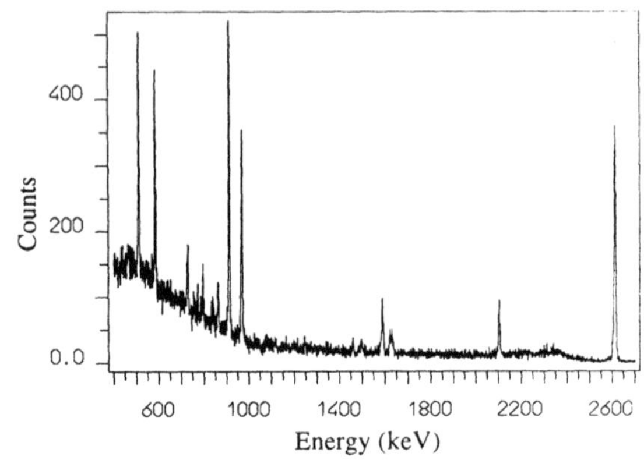

Figure 1. Milano group: sum spectrum of the 20 TeO_2 detectors exposed to a source of ^{232}Th.

The spectrum in the region above 2 MeV (Fig.2) shows only the line at 2615 keV (^{208}Tl), confirming the long term stability and reproducibility of the array, and no evidence for $0\nu\beta\beta$ decay. A lower limit of 8.7×10^{22} years on the half life was obtained at 90 % C.L. corresponding to effective neutrino mass limits ranging from 2.5 to 5.2 eV [19]. The experiment is currently continuing data taking.

The impressive results obtained with the 20

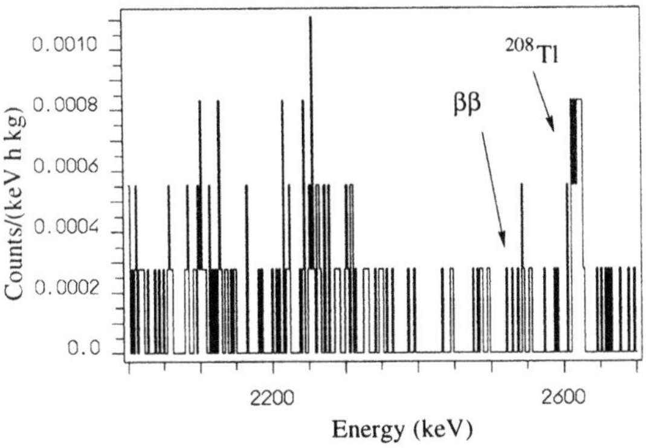

Figure 2. Background spectrum obtained by the Milano group with the 20 TeO_2 bolometer array.

bolometer array (reproducibility and long-term stability) have induced the Milano group to join other groups for the realization of a large scale experiment (CUORE) to search for neutrinoless Double Beta Decay, Cold Dark Matter and Solar Axions. It should consist in a close-packed array of 1000 TeO_2 bolometers, with a mass of 750 g each, for a total of 750 kg. The CUORE project (which stands for Cryogenic Underground Observatory for Rare Events and, in italian, means *heart*) is proposed by a still open collaboration, including at the moment: Milano, Gran Sasso and Firenze, Italy; Berkeley and South Carolina, USA; Leiden, the Netherlands; Neuchatel, Switzerland; Zaragoza, Spain. In a very preliminary approach, CUORE should consist of seventeen towers, each tower consisting of a stack of 15 modules. Each module, which is the smallest independent unit, should contain 4 crystals. The whole detector will be included in a 110 cm high, 75 cm diameter cylindrical volume, which will corresponds to the evacuated experimental space of the dilution refirgerator. The operation temperature should be around 10 mK. If needed, other materials could be easily studied in addition to or in place of the TeO_2 crystals. The total cost of CUORE should be around 8 M$, most of which for the crystals (6 M$), while the rest is mainly due to the refrigerator and the electronics. Since a single CUORE tower could be cooled down in the refrigerator presently housing the 20 element array, a test of the CUORE principle and a new very powerful DBD experiment, containing 5.7×10^{25} ^{130}Te nuclei could be carried out in a reasonably near future. The CUORE collaboration is proposing this project with the name "CUORICINO", which means *small CUORE*. Assuming, conservatively, for CUORE the same background and performances as in the 20 detector array (0.5 counts/(keV·kg·day)), an effective neutrino mass 5 y sensitivity of the order 0.2 eV can be estimated. However, this value could be lowered to 6×10^{-2} eV if a reasonable improvement of the present background by a factor 100 could be achieved by insisting on the radiopurity of the materials and on the self-shielding and granularity of the detector. A Monte Carlo evaluation of the background on the basis of the proposed preliminary structure and of the typical contamination levels of the employed materials (mainly copper and PTFE) is in progress. As far as WIMPs and axion searches are concerned, the CUORE potential will depend mainly on the threshold and the low energy background achievable. Two CUORE prototype bolometers (750 g TeO_2) were successfully operated in LNGS just during the conference.

3.3. Direct measurements of neutrino mass

The analysis of low energy beta spectra (e.g. ^{3}H or ^{187}Re), to search for possible non vanishing values of the electron antineutrino mass is an ideal application of thermal detectors. By allowing a true calorimetric measurement of the electronic spectrum, free from the systematics associated to molecular excited final states or to any mechanism which could imply any energy loss in the source, the thermal technique is complementary to the conventional spectrometer technique. Very good energy resolutions (few eV) can be in principle achieved, the only problem being the intrinsic slowness of low temperature calorimeters which could constrain statistical accuracy.

In the "source=detector" approach, two possibilities can be considered: either a detector realized with an absorber which naturally contains the beta active isotope, or one implanted or em-

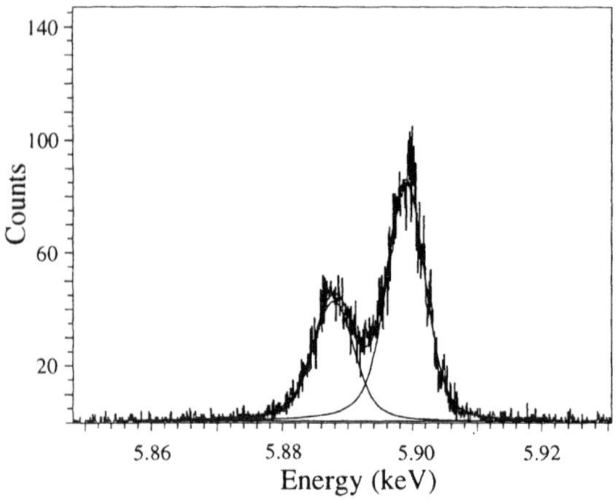

Figure 3. ^{55}Mn K_α doublet observed by the Milano group with a μ-bolometer consisting of a Tin absorber and a Ge NTD thermistor exposed to a ^{55}Fe source. The energy resolution is 5.8 eV FWHM after deconvolution of the natural line width [20].

bedded with the isotope under investigation. In the first case, Rhenium is the ideal candidate: ^{187}Re has in fact a natural isotopic abundance of 62.60 % and a very low end point ($\sim$ 2.5 keV). Assuming an energy resolution of 10 eV and a detector mass of a few mg of natural rhenium (corresponding to a counting rate of $\sim$ 10 Hz), a 5 eV sensitivity on the antineutrino mass in one year measurement could be achieved. Unfortunately pile-up effects constrain the maximum source rate to 0.1-1 Hz and a different approach based on the use of many similar detectors could be necessary in order to reach a good statistical accuracy. Due to its higher spectrum end point (18.7 keV), this is particularly compelling in the case of tritium, which represent the best candidate for an implanted LTD.

So far, only the Milano and Genoa groups are involved in such an investigation. In both cases the selected β-emitter is ^{187}Re.

Two similar detectors consisting of a Re metal absorber glued to an NTD Ge thermistor and differing only for the absorber mass (few hundred μg and $\sim$ 1 mg respectively) were developed by the Genoa group. Energy reolutions still far from the required ones ($\sim$ 30 and $\sim$ 75 eV FWHM respectively) were obtained. The Re spectrum obtained with the smaller detector was used to measure ^{187}Re end point and lifetime [21,9], while the second is currently used to study possible deformations of the β spectrum due to atomic effects on the emitted electron wavefunction (BEFS). To improve their detector performances this group has recently adopted a new technique based on the use of TES [9] instead of the usual Ge NTD. Preliminary results are encouraging and they hope to realize an actual Re detector in the near future. A different technique to search for a non-vanishing mass of the electron neutrino, based on the study of the X-ray spectra of EC decaying isotopes (e.g. ^{163}Ho), has been investigated by the same group. Also in this case, the preliminary results are encouraging but the statistical accuracy is still poor and could represent a strong limitation of the technique. No results on the neutrino mass has been yet quoted.

A different approach has been adopted by the Milano which, in collaboration with IRST (Istituto per la Ricerca Scientifica e Tecnologica, Trento, Italy) [9] is developing a technique to realize a large number of highly reproducible high resolution μ-calorimeters with Si thermometers.

Hundreds of thermistors exhibiting the same $R-T$ behaviour within the experimental accuracy has been so far realized. Recently, the use of very small Ge NTD thermistors has been also considered. Thermistor performances are studied by constructing detectors with Sn absorbers. Very good results have been obtained recently with both kind of thermistors: 13.5 eV and 5.8 eV FWHM (after deconvolution of the natural line width) with Si and Ge NTD sensors, respectively, at the 5.9 keV line of ^{55}Mn (Fig. 3). Concerning Rhenium detectors, dielectric compounds (e.g. $AgReO_4$) have been chosen by this group which is presently working on the absorbers purity. Preliminary results obtained with a 220 μg Re test detector with Si thermistor are encouraging (Fig. 4). As in the case of the Genoa group, the energy resolution ($\sim$56 eV at 5.9 keV) is still

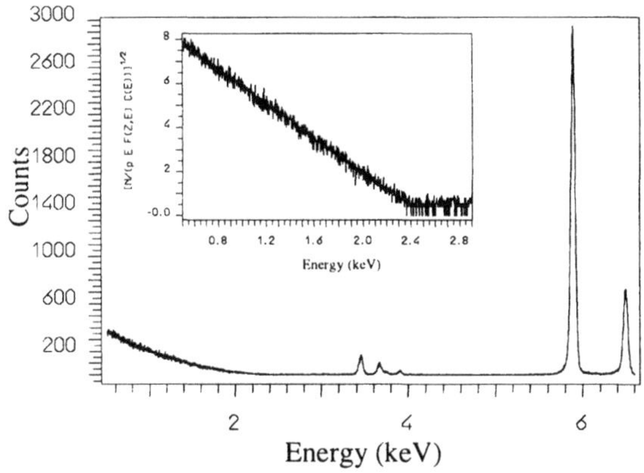

Figure 4. ^{187}Re spectrum collected by the Milano group with a 220 μg $AgReO_4$ absorber and a Si thermistor. The lines from external ^{55}Fe and Sn fluorescence sources are also apparent. The ^{187}Re Kurie plot is shown in the inset.

far from the required one. The realization of an array of 10 Re detectors with a total rate of $\sim$1 Hz, an energy resolution of 20 eV FWHM and an estimated sensitivity of 10 eV on m_ν is the main challenge of this group for the next future.

4. Conclusions

Low temperature thermal detectors are going to play a relevant role in Neutrino Physics. Due to their excellent energy resolution, wide choice of detecting materials and good efficiency in the detection of massive particles they have been proposed for a variety of applications. Actual experiments on $0\nu\beta\beta$ decay are already producing results and the technique looks promising also for future high sensitivity experiments on Neutrino Mass, Dark Matter and Solar Neutrinos.

REFERENCES

1. E. Fiorini and T.O. Niinikoski, *Nucl. Instr. Meth.* **224** (1984) 83.
2. S.H. Moseley et al., *J. Appl. Phys.* **59** (1984)1257.
3. For recent reviews on cryogenic detectors see D. Twerenbold, *Rep.Prog. Phys.* **59** (1996) 349 and N. Booth, B. Cabrera and E.Fiorini, *Ann. Rev. of Nucl.Sci.* **46** (1996) 471
4. B. Sadoulet, *These proceedings*
5. F. Pobell, *Matter and Methods at Low Temperatures*, (Springer-Verlag, Berlin-Heidelberg, 1992).
6. S. Vitale et al., *Proc. SPIE's 1992 International Symposium on Optical Applied Science and Engineering, 19-24 July 1992, San Diego CA, USA.*
7. A. Alessandrello et al., *Phys. Lett.* **B202** (1988) 611.
8. A. Alessandrello et al., *Phys. Lett.* **B420** (1998) 109.
9. For more details on thermal detectors and their applications, see: *Proc. VII Intern. Workshop on Low Temperature Detectors*, Munich, Germany, July 1997, *Ed. by S.Cooper.*
10. A. Alessandrello et al., *Astropart. Phys.* **3** (1995) 239.
11. J.N. Bahcall, *Astrophysics Preprint Series*, IASSNS-AST 93/41, Sep. 93, Princeton.
12. A.V.Kopylov et al., Int. Symp. Neutr. Cosmol., Baksan 1993, INRR Academ. Sci.
13. T.Kirsten, *These proceedings*
14. M. Moe and P. Vogel, *Annu. Rev. Nucl. Sci.* **44** (1994) 247, and references therein.
15. G.F. Dell'Antonio and E. Fiorini, *Suppl. Nuovo Cim.* **17** (1960) 132
16. H.V. Klapdor-Kleingrothaus, *These proceedings*
17. A. Alessandrello et al., *Phys. Lett.* **B285** (1992) 176, *Phys. Lett.* **B335** (1994) 519, *Nucl.Phys.B (Proc.Suppl.)* **48** (1996) 238.
18. A. Alessandrello et al., *Phys. Lett.* **B33** 156.
19. H.V. Klapdor-Kleingrothaus *Proc. of NANP-97, Dubna* , 1997.
20. P.L. Lee and S.I. Salem, *Phys. Rev.* **154** (1974) 2027.
21. E. Cosulich et al., *Phys. Lett.* **B295** (1992) 143.

N·H
ELSEVIER
Nuclear Physics B (Proc. Suppl.) 77 (1999) 376–385
NUCLEAR PHYSICS B
PROCEEDINGS
SUPPLEMENTS

Particle Physics Implications of Neutrinoless Double Beta Decay*

R. N. Mohapatra[a†]

[a]Department of Physics, University of Maryland, College Park, MD-20742, U. S. A.

Neutrinoless double beta decay is a sensitive probe of the patterns of neutrino masses and mixings if the neutrinos are Majorana particles as well as other new physics scenarios beyond the standard model. In this talk, the present experimental lower bound on the lifetime for $\beta\beta_{0\nu}$ is used to constrain the neutrino mixings and set limits on the parameters of the new physics scenarios such as the left-right symmetric models, R-parity violating SUSY models etc which lead to neutrinoless double beta decay. We then discuss proposed high precision searches for $\beta\beta_{0\nu}$ decay that can provide extremely valuable insight not only into the nature of neutrino mixings and masses but also put constraints on (or even rule out) new physics scenarios.

1. Introduction

In the standard electroweak model of Glashow, Weinberg and Salam, the absence of the right-handed neutrinos and the existence of an exact accidental global $B-L$ symmetry guarantees that the neutrinos are massless to all orders in perturbation theory. Any experimental evidence for a non-zero neutrino mass therefore constitutes evidence for new physics beyond the standard model and will be a major step towards a deeper understanding of new forces in nature[1]. Among the many experiments that are under way at this moment searching directly or indirectly (e.g. via neutrino oscillations) for neutrino masses, one of the most important ones is the search for neutrinoless double beta decay. This process is allowed only if the neutrino happens to be its own antiparticle (Majorana neutrino) as is implied by many extensions of the standard model. However, since $\beta\beta_{0\nu}$ decay changes lepton number (L_e) by two units any theory that contains interactions that violate electron lepton number L_e can in principle lead this process. This therefore reflects the tremendous versatility of $\beta\beta_{0\nu}$ decay as a probe of all kinds of new physics beyond the standard model. Indeed we will see that already very stringent constraints on new physics scenarios such as the left-right symmetric models with the see-saw mechanism[2] and supersymmetric models with R-parity violation[3], scales of possible compositeness of leptons etc are implied by the existing experimental limits[4] on this process. For a more detailed discussion of the theoretical situation than is possible here, see [5]. For an update of the experimental situation, both ongoing and in planning stage, see [6].

This talk is organized as follows: In section 2, I discuss the basic mechanisms for neutrinoless double beta decay ; in section 3, the implications of the present limits on the lifetime for neutrinoless double beta decay for neutrino mixings are discussed; in part section 4, I go on to discuss the kind of new physics scenarios that can be probed by $\beta\beta_{0\nu}$ decay and the constraints on the parameters of the new physics scenarios implied by present data.

2. Mechanisms for $\beta\beta_{0\nu}$ decay

As is wellknown, if the neutrino is its own antiparticle, the conventional four-Fermi interaction can lead to neutrinoless double beta decay via the diagram in Fig. 1. In physics scenarios beyond the standard models, if there are heavy Majorana fermions interacting with the electrons, diagrams similar to Fig. 1 with neutrino line replaced by the Majorana fermions can also lead to $\beta\beta_{0\nu}$ decay. Examples of such particles abound in literature: right-handed neutrino, photino, gluino to mention a few popular ones.

One could therefore give an arbitrary classi-

†Work supported by the National Science Foundation Grant No.PHY-9802551

PII S0920-5632(99)00447-8

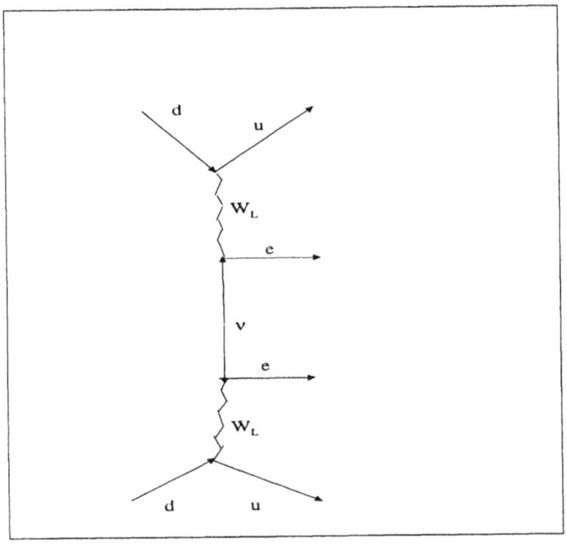

Figure 1. Feynman diagram involving neutrino majorana mass that contributes to $\beta\beta_{0\nu}$ decay.

fication of the mechanisms for $\beta\beta_{0\nu}$ decay into two kinds: (A) one class that involves the exchange of light neutrinos; and (B) the second class that involves heavy fermions or bosons. Furthermore, there are two distinct mechanisms for light neutrino exchange contributions: (a) helicity flip light neutrino mass mechanism and (b) helicity nonflip vector-vector or vector-scalar mechanism. In case (a), one can write the amplitude $A_{\beta\beta}$ for neutrinoless double beta decay to be:

$$A^{(m)}_{\beta\beta} \simeq \frac{G_F^2}{2}\langle\frac{m_\nu}{k^2}\rangle_{Nucl.} \quad (1)$$

whereas in case (b), it looks like:

$$A_{\beta\beta} \simeq \frac{G_F^2}{2}\langle\frac{\eta}{\gamma\cdot k}\rangle_{Nucl.} \quad (2)$$

To extract neutrino mass implications for neutrinoless double beta decay, we need to note the explicit form of $< m_\nu >$:

$$< m_\nu > = \Sigma_i U^2_{ei} m_i \quad (3)$$

where U_{ei} are the mixing matrix elements for the electron neutrino with the other neutrinos. Therefore a constraint on the $< m_\nu >$ can be converted into constraints on the neutrino mixings involving the first generation. Incidentally, one can also write $< m_\nu >= m_{ee}$ where m_{ee} is the ee entry of the neutrino mass matrix in the weak basis. Thus any theory which has zero entry in the ee location leads to vanishing neutrinoless double beta decay even if the neutrino is a Majorana particle.

It is important to remark that these kind of light neutrino exchange diagrams always lead to a long range neutrino potential inside the nucleons and therefore, crudely speaking the two nucleons "far" from each other can contribute in the double beta decay. This has important implications for the evaluation of the nuclear matrix element[7], an important subject we do not discuss here. We will instead use an effective momentum to parameterize the effect of the nuclear matrix element calculations (we will roughly choose $p_{eff} \approx 50$ MeV). The width for double beta decay amplitude is given by

$$\Gamma_{\beta\beta} \simeq \frac{Q^5|A|^2}{60\pi^3} \quad (4)$$

Here, Q is the available energy for the two electrons. Using the present most stringent limit on $\tau_{\beta\beta} \geq 1.1 \times 10^{25}$ years obtained for ^{76}Ge by the Heidelberg-Moscow group, one can obtain the upper limit on the width to be $\Gamma_{\beta\beta} \leq 3.477 \times 10^{-57}$ GeV; using Eq. (4), A for the light neutrino contribution, $Q \simeq 2$ MeV and $p_F \simeq 50$ MeV, one gets a rough upper limit of .7 eV for the neutrino mass. A more careful estimate leads to

$$< m_\nu > \leq .46\ eV \qquad \eta \leq 10^{-8} \quad (5)$$

(B): The second class of mechanisms consists of exchange of heavy particles which often arise in physics scenarios beyond the standard model. In the low energy limit, the effective Hamiltonian that leads to $\beta\beta_{0\nu}$ decay in these cases requires point interaction between nucleons; as a result, in general the nuclear matrix elements are expected to be smaller due to hard core repulsive nuclear potential; nevertheless, a lot of extremely useful information have been extracted about new physics where these mechanisms operate. Symbolically, such contributions can arise from effective Hamiltonians of the following type(we have suppressed all gamma matrices as well as color indices):

$$H^{(1)} = G_{eff}\ \bar{u}\Gamma d\bar{e}\Gamma F\ +\ h.c. \tag{6}$$

or

$$H^{(2)} = \lambda_\Delta \left(\frac{1}{M^3}\ \bar{u}\Gamma d\bar{u}\Gamma d\ +\ e^-e^-\right)\Delta^{++} + h.c. \tag{7}$$

Here F represents a neutral majorana fermion such as the right-handed neutrino (N)[8] or gluino $\tilde{G}$ or photino $\tilde{\gamma}$ and Δ^{++} represents a doubly charged scalar or vector particle. In the above equations, the coupling G_{eff} has dimension of M^{-2} and λ_Δ is dimensionless. The possibility of the doubly charged scalar contribution to $\beta\beta_{0\nu}$ was first noted in [9] and have been discussed subsequently in [10]. The contributions to neutrinoless double beta decay due to the above interactions lead to $\beta\beta_{0\nu}$ amplitudes of the form:

$$A^{(F)}_{\beta\beta} \simeq G^2_{eff}\frac{1}{M_F}(p^{eff})^3 \tag{8}$$

and

$$A^{\Delta}_{\beta\beta} \simeq \left(\frac{\lambda^2_\Delta}{M^3M^2_\Delta}\right)(p^{eff})^3 \tag{9}$$

Here again we have crudely replaced all nuclear effects by the effctive momentum parameter p^{eff}. If we choose $p^{eff} \simeq 50$ MeV, then the present lower limit on the lifetime for ^{76}Ge decay leads to a crude upper limit on the effective couplings as follows:

$$G_{eff} \leq 10^{-7}\left(\frac{M_F}{100\ GeV}\right)^{\frac{1}{2}} \tag{10}$$

and

$$\lambda_\Delta \leq 10^{-3}\left(\frac{M}{100\ GeV}\right)^{\frac{5}{2}} \tag{11}$$

In the second equation above, we have set $M = M_\Delta$. Note that these limits are rather stringent and therefore have the potential to provide useful constraints on the new physics scenarios that lead to such particles.

3. Implications for neutrino masses and mixings

This conference watched the history of neutrino physics take a remarkable new turn. Convincing evidence was presented by the Super-Kamiokande collaboration for the existence of neutrino oscillation of the atmospheric muon neutrinos to either ν_τ or a sterile neutrino. Using data both in the sub-GeV and multi-GeV energy range for the electron and the muon neutrinos as well as the zenith angle dependence of the muon data, the present fits at 90% confidence level seem to imply the following values for the oscillation parameters Δm^2 and $sin^2 2\theta$: $4\times 10^{-4} \leq \Delta m^2 \leq 5\times 10^{-3}$ eV^2 with $sin^2 2\theta$ between .8 to 1[11]. The possibility of $\nu_\mu - \nu_e$ oscillation as an explanation of the atmospheric anomaly seems to run into conflict with the recent CHOOZ [12] experiments. Neutrino oscillation also seems to be the only way to understand the deficit of the solar neutrinos[13]. The detailed oscillation mechanism in this case is however is unclear. The three possibilities are: a) Small-angle MSW[14], $\Delta m^2_{ei} \simeq 6\times 10^{-6} eV^2, sin^2 2\theta_{ei} \simeq 7\times 10^{-3}$; b) Large-angle MSW, $\Delta m^2_{ei} \simeq 9\times 10^{-6} eV^2, sin^2 2\theta_{ei} \simeq 0.6$; c) Vacuum oscillation, $\Delta m^2_{ei} \simeq 10^{-10} eV^2, sin^2 2\theta_{ei} \simeq 0.9$. The data on neutrino energy distribution presented at this conference indicates a preference towards vacuum oscillation rather than MSW mechanism. Turning to the laboratory experiments, the LSND[15] collaboration has presented evidence in favor of a possible oscillation of $\bar{\nu}_\mu \rightarrow \bar{\nu}_e$ as well as $\nu_\mu - \nu_e$. The preferred Δm^2 range seems to be $.24 \leq \Delta m^2_{e-\mu} \leq 10$ eV^2 with a mixing angle in the fea percent range. As already mentioned, $\beta\beta_{0\nu}$ gives only an upper bound of $< m_{\nu_e} > < .46$ eV.

Another effect of neutrino mass is in the arena of cosmology, where it not only effects whether the universe keeps expanding for ever or it eventually collapses onto itself, but it also determines the detailed manner in which structure formed in the early universe. This subject is in a constant state of flux due to new cosmological data coming in at a very rapid rate. But the idea that the present structure data may need a neutrino mass contribution to the dark matter is very much alive (see for instance Ref.[16] which seems to suggest that a total neutrino mass of 4-5 eV which contributes about 20% of the dark matter along with 70% cold dark matter and 10% baryon gives the best fit to the galaxy power spectrum data. This taken seriously would mean that $\Sigma_i m_{\nu_i} = 4 - 5$ eV).

With the above input information, if we stay within the minimal three neutrino picture, then the solar neutrino puzzle can be resolved by $\nu_e \rightarrow \nu_\mu$ oscillations and the atmospheric neutrino deficit by $\nu_\mu \rightarrow \nu_\tau$ oscillations and the LSND results cannot be accomodated. Note that these observables are controlled only by the mass square difference; on the other hand, the required hot dark matter implies that at least one or more of the neutrinos must have mass in the few eV range. It was pointed out[17] in 1993 that, in the minimal picture, this leads to a scenario, where all three neutrinos are nearly degenerate, with $m_{\nu_e} \approx 1.6$ eV. It is then clear that, in this case, in general there will be an observable amplitude for neutrinoless double beta decay mediated by the neutrino mass. In fact, if the limit on $\langle m_\nu \rangle$ is taken to be less than .47 eV as is implied for a certain choice of the nuclear matrix element, then the mixing must satisfy the constraint:

$$\Sigma_i U_{ei}^2 \simeq 0 \tag{12}$$

Since each of the elements in the above sum is complex, the U_{ei}^2 form the three sides of a triangle[18]. Then using the unitarity relation for the U matrix, it is clear that one must have $|U_{ei}| \leq 1/2$. On the other hand, the CHOOZ data for a general three neutrino oscillation picture implies that $4|U_{e3}|^2(1 - |U_{e3}|^2 \leq .2$. These two constraints then imply that $|U_{e3}| \leq .2$. This is indeed an interesting constraint and rules out (provided of course $\Delta m_{e3}^2 \geq 10^{-3}$ eV^2) a maximal mixing scenario for degenerate neutrinos that was proposed to reconcile sub-eV double beta decay neutrino mass limit i.e.[19]

$$U = \frac{1}{\sqrt{3}} \begin{pmatrix} 1 & \omega & \omega^2 \\ 1 & \omega^2 & \omega \\ 1 & 1 & 1 \end{pmatrix} \tag{13}$$

There is however another mixing pattern for the degenerate neutrino scenario which is consistent with both the CHOOZ experiment and the neutrinoless double beta decay bounds:

$$U = \begin{pmatrix} 1/\sqrt{2} & i/\sqrt{2} & 0 \\ 1/\sqrt{6} & -i/\sqrt{6} & -2/\sqrt{6} \\ 1/\sqrt{3} & -i/\sqrt{3} & 1/\sqrt{3} \end{pmatrix} \tag{14}$$

Other more general constraints for this case have been studied in several recent papers[20].

If we do not include the hot dark matter constraint, then there is no need to require that the neutrinos are degenerate in mass and one can live perfectly happily with a hierarchical pattern of neutrino masses as dictated by the simple type I seesaw formula. In that case, one can combine the atmospheric oscillation fits and the CHOOZ data to set an upper limit on $< m_\nu >$ equal to $\sqrt{\Delta m^2_{ATMOS}} sin^2\theta_{e\tau} \simeq .02$ eV[21]. Thus evidence for $< m_\nu >$ above this value would be an indication that either the neutrino mass pattern is not hierarchical or that the atmospheric neutrino puzzle involves transition between ν_μ and a sterile neutrino. Both of these are extremely valuable conclusions. The GENIUS proposal of the Heidelberg group[22] is expected to push the double beta decay limit to this level and could therefore test this conclusion.

4. Implications for physics beyond the standard model:

Let us now discuss the constraints implied by neutrinoless double beta decay searches on the new physics scenarios beyond the standard model. Let us first consider the the neutrino mass mechanism. Any theory which gives the electron neutrino a significant ($\simeq$ eV) Majorana mass or any other species (e.g. ν_μ or ν_τ) a large enough mass and mixing angle with the ν_e so that $U_{ei}^2 m_{\nu_i}$ is of order of an electron volt will make itself open to testability by the $\beta\beta_{0\nu}$ decay experiment. There are many theories with such expectations for neutrinos. Below I described two examples: (i) the singlet majoron model and (ii) the left-right symmetric model. Both these models are intimately connected with ways to understand the small neutrino mass in gauge theories.

4.1. The singlet majoron model:

This model[23] is the simplest extension of the standard model that provides a naturally small mass for the neutrinos by employing the the see-saw mechanism[24]. It extends the standard model by the addition of three right-handed neutrinos and the addition of a single complex Higgs field Δ which is an $SU(2)_L \times U(1)_Y$ singlet but with a lepton number +2. There is now a Dirac mass for the neutrinos and a Majorana mass for the right handed neutrinos proportional to the vacuum expectation value (vev) $\langle\Delta\rangle \equiv v_R$. This leads to a mass matrix for the neutrinos with the usual see-saw form:

$$M = \begin{pmatrix} 0 & m_D \\ m_D^T & f v_R \end{pmatrix} \tag{15}$$

This leads to both the light and heavy (right-handed) neutrinos being Majorana particles with the mutual mass relation being given by the see-saw formula:

$$m_{\nu_i} \simeq m_{iD}(M_{iR}^{-1})m_{iD}^T \tag{16}$$

where we have ignored all mixings and $M_{iR} \simeq f_{ii}v_R$ denote the masses of the heavy right-handed neutrinos . It is clear that the electron neutrino mass can be in the electron-volt range if the values of m_{1D} are chosen to be of similar order of magnitude to the electron mass. In fact, for $m_{1D} = m_e$, and $m_{1R} = 250$ GeV, one gets $m_{\nu_e} = 1$ eV which is the range of masses being probed by the ongoing and proposed $\beta\beta_{0\nu}$ experiments.

More importantly, this class of models leads to the new neutrinoless double beta decay process with majoron emission[25] which has a very different electron energy distribution than either 0ν or 2ν double beta decays. The relevant Feynman diagram is same that in Fig. 1 with a majoron line emanating from the light neutrino in the middle. The majoron coupling $g_{\nu\nu\chi}$ then replaces the neutrino mass in the $\beta\beta_{0\nu}$ amplitude. This observation has led to a considerable amount of experimental effort into searching for the majoron emitting double beta decay and limits at the level of $g_{\nu\nu\chi} \leq 10^{-5}$ are presently available.

A relevant question is whether majoron couplings at the level measurable are expected in reasonable extensions of the standard model. There have been extensive studies of this question and is beyond the scope of this review. But it is of interest to note that in the simplest singlet majoron model, one expects $g_{\nu\nu\chi} \simeq \Sigma_a m_{ea}^2 M_a^{-2} g_{aa\chi}$. In the absence of any mixings, this is proportional to m_{ν_1}/M_{N_1} which is expected to be of order 10^{-11} for an eV ν_e and 100 GeV for the $B-L$ breaking scale. However, if the ν_τ mass is in the MeV range as is allowed by LEP analysis, this coupling could easilly be in the 10^{-5} to 10^{-6} range which is clearly in the range accessible to experiments.

4.2. Left-right symmetric models:

Let us now consider the minimal left-right symmetric model with a see-saw mechanism for neutrino masses as described in [2]. Below, we provide a brief description of the structure of the model. The three generations of quark and lepton fields are denoted by $Q_a^T \equiv (u_a, d_a)$ and $\Psi_a^T \equiv (\nu_a,\ e_a)$ respectively, where $a = 1,\ 2,\ 3$ is the generation index. Under the gauge group $SU(2)_L \times SU(2)_R \times U(1)_{B-L}$, they are assumed to transform as $\Psi_{a\ L} \equiv (1/2,\ 0,\ -1)$ and $\Psi_{a\ R} \equiv (0,\ 1/2,\ -1)$ and similarly for the quarks denoted by $Q^T \equiv (u,\ d)$. In this model, there is a right-handed counterpart to the $W_L^\pm$ to be denoted by $W_R^\pm$. Their gauge interactions then lead to the following expanded structure for the charged weak currents in the model for one generation prior to symmetry breaking (for our discussion , the quark mixings and the higher generations are not very important; so we will ignore them in what follows.)

$$L_{wk} = \frac{g}{2\sqrt{2}}[W_{\mu L}^- J_L^\mu + L \to R] \tag{17}$$

where $J_L^\mu = (\bar{d}\gamma^\mu(1-\gamma_5)u + \bar{e}\gamma^\mu(1-\gamma_5)\nu_e)$

The Higgs sector of the model consists of the bi-doublet field $\phi \equiv (1/2,\ 1/2,\ 0)$ and triplet Higgs fields: $\Delta_L(1,\ 0,\ +2) \oplus \Delta_R(0,\ 1,\ +2)$

The Yukawa couplings for the lepton sector which are invariant under gauge and parity symmetry can be written as:

$$\mathcal{L}_Y = \overline{\Psi}_L h^\ell \phi \Psi_R + \overline{\Psi}_L \tilde{h}^\ell \tilde{\phi} \Psi_R + \Psi_L^T f \tau_2 \vec{\tau} \cdot \vec{\Delta}_L C^{-1} \Psi_L + L \to R + h.c. \quad (18)$$

where h, $\tilde{h}$ are hermitian matrices while f is a symmetric matrix in the generation space. Ψ and Q here denote the leptonic and quark doublets respectively.

The gauge symmetry is spontaneously broken by the vacuum expectation values: $< \Delta_R^0 >= v_R$; $< \Delta_L^0 >= 0$; and $< \phi >= \begin{pmatrix} \kappa & 0 \\ 0 & \kappa' \end{pmatrix}$. As usual, $< \phi >$ gives masses to the charged fermions and Dirac masses to the neutrinos whereas $< \Delta_R^0 >$ leads to the see-saw mechanism for the neutrinos in the standard way[2]. For one generation the see-saw matrix is in the form $m_\nu \simeq m_f^2/f v_R$ and leads as before to a light and a heavy state as discussed in the previous section. For our discussion here it is important to know the structure of the light and the heavy neutrino eigenstates:

$$\begin{aligned} \nu &\equiv \nu_e \ + \ \xi N_e \\ N &\equiv \ N_e \ - \ \xi \nu_e \end{aligned} \quad (19)$$

where $\xi \simeq \sqrt{m_{\nu_e}/m_N}$ and is therefore a small number. Substituting these eigenstates into the charged current Lagrangian, we see that the right-handed W_R interaction involves also the light neutrino with a small strength proportional to ξ. To second order in the gauge coupling g, the effective weak interaction Hamiltonian involving both the light and the heavy neutrino becomes:

$$\begin{aligned} H_{wk} = {} & \frac{G_F}{\sqrt{2}}(\overline{u}\gamma^\mu(1-\gamma_5)d[\overline{e}\gamma_\mu[(1-\gamma_5) \\ & + \xi(\frac{m_{W_L}^2}{m_{W_R}^2})(1+\gamma_5)]\nu + \xi\overline{e}(1-\gamma_5)N] \\ & + \frac{G_F}{\sqrt{2}}\left(\frac{m_{W_L}^2}{m_{W_R}^2}\right)(\overline{u}\gamma^\mu(1+\gamma_5)d\overline{e}\gamma_\mu(1+\gamma_5)N) \\ & + h.c. \end{aligned} \quad (20)$$

From Eq. (20), we see that there are several contributions to the $\beta\beta_{0\nu}$. Aside from the usual neutrino mass diagram (Fig.1), there is a contribution due to the wrong helicity admixture with $\eta \simeq \xi\left(\frac{m_{W_L}^2}{m_{W_R}^2}\right)$ and there are contributions arising from the exchange of heavy right-handed neutrinos. This last contribution is given by :

$$A_{\beta\beta}^{(R)} \simeq \frac{G_F^2}{2}\left(\frac{m_{W_L}^4}{m_{W_R}^4} + \xi^2\right)\frac{1}{m_N} \quad (21)$$

The present limits on neutrinoless double beta decay lifetime then imposes a correlated constraint on the parameters m_{W_R} and m_N[26]. If we combine the theoretical constraints of vacuum stability then, the present ^{76}Ge data provides a lower limit on the masses of the right handed neutrino (N_e) and the W_R of 1 TeV, which is a rather stringent constraint. We have of course assumed that the leptonic mixing angles are small so that there is no cancellation between the parameters.

Finally, the Higgs sector of the theory generates two types of contributions to $\beta\beta_{0\nu}$ decay. One arises from the coupling of the doubly charged Higgs boson to electrons (see Fig.2). The amplitude for the decay is same as in Eq. (6) except we have $\lambda_\Delta = f_{11}$ and

$$\frac{\lambda_\Delta}{M^3} = 2^{7/4} G_F^{3/2}\left(\frac{m_{W_L}}{M_{W_R}}\right)^3 \quad (22)$$

Using this expression, we find that the present ^{76}Ge data implies that (assuming $m_{W_R} \geq 1$ TeV)

$$M_{\Delta^{++}} \geq \sqrt{f_{11}} \ \ 80 GeV \quad (23)$$

A second type Higgs induced contribution arises from the mixing among the charged Higgs fields in ϕ and Δ_L which arise from the couplings in the Higgs potential, such as $\mathrm{Tr}(\Delta_L \phi \Delta_R^\dagger \phi^\dagger)$ after the full gauge symmetry is broken down to $U(1)_{em}$. Let us denote this mixing term by an angle θ. This will contribute to the four-Fermi interaction of the form given by the ϵ_1^{ee} term with

$$\epsilon_1^{ee} \simeq \frac{h_u f_{11} sin 2\theta}{4\sqrt{2} G_F M_{H^+}^2}, \quad (24)$$

where we have assumed that H^+ is the lighter of the two Higgs fields. We get $h_u f_{11} \sin 2\theta \leq 6 \times 10^{-9}(M_{H^+}/100\ GeV)^2$, which is quite a stringent constraint on the parameters of the theory. To appreciate this somewhat more, we point out that one expects $h_u \approx m_u/m_W \approx 5 \times 10^{-5}$ in which case, we get an upper limit for the coupling of the Higgs triplets to leptons $f_{11}\sin 2\theta \leq 10^{-4}$ (for $m_{H^+} = 100\ GeV$). Taking a reasonable choice of $\theta \sim M_{W_L}/M_{W_R} \sim 10^{-1}$ would correspond to a limit $f_{11} \leq 10^{-3}$. Limits on this parameters from analysis[27] of Bhabha scattering is only of order .2 or so for the same value of the Higgs mass.

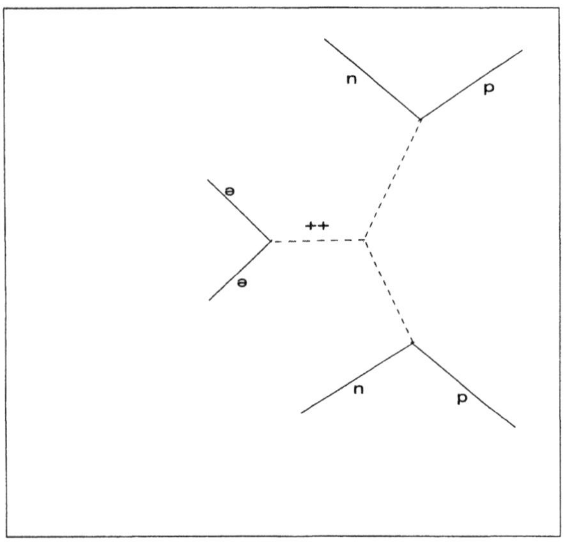

Figure 2. The Feynman diagram responsible for neutrinoless double beta decay due to the exchange of doubly charged Higgs bosons. The top and bottom solid lines are quark lines and the middle right solid lines are electron lines. The dashed lines are the scalar bosons with appropriate quantum numbers.

An interesting recent development is that once one supersymmetrizes the seesaw version of the left-right model just described, allowed values for the right handed scale get severely restricted by the requirement that the ground state of the tehory conserve electric charge. There are only two allowed domains for M_{W_R}: (i) if the ground state breaks R-parity, there is an upper limit on the W_R scale of about $\leq$ 10 TeV[28]. Since in this case, R-parity is spontaneously broken R-parity violating interactions conserve baryon number and the theory therefore is much improved in the sense of naturalness over the MSSM. What is interesting is that the GENIUS experiment can then completely scan the allowed range of this model. On the other hand, if R-parity is conserved, there must be a lower limit on M_{W_R} of about 10^{10} GeV[29]. In this case also there is a contribution to $\beta\beta_{0\nu}$ decay coming from the light doubly charged Higgs boson in the same manner described above[30]. This contribution scales like V_R^{-2} in the amplitude. Thus as the limits on neutrinoless double beta decay improve, at some point they will not only imply that the W_R mass is not only bigger than 10^{10} GeV or so; but they can also continue to improve this lower limit due to the contribution from the doubly charged Higgs boson whose mass is directly proportional to the square of v_R.

4.3. MSSM with R-parity violation:

The next class of theories we will consider is the supersymmetric stamdard model. As is well-known, the minimal supersymmetric standard model can have explicit[3] violation of the R-symmetry (defined by $(-1)^{3B+L+2S}$), leading to lepton number violating interactions in the low energy Lagrangian. The three possible types of couplings in the superpotential are :

$$W' = \lambda_{ijk} L_i L_j E_k^c + \lambda'_{ijk} L_i Q_j D_k^c + \lambda''_{ijk} U_i^c D_j^c D_k^c \; . \qquad (25)$$

Here L, Q stand for the lepton and quark doublet superfields, E^c for the lepton singlet superfield and U^c, D^c for the quark singlet superfields. i, j, k are the generation indices and we have $\lambda_{ijk} = -\lambda_{jik}$, $\lambda''_{ijk} = -\lambda''_{ikj}$. The $SU(2)$ and color indices in Eq. (24) are contracted as follows: $L_i Q_j D_k^c =$

$(\nu_i d_j^\alpha - e_i u_j^\alpha) D^c_{k\alpha}$, etc. The simultaneous presence of all three terms in Eq. (25) will imply rapid proton decay, which can be avoided by setting the $\lambda'' = 0$. In this case, baryon number remains an unbroken symmetry while lepton number is violated.

There are two types of to $\beta\beta_{0\nu}$ decay in this model. One class dominantly mediated by heavy gluino exchange[31] falls into the class of type II contributions discussed in the previous section. The dominant diagram of this class is ahown in Fig. 3. Detailed evaluation of the nuclear matrix element for this class of models has recently been carried out by Hirsch et. al.[32] and they have found that a very stringent bound on the following R-violating parameter can be given:

$$\lambda'_{111} \leq 4\times10^{-4}\left(\frac{m_{\tilde{q}}}{100GeV}\right)^2\left(\frac{m_{\tilde{g}}}{100GeV}\right)^{1/2} \quad (26)$$

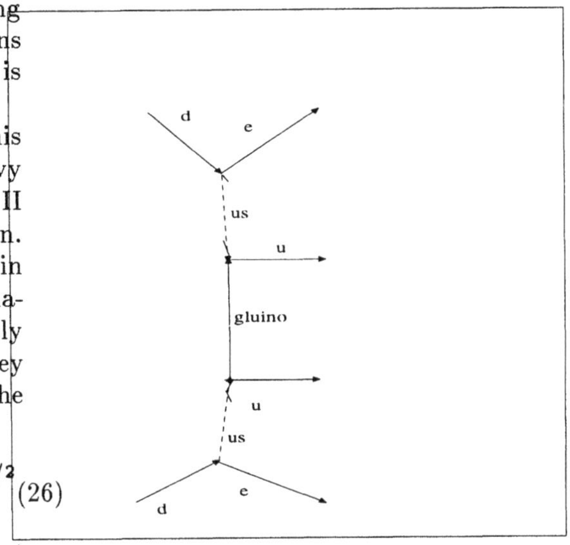

Figure 3. Gluino mediated contribution in MSSM with R-parity violation. *ds* stands for the down squark.

It has been recently pointed out by Faessler et al[32] that if one assumes the dominance of pion exchange in these processes, the limits λ'_{111} becomes more stringent by a factor of 2.

The second class of contributions fall into the light neutrino exchange vector-scalar type[33] and the dominant diagram of this type is shown in Fig.4.(where the exchanged scalar particles are the $\tilde{b} - \tilde{b}^c$ pair). This leads to a contribution to ϵ_2^{ee} given by

$$\epsilon_2^{ee} \simeq \left(\frac{(\lambda'_{113}\lambda'_{131})}{2\sqrt{2}G_F M^2_{\tilde{b}}}\right)\left(\frac{m_b}{M^2_{\tilde{b}^c}}\right) M' \quad (27)$$

where $M' = (\mu\tan\beta + A_b m_0)$.
Here A_b, m_0 are supersymmetry breaking parameters, while μ is the supersymmetric mass of the Higgs bosons. $\tan\beta$ is the ratio of the two Higgs vacuum expectation values and lies in the range $1 \leq \tan\beta \leq m_t/m_b \approx 60$. For the choice of all squark masses as well as μ and the SUSY breaking mass parameters being of order of 100 GeV, $A_b = 1$, $\tan\beta = 1$,the following bound on R-violating couplings is obtained:

$$\lambda'_{113}\lambda'_{131} \leq 3\times10^{-8} \quad (28)$$

This bound is a more stringent limit on this parameter than the existing ones. The present limits on these parameters are $\lambda'_{113} \leq 0.03, \lambda'_{131} \leq 0.26$, which shows that the bound derived here from $\beta\beta_{0\nu}$ is about five orders of magnitude more stringent on the product $\lambda'_{113}\lambda'_{131}$. If the exchanged scalar particles in Fig.9 are the $\tilde{s} - \tilde{s}^c$ pair, one obtains a limit

$$\lambda'_{121}\lambda_{112} \leq 1\times10^{-6} \quad (29)$$

which also is more stringent by about four orders of magnitude than the existing limits ($\lambda'_{121} \leq 0.26, \lambda'_{112} \leq 0.03$).

If the quarks and leptons are composite particles, it is natural to expect excited leptons which will interact with the electron via some effective interaction involving the W_L boson. If the excited neutrino is a majorana particle, then there will be contributions to $\beta\beta_{0\nu}$ decay mediated by the excited neutrinos (ν^*). The effctive interaction responsible for this is obtained from the primordial

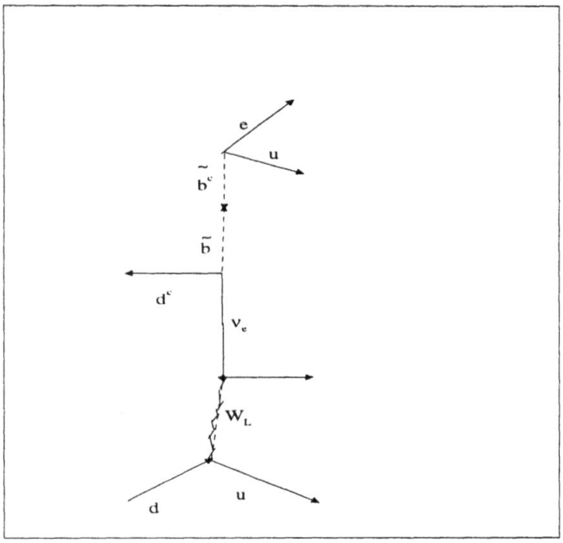

Figure 4. Vector-scalar contribution in MSSM with R-parity violation.

interaction:

$$H^{\nu^*}_{eff} = g\frac{\lambda_W^{(\nu^*)}}{m_{\nu^*}}\bar{e}\sigma^{\mu\nu}(\eta_L^*\nu_L^*+\eta_R^*\nu_R^*)W_{\mu\nu}+h.c. \quad (30)$$

Here L and R denote the left and right chirality states. This contribution falls into our type B heavy particle exchange category and has been studied in detail in two recent papers[34] and have led to the conclusion that it leads to a lower bound

$$m_{\nu^*} \geq 3.4 \times m_W \quad (31)$$

for $\lambda_W^{(\nu^*)} \geq 1$. This is a rather stringent bound on the compositeness scale.

In conclusion, neutrinoless double beta decay provides a very versatile way to probe scenarios of physics beyond the standard model. In this review, we have focussed only on the 0ν mode and briefly touched on the single majoron mode. Single and multi majoron modes which test for the possibility of lepton number being a spontaneously broken global symmetry have been extensively discussed in literature[35]. The 0ν mode acquires special interest in view of the recent discoveries in neutrino physics as well as certain SO(10) models predicting such spectra without contradicting the solar and atmospheric neutrino data.

REFERENCES

1. R.N. Mohapatra and P.B. Pal, "***Massive Neutrinos in Physics and Astrophysics***", Second Edition, World Scientific, Singapore, 1998.
2. R. N. Mohapatra and G. Senjanović, Phys. Rev. Lett. **44**, 912 (1980); Phys. Rev. **D23**, 165 (1981).
3. C. S. Aulakh and R. N. Mohapatra, Phys. Lett. **119B**, 136 (1983); F. Zwirner, Phys. Lett. **132B**, 103 (1983); L. Hall and M. Suzuki, Nucl. Phys. **B231**, 419 (1984); G. G. Ross and J. W. F. Valle, Phys. Lett. **B151**, 375 (1985).
4. H. Klapdor-Kleingrothaus, Prog. in Part. and Nucl. Phys., **32**, 261 (1994); A. Balysh et. al., Phys. Lett. (to appear).
5. R. N. Mohapatra, in ***Double Beta decay and Related Topics***, ed. H. Klapdor-Kleingrothaus and S. Stoica, World Scientific, 1995; p. 44; for earlier reviews see M. Doi, T. Kotani, E. Takasugi, Prog. Theor. Phys. Suppl. **83**, 1 (1985);H. Primakoff and S. P. Rosen, Rep. Prog. Phys. **22**, 121 (1959); W. C. Haxton and G. Stephenson, Prog. in Part. and Nucl. Phys. **12**, 409 (1984); H. Grotz and H. Klapdor, ***The Weak Interactions in Nuclear, Particle and Astrophysics***, Adam Hilger, Bristol, (1990); D. Caldwell, Nucl. Phys. Proc. Suppl. **B 13**, 547 (1990); M. Moe and P. Vogel, Ann. Rev. Nucl. Sc. **44**, 247 (1994).
6. H. Klapdor-Kleingrothaus, these proceedings; H. Ejiri, these proceedings; F. Avignone, talk at PASCOS98 (to appear in the proceedings).
7. see the articles by P. Vogel, K. Muto, S. Stoica and S. Suhonen in ***Double Beta Decay and Related Topics*** ed. H. Klapdor-Kleingrothaus and S. Stoica, World Scientific, 1995. For a recent review, see A. Faessler and F. Simkovic,

Tuebingen preprint (1998); H. Ejiri, these proceedings.
8. A. Halprin, P. Minkowski, S. P. Rosen and H. Primakoff, Phys. Rev. **D13**, 2567 (1976).
9. R.N. Mohapatra and J. Vergados, Phys. Rev. Lett. **47**, 1713 (1981).
10. J. Schecter and J.W.F. Valle, Phys. Rev. **D25**, 2951 (1982); W.C. Haxton, S.P. Rosen and G.J. Stephenson, *ibid.*, **D26**,1805 (1982); L. Wolfenstein, *ibid.*, **D26**, 2507 (1982).
11. T. Kajita, these proceedings.
12. CHOOZ collaboration, M Apolonio et al. hep-ex/9711002.
13. J. Bahcall, P. Krastev and A. Smirnov, hep-ph/9807216.
14. S. P. Mikheyev and A. Smirnov, Sov. J. Nucl. Phys. **42**, 913 (1985); L. Wolfenstein, Phys. Rev. **D 17**, 2369 (1978).
15. C. Athanassopoulos et al. Phys. Rev. Lett. **75**, 2650 (1995); LSND2 C. Athanassopoulos et al. Nucl-ex/9706006.
16. E. Gawiser and J. Silk, Astro-ph/9806197; Science, **280**, 1405 (1998).
17. D. Caldwell and R. N. Mohapatra, Phys. Rev. **D 48**, 3259 (1993); A. Joshipura, Z. Phys. **C 64**, 31 (1994).
18. F. Vissani, hep-ph/9708483.
19. R. N. Mohapatra and S. Nussinov, Phys. Lett. **B 346**, 75 (1995).
20. H. Minakata and O. Yasuda, Nucl. Phys. (to appear); hep-ph/9602386.
21. S. M. Bilenky, C. Giunti, C. W. Kim and M. Monteno, hep-ph/9711400.
22. H. Klapdor-Kleingrothaus, hep-ex/9802007 and these proceedings.
23. Y. Chikashige, R. N. Mohapatra and R. D. Peccei, Phys. Lett. **98B**, 265 (1981).
24. M. Gell-Mann, P. Ramond and R. Slansky, *in "Supergravity"*, Ed. D.Freedman et al. (North-Holland, Amsterdam, 1979); T. Yanagida, Prog. Th. Phys. **B135** (1978) 66; R.N. Mohapatra and G. Senjanović, Phys. Rev. Lett. **44** (1980) 912.
25. H. Georgi, S. L. Glashow and S. Nussinov, Nucl. Phys. **B 193**, 297 (1981).
26. R.N. Mohapatra, Phys. Rev. **D34**, 909 (1986).
27. M. Schwarz, Phys. Rev. **D40**, 1521 (1989); for a recent review, see F. Cuypers and S. Davidson, hep-ph/9609487; F. Cuypers and M. Raidal, hep-ph/9704224.
28. R. Kuchimanchi and R. N. Mohapatra, Phys. Rev. Lett. **75**, 3989 (1995).
29. Z. Chacko and R. N. Mohapatra, hep-ph/9712359; C. S. Aulakh, A. Melfo and G. Senjanović, hep-ph/9707258.
30. R. N. Mohapatra, Talk at PASCOS98, (1998).
31. R. N. Mohapatra, Phys. Rev. **D 34**, 3457 (1986).
32. M. Hirsch, H. Klapdor-Kleingrothaus and S. Kovalenko, Phys. Rev. Lett. **75**, 17 (1995); A. Faessler, S. Kovalenko, F. Simkovic and J. Schweiger, Phys. Rev. Lett. **78**, 183 (1997).
33. K. S. Babu and R. N. Mohapatra, Phys. Rev. Lett. **75**, 2276 (1995).
34. O. Panella and Y. N. Srivastava, College de France Preprint, LPC 94-39; E. Takasugi, hep-ph/9506379.
35. P. Bamert, C. Burgess and R. N. Mohapatra, Nucl. Phys. **B 449**, 25 (1995); R. N. Mohapatra and E. Takasugi, Phys. Lett. **B 211**, 192 (1988); For experimental study of these processes, see J. Hellmig, M. Hirsch, H. V. Klapdor-Klein-grothaus, B. Maier and H. Pas, in ***Double Beta Decay and Related Topics***, ed. H. V. Klapdor-Kleingrothaus and S. Stoica, World scientific (1995), p. 130.

Part 11

Dark Matter Search

ELSEVIER

Nuclear Physics B (Proc. Suppl.) 77 (1999) 389–397

NUCLEAR PHYSICS B
PROCEEDINGS
SUPPLEMENTS

Direct searches for dark matter

Bernard Sadoulet

Center for Particle Astrophysics
Lawrence Berkeley National Laboratory
and Physics Department, University of California, Berkeley
Berkeley, California 94720 USA

After briefly recalling the evidence which suggests that the dark matter pervading the universe is nonbaryonic, we review the present searches for the best motivated particle candidates: axions, light neutrinos and Weakly Interacting Massive Particles (WIMPs).

1. INTRODUCTION

In the last decade considerable additional evidence[1] has been gathered supporting the hypothesis that at least 90% of the mass in the universe is dark: it does not emit or absorb any form of electromagnetic radiation. Understanding this dark matter has become one of the more central problems in astronomy and cosmology. Once a subject of controversy among astronomers, its existence is now well established at a variety of scales.

The debate has shifted to measuring the amount of dark matter in the universe, studying its distribution and unraveling its nature. A central question is whether this dark matter is made of ordinary baryonic matter or is nonbaryonic. A number of cosmological observations, reviewed in section 2, indicate that it may indeed be nonbaryonic. This case is not undermined by the current observations of Massive Compact Halo Objects (MACHOs), as there is a fundamental ambiguity in the distance of the lensing objects.

Following our conclusion that the searches for nonbaryonic dark matter remain essential, we review the current detection efforts for axions (section 3), massive neutrinos (section 4), and Weakly Interacting Massive Particles (WIMPs – section 5).

2. THE CASE FOR NONBARYONIC DARK MATTER

2.1. Comparison of Ω and Ω_b

Figure 1 summarizes the current attempts of measuring the average density Ω of the universe in units of the critical density

$$\Omega = \frac{\rho}{\rho_c} \text{ with } \rho_c = 1.88 \times 10^{-26}\ h^2 \text{ kg m}^{-3}$$

where h is the Hubble expansion parameter in units of 100 km/s/Mpc (h = 0.65±0.1). Ω can be determined through an inventory of the masses of the various objects in the universe, for instance using the virial velocities in galaxy clusters. This intrinsically can give only a lower limit of Ω, as these methods only measure local density inhomogeneities. Dynamic methods attempt to relate the observed velocity deviations from the Hubble flow to the density concentrations and deduce from it an effective Ω, which unfortunately depends on the way the number density of galaxies tracks the mass density fluctuations. Cosmology tests can also be used to directly probe the geometry but, as this involves very distant objects, it is difficult to correct the measured quantities for evolution. This fundamental difficulty, which foiled the earlier attempts,[2] is still a cause for concern in the interpretation of high redshift supernovae.[3] Taken at face value, these exciting observations indicate that the universe is accelerating. They provide an

PII S0920-5632(99)00448-X

approximate measurement of the difference between the vacuum energy density and matter density, Ω_Λ-Ω_m. The sum between these quantities, Ω_Λ-Ω_m, can be obtained from the acoustic ("Doppler") peak in the microwave background power spectrum indicated by the Saskatoon and CAT data. Together these observations give Ω_m = 0.25 (+0.18-0.12, 95%CL interval).[4]

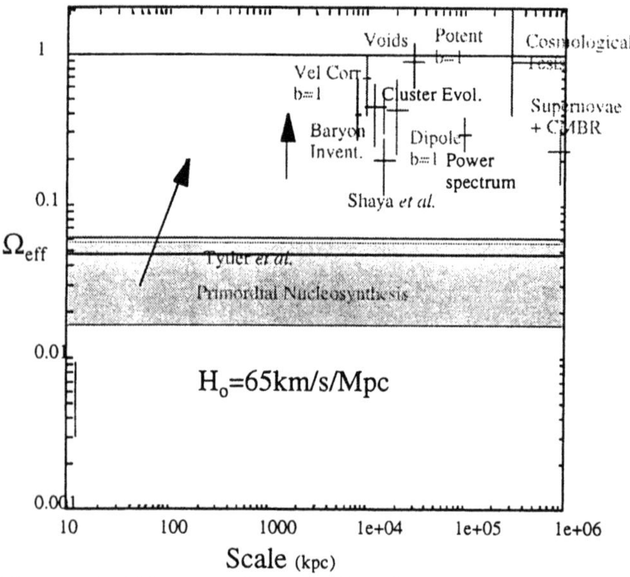

Figure 1. Effective Ω as a function of scale of measurement for H_0 = 65 km/s/Mpc. The bands give the Ω in baryons expected from primordial nucleosynthesis.

The combination of all these observations makes it rather convincing that dark matter does indeed exist, as the value obtained over large scales (≈ 0.3) is much greater than the contribution of stars (0.003-0.01).

It also provides a convincing argument for the nonbaryonic nature of dark matter. The shaded band displays the relatively narrow limits ($0.007 \lesssim \Omega_b h^2 \lesssim 0.024$) inferred from the observations of ^{4}He, D, ^{3}He and ^{7}Li in the very successful standard scenario of homogeneous primordial nucleosynthesis.[5] It is definitely below most measurements of Ω at large scale. If we believe the recent measurement of the D fraction in Lyman alpha systems by Tytler,[6] Ω_b may be close to the upper boundary of this band but our conclusion is not significantly affected.

2.2. Formation of the large-scale structure

A second argument for nonbaryonic dark matter is based on the fact that it provides the most natural explanation of the large-scale structure of the galaxies in the universe in terms of collapse of initial density fluctuations. They can be inferred from the COBE measurement of the temperature fluctuations of the cosmic microwave background. The deduced power spectrum of the (adiabatic) mass fluctuations at very large scale connects rather smoothly with the galaxy power spectrum measured at lower scale,[7] giving strong evidence for the formation of the observed structure by gravitational collapse. The observed spectral shape is natural with cold nonbaryonic dark matter but cannot be explained with baryons only, since they are locked in with the photons until recombination and cannot grow enough fluctuations to form the structure we see today.

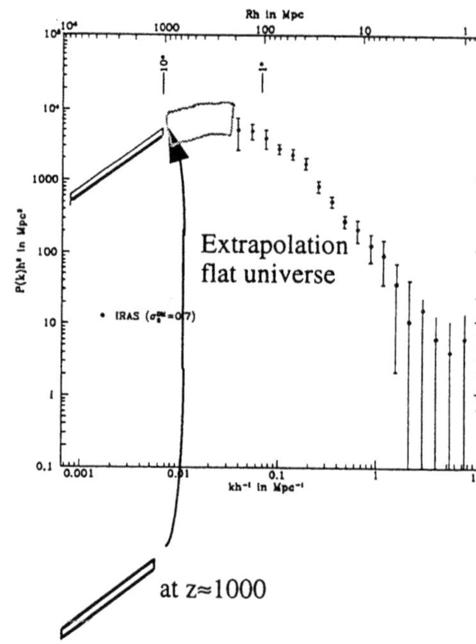

Figure 2. Measured power spectrum measured for IRAS galaxies and extrapolation of the COBE result assuming nonbaryonic dark matter and a flat universe (after Fisher et al., 1992). The contour in the middle gives a rough estimate of the power spectrum inferred from the measurement of the acoustic peak of the cosmic microwave background (after Scott et al., 1995).

Although their interpretation is more model-dependent, the recent measurements of temperature fluctuations at the degree scale of the cosmic microwave background appear to support this conclusion.[8] They are smoothly bridging the gap between the COBE extrapolation and large-scale structure.

2.3. Inefficiency of the formation of compact objects

A third general argument comes from the implausibility of hiding a large amount of baryons in the form of MACHOs. For instance, as the ratio of the mass in gas and stars to the total mass in clusters is of the order of 20%, this would require 80% of the initial gas to have condensed into invisible MACHOs. This is very difficult to understand within the standard scenarios of star formation. The same argument applies to galactic halos.

2.4. Impact of the MACHO observations

The intrinsic degeneracy arising in the interpretation of microlensing observations prevents the fascinating MACHO results from seriously undermining the case for nonbaryonic dark matter. As explained by Spiro in these proceedings,[9] the lensing duration is a degenerate function of the mass, distance and transverse velocity of the lens, and we do know where the lenses responsible for the observed events are. The location of the two observed double lenses in the host galaxies, the large mass implied by a halo distribution hypothesis, and the low event rate towards the Small Magellanic Cloud cast considerable doubt about the MACHOs forming the totality of the halo.

In any case, if a nonbaryonic component exists, it is difficult to prevent it from accreting (unless it is hot) and, even in the presence of MACHOs in the halos, it should constitute a significant portion of the halo and be present locally for detection. In fact, taking into account all kinematic information on the galaxy and the MACHO observations, the most likely density for a nonbaryonic component is close[10] to the canonical 0.3 GeV/cm^3 inferred from the velocity curves of our galaxy.[11]

2.5 Conclusion

In conclusion, it seems very difficult to construct a self-consistent cosmology without nonbaryonic dark matter and the MACHOs results do not so far undermine those arguments. Thus, it remains urgent to search for nonbaryonic dark matter.

A large number of candidates have been proposed over the years for such a nonbaryonic component. They range from shadow universes existing in some string models, strange quark nuggets formed at a first order quark-hadron phase transition,[12] Charged Massive Particles (CHAMPs),[13] and a long list of usually massive particles with very weak interactions. We should probably search first for particles that would also solve major questions in particle physics. According to this criterion, three candidates appear particularly well motivated.

3. SEARCHES FOR AXIONS

3.1. Cosmological axions

Axions are an example of relic particles produced out of thermal equilibrium, a case where we depend totally on the specific model considered to predict their abundance. These particles have been postulated[14] in order to dynamically prevent the violation of CP in strong interactions in the otherwise extremely successful theory of quantum chromodynamics. Of course, there is no guarantee that such particles exist, but the present laboratory and astrophysical limits on their parameters are such that if they exist, they would form a significant portion of cold dark matter.[15] Such low-mass cosmological axions could be detected by interactions with a magnetic field, which produce a faint microwave radiation detectable in a tunable cavity.[16] The first two searches[17] for cosmological axions performed a decade ago were missing a factor of 1000 in sensitivity.

This is no longer the case; Livermore, MIT, Florida and Chicago are currently performing an experiment which has published preliminary limits[18] and will reach (Fig. 3) a cosmologically interesting sensitivity at least for one generic type of axion (hadronic models[19]). By replacing their HEMT-based amplifiers by SQUID amplifiers, the collaboration hopes to improve their sensitivity down to the lowest couplings currently predicted (DFZ model[20]). Matsuka and his collaborators in Kyoto are developing a more ambitious scheme using Rydberg atoms, which are very sensitive photon detectors and should be able to directly reach the DFZ limit.

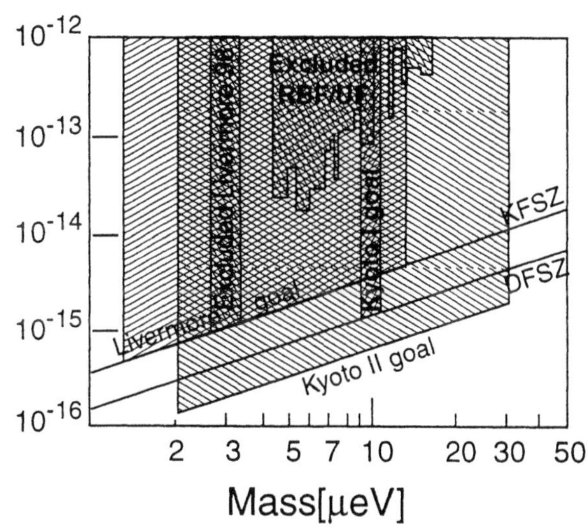

Figure 3. Expected sensitivity of the Livermore and Kyoto experiments. The lines labeled KSVZ and DFSZ refer to two generic species of axions. The shaded regions in the upper right are the previous experimental limits.

Although these experiments are very impressive, it should be noted that the decade of frequency (and therefore of mass) which can be explored with the present method is only one out of three which are presently allowed.

3.2. Solar axions

For large enough masses, axions can also be produced in the sun. Such axions would produce x-rays in germanium detectors by conversion in the field of the nucleus and it was suggested that a characteristic Braggs modulation could be observed. The Zaragoza and USC-PNL groups have searched for such effects, but the limits are still much above the required level of sensitivity.

4. LIGHT NEUTRINOS

4.1. Neutrinos in cosmology

Neutrinos of mass much smaller than 2 MeV/c fall in the generic category of particles which have been in thermal equilibrium in the early universe and decoupled when they were relativistic. Their current number density is approximately equal to that of the photons in the universe. The relic particle density is therefore directly related to its mass, and a neutrino species of 25 eV would give an Ω of the order of unity.[21] Note that neutrinos alone cannot lead to the observed large-scale structure, as fluctuations on scales greater than 40 h^{-1} Mpc are erased by neutrino streaming. They have to be mixed in with cold nonbaryonic dark matter[22] or seeded by topological defects. Moreover, because of phase space constraints, they cannot explain the dark matter halos observed around dwarf galaxies.[23]

Unfortunately, no good ideas exist of possible ways to detect cosmological neutrinos,[24] and one can only rely on the mass measurements of neutrinos in the laboratory through the study of beta spectra, neutrinoless double beta decay, and oscillation experiments.

4.2. Neutrino mass measurements

A large fraction of these proceedings is devoted to the neutrino mass measurements. For the sake of completeness, one may summarize the situation as follows: The direct mass measurement of the electron neutrino gives limits of 5 eV.[25] Model dependent limits of the order of 1 eV on the mass of Majorana neutrinos are given by neutrinoless double beta decay searches. The claim by the LSND group[26] for muon to electron neutrino oscillation with relatively large $\Delta m^2 \approx 6$ eV^2 oscillation is now challenged by the Karmen experiment.[27]

At this conference, the SuperKamiokande group has presented statistically significant results[28] demonstrating the disappearance of atmospheric muon neutrinos which points to an oscillation with Δm^2 of a few 10^{-3} eV^2 and a large mixing angle. A muon to electron neutrino oscillation is disfavored, both by Chooz[29] and internally by SuperKamiokande.

At the date of this writing, the situation of solar neutrinos is less clear. The combination of the chlorine, water Cerenkov and gallium experiments have now indicated for some time a depletion of solar neutrinos with respect to the standard solar model.[30] The most natural explanation is an MSW[31] or vacuum oscillation with Δm^2 of 10^{-6} eV^2 or 10^{-10} eV^2 respectively.[32] However, the distortion of the energy distribution observed by SuperKamiokande, which would be a direct confirmation independently of the solar model, is not fully understood. The essential measurements of neutral current events by SNO and of the ^{7}Be neutrino flux depletion by Borexino will not be available before 2000 and 2001 respectively.

If the oscillation interpretation of both atmospheric and solar neutrinos is correct, we are led to nearly degenerate masses of the three types of neutrinos (unless we have a light sterile neutrino), without any direct indication of the mass which may well be in the electron volt range if we do not invoke the see-saw mechanism. It is therefore important for cosmology to keep pushing the resolution of direct electron neutrino mass measurement.

5. WEAKLY INTERACTIVE PARTICLES

A generic class of candidates is constituted by particles which were in thermal equilibrium in the early universe and decoupled when they were non-relativistic. In this case, it can be shown that their present density is inversely proportional to their annihilation rate.[33] For these particles to have the critical density, this rate has to be roughly the value expected from weak interactions (if they have masses in the GeV/c^2 to TeV/c^2 range). This may be a numerical coincidence, or a precious hint that physics at the W and Z scale is important for the problem of dark matter. Inversely, physics at the W and Z^0 scale leads naturally to particles whose relic density is close to the critical density. In order to stabilize the mass of the vector intermediate bosons, one is led to assume the existence of new families of particles such as supersymmetry in the 100 GeV mass range. In particular, the lightest supersymmetric particle could well constitute the dark matter. This class of particles is usually called Weakly Interactive Massive Particles (WIMPs).

The most direct method to detect these WIMPs is by elastic scattering on a suitable target in the laboratory[34] (indirect methods are reviewed by B. Barish in this volume[35]). Elastic WIMP scattering would produce a roughly exponential spectrum with a mean energy dependent on their mass. The hope is to identify such a contribution in the differential energy spectrum measured by an ultra-low background detector, or at least to exclude cross sections that would lead to differential rates larger than observation.

5.1. Experimental challenges

In specific models such as supersymmetry, the knowledge of the order of magnitude of the annihilation cross section allows an estimation of their elastic scattering, taking into account the coherence over the nucleus. Typically, if scalar (or "spin independent") couplings dominate, the interaction rate of WIMPs from the halo is expected to be of the order of a few events per kilogram of target per week for large nuclei like germanium. We display in Figure 4, as the lower shaded region, the range of cross sections (rescaled to a proton target) expected[36] in grand unified theory inspired supersymmetric models, where scalar interactions usually dominate.

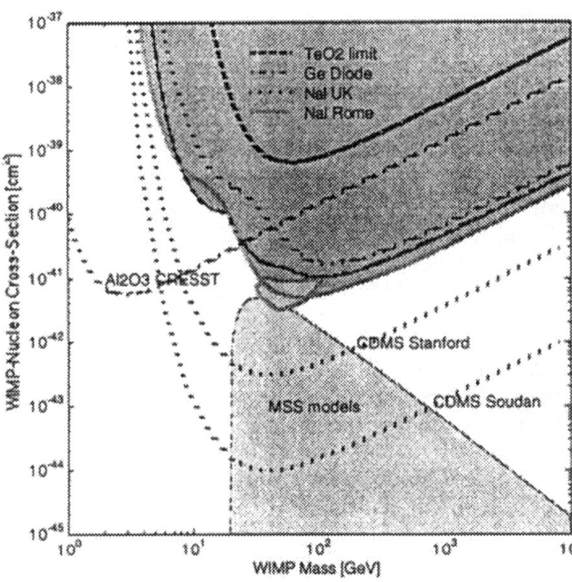

Figure 4. Current achieved limits for spin independent couplings as a function of the WIMP mass. This figure includes the results of the Rome[37] NaI, UK[38] NaI, Milan[39] TeO_2, Modane[40] Al_2O_3, and the Ge diode experiments: PNL-USC,[41] Oroville,[42] Neuchâtel-Caltech,[43] Heidelberg-Moscow,[44] and IGEX.[45] All the results have been converted to WIMP-nucleon cross sections assuming scalar interactions scaling as the square of the atomic number. The shaded region at the top is excluded by these experiments. The heart shaped in the middle is the region corresponding to the modulation signal claimed by Bernabei et al.[46] The shaded region at the bottom is the rate predicted by minimal supersymmetric models including the constraints from LEP and CDF.

The upper shaded regions summarize the current limits achieved with state of the art techniques for low radioactivity background. These limits barely skirt the supersymmetric region, although relaxing the unification assumptions enlarges it somewhat.[47]

Unfortunately, the expected rates can be very small for specific combinations of parameters where axial ("spin dependent") couplings dominate. In this case, the interaction takes place with the spin of the nucleus, which limits the number of possible targets, and the current limits are very far above the supersymmetry expectation.[36]

It is therefore essential to construct experiments with very low radioactive backgrounds or, even better, with active background rejection. The main tool for this purpose is to use the fact that WIMP interactions produce nuclear recoils, while the radioactive background is dominated by electron recoils (if neutrons are eliminated).

A second challenge faced by the experimentalist comes from the fact that the energy deposition is quite small, typically 10 keV for the mass range of interest. For detectors based only on ionization or scintillation light, this difficulty is compounded by the fact that the nuclear recoils are much less efficient in ionizing or giving light than electrons of the same energy. This increases the recoil energy threshold of such detectors, and one should be careful to distinguish between true and "electron-equivalent" energy, which may differ by a factor three (Ge) to twelve (I).

A third challenge is to find convincing signatures linking detected events to particles in the halo of the galaxy. The best one would be the measurement of the direction of the scattered nucleus,[48] a very difficult task. Short of this directionality signature, it is in principle possible to look for a change in the event rate and the spectrum of energy deposition with the time of the year.[49]

5.2. Prominent direct search strategies

In spite of these experimental challenges, low expected rates and low energy depositions, a number of experimental teams are actively attempting to directly detect WIMPs. A number of interesting attempts have been made to use mica which integrates for billions of years,[50] superheated microdots[51] which should be only sensitive to nuclear recoil, and low pressure time projection chambers which could give the directionality.[52] However, the main developments occurred along three main experimental strategies.

1. A first approach is to attempt to decrease the radioactive background as much as possible. Germanium is the detector of choice as it is very pure, and the first limits[41,42,43] were obtained by decreasing the threshold of double beta experiments. The most impressive results have been obtained by

the Heidelberg-Moscow group[44] with a background of 0.05 events/kg/day/electron-equivalent-keV around 20 keV (equivalent electron energy). This impressive performance comes from a careful screening of surrounding material, the large size of their crystal (2.5 kg), and signal shape discrimination. The IGEX and Baksan-USC-PNL[53] collaborations have achieved somewhat worse levels (0.25 events /kg/day/electron-equivalent-keV), but reached lower thresholds. The current combined exclusion plot is given in Fig. 4. GENIUS, an ambitious proposal[54] to immerse one ton of germanium detectors in an ultra-pure liquid nitrogen bath, pushes this strategy to the extreme.

However, this approach is fundamentally limited by the absence of discrimination against the radioactive background. Not only can this background not be partially rejected, but it also cannot be measured independently of the signal and subtracted. Once the background level is measured with sufficient statistical accuracy, the sensitivity of the experiment does not improve with exposure. In contrast, the combination of an active background rejection and subtraction allows a sensitivity increase as the square root of the target mass and the running time, until the subtraction becomes limited by systematics.[55]

2. A second approach has been to use large scintillators with pulse shape discrimination of nuclear and electronic recoils. The technique is simple and large masses can be assembled to search for modulation effects. The most impressive result so far has been obtained with NaI. The NaI groups[37,38] have published limits that claim to be slightly better than those obtained with conventional germanium detectors. However, these limits remain controversial, as they may not fully take into account systematics in the efficiency close to the threshold or in the rejection power from pulse shape discrimination. In any case, because sodium has a spin, these experiments so far give the best limits for spin dependent couplings. The Rome group has recently announced[46] a nearly three σ detection of a signal using the annual modulation expected for a WIMP spectrum (heart-shaped region in Fig. 4). This modulation signal represents less than 1% of the observed background and it is not yet clear that the systematics have been controlled at the required level. Overall, it is unlikely that NaI could make significant additional progress, as the small number of photoelectrons at the energies of interest and the lack of power of the pulse shape discrimination make it highly susceptible to systematics.

3. Therefore, more powerful discrimination methods need to be devised. Liquid xenon with simultaneous measurement of scintillation and ionization is a promising approach, albeit with relatively high thresholds, and not enough development so far to fully judge its potential. In contrast, the active development of novel "cryogenic" detectors based on the detection of phonons produced by particle interactions is beginning to bear fruit. In spite of the complexity of the very low temperature operation, four large setups are currently being routinely operated (Milano,[39] CDMS,[56] CRESST[57] and EDELWEISS[58]), with total detector mass ranging from 70 g to 7 kg.

For dark matter searches, this technology appears to have three advantages:

• It can lead to a much smaller threshold, as phonons measure the total energy of nuclear recoil without any loss. Already the performance of thermal phonon detectors in the laboratory exceeds that of ionization detectors. Berkeley[59] is now routinely getting a resolution of better than 900 eV and 450 eV FWHM in phonons and ionization respectively with 165 g detectors. The CRESST group has also demonstrated a FWHM of 235 eV at 1.5 keV in a 250 g crystal of sapphire. Four of these detectors are now installed in the CRESST experiment, which hopes to obtain without discrimination the limits shown in Fig. 4. Stanford[60] has recently shown that it is even possible to detect athermal phonons after very few bounces on the surface and get similar baseline resolution (1 keV).

• With the simultaneous measurement[61] of ionization and phonons in crystals of germanium or silicon, it is possible to distinguish between nuclear recoils and electron recoils. This approach is used by both the CDMS and the EDELWEISS collaborations. CDMS has demonstrated greater than 99% rejection with thermal and athermal phonon plus ionization technology down to 20 keV recoil energy (Figure 5 a and b). This allows them to reach at a shallower site an effective gamma contamination better than Heidelberg-Moscow, with much lower thresholds. Unfortunately, as often in such situations, a new background was uncovered: soft electrons incident on the surface suffer from ionization losses in a micron-thick dead layer and partially simulate nuclear recoils (Figure 5 c). This dead layer is due to back diffusion of the carriers and can be decreased by suitable modification of the contacts. Combining these improvements with better shielding, the team has recently be able to drastically reduce this problem and hopes to reach at its present Stanford site the limit displayed in Fig. 4. A second line indicates the expected sensitivity at a deep underground facility (the Soudan mine).

• A third advantage of phonon-mediated detectors is the greater amount of information obtained about very rare events. Already the simultaneous measurement of phonons and ionization gives two pieces of information instead of one, and allows a more efficient rejection of microphonics and spurious instrumental effects. The detailed measurement of out-of-equilibrium phonons is even more promising.

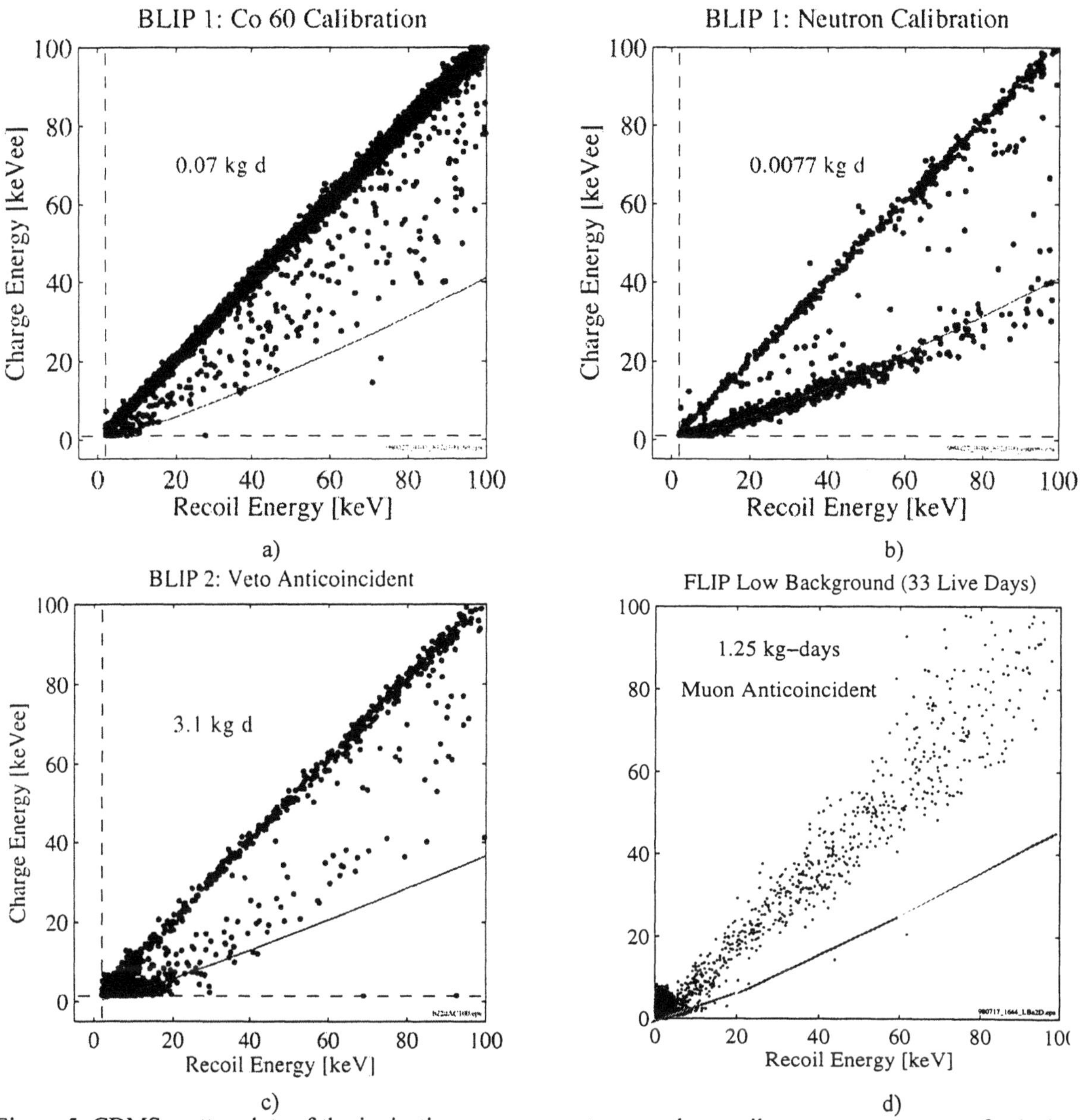

Figure 5. CDMS scatter plots of the ionization measurement versus the recoil energy measurement for (a, b, c) a 165 g Ge detector with thermal phonon readout (BLIP), and (d) a 100 g Si detector with athermal phonon sensing (FLIP) obtained at the Stanford Underground Facility Icebox. The ionization measurements are normalized to electron equivalent energy. Panels (a) and (b) show results of calibration runs with a ^{60}Co photon source (a) and a ^{252}Cf source producing neutrons (and photons). The line represents a fit to the region of nuclear recoil events. Panels (c) and (d) are obtained in low background running conditions. Note in (a) and (c) the soft electron component, intermediate between the diagonal photon line and the nuclear recoil line. In panel (c), after an athermal phonon signal rise time cut, only two events are left in the nuclear recoil region.

CDMS has recently demonstrated that geometrical fiducial cuts can be imposed using the phonon information and that the problematic surface electrons can be eliminated by a phonon rise time cut (Figure 5 d). In the long run, athermal phonons may allow a determination of the directionality for isotopically pure targets.

To summarize, cryogenic detectors are making fast progress and currently appear the best promise to explore a significant portion of the supersymmetric WIMP space in the next few years.

6. CONCLUSION

The case for nonbaryonic dark matter remains very strong and it is important to aggressively continue the current particle searches. An axion experiment is underway which should give us a definite answer about axions over a mass range of one order of magnitude (out of three that are still allowed).

Oscillation neutrino experiments are in progress, which cover the mass range of cosmological interest. The WIMPs search is very active with the installation of very large NaI scintillators, liquid xenon detectors, and phonon-mediated detectors that are beginning to be operational.

The combination of all these efforts may well give us the solution of a central puzzle of cosmology and astrophysics. It may even give us important information about particle physics, with perhaps the discovery of the long sought-for axions or supersymmetric particles.

ACKNOWLEDGMENTS

This work was supported by the Center for Particle Astrophysics, a National Science Foundation Science and Technology Center operated by the University of California under Cooperative Agreement no. AST-912005, and by the Department of Energy under the contract DE-AC03-76SF00098.

REFERENCES

1. See, e.g., the reviews by V. Trimble, Ann. Rev. Astron. Astrophys., 25 (1987) 425; J.R. Primack, D. Seckel and B. Sadoulet, Ann. Rev. Nucl. Part. Sci., 38 (1988) 751; S. Tremaine, Physics Today 45 (1992) 28.
2. A. Sandage, Physics Today, 34 (1990); E. Loh and E. Spillar, Astrophys. J., 303 (1986) 154; E. Loh and E. Spillar, Astrophys. J. Lett., 307 (1988) L1; E. Loh, Astrophys. J., 329 (1988) 24.
3. S. Perlmutter et al., Astrophys. J., 483 (1997) 565; S. Perlmutter et al., Nature, 391 (1998) 51; P.M. Garnavich et al., Astrophys. J. Lett., 493 (1998) L53; S. Perlmutter et al., Preprint 1998: Astro-ph 9812473.
4. G. Efstathiou et al., Preprint 1998: Astro-ph 981226.
5. J. Yang et al., Astrophys. J., 281 (1984) 493; K.A. Olive, D.N. Schramm, G. Steigman and T. Walker, Phys. Lett. B, 426 (1990); D.N. Schramm and M. Turner, Rev. of Modern Phys., 70 (1998) 303.
6. D. Tytler, X.-M. Fan and S. Burles, Nature, 381 (1996) 207.
7. C. Fisher, M. Davis, M.A. Strauss, A. Yahil et al., Astrophys. J., 389 (1992) 188.
8. D. Scott, J. Silk and M. White, Science, 268 (1995) 829
9. M. Spiro, in Neutrino 98, Proc. of the XVIII International Conference on Neutrino Physics and Astrophysics, Takayama, Japan, 4–9 June 1998, edited by Y. Suzuki and Y. Totsuka. This volume.
10. E.I. Gates, G. Gyuk, M.S. Turner, Phys. Rev. D, 53 (1996) 4138.
11. M. Fich and S. Tremaine, Ann. Rev. Astron. Astrophys., 29 (1991) 409.
12. E. Witten, Phys. Rev. D, 30 (1984) 272; A. De Rujula and S. Glashow, Nature, 312 (1984) 734.
13. A. De Rujula, S.L. Glashow and U. Sarid, Nucl. Phys. B, 333 (1990) 173.
14. R. Peccei and H. Quinn, Phys. Rev. Lett., 38 (1977) 1440.
15. M.S. Turner, Phys. Reports, 197 (1990) 167.
16. P. Sikivie, Phys. Rev. Lett., 51 (1983) 1415.
17. S. DePanfilis et al., Phys. Rev. Lett., 59 (1987) 839; S. DePanfilis et al., Phys. Rev. D., 40 (1989) 3153; C.A. Hagmann, University of Florida thesis (1990).
18. C. Hagmann et al., Phys. Rev. Lett., 80 (1998) 2043.
19. J.E. Kim, Phys. Rev. Lett., 43 (1979) 103; M.A. Shifman, A.I. Vainshtein and V.I. Zakharov, Nucl. Phys. B, 166 (1980) 493.
20. M. Dine, W. Fischler and M. Srednicki, Phys. Lett. B, 104 (1981) 199; A.P. Zhitniskii, Sov. J. Nucl. Phys., 31 (1980) 260.
21. R. Cowsik and J. McClelland, Astrophys. J., 180 (1973) 7.
22. M. Davis, F. Summers and D. Schlegel, Nature, 359 (1992) 393; D. Pogosyan and A.A. Starobinsky, Astrophys. J., 447 (1995) 465; J.R. Primack et al., Phys. Rev. Lett., 74 (1995) 2160; A. Klypin, R. Nolthenius and J.R. Primack, Astrophys. J., 474 (1997) 533.
23. S.D. Tremaine and J.E. Gunn, Phys. Rev. Lett., 42 (1979) 407; J. Madsen, Phys. Rev. D, 44 (1991) 999.
24. See, e.g., P.F. Smith and J.D. Lewin, Physics Reports, 187 (1990) 203.
25. V.M. Lobashev, in Neutrino 98, Proc. of the XVIII International Conference on Neutrino Physics and Astrophysics, Takayama, Japan, 4–9 June 1998, edited by Y. Suzuki and Y. Totsuka. This volume.
26. C. Athanassopoulos et al., Phys. Rev. Lett., 75 (1995) 2650; D.H. White et al., in Neutrino 98, Proc. of the XVIII International Conference on Neutrino Physics and Astrophysics, Takayama,

Japan, 4–9 June 1998, edited by Y. Suzuki and Y. Totsuka. This volume.
27. B. Zeitnitz et al., in Neutrino 98, Proc. of the XVIII International Conference on Neutrino Physics and Astrophysics, Takayama, Japan, 4–9 June 1998, edited by Y. Suzuki and Y. Totsuka. This volume.
28. T. Kajita, et al., in Neutrino 98, Proc. of the XVIII International Conference on Neutrino Physics and Astrophysics, Takayama, Japan, 4–9 June 1998, edited by Y. Suzuki and Y. Totsuka. This volume.
29. M. Apollonio et al., Phys. Lett. B, 420 (1998) 397.
30. J.N. Bahcall, S. Basu and M.H. Pinsonneault, 1998, astro-ph/9805135, to be published in Phys. Lett. B.
31. S.P. Mikheyev. and M.S. Smirnov, Nuovo Cim., 9C (1986) 17; L. Wolfenstein, Phys. Rev. D, 20 (1979) 2634.
32. N. Hata and P. Langacker, Phys. Rev. D, 56 (1997) 6107.
33. B. Lee and S. Weinberg, Phys. Rev. Lett., 39 (1977) 165; J. Silk and M. Srednicki, Phys. Rev. Lett., 53 (1984) 624.
34. M.W. Goodman and E. Witten, Phys. Rev. D, 31 (1985) 3059; J.R. Primack, D. Seckel and B. Sadoulet, Ann. Rev. Nucl. Part. Sci., 38 (1988) 751; J.D. Lewin and P.F. Smith, Astropart. Phys., 6 (1996) 87.
35. B. Barish et al., in Neutrino 98, Proc. of the XVIII International Conference on Neutrino Physics and Astrophysics, Takayama, Japan, 4–9 June 1998, edited by Y. Suzuki and Y. Totsuka. This volume.
36. G. Jungman, M. Kamionkowski and K. Griest, Phys. Rep., 267 (1996) 195.
37. R. Bernabei et al., Phys. Lett. B, 389 (1996) 757.
38. P.F. Smith et al., Phys. Lett. B, 379 (1996) 299; J.J. Quenby et al., Astropart. Phys., 5 (1996) 249.
39. A. Alessandrello et al., Nucl. Instr. and Methods, A370 (1996) 241.
40. A. de Bellefon, et al., Nucl. Instr. and Methods, A370 (1996) 230.
41. S.P. Ahlen et al., Phys. Lett. B, 195 (1987) 603.
42. D.O. Caldwell et al., Phys. Rev. Lett., 61 (1988) 510.
43. D. Reusser et al., Phys. Lett. B, 235 (1991) 143
44. M. Beck et al., Phys. Rev. Lett., 70 (1993) 2853; L. Baudis et al., Preprint 1998: hep-ex/9811045.
45. A. Morales, private communication (1997).
46. R. Bernabei et al., Rome II preprints, 1998, INFN/AE-8/23 and ROM2F/98/34.
47. A. Bottino et al., Phys. Lett. B, 402 (1997) 113.
48. D.N. Spergel, Phys. Rev. D, 37 (1988) 353.
49. A.K. Drukier, K. Freese and D.N. Spergel, Phys. Rev., 33 (1986) 3495; K. Freese, J. Frieman and A. Gould, Phys. Rev. D, 37 (1987) 3388.
50. D.P. Snowden-Ifft, E.S. Freeman and P.B. Price, Phys. Rev. Lett., 74 (1995) 4133.
51. J.I. Collar, Phys. Rev. D, 54 (1996) R1247.
52. K.N. Buckland, M.J. Lehner, G.E. Masekand and M. Mojaver, Phys. Rev. Lett., 73 (1994) 1067; M.J. Lehner, K.N. Buckland and G.E. Masek, Astropart. Phys., 8 (1997) 43.
53. A.A. Klimenko et al., JETP Lett. 67 (1998) 875.
54. J. Hellmig and H.V. Klapdor-Kleingrothaus, Zeit. fur Phys. A, 359 (1997) 351.
55. R.J. Gaitskell et al., Nucl. Phys. B (Proc. Suppl.), 51B (1996) 279.
56. S.W. Nam et al., Proc. of the VIIth International Workshop on Low Temperature Detectors, Munich, 1997, published by the Max Planck Institute of Physics, p. 217, and Web site: http://avmp01.mppmu.mpg.de/ltd7.
57. M. Sisti et al., Proc. of the VIIth International Workshop on Low Temperature Detectors, Munich, 1997, published by the Max Planck Institute of Physics, p. 232, and Web site: http://avmp01.mppmu.mpg.de/ltd7.
58. D. L'Hôte et al., Proc. of the VIIth International Workshop on Low Temperature Detectors, Munich, 1997, published by the Max Planck Institute of Physics, p. 237, and Web site: http://avmp01.mppmu.mpg.de/ltd7.
59. R.J. Gaitskell et al., Proc. of the VIIth International Workshop on Low Temperature Detectors, Munich, 1997, published by the Max Planck Institute of Physics, p. 221, and Web site: http://avmp01.mppmu.mpg.de/ltd7.
60. R.M. Clarke et al., Proc. of the VIIth International Workshop on Low Temperature Detectors, Munich, 1997, published by the Max Planck Institute of Physics, p. 229, and Web site: http://avmp01.mppmu.mpg.de/ltd7.
61. T. Shutt et al., Phys. Rev. Lett., 29 (1992) 3425. T. Shutt et al., Phys. Rev. Lett., 29 (1992) 3531.

ELSEVIER

Nuclear Physics B (Proc. Suppl.) 77 (1999) 398–401

NUCLEAR PHYSICS B
PROCEEDINGS
SUPPLEMENTS

Indirect Searches for Dark Matter

Barry C. Barish

California Institute of Technology Pasadena, CA 91125

The search for dark matter WIMPs is complementary to direct searches and accelerator searches. The status of present searches and prospects for the future are reviewed.

1. INTRODUCTION

The search for Weakly Interacting Massive Particles (WIMPs) is of fundamental importance to both particle physics and astrophysics[1]. In particle physics they could represent the much sought states associated with supersymmetric theories. In astrophysics they represent the favored candidate for the dark matter. These dual reasons provide very strong motivation to vigorously search for such particles.

Accelerator searches for WIMPS have been conducted at LEP and have produced important limits in the available supersymmetric parameter space[2]. Unfortunately, much of the parameter space is not covered at LEP. The alternate approaches are to detect WIMPs as dark matter particles, either using direct detection on the earth's surface or by indirect searches searching for effects of astrophysical WIMP interactions. These searches cover different regions of the parameter space.

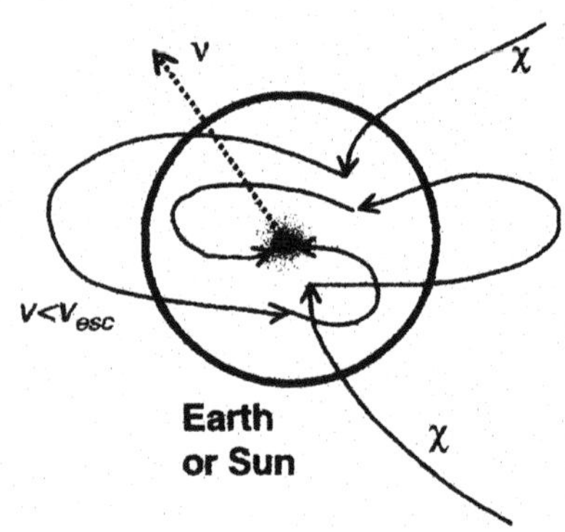

Figure 1. WIMP capture in the gravitational field of the sun or earth

WIMPs are the favored candidate to account for the local dark matter density of the galaxy ($\rho = 0.3$ Gev/cm^3). If so, these WIMPs have a Maxwell velocity distribution and typically are moving with galactic velocities of 300 km/sec. As illustrated in figure 1, the WIMPs can be captured in the gravitational field of the sun or the earth.

2. WIMP DETECTION SIGNATURES

Astrophysical WIMPs are attracted to the earth and sun by the gravitational fields, undergo collisions and thereby lose energy and are captured. Eventually through subsequent collisions and energy loss they migrate toward the central core where the WIMP density can become very large. This large flux makes it probable for WIMP-WIMP annihilation to occur at a detectable rate by observation of product neutrinos in large underground detectors.

The capture rates have been calculated by Gould[3] for the different supersymmetric WIMP candidates: sneutrino, Dirac ν, Majorana ν, Higgsino, Photino. etc. The most interesting supersymmetric dark matter candidate is the Neutralino

$$\chi = c_1\tilde{\gamma}^0 + c_2\tilde{Z}^0 + c_3\tilde{H}_1^0 + c_4\tilde{H}_2^0$$

In the minimal supersymmetric model, mass, cross section etc are determined by three parameters,

$$\tan\beta \equiv v_2 / v_1 , M_2 , \mu$$

(where $M_2 \to 0, \chi \to$ photino $\tilde{\gamma}^0$

and when $\mu \to 0, \chi \to$ Higgsino

$$\tilde{H}^0 = \sin\beta\, \tilde{H}_1^0 + \cos\beta\, \tilde{H}_2^0)$$

The annihilation phenomenology is similar to e^+e^-, except for different branching ratios.

$$\bar{\chi}\chi \to \bar{u}u, \bar{d}d, \bar{s}s, \bar{b}b, \bar{c}c, \bar{\tau}\tau, \bar{g}g, ... \text{ jets}$$

The annihilation products are produced in a very high density medium ($\rho \sim 13$ g/cm^3). The

0920-5632/99/$ – see front matter
PII S0920-5632(99)00449-1

mean free path for strong interactions at the center of the earth is only ~0.1 cm. This implies that particles with $\tau >> 10^{-11}$ sec will mostly interact before they decay and will not contribute high energy neutrinos. The phenomenology of production and decay of short lived states is required to predict the neutrino signal.

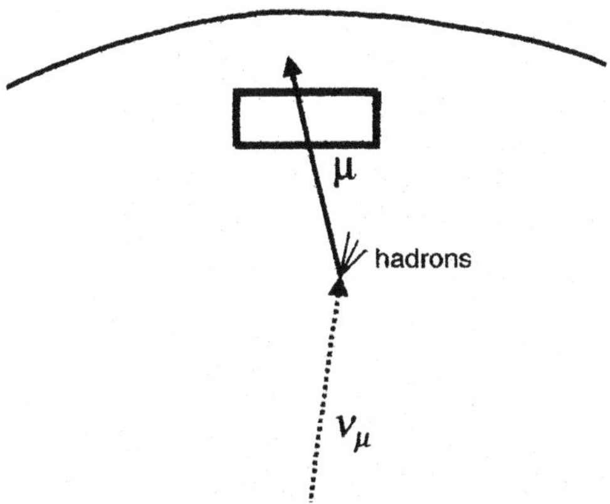

Figure 2. High energy muons from neutrino interactions point back to the sun or the earth's center.

3. EXPERIMENTAL RESULTS

The indirect WIMP search is conducted by looking for a signal indicating a source of high energy neutrinos in the direction of the sun or the center of the earth. For the sun the source is within the 0.5 degree angular size of the sun from the terrestrial detector. In practice, the detector angular distributions are determined by the angular resolution of the detector for neutrino detection. From the center of the earth, the distribution of WIMPs are within the core and the angular distribution for this source is considerably broader as illustrated in figure 3. It can be characterized by a source size of about ~14^0 (20 GeV/c)$^{0.5}$.

Searches have been performed and are continuing with the present generation of underground detectors. New results are presented at this conference by MACRO, for WIMPs from both the sun and the earth observing high energy neutrino interactions. Comparable results have been published by both Kamiokande[4] and Baksan[5].

The primary background events are from atmospheric neutrinos, which have yielded such important results at this conference in the search for neutrino oscillations. This background for these indirect searches for WIMPs, however, does not cluster in the direction of the sun or the center of the earth.

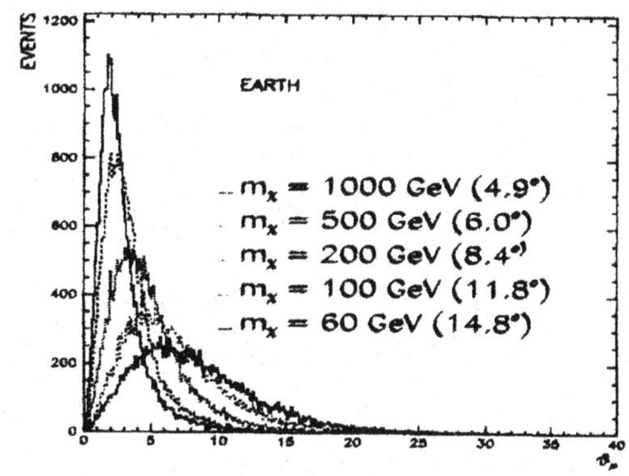

Figure 3. The angular source size for WIMPs from the center of the earth in MACRO, including the spread due to the angle between the muon and the neutrino and the detector angular resolution.

Uncertainties in the expected distributions of the atmospheric neutrino background due to the presence of possible neutrino oscillations is of concern in performing the WIMP analysis. It is important to point out that the observations for atmospheric neutrinos are in the direction that near the zenith, the number of events is lower than expectations. In the WIMP search we are searching for an excess, so that this means the limits that are given for WIMPs in this paper and in the literature are on the conservative side. The WIMP limits have been obtained using the expected level of background.

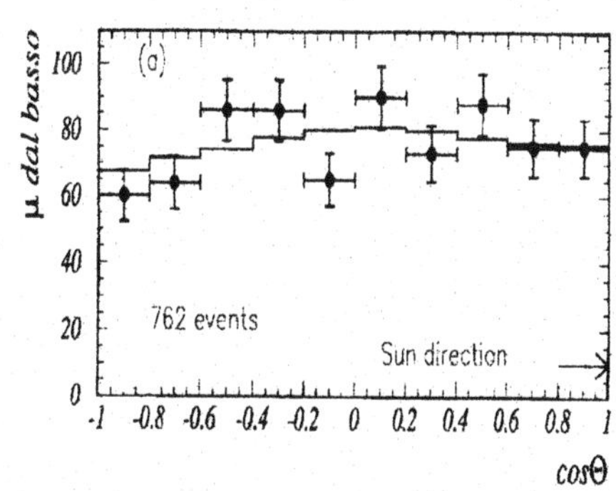

Figure 4. The angular distribution from 762 events of upward muons relative to the direction of the sun from the MACRO experiment.

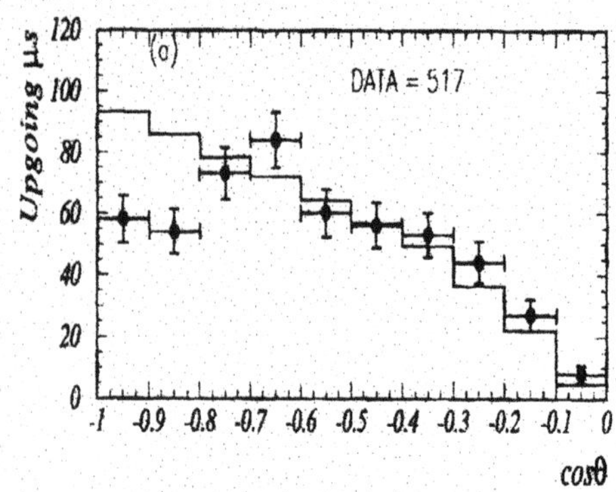

Figure 5. The angular distribution from 517 events of upward muons relative to the direction of the center of the earth from MACRO. For comparison the expectations from atmospheric neutrinos without neutrino oscillations is shown.

The angular distributions for the data from the MACRO experiment are given in figure 4 and 5 from the sun and the earth.
The observed distributions from the sun show no indication of an excess, and from the earth the data is actually lower than expected at the zenith as noted above.

Table 1: MACRO WIMP results

	EARTH			SUN		
Cone	Data	Backgr.	Flux Limit ($E_\mu > 1.5$ GeV) ($cm^{-2}\ s^{-1}$) Exp. ~ 2620 m^2 yr	Data	Backgr.	Flux Limit ($E_\mu > 2$ GeV) ($cm^{-2}\ s^{-1}$) Exp. ~ 890 m^2 yr
30°	76	123.6	2.28×10^{-14}	56	51.1	6.02×10^{-14}
24°	52	81.2	1.81×10^{-14}	33	33.0	3.85×10^{-14}
18°	32	46.6	1.46×10^{-14}	17	18.5	2.64×10^{-14}
15°	24	32.5	1.25×10^{-14}	11	13.0	2.12×10^{-14}
9°	10	11.9	7.54×10^{-15}	3	4.6	1.35×10^{-14}
6°	4	5.2	5.93×10^{-15}	2	2.0	1.38×10^{-14}
3°	0	1.3	3.72×10^{-15}	2	0.5	1.73×10^{-14}

Table 1: Selected and expected events and 90% C.L. muon flux limits for some of the 10 half-cones chosen around the core of the Earth and the Sun direction.

In Table I the new flux limits that have been presented to this conference from MACRO are given for both the sun and the earth.

No WIMP experiment searching for high energy neutrinos from the sun or center of the earth has presented positive evidence.

4. INTERPRETATION

The sun and the earth are complementary ways to search for WIMPs, because of the different compositions. For the earth, because of the heavy elements, the spin independent scalar interactions dominate and the sensitivity is enhanced at $M_N \sim M_X$. In contrast for the sun, because of the proton content, the spin dependent interactions are also important.

Using flux limits from the underground neutrino experiments, it is possible to rule out a portion of the available parameter space allowable in supersymmetric models, after taking into account the limits from LEP. Figure 6 shows the MACRO flux limit from the earth, along with model predictions by Bottino et al[6] varying the supersymmetric parameters. The points above the line are supersymmetric model parameters ruled out by the MACRO data.

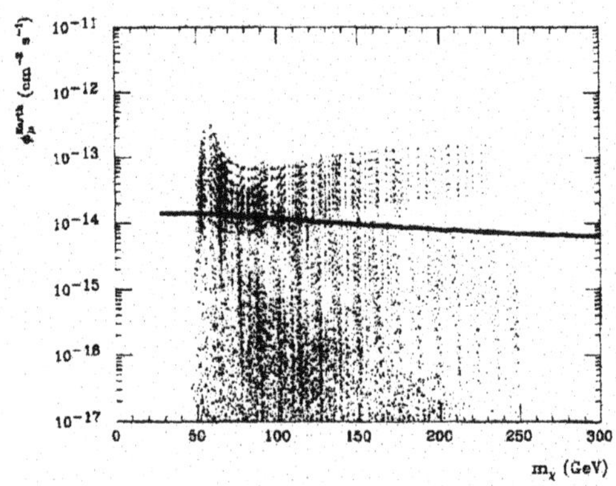

Figure 6. MACRO flux limits from the earth and supersymmetric model predictions of Bottino et al varying model parameters.

The search from the earth with direct detection has also ruled out some of the parameter space. In fact, it has been shown by Bergstrom[7] et al. that the models ruled out by the direct and indirect searches largely overlap. This is because the capture rate in the earth is determined by the same scalar interactions as for direct detection.

However the indirect search from the sun are sensitive to spin-dependent cross section and probe parameter space mostly not available to direct detection. The results for MACRO from the sun are shown in figure 7. The points above the line representing the experimental limit are ruled out.

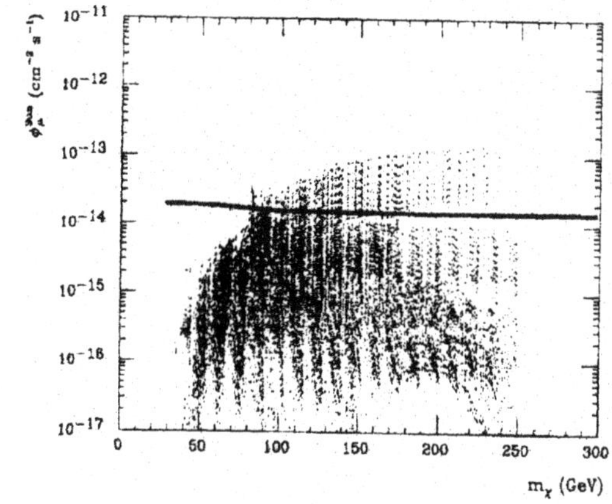

Figure 7. Limits from MACRO for neutrinos from the sun are shown, as well as supersymmetric model results from Bottino.

5. WIMP SEARCHES IN SPACE

Another possible way to make an indirect search for WIMPs by observing the products of annihilations as a signature is from annihilations in space. Two annihilation channels appear plausible, the neutralino annihilation into $\gamma\gamma$ and γZ final states, which give a signature of gamma rays with unique lines.

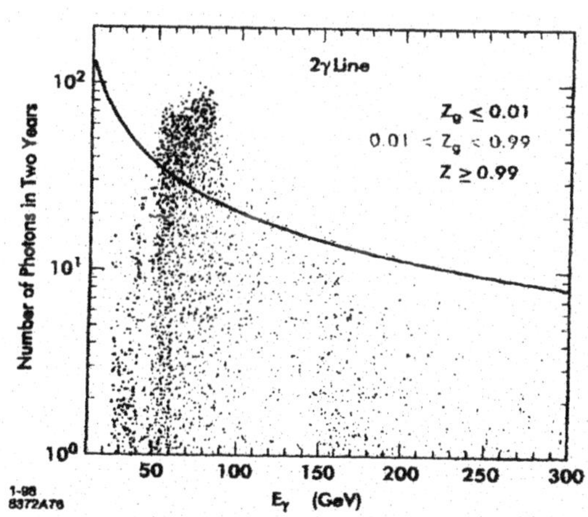

Figure 8. Projected sensitivity of GLAST for detection of high energy gamma ray lines from WIMP annihilations in space.

Bergstrom and Ullio have analyzed both these processes and these results have been compared to the projected sensitivities of the proposed GLAST mission. GLAST will be an improvement over EGRET, which has measured sources of high energy gamma rays. GLAST will have a large CsI calorimeter with energy resolution of $\sigma_E/E \sim 1.5\%$. This will allow searches for gamma-ray lines above 30 GeV.

Figure 8 shows the projections for GLAST after a two year run. A significant portion of the parameter space is accesible, especially that for higgsino-like neutralinos.

6. FUTURE PROSPECTS

The present generation of underground experiments, Kamiokande, Baksan and MACRO have found no evidence of a signal of from either the sun or the center of the earth. These limits on the neutrino flux exclude some supersymmetric models.

In the next decade there will be a new generation of direct detectors that should improve the sensitivity by a factor of 10 or more. There may also be large underwater or under ice neutrino detectors as large as 1 km^3 developed that can search for neutrinos from the sun and the center of the earth.

The implementation of larger and more sensitive direct search detectors should provide better information on spin independent interations. Thus, the future of indirect searches will be on the information from spin dependent interactions from the sun. These experiments are complementary to both the direct search experiments and those from accelerators.

REFERENCES

1. G. Jungman, M. Kamionkowski and K. Griest, *Phys. Rep.* **267** (1995) 195-373.
2. K. Ackerstaff et al, OPAL, CERN-PPE/97-083 to be published *Z. Phys. C*; D. Buskulic et al, ALEPH, *Z. Phys. C* **72** (1996) 549
3. A. Gould, *Astrophys. J.* **321** (1987) 877
4. M. Mori et al. (Kamiokande), *Phys.Rev. D* **48** (1993) 5505
5. M.M. Boliev et al. (Baksan), *Nucl. Phys. B* (Proc. Suppl.) **48** (1996) 83
6. Bottino et al *Astropart Phys.* **3** (1995) 63.
7. Bergstrom, Edsjjo and Gondolo *Phys. Rev. D* **55** (1977) 195 and extended to km^3 detectors in hep-ph/9806293, unpublished.

ELSEVIER

Nuclear Physics B (Proc. Suppl.) 77 (1999) 402–419

NUCLEAR PHYSICS B
PROCEEDINGS
SUPPLEMENTS

Baryonic Dark Matter

M. Spiro, E. Aubourg, and N. Palanque-Delabrouille
(DAPNIA/SPP, CEA-Saclay, 91191 Gif sur Yvette Cedex, France)

This paper discusses the existence of baryonic dark matter in the universe confronted with the fundamental parameters and observations in cosmology. The various possible baryonic candidates are discussed with special emphasis on the MACHOs, which would be dim compact halo objects clustered at galactic scales.

1. The Dark Matter Issue, H_0, Ω

There is a strong theoretical prejudice for $\Omega = 1$, *i.e.* that the Universe we live in is flat. One of the main reasons comes from the evolution of the cosmological density parameter Ω.

$$(\Omega - 1) \propto t^{2-2\alpha} \tag{1}$$

where α is $\frac{1}{2}$ for a radiation dominated universe and α is $\frac{2}{3}$ for a matter dominated universe. Therefore, $\Omega = 1$ is an equilibrium value, but should the initial value of Ω be only slightly different from 1, and the Universe would soon recollapse (if $\Omega > 1$) or get totally diluted (if $\Omega < 1$). However, the observations tend to show that today's value Ω_0 is of the order of unity (depending on the method of measurement used, $0.2 \leq \Omega_0 \lesssim 1$). Therefore, we can estimate the value of Ω at Planck's time, for example ($t_{Pl} = 5 \times 10^{-44}$ sec and today's Universe is about 15 billion years old) or at the time of freeze-out of the decay of neutrons into protons (allowing Nucleosynthesis to begin) *i.e.* $t = 1$ sec:

$$|\Omega_0 - 1| \leq \mathcal{O}(1)$$

$$|\Omega(t_{Pl}) - 1| \lesssim 10^{-60} \tag{2}$$

$$|\Omega(t_{\text{freeze-out}}) - 1| \lesssim 10^{-16} \tag{3}$$

which means that there was in the past an incredibly sharp tuning! Ω was much closer to 1 than it is today. An anthropic point of view is to say that if it were not so, we would not be here to observe it. But another solution is to admit that the Ω is *exactly* 1, the only value stable with time.

There is another hint to an $\Omega = 1$ Universe. Shortcomings of standard cosmology show evidence for various problems (fine tuning) in the simple big-bang scenario. Besides the flatness problem mentionned above , there is the horizon and homogeneity problem (how is it that regions of the universe which were not causally connected in the past have the same temperature) and the monopole problem (how is it that there are so few monopoles in the observable universe).

The theory of inflation (the original model was proposed by [52]) solves simultaneously all the problems mentioned above, by assuming there was an epoch, very early in the history of the Universe, when vacuum energy dominated. This means that the scale factor grows as $\exp^{Ht}$ and that Ω is brought infinitely close to unity whatever the initial conditions and provided that the vacuum energy drops to zero. This is however now disputed with the recent observations of distant Type Ia supernovae.

In the following sections, several experimental results will be presented, which allow to estimate the value of Ω, from the matter density in various structures (galaxies, clusters...).

1.1. "Visible" matter

Because the matter we see is concentrated in galaxies, the most straightforward way to get the mean density is to measure the luminosity density $\mathcal{L}$ of stars in galaxies and estimate the average mass-to-light ratio $< M/L >$. Then, assuming that the measured $< M/L >$ is representative of the Universe as a whole (an obviously questionable assumption), the mean *visible* matter density is given by the product $<\rho_{vis}> = <M/L> \times \mathcal{L}$. In blue light,[1] $\mathcal{L}_B =$

[1] The luminosities can be given in several bands (visible, blue or bolometric) which can yield as much as a 50%

0920-5632/99/$ – see front matter
PII S0920-5632(99)00451-X

$(1.6 \pm 0.2)h_0 \times 10^8\,L_\odot\,\mathrm{Mpc}^{-3}$ [34]. Using the value of $\rho_c == 2.8 \times 10^{11} h_0^2\,M_\odot\,\mathrm{Mpc}^{-3}$, the cosmological density parameter Ω and the mass-to-luminosity ratio M/L are then related by

$$\frac{M}{L_B} \sim 1500\,h\,\Omega\,\frac{M_\odot}{L_\odot} \tag{4}$$

¿From now on, mass-to-light ratios will always be given in solar units, *i.e.* in units of $M_\odot/L_\odot$.

Let's define Ω_{vis} as the value Ω_0 would have if the mass-to-light ratio of the Universe as a whole were similar to the mass-to-light ratio observed in stars, *i.e.* assuming $(M/L)_{Universe} \sim 3$. A combination of various recent observational results yields the following range on Ω_{vis} (using equation 4):

$$0.002\,h^{-1} \leq \Omega_{vis} \leq 0.006\,h^{-1} \tag{5}$$

1.2. Dynamical mass in galaxies

The most explicit evidence for galactic dark matter comes from the observation of the rotation curves (*i.e.* rotation velocity $v(R)$ as a function of the galactocentric radius R) of spiral galaxies. They can be obtained from measurements of the motion of peripheral stars (up to the edge of the "visible" component of the galaxy), and, further away, by studying the Doppler shift of emission lines (the 21-cm line of neutral hydrogen for instance). The inner part of the disk can also be traced using CO emission in molecular gas. Systematic studies on stellar velocities [65] gave results consistent with similar studies using HI regions [37].

Assuming that the mass in galaxies is spherically distributed, it is possible to relate the mass $M(R)$ within radius R to the rotation velocity $v(R)$:

$$\frac{G\,M(R)}{R} = v^2(R) \tag{6}$$

Figure 1 shows the discrepancies between the expextations from the visible mass distribution and the effective plateau in the rotation curve.

On average, within R_{lum}, the dynamical mass does not exceed its luminous counterpart by a large amount: $(M/L)_{dyn,\,R<R_{lum}} \sim 4$–6 compared to $(M/L)_{vis} \sim 1$–3. However, when including data out to about $\sim 3R_{lum}$, the observed *dynamical* mass-to-light ratio is of the order of $30h$ which is no longer consistent with the value for the *luminous* component. Since no turn-off of rotation curves has been observed so far, we can only give a lower bound of about 40 kpc, for most spiral galaxies, on the extent R_{halo} of the halo of the galaxy (beyond that radius, the column density of HI becomes too small to be detected). Extrapolating the behavior described above to the radius R_{halo} yields $(M/L)_{dyn} \simeq 70 R_{halo}/100\,\mathrm{kpc}$.

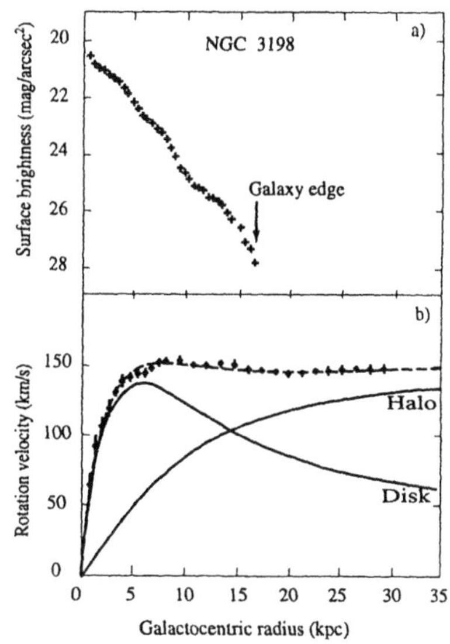

Figure 1. Rotation curve of NGC3198 [69].

This result can be converted into an expression for Ω:

$$\Omega_{halo} \sim 0.05 R_{halo}/100kpc \tag{7}$$

with R_{halo} between 40 kpc and 100 kpc

Because $\Omega_{halo} >> \Omega_{vis}$, this is a sure clue that dark matter exists on the scale of individual galaxies.

---

difference in the quoted value of the M/L ratio.

1.3. Mass on the scale of clusters of galaxies

• Derivation of mass from velocity dispersions

Evidence for dark matter on the scale of clusters ($\sim$ 10 Mpc) comes mainly from measurements of velocity dispersions and average separation of galaxies. Typically, the virial theorem is applied to the data, yielding M/L ratios near $100h$ (for small clusters) to keep them bound. Larger clusters have published mass-to-light ratios ranging from $30h$ to about $1000h$, but these values are often derived each in a different waveband (as mentioned before, this yields differences of factors of ~ 2), or with various choices of H_0 (although this factor cancels out when calculating Ω), or depending on whether the authors consider the virial mass ($2T + U = 0$) or the just-bound mass ($T + U = 0$). The above range would narrow down a lot when using uniform definitions to express the mass-to-light ratio. Typical ratios are of a few hundreds and dark matter must then dominate over the visible matter in clusters by about a factor of 10.

• Derivation of mass from X-ray data

Evidence for dark matter on the scale of clusters is confirmed by the gravitational mass derived from the X-ray emission of hot gas in the intra-cluster plasma. This method, however, rests on the assumption that the gas is in hydrostatic equilibrium in the cluster potential. Observed substructure in many clusters (meaning that they are fairly young), along with the fact that clusters are expected to relax on a time-scale comparable to the Hubble time (*i.e.* $\sim$ age of the Universe) imply that X-ray mass derivations are not without uncertainties. Yet, they should not be discarded. Recently, Bohringer and Neumann [35] showed that a typical rich cluster had the following mass composition: 2–7% in galaxies, 10–30% in gas and 60–85% in dark matter.

• Derivation of mass from strong gravitational lensing

Gravitational lensing provides further evidence for dark matter on these scales, and unlike the methods that have been described so far, it can also probe the total matter distribution. Galaxies behind the cluster are distorted by the cluster potential and some appear as giant arcs (highly distorted images with a very large amplification) or arclets (faint weakly lensed images). In the case of multiple images from a single source, each of them probes the potential in different points and it is then possible to recover *exactly* the matter distribution acting as a lens [54]! In the case of arclets, a statistical analysis can yield the same results. The main constraints come from the position, shape and orientation of the arcs and, when available, curvature radius, lengths or axis ratios...

Several gravitational lens systems have been observed with sufficient resolution to detect the various images of a single galaxy. In a few cases (3 images of the same source are needed), the lens system is over-constrained; the mass distribution can be computed and the position of other arclets predicted. For instance, from the observations of two arc systems with multiple images, Mellier *et al.* [55] could model the cluster lens ms2137–23 and predict the position of three arclets (counter images of the existing arcs) that were found to coincide with real objects of the field, with the right elongation and orientation. This gives strong confidence in the reliability of the results obtained, among which the mass-to-light ratio of about 100 for this cluster. Based on a similar analysis, the same group modeled the matter distribution in the cluster A370 and found a double-potential solution that reproduced very well the observed substructure in the main giant arc.

In both cases discussed above, the total matter, detected through the modeling of the gravitational lens, is strongly concentrated on the brightest giant galaxy seen in the optical (or brightest two galaxies as in A370), and seems to follow the light distribution. All cluster-lenses studied so far give mass-to-light ratios ≥ 100, confirming that clusters are dominated by a dark component.

To conclude on the total matter density derived from observation of strong gravitational lensing in clusters or from virial analysis, an average value of Ω on scales of $\sim$ 10 Mpc, assuming that

$(M/L)_{Universe} \sim (M/L)_{clusters}$, is:

$$\Omega_{cluster} \sim 0.2 \quad (8)$$

1.4. Peculiar velocities and large-scale dynamics

The large-scale dynamics of matter is another probe of the matter density of the Universe, here on cosmological scales.The POTENT procedure [48] constrains the cosmological density parameter Ω from the study of large-scale velocity flows.

The radial peculiar velocity u_i of individual galaxies is the difference between the total radial velocity of the galaxy as inferred from its redshift z_i ($v_i = cz_i$) and the Hubble velocity due to its distance r_i ($v_i = H_0 r_i$):

$$u_i = cz_i - H_0 r_i \quad (9)$$

Figure 2 shows the cosmological landscape and a map of both the 3-D velocity field (projected) and the recovered mass-density field.

Knowing the 3-D velocity field, it is possible to reconstruct the mass density fluctuation $\delta \equiv (\rho - \bar{\rho})/\rho$ where $\bar{\rho}$ is the average mass density (from the theory of GI):

$$\delta(\vec{r}) = ||I - f^{-1}\partial\vec{v}/\partial\vec{r}|| - 1 \quad (10)$$

where I is the identity matrix, f is a function of the cosmological density parameter Ω approximated by $f(\Omega) \approx \Omega^{0.6}$ and the bars denote the determinant. The mass density recovered by the POTENT analysis is consistent with both the galaxy density from redshift catalogs and the level of anisotropies detected by COBE, which brings further support to the method.

A comparison between the POTENT mass density field and the fields derived from galaxy redshift surveys (such as IRAS) provides an estimate of the ratio $\beta \equiv f(\Omega)/b$, which unfortunately relies on an assumed biasing factor b between galaxy and mass distributions. The biasing concept was introduced in 1984 so as to take into account a dark matter component by writing

$$\frac{\delta N_{gal}}{N_{gal}} = b\frac{\delta\rho}{\rho} \quad (11)$$

where N_{gal} is the number of galaxies and ρ the total density. It relies on the observation that clusters of galaxies are themselves much more clustered than galaxies are, which leads to believe

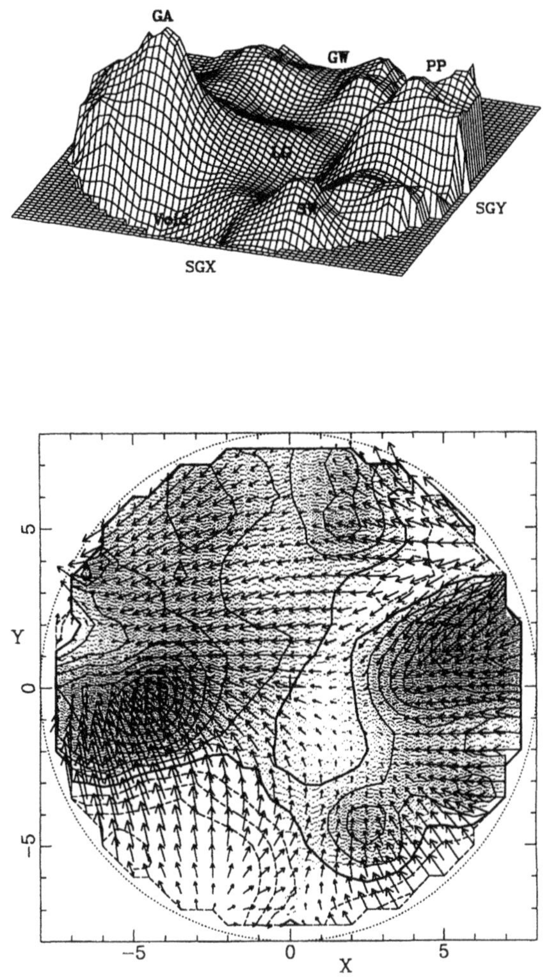

Figure 2. The fluctuation fields of velocity and mass-density in the super-galactic plane as recovered by POTENT from the velocities of ~ 3000 galaxies with $12\,h^{-1}$Mpc smoothing. The vectors shown are projections of the 3-D velocity field. Distances and velocities are in 1000 km/s (taking $H_0 = 1$). Contour spacing is 0.2 in δ, with the heavy contour marking $\delta = 0$ and dashed contours denoting negative fluctuations. The Local Group is at the origin, the Great Attractor is on the left, Perseus Pisces on the right and Coma at the top. On the 3-D plot (top), the height of the surface is proportional to δ [47].

that mass might be more smoothly distributed than light (*i.e.* galaxies) on scales $\sim 20\,h^{-1}$Mpc. Current estimates for the biasing factor are in the range $1 \leq b \leq 3$. A preliminary attempt to separate the Ω and b parameters from the POTENT analysis yields (as mentioned above, POTENT measures a value of $\Omega^{0.6}/b$ and not directly Ω)

$$\Omega > 0.46 \text{ (at 95\% confidence level) if } b > 0.5 \tag{12}$$

An interesting lower bound on Ω can be obtained from velocities in voids: the mass density fluctuation δ inferred from a given diverging velocity field becomes more negative as the cosmological density parameter Ω decreases (*cf.* equation 10), and could get smaller than -1 if Ω was taken too small, which is impossible since mass is non-negative. The POTENT data in voids is perfectly consistent with $\Omega \approx 1$ but values of $\Omega \approx 0.6$ already yield $\delta < -1$ in the deepest void observed in the sample! Values of $\Omega = 0.2$ and 0.3 are ruled out at the 2.9σ and 2.4σ levels in terms of the random error σ_δ [47].

In conclusion on large scale flows, using the POTENT analysis, strong evidence indicates that:

$$\Omega_{POTENT} > 0.2\,(0.3) \text{ at the } 2.9\sigma\,(2.4\sigma) \text{ level} \tag{13}$$

and data is consistent with:

$$\Omega_{POTENT} \approx 1 \tag{14}$$

1.5. Conclusions

The sketch of figure 3 summarizes all the observations mentioned here as well as the inflation expectation. As the scale increases, the value estimated for Ω gets closer and closer to the inflation prediction of $\Omega = 1$.. Whatever the method used to derive Ω out of large scale observations, it is clear that visible matter can only make up a small percentage of the Universe. There must be dark matter to explain the matter density observed.

2. Baryonic dark matter

Big Bang Nucleosynthesis begins when D can finally be formed, about 1 minute after the Big Bang. Light elements can then be produced, but only through 2-body reactions (because of the low

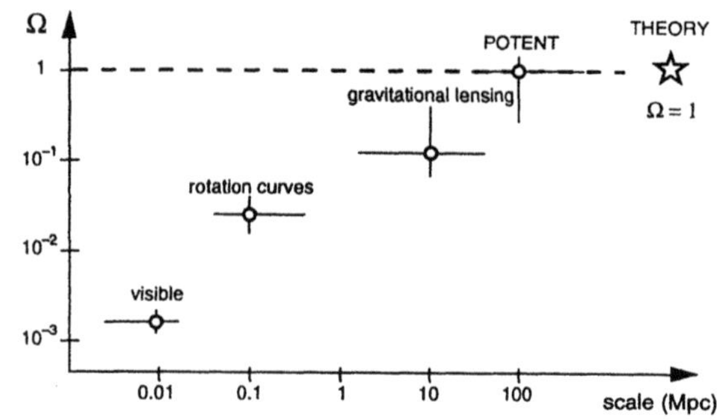

Figure 3. Values derived for the cosmological density parameter Ω as a function of the scale under study.

baryon density) such as:

$$\begin{aligned} n+p &\longrightarrow \mathrm{D} \\ \mathrm{D}+p &\longrightarrow {}^3\mathrm{He} \\ \mathrm{D}+n &\longrightarrow {}^3\mathrm{H} \\ \mathrm{D}+\mathrm{D} &\longrightarrow {}^4\mathrm{He} \\ &\dots \end{aligned} \tag{15}$$

A back-of-the-envelope estimate is sufficient to understand qualitatively the outcome of nucleosynthesis, and the dependence of the abundance of each nucleus formed on the baryon-to-photon ratio η.

This is the situation on Big Bang Nucleosynthesis as it stood until 1994. Figure 4 summarizes the theoretical predictions and the observational data presented above for the primordial abundances of the various light elements produced during BBN. The concordance range for the baryon-to-photon ratio is:

$$2.5 \times 10^{-10} \leq \eta \leq 6 \times 10^{-10} \tag{16}$$

A given value of η is related to the baryon density Ω_B since $\eta = (n_B/n_\gamma)$. The density of photons can be calculated from the temperature of the microwave background (which yields $\sim 411\,\mathrm{cm}^{-3}$) so the baryon density can be obtained by the relation:

$$\Omega_B = \frac{\rho_B}{\rho_c} = 3.66 \times 10^7\, \eta\, h^{-2} \qquad (17)$$

Standard nucleosynthesis therefore allows the following range on the baryon density:

$$0.009 \leq \Omega_B h^2 \leq 0.02 \qquad (18)$$

where $h = H_0/100\,\mathrm{km\,s^{-1}\,Mpc^{-1}}$ is in the range $[0.4 - 1]$ (*cf.* section 1).

2.1. Big Bang nucleosynthesis and the nature of dark matter

If we do not yet consider any of the recent D abundances obtained from quasar absorption systems, the predictions of big bang nucleosynthesis and observations agree well and yield the range given in equation 18 for the contribution of baryons to the cosmic density parameter. Considering the largest possible range of the "reduced" Hubble parameter h ($0.4 \leq h \leq 1.0$), this gives:

$$0.009 \leq \Omega_B \leq 0.14 \qquad (19)$$

A slightly larger interval can be obtained if one adds uncertainties due to the unsure understanding of the chemical evolution of the Universe since nucleosynthesis occurred, and to other, unaccounted for, possible systematics [44]:

$$1.6 \times 10^{-10} \leq \eta \leq 9.0 \times 10^{-10} \qquad (20)$$

Allowing for the largest range in h yields the following range for Ω_B:

$$\begin{aligned} 0.006h^{-2} &\leq \Omega_B \leq \ 0.03h^{-2} \\ \Longrightarrow 0.006 &\leq \Omega_B \leq \ 0.21 \end{aligned} \qquad (21)$$

This range needs to be compared to that obtained in section 1 for the amount of visible matter in the Universe (cf equation 5). Because of the h dependence of all these ranges, it is interesting to plot the results on an $H_0 - \Omega$ diagram as in figure 5 (adapted from [43]). There is no overlap between the ranges of Ω_{vis} and Ω_B, whatever the value of H_0. This underlines the existence of a large amount of *baryonic* dark matter. With the results of standard nucleosynthesis, at least 70% of the baryons in the Universe must be dark. Another interesting consequence of nucleosynthesis

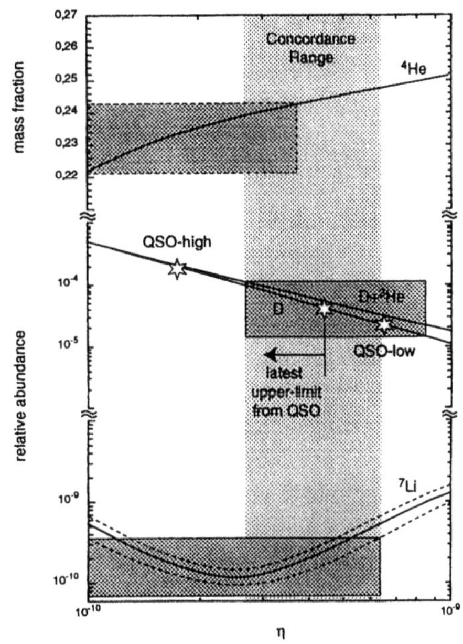

Figure 4. Confrontation between theoretical predictions from Big Bang Nucleosynthesis and observations. The dashed curves around the predicted ^{7}Li abundance indicate the 2σ error bars. The uncertainties are smaller for the other predicted abundances and therefore not shown. The gray boxes trace the regions allowed by each set of observations as quoted in the text. The ^{4}He box is dashed because of the very small dependence of Y_p on η: a small shift in the limit on Y_p induces a large change on the limit on η. The concordance range (shown in light gray) is $2.5 \times 10^{-10} \leq \eta \leq 6 \times 10^{-10}$, derived from the limits on D$+^3$He and on ^{7}Li. The two stars show the values of (D/H) as measured by [66] for the high value and by [77] for the low value.

is that for no value of the Hubble constant does the allowed range for the baryonic density overlap the $\Omega = 1$ line, and only for very small values of H_0 does it cross the limit for $\Omega_{clusters}$. This implies that if the results from POTENT are to be trusted, there must also be a tremendous amount of *non-baryonic* dark matter. Even on the scale of clusters, baryonic matter cannot account for all the mass detected.

Let us now consider the impact of the new deuterium detections on baryonic dark matter. Obviously, the low detection of deuterium leaves more than enough room for baryonic dark matter. If the trend toward low Deuterium detections continues, large amounts of baryonic dark matter are expected. None of these two detections, however, affect the need for non-baryonic dark matter.

An agreement on the primordial abundance of deuterium is therefore a major issue in order to conclude on the amount of baryonic dark matter in the Universe.

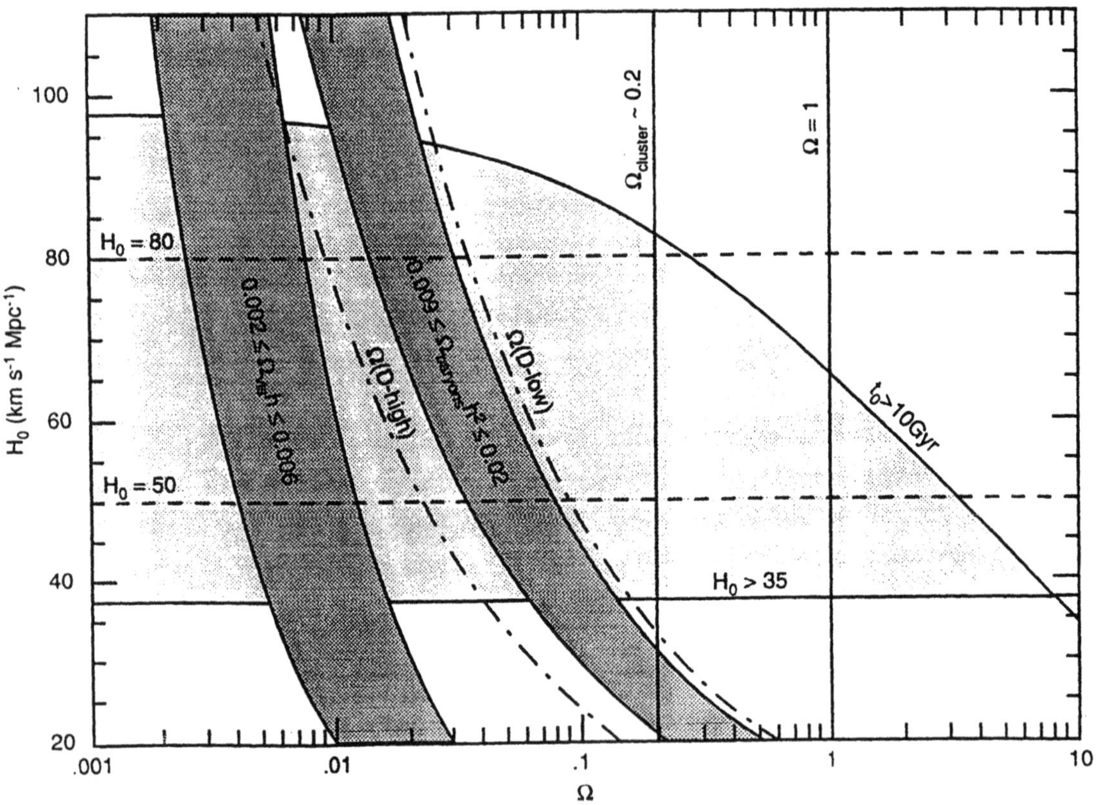

Figure 5. $H_0 - \Omega$ plot (assuming $\Lambda = 0$), indicating allowed (shaded in gray) and excluded regions. The fraction of visible matter in the Universe, Ω_{vis} is shown, along with the fraction of baryonic matter Ω_B resulting from nucleosynthesis, the average value of the fraction of matter in clusters of galaxies Ω_{clusters}, and the value $\Omega = 1$ theoretically preferred and obtained by the POTENT analysis. On each side of the region allowed by nucleosynthesis are shown the two recent deuterium detections (in mixed line). The two current bounds for the Hubble constant: $H_0 \geq 60\,\text{km}\,\text{s}^{-1}\,\text{Mpc}^{-1}$ and $H_0 \leq 70\,\text{km}\,\text{s}^{-1}\,\text{Mpc}^{-1}$ are plotted in dotted lines. Two more limits are indicated: the lower bound $H_0 \geq 35\,\text{km}\,\text{s}^{-1}\,\text{Mpc}^{-1}$ obtained from white dwarf stars and supernovae [28], and the lower bound on the age of the Universe $t_0 \geq 10\,\text{Gyr}$ obtained essentially from the age of globular clusters, under the reasonable assumption that the Universe can hardly be younger than its components (adapted from [43]).

2.2. Dark matter candidates

As we have discussed at length in the previous section, two types of dark matter are needed, baryonic and non-baryonic. Comparing the range for baryonic dark matter obtained from BBN (*cf.* equation 19) with the amount of matter on the scale of individual galaxies as deduced from their flat rotation curve, it is remarkable that these two ranges perfectly agree, making it reasonable to fill the halo of galaxies with baryonic dark matter. On larger scales, however, non-baryonic dark matter becomes inevitable (believing nucleosynthesis to be correct). The nature of dark matter has a large impact on several fields of physics: cosmology, astrophysics and particle physics. The formation of Large Scale Structures depends largely on whether most of the dark matter is *hot* or *cold*,[2] and the extension of the standard model of particle physics is based on the existence of new particles which must be accounted for as candidates for dark matter. I will mention separately candidates for baryonic and non-baryonic dark matter and present the current observational limits.

2.2.1. Baryonic candidates

One can consider a large range of baryonic candidates for dark matter, some of which are already ruled out by current observations.

- **Intergalactic baryonic dark matter**

  - **Gas**
    In clusters of galaxies, the ratio of the mass contained in form of hot gas to the mass contained in the form of stars is of the order of Ω_B/Ω_{star}. This is s strong incentive to look whether, on intergalactic scales, the presence of large quatities of cold or warm gas could be the main component of the baryonic dark matter.

Plasma associated with galaxies outside the great clusters makes a significant also still quite uncertain contribution to the baryon budget. The detected X-ray emission from plasma in groups in softer than for clusters. It is a reasonnable presumption that the absence of detections often represents a lower virial temperature than the absence of plasma.

In addition to the hot plasma which could be detected in X-rays, there might be a considerable amount of plasma in cool, thermally stable, photo-ionized clouds.

This is indeed what is seen at z¿=3 where the dominant baryonic mass component is in the Lyman-*alpha* forest gas detected by the trace neutral hydrogen in plasma [78].

- **Galactic baryonic dark matter**

  - **Gas?**
    The amount of the gas component in the disks or halos of galaxies is well known, and is far below that needed to obtain a flat rotation curve (*cf.* for instance figure 1). [62] However, it might be possible to put a significant mass of hydrogen in the form of fractal, cold molecular clouds [62]. It would not radiate (cold) and be diffuse enough not to be detected. This has been discussed and somewhat rejected from the isotropy of cosmic rays (Salati et al 96).

  - **Stellar remnants?**
    In the disks or haloes of galaxies, neutron stars and Black Holes have been considered as possible baryonic candidates for dark matter. However, old stars eject large quantities of metals (processed elements) in their environment which would be observed spectroscopically in molecular clouds or be part of the composition of younger stars. The study of the metallicity of population I stars (young stars) excludes neutron stars and

[2] *Hot* dark matter consists of particles whose velocity when they decouple from the expansion is still relativistic. Light neutrinos are the best examples of hot relics. Because of their free streaming, such particles are not expected to collapse on small scales and the first structures to form will therefore be fairly large (typically of the order of 40 Mpc for a 30 eV neutrino) and can only appear late in the history of the Universe (horizon size needs to be at least as large as the structure we want to form). In contrast, *cold* dark matter particles freeze-out when they are non-relativistic. Usual candidates are heavy neutrinos, WIMPS or axions. These particles collapse easily onto the matter over-densities and small objects form first.

stellar black holes from contributing significantly to the dark matter [39].

Very massive black holes ($m \gtrsim 10^3\,\mathrm{M}_\odot$), however, would swallow their nucleosynthetic products as the star collapses, with no release of heavy elements. But the precursors of very massive black holes would be highly luminous and because of cosmological redshift, they should be detectable, today, in the near-IR range [36]. The strong constraints on spectral distortion of the microwave background imposed by COBE essentially exclude this possibility.

A significant amount of stellar remnants would thus lead to excessive pollution and background light. Another solution must be sought for baryonic dark matter.

- **Low mass stars?**

Very low mass objects called *snowballs* have at one time been considered. They are non-degenerate hydrogen condensates of mass smaller than $0.01\,\mathrm{M}_\odot$. No nuclear reaction can occur in such light objects which makes them "invisible". However, several mass ranges have been excluded by plain observation of the low frequency of encounters of such objects (based on the number of craters on the moon or counts of interstellar comets). A very stringent limit has been set by [46] who claim that objects smaller than $\sim 10^{-7}\,\mathrm{M}_\odot$ would have evaporated by now.

Brown dwarfs are stars of mass comprised between $0.01\,\mathrm{M}_\odot$ and $0.08\,\mathrm{M}_\odot$. They are too light to produce energy by burning their hydrogen into ^{4}He, and can only ignite their deuterium (through $p + d \longrightarrow^3$ He). They are thus too faint to be detected directly and up to now, they still constitute a fairly good galactic dark matter candidate. A thorough review on brown dwarfs can be found in [38].

Red dwarfs are stars of mass between $0.08\,\mathrm{M}_\odot$ and $0.5\,\mathrm{M}_\odot$ which burn their hydrogen but remain very dim. They can be detected by their IR emission or proper motion measurements. Recently, using star counts in the Wide Field Camera of the Hubble Space Telescope, [30] have concluded that faint red dwarfs ($M_I > 10$)[3] contribute less than 6% of the unseen matter in the halo of the Galaxy and at most 15% of the mass of the disk. Using a different technique, [49] claim that faint red dwarfs account for $< 1\%$ of the Galactic dark halo for $M_I < 14$, and for $< 6\%$ for a limiting magnitude $M_I < 15$ at the 95% confidence level.

Finally, *white dwarfs* are stars in the mass range $\sim 0.3 - 1.4\,\mathrm{M}_\odot$ (residues of stars whose initial mass was less than $8\,\mathrm{M}_\odot$ or so) which abound in today's Universe. They are known to contribute as much as 20–30% of the stellar mass in old globular clusters [56]. They constitute the final state of low mass stars. The mass fraction of typical halos contributed by these objects is uncertain, but can be constrained in various ways. Based on Type Ia supernovae rates, [73] have set a tight upper limit on the fraction of white dwarfs in binary systems. Recent work by [42] assumes that galactic halos are baryonic, and shows that $\lesssim 10\%$ of the mass fraction of halos can then be in the form of white dwarfs. A higher mass fraction would make the halos so bright during the early phase in which the future white dwarfs are still on the main sequence that they would have been detected in deep (high redshift) surveys. Baryonic halos composed of an admixture of brown, red and white dwarfs seem more likely.

Amongst known candidates, very few are left to solve the baryonic dark matter problem: essentially cold molecular hydrogen clouds or low mass objects in the halo of galaxies (commonly called machos for "Massive Astrophysical Compact Halo Objects"). A promising technique to detect non-luminous low mass objects is through microlensing. The purpose of this thesis is to constrain the mass fraction of the Halo of our Galaxy

[3] I stands for the infrared band of the Johnson-Cousins system.

in machos using this technique and I refer to further chapters for more on this detection method.

2.2.2. Non-baryonic candidates

Baryonic dark matter could explain dark matter on the scale of individual galaxies (although, as we have seen, it looks improbable that halos be entirely composed of baryonic matter) but in no way can it explain the amount of dark matter on larger scales. Most of the non-baryonic candidates are motivated by particle physics, and several have been introduced so as to solve problems that are totally unrelated to cosmological dark matter. Most often considered are light neutrinos ($m < 1\,\mathrm{MeV}$), Weakly Interacting Massive Particles (or WIMPS) which include massive neutrinos and supersymmetric particles, and axions. A review on Particle Dark Matter was presented recently by [53].

3. Search for MACHO's with the gravitational microlensing technique

Very few possibilities remain for baryonic dark matter [40]. They are called Massive COmpact Halo Objects (MACHOs). These could be either low mass faint objects (0.1 to $0.3\,\mathrm{M}_\odot$), cold star remnants (white dwarfs of 0.3 to $0.7\,\mathrm{M}_\odot$, neutron stars of 1.3 to $1.7\,\mathrm{M}_\odot$), aborted stars of 10^{-7} to $0.08\,\mathrm{M}_\odot$ or even primordial black holes.

A way to detect these MACHOs is to look for the temporary brightening of a star that occurs when a MACHO passes next to its line of sight.

3.1. Gravitational microlensing

According to the principles of general relativity, the light rays from a source star are deflected by the presence of a massive deflector located near the line of sight between the star and the observer. This forms two distorted images of the source, as illustrated in figure 3.1.

In the particular configuration where all three objects are perfectly aligned, the two images merge into a ring, whose radius is called the Einstein radius R_E

$$R_E = \sqrt{\frac{4GM}{c^2} D_{OS}\, x(1-x)} \qquad (22)$$

where $x = D_{OD}/D_{OS}$ is the ratio of the distance observer-deflector to the distance observer-source and M the mass of the deflector. When probing the dark matter content of the Galactic halo, the source star is typically 60 kpc away (located in one of the Magellanic Clouds) from the observer and the lens typically a solar-mass object or lighter, so the angular separation between the images ($\sim 2R_E/D_{OD}$) is only of the order of the milliarcsecond. Given the limited resolution of optical telescopes with current technology, only the combined light intensity can therefore be recorded. The total magnification, however, is always greater than what the observer would receive from the source in the absence of the lens, which makes the latter detectable. As the lens moves in the halo with respect to the line-of-sight, the typical time scale Δt of a microlensing event is given by

$$\Delta t = \frac{R_E}{v_t} \simeq 90\sqrt{M/\mathrm{M}_\odot} \qquad (23)$$

where v_t is the transverse velocity of the lens which to first order can be taken as the rotation velocity of the galactic halo, ie 220 km/s.

The probability that a given star is amplified at a given time is very low, typically $5\,10^{-7}$. Millions of stars thus have to be monitored for years in order to ever be able to detect such a rare event. More details on the general principles of microlensing can be found in [13].

The only targets far enough to probe a large fraction of the galactic halo but close enough to resolve millions of stars are the large and

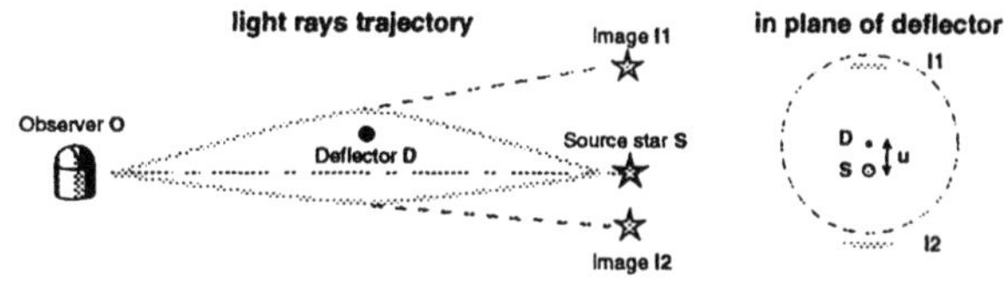

Figure 6. Deflection of light by a massive body D located near the line of sight between the observer O and the source star S. The dotted circle is the Einstein ring.

the small Magellanic Clouds (LMC and SMC respectively), observable from the southern hemisphere. Various experiments are involved in the survey of stars located in these two satellite galaxies: EROS 2 (French experiment observing in Chile), MACHO (American experiment observing in Australia) and OGLE 2 (Polish experiment observing in Chile). The following sections present the latest results obtained by these experiments on the contribution of dark compact objects to the mass of the Galactic halo.

Because the mass range to probe extends from planetary objects ($\sim 10^{-7}$ $M_\odot$) to stellar dark objects (a few solar masses), the event time-scales that the experiments have to be sensitive to vary from a few hours to a few months (see equation 23). Dedicated experiments and analyses are therefore performed separately to search for either small or large mass deflectors.

3.2. Limits on contribution of small mass objects

To search for planetary mass dark matter in the galactic halo, the EROS 1 CCD experiment, on the one hand, monitored 150 thousand stars in the LMC, with a high efficiency of $\sim$ 80%, thanks to a very good time sampling. On the other hand, the MACHO experiment monitored 8.6 million stars thanks to a large coverage of the LMC, but with only a $\sim$ 1% efficiency to short time-scale events since their observational strategy was optimized for long time-scale events. Neither of the two experiments found any such event, which allowed them to set quite stringent limits on the maximum contribution of small mass objects to the dark mass of the Galactic halo [16,5]. Figure 7 shows the limits obtained by either project, under the hypothesis of a "standard" isotropic and isothermal spherical halo entirely filled of dark compact objects all having the same mass M which is indicated as the abscissa of the graph. The vertical axis represents the maximum halo mass fraction of the halo that could be composed of objects of a given mass M.

As explained above, EROS and MACHO have chosen very different analysis techniques, and there is little overlap in exposure for the two projects. Combining both sets of data after

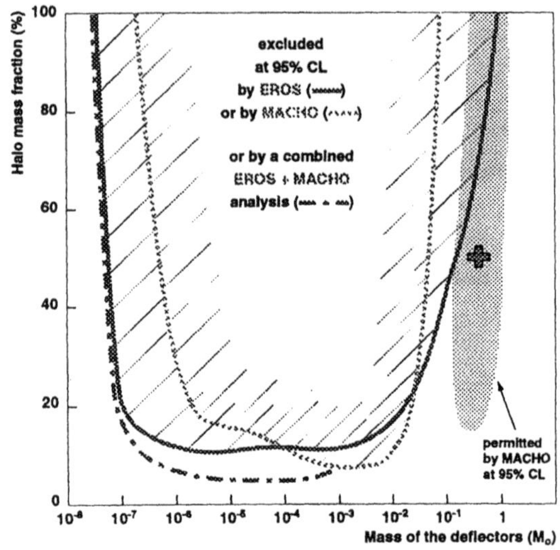

Figure 7. Exclusion diagram showing the 95% CL limits as obtained by EROS (full curve), by MACHO (dotted curve) and by a combined analysis of EROS + MACHO data (dot-dashed curve). Also shown is the 95% CL region compatible with the 6 events detected by the MACHO two year analysis.

removal of this small overlap thus yields even stronger limits [1]. The new exclusion diagram obtained is also illustrated in figure 7. It can be seen that not more than ~10% of the halo can be composed of objects in the mass range $[10^{-7} - 0.02]$ $M_\odot$, at the 95% confidence level. Because we are using δ-function mass distributions and since the limit is quite flat in the mass interval mentioned above, any mass function that peaks in this range is also excluded at the same confidence level.

3.3. Contribution of high mass objects

3.3.1. Present results toward the LMC

Both the EROS 1 experiment (using photographic plates) and the MACHO experiment (using CCD's) have searched and found long time-scale microlensing candidates. A total of 10 events have been detected, two by EROS [7] and eight by MACHO [6], although one of them is slightly asymmetric and thus often disregarded as a microlensing candidate and another is an LMC binary event where the lens is most probably located in the LMC itself.

The typical Einstein radius crossing time associated to the LMC events is of the order of 40 days, which implies a surprisingly high most probable mass for the lenses: $\sim 0.5\ M_\odot$. This mass is much larger than the upper limit of $0.08\ M_\odot$ for brown dwarfs, and the lenses could be interpreted as, instead, white dwarfs or black holes. The optical depth implied by the mean duration of the events is compatible with about half that required to account for the dynamical mass of the dark halo. Such an interpretation, however, is not quite accepted among astronomers: by observing younger galaxies where we could detect the light due to a significant white dwarf component in the halo, a limit of 10% has been set on their contribution [8]; furthermore, if indeed half of the dark halos of galaxies consisted of such stellar remnants, we should observe an enrichment of the interstellar medium in Helium, which we do not. There is thus no consensus yet as to the nature of the deflectors causing the observed events.

The 95% CL region allowed by the MACHO experiment due to the detection of 6 events (*ie* disregarding the LMC binary and the asymmetric event) is illustrated in figure 7.

3.3.2. Future data

The interpretation of the present data is ambiguous, and huge error bars remain on both the most probable mass of the deflectors and the halo fraction in compact objects. More statistics is thus required, and several experiments are accumulating data to answer the questions of the presence or not and the nature of dark compact objects in the halo of the Galaxy:

- The EROS 2 experiment is now taking data with a completely redesigned setup and a new strategy. Using a wide field CCD camera (data taken in two colors simultaneously, with in each color a 1 square degree mosaic consisting of eight 2048×2048 CCD's), EROS covered 66 deg^2 on the LMC during the first year of observation (August 1996 - May 1997) and a total of 88 deg^2 the second year. The exposure times and time sampling are adapted to a search for long time-scale events.

- The MACHO experiment is presently analyzing four years of data on 15 deg^2 in the LMC, which means an increase of a factor of 2 in time scale and 1.4 in area. Preliminary results indicate 6 new events with time-scales ranging between 15 and 110 days with an average of about 50 days. This would confirm a high mass for halo deflectors if the lenses are indeed located in the halo of the Galaxy.

- The OGLE 2 experiment uses an upgraded setup and started taking data in summer 1997. They now also cover fields in the Large and the Small Magellanic Clouds (their previous strategy concerned only fields toward the Galactic Center, thus probing disk Dark Matter and not halo Dark Matter).

3.4. Highlights toward the SMC

The Small Magellanic Cloud gives a new line-of-sight through the Milky Way halo and a new population of source stars. The use of various lines-of-sight is very important since a comparison of the event rates is a powerful tool for discriminating between various shapes for the dark halo [17,10]. In addition, this allows for discrimination between various theories for the populations responsible for the LMC lenses [20].

The EROS and MACHO experiments (and more recently the OGLE experiment) thus monitor stars in the SMC to search for microlensing events. EROS recently published the first analysis on SMC data, whose results are presented hereafter.

3.4.1. First analysis of SMC data

The EROS 2 experiment covers the densest 10 square degrees of the Small Magellanic Cloud. On these 10 fields, a total of 5.3 millions light curves were built and subjected to a series of selection criteria and rejection cuts to isolate microlensing candidates [15]. Ten light curves passed all cuts and were inspected individually. Several correspond to physical processes other than microlensing (one of them, for instance, is the light curve of a nova that exploded in the SMC), and only one of the candidates passes this final visual inspection. This candidate was also seen and detected online by MACHO. Its light curve is shown in figure 8. Once corrected for blending — because of the high stellar density of the fields monitored in microlensing surveys, the flux of each reconstructed star generally results from the superposition of the fluxes of many source stars— the event light curve is well fitted by that of a microlensing event with an Einstein radius crossing time of 123 days, a maximum magnification of 2.6 occurring on January 11, 1997 and a χ^2/d.o.f. = 332/217 = 1.5. The best microlensing fit is for 70% of the monitored flux being amplified and 30% being the contribution of blending unamplified light.

The source star being very bright, the value of the reduced χ^2 of the fit is surprisingly high. A search for periodicity was therefore performed on the light curve, and a modulation was detected, with an amplitude of 2.5% and a period of 5.2 days. Fitting again the candidate light curve for microlensing allowing for a periodic modulation yields much more satisfactory residuals than before: χ^2/d.o.f. = 199/214. This strongly supports the microlensing interpretation of the observed magnification. The modulation detected was later confirmed by the OGLE experiment, on their own data taken after the event occurred (first points in June 1997). They also confirm the value of the blending coefficient of the source star since their new camera allows the separation of the two components of the blend and thus the individual measure of each of the two fluxes.

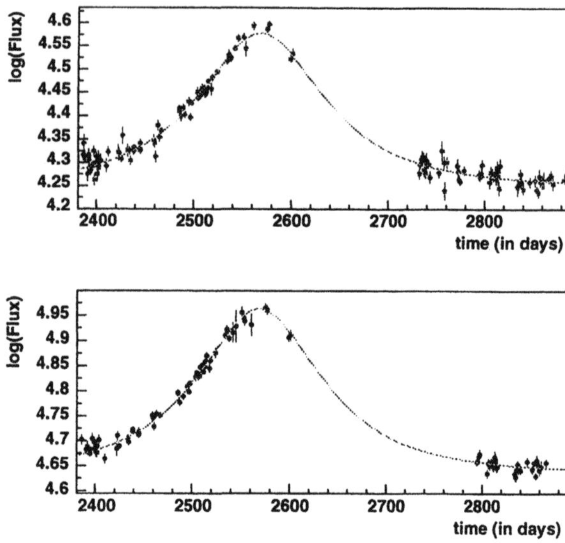

Figure 8. Light curve of microlensing candidate SMC #1, with a microlensing fit including blending superimposed. Time is in days since Jan. 0, 1990 (Julian date 2,447,891.5). Red light curve on top, blue on the bottom.

3.4.2. Estimate of Halo fraction in compact objects and lens mass

The time-scale of the observed event allows one to estimate the fraction of the halo that can be composed of dark compact objects generating microlensing events, independently of their mass. Assuming that the deflector is in the halo of the Milky Way, and considering a standard halo model (*ie* an isotropic and isothermal spherical halo), the EROS experiment estimated that the detected event is compatible with about 50% of the mass of the halo in dark compact objects. This fraction can vary by as much as a factor of two when considering other halo models (flattened halos for instance, or thinner halos and thicker disks so as to reproduce the rotation curve of the Galaxy but have less mass in the dark halo component).

Using a likelihood analysis also based on the time-scale of the detected event, the most probable mass of the deflector generating the event can be estimated. Under the assumption of a standard halo composed of dark compact objects having a single mass M (ie the mass function is supposed to be a Dirac-function), the most probable mass of the Halo deflector, given with 1σ

error bars, is:

$$M = 2.6^{+8.2}_{-2.3}\,\mathrm{M}_\odot \qquad (24)$$

The event has the highest time-scale observed so far, and consequently the highest most probable mass. Only a neutron star or a black hole could be that massive and yet be dark. It is even harder than for the LMC events to explain how the halo of the Galaxy could be filled (even partially) with such heavy dark objects. Other interpretations therefore have to be looked at seriously.

3.4.3. Interpretation as SMC self lensing

The very long time-scale of the observed event suggests that it could show measurable distortions in its light curve due to the motion of the Earth around the Sun (the parallax effect: [11]), provided that the Einstein radius projected onto the plane of the Earth is not much larger than the Earth orbital radius. The first detection of parallax in a gravitational microlensing event was observed by [4].

No evidence for distortion due to parallax is detectable on the light curve, implying either a very massive deflector with a very large Einstein radius, or a deflector near the source. The absence of parallax detection implies the following relation between the mass of the deflector and its distance to the observer, at the 95% CL:

$$\frac{M}{\mathrm{M}_\odot} \times \frac{x}{1-x} > 0.7 \qquad (25)$$

where, as previously, $x = D_{OD}/D_{OS}$.

If the deflector is in the **halo of the Galaxy** (assuming a standard halo, $x < 0.66$ at the 95% CL) this yields a lower limit on the mass of the deflector: $M > 0.6\,\mathrm{M}_\odot$.

If the deflector is located in the **SMC itself**, $1-x \sim 1/10$ and the mass of the deflector is then $M \sim 0.1\mathrm{M}_\odot$, typical of a brown dwarf or faint star in the SMC.

To validate a possible SMC self lensing interpretation of the first event detected toward this new line-of-sight, it is necessary to check whether the SMC stellar population could provide such an event, in terms of duration and optical depth (probability that at a given time, a given star be magnified).

Various authors have suggested that the SMC is quite elongated along the line-of-sight, with a depth varying from a few kpc to as much as 20 kpc, depending on the region under study. We will approximate the SMC density profile by a prolate ellipsoid:

$$\rho = \frac{\Sigma_0}{2h}\, e^{-|z|/h}\, e^{-r/r_d} \qquad (26)$$

where z is along the line-of-sight and r is transverse to the line-of-sight. The depth h will be a free parameter, allowed to vary between 2.5 and 7.5 kpc. The values of the various parameters of the model are fit to the isophote levels of the cloud (which yields $\Sigma_0 \simeq 400\,\mathrm{M}_\odot\mathrm{pc}^{-2}$ and $r_d = 0.5$ kpc) and considering a mass-to-light ratio of 3 $\mathrm{M}_\odot/\mathrm{L}_\odot$ (typical of the values measured in the disk of the Milky Way). For $h = 2.5$, 5.0 or 7.5 kpc, the predicted optical depths are $\tau = 1.0\;10^{-7}$, $1.7\;10^{-7}$ or $1.8\;10^{-7}$ respectively, to be compared with the experimental optical depth of $3.3\;10^{-7}$. Considering the very limited statistics we have, the model is consistent with the observations.

Finally, considering a velocity dispersion of 30 km/s in the SMC (and the 123 days time-scale of the event), the mass M of the deflector can be estimated according to an assumed distance between the source star and the lens. If the deflector is 5 kpc (resp. 2.5 kpc) from the source, its mass is $M \sim 0.1\,\mathrm{M}_\odot$ (resp. $0.2\,\mathrm{M}_\odot$), compatible with the results obtained from the parallax analysis.

An SMC self-lensing interpretation of the first microlensing event detected toward this new line-of-sight is thus quite plausible.

3.4.4. A Binary Lens towards the SMC

After the detection of this first SMC event, the MACHO collaboration alerted the microlensing community of an ongoing microlensing event (IAU circular 6935), which was later identified as a binary source event. In that case, the variation of the amplification is no longer simple: the gravitational potential of the double lens gives rise to cautic lines. When the source star crosses such a caustic, the amplification becomes singular. It is thus possible to resolve the finite size of the star by measuring the duration of the caustic crossing.

The measurements obtained on this microlensing event allowed to predict the date of the second caustic crossing, June 18, 1998. All microlensing collaborations took data this night. Among them, the PLANET collaboration obtained well sampled data at the time of the maximum, and the EROS collaboration equivalent data at the end of the caustic crossing (see figure 9).

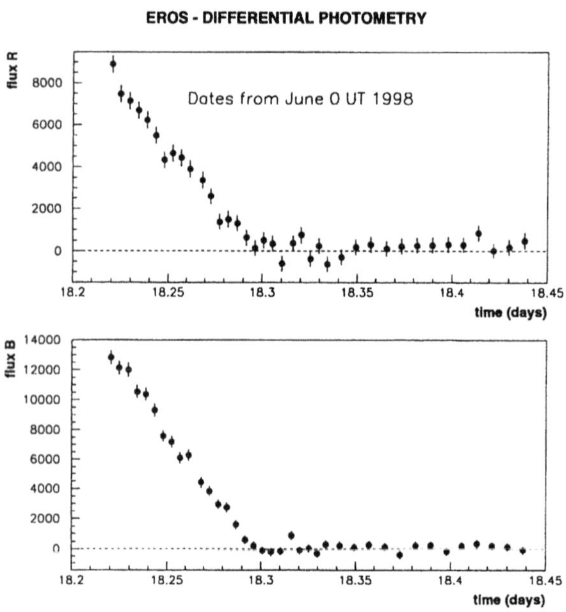

Figure 9. Differential photometry of EROS data taken on 18 June 1998. R data on top, B data on bottom, in ADU.

Both data sets allowed to put constraints on the duration of the caustic crossing [2,3]. Combining this result with an estimate of the size of the source star, it was then possible to put limits on the proper motion of the lens. The most plausible interpretation for this event is that the lens lies in the SMC itself: only 7% of the halo population has a proper motion compatible with the one measured.

3.5. Microlensing conclusions

Two targets have been explored so far, in the search for dark matter in the halo of the Milky Way. They are the Large and the Small Magellanic Clouds.

The LMC data collected by the MACHO and the EROS experiments have allowed them to exclude any major contribution to the dark mass of the halo from compact objects in the mass range $10^{-7}\,M_{\odot}-0.02\,M_{\odot}$. Eight events compatible with microlensing by halo lenses were detected, with an average time-scale of 40 days, which could be interpreted as about 50% of the halo dark matter in the form of $\sim 0.5M_{\odot}$ objects. A huge controversy remains as to the nature of these objects.

The SMC data has yielded one event found during the analysis of the first year of data, and one binary event detected online by the MACHO group. The first event has the longest time-scale observed so far: 123 days. If the lens causing the event is assumed to be in the halo of the Galaxy then its most probable mass is $2.6M_{\odot}$, with a lower limit of $0.6M_{\odot}$ coming from parallax analysis. Such a high mass is very hard to explain. A more plausible explanation is to assume that the lens and the source star are both located in the SMC. For the second event, caustic crossing time indicates clearly that the lens is in the SMC. The mass of the deflectors would then be typical of that of a faint star in the SMC, and the experimental optical depth compatible with a "thick disk" model of the SMC.

The status of microlensing experiments and their implications on the galactic structure can be summarized in a few words. With about 100 microlensing events detected toward the Galactic Center, nearly 10 toward the LMC, 1 toward the SMC and none yet toward the Andromeda Galaxy (M31), there is strong evidence for the existence of a bar in the bulge and for lenses residing either in the halo of our Galaxy or in the LMC/SMC themselves. The main question now raised by these results is to determine where the lenses generating the detected events belong. Are they halo objects or intrinsic to either cloud?

More statistics is still being accumulated. The MACHO experiment will run until 1999, EROS 2 plans to run until 2002 and OGLE 2 is just starting to take new data. The answer to this problem can then come from at least three possible studies:

- The comparison of the time-scales of the events toward the LMC and the SMC. They are expected to be similar if the deflectors are in the Galactic halo but different (due to the different velocity dispersions) if they are intrinsic to each cloud.

- The analysis of the spatial distribution of LMC events. The events are expected to be distributed evenly over the entire cloud if the lenses belong to the Galactic halo, while they should follow the stellar density of the LMC if they are LMC stars themselves.

- Finally, because the disk of the Andromeda Galaxy is slanted with respect to the line-of-sight, different fractions of its halo will be probed according to which end of the disk is being monitored; this will yield a larger number of events on the far side than on the near side, for microlensing events produced by deflectors in the halo of M31. Several experiments are exploring this line of sight and the first results are expected soon.

4. Conclusions

As discussed in this paper, two types of dark matter are needed, baryonic and non-baryonic. For baryonic dark matter, there are two basic candidates: intergalactic gas or MACHOs in galatic halos. There are indeed some observations through microlensing effects which tend to support the idea that MACHOs could be the main candidate for baryonic dark matter. However the high masses inferred from the microlensing events (0.1 to 1 solar mass) makes this hypothesis very puzzling and the subject of baryonic dark matter clearly awaits further observations.

REFERENCES

1. Afonso, C. et al. (EROS and MACHO coll.), 1998, ApJ, 199, L12.
2. Afonso, C. et al. (EROS coll.), 1998, A&A, 337, L17.
3. Albrow, M.D. et al. (PLANET coll.), 1998, astro-ph/9807086.
4. Alcock, C. et al. (MACHO coll.), 1995, ApJ, 454, L125.
5. Alcock, C. et al. (MACHO coll.), 1996, ApJ, 471, p. 774.
6. Alcock, C. et al. (MACHO coll.), 1997, ApJ, 490, p. 59.
7. Ansari, R. et al. (EROS coll.), 1996, A&A, 314, p. 94.
8. Charlot, S. and Silk, J., 1995, ApJ, 445, p. 124.
9. Copi, C. and Schramm, D. 1996, Comments Nucl. Part. Phys., 22, 1.
10. Frieman, J. and Scoccimarro, R., 1994, ApJ, 431, L23.
11. Gould A., 1992, ApJ, 392, p. 442.
12. Paczyński, B., 1986, ApJ, 304, p. 1.
13. Paczyński, B., 1996, Ann. Rev. Astron. Astrophys., 34.
14. Palanque-Delabrouille N., 1997, PhD thesis, University of Chicago and Université de Paris 7.
15. Palanque-Delabrouille, N. et al (EROS coll.), 1998, A&A, 332, p. 1.
16. Renault, C. et al (EROS coll.), 1997, A&A, 324, L69.
17. Sackett, P. and Gould, A., 1993, ApJ, 419, p. 648.
18. Salati, P. et al, 1996, A&A, 313, p. 1.
19. de Vaucouleurs, G. and Freeman, K.C., 1970, *in* Galaxies, p. 163.
20. Zhao, H., 1997, submitted to ApJ (astro-ph/9703097).
21. C. Alard, S. Mao, and J. Guibert. *Astronomy and Astrophysics*, 300:L17, 1995.
22. C. Alcock et al. (macho collaboration). *Nature*, 365:621, 1993.
23. C. Alcock et al. *Phys. Rev. Lett.*, 74:2867, 1995.
24. C. Alcock et al. (macho collaboration). *Astrophysical Journal*, 454:L125, 1995.

25. C. Alcock et al. (macho collaboration). *Astrophysical Journal*, 461:84, 1996.
26. R. Ansari et al. (eros collaboration). *Astronomy and Astrophysics*, 299:L21, 1995.
27. R. Ansari et al. (eros collaboration). *Astronomy and Astrophysics*, 314:94, 1996.
28. W. Arnett et al. *Nature*, 314:337, 1985.
29. E. Aubourg et al. (eros collaboration). *Nature*, 365:623, 1993.
30. J. N. Bahcall et al. *Astrophysical Journal*, 435:L51, 1994.
31. P. Baillon et al. *Astronomy and Astrophysics*, 277:1, 1993.
32. J. Beaulieu, D. Sasselov, C. Renault, et al. The effect of metallicity on the cepheid distance scale and its implication for the hubble constant determination. *Astronomy and Astrophysics Letters*, 318:L47, 1997.
33. J. Berger et al. *Astron. Astrophys. Suppl. Ser.*, 87, 1991.
34. B. Binggeli, A. Sandage, and G. A. Tammann. *Annu. Rev. Astron. Astrophys.*, 26:509, 1988.
35. H. Bohringer and D. Neumann. In J. Trân Thanh Vân, editor, *Rencontres de Moriond*. Editions FRONTIERES, 1995.
36. J. Bond et al. *Astrophysical Journal*, 306:428, 1986.
37. A. Bosma. *Astronomical Journal*, 86:1791, 1981.
38. A. Burrows and J. Liebert. *Reviews of Modern Physics*, 65, No.2:301, 1993.
39. B. Carr. *Annu. Rev. Astron. Astrophys.*, 32:531, 1994.
40. R. Carswell et al. *Mon. Not. R. Astron. Soc.*, 268:L1, 1994.
41. F. Cavalier. *Recherche de naines brunes dans le halo galactique par effet de lentille gravitationnelle Analyse des données photographiques de l'expérience EROS.* PhD thesis, Université Paris 11 Orsay, 1994.
42. S. Charlot and J. Silk. *Astrophysical Journal*, 445:124, 1995.
43. C. Copi and D. Schramm. *Comments Nucl. Part. Phys.*, 22:1, 1996.
44. C. Copi, D. Schramm, and M. Turner. *Science*, 267:192, 1995.
45. A. De Rújula et al. *CERN*, TH.5787/91, 1991.
46. A. De Rújula et al. *Astronomy and Astrophysics*, 254:99, 1992.
47. A. Deckel. *Annu. Rev. Astron. Astrophys.*, 32, 1994.
48. A. Deckel et al. *Astrophysical Journal*, 412:1, 1993.
49. C. Flynn, A. Gould, and J. Bahcall. *Astrophysical Journal Letters*, astro-ph/9603035, 1996.
50. W. Freedman et al. *Nature*, 371:757, 1994.
51. K. Griest. *Astrophysical Journal*, 366:412, 1991.
52. A. Guth. *Physical Review D*, 23:347, 1981.
53. M. Kamionkowski. In J. Trân Thanh Vân, editor, *Rencontres de Blois*. Editions FRONTIERES, 1996.
54. J. Kneib. *Ph.D. Thesis*. PhD thesis, Université Paul Sabatier, Toulouse, 1993.
55. Y. Mellier et al. *Astrophysical Journal*, 407:33, 1993.
56. G. Meylan. *Astronomy and Astrophysics*, 214:106, 1989.
57. B. Paczyński. *Astrophysical Journal*, 304:1, 1986.
58. B. Paczyński. *preprint*, astro-ph 9411004, 1994.
59. B. Paczyński et al. *Ap. J. Lett*, 435, 1994.
60. N. Palanque-Delabrouille. *Research on Galactic Dark Matter Implied by Gravitational Microlensing.* PhD thesis, Université Paris 7 and University of Chicago, 1997.
61. N. Palanque-Delabrouille et al. (eros collaboration). *Astronomy and Astrophysics*, astro-ph/9710194, 1997.
62. D. Pfenniger and F. Combes. *Astronomy and Astrophysics*, 285:94, 1994.
63. C. Renault. *Recherche de matière noire galactique par effet de microlentille gravitationnelle (sous forme d'objets compacts de faible masse).* PhD thesis, Université Paris 7, 1996.
64. C. Renault et al. (eros collaboration). *Astronomy and Astrophysics*, 324:L69, 1997.
65. V. C. Rubin, K. Ford, and N. Thonnard. *Astrophysical Journal*, 238:471, 1980.
66. M. Rugers and C. Hogan. *Astrophysical Journal Letters*, 459:1, 1996.

67. A. Saha et al. *Astrophysical Journal*, 438:8, 1995.
68. A. Saha et al. *Astrophysical Journal*, S107:693, 1996.
69. R. Sancisi and T. van Albada. In *Dark Matter in the Universe*, page 67, 1987.
70. A. Sandage et al. *Astrophysical Journal*, 423:L13, 1994.
71. A. Sandage et al. *Astrophysical Journal*, 460:L15, 1994.
72. P. Schechter et al. *Pub. Astr. Soc. Pacif.*, 105:1342, 1993.
73. T. Smecker and R. Wyse. *Astrophysical Journal*, 372:448, 1991.
74. C. Stubbs et al. *SPIE Proc.*, 1900:192, 1993.
75. N. R. Tanvir et al. *Nature*, 377:27, 1995.
76. V. Trimble. *Annu. Rev. Astron. Astrophys.*, page 425, 1987.
77. D. Tytler, X. Fan, and S. Burles. *preprint*, astro-ph/9603069, 1996.
78. M. Fukujita et al. *preprint*, astro-ph/9712020, 1997.

NUCLEAR PHYSICS B PROCEEDINGS SUPPLEMENTS

ELSEVIER Nuclear Physics B (Proc. Suppl.) 77 (1999) 420–426

Theoretical overview: emphasis on neutrinos

David O. Caldwell[a]*

[a]Institute for Nuclear and Particle Astrophysics and Cosmology and Physics Department, University of California, Santa Barbara, CA 93106-9530, U.S.A.

The missing mass of the universe is likely to consist of at least three components: baryons, neutrinos, and a particle which was weakly interacting and slow moving ("cold") in the early universe. It is doubtful that the baryonic candidates observed by microlensing could be an appreciable part of our local dark matter halo. The dominant cold component might be the supersymmetric neutralino, and if so, recent accelerator constraints making this more gaugino-like should make direct detection of the dark matter particle easier. Despite recent observations favoring a low-density universe, such a model (even with the addition of a cosmological constant) does not fit universe structure. The only model which does has a critical density, 20% neutrinos, 10% baryons, and 70% cold dark matter. The data are fit better if the neutrino dark matter is shared between two neutrino species (ν_μ and ν_τ) rather than one, a mass pattern which also explains the solar and atmospheric neutrino deficits and the LSND experiment. It is shown here that KARMEN and other experiments do not conflict with the LSND results in the appropriate mass region. Further support for this mass pattern is provided by the need for a sterile neutrino to rescue heavy-element nucleosynthesis in supernovae.

1. INTRODUCTION

While the amount of dark matter may be less than indicated until recently, it is ever more likely that dark matter has a mix of ingredients. After a brief discussion of the contentious issue of dark matter density, there follows mention of some new developments regarding baryonic and cold dark matter. Because the baryonic and cold constituents of dark matter are discussed in these Proceedings by Spiro and Sadoulet, respectively, the emphasis of this report will be on a neutrino component of the missing mass of the universe.

There is now some evidence from the first Doppler peak observed in the cosmic microwave background radiation that the total energy density of the universe is the critical value; i.e., $\Omega = 1$, and the universe will expand forever at an ever decreasing rate. Such a flat universe has the only time-stable value of density and is expected in all but very contrived models of an early era of exponential expansion, or "inflation". It has usually been assumed that $\Omega = \Omega_m$; that is, the energy density is the matter density. Recent evidence points to $0.3 \le \Omega_m \le 0.6$, however, based on a variety of observations: high-redshift supernovae type Ia, evolution of galactic clusters, high baryon content of clusters, lensing arcs in clusters, and dynamical estimates from infrared galaxy surveys. On this basis it has become popular to assume that $\Omega_m \approx 0.3$, but $\Omega = 1$ through the addition of a vacuum energy density, usually designated as a cosmological constant, Λ. The model with $\Omega_m = 0.3$, $\Omega_\Lambda = 0.7$ is in trouble with recent determinations of the age of the universe. More generally, in Section 4 it will be shown that either a low-density universe or a critical density universe with a cosmological constant certainly does not fit universe structure as measured over three orders of magnitude in distance scale by the cosmic microwave background and galaxy surveys. The only model which fits these extensive data is one having $\Omega_m = 1$, of which $\Omega_\nu = 0.2$ is in neutrinos, and $\Omega_b = 0.1$ in baryons, with the main component being cold dark matter. Clearly there is a serious conflict among these observations, and one can hope that out of conflict comes important progress, as is so often the case.

*Supported in part by the U.S. Department of Energy.

PII S0920-5632(99)00456-9

2. BARYONIC DARK MATTER

While the MACHO, EROS, and OGLE groups have observed a large number of microlensed events, it is dubious that these represent halo dark matter. In fact, it may be the most important result from the MACHO and EROS experiments so far that they have excluded as being more than 20% of the halo dark matter all compact objects in the range of 5×10^{-7} to 0.02 solar masses. Hydrogen-burning stars (> 0.08 solar masses) are excluded by direct observations, apparently leaving brown dwarf stars as the most likely objects being lensed. Because the IMF (initial mass function, or distribution of stellar masses) appears to be universal, it is very unlikely that there could be enough brown dwarfs to make up a substantial portion of the halo mass, and very few such stars have been seen, even nearby.

Although not strictly determined by the microlensing measurements, the average mass found is likely to be about half a solar mass, a totally unexpected result. This poses further difficulties, since it would seem to require the objects be white dwarfs or primordial black holes (PBH). The former would have to be very cold, hence old, not to have been seen, and their age would probably have to exceed that of the universe. Furthermore, the production of the many white dwarfs needed for halo dark matter would have produced more heavy elements than are observed. The PBH are possible, although there is an argument involving tidal friction claiming to exclude them. PBH annihilation would be a spectacular event and has not been seen.

Baryonic dark matter must exist, since the amount predicted by big bang nucleosynthesis far exceeds that which has been observed. Recent observations suggest that nearly all of the missing baryons are in the form of hydrogen gas, however, leaving too little to populate the halo of our galaxy with compact objects. This limitation would not affect the possibility of a PBH halo, though.

With all proposed candidates unlikely at best, what could be the source of the microlensed events? Since the light sources used for lensing are the stars of the Large Magellanic Cloud (LMC), it is possible that some lensing objects are either in the LMC also, or in a stream of stars pulled out of the LMC by our galaxy. This star population, which has been predicted [1] and possibly observed [2], would then not be representative of the dark matter halo.

3. COLD DARK MATTER

Baryonic dark matter does not appear to constitute an appreciable part of the local dark matter. Furthermore, in searching for cold dark matter it is only its local density which matters, and that is determined by properties of the Milky Way, especially rotation rates, and has nothing to do with determinations of Ω_m. Thus a search for this major component of dark matter is not jeopardized by recent developments.

Indeed, the most important recent development in the search for the theoretically best motivated WIMP (Weakly Interacting Massive Particle) candidate for cold dark matter makes its detection by scattering from detector nuclei more likely. This candidate is the still undiscovered supersymmetric neutralino, some mixture of gaugino and higgsino particles. For such direct detection, a neutralino which is mainly gaugino has a much bigger cross section than does one which is mainly higgsino. For accelerator searches for neutralinos, and also for indirect detection of neutralino dark matter by observing products of their annihilation in the earth or the sun, the mainly higgsino type is preferred. Thus recent accelerator limits, particularly from LEP, especially constrain the parameter space in which the neutralino would be mainly higgsino. A recent analysis [3] shows that there remains about two orders of magnitude more parameter space for a gaugino neutralino dark matter particle than for a higgsino one.

This analysis [3] also shows that a higgsino dark matter particle needs to be more massive than 73 GeV, whereas a gaugino candidate could be much lighter. If the restriction is used that there is scalar mass universality in order to reduce the number of parameters, then neutralinos of even the gaugino type need to be heavier than 42 GeV. These results include LEP data up to 183 GeV.

Another theoretical development is a proliferation of dark matter candidates, especially those which are much more massive than the usual WIMPs. So far there has not been much experimental response to these proposals, which are too numerous to go into here. It is likely that the neutralino search, now going on in over two dozen experiments, will have to play out before some of these quite different and specific searches are attempted.

4. NEUTRINO DARK MATTER

As mentioned in the Introduction, the only model (CHDM) which fits measurements of universe structure is one having $\Omega_m = 1$, $\Omega_\nu = 0.2$, and $\Omega_b = 0.1$ (not a critical number), with cold dark matter making up the remainder. This is work of Gawiser and Silk [4], who used all published data from the cosmic microwave background and galaxy surveys covering three orders of magnitude in distance scales. They compared the data with ten models of universe structure, but of concern here are only three of these, CHDM, an open universe model (OCDM) having $\Omega_m = 0.5$, and one (ΛCDM) having $\Omega_m = 0.5$ and $\Omega_\Lambda = 0.5$. In the latter two cases the parameters were varied to get the best fits, resulting in $\Omega_m = 0.5$ with $\Omega_b = 0.05$ and the rest in cold dark matter. The probabilities of the fits were CHDM = 0.09, OCDM = 2.9×10^{-5}, and ΛCDM = 1.1×10^{-5}. If one dubious set of data is removed, the APM cluster survey (which disagrees with galaxy power spectra), these probabilities become CHDM = 0.34, OCDM = 6.7×10^{-4}, and ΛCDM = 4.3×10^{-4}.

Had it been possible to extend the fit to even smaller scales, the discrepancy between CHDM and the others would have been even greater, but this is the non-linear regime requiring simulations. The CHDM model with two neutrinos gives an excellent fit [5] to the data at this extended scale, whereas the others deviate even more strongly than in the linear region.

In principle the needed neutrino mass for dark matter could come from one, two, or three neutrinos, but a fourth one would sufficiently alter the universe expansion rate at the era of nucleosynthesis to spoil the agreement between calculations and observed abundances of light elements. There is now quite good concordance between the ^{4}He abundance values [6] and the primordial D/H ratio [7], reinstating the three-neutrino limit which has been in question recently.

If neutrino dark matter were due to one neutrino, that would presumably be ν_τ, and this would be ruled out if, as fits the Super-Kamiokande data [8] best, the atmospheric anomalous ν_μ/ν_e ratio is due to $\nu_\mu \to \nu_\tau$, since the mass-squared difference required is $\Delta m^2_{\mu\tau} \sim 10^{-3}$ eV^2, and the needed neutrino mass is $94\Omega_\nu h^2 \sim 5$ eV, where h is the Hubble constant in units of 100km·s^{-1}·Mpc^{-1}. Other processes to explain the atmospheric results are very unlikely: $\nu_\mu \to \nu_e$ does not fit the Super-Kamiokande angular distributions, and the CHOOZ ν_e disappearance experiment [9] rules out almost all the parameter space anyway; $\nu_\mu \to \nu_s$ (a sterile neutrino) now would appear to have a problem with the nucleosynthesis limit, since near maximal mixing is required.

A three-neutrino scheme could have $\nu_\mu \to \nu_\tau$ for the atmospheric case, $\nu_\mu \to \nu_e$ (with $\Delta m^2_{\nu e} \lesssim 10^{-5}$ eV^2) for the solar ν_e deficit, and the three nearly mass degenerate neutrinos could give the needed dark matter. When this was first suggested [10] there was a possible problem with neutrinoless double beta decay. While limits on that have improved, theoretical ways have been found to ameliorate the problem. If LSND is correct, however, this scheme is certainly ruled out.

That leaves two-neutrino dark matter. This scheme [10] requires four neutrinos, with the solar deficit explained by $\nu_e \to \nu_s$, with both neutrinos quite light, the atmospheric effect due to $\nu_\mu \to \nu_\tau$, which share the dark matter role, and the LSND $\nu_\mu \to \nu_e$ demonstrating the mass difference between these two nearly mass-degenerate doublets. Note that the solar $\nu_e \to \nu_s$ is for the small mixing angle (or vacuum oscillation), so ν_s does not affect nucleosynthesis. The original motivation for this mass pattern preceded LSND and was simply to provide some hot dark matter, given the solar and atmospheric phenomena. If LSND is correct, it becomes the unique pattern.

This neutrino scheme was the basis for simulations [11] which showed that two-neutrino dark matter fits observations better than the one-neutrino variety. The latter produces several problems at a distance scale of the order of $10h^{-1}$ Mpc, particularly overproducing clusters of galaxies. Whether the ~ 5 eV of neutrino mass is in the form of one neutrino species or two makes no difference at very large or very small scales, but at $\sim 10h^{-1}$ Mpc the larger free streaming length of $\sim 5/2$ eV neutrinos washes out density fluctuations and hence lowers the abundance of galactic clusters. In every aspect of simulations done subsequently the two-neutrino dark matter has given the best results. For example, a single neutrino species, as well as low Ω_m models, overproduce void regions between galaxies, whereas the two-neutrino model agrees well with observations [12].

5. THE LSND EXPERIMENT AND DARK MATTER

Whether the LSND experiment is observing neutrino oscillations is the crucial issue for choosing among the neutrino mass patterns. At Neutrino '98 new results from the KARMEN experiment [13] were presented which led many people to believe that the LSND results were contradicted. This erroneous conclusion resulted from the way the mass-squared difference (Δm^2) vs. mixing-angle ($\sin^2 2\theta$) plots were presented, and that issue needs to be clarified here.

In its 1996 publication [14], LSND claimed a signal in $\bar{\nu}\mu \to \bar{\nu}_e$ on the basis that 22 events of the type $\bar{\nu}_e p \to e^+ n$ were seen, using a stringent criterion to reduce accidental coincidences between e^- or e^+ and γ rays mimicking the 2.2-MeV γ from $np \to d\gamma$, whereas only 4.6 ± 0.06 events were expected. The probability of this being a fluctuation is 4×10^{-8}. Note especially that these data were restricted to the energy range 36 to 60 MeV to stay below the $\bar{\nu}_\mu$ endpoint and to stay above the region where backgrounds are high due to the $\nu_e{}^{12}\text{C} \to e^- X$ reaction. In plotting Δm^2 vs. $\sin^2 2\theta$, however, events down to 20 MeV were used to increase the range of E/L, the ratio of the neutrino's energy to its distance from the target to detection. This was done because the plot employed was intended to show the favored regions of Δm^2, and all information about each event was used. A likelihood analysis was used, and the contours shown in Fig. 1 are at 2.3 and 4.5 log-likelihood units from the maximum. If this were a Gaussian likelihood distribution, which it is not (its integral being infinite), the contours would correspond to 90% and 99% likelihood levels, but in addition they have been smeared to account for possible systematic errors. Those contours have been widely misinterpreted as confidence levels—which they certainly are not—because they were plotted along with confidence-level limits from other experiments. This confusion of comparing likelihood levels for the LSND data with confidence levels from KARMEN is exacerbated by using the 20–36 MeV region for the LSND data. This higher background range makes some difference for the 1993–5 data, but an apreciable difference for the parasitic 1996–7 runs, which were at a low event rate, decreasing the ratio of signal/background events. This distorts the energy spectrum, making the higher Δm^2 values desirable for dark matter appear less likely.

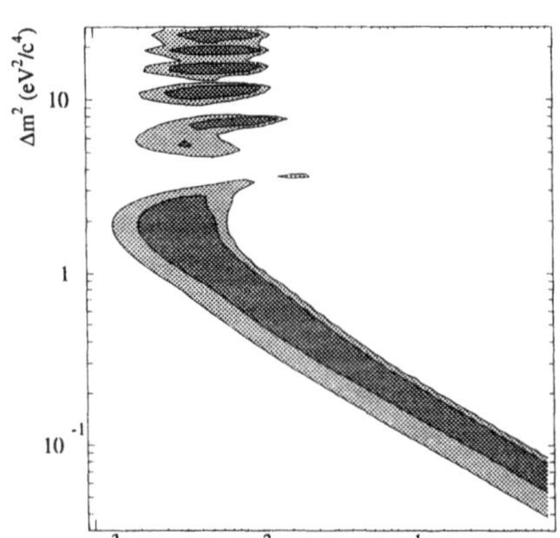

Figure 1. Mass-squared difference (Δm^2) vs. degree of mixing ($\sin^2 2\theta$) for a $\bar{\nu}_\mu \to \bar{\nu}_e$ explanation of the LSND beam-excess data. Shown are regions of Δm^2 favored using the energy (from 20 to 60 MeV) and distance from the source of each event.

The 1993–7 likelihood plot was compared at Neutrino '98 with KARMEN results which used the new "unified procedure" [15] for confidence levels. Because KARMEN saw no events, a limit based on that looked as if LSND were ruled out. KARMEN expected to see 2.88 events, and a "sensitivity" contour corresponding to that event rate does not exclude very much of the LSND parameter space.

A fairer comparison of the two experiments is to use the same procedure for each, so here Bayesian confidence levels are employed. Because no attempt is made to use E/L to further constrain Δm^2, this is not the correct way to determine favored regions of Δm^2. The effect of excluding the heavily contaminated 20–36 MeV region can be seen clearly from a comparison of Figs. 2 and 3. Figure 2, made using data with e^+ energy between 20 and 60 MeV, seems to show that other experiments exclude most of the LSND region, whereas Fig. 3, which uses the cleaner data with e^+ energy between 36 and 60 MeV, shows that there is a wide range of Δm^2 not in contradiction with other experiments. Also shown in Fig. 3 is the LSND $\nu_\mu \to \nu_e$ result [16] which although quite broad tends to favor higher Δm^2 values.

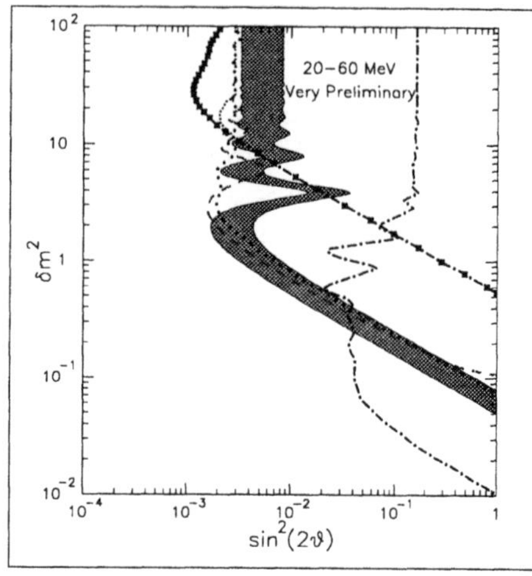

Figure 2. LSND upper and lower 90% Bayesian confidence limits for $\bar{\nu}_\mu \to \bar{\nu}_e$ oscillations using 1993-1997 data with $20 < E_e < 60$ MeV. Also shown is the 90% Bayesian confidence upper limit from Karmen as of summer, 1998 (dashed), as well as upper limits from NOMAD (dash-dot-star), E776 (dotted), and Bugey (dash-dot).

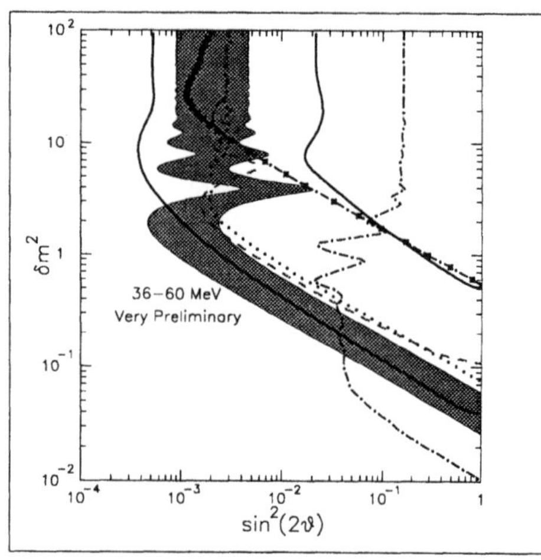

Figure 3. Same as Figure 2, but from $36 < E_e < 60$ MeV data. The added curves show the 80% confidence band for the LSND $\nu_\mu \to \nu_e$ result.

6. SUPPORTING INFORMATION FROM SUPERNOVA NUCLEOSYNTHESIS

If LSND is correct, a sterile neutrino is required, and two-neutrino dark matter is established. Any independent information favoring a sterile neutrino would support this four-neutrino scheme and the two-neutrino dark matter. Such information can come from that neutrino laboratory, the supernova.

While the $\Delta m^2_{e\mu} \sim 6$ eV^2 value is desirable for two-neutrino dark matter, it apparently would cause a conflict with the production of heavy elements in supernovae. This r-process of rapid neutron capture occurs in the outer neutrino-heated ejecta of Type II supernovae. The existence of this process would seem to place a limit on the mixing of ν_μ and ν_e because energetic ν_μ ($\langle E \rangle \approx 25$ MeV) coming from deep in the supernova core could convert via an MSW transition to ν_e inside the region of the r-process, producing ν_e of much higher energy than the thermal ν_e ($\langle E\rangle \approx 11$ MeV). The latter, because of their charge-current interactions, emerge from farther out in the supernova where it is cooler. Since the cross section for $\nu_e n \to e^- p$ rises as the square of the energy, these converted energetic ν_e would deplete neutrons, stopping the r-process. Calculations [17] of this effect limit $\sin^2 2\theta$ for $\nu_\mu \to \nu_e$ to $\lesssim 10^{-4}$ for $\Delta m^2_{e\mu} \gtrsim 2$ eV^2, in conflict with compatibility between the LSND result and a neutrino component of dark matter.

The sterile neutrino, however, can not only solve this problem, but also rescue the r-process itself. While recent simulations have found the r-process region to be insufficiently neutron rich, very recent realization of the full effect of α-particle formation has created a disaster for the r-process [18]. The initial difficulty of too low entropy (i.e., too few neutrons per seed nucleus, like iron) has now been drastically exacerbated by calculations [18] of the sequence in which all available protons swallow up neutrons to form α particles, following which $\nu_e n \to e^- p$ reactions create more protons, creating more α particles, and so on. The depletion of neutrons by making α particles and by $\nu_e n \to e^- p$ rapidly shuts off the r-process, and essentially no nuclei above $A = 95$ are produced.

The sterile neutrino would produce two effects [19]. First, there is a zone, outside the neutrinosphere (where neutrinos can readily escape) but inside the $\nu_\mu \to \nu_e$ MSW ("LSND") region, where the ν_μ interaction potential goes to zero, so a $\nu_\mu \to \nu_s$ transition can occur nearby, depleting the dangerous high-energy ν_μ population. Second, because of this ν_μ reduction, the dominant process in the MSW region reverses, becoming $\nu_e \to \nu_\mu$, dropping the ν_e flux going into the r-process region, hence reducing $\nu_e n \to e^- p$ reactions and allowing the region to be sufficiently neutron rich. This rescuing scenario—the only robust one which has been found after many attempts—works even better if the MSW region is inside the radius at which the weak interactions freeze out, which is certainly the case if Δm^2 is as large as 6 eV^2.

7. CONCLUSIONS

The search for cold dark matter is not impeded by the evidence for a low-density ($\Omega_m < 1$) universe, or by the existence of microlensed objects (MACHOs). Indeed, accelerator limits greatly reducing the parameter space for higgsinos make the likelihood greater for the direct detection of neutralino dark matter, since it would of the higher cross section gaugino type, if such supersymmetric particles exist. The nature of the MACHOs is an intriguing mystery, but it is unlikely that they constitute an appreciable part of our local dark matter halo.

A neutrino component of dark matter appears very probable, both from the astrophysics and particle physics standpoints. Despite the evidence for $\Omega_m < 1$, the one model which fits universe structure has $\Omega_m = 1$, with 20% neutrinos and most of the rest as cold dark matter. Open universe and low-density models with a cosmological constant give extremely bad fits. This conflict should be the source of future progress, but since there are $10^2/\text{cm}^3$ of neutrinos of each active species left over from the early universe, the ultimate answer on neutrino dark matter will come from determinations of neutrino mass. While the solar and atmospheric evidences for neutrino mass are important, the crucial issue is the much larger mass-squared difference observed by the LSND experiment. In the mass region needed for dark matter, no other experiment excludes the LSND result, if data from the different experiments are compared using the same procedures.

The resulting mass pattern, $\nu_e \to \nu_s$ for solar, $\nu_\mu \to \nu_\tau$ for atmospheric, and $\nu_\mu \to \nu_e$ for LSND, requires a sterile neutrino and provides two-neutrino (ν_μ and ν_τ) dark matter. This form of dark matter fits observational data better than the one-neutrino variety. Furthermore, the sterile neutrino appears to be necessary to rescue the production of heavy elements by supernovae. Thus evidence is increasing for this four-neutrino mass pattern.

ACKNOWLEDGMENTS

I wish to thank Steven Yellin for producing Figures 2 and 3.

REFERENCES

1. H.-S. Zhao, astro-ph/9703097.
2. D. Zaritsky, private communication.
3. J. Ellis *et al.*, hep-ph/9801445.
4. E. Gawiser and J. Silk, *Science* **280** (1988) 1405.
5. J.R. Primack and A. Klypin *Nucl. Phys. (Proc. Suppl.)* **51B** (1996) 30.
6. Y.I. Izotov and T.X. Thuan, *Astrophys. J.* **500** (1998) 188.
7. S. Burles and D. Tytler, *Astrophys. J.* **499** (1998) 699.
8. Y. Fukuda *et al.*, hep-ex/9807003 and these Proceedings.
9. M. Apollonio, *Phys. Lett.* **B420** (1998) 397.
10. D.O. Caldwell, *Perspectives in Neutrinos, Atomic Physics and Gravitation*, Editions Frontières, Gif-sur-Yvette, France, 1993, p. 187; D.O. Caldwell and R.N. Mohapatra, *Phys. Rev.* **D48** (1993) 3259.
11. J.R. Primack, J. Holtzman, A. Klypin, and D.O. Caldwell, *Phys. Rev. Lett.* **74** (1995) 2160.
12. S. Ghigna *et al.*, *Astrophys. J.* **479** (1997) 580;*ibid.* **437** (1994) L71;J.R. Primack, astro-ph/9707285.
13. B. Zeitnitz, these Proceedings.
14. C. Athanassopoulos *et al.*, *Phys. Rev. Lett.* **77** (1996) 3082; *Phys. Rev.* **C54** (1996) 2685.
15. G.J. Feldman and R.D. Cousins, *Phys. Rev.* **D57** (1998) 3873.
16. C. Athanassopoulos *et al.*, *Phys. Rev. Lett.* **81** (1998) 1774.
17. Y.-Z. Qian *et al.*, *Phys. Rev. Lett.* **71** (1993) 1965; Y.-Z. Qian and G.M. Fuller, *Phys. Rev.* **D51** (1995) 1479; G. Sigl, *Phys. Rev.* **D51** (1995) 4035.
18. G.M. Fuller and B.S. Myer, *Astrophys. J.* **453** (1995) 202;B.S. Myer, G.C. McLaughlin, and G.M. Fuller (in preparation, 1998).
19. D.O. Caldwell, G.M. Fuller, and Y.-Z. Qian (to be submitted to *Phys. Rev. Lett.* 1998).

Part 12

Neutrinos in Cosmology and Astrophysics

ELSEVIER

Nuclear Physics B (Proc. Suppl.) 77 (1999) 429–434

NUCLEAR PHYSICS B
PROCEEDINGS
SUPPLEMENTS

Supernova Neutrinos: Review

Hollis E. Dalhed, James R. Wilson, and Ronald W. Mayle

University of California,
Lawrence Livermore National Laboratory,
7000 East Avenue, Mail Stop L-015
Livermore, CA 94550

This review discusses the role of neutrinos in the dynamics of supernovae. Neutrinos not only provide an essential probe into the core collapse mechanism, but also probably play an active role in the explosion mechanism. The time history of the neutrino fluences and their spectral shapes are affected by such physics as the nuclear equation of state and convection modeling. In turn, neutrinos play an active role in depositing energy in the region behind the outgoing shock wave and in the potential formation of heavy nuclei via r-process nucleosynthesis.

1. Introduction

Neutrinos play a critical role in our understanding of supernova explosions. The time history and spectral details of the neutrino burst provide information on the dynamics of the core collapse and subsequent neutron star formation and evolution. These details are dependent on our understanding of the nuclear equation of state and its impact on core collapse and evolution. Neutrinos may also be critically responsible for the explosion of Type II supernovae through the "delayed explosion" mechanism [1]. Convection is believed to play an important role in supernova dynamics, and detailed observation of the neutrino signal might help in our understanding of this process. Additionally the late time neutrino flux, perhaps modified by oscillations, is believed to play a critical role in the formation of heavy r-process nuclei. We will discuss each of these aspects of supernova neutrinos in this review.

1.1. Type II supernovae

Type II supernovae result from the gravitational collapse of progenitor stars which were approximately 10 to 30 $M_\odot$ on the main sequence. When the central iron core reaches ~ 1.5 $M_\odot$, it gravitationally collapses, leaving behind the remaining mass in the surrounding envelope or mantel. An outgoing shock wave is generated when the core reaches supernuclear density and rebounds. This shock wave must traverse the core and overlying material, disintegrating nuclei and providing sufficient kinetic energy to eject the mantel.

1.2. Neutrino creation and emission

At the start of collapse, when densities in the core rise above approximately 10^9 g/cm^3, electrons are captured on nuclei and the resulting electron neutrinos freely escape the collapsing star. Eventually the core density becomes high enough to trap neutrinos with the resulting chemical equilibrium leading to the diffusive emission of all flavors of neutrinos. Almost all of the gravitational binding energy will eventually be emitted in the form of neutrinos.

1.2. Delayed explosion mechanism

The final gravitational binding energy of the neutron star is very roughly 2-3 $\times 10^{53}$ ergs. About one or two percent of this energy goes into the kinetic energy of the ejecta, while the vast majority is carried off by the neutrinos. The shock wave, in exiting the mantel, expends some of its energy disintegrating nuclei. If the shock is strong enough, it will continue completely through the mantel, resulting in a "prompt" explosion. However, if the shock is not sufficiently strong, it will stall into an

PII S0920-5632(99)00458-2

accretion shock and the explosion will fail. It is possible that some of the neutrino energy flooding out of the core is coupled into the matter behind the stalled shock, heating this matter and eventually providing enough ram pressure to force the shock to continue through the remainder of the mantel. This is the "delayed explosion" mechanism.

1.3. Supernova synergism

Whether the Type II explosion mechanism is prompt or delayed, details of the evolution certainly influence and are in turn are influenced by neutrinos. In the remainder of this review we will look at several key areas where this synergism is very important to our understanding of supernova explosions and neutrinos.

2. NUCLEAR EQUATION OF STATE

The nuclear equation of state determines the strength of the shock wave resulting from the bounce and rebound of the core. A "soft" equation of state allows deeper penetration of the bounce and stronger resulting shock wave. Additionally the nuclear equation of state determines matter conditions in the neutrino-emitting region of the core, thereby determining spectral and temporal details of the neutrino flux.

The nuclear equation of state used in our evolution code [2] is augmented by the inclusion of kaon and pion degrees of freedom [3]. At sufficiently high densities, reactions such as $p \rightarrow n + \pi^+$ and $n \rightarrow p + \pi^-$ occur. The existence of these excitations serves as an additional degree of freedom in determining the energy and pressure of the nuclear matter. The effect of including the extra degrees of freedom in the equation of state is shown in Figure 1. For a density approximately six times nuclear and for a ratio of free electrons to baryon number ($\bar{z}$) of 0.25, Figure 1 plots temperature versus internal energy for an equation of state including pion and kaon excitations and one without. The values of density and $\bar{z}$ used are quite representative of conditions attained during supernova core collapse. Especially near degeneracy the increased temperature due to the shift of

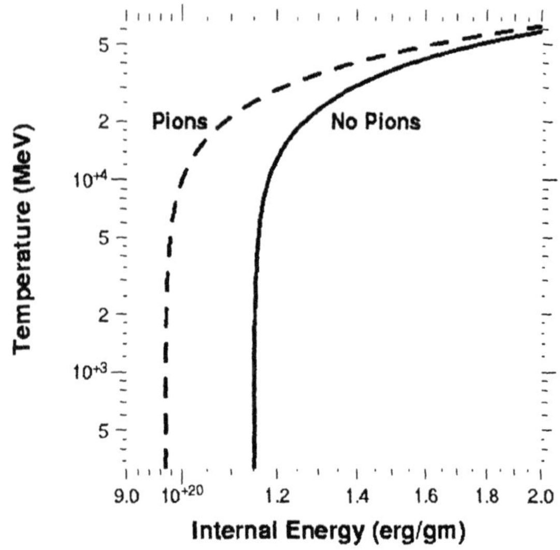

Figure 1. Temperature versus internal energy for an equation of state including pion and kaon excitations (dashed curve) and one without (solid curve). The density is approximately six times nuclear and the ratio of free electrons to baryon number is 0.25.

chemical potential caused by the pion and kaon excitations is quite pronounced.

Although they are not currently included in our equation of state, similar important effects due to quark/hadron or strange matter phase transitions and collective nuclear effects are currently being implemented.

3. CONVECTION

One of the most intense and interesting areas of supernova modeling centers on the role of convection in the explosion mechanism. That unstable regions exist within the star during collapse is quite clear. It is also quite clear that the ability to model this inherently three dimensional problem with the full set of physics models necessary to determine success or failure of the explosion modeling is still some time in the future. Nonetheless considerable qualitative if not quantitative progress can be made using one dimensional mixing length theory to explore this important phenomenon.

3.1. Stability criterion

The unstable buoyancy of displaced fluid elements drives convective instabilities. The general criterion for this instability [4] is given by

$$\frac{T}{C_p}\frac{\partial\rho}{\partial T}\bigg|_{p,Y_e}\lambda\frac{ds}{dr}+\frac{\partial\rho}{\partial Y_e}\bigg|_{p,s}\lambda\frac{dY_e}{dr}\geq 0 \qquad (1)$$

where λ is a mixing length to be described later, s is the entropy per baryon, and Y_e is the lepton fraction. Normally the derivatives of density with respect to temperature and lepton fraction are both negative. Thus Equation (1) indicates that instabilities may be driven by negative gradients in the entropy and/or the lepton fraction Y_e.

3.2. Instabilities in a supernova

Convective instabilities occur in a supernova both below the neutrinosphere in the core of the nascent neutron star, and in the high entropy region above the neutrinosphere but below the outgoing shock wave. Normally in the region inside the core of the neutron star the gradient of entropy is positive and thus stabilizing while the gradient of lepton number is strongly negative and thus destabilizing. In analogy to the well-known phenomenon of the instability of warm salt water overlying cold fresh water, this instability has been called "neutron finger" convection [4]. In the high entropy region outside of the neutrinosphere, normally the entropy gradient is negative while the lepton fraction gradient can be quite convoluted. In either case, we model the effects of convection using mixing length theory, with the mixing length being determined by Equation 1 with the appropriate gradients.

3.3. Mixing length theory

Linear analysis [5] of the type of instability we are discussing gives the wavelength of the most unstable mode as

$$\lambda_F=\left(\frac{\kappa_T\eta}{g\alpha d\ln T/dr}\right)^{1/4} \qquad (2)$$

where the various constants are defined in [5]. Inserting representative values for a supernova gives $\lambda_F=2\times10^5$ cm, which is quite in line with values obtained in two dimension calculations [6] and with values used in our calculations.

The prototypical evolution equations for convection are represented by

$$\rho\frac{\partial X}{\partial t}=\nabla\bullet(\rho D\nabla X) \qquad (3)$$

where X represents lepton fraction, internal energy, or mass fractions of the various nuclei present. The diffusion coefficient is $D=\frac{1}{3}\upsilon_c\lambda$ where $\frac{1}{2}\rho\upsilon_c^2=g\delta\delta\rho\lambda$ with $\delta\delta\rho=\left(\frac{d\rho}{dr}-\frac{\Delta\rho}{\lambda}\right)\lambda$ and $g=-\frac{1}{\rho}\frac{dP}{dr}=\frac{P}{\rho\lambda_P}$. Now using $\Delta\rho=\lambda\frac{\partial\rho}{\partial r}$ from Equation (1), the convective velocity is determined. With an appropriate choice of the mixing length λ guided by Equation (2), the diffusion coefficients D are determined and the evolution equations are solved.

3.4. Neutrinos, explosion, and convection

Arguably convection plays a key role not only in the supernova explosion but also in the nature of the emitted neutrino flux. The physical effect of convection is of course the turbulent mixing of various parts of the star on fairly rapid time scales. In the central core, hot material from near the center is convected nearer the neutrinosphere, impacting conditions in the region where neutrinos are emitted. Convection outside the neutrinosphere will likewise mix material in the critical high entropy region behind the outgoing shock.

Figure (2a) shows trajectories of various mass points in a successful modeling of a 20 $M_\odot$ star collapse and explosion. By 500 milliseconds after bounce the formation of the high entropy bubble is becoming apparent, and eventually almost all of the mass above this region will be ejected. By contrast, Figure (2b) shows an identical calculation except that no convection modeling is done. This clearly shows that no high entropy bubble will form, and in this calculation the matter will continue to accrete onto the nascent neutron star until a black hole is formed.

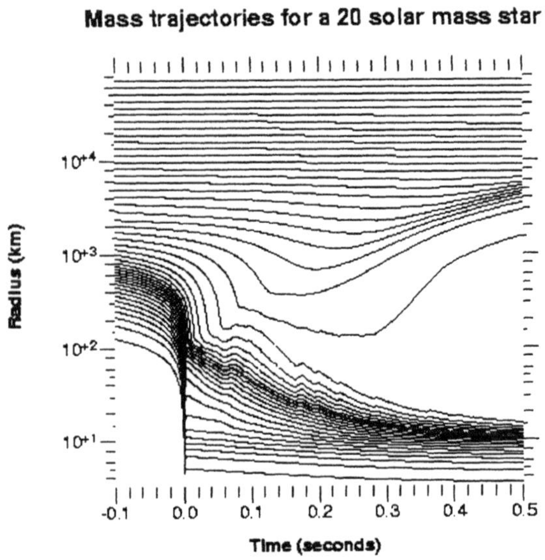

Figure 2a. Mass trajectories for the collapse and explosion of a 20 $M_\odot$ star with convection modeling. The heavy line depicts the approximate location of the neutrinosphere.

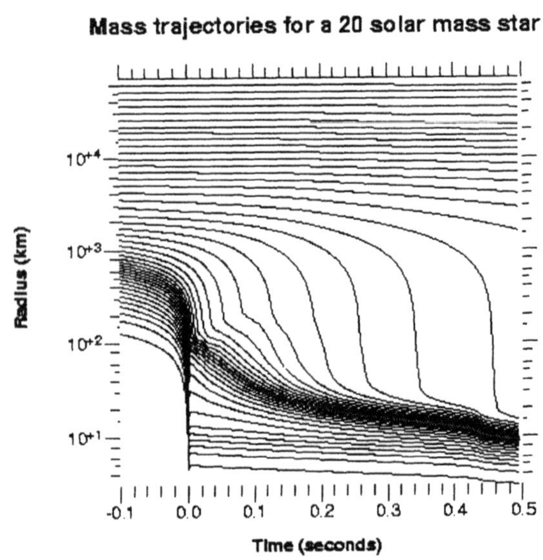

Figure 2b. Mass trajectories for the collapse of a a 20 $M_\odot$ star without convection modeling.

Convection will also affect the neutrino flux in the high entropy region behind the shock, affecting neutrino heating and possibly neutrino influence in the late-time production of r-process elements. Figure 3 shows the electron neutrino luminosity in this region at a time of 300 milliseconds after bounce. The curves correspond to different values of the mixing length, where the value of the parameter indicates the ratio of the mixing length parameter λ to the length of the region determined to be unstable.

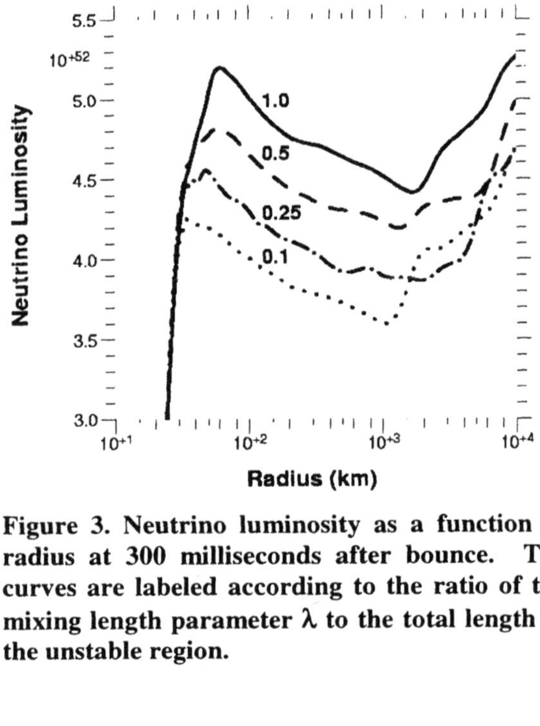

Figure 3. Neutrino luminosity as a function of radius at 300 milliseconds after bounce. The curves are labeled according to the ratio of the mixing length parameter λ to the total length of the unstable region.

Neutrino luminosity differences in the region depicted in Figure 3 may be important to whether neutrino heating can reenergize the stalled shock or not. These differences may or may not be important to r-process nucleosynthesis, which occurs during approximately 10 to 20 seconds after bounce. This is currently being investigated.

4. OBSERVABLE CHARACTERISTICS OF THE NEUTRINO SIGNAL

The observed neutrino signal from a supernova provides invaluable information into many facets of the physics of the explosion. Both the time history and the spectral shape of the luminosity provide details that will aid in the computational modeling.

4.1 Time history of the neutrino signal

Shown in Figure 4 are the time histories of the luminosities of the electron neutrinos, the antielectron neutrinos, and the combined luminosities of the mu and tau neutrinos and their antipartners. (In the calculation the mu, tau, antimu and antitau neutrinos are treated as a single distribution since they are dynamically indistinguishable.) The very early peak of the ν_e luminosity due to the initial capture of electrons on nuclei to start the collapse lasts for several milliseconds. The $\overline{\nu}_e$ luminosity rises quickly and

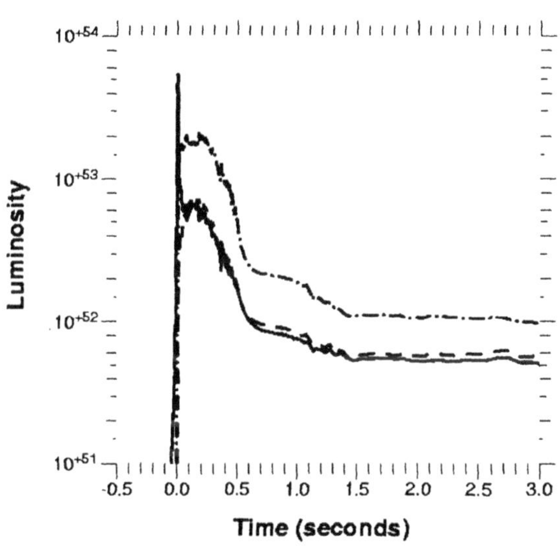

Figure 4. Time histories of the ν_e (solid), $\overline{\nu}_e$ (dash), and $\left(\nu_\mu+\overline{\nu}_\mu+\nu_\tau+\overline{\nu}_\tau\right)$ (dashdot) luminosities. The luminosities are calculated at a distance of 300 km from the neutrinosphere of a 20 $M_\odot$ star.

follows the ν_e luminosity thereafter. The combined $\left(\nu_\mu+\overline{\nu}_\mu+\nu_\tau+\overline{\nu}_\tau\right)$ luminosities also follow a quite similar shape as the $\overline{\nu}_e$ luminosity but with a weight of four. Detailed observation of the shapes of these time histories might provide critical information about the nature of the explosion mechanism, whether prompt or delayed.

4.2 Spectral shape of the neutrino signal

Because of the enormous densities in the nascent neutron star core with consequently high chemical potentials, the spectra of the emitted neutrinos are not Fermi-Dirac. This is shown in Figure 5, where it can be seen that for many seconds past bounce, the spectra are depleted relative to Fermi-Dirac at both low and high energies.

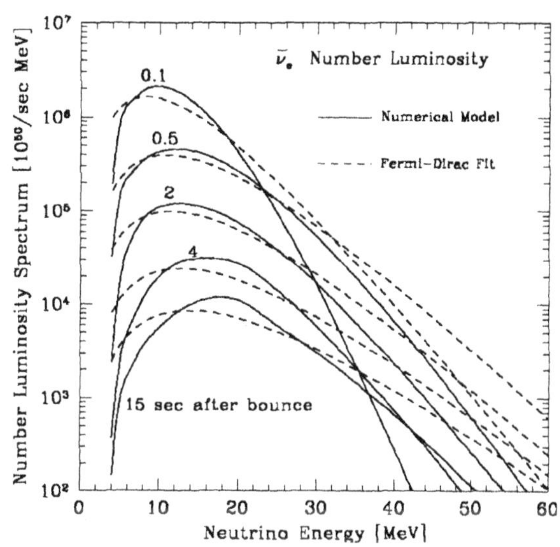

Figure 5. Spectra of the $\overline{\nu}_e$ flux at various times past bounce. Overlayed on each spectrum is the corresponding Fermi-Dirac distribution, showing the departures from equilibrium of the $\overline{\nu}_e$ spectra. The calculations are at a distance of 300 km from the neutrinosphere of a 20 $M_\odot$ star.

Additionally, the time histories of the average energies of the various neutrino species differ. This is largely due to opacity differences for the various species in the neutrino-emitting region of the star. Generally, the average ν_e energy settles in time near 12 MeV, the average $\bar{\nu}_e$ energy settles near 15 MeV, and the average $\nu_{\mu,\tau} / \bar{\nu}_{\mu,\tau}$ settles near 25 MeV. Knowledge of these averages and detail of the spectra provide valuable information about conditions at the neutrinosphere.

5. R-PROCESS NUCLEOSYNTHESIS

Viability of the delayed neutrino heating explosion mechanism is extremely appealing for Type II supernovae being considered as the site of the production of neutron-rich heavy elements. Constraints on the necessary ingredients of high entropy ($S/k_b > 500$), small lepton fraction ($Y_e <$ 0.45), and the time necessary to produce the observed amount of ejecta strongly suggest conditions observed in the delayed explosion mechanism. Production of the r-process nuclei depends sensitively on details of time history and spectra of the neutrino species. Getting this production correct is perhaps one of the most difficult aspects of modeling supernova neutrinos.

5.1 General relativistic effects

It is becoming clear that general relativistic effects will probably play a significant role in the correct determination of the neutrino fluxes. At times of interest for r-process, red shifts at the neutrinosphere are approaching 1.5 and neutrino trajectories are strongly bent. The role of the red shift and gravitational bending on annihilation is currently being reinvestigated and is expected to be a non-negligible effect on the r-process.

5.2 Neutrino oscillations

Neutrino oscillations will have an impact on r-process nucleosynthesis. Calculations indicate that conditions in the high entropy bubble behind the shock, believed to be the r-process site, are right for level crossings for a wide range of δm^2. Effects due to $\nu_e \Leftrightarrow \nu_{\mu,\tau}$ oscillations have modest impact on Ye, while $\bar{\nu}_e \Leftrightarrow \bar{\nu}_{\mu,\tau}$ oscillations have a profound effect. We are continuing to investigate effects due to oscillations, and are beginning to include effects due to oscillations involving a sterile neutrino.

6. CONCLUSIONS

Neutrinos play not only a critical diagnostic role in supernova evolution but also play a dynamic role as well. The interplay of neutrino formation, transport and interaction with other physical phenomena is quite complex. Full understanding of the role of neutrinos in the dynamics of the explosion mechanism, r-process formation, and as a probe of conditions in the nascent neutron star is still rapidly developing. Rich new areas of relevant physics, such as the possibility of neutrino-plasmon coupling [7], are just unfolding. Development of multidimensional computational modeling, new theoretical advances, and hopefully spectacular observational data make this an incredibly exciting area of research.

This work was performed under the auspices of the U.S. Department of Energy by Lawrence Livermore National Laboratory under Contract No. W-7405-ENG-48.

REFERENCES

1. J.R. Wilson and R.W. Mayle, in The Nuclear Equation of State, NATO ASI Series Volume 216A, W. Greiner and H. Stöcker (eds.), Plenum Press, New York and London (1989), 731-750.
2. T.L. McAbee and J.R. Wilson, Nucl. Phys. **A576** (1994) 626.
3. R.W. Mayle, M. Tavani, and J.R. Wilson, ApJ **418** (1993), 398.
4. J.R. Wilson and R.W. Mayle, Phys. Rep. **163** Nos. 1-3 (1988) 63.
5. J.P. Cox and R.T. Giuli, Principles of Stellar Structure (Gordon and Breach, New York, 1968) Ch. 14.
6. D.S. Miller, J.R. Wilson, and R.W. Mayle, ApJ 415 (1993) 278.
7. L.O.Silva, R. Bingham, J.M. Dawson, and W.B. Mori, LANL preprint physics/9807049.

NUCLEAR PHYSICS B
PROCEEDINGS
SUPPLEMENTS

ELSEVIER

Nuclear Physics B (Proc. Suppl.) 77 (1999) 435–439

Future supernova neutrino detection

W. Fulgione

Istituto di Cosmogeofisica del CNR-Torino
Istituto Nazionale di Fisica Nucleare Sezione di Torino

The responses of different neutrino detectors to the signal expected from a gravitational stellar collapse in our Galaxy are analysed and compared in this paper. Energetic and temporal characteristics of the detected neutrino signal are studied in order to investigate their information content concerning the stellar collapse and neutrino properties.
Some consequences on ν physics and astrophysics are discussed.

1. Introduction

Neutrino bursts from gravitational collapse in our Galaxy or from the Large and Small Magellanic Clouds can be detected today by a number of neutrino telescopes around the world.

The total detector mass is about 4 times the active mass at the time of SN 1987A but with a great improvement of detector sensitivities (expecially due to the lower energy thresholds).

A list of running topical SN ν telescopes is shown in Tab.1 with their effective sensitive mass. Furthermore the Sudbury Neutrino Observatory (SNO) has been included in the discussion because of its peculiarity and because it will be operational in the next future.

Table 1
Supernova Neutrino Telescopes

water $\check{C}$ 34100 t	scintillators 1900 t	h. water $\check{C}$ 1000 t
SK (32500)	LVD (670)	SNO (1000)
SNO (1600)	MACRO (560)	
	Baksan (330)	
	LSND (150)	
	ASD (100)	
	LSD (90)	

In this work, the detector capabilities to distinguish the SN signal from the background will not be discussed. Each detector by itself or with the help of the time coincidence with other detectors is supposed to be able to disentangle the ν signal from a gravitational collapse occurring in our Galaxy ($20kpc$) or in the Magellanic Clouds ($50kpc$) [1].

In the following, we will discuss the spectral characteristics and the time profile of the detected signal, which carry information on the collapse dynamics, explosion mechanism, properties of protoneutron star matter and neutrinos.

The number of expected events in different targets and for various processes from a SN at $10kpc$ are shown in Tab.2, and they will be used to give numerical examples. They are scaled from [2] or calculated on the basis of that model.

Different liquid scintillator detectors (LS) have been considered as a one, with a total mass of $1.9kt$. In order to have an uniform data set, in occasion of the next SN signal, an effort from each detector team will be needed to normalize the measurements on efficiency, thresholds and energy resolution.

2. ν spectra

The knowledge of the different emission temperatures of ν flavours gives us some indication on the status of matter at the source (neutronization), and the time variation of the average ν energy could be diagnostic of the explosion mechanism. The hardening of ν spectrum is expected just after the core bounce in the prompt mechanism while it would be gradual and delayed in

0920-5632/99/$ – see front matter
PII S0920-5632(99)00460-0

the delayed one.

2.1. $\bar{\nu}_e$ and ν_e

All detectors listed in Tab.2 are sensitive to $\bar{\nu}_e$ and can provide the time integrated $\bar{\nu}_e$ spectrum from which is possible to determine $< T_{\bar{\nu}_e} >$. A very detailed spectrum is expected from the SK data due to the huge number of events (> 5500) from the inverse β decay:

$$\bar{\nu}_e p \rightarrow n e^+ \quad (1)$$

In water Čerenkov detectors, (and in the "standard" SN ν model), the contribution to the total number of events from other interaction channels is less than 5%, as shown in Tab.2.

Moreover, SK will allow to study the SN explosion mechanism by the time variation of the average $\bar{\nu}_e$ energy during the different phases of the emission[3].

Anyway, the low energy region ($E_d < 10MeV$) of the detected spectrum in water Čerenkov is disturbed by the presence of low energy $\gamma's$ from the n.c. excitation of ^{16}O [4]:

$$\nu_i^{16}O \rightarrow \nu_i' \gamma X \quad (2)$$

and of electrons from:

$$\nu_i e^- \rightarrow \nu_i' e^- \quad (3)$$

The background due to such interactions in the energy range $5 \leq E_d \leq 10MeV$, for high emission temperatures, could be about two times the $\bar{\nu}_e p$ signal [5]. As pointed out in [3] a poor knowledge of the $\bar{\nu}_e$ spectrum at low energies (where the deviation from the standard Fermi-Dirac distribution could be strong) induces errors in determining the total luminosity and the average $\bar{\nu}_e$ energy.

LS, because of their capability to recognize $\bar{\nu}_e$ interactions by the n capture signature, can provide an unbiassed $\bar{\nu}_e$ spectrum from (1). Moreover they can in principle work at lower detection threshold, because of their light production and because the observed energy in the $\bar{\nu}_e p$ reaction is boosted by the two γ's from e^+e^- annihilation ($E_d = E_{\bar{\nu}_e} - 1.8MeV + 2m_e c^2$).

If the energy spectrum of ν's emitted by the next SN explosion would be softer than expected at present in the most accredited scenario, or if the collapsing star gave up into a black hole (short duration $1 \div 2$ sec, quick turn off, high luminosity and soft spectra are predicted for the ν burst in this case) the detector sensitivity to low energies would become of utmost importance.

Finally, concerning ν_e, SNO has a unique sensitivity. The c.c. reaction with deuterium:

$$\nu_e d \rightarrow p p e^- \quad (4)$$

is responsible for about 20% of the total signal from a SN in heavy water detectors while the ν_e contribution in liquid scintillators and water detectors is of the order of per cent.

Sources of background for such reaction are: unrecognized c.c. $\bar{\nu}_e d$ interactions $[80\% \cdot (1-\epsilon_n)^2]$ where ϵ_n is the n detection efficiency, and $\nu_i e^-$ scattering [10%].

2.2. ν_x ($\nu_x = \nu_\mu, \nu_\tau, \bar{\nu}_\mu, \bar{\nu}_\tau$)

While ν_e and $\bar{\nu}_e$'s are absorbed and emitted in c.c. mediated reactions with nuclei, μ and τ neutrinos interact with the protoneutron star matter only via neutral currents. Therefore ν_μ and ν_τ decouple from the stellar gas deeper inside the star and are emitted with much higher characteristic spectral temperatures.

The ν_x energy cannot be directly measured in most of the reactions of Tab.2, therefore the ν_x emission temperature can be only extracted from the observed number of events (N_{ν_x}).

Heavy water Čerenkov detectors are the most sensitive to ν_x. More than 45% of the signal in SNO D_2O detector is expected to be due to μ and τ neutrinos and antineutrinos via the neutral current breakup reactions of deuterium (about 200 events for a SN at $10kpc$):

$$\nu_i d \rightarrow \nu_i' n p \quad (5)$$

Sources of background, during a ν burst from a SN collapse, are: ν_e ($\bar{\nu}_e$) contribution to the breakup reaction that, even if partially suppressed by the energy dependence of the cross section and their softer spectra, can be estimated as $\approx 35\%$ of the ν_x signal and unrecognized $\bar{\nu}_e d$ reactions with deuterium $\bar{\nu}_e d \rightarrow n n e^+$ ($60\%(1-\epsilon_{e^+})$ where ϵ_{e^+} is the e^+ detection efficiency).

Table 2
Expected number of events for the different processes in various targets ($\nu_i = all$, $\nu_x = \nu_\mu, \nu_\tau, \bar{\nu}_\mu, \bar{\nu}_\tau$)

	heavy w. $\check{C}$ 1000	**water $\check{C}$** 34100	**liquid scint.** 1900	
$\bar{\nu}_e p \rightarrow n e^+$	–	5658	365	$E_\nu^{tot} = 2.9 \cdot 10^{53} erg$
$\nu_i(\nu_x)e^- \rightarrow \nu_i'(\nu_x')e^-$	8(2)	153(54)	12(3)	
$\nu_e^{16}O \rightarrow^{16} Fe^-$	1	39	–	$< E_{\nu_e} >= 9.9MeV$
$\bar{\nu}_e^{16}O \rightarrow^{16} Ne^+$	1	49	–	
$\nu_e^{12}C \rightarrow^{12} Ne^-$	–	–	1	$< E_{\bar{\nu}_e} >= 11.6MeV$
$\bar{\nu}_e^{12}C \rightarrow^{12} Be^+$	–	–	1	
$\nu_x^{16}O \rightarrow \nu_x'\gamma X$	–	≈ 40	–	$< E_{\nu_x} >= 15.4MeV$
$\nu_i(\nu_x)^{12}C \rightarrow \nu_i'(\nu_x')\gamma^{12}C$	–	–	21(17)	
$\nu_e d \rightarrow ppe^-$	82	–	–	$\frac{dN_\nu}{dE} = \frac{E^2}{exp(\frac{E}{T_{\nu_i}} - \eta_{\nu_i})+1}$
$\bar{\nu}_e d \rightarrow nne^+$	67	–	–	
$\nu_i(\nu_x)d \rightarrow \nu_i'(\nu_x')np$	272(200)	–	–	
total	≈ 430	≈ 5950	≈ 400	

In spite of its little neutral current response SK is expected to register about $50 + 40$ events from ν_x interactions (3) and (2). The $\nu_x e^-$ scatterings (≈ 50 expected events) can be selected by their angular signature, and keep some information of the incident ν_x energy. Anyway they suffer the contamination from: $\overset{(-)}{\nu_e} e^-$, which are a factor ≈ 2 more frequent because of the c.c. contribution to the ν_e scattering cross section, and $\bar{\nu}_e p$ (≈ 400 events) in the 30° cone around the SN direction.

The number of n.c. interactions with ^{16}O, on the other hand, is very sensitive to the high energy tail of the ν spectra, so in principle it is sensitive to the temperature T_{ν_x}. Nevertheless the energy resolution needed to disentangle these reactions from the continuum of $\bar{\nu}_e p$, in the energy region close to the detector threshold, could be unrealistic.

The n.c. interactions with ^{12}C in scintillators have a clear signature because γ's from ^{12}C deexcitation are monoenergetic:

$$\nu_i^{12}C \rightarrow^{12} C\nu_i'\gamma(E_\gamma = 15.11MeV) \qquad (6)$$

The n detection efficiency and average n capture time play an important rule to disentangle, and reject, the strongest source of background, for this reaction, namely $\bar{\nu}_e p$.

The sensitivity of LS to the ratio $T_{\nu_x}/T_{\bar{\nu}_e}$ can be seen in Fig.1 where the ν burst from a SN collapse is represented by the two observables: $< E_d >$ (the average detected energy from $\bar{\nu}_e p$ scattering) and the ratio of n.c. interactions with carbon nuclei on the total ($N_{n.c.}/N_{tot}$) [6].

Different parameters of the ν energy spectra have been tested: ($1.5 \le T_{\nu_e} \le 3.5MeV; 0 \le \eta \le 4$). The source distance and the total emitted energy only determine the errors: in the plot statistical errors for a SN at $10kpc$ are shown (the average detected energy can be determined to about 5%). The result depends on the energy partition between neutrino flavours and on the efficiency to detect $15.11MeV$ γ quanta (ϵ_γ). Energy equipartition and $\epsilon_\gamma = 1.$ have been assumed here [7].

Even if the expected number of n.c. interactions with ^{12}C in LS is a factor 20 less than the number of $\nu_x d$ interactions in an heavy water detector of the same mass, live LS detectors are sensitive to the ratio $T_{\nu_x}/T_{\bar{\nu}_e}$.

3. ν luminosity

The first detectable neutrino signal from a gravitational stellar collapse is the shock break-

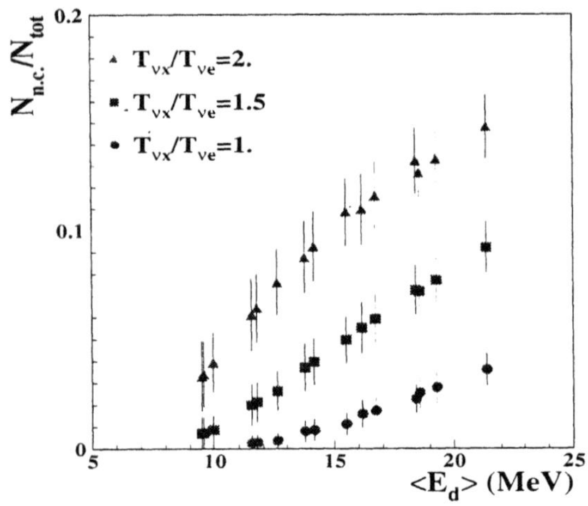

Figure 1. Ratio of $\nu_i^{12}C$ events on the total vs. the average detected energy from $\bar{\nu}_e p$ scattering in LS. Errors are statistical for $D = 10kpc$.

out ν_e flash. Estimates of the total number of ν_e events during this phase in LS, SNO and SK are $\approx 1, 10, 10$. The heavy water Čerenkov detector high sensitivity to ν_e through reaction (5) has been mentioned, while it is matter of controversial whether SK could disentangle ν_e interactions from the background due to the rising contribution of $\bar{\nu}_e p$ scattering [3].

Another prediction of the current theory is the rapid turn on of the $\bar{\nu}_e$ emission. The event rate due to $\bar{\nu}_e$ interactions should be about $0.3 events \cdot ton^{-1} \cdot sec^{-1}$ after 50-100 ms from the breakout flash, this means that the triangulation tecnique could be applied succesfully to point back to the SN [2].

Other features in this first phase of the ν emission, as the possible modulation of the luminosity after the bounce, could be studied by each detector because more than 50% of the total signal is expected to lay in the first second. During this time interval experimental difficulties could arise due to the high event rate (detector dead time and buffer's capability).

The next cooling phase, characterized by a slow decrease of the luminosity and of the average ν energy, ends, after a few tens of seconds, when the star becomes transparent to ν radiation. The possibility to follow the entire process up to the turning off of the ν signal depends on the masses, the backgrouds (energy and rate) and the energy thresholds of detectors. SK could have the necessary sensitivity.

We want to mention here the ultrahigh-energy ν observatory AMANDA which is equipped to monitor the single photomultiplier (at present about 300) counting rate every half a second [8]. The amount of ice under observation could in future make this detector the most suitable to study the time profile of the ν emission. Ice properties, which govern the AMANDA performances, and counting rate fluctuations will determine the actual sensitivity of this detector, although an higher sampling rate would be advisable.

4. ν mass and mixing

A finite ν mass would cause a neutrino signal time spreading, the delay of arrival times depending on the ν energy as:

$$\Delta_t = 5.15 \cdot \frac{D}{10kpc} \cdot (\frac{m_\nu c^2}{1eV})^2 (\frac{E_\nu}{10MeV})^{-2} msec \quad (7)$$

Different strategies have been proposed in these years to measure or constrain the masses of different flavours from the next SN signal.

Concerning electron neutrinos [present limit $m_{\nu_e} < 10 \div 15eV$] a mass just above a few eV almost completely erases the breakout burst, demodulates the luminosity pulsations and smooths the rise time of the $\bar{\nu}_e$ signal.

If rapid features (ν_e breakout burst or pulsations on the $\bar{\nu}_e$ luminosity or a rapid rise time) are seen, the ν_e mass must be less than few eV. But if these features will be absent, any conclusion on ν_e mass independently on SN models would be difficult.

A possible tool is the correlation between energy and arrival time on an event by event basis. Reconstructing the smeared pulsations of the signal by a specific value of $m_{\nu_e}\sqrt{D}$ [2] one could obtain a model independent measurement of m_{ν_e}

(but dependent on the knowledge of the source distance D). Alternatively, if the $\bar{\nu}_e$ spectrum at the source is constant in the first tens of milliseconds, the sensitivity $m_{\nu_e} \leq 5eV$ can be achieved by studying the time-energy correlation in the first 100-300 SK events [9].

For what concerns ν_μ and ν_τ [$m_{\nu_\mu} < 170keV$, $m_{\nu_\tau} < 24MeV$] the two components cannot be separated because they are experimentally indistinguishable.

The heavy water Čerenkov detector gives us some chances to study the ν_x event rate time profile, depending on the capability to disentangle deuterium breakup reactions and to model the ν_e and $\bar{\nu}_e$ contribution to n.c. in heavy water from the c.c. signal in other detectors.

A fast rise time or any rapid feature in the ν_x signal would constrain m_{ν_μ} and m_{ν_τ} to cosmologically interesting values in a model independent way.

On the other hand, in the hypotesis that the ν_x luminosity reflects the $\bar{\nu}_e$ one, it is possible to study m_{ν_μ} and m_{ν_τ} down to $30 \div 150eV$ by comparing the time profile of the $\bar{\nu}_e p$ signal and the signal induced by ν_x, by using the νe^- scattering in SK [10] or the $\nu^{16}O$ in SK [5] or the $\nu^{12}C$ in LS [11] or the νd in SNO [12].

Finally a possible different way, suggested in [13], consists on the detection of the presupernova MeV γ-burst induced by ν interactions with the stellar envelope. In this case, if the γ intensity would be high enough, the time delay between ν's and photons could be directly measured avoiding any dependence on the unknown details of the SN model.

The effect of ν mixing on the signal from the SN collapse has been discussed by many authors, we will limit ourself to notice some salient aspects.

The ν emission can be divided into two phases: in the first one (the shock breackout flash) mostly ν_e are emitted, in the latter all ν flavours are emitted with about the same total energy per flavour.

During the first phase the SN experiment is equivalent to the solar ν experiment with a minimum detectable $\Delta m^2 \geq 10^{-19}eV^2$ for $D = 10kpc$. SNO can detect the shock breakout ν_e flash in the c.c and n.c. channels. In the case of $\nu_e \rightarrow \nu_x$ oscillations, the ratio between the number of events in the two channels would change signalling flavour oscillations up to values of Δm^2 otherwise undetectable.

In the second phase, ν mixing induces a merging of the different spectra: $\nu_\mu \rightarrow \nu_\tau$ oscillations do not induce detectable effects (at least for $m_{\nu_\tau} \ll 10MeV$) because the two fluxes are expected to be equal, while $\bar{\nu}_e \rightarrow \bar{\nu}_x$ oscillations could induce a detectable distortion of the $\bar{\nu}_e$ spectrum.

5. Conclusion

The neutrino burst from the next galactic stellar collapse will be observed by a number of different massive, underground neutrino telescopes. These detectors provide temporal, energetic, angular and flavour information with an unprecedented sensitivity.

The possibility to obtain the most complete information and especially to check its consistency will be guaranteed by the comparison among the different detector responses.

REFERENCES

1. see for example: M.Aglietta et al. N.Cim.A 105 1992 1793; Y.Oyama et al. NIM A 340 1994 612; M.Ambrosio et al. Astrop.Phys. 8 1998 123; W.Fulgione et al. NIM A 368 1996 512; V.G.Ryasny NIM A 374 1996 377
2. A.Burrows et al. Phys.Rev. D45 1992 3361
3. T.Totani et al. ApJ 496 1998 2
4. K.Langanke et al. Phys.Rev.Lett. 76 1996 2629
5. J.F.Beacom and P.Vogel hep-ph/9802424
6. E.Kemp Ph.D.thesis, Universidade Estadual de Campinas (BR) 1998
7. P.Antonioli et al. NIM A 309 1991 569
8. P.B.Price et al. Proc. 32nd Rencontres de Moriond 1997
9. T.Totani Phys.Rev.Lett. 80 1998 2039
10. G.Fiorentini and C.Acerbi Astrop.Phys. 7 1997 245
11. O.G.Ryazhskaya et al. 23rd ICRC HE 5.1.4
12. J.F.Beacom and P.Vogel hep-ph/9806311
13. G.S.Bisnovati-Kogan et al. Astr.Space Sci. 35 23 1975; O.G.Ryazhskaya INFN/AE-98/10

Nuclear Physics B (Proc. Suppl.) 77 (1999) 440–444

NUCLEAR PHYSICS B
PROCEEDINGS
SUPPLEMENTS

Pulsar Velocities without Neutrino Mass

D. Grasso[a], H. Nunokawa[b] and J. W. F. Valle[a]

[a]Instituto de Física Corpuscular - IFIC/CSIC, Departamento de Física Teòrica
Universitat de València, 46100 Burjassot, València, Spain

[b]Instituto de Física Gleb Wataghin, Universidade Estadual de Campinas
13083-970 Campinas, SP Brazil

We show that pulsar velocities may arise from anisotropic neutrino emission induced by resonant conversions of massless neutrinos in the presence of a strong magnetic field. The main ingredient is a small weak universality violation and neither neutrino masses nor magnetic moments are required.

1. Introduction

One of the most challenging problems in modern astrophysics is to find a consistent explanation for the high velocity of pulsars. Observations [1] show that these velocities range from zero up to 900 km/s with a mean value of 450 ± 50 km/s. An attractive possibility is that pulsar motion arises from an asymmetric neutrino emission during the supernova (SN) explosion. In fact, neutrinos carry more than 99% of the new-born proto-neutron star's gravitational binding energy so that even a 1% asymmetry in the neutrino emission could generate the observed pulsar velocities. To find the origin of such asymmetry is, however, not a minor task. Although several possible realizations of such asymmetric neutrino emission in the framework of the Standard Model (SM) of particle physics have been already explored [2] a conclusive solution of the problem is still lacking, and there is some motivation for looking also at solutions that involve physics beyond the SM.

2. Pulsar Kick from Neutrino Oscillation

Recently, several neutrino conversion mechanisms in matter have been invoked as a possible engine for powering pulsar motion. Kusenko and Segré proposed a mechanism [3] based on MSW conversions [4]. The idea is based on the observation that the strong magnetic field present during a SN explosion gives rise to some angular dependence of the matter induced neutrino potentials [5]. As a consequence, in the presence of non-vanishing ν_τ mass and mixing the resonance sphere for the $\nu_e - \nu_\tau$ conversions is distorted. If the resonance surface lies between the ν_τ and ν_e neutrino spheres, such a distortion would induce a temperature anisotropy in the flux of the escaping tau-neutrinos produced by the conversions, hence a recoil kick of the proto-neutron star. In order to account for the observed pulsar velocities the required strength of the dipolar component of the magnetic field between the two neutrino spheres must exceed 10^{15} Gauss [6] or even larger [7]. Another crucial ingredient in this mechanism is the neutrino squared mass difference, $\Delta m^2 \gtrsim 10^4 \text{eV}^2$, which leads to $m_{\nu_\tau} \gtrsim 100$ eV or so, assuming a negligible ν_e mass. This is necessary in order for the resonance surface to be located between the two neutrino-spheres. However, we note that such requirement is at odds with cosmological bounds on neutrinos masses unless the τ-neutrino is unstable.

Akhmedov, Lanza and Sciama [8] proposed a similar pulsar acceleration mechanism based on neutrino spin-flavour precession (SFP) [9] which is resonantly enhanced by matter [10] (RSFP). The lowest magnetic field strength required is $B > 2 \times 10^{16}$ Gauss, as long as the neutrino magnetic moment exceeds $\mu_\nu \gtrsim 10^{-15}\ \mu_B$.

In this work we investigate the relevance, for pulsar motion, of a different kind of neutrino conversion mechanism, not requiring any neu-

0920-5632/99/$ – see front matter
PII S0920-5632(99)00462-4

trino mass nor magnetic moment, proposed in ref. [11]. In contrast to the physics of the solar neutrino problem, this new conversion mechanism was shown to be potentially relevant for supernova physics [12].

3. The Particle Physics Model

The simplest underlying particle physics model that realizes this new conversion mechanism postulates the existence of two new $SU(2) \otimes U(1)$ singlet leptons for each generation of leptons, in such a way that lepton number symmetry is exact in the Lagrangian [13]. These extra states can arise in various extensions of the SM, such as superstring models [14]. However, the model is very interesting on its own right, both conceptually as well as phenomenologically [15]. As a result of the postulated lepton number symmetry neutrinos remain massless to all orders of perturbation even after the gauge symmetry breaking. However, unlike the situation in the SM there is a non-trivial Kobayashi-Maskawa-like mixing in the weak leptonic charged current [16]. The simplest such scheme contains three two-component gauge singlet neutral leptons S added to the three right-handed neutrino components ν^c present in SO(10). For definiteness we consider this model at the $SU(2) \otimes U(1)$ level. The assumed lepton number conservation leads to a neutral mass matrix with the following texture in the basis (ν, ν^c, S):

$$\begin{pmatrix} 0 & D & 0 \\ D^T & 0 & M \\ 0 & M^T & 0 \end{pmatrix}, \tag{1}$$

where the Dirac matrix D describes the weak doublet ν ν^c mass term, whereas M connects the singlet states ν^c and S. It is easy to see that, the three conventional neutrinos remain massless, while the other six neutral 2-component leptons combine into three heavy Dirac fermions. This model offers a viable alternative to the *see-saw* model.

The resulting charged-current Lagrangian in the massless-neutrino sector is [11],

$$\mathcal{L}_{CC} = \frac{ig}{\sqrt{2}} W_\mu \bar{e}_{aL} \gamma_\mu K_{ai} \nu_{iL} + \text{h.c.} , \tag{2}$$

where $a = e, \mu, \tau$, $i = 1, 2, 3$ and the mixing matrix K is given as $K = R\mathcal{N}$, [11,16]. For definiteness and simplicity we consider the case of two neutrinos, for which

$$R = \begin{pmatrix} \cos\theta & \sin\theta \\ -\sin\theta & \cos\theta \end{pmatrix}, \tag{3}$$

and where $\mathcal{N} = \text{diag}(\mathcal{N}_1, \mathcal{N}_{2,3})$. Note that K is not unitary, since it is a sub-matrix of the full rectangular matrix including also the heavy states [16]. The two neutrino flavors are effectively nonorthogonal, i.e., $\langle \nu_e | \nu_{\mu,\tau} \rangle \equiv -\sin\theta\cos\theta(\mathcal{N}_1^2 - \mathcal{N}_{2,3}^2)$. The non-diagonal elements of the matrix K cannot be rotated away through a redefinition of the massless-neutrino fields. In this way a non-trivial mixing arises among the massless neutrinos.

The corresponding form of the neutral-current Lagrangian is [11],

$$\mathcal{L}_{NC} = \frac{ig}{2\cos\theta_W} \mathcal{N}_i^2 Z_\mu \bar{\nu}_{iL} \gamma_\mu \nu_{iL} , \tag{4}$$

where $\mathcal{N}^2 = K^\dagger K$ so that the neutral current is diagonal, but i-dependent, signaling the violation of weak universality. It is also convenient to define $\mathcal{N}_i^2 \equiv (1 + h_i^2)^{-1}$, $i = 1, 2(3)$, where the h_i parameters reflect the deviation from the *standard* neutrino coupling. Since no oscillations between two strictly massless neutrinos can develop in vacuum, it follows that laboratory limits on the leptonic mixing angle θ are very weak. However, it will be sufficient for our purposes to assume that the mixing angle θ is very small. In this way we have $\nu_i \sim \nu_a$ $[a = e, \mu(\tau)]$, so that $h_i^2 \sim h_a^2$. The parameters h_i^2 are also constrained experimentally. For the third generation one can still allow h_τ^2 values in the range of a few percent [17], whereas the constraints on h_e^2 and h_μ^2 are more stringent. For this reason we consider from now on the case of $\nu_e \leftrightarrow \nu_\tau$ conversions, for which universality violation can reach 10^{-2} or so [17].

4. Massless Neutrino Conversion in a Strong Magnetic Field

Within the framework described in the previous section, it has been shown that neutrino can under go resonant conversion even if neutrino mass is strictly zero [11]. The resonance condition is given by,

$$Y_e' = \eta Y_n', \tag{5}$$

where

$$\eta \equiv \frac{1}{2}\left(h_\tau^2 - h_e^2\right) , \tag{6}$$

measures the violation of lepton universality and,

$$Y_e' \equiv Y_e - Y_e^0 \cos\phi \qquad Y_n' \equiv Y_n + Y_e^0 \cos\phi . \tag{7}$$

Here $Y_e \equiv \frac{n_e}{n_p+n_n}$ and $Y_n = 1 - Y_e$ are the electron and neutron number per baryon, respectively. The terms proportional to $\cos\phi$ in eq. (7) arise due to the presence of the magnetic field. The main effect of the magnetic field is to polarize the electrons inducing an angular dependence in the axial part of the potentials [5] that affect the entries of the neutrino evolution Hamiltonian described in [11,12]. Since the net spin carried by electrons in higher Landau levels vanishes, there is a cancellation of their contributions to the axial part of the neutrino interaction Hamiltonian [5]. As a result only electrons in the lowest Landau level will contribute to the axial term and hence to the asymmetric neutrino emission. This is parametrized through effective electron and neutron fractions Y_e' and Y_n' which depend on the angle ϕ between the neutrino propagation direction and $\vec{B}$. In eq. (7) the superscript zero refers to the contribution from the electrons in the lowest Landau level. The resonance condition (5) can be rewritten as,

$$Y_e\left(1 - \lambda_e \cos\phi\right) = \eta Y_n \left(1 + \lambda_e \frac{Y_e}{Y_n} \cos\phi\right) , \tag{8}$$

where $\lambda_e \equiv Y_e^0/Y_e$ is the fraction of the electrons in the lowest Landau level with respect to the total electron density, which is the measure of the electron polarization due to the magnetic field.

In order to establish that massless neutrino conversions can play a role explaining the origin of pulsar velocities, we need to verify that the resonance condition eq. (5) can indeed be fulfilled in a SN environment between the e and the τ neutrino-spheres. The mean resonance position is obtained by averaging eq. (8) over ϕ, giving

$$Y_e = \eta Y_n . \tag{9}$$

Apart from small **B** induced corrections to Y_e, the condition eq. (9) coincides with the free-field resonance condition given in ref. [12] which allows us to apply here some of the arguments used there. We see that, for experimentally allowed values of $\eta \lesssim 10^{-2}$, the condition in eq. (9) can be fulfilled if $Y_e \lesssim 10^{-2}$. This is indeed possible close to the neutrino-spheres as a consequence of the strong deleptonization taking place in that region during the Kelvin-Helmholtz cooling phase. A rough estimate of the value of Y_e between the neutrino-spheres can be found in ref. [12] in which the order of minimum values of Y_e is found to be $\sim 10^{-2}$, in agreement with numerical SN models [18]. Therefore, we can expect resonant massless-neutrino conversion to occur between the two neutrino-spheres for a range of values of the parameter η which is not experimentally excluded.

The adiabaticity of resonant massless-neutrino conversions can be easily verified by looking at the probability for $\nu_e \leftrightarrow \nu_\tau$ and $\bar{\nu}_e \leftrightarrow \bar{\nu}_\tau$ conversions, given in ref. [12]. One can show that from eq. (19) in ref. [12] that the adiabatic conversion, i.e., $P(\nu_e \leftrightarrow \nu_\tau) \sim 1$, in the region between the ν_τ and ν_e neutrinospheres, is realized so long as $\sin^2 2\theta \gtrsim 10^{-7}$.

5. Asymmetric Neutrino Emission

The anisotropy in the total momentum of the escaping neutrinos can be computed as in [3,7] starting from the resonance condition eq. (8). Since, as we discussed $Y_e \ll Y_n$ in the region of our interest, we neglect the second term on the right side of eq. (8). Following [3] we parametrize the resonance surface equation as follows:

$$r(\phi) = r_0 + \delta \cos\phi, \tag{10}$$

where r_0 is the radius of the free-field resonance sphere defined by eq. (9). The displacement δ is found by subtracting the resonance condition computed for $\phi = \pi$ with the same computed in $\phi = 0$. We find

$$Y_e(r+\delta) - Y_e(r-\delta) \simeq \frac{dY_e}{dr} 2\delta = 2\lambda_e Y_e(r_0) \;, \quad (11)$$

which can be rewritten as,

$$\delta = \lambda_e h_{Y_e}(r_0) \;, \quad (12)$$

where

$$h_{Y_e} \equiv \left(\frac{d \ln Y_e}{dr} \right)^{-1} . \quad (13)$$

The deformation of the resonance sphere implies an angle dependence in the temperature of the escaping ν_τ 's, hence to an asymmetry in the momentum that these neutrinos carry away from the SN. This is given by [7]

$$\frac{\Delta k}{k} \approx \frac{1}{6} \frac{\int_0^\pi F_\nu \cos\phi \sin\phi d\phi}{\int_0^\pi F_\nu \sin\phi d\phi} \simeq \frac{2}{9} \delta h_T^{-1} \quad (14)$$

where the temperature variation scale height h_T is defined in analogy to the definition of h_{Y_e} given in eq. (13). The factor 1/6 in eq. (14) accounts for the fact that, out of the six neutrino and anti-neutrino species, we are assuming that only the ν_τ carries a momentum anisotropy. We note, however, that an extra factor of two may be gained in eq. (14) if, at the same time, $\bar{\nu}_\tau$ emission suffers resonant conversion between $\bar{\nu}_e$ and $\bar{\nu}_\tau$ spheres. This is indeed possible as the resonance condition eq. (5) for $\nu_e - \nu_\tau$ and $\bar{\nu}_e - \bar{\nu}_\tau$ conversions coincide, which is a characteristic feature of this mechanism [11], in sharp contrast with the case of MSW conversions [4], where either neutrinos or anti-neutrinos can resonantly convert, but not both.

By substituting eq. (12) in eq. (14) we finally get

$$\frac{\Delta k}{k} \approx \frac{2}{9} \frac{h_{Y_e}}{h_T} \lambda_e \;. \quad (15)$$

Numerical SN simulations [19] typically give $h_{Y_e}/h_T \approx 1$ between the two neutrino-spheres. Hence, we see from eq. (15) that in order for massless neutrino resonant conversions to account for the observed pulsar velocities one needs $\lambda_e \approx 5 \times 10^{-2}$ in the region between the two neutrino-spheres. Using the expression for the polarization λ_e given in [20] we determine the required

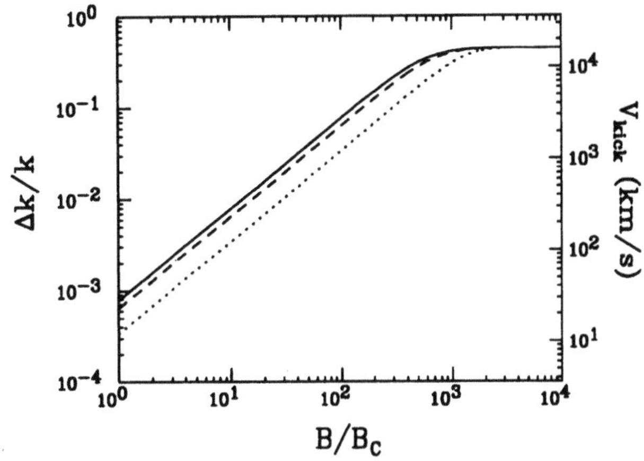

Figure 1. Magnitude of kick velocity versus magnetic field, in units of the critical field $B_c = 4.4 \times 10^{13}$ Gauss.

value of the magnetic fields strength as $\approx 10^{15}$ Gauss, which seems reasonable from the astrophysics point of view [21], and lower than required in the MSW [6] and RSFP [8] neutrino conversion mechanisms.

This is illustrated in Fig. 1 for three different choices of ρY_e at resonance consistent with SN models. The three lines correspond to ρY_e values (in g/c.c.) 5×10^9 (solid), 1×10^{10} (dashed) and 5×10^{10} (dotted) at resonance. In Fig. 1 we assume that both ν_τ and $\bar{\nu}_\tau$ carry a momentum anisotropy. The kick velocity values given in the right ordinate assume a pulsar mass of 1.4 $M_\odot$ and a total energy released by all neutrino species of 3×10^{53} erg.

6. Conclusions

We have proposed a viable scheme for generating pulsar velocities from anisotropic neutrino emission induced by resonant conversions of massless neutrinos in the presence of a strong magnetic field. Our proposal rests on the idea that there is a small violation of universality in the weak interactions and neither neutrino masses nor magnetic moments are needed. Although the mechanism we proposed makes use of neutrino non-orthonormality instead of neutrino mass, it should be clear that neutrino masses could be present. Though too small to play a role in pulsars they could explain, for example, the observed deficit of solar neutrinos. The particle physics model we have presented admits a simple extension [22] in which there is a small non-zero μSS entry in the mass matrix. For μ in the eV range one easily gets neutrino masses in the $\eta^2\mu \sim 10^{-3}\ eV$ range, as required for the understanding of the solar neutrino data.

Slightly more detailed discussion of the work presented here is found in ref. [23].

We finally note that, very recently, Raffelt and Janka [24] have claimed that the kick effect in in this type of scenario is vastly overestimated because the temperature variation over the deformed neutrinosphere is not an adequate measure for the anisotropy of the neutrino emission, and have concluded that the required magnetic field must be, at least, more than one order of magnitude larger.

Acknowledgments

This work was supported by DGICYT grant PB95-1077, CICYT-INFN, European Union TMR network ERBFMRXCT960090. H. N. was supported by a FAPESP fellowship. We thank H.-T. Janka, A. Rossi, A. Yu. Smirnov and H. Suzuki for useful correspondence and discussions.

REFERENCES

1. A.G. Lyne and D.R. Lorimer, Nature **369** (1994) 127.
2. N.N. Chugai, Sov. Astron. Lett.**10**, 87, 1984; A. Vilenkin, Astrophys. J. **451** (1995) 700.
3. A. Kusenko, G. Segrè, Phys. Rev. Lett. **77** (1996) 4872.
4. M. Mikheyev, A. Smirnov, Sov. J. Nucl. Phys. **42** (1986) 913; L. Wolfenstein, Phys. Rev. **D17** (1978) 2369; Phys. Rev. **D20** (1979) 2634.
5. J.C.D'Olivo, J.F.Nieves and P.B. Pal, Phys. Rev. **D40** (1989) 3679; V.B. Semikoz, J.W.F. Valle, Nucl. Phys. **B425** (1994) 65; Nucl. Phys. **485** (1997) 585(E); J.C.D'Olivo, J.F.Nieves, Phys. Lett. **B383** (1996) 87; P. Elmfors, D. Grasso, G. Raffelt, Nucl. Phys. **B479** (1996) 3.
6. A. Kusenko, G. Segrè, Phys. Rev. Lett. **79** (1997) 2751.
7. Y.Z. Qian, Phys. Rev. Lett. **79** (1997) 2750.
8. E.Kh. Akhmedov, A. Lanza and D.W. Sciama, Phys. Rev. **D56** (1997) 6117.
9. J. Schechter, J.W.F. Valle, Phys. Rev. **D24** (1981) 1883; Phys. Rev. **D25** (1982) 283.
10. E.Kh. Akhmedov, Phys. Lett. **B213** (1988) 64; C. S. Lim and W. Marciano, Phys. Rev. **D37** (1988) 1368.
11. J.W.F. Valle, Phys. Lett. **B199** (1987) 432.
12. H. Nunokawa, Y.Z. Qian, A. Rossi and J.W.F. Valle, Phys. Rev. **D54** (1996) 4356.
13. R. Mohapatra and J. W. F. Valle, Phys. Rev. **D34** (1986) 1642; D. Wyler and L. Wolfenstein, Nucl. Phys. **B218** (1983) 205.
14. E. Witten, Nucl. Phys. **B268** (1986) 79.
15. J. W. F. Valle, hep-ph/9603307; Prog. Part. Nucl. Phys. **26** (1991) 91.
16. J. Schechter and J. W. F. Valle, Phys. Rev. **D22** (1980) 2227.
17. M. Gronau, C. N. Leung, J. L. Rosner, Phys. Rev. **D29** (1984) 2539; P. Langacker, D. London, Phys. Rev. **D38** (1988) 886 and Phys. Rev. **D38** (1988) 907; A. Ilakovac, A. Pilaftsis, Nucl. Phys. **B437** (1995) 491; E. Nardi, E. Roulet, D. Tommasini, Phys. Lett. **B344** (1995) 225.
18. See for e.g., H. Suzuki, *Physics and Astrophysics of Neutrinos*, ed. by M. Fukugita and A. Suzuki, p. 763 (Springer-Verlag, Tokyo, 1994); J. R. Wilson and R. Mayle, in *Advanced Study Institute on the Nuclear Equation of State, Peñiscola, Spain, NATO ASI Series B, Vol216A*, edited by Walter Greiner and Horst Stöcker, p. 731, (New York, Plenum Press, 1989); H.-Th. Janka and W. Keil, astro-ph/9709210.
19. H. Susuki, private communication.
20. H. Nunokawa, V.B. Semikoz, A.Yu. Smirnov and J.W.F. Valle, Nucl. Phys. **B501** (1997) 17.
21. C. Thompson, R. Duncan, Astrophys. J. **408** (1993) 194.
22. M. C. Gonzalez-Garcia, J. W. F. Valle, Phys. Lett. **B216** (1989) 360.
23. D. Grasso, H. Nunokawa and J. W. F. Valle, Phys. Rev. Lett. **81** (1998) 2412.
24. G. Raffelt and H.-T. Janka, astro-ph/9808099.

NUCLEAR PHYSICS B
PROCEEDINGS
SUPPLEMENTS

ELSEVIER Nuclear Physics B (Proc. Suppl.) 77 (1999) 445–449

The neutrino ground state in a neutron star

Ken Kiers[a*†] and Michel H.G. Tytgat[b]

[a]High Energy Theory, Department of Physics
Brookhaven National Laboratory, Upton, NY 11973-5000, USA

[b]Service de Physique Théorique, Université Libre de Bruxelles, CP225
Bd du Triomphe, 1050 Bruxelles, Belgium

We address a recent claim that the stability of neutron stars implies a lower bound on the mass of the neutrino. We argue that the result obtained by some previous authors is due to an improper summation of an infrared-sensitive series and that a non-perturbative "resummation" of the series yields a finite and well-behaved result. The stability of neutron stars thus gives no lower bound on the mass of the neutrino.

1. INTRODUCTION

In this talk we present a calculation of the interaction energy due to multi-body neutrino exchange in a neutron star [1].

Consider the series shown in Fig. 1, which represents the "self-energy" of a neutron star due to neutrino exchange. In this figure the crosses represent insertions of the neutron density. It is not difficult to see that the term in this series with k insertions of the neutron density scales approximately as [2]

$$W^{(k)} \sim \frac{C_k}{R}\left(\frac{G_F N}{R^2}\right)^k, \tag{1}$$

where $N \sim 10^{57}$ is the total number of neutrons in the star, $R \sim 10$ km is the radius of the star and C_k is a dimensionless numerical coefficient. The thing which is perhaps surprising in this expansion is that the "expansion parameter" $G_F N/R^2$ is of order 10^{12}. A direct summation of the terms in this series (but truncated after N terms) yields an enormous value for the interaction energy [2]. This result has led some previous authors to claim that neutrinos must have a mass of at least 0.4 eV in order to allow neutron stars to exist as stable objects [2–4].

Our approach to this problem is quite different. We contend that the series represented in Fig. 1 is actually infrared-divergent and must be "resummed" non-perturbatively in order to yield a sensible result. This approach was previously advocated in Ref. [5]. We have performed the required resummation and find that the apparent infrared divergence is an artifact of the expansion and that the interaction energy is finite and well-behaved [6]. There is thus no lower bound on the mass of the neutrino. Along the way we also encounter some interesting physics. We demonstrate, for example, that the ground state of the system actually contains a non-zero neutrino number – a result which was previously anticipated in Ref. [7]. In our simple model for the density of the neutron star it is straightforward to calculate both the energy and the neutrino number of the ground state.

Before describing the calculation in more de-

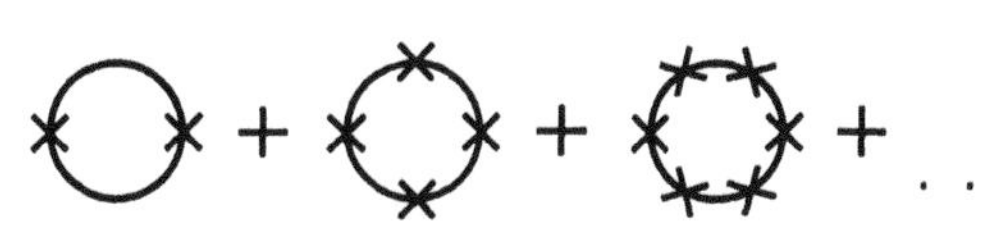

Figure 1. Perturbative expansion of the shift in the neutrino ground state energy due to the presence of a neutron star. Solid lines represent neutrino propagators and crosses represent insertions of the neutron density.

*Talk presented at "NEUTRINO 98", Takayama, Japan, June 4-9, 1998. This contribution to the proceedings is excerpted from Ref. [1].
†Address after September, 1998: Physics Department, Taylor University, 236 West Reade Ave., Upland, IN 46989, USA

PII S0920-5632(99)00464-8

tail, let us note that there have been several groups which have examined various aspects of this problem [5,8–12]. In particular, Arafune and Mimura [12] have confirmed our asymptotic result using an analytical approximation.

2. NEUTRINO GROUND STATE

2.1. Preliminaries

Our goal is to calculate the shift in the neutrino ground state energy due to the presence of the star. This energy shift may be defined in terms of the neutrino Hamiltonian $H_{(0)}$ in the presence (absence) of the star as follows [13]:

$$W = \langle\hat{0}|H|\hat{0}\rangle - \langle 0|H_0|0\rangle. \tag{2}$$

Here $|\hat{0}\rangle$ denotes the neutrino ground state in the presence of the star, while $|0\rangle$ denotes the usual matter-free vacuum state. As we have already alluded, the state $|\hat{0}\rangle$ contains in general a non-zero neutrino number (i.e., it is "charged"). Note that the expression in Eq. (2) is a formal, ultraviolet-divergent quantity which needs to be renormalized. This renormalization may be done quite easily using the usual techniques.

In order to proceed, it is convenient to introduce an effective Lagrangian for the neutrino field. After integrating out all of the other particles in the theory, one obtains [14,5,15]

$$\mathcal{L}_{\text{eff}} = \overline{\psi}_L \left[i\partial\!\!\!/ + \alpha\gamma^0 \right] \psi_L \tag{3}$$

where $\psi_L = \frac{1}{2}(1-\gamma_5)\psi$, and where

$$\alpha(\vec{x}) = G_F \rho_n(\vec{x})/\sqrt{2} \sim 20 \text{ eV} \tag{4}$$

is the electroweak potential induced by the finite neutron density ($\rho_n \approx 0.4$ fm^{-3} in a typical neutron star). This potential is identical to the one which is usually considered in the well-known Mikheyev-Smirnov-Wolfenstein (MSW) effect [16]. Note that the potential term in (3) resembles a position-dependent chemical potential, so that it is not at all surprising that the ground state of the system has a non-zero neutrino number.

It is straightforward to derive the following perturbative expansion for W in terms of the potential $\alpha(\vec{x})$ and the neutrino propagator $G_0(\vec{x},\vec{x}';\omega)$ [1]:

$$W = \frac{1}{2\pi i}\sum_{k=1}^{\infty}\frac{1}{k}\int_C d\omega\, \text{Tr}_{\mathbf{x}}\left[\alpha G_0(\omega)\right]^k \tag{5}$$

$$\equiv \sum_k W^{(k)}. \tag{6}$$

All of the odd terms in this expansion disappear so that this series corresponds precisely to that which is represented diagrammatically in Fig. 1. One aspect of the perturbative expansion which is useful is that it neatly isolates the ultraviolet divergence in W. In fact, the only ultraviolet divergent term in (6) is that with $k = 2$. This term is related to the vacuum polarization of the Z in the complete theory. A final note concerning this expansion is that while the terms with $k \geq 4$ are separately finite, their sum is infrared-sensitive and is in fact ill-defined for "large" stars [$\alpha R \gg \mathcal{O}(1)$]. The non-perturbative "resummation" of these terms is the main goal of our calculation.

2.2. Comparison point

It is useful to compare the results which we will describe with results which were obtained previously in the literature. The perturbative expansion of W has been considered in Refs. [2–4], where it was found that already by the eighth term in the expansion the interaction energy exceeded the gravitational binding energy of the star. After summing up the entire series [which in the approach of Refs. [2–4] was actually a truncated sum, not an infinite series as we have in Eq. (6)], it was found that the interaction energy exceeded the rest mass of the universe. It was argued that the only way to regulate the sum was to give all neutrino flavours a mass of at least 0.4 eV.

Our philosophy in this matter is that the perturbative expansion is simply outside of its radius of convergence when $\alpha R \gg \mathcal{O}(1)$ and that the series needs to be resummed using a non-perturbative approach [5]. We cannot consider a realistic neutron star (i.e., with $\alpha R \sim G_F N/R^2 \sim 10^{12}$) using our numerical approach, but it is actually sufficient to restrict our analysis to $\alpha R \leq \mathcal{O}(100)$, since for $\alpha R \sim \mathcal{O}(1)$ we already observe a "cross-over" to the non-perturbative regime. A clear signal of this cross-over is that the ground state obtains a non-zero charge. (The charge of the ground state is exactly zero to any finite order in perturbation theory.)

Consider the comparison point $\alpha R = 20$. This point could correspond to a tiny "star" with a realistic neutron density ($\alpha = 20$ eV), but with a tiny radius ($R = 2 \times 10^{-5}$ cm). In this case the truncated sum in Ref. [3] gives $\sum_{k=4}^{N} W^{(k)} \sim 10^{66}$ eV. By way of comparison, our non-perturbative resummation gives $\sum_{k=4}^{N} W^{(k)} \sim -2.3$ keV.

2.3. Phase shift formulas

An exact, non-perturbative expression for the interaction energy W is given by the following expression due to Schwinger [13]

$$W = \frac{1}{2\pi}\sum_{l=0}^{\infty}(2l+2)\int_0^{\infty} d\omega\ [\delta_l(\omega) + \delta_l(-\omega)]\,. \quad (7)$$

Here $\delta_l(\omega)$ is the scattering phase shift for a neutrino incident on the star and l labels the orbital angular momentum. A similar expression may be obtained for the "charge" of the ground state:

$$q = -\frac{1}{2\pi}\sum_{l=0}^{\infty}(2l+2) \times \int_0^{\infty} d\omega\ \left[\frac{\partial}{\partial\omega}\delta_l(\omega) - \frac{\partial}{\partial\omega}\delta_l(-\omega)\right]. \quad (8)$$

The factor $(2l+2) \equiv (2j+1)$ is the degeneracy factor for a given energy ω and total angular momentum j.

The beauty of the above expressions for W and q are that they are valid for any value of αR. Note, however, that Eqs. (7) and (8) are still formal ultraviolet-divergent expressions which need to be renormalized.

3. NUMERICAL EVALUATION

Let us first choose a density profile for the neutron star. Since the quantities which are of interest to us (i.e., W and q) are only sensitive to the gross features of the star, it is convenient to choose a very simple – albeit unrealistic – density profile:

$$\alpha(r) = \alpha\theta(R - r). \quad (9)$$

Recall that $\alpha = G_F\rho_n(\vec{x})/\sqrt{2} \sim 20$ eV in a neutron star. In this very simple model it is straightforward to obtain closed expressions for the scattering phase shifts, which simplifies our numerical work considerably.

We may now renormalize W and q. Since our model is renormalizable the ultraviolet divergences in W and q are confined to the first few terms in the perturbative expansion. These terms may be isolated by Taylor-expanding the phase shift formulas in αR. [Note that by Taylor-expanding in αR we recover the perturbative expansion defined in Eq. (5).] The procedure is then as follows: (i) Taylor expand W and q in order to isolate the divergent terms; (ii) subtract out the divergent terms; (iii) regularize and renormalize the divergent terms using conventional methods; (iv) add the finite, renormalized terms back in. This procedure yields the following expression for the renormalized energy:

$$W_{\rm ren} = W_{\rm ren}^{(2)} + W^{(4+)}, \quad (10)$$

where

$$W^{(4+)} = \frac{1}{2\pi}\sum_{l=0}^{\infty}(2l+2)\int_0^{\infty} d\omega \left[\overline{\delta}_l(\omega) + \overline{\delta}_l(-\omega)\right], \quad (11)$$

and where $\overline{\delta}_l(\omega) \equiv \delta_l(\omega) - \delta_l^{(2)}(\omega)$, with $\delta_l^{(2)}(\omega)$ being the second term in the Taylor expansion of $\delta_l(\omega)$. $W_{\rm ren}^{(2)}$ is essentially the vacuum polarization of the Z in the full theory (but convoluted over the neutron star) and corresponds to the first term in the diagrammatic expansion in Fig. 1. This term has been discussed in detail in Ref. [1] and will not be considered further here. The second term in Eq. (10), $W^{(4+)}$, is of particular interest to us since it represents the non-perturbative "resummation" of the terms with four or more insertions of the neutron density in the diagrammatic expansion shown in Fig. 1. As we have noted, a direct summation of these terms does not lead to a sensible result when $\alpha R > \mathcal{O}(1)$. The resummed result in Eq. (11), however, is always well-defined and leads to a well-behaved and sensible result.

We may similarly obtain an expression for the renormalized charge:

$$q_{\rm ren} = \frac{1}{2\pi}\sum_{l=0}^{\infty}(2l+2)\left[\delta_l(0^+) - \delta_l(0^-)\right]. \qquad (12)$$

In order to calculate the charge, then, we need only know the scattering phase shifts at the origin.

3.1. Small αR

For small αR the "resummed" energy, $W^{(4+)}$, is well-approximated by the leading term in the Born expansion. We have checked this explicitly by calculating $W^{(4+)}$ directly using the phase shift formula and by Taylor-expanding the phase shifts to fourth order and (analytically) integrating the resulting expressions. The result which we obtain is

$$W^{(4+)} \simeq W^{(4)} \simeq -0.00115\frac{(\alpha R)^4}{R}. \qquad (13)$$

For small αR the phase shifts are zero at the origin, so that the ground state of the system is uncharged [see Eq. (12)].

3.2. Larger αR

As αR is increased, there is a critical point beyond which it becomes energetically favourable for the neutrinos to "condense"; thus, for $\alpha R > \alpha R|_{\rm crit}$, the ground state of the system carries a non-zero neutrino number. There are in fact an infinite number of points at which the charge of the ground state changes discontinuously as αR is increased. In our simple model for the density profile of the neutron star there is a correspondingly simple condition for a new charge to be added to the ground state:

$$j_l(\alpha R) = 0. \qquad (14)$$

In the quantum mechanical scattering problem, these values of αR correspond to the points at which a resonance crosses from the positive to the negative energy continuum [1].

Figure 2 shows a plot of the renormalized charge as a function of αR. The solid curve gives the exact charge, which has periodic jumps, and the dashed curve gives the charge expected in the large volume limit for a system with chemical potential $\mu = \alpha$: $q_{\rm cond} = 2(\alpha R)^3/(9\pi)$. Clearly, as αR gets large the exact result tends to this limit. The critical point – at which the ground state becomes charged and beyond which we should not trust the perturbative expansion – is seen to occur at $\alpha R = \pi$ in this model. This is the first zero of $j_0(\alpha R)$.

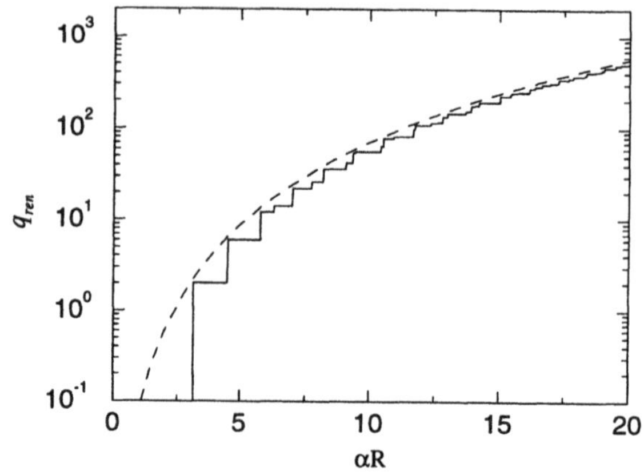

Figure 2. Plot of the charge as a function of αR. The solid curve gives the exact charge and the dashed curve gives the charge expected for a system with chemical potential $\mu = \alpha$.

Figure 3 shows plots of both the charge and the energy as functions of αR. Both the charge and the energy are normalized to the values expected for a "condensate" in the large-volume limit. For small αR the charge is zero and the energy is well-described by the leading term in the Born expansion [which is shown by the lower dashed line in Fig. 3(b)]. At $\alpha R = \pi$ the ground state becomes charged and perturbation theory breaks down. As αR is increased, both the charge and the energy tend toward the values expected for a large system with chemical potential $\mu = \alpha$. [Note that $W_{\rm cond} = -\alpha^4 R^3/(18\pi)$.] The asymptotic trend apparent in Fig. 3 has recently been confirmed by Arafune and Mimura using an analytical approximation [12].

The comparison point which was singled out above ($\alpha R = 20$) is included in the points shown in Fig. 3. A previous (perturbative) calculation of the energy associated with this point yielded an enormous value [3], while our non-perturbative

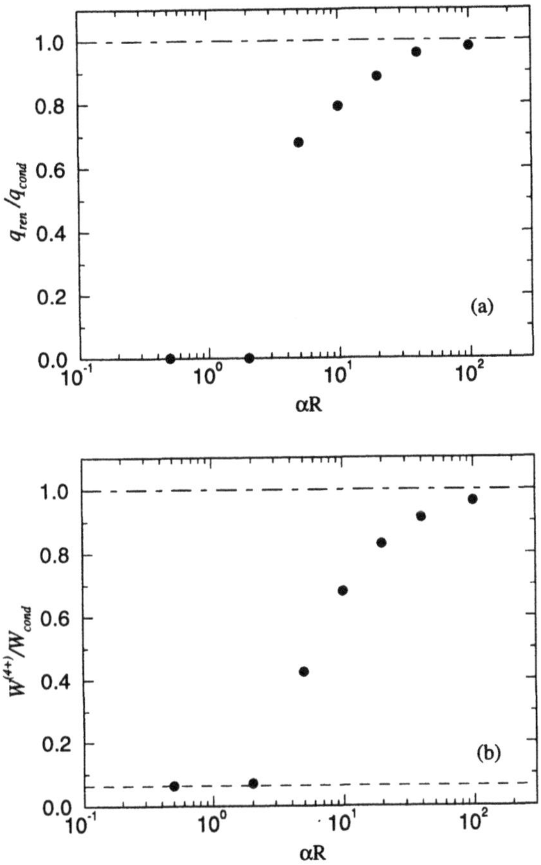

Figure 3. Normalized plots of (a) the charge and (b) the energy as a function of αR. The dots give the results of our exact (non-perturbative) calculations.

approach yields a very small and innocuous value.

4. CONCLUSIONS

It has been argued in Refs. [2–4] that the interaction energy due to the exchange of massless or very light neutrinos in a neutron star would be enormous and would destabilize the star. We have addressed this claim by performing an explicit non-perturbative calculation of the interaction energy. We find that, once properly resummed, the interaction energy is a finite and well-behaved quantity which is far too small to have any effect on the fate of a neutron star. Thus, contrary to previous claims, the stability of neutron stars places no lower bound on the mass of the neutrino.

We are indebted to R. Jaffe for many insightful and helpful conversations during the course of this work. This research was supported in part by the U.S. Department of Energy under contract number DE-AC02-76CH00016. K.K. is also supported in part by the Natural Sciences and Engineering Research Council of Canada. K.K. also wishes to thank the conference organizers for partial support.

REFERENCES

1. K. Kiers and M.H.G. Tytgat, Phys. Rev. D **57**, 5970 (1998).
2. E. Fischbach, Ann. of Phys. **247**, 213 (1996).
3. B. Woodahl, M. Parry, S.-J. Tu and E. Fischbach, hep-ph/9709334.
4. E. Fischbach and B. Woodahl, hep-ph/9801387.
5. As. Abada, M.B. Gavela and O. Pène, Phys. Lett. B **387**, 315 (1996).
6. In order to simplify our calculation, we consider only the case $m_\nu = 0$, but by extension our result is also valid for very light – but not strictly massless – neutrinos.
7. A. Loeb, Phys. Rev. Lett. **64** (1990) 115.
8. A. Y. Smirnov and F. Vissani, hep-ph/9604443.
9. As. Abada, O. Pène and J. Rodriguez-Quintero, hep-ph/9712266.
10. M. Kachelriess, hep-ph/9712363.
11. As. Abada, O. Pène and J. Rodriguez-Quintero, hep-ph/9802393.
12. J. Arafune and Y. Mimura, hep-ph/9805395.
13. J. Schwinger, Phys. Rev. **94**, 1362 (1954).
14. P.D. Mannheim, Phys. Rev. **D 37**, 1935 (1988).
15. K. Kiers and N. Weiss, Phys. Rev. **D56**, 5776 (1997).
16. L. Wolfenstein, Phys. Rev. D **17** (1978) 2369; Phys. Rev. D **20** (1979) 2634;
S.P. Mikheyev and A.Yu. Smirnov, Yad. Fiz. **42** (1985) 1441 [Sov. J. Nucl. Phys. **42** (1985) 913]; Il Nuovo Cimento C **9** (1986) 17.

ELSEVIER

Nuclear Physics B (Proc. Suppl.) 77 (1999) 450–455

NUCLEAR PHYSICS B
PROCEEDINGS
SUPPLEMENTS

Neutrino Mass and Baryon Asymmetry*

H. Murayama[ab]

[a]Department of Physics, University of California, Berkeley, CA 94720

[b]Theoretical Physics Group, Lawrence Berkeley National Laboratory, Berkeley, CA 94720

The atmospheric neutrino data reported at this meeting strongly suggests neutrino oscillation, while the situation with the solar neutrino data is becoming more and more convincing as well. Both suggest tiny but finite neutrino masses. I first review why such neutrino masses are closely related to baryon-number violation in the early Universe. I also present a simple scheme of baryogensis, the Minimal Supersymmetric Leptogenesis, which generates the baryon asymmetry of desired order of magnitude with the mass range suggested by atmospheric and solar neutrino data, but not with the Hot Dark Matter (HDM) mass range. Finally I point out a dark horse HDM candidate (hadronic axion) which may be testable by SuperKamiokande with a galactic supernova.

1. Introduction

Just like everybody else attending this conference, I was impressed by the results on atmospheric neutrino by SuperKamiokande collaboration [1]. Even though the evidence for ν_μ oscillation needs to be established by other analyses and experiments, it made a very strong case for the finite mass of neutrinos. The situation with solar neutrino data is becoming more and more convincing as well; the data again suggests tiny but finite neutrino masses.

There is a strong prejudice among theorists: tiny but finite masses of neutrinos must be Majorana-type and hence imply lepton-number (L) violation. When further combined with the known effect of the standard model anomaly, lepton-number violation implies baryon-number (B) violation at high temperatures. Therefore, the origin of neutrino masses itself can even be the source of the cosmological baryon asymmetry: an idea called "leptogenesis." The range of neutrino masses suggested by the latest data, $m_\nu \sim 0.03$ eV from atmospheric neutrino data or $m_\nu \sim 0.003$ eV from solar neutrino data is a good match to the idea of the Minimal Supersymmetric Leptogenesis [2].

*This work was supported in part by the U.S. Department of Energy under Contracts DE-AC03-76SF00098, in part by the National Science Foundation under grant PHY-95-14797, and also by Alfred P. Sloan Foundation.

2. m_ν and $\not{B}$

The finite mass of the neutrino can come from two possible sources: Dirac and Majorana mass terms. Led by the principle of minimality, we do not want to introduce additional particles to our discussions, such as right-handed neutrinos, even though they may well exist at some high energy scale. The minimalist approach to introduce the finite neutrino masses to the Standard Model is the following non-renormalizable term in the Lagrangian:

$$\mathcal{L} = \frac{m_\nu}{2v^2}(LH)(LH) = \frac{1}{2M}(LH)(LH). \quad (1)$$

By substituting the vacuum expectation values for the Higgs boson, this term generates neutrino masses, which are necessarily of the Majorana-type and violate the L by two units. Here, the mass scale M refers to physics which generates this L-violating interaction.

On the other hand, at high temperatures $T \gtrsim 100$ GeV in the early Universe, W- and Z-bosons were massless. Then "electric" and "magnetic" fields of W and Z fields were fluctuating in the hot plasma similar to the electromagnetic fields. Certain changes in W fields is known to result in a change of the B and L by one unit, $\Delta B = \Delta L = 1$ keeping $B - L$ unchanged. This is a violation of $B + L$ number and is a consequence of the electroweak anomaly effect, or

PII S0920-5632(99)00466-1

sometimes called sphaleron effect, which I will explain shortly below. The most important point is that, as explained above, the Majorana neutrino mass violates the L, and hence the co-existence of the Majorana neutrino mass and the electroweak anomaly effect which violates $B + L$ actually violates B. Therefore, the total baryon number of the Universe was not constant in the early Universe. This can work in both ways: the effect can either erase the pre-existing baryon asymmetry or can create the baryon asymmetry if there is a deviation from thermal equlibrium and some form of CP violation.

Here is a cartoon-level explanation of the electroweak anomaly effect. In the background of the fluctuating W-electric and W-magnetic fields, we look at the spectrum of the quarks and leptons in their Dirac equations. The spectrum may look like the one shown in Fig. 1, where all negative energy states are filled in the "Dirac sea," while the positive energy states are vacant. Due to the fluctuating background W-field, all energy eigenvalues move up and down constantly. Once in a while, the fluctuation of the W-field can happen in a way such that all energy eigenvalues are shifted upwards by one unit. Then the resulting state again has all the negative energy states filled, most of the positive energy states vacant, but the lowest positive energy state is now filled! For an observer, this would appear as a creation of one particle. Since all quarks and leptons couple to the W-field in the exactly the same manner (quark-lepton universality of the weak interaction), if one lepton is created this way, one quark is also created the same way for each color. This results in $\Delta L = 1$, $\Delta Q = N_c = 3$. Since quarks carry baryon number $B = 1/3$, $\Delta B = 1$ and hence $B - L$ is conserved. Even though this explanation may appear just a cartoon, this is actually quite rigorous.

Now that we learnt that the baryon number is not conserved in the early Universe with finite neutrino masses, a consequence of this is that all pre-existing baryon asymmetry is washed out by the temperature

$$T \gtrsim 2.5 \times 10^{14}\ \text{GeV} \left(\frac{10^{-3}\ \text{eV}^2}{m_\nu^2} \right). \qquad (2)$$

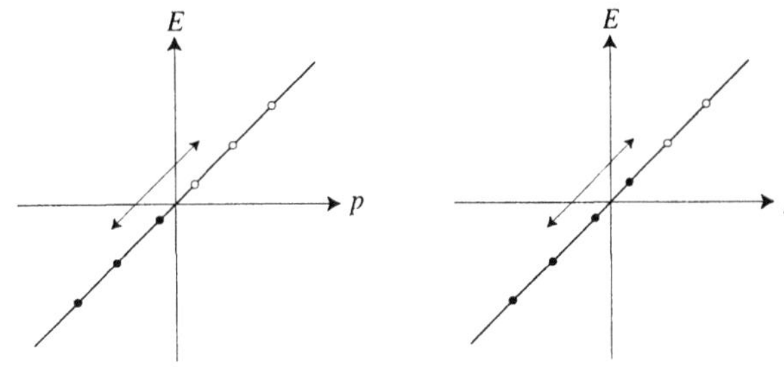

Figure 1. The energy levels of quarks, leptons in the fluctuating W-field background. All the negative energy states are filled in the Dirac sea originally in the left figure. The fluctuation of the W-field shifts all states up and down. As a result of the fluctuation, the spectrum can smoothly go over to the right-figure which is observed as a state with one particle created.

Therefore, the neutrino mass suggested by the atmospheric neutrino data reported at this conference requires the baryogenesis below this temperature, *i.e.*, well below the GUT-scale $\sim 2 \times 10^{16}$ GeV.

There is another reason why we want baryogensis to happen at a not-too-high temperature. Because the small neutrino mass $m_\nu \sim \sqrt{10^{-3}\ \text{eV}^2}$ suggests physics of high mass scale $M \sim 10^{14}$ GeV which is much higer than the electroweak scale $v \sim 174$ GeV, we need a mechanism to stabilize the hierarchy $v \ll M$. The most natural direction is to have supersymmetry. However, once supersymmetry is introduced, the superpartner of the graviton G, spin 3/2 gravitino $\tilde{G}$, causes a cosmological problem. The gravitino can be produced copiously in the early Universe (*e.g.*, $gg \to \tilde{g}\tilde{G}$). But gravitino has a long life time $\tau_{3/2} \simeq M_{Pl}^2/m_{\tilde{G}}^3$, and its decay product always contains the lightest supersymmetric particle (LSP) such as photino (*e.g.*, $\tilde{G} \to \gamma\tilde{\gamma}$).[2] The more gravitinos produced, the more photinos remain, and their energy density exceed the critical density. This fact limits the total production of

[2]Actually, current collider limits do not allow a pure photino to be the LSP. We use this notation just for simplicity.

gravitinos from above. Since the gravitino production rate is roughly proportional to the temperature, we find that the reheating temperature after the inflation T_{RH} should not exceed a certain limit, given by [3]

$$T_{RH} \leq 1 \times 10^{12} \text{ GeV} \left(\frac{h_0}{0.8}\right)^2 \left(\frac{20 \text{ GeV}}{m_{\tilde{\gamma}}}\right). \quad (3)$$

This is a valid formula for a relatively heavy gravitino ($m_{\tilde{G}} \gtrsim 10$ GeV).[3] This argument again suggests a need for baryogenesis not at a too-high temperature such as the GUT-scale.

Given these constraints, a simple and attractive possibility of baryogenesis is the following [4]. At a very high temperature above 10^{12} GeV, there may have been a baryon asymmetry in the Universe, but it is either washed out by the electroweak anomaly effect, or inflated away. The cosmological inflation resets the baryon asymmetry and the Universe starts anew at a reheating temperature of 10^{12} GeV. At a lower temperature (but still above the electroweak phase transition), somehow lepton asymmetry is generated. The mechanism of generating the lepton asymmetry is of course model-dependent, but can well be intimately related to the origin of the neutrino masses. Because of the electroweak anomaly effect, the generated lepton asymmetry is partially converted to the baryon asymmetry. Finally after the electroweak phase transition, the electroweak anomaly effect becomes unimportant and the baryon number becomes a conserved quantity. This is an idea called "leptogenesis." Since the B-violation enters only through the electroweak anomaly effect, the constraints from the proton decay can be easily evaded despite the energy scale of the baryogenesis lower than the GUT-scale. The original idea of leptogenesis employed the CP-violating decay of the right-handed neutrinos [4], which generate the L-violating operator (1). Even though this is still a possibility [5], it requires that the right-handed neutrinos to be produced after reheating, which is a non-trivial constraint on such a model. Below, I discuss a simpler possibility which does not depend on the details of the origin of neutrino masses.

[3]Lighter gravitino may decay very late and can potentially screw up the Big-Bang Nucleosynthesis. We do not consider this possibility here.

3. The Minimal Supersymmetric Leptogenesis

I now present a simpmle mechanism for leptogenesis based on the L-violating operator (1) which is needed to generate the neutrino mass and supersymmetry which is need to stabilize the hierarchy. There is only one assumption in the mechanism which I will state explicitly below. Then this is enough to generate the correct size of the baryon asymmetry with a range of neutrino masses suggested by atmospheric and solar neutrino data.

I start with an interesting peculiarity in the supersymmetric standard model. The superpartner of the lepton doublet, slepton doublet $\tilde{L}$, has exactly the opposite quantum numbers from the Higgs doublet H_u: both are electroweak doublets and the hypercharges are $-1/2$, $+1/2$, respectively. Because of the opposite quantum numbers and the specific form of supersymmetric Lagrangian, the scalar potential for them is proportional to $V \sim (|\tilde{L}|^2 - |H_u|^2)^2$. Therefore, the scalar potential is flat along the direction $\sqrt{2}\phi \equiv |\tilde{L}| = |H_u|$. Given the L-violating operator (1) needed for the neutrino mass, the superpotential is given by $W = \frac{1}{8M}\phi^4$, and the scalar potential by

$$V \sim m^2|\phi|^2 + \frac{m}{8M}(\phi^4 + \phi^{*4}) + \frac{|\phi|^6}{4M^2}. \quad (4)$$

Here, $m \sim 100$ GeV describes the effect of supersymmetry breaking. Note the symmetry of the scalar potential under the phase change $\phi \to e^{i\theta}\phi$ except the second term. This phase invariance is the lepton-number symmetry, which is violated by the second term.

Here comes the assumption: the fields along this direction ϕ had acquired a large amplitude ($\gtrsim 10^{13}$ GeV) with a random phase by the end of the inflation. Then the leptogenesis follows automatically. This assumption can be justified within various contexts, such as quantum fluctuation in de Sitter background, intial conditions motivated by chaotic inflation idea, negative mass

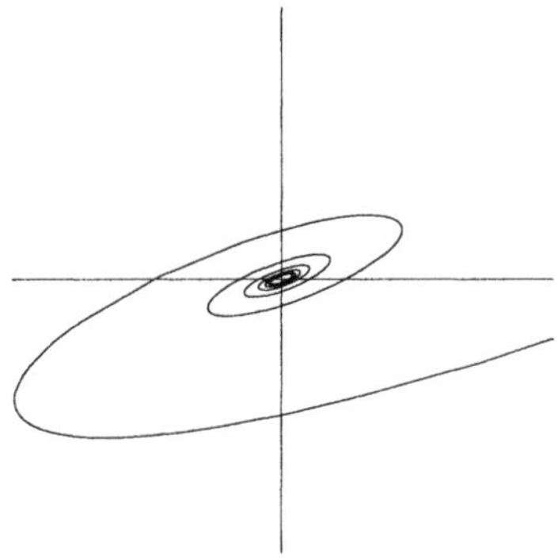

Figure 2. A motion of the flat direction on the complex plane. In this plot, the "ridges" lie along the horizontal and vertical axes, while the "valleys" along the diagonals.

squared during the inflation, and no-scale type supergravity during the inflation. It turns out that all of these different contexts predict more-or-less the same order of magnitude for the amplitude of the flat direction at the end of the inflation of order $\phi \sim 10^{14}$ GeV.

Once the flat direction ϕ has a large amplitude, one can picture the motion of the scalar field as a ball which is released on the slop of the potential (4). The potential is close to axially symmetric (phase invariance) but with small "valleys" and "ridges" due to the second term in (4). Because I assumed a random phase for the intial value of ϕ, the initial value is generically not exactly in the "valley." The downward slope from the "ridge" kicks the ball to give an "angular momentum" and the ball begins to rotate around the origin while its radius decreases due to the expansion of the Universe (see Fig. 2). The acquired angular momentum corresponds to the lepton number of the slepton field. What violates CP in this process is the initial random phase of the field ϕ.

Eventually the flat direction ϕ decays into relativistic particles with a finite lepton number. The electroweak anomaly effect partially converts the lepton number to the baryon number, and we find the baryon-to-photon ratio to be

$$\eta = \frac{n_B}{n_\gamma} = 2.46 Y_L$$
$$\simeq \quad 1 \times 10^{-10} \left(\frac{\sqrt{10^{-3}\ \mathrm{eV}^2}}{m_\nu} \right) \left(\frac{T_{RH}}{10^{12}\ \mathrm{GeV}} \right) \quad (5)$$

It is interesting that the range of neutrino masses suggested by atmospheric neutrino data $m_\nu^2 \sim 5 \times 10^{-4}$–$6 \times 10^{-3}$ eV^2 and solar neutrino data $m_\nu^2 \sim 10^{-5}$ eV^2 is in the right ballpark with the baryon-to-photon ratio required in the Big-Bang Nucleosynthesis $\eta = 4$–7×10^{-10} if the reheating temperature is high enough. And such a high reheating temperature implies a relatively high gravitino production rate which can give the correct order of magnitude for the LSP abudance from its decay product as discussed earlier (see Eq. (3)). This argument, however, clearly disfavors neutrino mass in the hot dark matter range $m_\nu \sim 4$ eV.[4]

4. Dark Horse HDM Candidate

Since the Minimal Supersymmetric Leptogenesis I have described disfavors neutrinos as hot dark matter candidate, it is a natural question if there could be any other possible candidates forthe HDM particle. As emphasized by Caldwell at this meeting [6], having a right mixture of the CDM and HDM gives the best fit to the current data of the CMBR anisotropy and large-scale structures. I'd like to point out a dark horse candidate to the HDM particle: a hadronic (or KSVZ) axion [7].

Hadronic axion is a special type of axion which does not couple to electron at the tree-level. This allows a relatively low decay constant despite the strong constraint on axion-electron coupling from white dwarfs. The remaining constraints are from axion-nucleon coupling from the duration of neutrino burst observed by Kamiokande and IMB from SN1987A, which leaves a window in the ax-

[4]Recall that the hot dark matter combined with atmospheric and solar neutrino data requires almost degenerate neutrinos of all three generations at 4 eV range and hence *none* of the neutrinos is as light as required in the leptogenesis scheme discussed here which requires at least one of the neutrinos to be in this range.

ion decay constant $3 \times 10^5 < f_a < 2 \times 10^6$ GeV not excluded. Finally, the axion-photon coupling cannot be too large to affect the constraints from red giants and the telescope search for UV rays from axion decay. However, the axion-photon coupling has two competing contributions which can in principle cancel with each other. If the axion-photon coupling is sufficiently suppressed, these constraints also leave the window not excluded (see Fig. 3). I refer to the presentation by Raffelt at this meeting [8] for further details. What had been not noticed is that the axion mass in this window is precily the correct one for the HDM, around 10 eV.[5]

An interesting possibility to close this window completely, and hence exclude the possibility of axionic HDM, is to study a galactic supernova with SuperKamiokande detector. The axion burst from the supernova would excite ^{16}O nuclei via M1 transition which leads to multiple photon signature from deexcitation.

5. Conclusion

The main point in the talk was that the finite neutrino mass, if Majorana-type as believed from theoretical prejudices, implies baryon-number non-conservation in the early Universe. Therefore, it is not unreasonable to expect that the neutrino mass has an intimate connection to the origin of the baryon asymmetry in the Universe. I presented one minimal scheme for such as possibility, the Minimal Supersymmetric Leptogenesis, which generates the required amount of the baryon asymmetry with neutrino mass in the range suggested by atmospheric and solar neutrino data. This scheme disfavors neutrino mass in the HDM range; an alternative candidate for the HDM is a hadronic axion, which can be studied with a possible axion burst from a supernova by SuperKamiokande experiment.

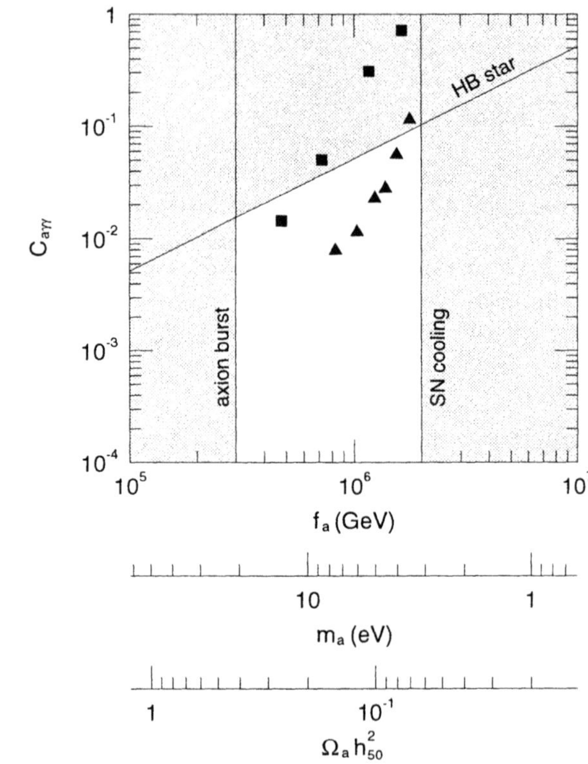

Figure 3. The constraint on the axion-photon coupling in the hadronic axion window. The axion mass of $m_a \sim 10$ eV gives the desired HDM density $\Omega_a \sim 0.2$ [7].

Acknowledgements. I thank Takeo Moroi who collaborated with me on most of the works described here. This work was supported in part by the U.S. Department of Energy under Contracts DE-AC03-76SF00098, in part by the National Science Foundation under grant PHY-95-14797, and also by Alfred P. Sloan Foundation.

[5]The hadronic axion abundance is not much smaller than neutrino abundance.

REFERENCES

1. T. Kajita, this proceedings.
2. T. Moroi and H. Murayama, in preparation.
3. T. Moroi, Ph.D. Thesis, hep-ph/9503210.
4. M. Fukugita and T. Yanagida, *Phys. Lett.* **174B**, 45 (1986).
5. See, *e.g.*, Q. Shafi, this proceedings.
6. D. Caldwell, this proceedings.
7. T. Moroi and H. Murayama, hep-ph/9804291, to appear in *Phys. Lett.* **B**.
8. G. Raffelt, this proceedings.

NUCLEAR PHYSICS B
PROCEEDINGS
SUPPLEMENTS

ELSEVIER

Nuclear Physics B (Proc. Suppl.) 77 (1999) 456–461

Axion Hunting at the Turn of the Millenium

Georg Raffelt

Max-Planck-Institut für Physik, Föhringer Ring 6, 80805 München, Germany

The status of several current and proposed experiments to search for galactic dark-matter and solar axions is reviewed in the light of astrophysical and cosmological limits on the Peccei-Quinn scale.

1. INTRODUCTION

Twenty years after their inception [1], axions [2] remain a popular solution to the strong CP problem as well as a candidate for the cold dark matter of the universe. As a research topic, axions and related issues seem to have retained much of their appeal (Fig. 1), and indeed 1998 could yet become the year with the largest number of axion research papers ever. All that is missing in this flurry of activities is the appearance of the main character, the axion itself, which thus far has eluded all attempts at a discovery.

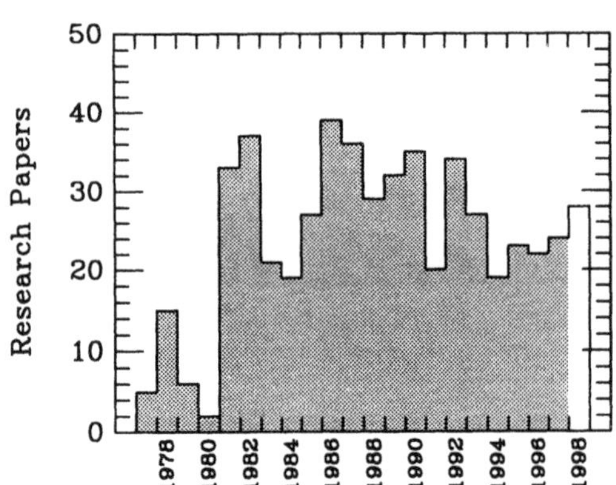

Figure 1. Original research papers in the SPIRES HEP data base at SLAC [3] with *axion*, *axino*, or *Peccei-Quinn* in their title. The date is when a preprint was recorded, not the year of publication in a journal. The papers of 1977 include those of Ref. [1] which do not use our keywords, and the ones for 1998 are only up to the end of June.

In the framework of "invisible axion" models where the scale f_a at which the Peccei-Quinn symmetry is spontaneously broken could be arbitrarily large, and where axions therefore could be arbitrarily weakly interacting, previous experiments really did not stand a plausible chance of finding these particles. Therefore, the most exciting recent development is that there are now two full-scale search experiments for galactic dark-matter axions in operation which do have a realistic discovery potential. Also, there is a surprising amount of activity around the search for solar axions which is beginning to become competitive with astrophysical limits. The chance of an actual discovery, however, appears more remote than in the search for galactic axions.

2. ASTROPHYSICAL LIMITS

Axion models are characterized by the Peccei-Quinn scale f_a, or equivalently by the axion mass $m_a = 0.60$ eV $(10^7\ \text{GeV}/f_a)$. Several astrophysical lower limits on f_a (Fig. 2) are based on the requirement that the axionic energy loss of stars, notably globular-cluster stars or the core of supernova (SN) 1987A, is not in conflict with certain observed properties of these objects [4,5]. These limits imply $m_a \lesssim 10^{-2}$ eV or $f_a \gtrsim 10^9$ GeV, indicating that axions, if they exist, are both extremely light and very weakly interacting.

These limits on the axion mass are indirectly derived from limits on the coupling strength to photons (globular cluster stars) and nucleons (SN 1987A). The axionic two-photon interaction is $\mathcal{L}_{\text{int}} = g_{a\gamma}\mathbf{E}\cdot\mathbf{B}\,a$, where

$$g_{a\gamma} = -\frac{3\alpha}{8\pi f_a}\,\xi = -\frac{m_a/\text{eV}}{0.69\times 10^{10}\ \text{GeV}}\,\xi \qquad (1)$$

PII S0920-5632(99)00468-5

with

$$\xi \equiv \frac{4}{3}\left(\frac{E}{N} - 1.92 \pm 0.08\right). \qquad (2)$$

E/N is a model-dependent ratio of small integers. In the DFSZ model or GUT models one has $E/N = 8/3$, corresponding to $\xi \approx 1$, and it is this case for which the globular-cluster limit

$$g_{a\gamma} \lesssim 0.6 \times 10^{-10}\ \mathrm{GeV}^{-1} \qquad (3)$$

is shown in Fig. 2 as a limit on the axion mass, $m_a \lesssim 0.4$ eV. The axion-photon coupling for a variety of models has recently been compiled [6]; often-discussed cases are $E/N = 8/3$ (DFSZ) or $E/N = 0$ (KSVZ).

However, models with $E/N = 2$ can be constructed, allowing for a near or complete cancellation of $g_{a\gamma}$. In this case there is no globular-cluster limit on m_a or f_a so that there is a small window for m_a near 10 eV. It was recently shown [7] that in this range axions could be a cosmological hot dark matter component which certain structure-formation arguments suggest in addition to the main cold dark matter.

In the early universe, axions are thermalized if $f_a \lesssim 10^8\,\mathrm{GeV}$ [8], a region excluded by the stellar-evolution limits except for the special case $E/N = 2$. If inflation occurred after the Peccei-Quinn symmetry breaking or if $T_{\mathrm{reheat}} < f_a$, the "misalignment mechanism" [9] leads to a contribution to the cosmic critical density of $\Omega_a h^2 \approx 1.9 \times 3^{\pm 1}\,(1\,\mu\mathrm{eV}/m_a)^{1.175}\,\Theta_{\mathrm{i}}^2 F(\Theta_{\mathrm{i}})$ where h is the Hubble constant in units of $100\,\mathrm{km\,s^{-1}\,Mpc^{-1}}$. The function $F(\Theta)$ with $F(0) = 1$ and $F(\pi) = \infty$ accounts for anharmonic corrections to the axion potential. Because the initial misalignment angle Θ_{i} can be very small or very close to π, there is no real prediction for the mass of dark-matter axions even though one would expect $\Theta_{\mathrm{i}}^2 F(\Theta_{\mathrm{i}}) \sim 1$.

A possible fine-tuning of Θ_{i} is limited by inflation-induced quantum fluctuations which in turn lead to temperature fluctuations of the cosmic microwave background [10,11]. In a broad class of inflationary models one thus finds an upper limit to m_a where axions could be the dark matter. According to the most recent discussion [11] it is about 10^{-3} eV (Fig. 2).

If inflation did not occur at all or if it occurred before the Peccei-Quinn symmetry breaking with $T_{\mathrm{reheat}} > f_a$, cosmic axion strings form by the Kibble mechanism [12]. Their motion is damped

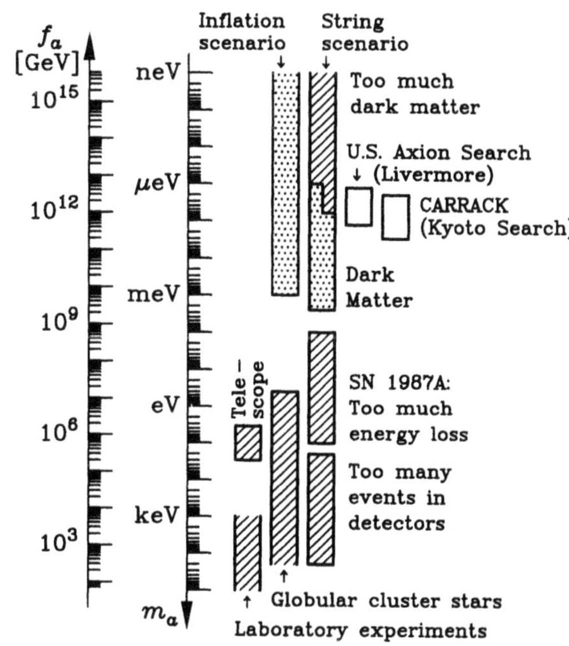

Figure 2. Astrophysical and cosmological exclusion regions (hatched) for the axion mass m_a or equivalently, the Peccei-Quinn scale f_a. An "open end" of an exclusion bar means that it represents a rough estimate; its exact location has not been established or it depends on detailed model assumptions. The globular cluster limit depends on the axion-photon coupling; it was assumed that $E/N = 8/3$ as in GUT models or the DFSZ model. The SN 1987A limits depend on the axion-nucleon couplings; the shown case corresponds to the KSVZ model and approximately to the DFSZ model. The dotted "inclusion regions" indicate where axions could plausibly be the cosmic dark matter. Most of the allowed range in the inflation scenario requires fine-tuned initial conditions. In the string scenario the plausible dark-matter range is controversial as indicated by the step in the low-mass end of the "inclusion bar."

primarily by axion emission rather than gravitational waves. After axions acquire a mass at the QCD phase transition they quickly become nonrelativistic and thus form a cold dark matter component. The axion density such produced is similar to that from the misalignment mechanism for $\Theta_i = \mathcal{O}(1)$, but in detail the calculations are difficult and somewhat controversial between two groups of authors [13,14]. Taking into account the uncertainty in various cosmological parameters one arrives at a plausible range for dark-matter axions as indicated in Fig. 2.

3. DARK MATTER SEARCH

If axions are the galactic dark matter one can search for them in the laboratory. The detection principle is analogous to the Primakoff process for neutral pions, i.e. the two-photon vertex allows for axion transitions into photons in the presence of an external electromagnetic field (Fig. 3). Dark matter axions would have a mass in the μeV to meV range. As they are bound to the galaxy their velocity dispersion is of order the galactic virial velocity of around $10^{-3}c$ so that their kinetic energy is extremely small relative to their rest mass. Noting that a frequency of 1 GHz corresponds to 4 μeV, the Primakoff conversion produces microwaves. Galactic axions are nonrelativistic while the resulting photons are massless so that the conversion involves a huge momentum mismatch which can be overcome by looking for the appearance of excitations of a microwave cavity rather than for free photons.

An axion search experiment thus consists of a high-Q microwave resonator placed in a strong external magnetic field ("axion haloscope" [15]).

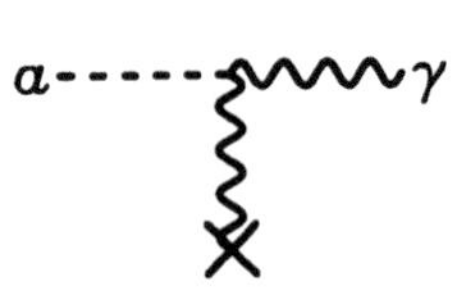

Figure 3. Primakoff conversion of axions into photons in an external electromagnetic field.

The microwave power output of such a detector on resonance is [15,16]

$$\begin{aligned} P &\approx 0.4\times 10^{-22}\,\text{Watts}\left(\frac{V}{0.2\,\text{m}^3}\right) \\ &\times \left(\frac{B}{7.7\ \text{Tesla}}\right)^2\left(\frac{C}{0.65}\right)\left(\frac{Q}{10^5}\right) \\ &\times \left(\frac{\rho_a}{300\ \text{MeV cm}^{-3}}\right)\left(\frac{m_a}{1\ \mu\text{eV}}\right), \end{aligned} \tag{4}$$

where V is the cavity volume, B the applied magnetic field, C a mode-dependent form factor which is largest for the fundamental T_{010} mode, Q the loaded quality factor, and ρ_a the local galactic axion density. If m_a were known it would be easy to detect galactic axions with this method—one may verify or reject a tentative signal by varying, for example, the applied magnetic field strength. Therefore, it would be hard to mistake a background signal for dark-matter axions. The problem is, of course, that m_a is not known so that one needs a tunable cavity, stepping its resonance through as large a frequency range as possible and to look for the appearance of microwave power beyond thermal and amplifier noise.

Two pilot experiments of this sort [17,18] have excluded the range of axion masses and coupling strengths indicated in Fig. 4. For a standard local halo density of about 300 MeV cm^{-3} they were not sensitive enough to reach realistic axion models. Two current experiments with larger cavities, however, have the requisite sensitivity.

The U.S. Axion Search [19] uses conventional microwave amplifiers (HEMTs) which limit the useful cavity temperature to about 1.4 K. A first exclusion slice has been reported [19]—see Fig. 4 where the ultimate search goal is also shown. In a next-generation experiment one would use SQUID amplifiers, increasing the sensitivity to encompass more weakly coupled axion models.

The Kyoto experiment CARRACK [20], on the other hand, uses a completely novel detection technique, based on the excitation of a beam of Rydberg atoms which passes through the cavity. This is essentially a counting method for microwaves which does not require a (noisy) amplifier so that one can go to much lower physical cavity temperatures. This enhances the sensitivity and also allows one to use smaller cavity volumes and thus to search for larger axion masses. With the current setup a narrow slice of axion masses is to be searched (Fig. 4), while a new apparatus currently under construction will allow for the coverage of a much broader mass range.

The search goals of these second-generation experiments covers the lower range of plausible axion masses in the framework of the cosmological string-scenario of primordial axion production, and a significant portion of the plausible mass range in the inflation scenario if one does not wish to appeal to fine-tuned initial conditions of the axion field (Fig. 2). If these experiments fail to turn up axions, it would be extremely important to extend the experimental search into a regime of larger masses toward the meV scale. This would require new detection methods.

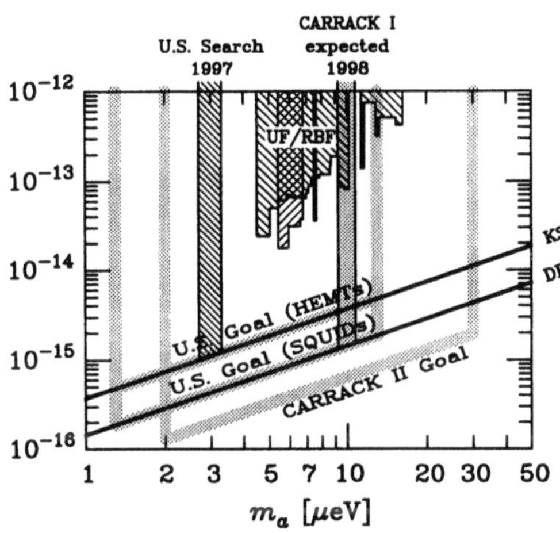

Figure 4. Limits on galactic dark matter axions from the University of Florida (UF) [17] and the Rochester-Brookhaven-Fermilab (RBF) [18] pilot experiments and the recent limit from the U.S. Axion Search [19]. Also shown are the search goals for the U.S. experiment employing HEMTs for microwave detection, for a next generation experiment using SQUIDs, the 1998 search goal for CARRACK I (Kyoto) and for CARRACK II, both using Rydberg atoms.

4. SOLAR AXIONS

4.1. Helioscope Method

Another classic way to search for axions is to use the Sun as a source and to attempt an experimental detection of this flux. Unfortunately, the experimental sensitivity typically lies in an f_a range which is already excluded by the stellar-evolution limits of Fig. 2 so that one needs to appeal to large systematic uncertainties of the astrophysical bounds in order to hope for a positive detection. On the other hand, such experiments can provide independent limits on the parameters of axions and similar particles even if the chances for a positive detection seem slim.

In the so-called "helioscope" method [15,21] one again uses the Primakoff effect (Fig. 3) by pointing a long and strong dipole magnet toward the Sun. The axions produced in the hot interior of the Sun would have typical energies of a few keV and would thus convert into x-rays which can then be picked up by a detector at the downstream end of the magnet. A pioneering experiment was conducted several years ago [22], but detecting axions would have required a flux larger than what is compatibel with the solar age.

Recently, first results were reported from the Tokyo axion helioscope where a dipole magnet was gimballed like a telescope so that it could follow the Sun and thus reach a much larger exposure time [23]. The limit on the axion-photon coupling of $g_{a\gamma} \lesssim 6 \times 10^{-10}\ \mathrm{GeV}^{-1}$ is less restrictive than the globular-cluster limit of Eq. (3), but more restrictive than the solar-age limit of $25 \times 10^{-10}\ \mathrm{GeV}^{-1}$ [24], and also more restrictive than a recent solar limit of about $10 \times 10^{-10}\ \mathrm{GeV}^{-1}$ which is based on helioseismological sound-speed profiles of the Sun [25].

Another helioscope project with a gimballed dipole magnet was begun in Novosibirsk several years ago [26], but its current status has not been reported for some time. A very intruiging project at CERN would use a decommissioned LHC test magnet that could be mounted on a turning platform to achieve reasonably long times of alignment with the Sun [27]. With this setup one would begin to compete with the globular cluster limit of Eq. (3).

The helioscope approach is bedevilled by the same problem which requires the use of a resonant cavity in the galactic axion search, viz. the momentum mismatch between (massive) axions and (massless) photons in the Primakoff process. For example, the above limit of the Tokyo helioscope applies only for $m_a \lesssim 0.03$ eV, implying that the "axion-line"—the relationship between $g_{a\gamma}$ and m_a of Eq. (1)—is not even touched, i.e. the limit applies only to particles which for a given $g_{a\gamma}$ have a smaller mass than true axions.

In a next step one will fill the transition region with a pressurized gas, giving the photon a dispersive mass in order to overcome the momentum mismatch [21]. As in the cavity experiments, this is a resonant method (the match is only good for a small range of axion masses) so that one needs to take many runs with varying gas pressure to cover a broad m_a range. In this way it is hoped to eventually cut across the axion line. The same approach would have to be used for the proposed CERN helioscope.

4.2. Bragg Diffraction

An alternative method to overcome the momentum-mismatch problem in the Primakoff process is to use an inhomogeneous external electromagnetic field which has strong Fourier components for the required momentum transfer. It has been suggested to use the strong electric fields of a crystal lattice for this purpose [28]. In practice one can use germanium detectors which were originally built to search for neutrinoless double-beta decay and for WIMP dark matter. The Ge crystal serves simultaneously as a "transition agent" between solar axions and x-rays and as an x-ray detector. The beauty of this method is that one can piggy-back on the existing Ge experiments, provided one determines the absolute orientations of the crystal axes relative to the Sun because the expected conversion rate depends on the lattice orientation in analogy to Bragg diffraction.

A first limit produced by the SOLAX Collaboration [29] of $g_{a\gamma} \lesssim 30 \times 10^{-10}$ GeV^{-1} is not yet self-consistent as the properties of the Sun already require $g_{a\gamma} \lesssim 10 \times 10^{-10}$ GeV^{-1}. However, the limit easily cuts across the axion line (it applies for $m_a \lesssim 1$ keV), and no doubt it can be significantly improved as $\beta\beta$ and WIMP search experiments grow in size and exposure time.

4.3. Mössbauer Absorption

If axions essentially decouple from photons for $E/N = 2$ models, and if they also do not couple to electrons at tree level, there is a small window of allowed axion masses in the neighborhood of 10 eV (Fig. 2). One can search for axions in this range by appealing only to their coupling to nucleons. The Sun would emit a nearly monochromatic 14.4 keV axion line from thermal transitions between the first excited and ground state of ^{57}Fe which is quite abundant in the Sun. In the laboratory one can then search for the axion absorption process which would give rise to x-rays as ^{57}Fe de-excites [30]. Of course, the Doppler broadening of the line in the Sun of about 5 eV is much larger than the natural line width of order 10 neV so that the Mössbauer absorber in the laboratory picks up only a small fraction of the total flux. Even so it may be possible to detect or significantly constrain solar axions in an experiment which is now in preparation in Tokyo [31]. A recent pilot experiment by another group did not have enough sensitivity to find axions in the above window [32].

5. SUMMARY

A surprisingly large number of experiments to search for solar and galactic dark-matter axions have recently emerged. The U.S. Axion Search as well as the Kyoto experiment CARRACK have now reached a sensitivity where they could realistically detect galactic dark matter axions, surely an important step because the role of axions as an alternative to supersymmetric particles as a cold dark matter candidate is perhaps the most important aspect of the continuing interest in axion physics. As it stands, axion dark matter could well show up before the millenium ends!

ACKNOWLEDGMENTS

Partial support by the Deutsche Forschungsgemeinschaft under grant No. SFB-375 is acknowledged.

REFERENCES

1. R.D. Peccei and H.R. Quinn, Phys. Rev. Lett. **38**, 1440 (1977); Phys. Rev. D **16**, 1791 (1977). S. Weinberg, Phys. Rev. Lett. **40**, 223 (1978). F. Wilczek, ibid. 279.
2. J.E. Kim, Phys. Rept. **150**, 1 (1987). H.-Y. Cheng, Phys. Rept. **158**, 1 (1988). M.S. Turner, Phys. Rept. **197**, 67 (1990). G.G. Raffelt, Phys. Rept. **198**, 1 (1990).
3. http://www-spires.slac.stanford.edu/find /spires.html
4. G. Raffelt, *Stars as Laboratories for Fundamental Physics* (University of Chicago Press, Chicago, 1996).
5. G. Raffelt, Mini Review in *The Review of Particle Physics*, C. Caso et al., The European Physical Journal **C3**, 1 (1998). See also http://pdg.lbl.gov/
6. J.E. Kim, hep-ph/9802220.
7. T. Moroi and H. Murayama, hep-ph/9804291.
8. M. Turner, Phys. Rev. Lett. **59**, 2489 (1987).
9. J. Preskill, M. Wise and F. Wilczek, Phys. Lett. B **120**, 127 (1983). L. Abbott and P. Sikivie, ibid. 133. M. Dine and W. Fischler, ibid. 137. M.S. Turner, Phys. Rev. D **33**, 889 (1986).
10. D.H. Lyth, Phys. Lett. B **236**, 408 (1990). M.S. Turner and F. Wilczek, Phys. Rev. Lett. **66**, 5 (1991). A. Linde, Phys. Lett. B **259**, 38 (1991).
11. E.P.S. Shellard and R.A. Battye, "Inflationary axion cosmology revisited", in preparation (1998); the main results can be found in: E.P.S. Shellard and R.A. Battye, astro-ph/9802216.
12. R.L. Davis, Phys. Lett. B **180**, 225 (1986). R.L. Davis and E.P.S. Shellard, Nucl. Phys. B **324**, 167 (1989).
13. R.A. Battye and E.P.S. Shellard, Nucl. Phys. B **423**, 260 (1994); Phys. Rev. Lett. **73**, 2954 (1994); (E) ibid. **76**, 2203 (1996); astro-ph/9706014, to be published in: Proc. Dark Matter 96, Heidelberg, ed. by H.V. Klapdor-Kleingrothaus and Y. Ramacher.
14. D. Harari and P. Sikivie, Phys. Lett. B **195**, 361 (1987). C. Hagmann and P. Sikivie, Nucl. Phys. B **363**, 247 (1991).
15. P. Sikivie, Phys. Rev. Lett. **51**, 1415 (1983); (E) ibid. **52**, 695 (1984).
16. L. Krauss, J. Moody, F. Wilczek and D. Morris, Phys. Rev. Lett. **55**, 1797 (1985).
17. C. Hagmann, P. Sikivie, N.S. Sullivan, D.B. Tanner, Phys. Rev. D **42**, 1297 (1990).
18. W.U. Wuensch et al., Phys. Rev. D **40**, 3153 (1989).
19. C. Hagmann et al., Phys. Rev. Lett. **80**, 2043 (1998).
20. S. Matsuki, I. Ogawa and K. Yamamoto, Phys. Lett. B **336**, 573 (1994). I. Ogawa, S. Matsuki and K. Yamamoto, Phys. Rev. D **53**, R1740 (1996). S. Matsuki et al., Nucl. Phys. B (Proc. Suppl.) **51B**, 213 (1996).
21. K. van Bibber, D. Morris, P. McIntyre, and G. Raffelt, Phys. Rev. D **39**, 2089 (1989).
22. D.M. Lazarus et al., Phys. Rev. Lett. **69**, 2333 (1992).
23. S. Moriyama et al., hep-ex/9805026, submitted to Phys. Lett. B. S. Moriyama, *Direct Search for Solar Axions by Using Strong Magnetic Field and X-Ray Detectors*, Ph.D. Thesis (in English), University of Tokyo, 1998.
24. G. Raffelt and D. Dearborn Phys. Rev. D **36**, 2211 (1987).
25. H. Schlattl, A. Weiss and G. Raffelt, in preparation, 1998.
26. P.V. Vorob'ev and I.V. Kolokolov, astro-ph/9501042, unpublished.
27. K. Zioutas et al., astro-ph/9801176, submitted to Nucl. Instrum. Meth. A.
28. W. Buchmüller and F. Hoogeveen, Phys. Lett. B **237**, 278 (1990). E.A. Paschos and K. Zioutas, Phys. Lett. B **323**, 367 (1994). R.J. Creswick et al., hep-ph/9708210, submitted to Phys. Lett. B.
29. F.T. Avignone III et al. (SOLAX Collaboration), astro-ph/9708008, submitted to Phys. Lett. B. A.O. Gattone et al. (SOLAX Collaboration), astro-ph/9712308, Proc. TAUP 97.
30. S. Moriyama, Phys. Rev. Lett. **75**, 3222 (1995).
31. S. Moriyama and M. Minowa, private communication.
32. M. Krčmar et al., nucl-ex/9801005.

Part 13

Ultra-High Energy Neutrinos

ELSEVIER

Nuclear Physics B (Proc. Suppl.) 77 (1999) 465–473

NUCLEAR PHYSICS B
PROCEEDINGS
SUPPLEMENTS

High energy neutrino astrophysics

R.J. Protheroe[a]

[a]Department of Physics and Mathematical Physics
The University of Adelaide, Adelaide, Australia 5005

I give a brief discussion of possible sources of high energy neutrinos of astrophysical origin over the energy range from $\sim 10^{12}$ eV to $\sim 10^{25}$ eV. In particular I shall review predictions of the diffuse neutrino intensity. Neutrinos from interactions of galactic cosmic rays with interstellar matter are guaranteed, and the intensity can be reliably predicted to within a factor of 2. Somewhat less certain are intensities in the same energy range from cosmic rays escaping from normal galaxies or active galactic nuclei (AGN) and interacting with intracluster gas. At higher energies, neutrinos will definitely be produced by interactions of extragalactic cosmic rays with the microwave background. With the discovery that gamma ray bursts (GRB) are extragalactic, and therefore probably the most energetic phenomena in the Universe, it seems likely that they will be copious sources of high energy neutrinos. Other sources, such as AGN and topological defects, are more speculative. However, searches for neutrinos from all of these potential sources should be made because their detection would have important implications for high energy astrophysics and cosmology.

1. INTRODUCTION

The technique for constructing a large area (in excess of 10^4 m^2) neutrino telescope has been known for more than two decades [1]. The pioneering work of the DUMAND Collaboration led to the development of techniques to instrument a large volume of water in a deep ocean trench with strings of photomultipliers to detect Cherenkov light from neutrino-induced muons [2]. Locations deep in the ocean shield the detectors from cosmic ray muons. The second generation of high energy neutrino telescope such as AMANDA [3] located deep in the polar ice cap at the South Pole, and NT 200 in operation in Lake Baikal, Siberia [4], have demonstrated the feasibility of constructing large area experiments for high energy neutrino astronomy. The next generation telescopes, such as the planned extension of AMANDA, ICECUBE [5], and ANTARES [6], may have effective areas of 0.1 km^3, or larger, and be sufficiently sensitive to detect bursts of neutrinos from extragalactic objects and to map out the spectrum of the diffuse high energy neutrino background. In this paper I focus on possible astrophysical sources of neutrinos contributing to the diffuse high energy neutrino background from $\sim$ 1 TeV to the GUT scale.

2. COSMIC RAY INTERACTIONS WITH MATTER

There will definitely exist a diffuse galactic neutrino background due to interactions of the galactic cosmic rays with interstellar matter. The spectrum of cosmic rays is reasonably well known, as is the matter distribution in our galaxy. Estimates of the neutrino intensity have been made by Silberberg and Shapiro [7], Stecker [8], Domokos et al. [9], Berezinsky et al. [10], and Ingelman and Thunman [11], and the more recent predictions are shown in Fig. 1. The differences of about a factor of 2 between the predictions are accountable in terms of the slightly different models of the interstellar matter density, and cosmic ray spectrum and composition used. Also shown is the atmospheric neutrino background as estimated by Lipari [12]. In addition, there will be a very uncertain background (not plotted) due to charm production (see refs. [13,14] for a survey of predictions).

Somewhat less certain is the flux of neutrinos from clusters of galaxies. This is produced by *pp* interactions of high energy cosmic rays with intracluster gas. Berezinsky et al. [15] have made predictions of this, and I show in Fig. 2 their estimates of the diffuse neutrino intensity due

0920-5632/99/$ – see front matter
PII S0920-5632(99)00469-7

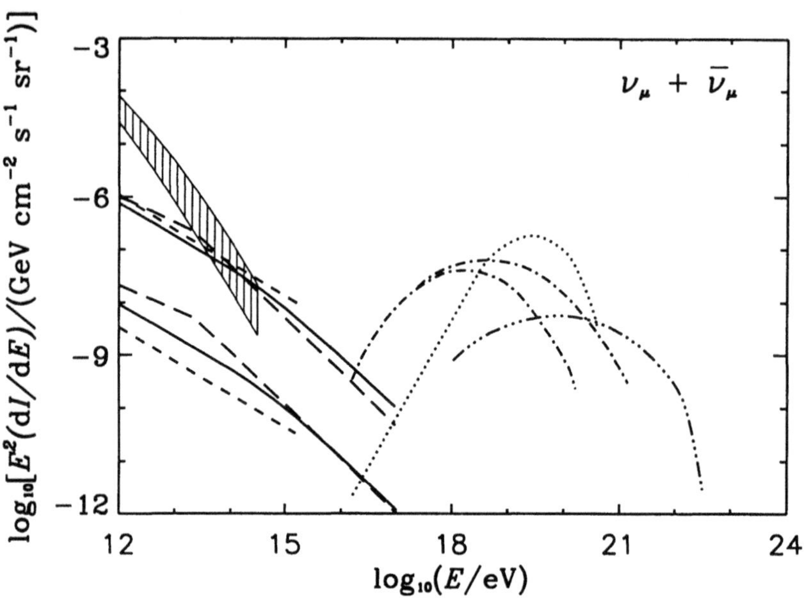

Figure 1. Neutrinos from cosmic ray interactions with the interstellar medium (upper curves for $\ell = 0°$, $b = 0°$, lower curves for $b = 90°$): — — — Domokos et al. [9]; - - - - Berezinsky et al. [10]; ——— Ingelman and Thunman [11]. The band with vertical hatching shows the range of atmosheric neutrino background [12] as the zenith angle changes from 90° (highest) to 0° (lowest). Neutrinos from cosmic ray interactions with the microwave background: — · — · — · — Protheroe and Johnson [23] for $E_{\rm max} = 3 \times 10^{20}$ eV and 3×10^{21} eV; ······ Hill and Schramm [24]; — · · · — · · · — · · · — assuming the highest energy cosmic rays are due to GRB according to Lee [25].

to interactions of cosmic rays produced by normal galaxies and AGN together with an upper limit based on assuming the observed γ-ray background results from π^0 production. Later estimates by Colafrancesco and Blasi [16] are also shown.

3. COSMIC RAY INTERACTIONS WITH RADIATION

Moving to higher energies, cosmic rays above $\sim 10^{20}$ eV will interact with photons of the cosmic microwave background radiation (CMBR) [17,18]. Again, we know that both ingredients exist (the highest energy cosmic ray detected has an energy of 3×10^{20} eV [19], and at least 6 cosmic rays have been detected above 10^{20} eV by the AGASA array [20]), and so pion photoproduction at these energies will occur, resulting in a diffuse neutrino background (Stecker [8]). However, the intensity in this case is model-dependent because it is not certain precisely what the origin of the highest energy cosmic rays is, and whether in fact they are extragalactic, although this seems very probable (see [21] for a discussion of the highest energy cosmic rays). One of the most likely explanations of the highest energy cosmic rays is acceleration in Fanaroff-Riley Class II radio galaxies as suggested by Rachen and Biermann [22]. Protheroe and Johnson [23] have repeated Rachen and Biermann's calculation in order to calculate the flux of diffuse neutrinos and γ-rays which would accompany the UHE cosmic rays, and their result has been added to Fig. 1. Any model in which the cosmic rays above 10^{20} eV are of extragalactic origin will predict a high energy diffuse neutrino intensity probably within an order of magnitude of this at 10^{19} eV. For example, I show an earlier estimate by Hill and Schramm [24]. Also shown is an estimate by Lee [25] of the

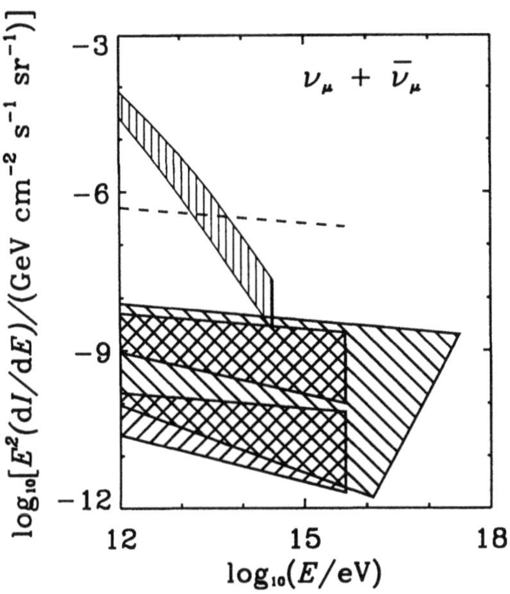

Figure 2. Neutrinos from cosmic ray interactions in clusters of galaxies (Berezinsky et al. [15]): lower hatched area – normal galaxies; upper hatched area – AGN; dashed line – upper bound from γ-ray data. Large hatched region – comic ray interactions with intergalactic medium in clusters of galaxies [16]

diffuse neutrino intensity estimated in a model in which the highest energy cosmic rays have their origin in sources of gamma ray bursts.

4. GAMMA RAY BURSTS

Gamma ray bursts (GRB) are observed to have non-thermal spectra with photon energies extending to MeV energies and above. Recent identification of GRB with galaxies at large redshifts (e.g. GRB 971214 at $z = 3.42$ [26]) show that the energy output in γ-rays alone from these objects can be as high as 3×10^{53} erg if the emission is isotropic, making these the most energetic events in the Universe. GRB 980425 has been identified with an unusual supernova in ESO 184-G82 at a redshift of $z = 0.0085$ implying an energy output of 10^{52} erg [27]. These high energy outputs, combined with the short duration and rapid variability on time-scales of milliseconds, require highly relativistic motion to allow the MeV photons to escape without severe photon-photon pair production losses. The energy sources of GRB may be neutron star mergers with neutron stars or with black holes, collapsars associated with supernova explosions of very massive stars, hyper-accreting black holes, hypernovae, etc. (see [28,29] for references to these models).

The relativistic fireball model of GRB (Meszaros and Rees [30]) provides the framework for estimation of neutrino fluxes from GRB. A relativistic fireball sweeps up mass and magnetic field, and electrons are energized by shock acceleration and produce the MeV γ-rays by synchrotron radiation. Protons will also be accelerated, and may interact with the MeV γ-rays producing neutrinos via pion photoproduction and subsequent decay at energies above $\sim 10^{14}$ eV [31,33]. Acceleration of protons may also take place to energies above 10^{19} eV, producing a burst of neutrinos at these energies by the same process [34]. These energetic protons may escape from the host galaxy to become the highest energy cosmic rays [35,36]. Additional neutrinos due to interactions of the highest energy cosmic rays with the CMBR will be produced as discussed in the previous section.

For a sufficiently intense GRB, it may be possible to identify neutrinos from individual GRB. Integrating over all GRB in the Universe, Waxman and Bahcall [31,32] have predicted the diffuse neutrino intensity, and this has been plotted in Fig. 3 with a steepening at 10^{16} eV, and with a continuation to higher energies as suggested by Vietri [34].

5. ACTIVE GALACTIC NUCLEI

The 2nd EGRET catalog of high-energy γ-ray sources [37] contains over 40 high confidence identifications of AGN, and all appear to be blazars (radio-loud AGN having emission from a relativistic jet closely aligned to our line of sight). Since the publication of the 2nd catalog, the number of blazars detected by EGRET has increased to nearly 70 (see refs. [38,39] for reviews). TeV emission has been observed from three blazars, the BL Lac Objects Mrk 421, Mrk 501 and 1ES 2344+514 [40]. Clearly, the γ-ray emission is as-

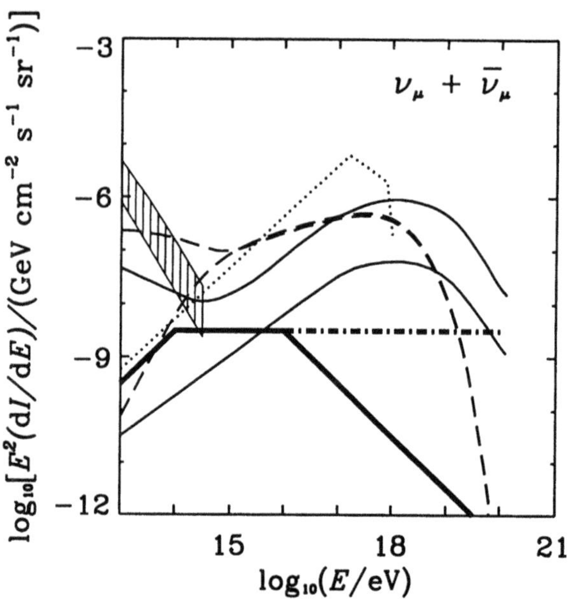

Figure 3. Diffuse neutrinos from GRB: Thick solid line Waxman & Bahcall [31], with extension to higher energies suggested by Vietri [34] (thick chain line). Diffuse $p\gamma$ (lower) and $pp + p\gamma$ (upper) neutrinos from blazars: ——— Mannheim [48] ($p\gamma$) Model A (lower curve), ($p\gamma$+pp) Model B (upper curve); - - - - - Protheroe [46] ×25% - see text; $\cdots\cdots$ $p\gamma$ Halzen & Zas [49].

sociated with AGN jets. Blazars appear also to be able to explain about 25% of the diffuse γ-ray emission [41], and models where γ-ray emission does not originate in the jet are unlikely to contribute significantly to the diffuse γ-ray (and neutrino) intensity (see Protheroe and Szabo [42] and references therein for predictions for non-blazar AGN). Several of the EGRET AGN show γ-ray variability with time scales of $\sim$ 1 day [43] at GeV energies, and variability on time scales of $\sim$ 1 hour or less [44,45] has been observed at $\sim$ 1 TeV for some BL Lacs. These variability timescales place important constraints on the models, and not all models developed so far are consistent with this. I shall survey the neutrino emission predicted in blazar models irrespective of this, assuming they may be made to accommodate the latest variability measurements.

Most theoretical work on γ-ray emission in AGN jets involved electron acceleration and inverse Compton scattering, and these models will predict no neutrinos. In proton blazar models, protons are accelerated instead of, or as well as, electrons. In this case interactions of protons with matter or radiation would lead to neutrino production. In some of the proton blazar models energetic protons interact with radiation via pion photoproduction (see e.g. [21] for references and a discussion of $p\gamma$ interactions). This radiation may be reprocessed or direct accretion disk radiation [46], or may be produced locally, for example, by synchrotron radiation by electrons accelerated along with the protons [47,48]. Pair synchrotron cascades initiated by photons and electrons resulting from pion decay give rise to the emerging spectra, and this also leads to quite acceptable fits to the observed spectra. These models can produce neutrinos and also higher energy radiation than electron models because protons have a much lower synchrotron energy loss rate than electrons for a given magnetic environment. In both classes of model, shock acceleration has been suggested as the likely acceleration mechanism (see [21] for references).

By appropriately integrating over redshift and luminosity in an expanding universe, using a luminosity function (number density of objects per unit of luminosity) appropriate to blazars, and using the proton blazar models to model the γ ray and neutrino spectra one can estimate the diffuse neutrino background expected from blazars. In Fig. 3 I have added intensities of $(\nu_\mu + \bar{\nu}_\mu)$ predicted in proton blazar models by Mannheim [48], Protheroe [46] (×0.25 as only $\sim$25% of γ-ray background is due to AGN [41] – original calculation assumed 100%) and Halzen and Zas [49]. For some of these models expected muon rates have been calculated [50–52].

6. TOPOLOGICAL DEFECTS

Finally, I discuss perhaps the most uncertain of the components of the diffuse high energy neutrino background, that due to topological defects (TD). In a series of papers [53–56], TD have been suggested as an alternative explanation of

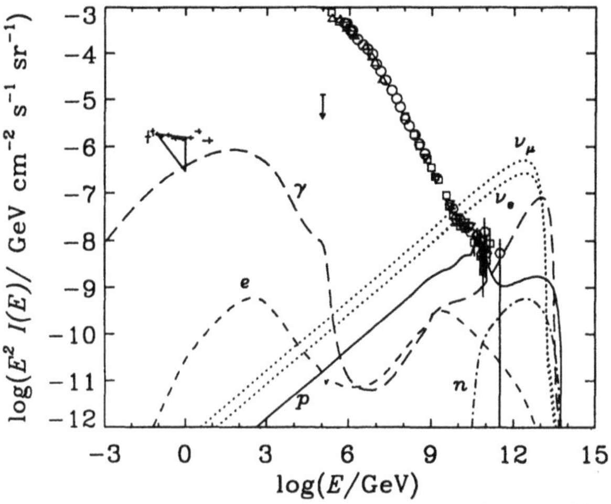

Figure 4. The result of [58] for $M_X c^2 = 10^{14.1}$ GeV, a magnetic field of 10^{-9} gauss, and $p = 2$, normalized the spectrum of "observable particles" (nucleons, photons, electrons) to the 3×10^{11} GeV data [19]). $H_0 = 75$ km $\mathrm{s^{-1}}$ $\mathrm{Mpc^{-1}}$ and $q_0 = 0.5$ are assumed.

the highest energy cosmic rays. In this scenario, the observed cosmic rays are a result of top-down cascading, from somewhat below (depending on theory) the GUT scale energy of $\sim 10^{16}$ GeV [57], down to 10^{11} GeV and lower energies. These models put out much of the energy in a very flat spectrum of neutrinos, photons and electrons extending up to the mass of the "X–particles" emitted.

Protheroe and Stanev [58] argue that these models appear to be ruled out by the GeV γ-ray intensity produced in cascades initiated by X-particle decay for GUT scale X-particle masses. The γ-rays result primarily from synchrotron radiation of cascade electrons in the extragalactic magnetic field. Fig. 4, taken from ref. [58], shows the neutrino emission for a set of TD model parameters just ruled out according to Protheroe and Stanev [58] for a magnetic field of 10^{-9} G and X-particle mass of 1.3×10^{14} GeV. Clearly for such magnetic fields and higher X-particle masses (e.g. GUT scale), TD cannot explain the highest energy cosmic rays. Indeed there is evidence to suggest that magnetic fields between galaxies in clusters could be as high as 10^{-6} G [59]. However, for lower magnetic fields and/or lower X-particle masses the TD models might explain the highest energy cosmic rays without exceeding the GeV γ-ray limit. For example, Sigl et al. [60] show that a TD origin is not ruled out if the extragalactic field is as low as 10^{-12} G, and Birkel & Sarkar [61] adopt an X-particle mass of 10^{12} GeV. Yoshida et al. [62] investigate various TD scenarios with GUT scale masses, and their predicted neutrino fluxes are generally higher than those of Sigl et al. [60], but such high neutrino intensities are likely to be excluded because the γ-rays, due to cascading even in a 10^{-12} G field, would probably exceed the GeV flux. The intensities are compared in Fig. 5. A novel feature of the work of Yoshida et al. [62] is the inclusion of interactions of high energy neutrinos with the 1.9 K cosmic neutrino background, and this can be important at the very highest energies.

I emphasize that the predictions summarized in Fig. 5 are *not* absolute predictions, but the intensity of γ-rays and nucleons in the resulting cascade is normalized in some way to the highest energy cosmic ray data. It is my opinion that GUT scale TD models are neither necessary nor able to explain the highest energy cosmic rays without violating the GeV γ-ray flux observed. The predicted neutrino intensities are therefore *extremely* uncertain. Nevertheless, it is important to search for such emission because, if it is found, it would overturn our current thinking on the origin of the highest energy cosmic rays and, perhaps more importantly, our understanding of the Universe itself.

7. DISCUSSION

Very recently, Waxman and Bahcall [32] have used some arguments based on the observed cosmic ray spectrum to obtain an upper bound to high energy neutrinos from astrophysical sources. Their argument hinges on sources of astrophysical neutrinos being sources of the highest energy cosmic rays which happen also to produce neutrinos by $p\gamma$ interactions. Hence, except for sources with a very high optical depth for protons, the maximum neutrino intensity will be about 10% of

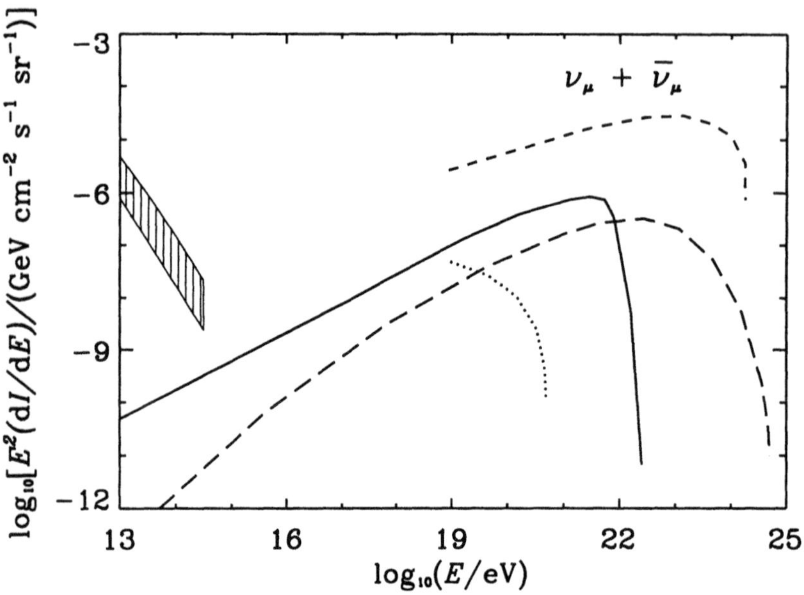

Figure 5. Neutrinos from topological defects: —— TD model just ruled out according to Protheroe & Stanev [58] ($10^{14.1}$ GeV, 10^{-9} gauss); — — — — TD model just allowed according to Sigl et al. [60] (2×10^{16} GeV, 10^{-12} gauss); ······ Birkel & Sarkar [61] (10^{12} GeV, 0 gauss); - - - - - Yoshida et al.[62] (10^{16} GeV, 0 gauss).

the extragalactic ($\sim E^{-2}$) component of the highest energy cosmic rays. Examining AGN models, they find that predictions for proton blazar models exceed their bound.

In the case of AGN, they also suggest that the optical depth to $p\gamma$ at $\sim 10^{19}$ eV must be much less than 1 (to enable TeV γ-rays to escape without significant $\gamma\gamma$ pair production losses), with the consequence that the ultra high energy cosmic ray production far exceeds the ultra high energy neutrino production, pushing the neutrino upper bound even lower.

TeV γ-rays, however, have so far only been seen from 3 blazars, and it is by no means certain that TeV γ-rays are emitted by all blazars and so high $p\gamma$ optical depths are not necessarily ruled out (note, however, that the infrared background limits how far away one can observe objects at TeV energies [63]). Also, in at least one of the proton blazar models [46] the optical depth of protons to $p\gamma$ at 10^{19} eV is high because the proton directions are isotropic in the jet frame whereas the radiation field is highly anisotropic, coming from near the base of the jet, and the photons cascade down to TeV energies where the $\gamma\gamma$ optical depth along the jet direction is low because of the radiation being anisotropic. Admitedly, neutrons are produced in a fraction of $p\gamma$ interactions, and the neutrons escape as cosmic rays, and so the effective optical depth for nucleons can not exceed ~ 1 by much, and so it is probable that this proton blazar model is ruled out.

The main argument relating the neutrino upper bound to the observed ultra high energy cosmic ray flux relies on the cosmic rays of energy 10^{19} eV being able to reach Earth from AGN during the Hubble time. There is evidence to suggest that magnetic fields between galaxies in cores of clusters (the most likely place to find an an AGN) could be as high as 10^{-6} G [59]. With such high magnetic fields it is not obvious that 10^{19} eV protons will reach us from most AGN contributing high energy neutrinos.

Thus, I believe that the "upper bound" is model dependent, and that its calculation is complicated by cosmic ray propagation effects. While

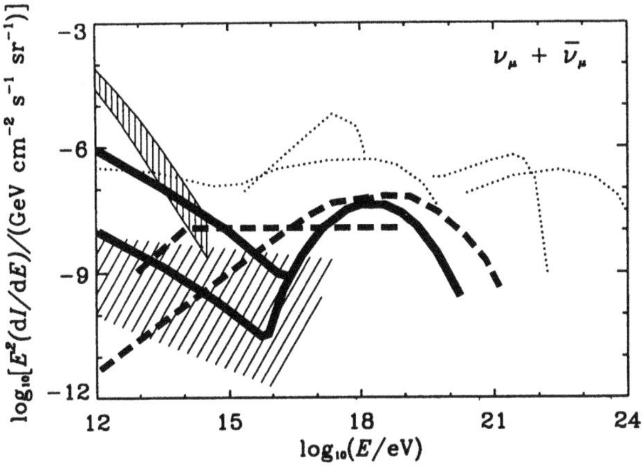

Figure 6. Grand Unified Neutrino Spectrum – a personal opinion about the predicted neutrino intensities: thick solid lines – certain; long dashed lines – almost certain; short dashed lines – speculative; dotted lines – highly speculative.

I would certainly classify the higher AGN fluxes as speculative, or highly speculative, I believe the lower ones are not ruled out by the argument of Waxman and Bahcall. Nevertheless, the work of Waxman and Bahcall is very important in reminding us that for any model used to predict high energy neutrino fluxes we must check that it does not overproduce cosmic rays.

Plotting a representative sample of the diffuse flux predictions from Figs. 1, 2, 3 and 5 in the same figure one has a "grand unified neutrino spectrum" (with apologies to Ressell and Turner [64]). This is shown in Fig. 6 where I have labelled the various curves as "speculative", "highly speculative", "certain" or "almost certain". These labels reflect my own personal opinion or prejudice and should not be taken too seriously – other opinions are equally valid.

8. PROSPECTS FOR OBSERVATION

With the construction in the relatively short term of 0.1 km^2 neutrino telescopes, and in the longer term of 1 km^2 detectors, it is useful to estimate the signals expected due to various possible neutrino intensities. At high energies, electron neutrinos may also be detected through the resulting cascade, and this is particularly important when looking for horizontal air showers, for example with the proposed AUGER detectors [62]. Several estimates of event rates have been made for various energy thresholds, or for horizontal air showers due to neutrino interactions (including ν_e and $\bar{\nu}_e$, see e.g. Gandhi et al. [52]).

To illustrate how the $(\nu_\mu + \bar{\nu}_\mu)$ signals expected from different astrophysical neutrino spectra would be detected by telescopes with different energy thresholds, I have made approximate estimates of the event rates *as a function of minimum muon energy* using the $P_{\nu\to\mu}(E_\nu, E_\mu^{\min})$ function given in Fig. 2 of ref. [13] for $E_\mu^{\min} = 1$ GeV, modified for other $E_\mu^{\min}$ values in a way consistent with that given for $E_\mu^{\min} = 1$ TeV. The effects of shadowing for vertically upward-going neutrinos have been included using the shadow factor $S(E_\nu)$ given in Fig. 20 of ref. [51]. I have estimated the expected neutrino induced muon signal for four representative neutrino intensities. The vertically upward-going and horizontal muon signals are shown separately for each case in Fig. 7 together with the atmospheric neutrino induced muon signals for the two directions. As can be seen, the highest signals would be due to the proton blazar models, with several events per year expected in a 0.1 km^2 detector. However, one should be cautious as these intensities are somewhat speculative (as discussed earlier). Detection of muon signals in one year from the other intensities estimated would be marginal for a 0.1 km^2 detector, but achievable with a 1 km^2 detector. Detection of transient neutrino signals, correlated with observations of the same source in photons (e.g. GRB, AGN) should therefore be the goal of high energy neutrino astronomy in the short term.

One should consider the consequences for astrophysical neutrinos of the discovery of the oscillation of atmospheric ν_μ, probably into ν_τ, by Super-Kamiokande [65] with an oscillation length of $\lambda_{\rm osc} \sim 10^3(E/\mathrm{GeV})$ km. On an astrophysical scale, the oscillation length $\lambda_{\rm osc} \sim 3\times10^{-11}(E/\mathrm{TeV})$ kpc is very small, and integrating contributions to the neutrino intensity over

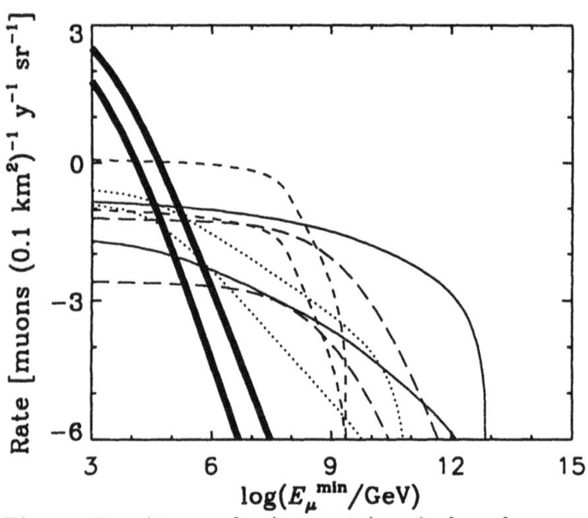

Figure 7. Atmospheric neutrino induced muon signal (thick solid lines) and putative astrophysical neutrino induced muon signals expected for the following $(\nu_\mu + \bar{\nu}_\mu)$ intensities: GRB intensity [31] extended to 10^{20} eV (dotted curves); $p\gamma$ proton blazar [48] (short dashed curves); TD model just ruled out according to ref. [58] (solid curves); interactions of the highest energy cosmic rays with the CMBR [23] for $E_{\max} = 3 \times 10^{21}$ eV (long dashed curves). Upper curves show horizontal signals, lower curves show vertical (upward) signals.

astrophysical dimensions one would naively expect the $(\nu_\mu + \bar{\nu}_\mu)$ flux to be 50% lower (assuming $\sin^2 2\theta = 1$), and to be accompanied by a similar $(\nu_\tau + \bar{\nu}_\tau)$ flux. The unique signature for detection of tau neutrinos has been discussed in ref. [66].

Neutrino astronomy is developing during an era in which exciting discoveries are being made in other areas of high energy astrophysics. These include detection of rapidly varying TeV γ-ray signals from AGN, discovery that GRB are extragalactic and probably the most energetic phenomena ocurring in the Universe today, and detection at Earth of cosmic rays with energies well above 10^{20} eV opening the question of whether their origin is through particle acceleration at radio galaxies or GRB, or from topological defects left over from the big bang. Hadronic processes may have a role in all these phenomena, and searching for high energy neutrinos may lead to greater understanding of the highest energy phenomena in the Universe. Clearly, for this to happen in parallel with the other observations rapid development of neutrino telescopes with sensitive areas of ~ 1 km^2 or larger, operating over a wide range neutrino energies, is essential.

ACKNOWLEDGMENTS

I thank A. Mücke and Q. Luo for reading the manuscript, and J. Bahcall and E. Waxman for helpful comments. This research is supported by a grant from the Australian Research Council.

REFERENCES

1. V.S. Berezinskiĭ and G.T. Zatsepin, *Sov. Phys. Usp.* **20** (1977) 361.
2. J.G. Learned, in "Frontiers of Neutrino Astrophysics", eds. Y. Suzuki and K. Nakamura, (Universal Acad. Press, Tokyo, 1993) p. 341.
3. P.C. Mock *et al.*, in 24th Int. Cosmic Ray Conf. (Rome), **1** (1995) 758.
4. V.A. Balkanov et al., in Proc. of 5th Int. Workshop on Topics in Astroparticle and Underground Physics (LNGS INFN, Assergi, September 7-11, 1997)
5. X. Shi, G.M. Fuller, F. Halzen, submitted to *Phys. Rev. Lett.* (1998) astro-ph/9805242
6. F. Blanc et al., ANTARES proposal (1997) astro-ph/9707136
7. R. Silberberg and M.M. Shapiro, "Proc. of the 15th Int. Cosmic Ray Conf. (Plovdiv)", vol. 6, p. 237 (1977).
8. F.W. Stecker, *Ap. J.* **228** (1979) 919.
9. G. Domokos et al., *J. Phys. G.: Nucl. Part. Phys.* **19** (1993) 899.
10. V.S. Berezinsky, T.K. Gaisser, F. Halzen and T. Stanev, *Astroparticle Phys.* **1** (1993) 281.
11. G. Ingelman and M. Thunman, preprint (1996) hep-ph/9604286.
12. P. Lipari, *Astroparticle Phys.* **1** (1993) 195.
13. T.K. Gaisser, F. Halzen, and T. Stanev, *Phys. Rep.*, **258**, (1995) 173.
14. T.K. Gaisser, Talk given at the OECD Megascience Forum Workshop, Taormina, Sicily, 22/23 May, 1997. astro-ph/9707283

15. V.S. Berezinsky, P. Blasi and V.S. Ptuskin, preprint (1996) astro-ph/9609048.
16. S. Colafrancesco and P. Blasi, *Astroparticle Phys.* in press (1998) astro-ph/9804262
17. K. Greisen, *Phys. Rev. Lett.* **16** (1966) 748.
18. G.T. Zatsepin and V.A. Kuz'min, *JETP Lett.* **4** (1966) 78.
19. D.J. Bird *et al.*, *Ap. J.* **441** (1995) 144.
20. M. Takeda et a., *Phys. Rev. Lett.* **81** (1998) 1163.
21. R.J. Protheroe, in "Towards the Millennium in Astrophysics: Problems and Prospects", Erice 1996, eds. M.M. Shapiro and J.P. Wefel (World Scientific, Singapore), in press (1998). astro-ph/9612213
22. J.P. Rachen and P.L. Biermann, *Astron. Astrophys.* **272** (1993) 161.
23. R.J. Protheroe and P.A. Johnson, *Astroparticle Phys.* **4** (1995) 253; erratum **5** (1996) 215.
24. C.T. Hill and D.N. Schramm *Phys. Rev. D* **31** (1985) 564
25. S. Lee, preprint (1996) astro-ph/9604098.
26. S.R. Kulkarni et al., *Nature*, **393** (1998) 35
27. T.J. Galama, et al., *Nature*, in press (1998) astro-ph/9806175
28. R. Popham, S.E. Woosley, and C. Fryer, *Ap. J.*, submitted (1998) astro-ph/9807028
29. K. Iwamoto, et al., *Nature*, in press (1998) astro-ph/9806382
30. P. Meszaros and M. Rees, *Mon. Not. R. Astr. Soc.* **269** (1994) 41P
31. E. Waxman and J. Bahcall, *Phys. Rev. Lett.* **78** (1997) 2292.
32. E. Waxman and J. Bahcall, submitted to *Phys. Rev. D* (1998) hep-ph/9807282
33. J.P. Rachen and P. Meszaros, *Phys. Rev. D.* submitted (1998) astro-ph/9802280
34. M. Vietri, *Phys. Rev. Lett.* **80** (1998) 3690
35. E. Waxman, *Phys. Rev. Lett.* **75** (1995) 386
36. M. Vietri, *Ap. J.*, *453* (1995) 883.
37. D.J. Thompson et al., *Ap. J. Suppl.*, **101** (1995) 259.
38. C. von Montigny et al., *Ap. J.*, **440** (1995) 525.
39. R. Mukherjee et al., *Ap. J.*, **490** (1997) 116.
40. M. Catanese, *Astrophys. J.* **501** (1998) 616
41. J. Chiang and R. Mukherjee, *Ap. J.*, **496** (1998) 752.
42. A.P. Szabo and R.J. Protheroe, *Astroparticle Phys.* **2** (1994) 375
43. D.A. Kniffen et al., *Ap. J.*, **411** (1993) 133.
44. J.A. Gaidos et al., *Nature* **383** (1996) 319.
45. R.J. Protheroe et al., in "Invited, Rapproteur, and Highlight Papers of the 25th Int. Cosmic Ray Conf. (Durban)", eds. M.S. Potgieter et al., pub. World Scentific (Singapore), p. 317 (1998). astro-ph/9710118
46. R.J. Protheroe, in Proc. IAU Colloq. 163, *Accretion Phenomena and Related Outflows*, ed. D. Wickramasinghe et al., ASP Conf. series, Vol. 121, pp 585–588 (1997) astro-ph/9607165
47. K. Mannheim and P.L. Biermann, *Astron. Astrophys.* **221**, (1989) 211.
48. K. Mannheim, *Astropart. Phys.* **3** (1995) 295.
49. F. Halzen, E. Zas, *Ap. J.* **488** (1997) 669
50. G.C. Hill, *Astropart. Phys.* **6** (1997) 215
51. R. Gandhi et al., *Astroprt. Phy.* **5** (1996) 81.
52. R. Gandhi et al., preprint (1998) hep-ph/9807264
53. Hill, C.T., *Nucl. Phys. B* **224** 469 (1983)
54. F.A. Aharonian, P. Bhattacharjee, and D.N. Schramm, *Phys. Rev. D* **46** 4188 (1992)
55. P. Bhattacharjee, C.T. Hill and D.N. Schramm, *Phys. Rev. Lett.* **69** 567 (1992).
56. A.J. Gill and T.W.B. Kibble, *Phys. Rev. D* **50** 3660 (1994).
57. U. Amaldi, W. de Boer and H. Fürstenau, *Phys. Lett.*, **B260** 447 (1991).
58. R.J. Protheroe and T. Stanev, *Phys. Rev. Lett.* **77** (1996) 3708.
59. P.P. Kronberg, *Rep. Prog. Phys.* **57** (1994) 325.
60. G. Sigl, S. Lee, D. Schramm and P. Coppi, *Phys. Lett. B* **392** (1997) 129.
61. M. Birkel and S. Sarkar, preprint (1998) hep-ph/9804285
62. S. Yoshida, H. Dai, C.C.H. Jui, P. Sommers, *Astropart. Phys.* **479** (1997) 547.
63. F.W. Stecker and O.C. De Jager, *Astron. Astrophys.* in press (1998) astro-ph/980419
64. T. Ressell & M. Turner, *Comments in Astrophysics* **14** (1991) 323.
65. Y. Fukuda et al. *Phys. Rev. Lett.* **81** (1998) 1562
66. J.G. Learned and S. Pakvasa, *Astropart. Phys.* **3** (1995) 267.

ELSEVIER

Nuclear Physics B (Proc. Suppl.) 77 (1999) 474–485

NUCLEAR PHYSICS B
PROCEEDINGS
SUPPLEMENTS

The AMANDA Neutrino Telescope*

The AMANDA Collaboration

E.C. Andrés[6,11] P. Askebjer[6] S.W. Barwick[2] R.C. Bay[1] L. Bergström[6] A. Biron[9] J. Booth[2] O. Botner[7] A. Bouchta[6] S. Carius[8] M. Carlson[4] W. Chinowsky[10] D. Chirkin[1] J. Conrad[7] C.G.S. Costa[4] D. Cowen[3] E. Dalberg[6] T. DeYoung[4] J. Edsjö[6] P. Ekström[6] A. Goobar[6] L. Gray[4] A. Hallgren[7] F. Halzen[4] R. Hardtke[4] S. Hart[11] Y. He[1] C. P. de los Heros[7] G. Hill[4,11] P.O. Hulth[6] S. Hundertmark[9] J. Jacobsen[4] A. Jones[11] V. Kandhadai[4] A. Karle[4] J. Kim[2] H. Leich[9] M. Leuthold[9] P. Lindahl[8] I. Liubarsky[4] P. Loaiza[7] D. Lowder[1] P. Marciniewski[7] T.C. Miller[5] P. Miocinovic[1] P.C. Mock[2] R. Morse[4] M. Newcomer[3] P. Niessen[9] D. Nygren[10] R. Porrata[2] D. Potter[11] P.B. Price[1] G. Przybylski[10] W. Rhode[1] S. Richter[11] J. Rodriguez[6] P. Romenesko[4] D. Ross[2] H. Rubinstein[7] T. Schmidt[9] E. Schneider[2] R. Schwarz[11] U. Schwendicke[9] G. Smoot[10] M. Solarz[1] V. Sorin[6] C. Spiering[9] P. Steffen[9] R. Stokstad[10] O. Streicher[9] I. Taboada[3] T. Thon[9] S. Tilav[4] C. Walck[6] C.H. Wiebusch[9] R. Wischnewski[9] K. Woschnagg[1] W. Wu[2] G. Yodh[2] S. Young[2]

[1]University of California, Berkeley, USA
[2]University of California, Irvine, USA
[3]University of Pennsylvania, USA
[4]University of Wisconsin, Madison, USA
[5]Bartol Research Institute,USA
[6]Stockholm University, Sweden
[7]University of Uppsala, Sweden
[8]Kalmar University, Sweden
[9]DESY – Institute for High Energy Physics, Germany
[10]Lawrence Berkeley National Laboratory, USA
[11]South Pole Winter-Overs, Antarctica

*Presented by F. Halzen, Physics Department, University of Wisconsin, Madison, WI 53706

With an effective telescope area of order 10^4 m^2 for TeV neutrinos, a threshold near ~50 GeV and a pointing accuracy of 2.5 degrees per muon track, the AMANDA detector represents the first of a new generation of high energy neutrino telescopes, reaching a scale envisaged over 25 years ago. We describe early results on the calibration of natural deep ice as a particle detector as well as on AMANDA's performance as a neutrino telescope.

1. INTRODUCTION AND SUMMARY

The Antarctic Muon and Neutrino Detector Array AMANDA is a multi-purpose instrument; its science missions cover particle physics, astronomy and astrophysics, cosmology and cosmic ray physics[1]. Its deployment creates new opportunities for glaciology[2]. The first-generation detector is designed to reach a relatively large telescope area and detection volume for a neutrino threshold not higher than 100 GeV. This relatively low threshold permits calibration of the novel instrument on the known flux of atmospheric neutrinos. Its architecture has been optimized for reconstructing the Cherenkov light front radiated by up-going, neutrino-induced muons which must be identified in a background of down-going, cosmic ray muons which are more than 10^5 times more frequent for a depth of 1–2 kilometer.

The status of the AMANDA project can be summarized as follows:

- Construction of the first generation AMANDA detector[3] was completed in the austral summer 96–97. It consists of

0920-5632/99/$ – see front matter
PII S0920-5632(99)00470-3

300 optical modules deployed at a depth of 1500–2000 m; see Fig. 1. An optical module (OM) consists of an 8 inch photomultiplier tube and nothing else. OM's have only failed during deployment, at a rate of less than 3 percent.

- Data taken with 80 OM's, deployed one year earlier in order to verify the optical properties of the deep ice, have been analysed. We will present the results here. This partially deployed detector will be referred to as AMANDA-B4. Reconstructed up-going muons are found at a rate consistent with the expected flux of atmospheric neutrinos. The exercise shows that calibration of the full detector on atmospheric neutrinos of approximately 100 GeV energy and above, is possible as we will show further on.

- First calibration of the full detector is now completed and analysis of the first year of data is in progress. Preliminary results based on the analysis of 1 month of data confirm the performance of the detector derived from the analysis of AMANDA-B4 data. Events reconstructed as going upwards, like the one shown in Fig. 2, are found, as expected.

As part of a research and development effort preparatory to developing a kilometer-scale neutrino detector, we have deployed 3 strings, instrumented with 42 OMs between 1.3 and 2.4 kilometers; see Fig. 1. The strings deviated from vertical by less than 1 m over 2.4 km; see Fig. 3. They also form part of an intermediate detector, AMANDA II, which will extend the present telescope by approximately an order of magnitude in affective area for TeV energies. It will be completed in 99-00 with the addition of eight more strings. The analogue signals made by photoelectron pulses in the new OMs are transferred to the surface over both twisted pair and fiber optic cables. The relative sharpness of the pulses at the surface is compared in Fig. 4. Also, bright light sources surrounding a pair of TV cameras were lowered into the last hole. The resulting images visually confirm the exceptional clarity of the ice inferred from previous indirect measurements.

After a brief review of our results on the optics of the ice, we will discuss muon track reconstruction and the status of the calibration of the detector on the flux of atmospheric neutrinos. We will conclude with a brief description of the data analysis of the first year of a data taken with the completed detector.

2. OPTICS OF DEEP ICE

As anticipated from transparency measurements performed with shallow strings above 1 km depth[2] (see Fig. 1), ice is bubble-free at 1400–1500 meters and below. The performance of the AMANDA detector is encapsulated in the event shown in Fig. 5. Coincident events between AMANDA-B4 and the four shallow strings have been triggered at a rate of 0.1 Hz. Every 10 seconds a cosmic ray muon is tracked over 1.2 kilometers. The contrast in detector response between the strings near 1 and 2 km depths is striking: while the Cherenkov photons diffuse on remnant bubbles in the shallow ice, a straight track with velocity c is registered in the deeper ice. The optical quality of the deep ice can be assessed by viewing the OM signals from a single muon triggering 2 strings separated by 79.5 m; see Fig. 5b. The separation of the photons along the Cherenkov cone is well over 100 m, yet, despite some evidence of scattering, the speed-of-light propagation of the track can be readily identified.

The optical properties of the ice are quantified by studying the propagation in the ice of pulses of laser light of nanosecond duration. The arrival times of the photons after 20 m and 40 m are shown in Fig. 6 for the shallow and deep ice[4]. The distributions have been normalized to equal areas; in reality, the probability that a photon travels 70 m in the deep ice is $\sim 10^7$ times larger. These critical results have been verified by the deployment of nitrogen lasers, pulsed LED's and DC lamps in the deep ice; see Table 1. We have established that ice is an adequate medium to do neutrino astronomy. A comparison of the optical properties of ice, lake and ocean detectors is summarized in Table 2.

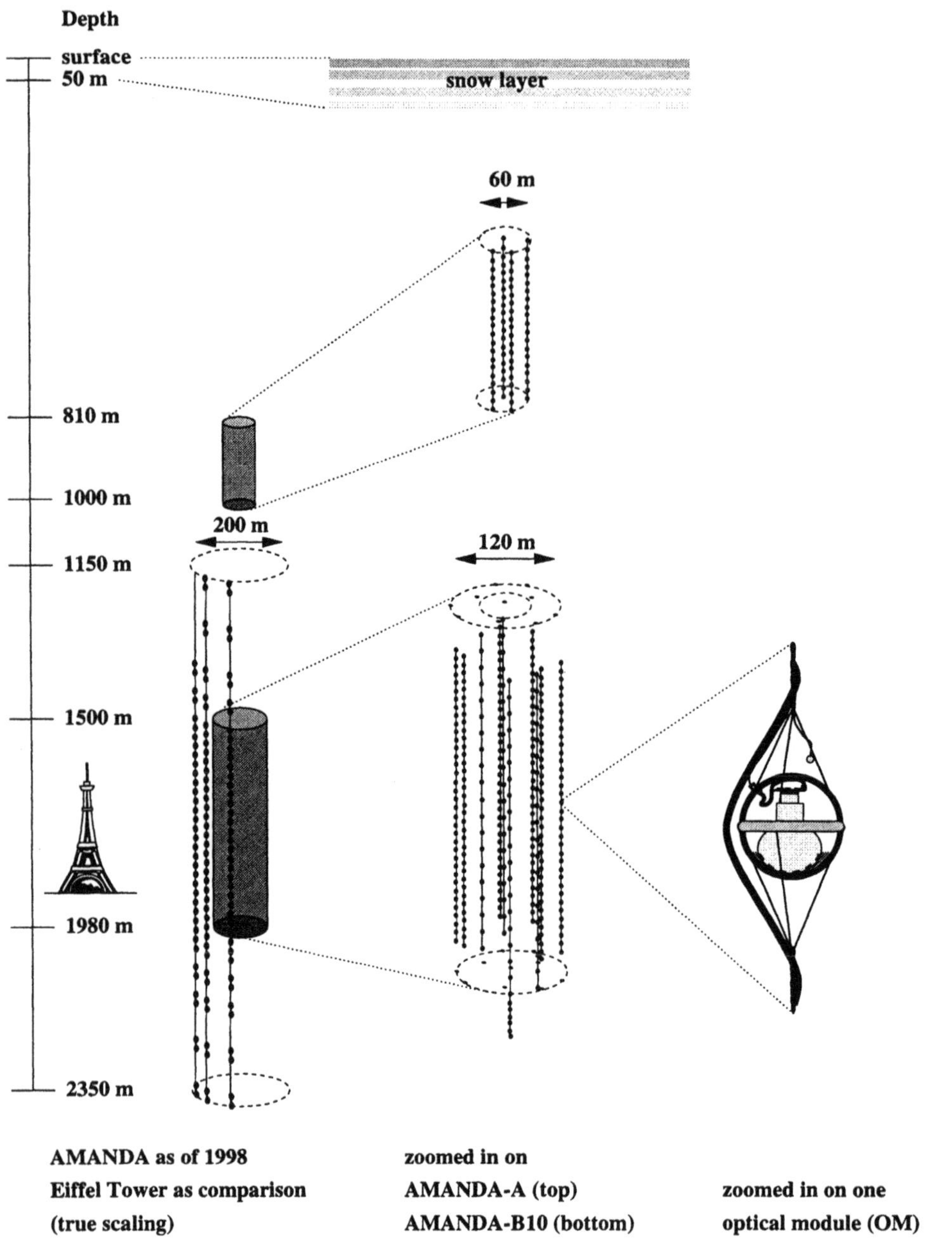

Figure 1. The Antarctic Muon And Neutrino Detector Array (AMANDA).

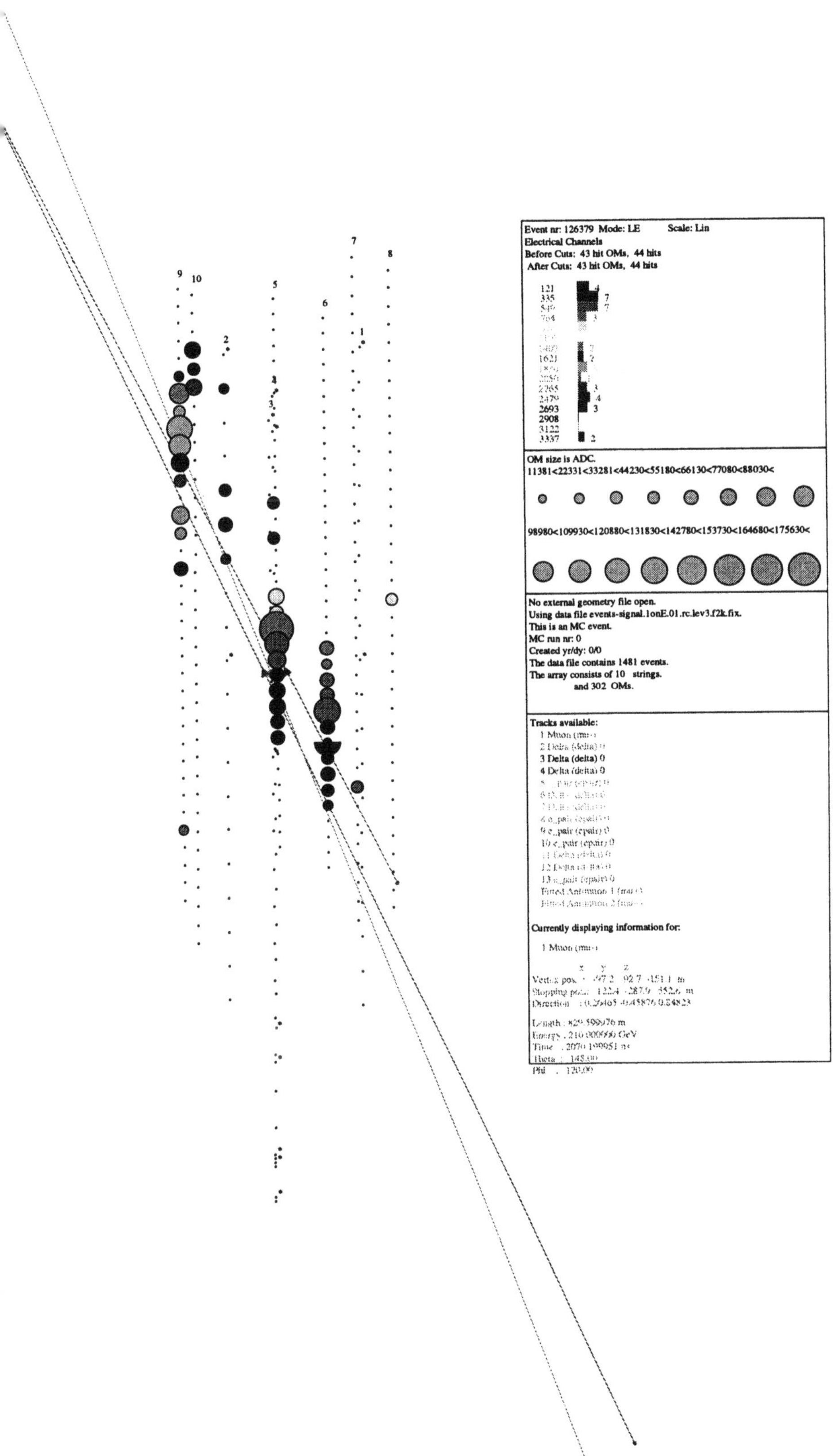

Figure 2. Monte Carlo simulation of 100 GeV neutrino-induced muon track recorded in the completed AMANDA detector.

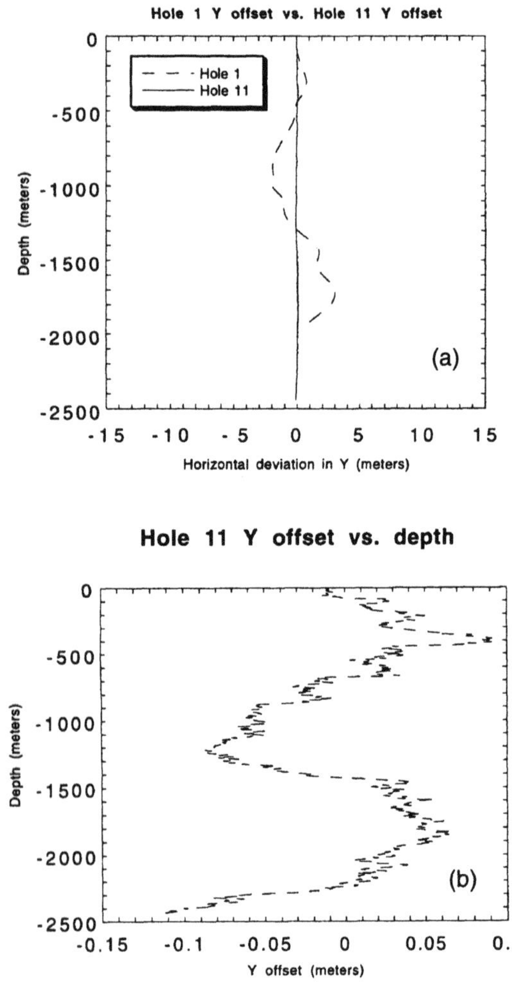

Figure 3. a) Progress in drilling: telemetry data from the drill compare the deviation from vertical. b) Excursions transverse to the vertical direction are smaller than 1 m over 2.4 km.

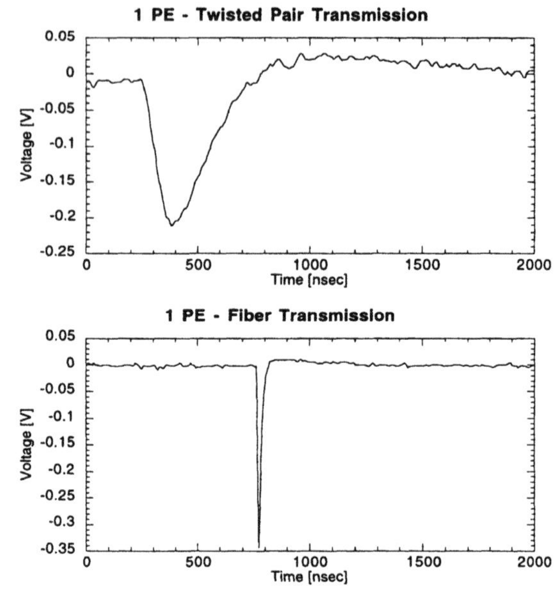

Figure 4. Time profile of *the same* single photoelectron pulse after transmission over a twisted pair (top) and fiber optic cable (bottom). The signal is from an optical module deployed at a depth of 1.7 km for strings 1 and 13.

Inter-string laser shots are also used to determine the geometry of the detector. In conjunction with telemetry from the drill, the OMs have been positioned with an absolute precision of better than 1 meter. Mapping the detector has been by far the most challenging aspect of the calibration of this novel instrument. A precise knowledge of the location of the optical sensors is crucial for track reconstruction. Therefore, two completely independent methods were developed for the final determination of the geometry. One method makes use of drill data. A variety of sensors are installed in the drill to determine its speed and direction during drilling. Every second a data string is transmitted to the control system and recorded. The analysis of this data provides the first information about the string position. The depth of the string is independently determined with pressure sensors. The final positioning of the strings is done with a laser calibration system. Laser pulses (532 nm) are transmitted with optical fibers to every optical module on strings 1–4, and to every second module on strings 5–10. After the timing calibration is completed, the laser calibration provides time of flight measurements to determine the distances between strings and a check on possible vertical offsets. More

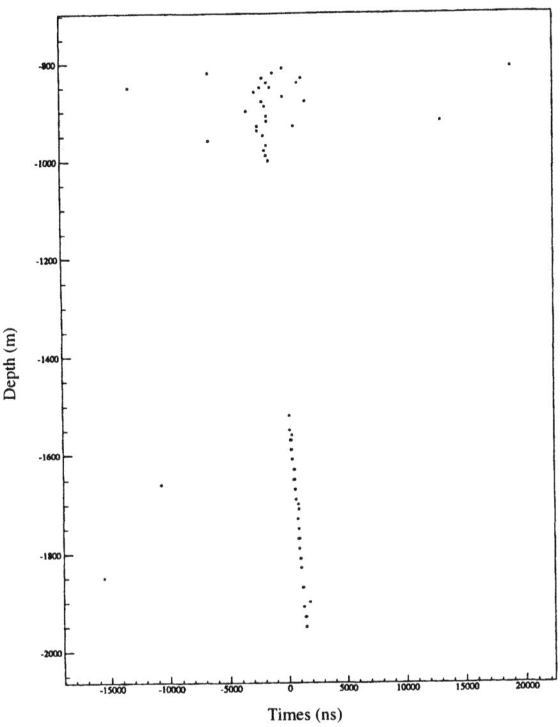

Figure 5a. Cosmic ray muon track triggered by both shallow and deep AMANDA OM's. Trigger times of the optical modules are shown as a function of depth. The diagram shows the diffusion of the track by bubbles above 1 km depth. Early and late hits, not associated with the track, are photomultiplier noise.

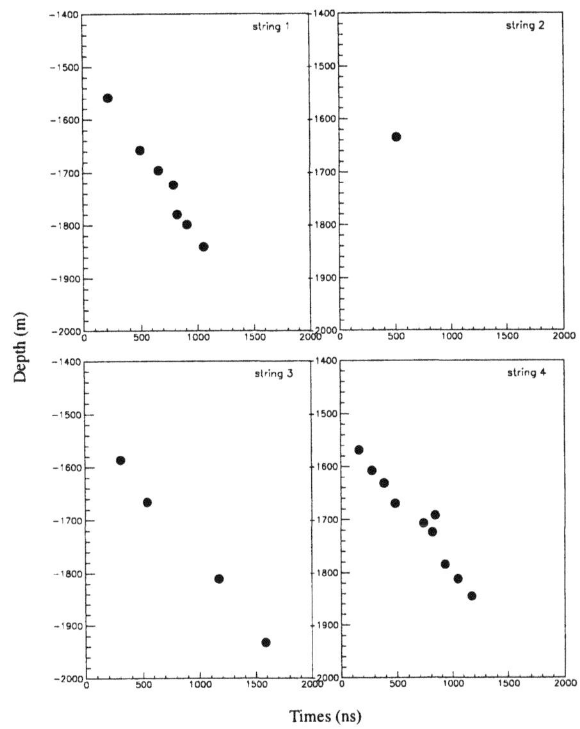

Figure 5b. Cosmic ray muon track triggered by both shallow and deep AMANDA OM's. Trigger times are shown separately for each string in the deep detector. In this event the muon mostly triggers OM's on strings 1 and 2, which are separated by 77.5 m.

than a hundred laser runs provide a large data base, both to determine the geometry and to verify the timing calibration. Figure 7 shows laser data from string 8 recorded on string 7, with the results from a global fit to data from all 10 strings plotted as a solid line. The vertical offset between the strings from pressure sensor data was found to be 0.9 m and the distance between them has been determined to 29.9 m. The errors given in the figure are the statistical errors from the global fit. The position error of the optical sensors is less than 1 m, thus matching the time resolution of the sensors.

3. RECONSTRUCTION OF MUON TRACKS

The AMANDA detector was antecedently proposed on the premise that inferior properties of ice as a particle detector with respect to water could be compensated by additional optical modules. The technique was supposed to be a factor 5$\sim$10 more cost-effective and, therefore, competitive. The design was based on then current information[5] that the absorption length at 370 nm, the wavelength where photomultipliers are maximally efficient, had been measured to be 8 m. The strategy would have been to use a large num-

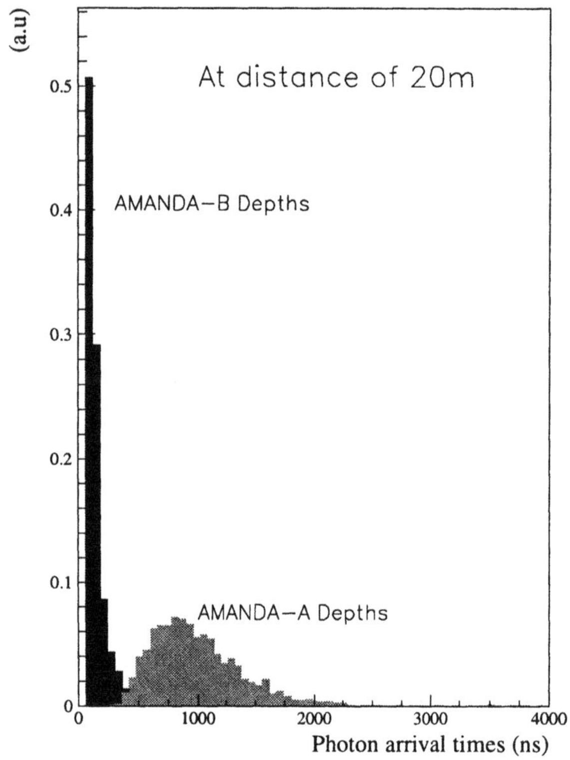

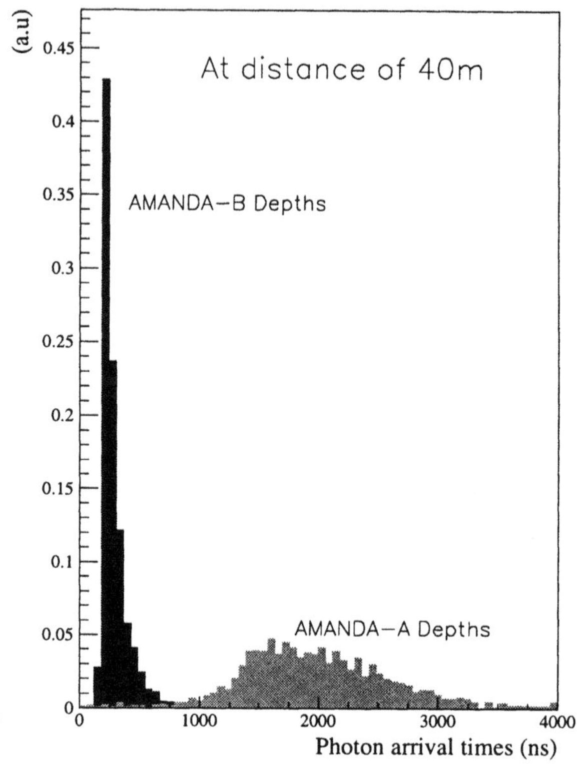

Figure 6. Propagation of 510 nm photons indicate bubble-free ice below 1500 m, in contrast to ice with some remnant bubbles above 1.4 km.

ber of closely spaced OM's to overcome the short absorption length. Muon tracks triggering 6 or more OM's were reconstructed with degree accuracy. Taking data with a simple majority trigger of 6 OM's or more, at 100 Hz yielded an average effective area of 10^4 m^2, somewhat smaller for atmospheric neutrinos and significantly larger for the high energy signals.

The reality is that the absorption length is 100 m or more, depending on depth[2]. With such a large absorption length, scattering becomes a critical issue. The scattering length is 25–30 m (preliminary; this number represents an average value which may include the combined effects of deep ice and the refrozen ice disturbed by the hot water drilling). Because of the large absorption length, OM spacings are now similar, actually larger, than those of proposed water detectors. A typical event triggers 20 OM's, not 6. Of these more than 5 photons are, on average, "not scattered." They are referred to as direct photons, i.e. photons which arrive within time residuals of $[-15; 25]$ ns relative to the calculated time it takes for unscattered Cherenkov photons to reach the OM from the reconstructed muon track. The choice of residual reflects the present resolution of our time measurements and allows for delays of slightly scattered photons. In the end, reconstruction is therefore as before, although additional information can be extracted from scattered photons by minimizing a likelihood function which matches their observed and expected delays[6].

The method is illustrated with AMANDA-B4 data in Fig. 8, where the measured arrival directions of background cosmic ray muon tracks, re-

Table 1
Complementary tools used in the determination of the optical properties of in-situ South Pole ice.

- surface YAG laser (410–600 nm) connected by fiber optic to ~ 300 diffuser balls
- 5 N_2 lasers (337 nm) between 1300–2300 m
- pulsed LEDs (390, 450 nm)
- DC lamps
- DC beacons
- multiple radio antennas (150–300 m)
- 2 TV cameras to 2400 m

Table 2
Optical properties of South Pole ice at 1750 m, Lake Baikal water at 1 km, and the range of results from measurements in ocean water below 4 km.

	(1700 m) AMANDA	BAIKAL	OCEAN
attenuation	~ 30 m	~ 20 m	50–55 m
absorption	105 ± 10 m	20 m	—
(refers to peak value)	335–400 nm	470 nm*	470 nm*
scattering length	24 ± 2 m	150–300 m	—

*smaller for bluer wavelengths

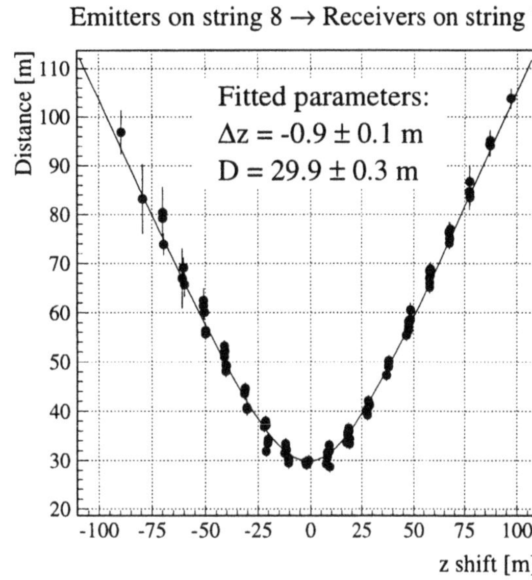

Figure 7. Shots of laserlight determine the distance and the vertical offset between strings.

constructed with 5 or more unscattered photons, are confronted with their known angular distribution. There is an additional cut in Fig. 8 which requires that the track, reconstructed from timing information, actually traces the spatial positions of the OM's in the trigger. The power of this cut, especially for events recorded with only 4 strings, is very revealing. In a kilometer-scale detector, geometrical track reconstruction using only the positions of triggered OM's is sufficient to achieve degree accuracy in zenith angle. We conclude from Fig. 8 that the agreement between data and Monte Carlo simulation is adequate. Less than one in 10^5 tracks is misreconstructed as originating below the detector[4].

Visual inspection reveals that the misreconstructed tracks are mostly showers, radiated by muons or initiated by electron neutrinos, which are misreconstructed as up-going tracks of muon neutrino origin. They can be readily identified on the basis of the characteristic nearly isotropic distribution of the OM amplitudes, and by the fact that the direct hits occur over a short distance near the origin of the shower, rather than spread over a longer muon track.

We have verified the angular resolution of AMANDA-B4 by reconstructing muon tracks registered in coincidence with a surface air shower array SPASE[7]. Figure 9 demonstrates that the zenith angle distribution of the coincident SPASE-AMANDA cosmic ray beam reconstructed by the surface array is quantitatively reproduced by reconstruction of the muons in AMANDA.

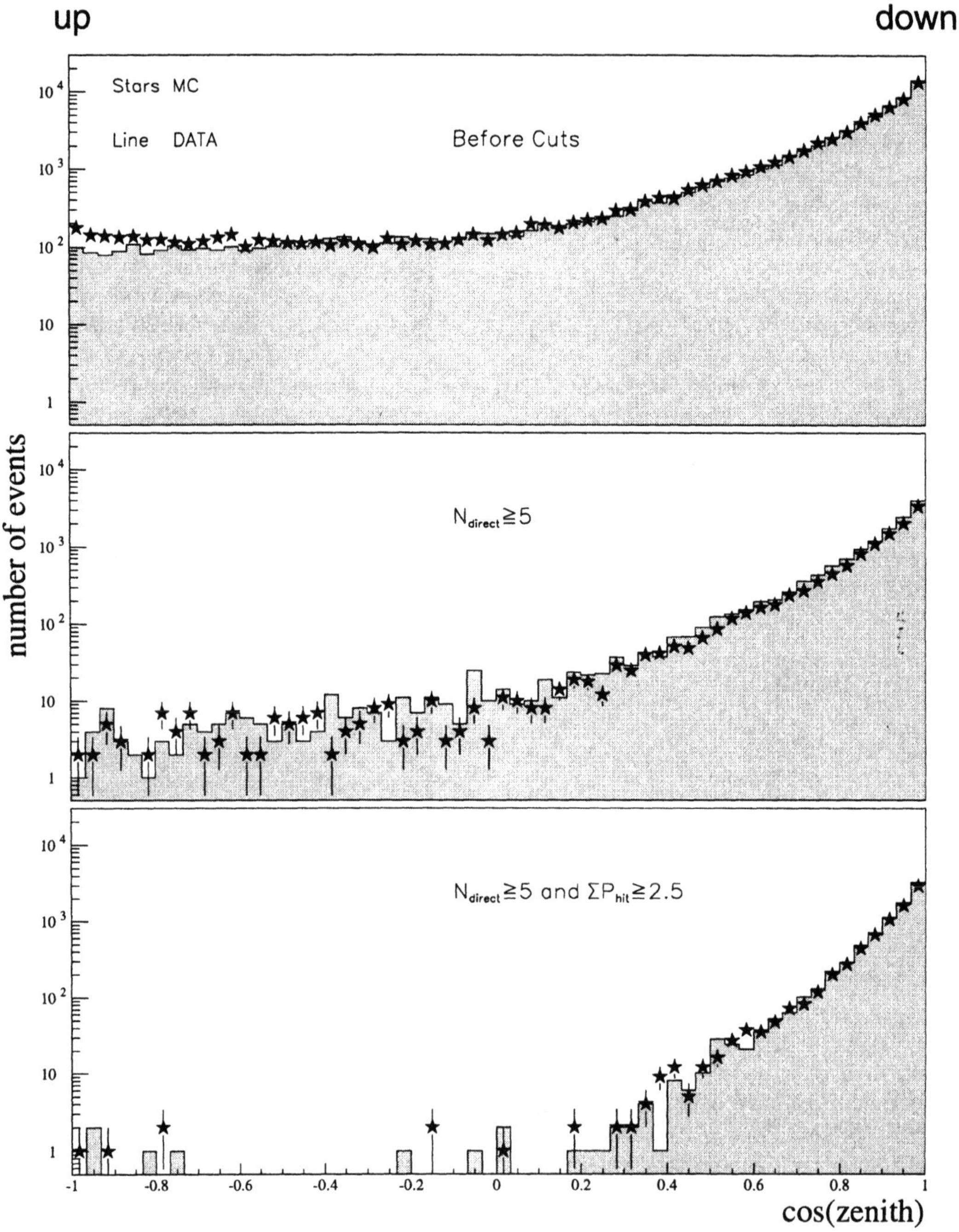

Figure 8. Reconstructed zenith angle distribution of muons: data and Monte Carlo. The relative normalization has not been adjusted at any level. The plot demonstrates a rejection of cosmic ray muons at a level of 10^{-5} with only 80 OMs.

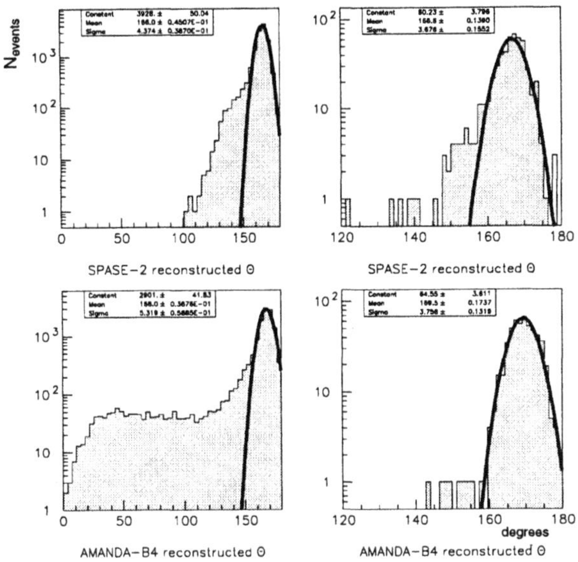

Figure 9. Zenith angle distributions of cosmic rays triggering AMANDA and the surface air shower array SPASE. Reconstruction by AMANDA of underground muons agrees with the reconstruction of the air shower direction using the scintillator array, and with Monte Carlo simulation. The events are selected requiring signals on 2 or more strings (left), and 5 or more direct photons (right).

Monte Carlo simulation, based on the AMANDA-B4 reconstruction, predicts that AMANDA-B10 is a 10^4 m^2 detector for TeV muons, with 2.5 degrees mean angular resolution per track[6]. The effective area is less for atmospheric neutrinos, but in excess of 0.1 km^2 for PeV neutrinos.

4. CALIBRATION ON ATMOSPHERIC NEUTRINOS

Because of the novel technique, the collaboration has maintained 3 fully independent Monte Carlo programs simulating the signals, the detector medium and the detector itself. They quantitatively reproduce the response of the detector to cosmic ray muons: the trigger rate and the amplitude and arrival times of Cherenkov photons for each OM[4]. For reconstruction, 2 independent routines and 3 neural nets are available.

Understanding the performance of the instrument near threshold requires a detailed calibration of the detector which is still in progress. Although is not critical for operating the detector as a high energy neutrino telescope, it is for the detection of the flux of atmospheric neutrinos which falls sharply with energy. As a first calibration we have attempted to identify gold-plated events which are contained in the detector (within the instrumented volume and within 20° of vertical) and which have a track-length in excess of 100 m ($E_\mu > 20$ GeV). Calculation of their rate is straightforward (see Table 3) except for the evaluation of the efficiency of the cut requiring 6 or more, direct photons with residuals in the interval $[-15, +15]$ ns. Monte Carlo simulation gives 5%[8]. The narrow, long AMANDA-B4 detector (which constitutes the 4 inner strings of AMANDA-B10) thus achieves optimal efficiency for tracks which travel vertically upwards through the detector. Because of edge effects, the efficiency, which is of course a very strong function of detector size, is only a few percent after final cuts, even near the vertical direction. The bottom line is that we expect a few events per year satisfying the cuts imposed; see Table 3.

We reconstructed 6 months of filtered AMANDA-B4 events subject to the conditions that 8 OMs report a signal in a time window of 2 microseconds. The two events, shown in Fig. 10, satisfy the cuts outlined in Table 3. Their properties are summarized in Table 4. They have been used to study the capability of AMANDA to search for neutrinos resulting from the annihilation of dark matter particles gravitationally trapped at the center of the earth[8].

We conclude that tracks reconstructed as up-going are found at a rate consistent with the expectation that they are induced by atmospheric neutrinos. The event rates are too low to attempt a detailed calibration of the technique. The result is nevertheless encouraging because such events occur at the rate of about 1 per day in the full detector; see Table 3. Calibration of the full detector on atmospheric neutrinos should be feasible. This work is in progress and preliminary

Table 3
Predicted atmospheric neutrino rate for events with i) track-length in excess of 100 m, ii) contained in the instrumented volume of the detector, iii) close to vertical direction, and iv) 6 or more direct hits. The results of AMANDA-B4 are contrasted with the anticipated rate for AMANDA-B10.

- close to vertical
- muon track > 100 m

$$\#\ \mu\text{'s} = 375 \left[\frac{E_\mu}{20\ \text{GeV}}\right]^{-0.5} \quad (10^4\ \text{m}^2\ \text{sr yr})^{-1}$$

		event rate		event rate
radius	35 m	144 $\text{yr}^{-1}\ \text{sr}^{-1}$	> 60 m	424 $\text{yr}^{-1}\ \text{sr}^{-1}$
$\Delta\theta$ from vertical	20°	70 yr^{-1}	> 45°	1046 yr^{-1}
efficiency ($N_{\text{dir}} \geq 6$)	5%	3.5 yr^{-1}	> 10%	105 yr^{-1}
rate	found 2 in 6 months		> 0.3 per day	
	80 OMs		300 OMs	

Table 4
Characteristics of the two events reconstructed as up-going muons.

Event ID#	4706879	8427905
α [m/ns]	0.19	0.37
Length [m]	295	182
Closest approach [m]	2.53	1.23
θ_{rec}[°]	14.1	4.6
ϕ_{rec}[°]	92.0	348.7
Likelihood/OM	5.9	4.2
OM multiplicity	14	8
String multiplicity	4	2

results based on 1 month of data are consistent with the performance of AMANDA as deduced from the AMANDA-B4 analysis.

Table 5
Summary of the filtering of the 1997 data collected with the completed detector.

Filter efficiency for **atmospheric neutrinos: 80%** (Monte Carlo estimate)

Filter efficiency for **data: 10%**

Other special event categories:
- Coincidence events (AMANDA-SPASE, AMANDA-GASP)
- High-multiplicity ("big") events
- Events coincident with known GRBs, Δt in (−1, +5) minutes
- Events consistent with high-energy EM cascades

Filter output summary:

Initial dataset:	500 GB
Filtered data:	53.5
Cascade data:	15.9
SPASE data:	28.2
"Big" events:	11.9
GRB data:	2.4
GASP data:	0.6
Run logs:	0.01

Filtering completed May 1998

5. DATA ANALYSIS

Even with incomplete calibration, the detector can be operated as a high energy telescope. Events of PeV energy, predicted from such sources as gamma ray bursts and active galactic nuclei, are less challenging to identify than threshold atmospheric neutrinos. Our analysis procedure of the 1997 data collected with the completed AMANDA detector is sketched in Table 5. The 100 Hz AMANDA-B10 trigger has generated a data set of 500 GigaBytes which has been reduced by a factor 10 by removing muon tracks that are clearly identified as down-

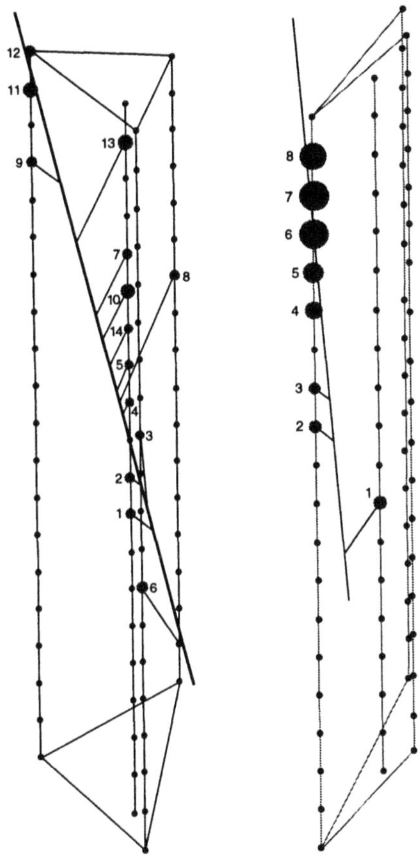

Figure 10. Events reconstructed as up-going satisfying the constraints of Table 3.

going cosmic ray background events. This filtering required 1800 hours of Cray T3E time at NERSC/LBL. While it filters 65% of the background, a Monte Carlo estimate is that 80% of the atmospheric neutrino signal is retained. The filtered data set of only 500 GigaBytes can be analysed at the collaborating home institutions. Special filters also extracted events with the characteristics of large electromagnetic showers, events where more than 100 OMs report, events in coincidence with the SPASE air shower array and the GASP atmospheric Cherenkov telescope, and events within $(-1, +5)$ minutes of a gamma ray burst. Analysis of all categories of events is in progress.

AMANDA has also been operating as a burst detector of MeV neutrinos with, for instance, the capability of detecting galactic supernovae.

ACKNOWLEDGEMENTS

The AMANDA collaboration is indebted to the Polar Ice Coring Office and to Bruce Koci for the successful drilling operations, and to the National Science Foundation (USA), the Swedish National Research Council, the K.A. Wallenberg Foundation and the Swedish Polar Research Secretariat. F.H. is supported in part by the U.S. Department of Energy under Grant No. DE-FG02-95ER40896 and in part by the University of Wisconsin Research Committee with funds granted by the Wisconsin Alumni Research Foundation.

REFERENCES

1. For a review, see T. K. Gaisser, F. Halzen and T. Stanev, *Phys. Rep.* **258**(3), 173 (1995); R. Gandhi, C. Quigg, M. H. Reno and I. Sarcevic, *Astropart. Phys.*, **5**, 81 (1996).
2. The AMANDA collaboration, *Science* **267**, 1147 (1995).
3. S. W. Barwick *et al.*, *The status of the AMANDA high-energy neutrino detector*, in Proceedings of the 25th International Cosmic Ray Conference, Durban, South Africa (1997).
4. S. Tilav *et al.*, *First look at AMANDA-B data*, in Proceedings of the 25th International Cosmic Ray Conference, Durban, South Africa (1997).
5. S. W. Barwick *et al*, *Proceedings of the 22nd International Cosmic Ray Conference*, Dublin (Dublin Institute for Advanced Studies, 1991), Vol. 4, p. 658.
6. C. Wiebusch *et al.*, *Muon reconstruction with AMANDA-B*, in Proceedings of the 25th International Cosmic Ray Conference, Durban, South Africa (1997).
7. T. Miller *et al.*, *Analysis of SPASE-AMANDA coincidence events*, in Proceedings of the 25th International Cosmic Ray Conference, Durban, South Africa (1997).
8. R. Bay *et al*, The AMANDA collaboration, Physics Reports **306**, to be published; A. Bouchta, University of Stockholm, PhD thesis (1998)

NH ELSEVIER

Nuclear Physics B (Proc. Suppl.) 77 (1999) 486–491

NUCLEAR PHYSICS B
PROCEEDINGS
SUPPLEMENTS

The Lake Baikal Experiment

V.A.Balkanov[a], I.A.Belolaptikov[g], L.B.Bezrukov[a], N.M.Budnev[b], A.G.Chensky[b], I.A.Danilchenko[a], Zh.-A.M.Djilkibaev[a], G.V.Domogatsky[a], A.A.Doroshenko[a], S.V.Fialkovsky[d], O.N.Gaponenko[a], A.A.Garus[a], T.I.Gress[a], A.M.Klabukov[a], A.I.Klimov[f], S.I.Klimushin[1], A.P.Koshechkin[a], E.V.Kuznetzov[a], V.F.Kulepov[d], L.A.Kuzmichev[c], S.V.Lovtzov[b], B.K.Lubsandorzhiev[a], M.B.Milenin[d], R.R.Mirgazov[b], A.V.Moroz[b], N.I.Moseiko[c], V.A.Netikov[a], E.A.Osipova[c], A.I.Panfilov[a], Yu.V.Parfenov[b], A.A.Pavlov[b], E.N.Pliskovsky[a], P.G.Pohil[a], E.G.Popova[c], M.I.Rozanov[e], V.Yu.Rubzov[b], I.A.Sokalski[a], CH.Spiering[h], O.Streicher[h], B.A.Tarashansky[b], T.Thon[h], R.V.Vasiljev, R.Wischnewski[h], I.V.Yashin[c].

[a] Institute for Nuclear Research, Moscow, Russia
[b] Irkutsk State University, Irkutsk, Russia
[c] Institute of Nuclear Physics, MSU, Moscow, Russia
[d] Nizhni Novgorod State Technical University, Nizhni Novgorod, Russia
[e] St.Petersburg State Marine Technical University, St.Petersburg, Russia
[f] Kurchatov Institute, Moscow, Russia
[g] Joint Institute for Nuclear Research, Dubna,Russia
[h] DESY-IfH, Berlin/Zeuthen, Germany

presented by G.V.Domogatsky

We review the present status of the Baikal Neutrino Project. The construction and performance of the large deep underwater Cherenkov detector NT-200 with 192 PMTs [1], which is currently taking data in Lake Baikal, are described. Some results from intermediate detector stages are presented.

1. Detector and Site

The Baikal Neutrino Telescope is deployed in Lake Baikal, Siberia, 3.6 km from shore at a depth of 1.1 km. At this depth, the maximum light absorbtion length for wavelengths between 470 and 500 nm is 20 ± 2 m. Scattering is strongly forward peaked $\langle \cos\theta \rangle \approx (0.85 - 0.95)$, with a scattering length about 15–30 m.

NT-200, the medium-term goal of the collaboration [2], was put into operation at April 6th, 1998 and consists of 192 optical modules (OMs) – see fig.1. An umbrella-like frame carries 8 strings, each with 24 pairwise arranged OMs. Three underwater electrical cables connect the detector with the shore station. Deployment of all detector components is carried out during 5–7 weeks in late winter when the lake is covered by thick ice.

In April 1993, the first part of *NT-200*, the detector *NT-36* with 36 OMs at 3 strings, was put into operation and took data up to March 1995. A 72-OM array, *NT-72*, run in 1995-96. In 1996 it was replaced by the four-string array *NT-96*. *NT-144*, a six-string array with 144 OMs, was taking data in 1997-98.

Summed over 840 days effective lifetime, $4.6 \cdot 10^8$ muon events have been collected with *NT-36, -72, -96, -144*.

The OMs are grouped in pairs along the strings. They contain 37-cm diameter *QUASAR* PMTs which have been developed specially for our project [1–3]. The two PMTs of a pair are switched in coincidence in order to suppress background from bioluminescence and PMT noise. A pair defines a *channel*.

A *muon-trigger* is formed by the requirement of $\geq N$ *hits* (with *hit* referring to a channel) within 500 ns. N is typically set to 3 or 4. For such events, amplitude and time of all fired channels are digitized and sent to shore. A separate *monopole trigger* system searches for clusters of

PII S0920-5632(99)00475-2

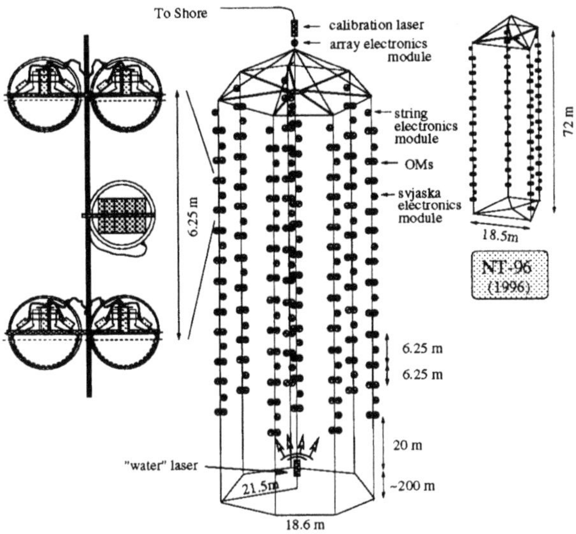

Figure 1. Schematic view of the Baikal Telescope *NT-200*. The array is time-calibrated by two nitrogen lasers. The one (fiber laser) is mounted just above the array. Its light is guided via optical fibers to each OM pair. The other (water laser) is arranged 20 m below the array. Its light propagates directly through the water. The expansion left-hand shows 2 pairs of optical modules ("svjaska") with the svjaska electronics module, which houses parts of the read-out and control electronics. Top right, the 1996 array *NT-96* is sketched.

sequential hits in individual channels which are characteristic for the passage of slowly moving, bright objects like GUT monopoles.

In the initial project of *NT-200* [1,2], the optical modules were directed alternating upward and downward (fig.1). However, from the experience with *NT-36* and *NT-72* we have found that the sensivity of uplooking OMs decreases due to sedimentation by 50% after 150 days. Hence starting from *NT-96*, the orientation of OMs has been changed: only OMs from two layers of the array (the second and eleventh) look upward, and all others look downward. Nevertheless we possibly come back toward a symmetrical structure if the problems with sedimentation will be solved.

2. METHODICAL INVESTIGATIONS

2.1. "Shadow" of the shore in muons

NT-200 is placed at a distance of 3.6 km to the nearby shore of the lake. The opposite shore is about 30 km away. This asymmetry opens the possibility to investigate the influence of the close shore to the azimuth distribution under large zenith angles, where reconstruction for the comparatively "thin" *NT-96* is most critical. A sharp decrease of the muon intensity at zenith angles of 70^0-90^0 is expected. The comparison of the experimental muon angular distribution with MC calculations gives us an estimation of the accuracy of the reconstruction error close to the horizontal direction. Indeed, the *NT-96* data show a pronounced dip of the muon flux in the direction of the shore and for zenith angles larger than 70^0 – in very good agreement with calculations which take into the effect of the shore.

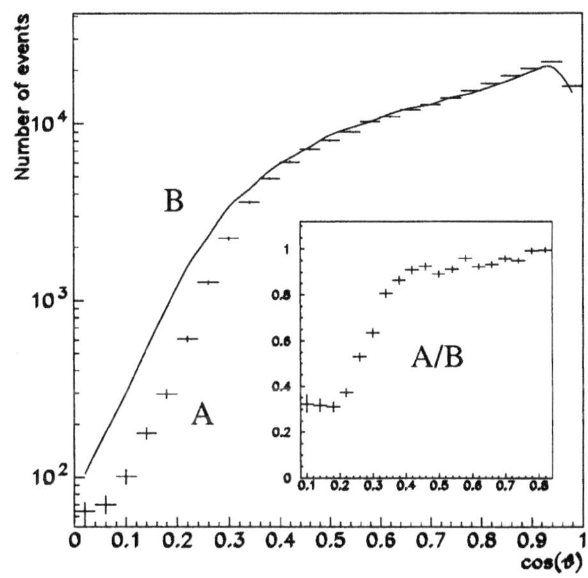

Figure 2. Atmospheric muons (vs zenith angle θ) as it is measured in the direction to the nearest point of the shore(A) and in opposite one(B) - "open" water. The small picture shows the ratio A to B.

2.2. Coincidence operation of underwater telescope and Cherenkov EAS array

During the expedition 1998, a Cherenkov air shower array consisting of four QUASAR tubes was deployed on the ice, just above the underwater telescope, in order to check the angular resolution of the latter. The angular error in the determination of the EAS direction is less than 1^0. This fact allows us to estimate the muon track reconstruction error of the underwater detector at angles close to the vertical (0^0-30^0). The total number of coincidence events is only 450 since *NT-200* started operation at the end of a moonless period. We presently analyze this data.

3. SOME PHYSICS RESULTS

Earlier we have discussed the main results obtained with the first small detector *NT-36* - investigation of atmospheric muon flux, searching for nearly vertically upward moving muons and searching for slowly moving GUT monopoles [2,4,5]. Below we present selected results obtained with *NT-96*.

3.1. Identification of nearly vertically upward moving muons

Different to the standard analysis [2], the method presented in this section relies on the application of a series of cuts which are tailored to the response of the telescope to nearly vertically upward moving muons [4,6]. The cuts remove muon events far away from the opposite zenith as well as background events which are mostly due to pair and bremsstrahlung showers below the array and to naked downward moving atmospheric muons with zenith angles close to the horizon ($\theta > 60°$). The candidates identified by the cuts are afterwards fitted in order to determine the zenith angle.

We included all events with $\geq$4 hits along at least one of all hit strings. To this sample, a series of 6 cuts is applied. Firstly, the time differences of hit channels along each individual string have to be compatible with a particle close to the opposite zenith (1). The event length should be large enough (2), the maximum recorded amplitude should not exceed a certain value (3) and the center of gravity of hit channels should not be close to the detector bottom (4). The latter two cuts reject efficiently brems showers from downward muons. Finally, also time differences of hits along *different* strings have to correspond to a nearly vertical muon (5) and the time difference between top and bottom hit in an event has to be larger than a minimum value (6).

The effective area for muons moving close to opposite zenith and fulfilling all cuts exceeds 1000 m^2.

Within 70 days of effective data taking, $8.4 \cdot 10^7$ events with the muon trigger $N_{hit} \geq 4$ have been selected.

Table 1
The expected number of atmospheric neutrino events and background events, and the observed number of events after cuts *1–6*.

after cut N^o →	1	2	3	4	5	6
atm. ν, MC	11.2	5.5	4.9	4.1	3.8	3.5
background, MC	7106	56	41	16	1.1	0.2
experiment	8608	87	66	28	5	4

Table 1 summarizes the number of events from all 3 event samples (MC signal and background, and experiment) which survive the subsequent cuts. After applying all cuts, four events were selected as neutrino candidates, compared to 3.5 expected from MC. One of the four events has 19 hit channels on four strings and was selected as neutrino candidate by the standard analysis too. The zenith angular distribution of these four neutrino candidates is shown in the inner box of fig.3.

Regarding the four detected events as being due to atmospheric neutrinos, one can derive an the upper limit on the flux of muons from the center of the Earth due to annihilation of neutralinos - the favored candidate for cold dark matter.

The limits on the excess muon flux obtained with underground experiments [7–9] and *NT-96* are shown in Table 2. The limits obtained with *NT-96* are 4–7 times worse then the best underground limits since the data collecting time of *NT-96* was only $\approx$ 70 days.

Table 2
90% C.L. upper limits on the muon flux from the center of the Earth for four regions of zenith angles obtained in different experiments

Zenith angles	Flux limit ($10^{-14} \cdot (cm^2\ sec)^{-1}$)			
	NT-96 $> 10GeV$	*Baksan* $> 1GeV$	*MACRO* $> 1.5GeV$	*Kam-de* $> 3GeV$
$\geq 150°$	11.0	2.1	2.67	4.0
$\geq 155°$	9.3	3.2	2.14	4.8
$\geq 160°$	5.9 – 7.7	2.4	1.72	3.4
$\geq 165°$	4.8	1.6	1.44	3.3

This result, however, illustrates the capability of underwater experiments with respect to the search for muons due to neutralino annihilation in the center of the Earth.

3.2. Selection of neutrino events over a large solid angle

The signature of neutrino induced events is a muon crossing the detector from below. With the flux of downward muons exceeding that of upward muons from atmospheric neutrino interactions by about 6 orders of magnitude, a careful reconstruction is of prime importance.

In contrast to first stages of the detector (*NT-36* [4]), *NT-96* can be considered as a real neutrino telescope for a wide region in zenith angle θ. After the reconstruction of all events with ≥ 9 hits at ≥ 3 strings (trigger *9/3*), quality cuts have been applied in order to reject fake events. Furthermore, in order to guarantee a minimum lever arm for track fitting, events with a projection of the most distant channels on the track (Z_{dist}) less than 35 meters have been rejected. Due to the small transversal dimensions of *NT-96*, this cut excludes zenith angles close to the horizon.

The efficiency of the procedure has been tested with a sample of $1.8 \cdot 10^6$ MC-generated atmospheric muons, and with MC-generated upward muons due to atmospheric neutrinos. It turns out that the signal to noise ratio is > 1 for this sample.

The reconstructed angular distribution of $2 \cdot 10^7$ events taken with *NT-96* in April/September 1996 – after all cuts – is shown in fig.3.

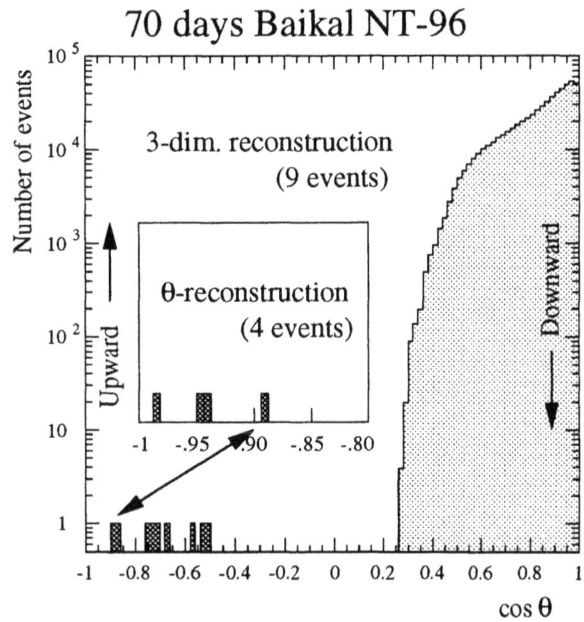

Figure 3. Experimental angular distribution of events satisfying trigger *9/3*, all final quality cuts and the limit on Z_{dist} (see text). The subpicture shows the events selected by using the method described in subsection 3.1. The event found by both algorithms is marked by the arrow.

From 70 days of *NT-96* data, 12 neutrino candidates have been found. Nine of them have been fully reconstructed. Three nearly upward vertical tracks (see subsection 3.1) hit only 2 strings and give a clear zenith angle but ambiguities in the azimuth angle – similar to the two events from *NT-36* [2]. This is in good agreement with MC expectations.

3.3. Search for fast monopoles ($\beta > 0.75$)

Fast bare monopoles with unit magnetic Dirac charge and velocities greater than the Cherenkov threshold in water ($\beta = v/c > 0.75$) are promising survey objects for underwater neutrino telescopes. For a given velocity β the monopole Cherenkov radiation exceeds that of a relativistic muon by a factor $(gn/e)^2 = 8.3 \cdot 10^3$ ($n = 1.33$

- index of refraction for water) [10,11]. Therefore fast monopoles with $\beta \geq 0.8$ can be detected up to distances 55 m ÷ 85 m which corresponds to effective areas of $(1–3)\cdot 10^4$ m^2.

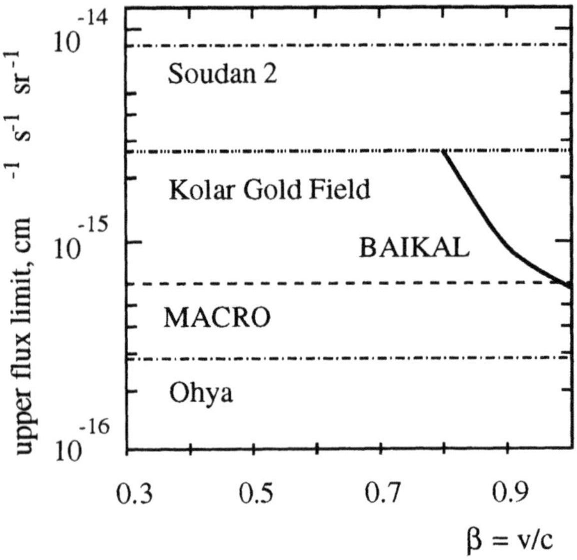

Figure 4. The 90% C.L. Baikal upper limit for an isotropic flux of bare magnetic monopoles compared with other published limits.

The natural way for fast monopole detection is based on the selection of events with high multiplicity of hits. In order to reduce the background from downward atmospheric muons we restrict ourself to monopoles coming from the lower hemisphere.

Two independent approaches have been used for selection of upward monopole candidates from the 70 days of *NT-96* data. The first one is similar to the method which was applied to upward moving muons (see subsection 3.1), with an additional cut $N_{hit} > 25$ on the hit multiplicity. The second one cuts on the value of space-time correlation, followed by a cut $N_{hit} > 35$ on the hit multiplicity. The upper limits on the monopole flux obtained with the two different methods coincide within errors.

The same type of analysis was applied to the data taken during 0.42 years lifetime with the neutrino telescope *NT-36* [12].

The combined 90% C.L. upper limit obtained by the Baikal experiment for an isotropic flux of bare fast magnetic monopoles is shown in fig.4, together with the best limits from underground experiments Soudan2, KGF, MACRO and Ohya [13–16] in Fig.4.

4. Conclusions and Outlook

The Baikal detector is well understood, and first atmospheric neutrinos have been identified. Also muon spectra have been measured, and limits on the fluxes of magnetic monopoles as well as of neutrinos from WIMP annihilation in the center of the Earth have been derived.

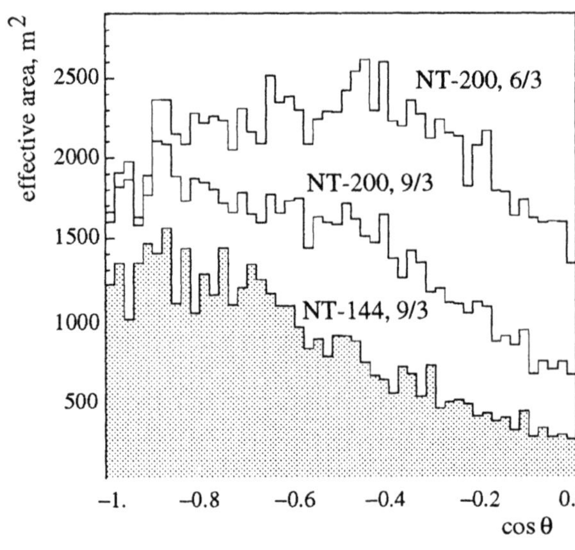

Figure 5.
The effective area of detectors vs zenith angle for events under trigger *9/3*(*6/3*) and at the condition $Z_{dist} > 35\,m$ (see text). Results are shown for *NT-200* and *NT-144*.

In the following years, *NT-200* will be operated as a neutrino telescope with an effective area between 1000 and 5000 m^2, depending on the energy. It will investigate atmospheric neutrino spectra above 10 GeV (about 1 atmospheric neutrino per day). Fig.5 shows the effective area for atmospheric neutrinos as a function of the zenith angle. We give the area after a cut on the event length and for two software triggers (*6/3* and *9/3*). Also shown is the area for the smaller detector *NT-144* operated in 1997.

Presumably still too small to detect neutrinos from AGN and other extraterrestrial sources, *NT-200* can be used to search for neutrinos from WIMP annihilation and for magnetic monopoles. It will also be a unique environmental laboratory to study water processes in Lake Baikal.

Apart from its own goals, *NT-200* is regarded to be a prototype for the development a telescope of next generation with an effective area of 50,000 to 100,000 m^2. The basic design of such a detector is under discussion at present.

This work was supported by the Russian Ministry of Research,the German Ministry of Education and Research and the Russian Fund of Fundamental Research (grants 96-02-17308, 97-02-17935, 97-02-31010, 97-02-96589 and 97-05-96466).

REFERENCES

1. I.A.Sokalski and Ch.Spiering (eds.) 1992 *The Baikal Neutrino Telescope NT-200, BAIKAL 92-03*
2. I.A.Belolaptikov *et al.*, *Astroparticle Physics* 7 (1997) 263.
3. R.I.Bagduev *et al.* 1998 *Preprint DESY* 98-091, July 1998
4. L.B.Bezrukov *et al.*, *Proc. of the 2nd Workshop on the Dark Side of the Universe*, 221 (Rome, 1995) (astro-ph/9601161)
5. L.B.Bezrukov *et al.*, *Proc. of the 2nd Workshop on the Dark Side of the Universe*, 221 (Rome, 1995) (astro-ph/9601160)
6. V.A.Balkanov *et al.* 1998 *Preprint INR 0972/98* (in russian)
7. M.M.Boliev *et al.* 1996 *Nucl.Phys.* (Proc. Suppl.) **48** 83
8. T.Montaruli *et al.* 1997 *Proc. 25-th ICRC* Durban–South Africa, vol.7, 185
9. M.Mori *et al.* 1993 *Phys. Rev.* **D48** 5505
10. I.M.Frank 1988 *Vavilov-Cherenkov Radiation* (Moscow: Nauka) 192 (in russian)
11. D.A.Kirzhnits and V.V.Losjakov 1985 *Pis'ma Zh. Eksp.Theor.Fz.* **42** 226
12. V.A.Balkanov *et al.* 1998 Preprint INR (Moscow: INR) (in russian)
13. S.Orito *et al.* 1992 *Phys. Rev. Lett.* **66** 1951.
14. M.Ambrosio *et al.* 1998 *MACRO Preprint* MACRO/PUB 98/3
15. H.Adarkar *et al.* 1990 *Proc. 21st ICRC. Adelaide.* 95
16. J.L.Thorn *et al.* 1992 *Phys. Rev.* **D46** 4846

ELSEVIER

Nuclear Physics B (Proc. Suppl.) 77 (1999) 492–497

NUCLEAR PHYSICS B
PROCEEDINGS
SUPPLEMENTS

Neutrino telescopes under the ocean: The case for ANTARES

L. Moscoso[a*]

[a]DSM/DAPNIA/SPP, CEA/Saclay,
91191 Gif-Sur-Yvette CEDEX, France

Neutrino telescopes offer an alternative way to explore the Universe. Several projects are in operation or under construction. A detector under the ocean is very promising because of the very accurate angular resolution that it provides. The ANTARES project is intended to demonstrate the feasibility of such a detector.

1. INTRODUCTION

High energy cosmic neutrinos should provide a new means to explore the sky. Due to the weakness of their interaction with matter and the absence of their interaction with the electromagnetic radiations, neutrinos can travel in the Universe without being absorbed. On the contrary, because photons interact with matter, with the infra-red (IR) radiation and with the cosmic microwave background (CMB), the Universe is opaque for high energy gamma rays.

Charged cosmic rays such as protons are deflected by the galactic and extra-galactic magnetic fields. So, only UHE protons, above 10^{20} eV, are rigid enough to point back to their source. Nevertheless, at these extreme energies protons also suffer from interactions with IR radiations and with the CMB which limit their free path-length to about 50 Mpc. At greater distances protons continuously lose energy and become less and less rigid.

Therefore, it appears that the only way to explore the Universe in the high energy range and at great distances is to detect neutrinos.

2. DETECTION OF HIGH ENERGY NEUTRINOS

High energy muon neutrinos can be detected by searching for long-range muons produced in charged current exchange interactions of neutrinos with the matter surrounding the detector. Due to the increase of the νN cross-section with the neutrino energy and to the increase of the muon path-length with the muon energy, the probability to detect a muonic neutrino aimed towards the detector is an increasing function of the neutrino energy. This means that high energy neutrinos will be statistically enhanced. Moreover, the angle between the neutrino and the produced muon is very small for high energy neutrinos. So the direction of the parent neutrino is well determined.

Despite its increase with the neutrino energy, the νN cross-section remains small and, moreover, the neutrino flux is expected to be a decreasing function of the neutrino energy with a differential $\simeq E^{-2}$ behaviour. The detector area must be large enough to provide sufficient sensitivity to detect cosmic sources over the widest possible solid angle on a reasonable time scale. For this reason a volume of detection of 1 km^3 is needed.

At the surface of the Earth the main source of background is the flux of particles produced in the cascades initiated in the atmosphere by primary cosmic rays. The major component is downward-going muons which can be rejected by selecting only upward-going particles. In order to suppress particles produced in the back-scattering of atmospheric muons, the detector must be well shielded. The remaining source of physical background is the flux of upward-going neutrinos produced in the atmospheric cascades.

The most economic way to realize a km-scale well-shielded detector is to build a 3-D array of optical modules in the deep ocean or in polar ice. In these media high energy muons crossing

*On behalf of the ANTARES collaboration

PII S0920-5632(99)00476-4

the detector produce Čerenkov light at an angle $\theta_C \simeq 43°$. The reconstruction of the muon direction is performed by using the information on the arrival times of the photons recorded by the optical modules.

3. THE ANTARES PROJECT

The main advantages of deploying the detector in the ocean compared to ice are the long scattering length and low scattering angle of light in water, the possibility to deploy on different sites located at different latitudes and the possibility to find sites at great depth. Nevertheless, several questions must be answered. In particular tests are needed to master the deployment and connection operations in the deep sea. Moreover the environmental parameters like the optical background rate, the bio-fouling rate and the water transparency must be measured. The ANTARES project [1] has two goals:

1. Realization of apparatuses capable of measuring environmental parameters such as optical background, bio-fouling and water transparency;
2. Construction and deployment of a 3-D prototype ("Demonstrator") scalable to a cubic kilometer detector.

These operations are performed off Toulon (France), 30 km from the shore at 2350 m depth. When the feasibility of a large detector has been demonstrated, further steps towards a cubic kilometer will be proposed.

3.1. The demonstrator

The deployment of a large network of optical modules in the deep ocean is one of the major problems to be solved. A very detailed programme has been defined in the ANTARES project to proceed by steps in order to ensure the good quality of the procedures and to reduce causes of failures.

The final prototype will consist of three strings, equipped with about 30 optical modules each, electrically interconnected through a junction box which will be linked to the shore via an electro-optical cable. The signals delivered by each optical module will be transmitted to the shore station through the optical fibers of the electro-optical cable.

The 40 km long electro-optical cable was successfully deployed in May 98. Mechanical tests of the first string started in July 98 and finished in September. Connection-deconnection tests are foreseen for the end of 98. A first string equipped with 8 optical modules, electronics, and positioning and slow-control systems will be deployed by the end of 98 and connected to the electro-optical cable. The first fully equipped string is foreseen for 99.

3.2. Environmental parameters

The site where the final very large detector will be installed must have very good properties from all points of view. The ANTARES collaboration is constructing a system of autonomous detectors capable of measuring the sea quality at any depth down to 4 000 m in order to chose the site and the characteristics of the final detector.

3.2.1. Optical background

The optical background has two main components: a continuous component due to Čerenkov light produced by electrons emitted in the β decay of the ^{40}K present in the sea water and a variable component due to the bio-luminescence emitted by bacteria and fishes.

Several deployments of 350 m long strings at 2 350 m depth have been performed. Each string supports several PMTs (1-3) and other monitors such as current-meters, thermometers, compasses and tilt-meters. Each string is equipped with a data-logger, an acoustical modem to transmit data to the boat, and a system of releases for recovery. The power is supplied by lithium batteries. Figure 1 shows an example of the variations of the counting rate recorded with our system during a period of medium activity. The pulse height threshold used to measure the counting rate was set to a value corresponding to 1/3 of the mean amplitude for a single photo-electron. The figure clearly shows the bio-luminescent activity consisting of bursts of light lasting a few seconds. The continuous background due to the ^{40}K decay is also visible with a frequency of $\approx$ 40 kHz. The

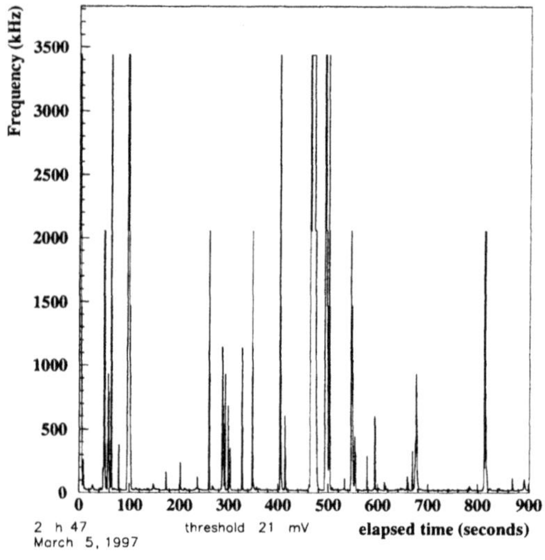

Figure 1. Variations of the counting rate with time.

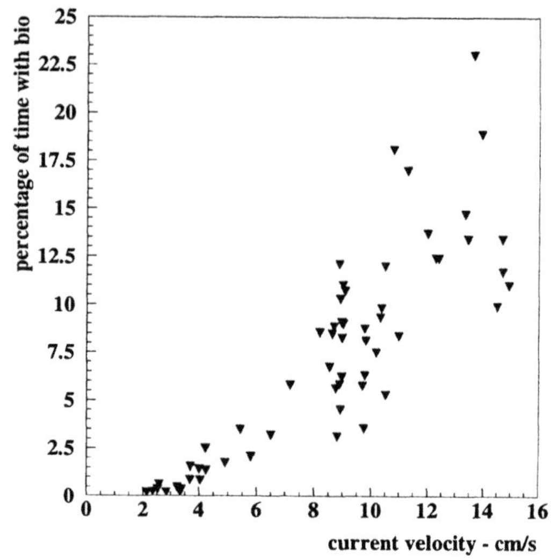

Figure 2. Percentage of time where the counting rate exceeded by 15% the ^{40}K background vs the current speed.

fraction of time during which the counting rate exceeded the ^{40}K background by at least 15% was found to be strongly correlated to the current speed, as shown in figure 2.

The measurement performed simultaneously with two PMTs 40 m apart showed a very little time correlation, suggesting that the size of the region where the bio-luminescence is active is generally less than 40 m.

3.2.2. Bio-fouling

The measurement of the bio-fouling of the glass sphere which will house the PMT is a long term operation. It has been performed twice for periods of 3 and 8 months. The system was a string equipped with two glass spheres; one of them (A) housed a system of light diodes (LED) which were continuously monitored, and the other one (B) housed several PIN diodes located at different latitudes of the sphere. During the first deployment the frame supporting the spheres was oriented vertically with the sphere A in the upper position emitting light downwards towards the sphere B.

The result of this measurement, of 3 months duration, is shown in figure 3.

The counting rate of upward-facing PIN decreased by 60% in 3 months while the rates of the PINs located at other latitudes decreased less. The counting rates are highly discontinuous and increase suddenly in the presence of high water current velocity. This suggests that *i)* the bio-fouling covered only a small angular region on the top of the glass sphere and *ii)* the bio-fouling was partially removed by the water current probably because it was not strongly glued on the glass.

A second measurement was performed after orienting the frame horizontally. The result of this 8-month-long measurement is shown in figure 4. Taking into account that the LED sphere suffers the same fouling as the PIN sphere, one can estimate that the horizontal region of the glass sphere loses only about 1.5% of transparency in one year.

3.2.3. Water transparency

The light attenuation as a function of the distance from the source was measured using a 350 m long mooring string incorporating a 33 m long

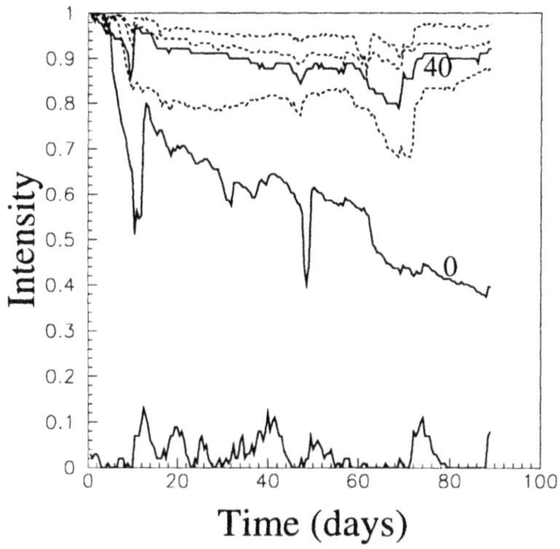

Figure 3. Evolution of the PIN diode counting rates measured during the first period of 3 months with the frame supporting the spheres in the vertical position. The upper curves depict the counting rates of PIN diodes at different latitudes. Full lines are for PIN diodes at zenith angles of 0° (top of the sphere) and 40° as indicated by the labels. Dashed lines are for PIN diodes at 20°. The intensities are normalized to the first measurement. The curve on the bottom of the figure shows the variation of the water current velocity in m/s.

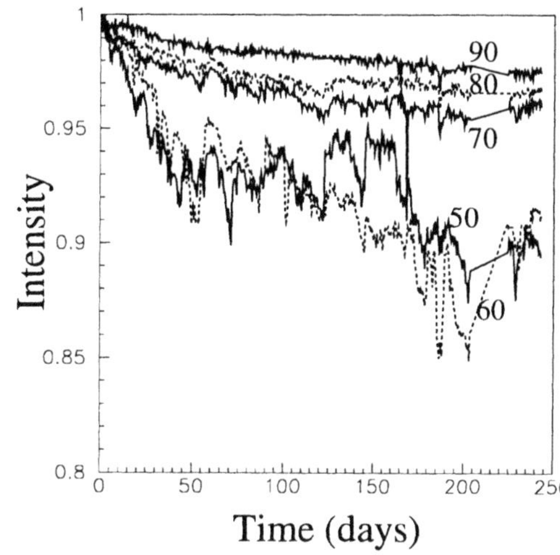

Figure 4. Evolution of the PIN diode counting rates measured during the second period of 8 months with the frame supporting the spheres in the horizontal position. Different curves depict the counting rates of PIN diodes at different latitudes. The zenith angle (degrees) of the position of the PIN diode is indicated by the label on each curve. The intensities are normalized to the first measurement. Note that the vertical scale is different from that of figure 3.

rigid structure holding an optical module at one end facing a motorized trolley carrying a light source along the structure. Continuous light sources (LEDs) emitting at different wavelengths are used. The measured attenuation length is 39±3 m for a wavelength of 466 nm. In July 98 a pulsed light source was used to disentangle the contributions from light absorption and from light scattering. The analysis is in progress.

3.3. Optical modules

Different large photo-cathode photo-multipliers are being tested (EMI and Hamamatsu 8" tubes). Larger tubes (10" and 11") are also foreseen. A dark box equipped with a mechanical system allowing a blue LED to scan the photo-cathode area is used to measure the response of the phototubes as a function of the position of the light spot. A water tank is used to study the overall response of the optical module to the Čerenkov light emitted by cosmic muons in water.

3.4. Data transmission

The optical module signals will be transmitted to the shore through the optical fibers of the electro-optical cable. Analog data transmission will be used first and replaced later by digital transmission of data from a signal sampling device based on an ASIC chip currently under development. The electro-optical cable, deployed in May 98, is equipped with four mono-mode fibers. The measured attenuation is 0.33 dB/km at a wavelength of 1 310 nm.

3.5. Positioning and slow control

The knowledge of the relative position of the optical modules should match the size of the photo-cathodes of the PMTs (20 cm). This will be achieved by sonar triangulation between an external base and acoustic detectors along the strings. Adjustment of the PMT voltages, recording of the environmental parameters, and measurement of the detector geometry will be managed by the slow-control system. These monitoring data will be transmitted to the shore through the electro-optical cable.

3.6. Software

A software package has been developed to simulate the neutrino interaction in the medium surrounding the detector, the muon tracking, the Čerenkov light emission and the detector response. The optical background and the distortions of the detector by the water currents are also simulated.

4. TOWARDS A CUBIC KILOMETER DETECTOR

The development of a cubic kilometer detector is a very complex challenge which must be reached by steps. The ANTARES collaboration envisages an intermediate stage consisting of a detector of 0.1 km^2 made of a network of about 1000 PMTs. The software packages developed for the ANTARES project have been used to estimate the performance expected for this intermediate stage. Different detector layouts have been considered, all made of 15 strings equipped with a total of about 1000 PMTs. The results are quite similar for the different layouts.

The reconstruction algorithm applied to simulated events shows that the angular resolution for reconstructed tracks will be better than 0.2°. Moreover, for muons with energies that trigger a km-scale detector (above 1-10 TeV), the angle between the muon and the parent neutrino is smaller than 0.1°. Figure 5 shows the distribution of the angle between the direction of the muon entering the detector and the direction of the reconstructed track for muons induced by simulated neutrinos with a E^{-2} energy spectrum.

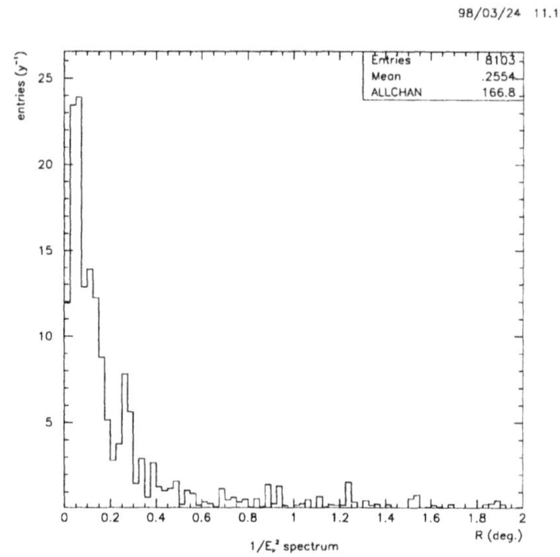

Figure 5. Distribution of the angle between the direction of the muon entering the detector and the direction of the reconstructed track for muons induced by simulated neutrinos with a E^{-2} differential energy spectrum.

5. EXTRAPOLATION OF THE EGRET CATALOGUE

Powerful extra-galactic objects such as AGNs or topological defects could contribute to a diffuse flux of high energy neutrinos significantly larger than the flux of atmospheric neutrinos at high energy [3,4].

Moreover, due to the extremely low angle between the muon and the parent neutrino and to the good quality of the muon direction measurement, the atmospheric neutrino background contaminating each individual source can be reduced to a very low level by selecting very small angular regions of the sky. In that way, a signal of only a few events could be significant. The sensitivity of the detector to muon neutrinos can be estimated from measured low-energy gamma-ray fluxes by assuming that *i)* the low energy gamma-rays are of hadronic origin and *ii)* the emitted gamma-rays have a differential energy spectrum E^{-2}.

With these assumptions the muon neutrino flux is about 40% of the flux of gammas at the production source. Using the 2nd EGRET catalog [2] for sources measured during the P12 period, the derived neutrino flux has been extrapolated to the energies where the neutrino detector is sensitive.

Although no individual extra-galactic source can be detected with an exposure of $0.1\,\mathrm{km}^2$·year, a statistically significant effect can be detected by adding the contributions of all the extra-galactic sources. A calculation made for the 43 identified AGN gives a total number of 8-67 events (depending on the value of the differential spectral index used: respectively 2.2 and 2) to be compared to a total background of 2.7 events.

A possible scenario would be to start with a $0.1\,\mathrm{km}^2$ detector made of about 15 strings equipped with about 1000 optical modules. If a statistical enhancement correlated with the positions of the AGN sources could be detected on a time scale of the order of one year this would motivate the construction of the cubic kilometer detector.

6. CONCLUSIONS

The sky survey with high energy neutrinos is essential in order to obtain new information complementary to that obtained from low energy gamma-rays and short distance high energy gamma-rays and ultra high energy protons. This requires a detector at the kilometer scale which can only be developed in stages. The construction of a deep ocean prototype and a programme of measurement of environmental parameters is the first step necessary to demonstrate the feasibility of such a detector.

Studies performed by the ANTARES collaboration indicate that a detector in the ocean can determine the origin of each event with a very fine angular resolution. This is a major advantage of the underwater technique, making it possible to reduce the background from known point sources to very low level.

REFERENCES

1. The ANTARES proposal and contributions to conferences by the ANTARES Collaboration can be found in the ANTARES Web pages: http://antares.in2p3.fr/antares/
2. The Second EGRET Catalog of High-Energy Gamma-Ray Sources, http://cossc.gsfc.nasa.gov/cossc/egret/egret catalog/cattex.html
3. F. W. Stecker, C. Done, M. H. Salamon and P. Sommers, Phys. Rev. Lett. 66 (1991) 2697 and Phys. Rev. Lett. 69 (1992) 2738(E); L. Nellen, K. Mannheim and P. L. Biermann Phys. Rev. D 47 (1994) 5270; A. P. Szabo and R. J. Protheroe, Astropart.Phys. 2 (1994) 375; K. Mannheim, Astropart.Phys. 3 (1995) 295; R. J. Protheroe, High Energy Neutrinos from Blazars, Accretion Phenomena and Related Outflows, IAU Colloq. 163, 1996, ADP-AT-96-7.
4. P. Bhattacharjee, Ch. T. Hill and D. N. Schramm, Phys. Rev. Lett. 69 (1992) 567; G. Sigl, S. Lee, D. N. Schramm and P. Coppi, Phys. Lett. B 392 (1997) 129.

ELSEVIER

Nuclear Physics B (Proc. Suppl.) 77 (1999) 498–508

NUCLEAR PHYSICS B
PROCEEDINGS
SUPPLEMENTS

Extremely high energy cosmic rays and neutrinos

James W. Cronin[a]*

[a]Enrico Fermi Institute, University of Chicago,
5640 S. Ellis Ave., Chicago, IL, 60637, USA

The evidence for the existence of cosmic rays with energies in excess of 10^{20} eV is now overwhelming. There is so far no indication of the GZK cutoff in the energy spectrum at 5×10^{19} eV. This conclusion is not firm for lack of statistics. A cutoff would be expected if the sources of the cosmic rays were distributed uniformly through out the cosmos. The sources of cosmic rays with energy above the GZK cutoff must be at a distance ≤ 100 Mpc, and if they are protons they are very likely to point to these sources. There are no easy explanations how known astrophysical objects can accelerate protons (or atomic nuclei) to these energies. This difficulty has led to speculation that there may be exotic sources such as topological defects which produce these energetic cosmic rays directly along with a copious supply of neutrinos of similar energy. The fluxes of these cosmic rays is very low and large instruments are required to observe them even with modest statistics. One such instrument, the Pierre Auger Observatory, is described in some detail. It is designed for all-sky coverage and the construction of its southern site will begin in Argentina in 1999. This instrument has the capability to detect neutrinos with energy $\geq 10^{18}$ eV and for some predictions neutrinos would actually be observed.

1. The cosmic ray energy spectrum above 10^{18} eV

In recent years the interest in extremely high energy cosmic rays (EHECR), those with energy $\geq 10^{18}$ eV (EeV), has revived because of a number of discoveries. Therefore there are many excellent reviews, books, and conference proceedings to which the reader is referred [1]. The energy spectrum of cosmic rays is quite well measured up to 10^{19} eV. Above the knee (3×10^{15} eV) it falls as a power law in energy, $dN/dE \sim E^{-\alpha}$, with an index α=3.

Above 10^{17} eV the cosmic ray spectrum has significant structure, which is displayed in Fig 1, where the differential spectrum has been multiplied by E^3 to better display the observed structures. These data are the contribution of four experiments which have operated over the past 20 years. These experiments observe the cosmic rays indirectly by means of the air showers they produce. They are from the Haverah Park surface array in England [2], the Yakutsk surface array in Siberia [3], the Fly's Eye fluorescence detector in Utah [4], and the AGASA surface array in Japan [5]. About the time of this conference the AGASA group reported the results of seven years of operation with their 100 km^2 array, giving the largest exposure from a single detector ever reported [6]. Before plotting, the energy scale of each experiment has been adjusted by amounts $\leq 20\%$ to show most clearly the common features. The method of energy determination in each of these experiments is quite different, and the fact that they agree within 20% is remarkable.

The spectrum continues with an index of 3.0 until about 5×10^{17} eV where it steepens with an index of about 3.3. Above an energy of 10^{18} eV it is difficult for the galaxy to contain even iron nuclei and galactic accelerators that can produce such energies cannot be imagined. If cosmic rays at these energies continue to be produced in the galaxy, they should show a strong anisotropy which correlates with the distribution of matter in our galaxy. Recently the AGASA group has observed a possible correlation with the galactic center and a spiral arm in the Cygnus region in a narrow energy range around 10^{18} eV [20]. Above this energy such a correlation is not observed probably due to lack of statistics. Above

*present address, Department of Physics, High-Energy Astrophysics Institute, University of Utah, 201 JFB, Salt Lake City, UT, 84112, USA

PII S0920-5632(99)00484-3

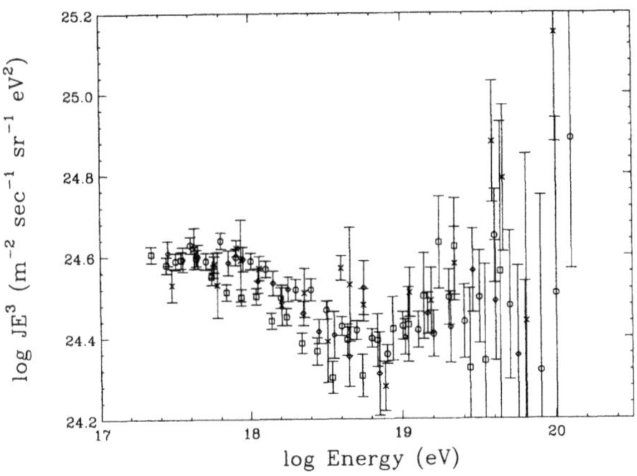

Figure 1. Upper end of the cosmic ray spectrum. Haverah Park [2] points (crosses) serve as a reference. Yakutsk [3] points [diamonds] have been reduced by 20%. Fly's Eye [3] points (squares) have been increased by 10%. Agasa [4] points (circles) have been reduced by 10%.

5×10^{18} eV the spectrum hardens to a spectral index of 2.7. This hardening of the spectrum may be due to a new component that is extragalactic.

The composition of the cosmic rays is notoriously difficult to measure with the indirect air shower methods. Such evidence as does exist suggests that the composition is moving towards a lower mean atomic number as the energy increases from 10^{17} eV to 10^{19} eV [8].

2. The Difficulty of Acceleration

Above 10^{19} eV the precision of the spectrum measurement suffers from lack of statistics. There have been about 60 events recorded with energy greater than 5×10^{19} eV. Yet it is above this energy that the scientific mystery is the greatest. There is little understanding how known astrophysical objects can produce particles of such energy. At the most primitive level, a necessary condition for the acceleration of a proton to an energy of 10^{20} eV requires that the product of the magnetic field B and the size of the region R be much larger than 3×10^{17} gauss-cm. This value is appropriate for a perfect accelerator such as might be scaled up from the Tevatron at Fermilab. The Tevatron has a BR=3×10^9 gauss-cm and accelerates protons to 10^{12} eV. The possibility of acceleration of cosmic rays to energies above 10^{19} eV seems difficult and the literature is filled with speculations. Two reviews which discuss the basic requirements are given by Greisen [9] and Hillas [10]. While these were written some time ago, they are excellent in outlining the basic problem of cosmic ray acceleration. Biermann [11] has recently reviewed all the ideas offered to achieve these high energies. Hillas in his outstanding review of 1984 presented a plot which graphically shows the difficulty of cosmic ray acceleration to 10^{20} eV. Figure 2 is a reproduction of his figure. Plotted are the size and strength of possible acceleration sites. The upper limit on the energy is given by;

$$E_{18} \leq 0.5\beta Z B_{\mu g} L_{kpc}.$$

Here the E_{18} is the maximum energy measured in units of 10^{18} eV. L_{kpc} is the size of the accelerating region in units of kilo-parsec, and $B_{\mu g}$ is the magnetic field in μgauss. The factor β was introduced by Greisen to account for the fact that the effective magnetic field in the accelerator analogy is much less than the ambient field. The factor β in the Hillas discussion is the velocity of the shock wave (relative to c) which provides the acceleration. Lines corresponding to a 10^{20} eV proton with β=1 and 1/300 are plotted. A line is also plotted for iron nuclei (β=1). With Z=26, iron is in principle easier to accelerate, but in a realistic situation it is difficult to avoid the disintegration of the nucleus during the acceleration process. Real proton accelerators should lie *well* above the solid line. The figure is also relevant for "one shot" acceleration as it represents the emf induced in a conducter of length L moving with a velocity β through a uniform magnetic field B. Synchrotron energy loss is also important. For protons the synchrotron loss rate at 10^{20} eV requires that the magnetic field be less than 0.1 gauss for slow acceleration (the accelerator analogy)[9]. The conclusion from this figure is that the acceleration of cosmic rays to 10^{20} eV is not a simple matter. Because of this difficulty some authors have seriously postulated that the cosmic rays are not accelerated but are directly produced by "top down" processes. Defects in the fabric of space-time can have huge energy content, and can release this energy in the form of high energy cosmic rays [12].

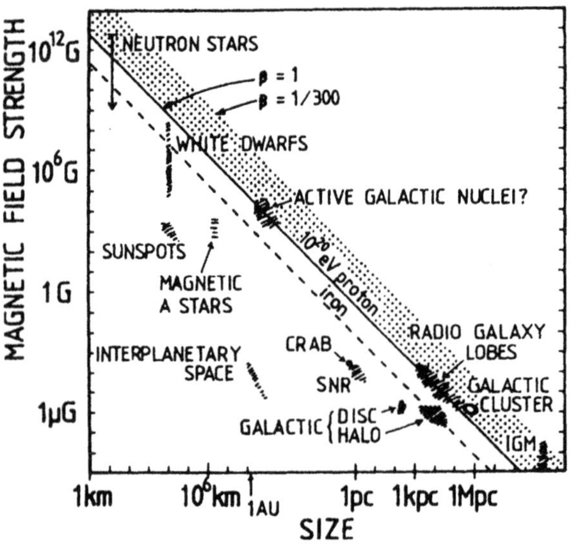

Figure 2. Hillas plot[10]. Size and magnetic field of possible sites of acceleration. Objects below the solid line cannot accelerate protons to 10^{20} eV.

3. Nature's Diagnostic Tools

There are some natural diagnostic tools which make the analysis of the cosmic rays above 5×10^{19} eV easier than at lower energies. The first of these is the 2.7K Cosmic Background Radiation (CBR). Greisen [13] and Zatsepin and Kuz'min [14] pointed out that protons, photons, and nuclei all interact strongly with this radiation (GZK effect). As an example a collision of a proton of 10^{20} eV colliding with a CBR photon of 10^{-3} eV produces several hundred MeV in the center of mass system. The cross section for pion production is quite large so that collisions are quite likely, resulting in a loss of energy for the primary proton. Almost independent of the initial energy of a proton, it will be found with less than 10^{20} eV after propagating through a distance of 100 Mpc (3×10^8 light years). Thus the observation of a cosmic ray proton with energy greater than 10^{20} eV implies that its distance of travel is less than 100 Mpc and that its initial energy at its source had to have been much greater. This distance corresponds to a red shift of 0.025 and is small compared to the size of the universe. Similar arguments can be made for nuclei or photons in the energy range considered. There are a limited number of possible sources which fit the Hillas criteria (Fig. 2) within a volume of radius 100 Mpc about the earth. The fact that the cosmic rays, if protons, will be little deflected by galactic and extragalactic magnetic fields serves as a second diagnostic tool [15]. The deflection of protons of energy 5×10^{19} eV by the galactic magnetic field (~ 2 μgauss) and the intergalactic magnetc fields ($\leq 10^{-9}$ gauss) is only a few degrees, so that above 5×10^{19} eV it is possible that the cosmic rays will point to their sources. We thus approach an astronomy where the "light" is cosmic rays and the sources visible are less than 100 Mpc away.

4. Cosmic Ray Astronomy

The energy 5×10^{19} eV represents a lower limit for which the notion of an astronomy of charged particles from "local" sources can be applied. The GZK effect enhances the number of events from sources within a distance of 100 Mpc. Of these events, two particularly stand out with energies reported to be 2×10^{20} eV by the AGASA experiment [16] and 3×10^{20} eV by the Fly's Eye experiment [17]. More recently a total of six events with energy $\geq$ 10^{20} eV have been reported by the AGASA experiment [6]. For all these events the probable distance to the source is less than 50 Mpc.

The events above 5×10^{19} eV are too few to derive a spectral index. It is not clear that a single spectrum is even the proper way to characterize these events. Since they must come from "nearby" the actual number of sources may not form an effective continuum in space, so the spectrum observed may vary with direction. The distribution of matter within 100 Mpc is not uniformly distributed over the sky. It is probably more fruitful to take an astronomical approach and plot the arrival directions of these events on the sky in galactic coordinates.

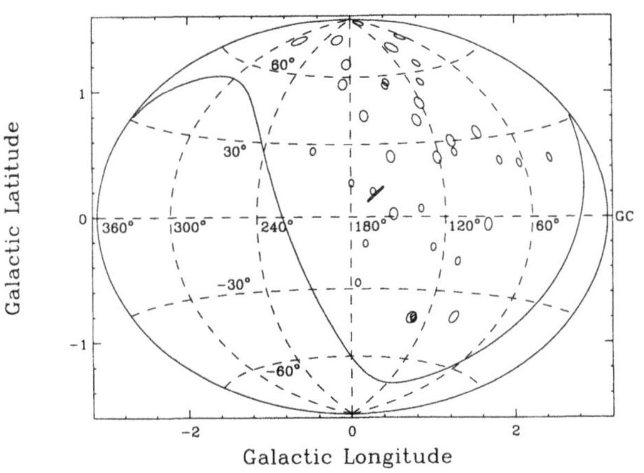

Figure 3. Plot in galactic coordinates of arrival directions of cosmic rays with energy $\geq 5 \times 10^{19}$ eV. Large symbols, Haverah Park [18]; small symbols, AGASA [19]; rectangle, error box (2σ) for the Fly's Eye event with energy 3×10^{20} eV [17]. The size of the symbols indicate the resolution of the experiments (63% of the events within the symbol). The empty region bounded by the solid line is the part of the sky not seen by the experiments which are located in the Northern Hemisphere.

Arrival direction data are available for the Haverah Park experiment [18] and the AGASA experiment [19] as is the arrival direction for the large Fly's Eye event [17]. In Fig. 3 we plot the arrival directions of 20 AGASA events and 16 Haverah Park events. The size of the symbols corresponds to the angular resolution. In addition, the error box for the most energetic event recorded by the Flys Eye experiment is plotted. What is remarkable in this figure is the number of coincidences of cosmic rays coming from the same direction in the sky. Of 20 events reported by AGASA, there are two pairs. The probability of a chance concidence for this is about 2%. The addition of the Haverah Park events shows a coincidence with one of the AGASA pairs. And the Fly's Eye event coincides with one of the AGASA events. It is not possible to estimate properly the chance probability for these overlaps, but the possibility that they may be real should not be ignored. The triple coincidence contains the AGASA event of 2×10^{20} eV, the Haverah Park event of about 1×10^{20} eV, and an AGASA event of 5×10^{19} eV. The Fly's Eye event of 3×10^{20} eV is in coincidence with an AGASA event of 6×10^{19} eV. The third pair contains AGASA events of 6×10^{19} eV and 8×10^{19} eV respectively.

The triple coincidence is particularly interesting if it is not the result of pure chance. It contains cosmic rays separated by a factor of four in energy which have not been separated in space by more than a few degrees. This is an encouraging prospect for future experiments where, with many more events, one may observe point sources, clusters, and larger scale anisotropies in the sky. The crucial questions will be: Does the distribution of cosmic rays in the sky follow the distribution of matter within our galaxy or the distribution of "nearby" extragalactic matter, or is there no relation to the distribution of matter? Are there point sources or very tight clusters? What is the energy distribution of events from these clusters? Are these clusters associated with specific astrophysical objects? If there is no spatial modulation or no correlation with observed matter, what is the spectrum? This situation would imply an entirely different class of sources which are visible only in the "light" of cosmic rays with energy $\geq 5 \times 10^{19}$ eV. Of course there may be a combination of these possibilities. If even crude data on primary composition is available, it can be divided into catagories of light and heavy components which may have different distributions. Crucial to these considerations is uniform exposure over the whole sky. And a final and perhaps most fundamental question is: Is there an end to the cosmic ray spectrum?

5. The Pièrre Auger Observatory

The discussion so far makes clear that the EHECR's are a mystery that will require even larger detectors than the present AGASA detector with an aperture of $\sim$ 125 km^2-sr, and the High Resolution Fly's Eye (HiRes) with anticipated time-averaged apertures of 300 km^2-sr at 10^{19} eV and 1000 km^2-sr at 10^{20} eV. The rate of cosmic rays with energy $\geq$ 5 $\times$ 10^{19} eV is about 4 km^{-2}-sr^{-1}-century^{-1} and 1 km^{-2}-sr^{-1}-century^{-1} for energies $\geq$ 10^{20} eV. The HiRes experiment is expected to begin operation in 1999.

We will describe a more ambitious approach to the problem which is the Pièrre Auger Observatory named after the French physicist who discovered extensive air showers [21]. In 1938 Auger demonstrated that particles were arriving from outside the earth with energies $\geq 10^{15}$ eV[22].

The Auger project is a comprehensive experiment designed to study cosmic rays with energy $\geq$ 10^{19} eV with the least posible bias concerning theories of their origin. Since the cosmic rays are likely to point to the sources a comprehensive study requires that the entire celestial sphere be observed. The assumption that a single site is sufficient violates the principle stated above.

Two instruments will be built at mid-latitude sites in the southern and northern hemispheres. Each instrument will observe cosmic rays at zenith angles up to 60°, so that as the earth turns the whole sky is nearly uniformly observed. Each instrument is a hybrid consisting of a surface array to measure the lateral distribution of the shower particles on the ground, and a fluorescence detector to measure the longitudinal development of the shower. The configuration of the fluorescence detectors is such that when conditions permit their operation (dark moonless nights) they will register $\geq$ 90% of the showers which trigger the surface array. Approximately 10% of the showers will be observed by both detectors. This subset of events will permit a cross check of the energy and provide the maximum possible information on the composition of the primary.

For further details we refer the interested reader to ref 21 where the Design Report and more than 200 technical reports can be found.

An international collaboration consisting of 19 countries will pool resources to build two cosmic ray observatories each with an aperture of 7000 km^2-sr for energies $\geq$ 10^{19} eV. The observatories are to be built in Mendoza Provence, Argentina (35.2° S, 69.2° W, altitude 1400m) and the state of Utah in the United States (39.1°, 112.6° W, altitude 1400m).

Each surface array consists of 1600 tanks of water, 3.6m in diameter and 1.2m deep. The tanks are spaced on a triangular grid with nearest neighbor distances of 1.5 km. They detect the shower particles by the Čerenkov light produced. The tanks are lined with an efficient diffuse reflector (tyvec) and three photomultipliers are mounted on top of the tank. The response of the tank is proportional to the amount of Čerenkov light produced independent of its angle. The water detector exploits the properties of the shower far from the core. Far from the core ($\sim$ $1km$) the shower consists principally of photons and about 1/10th as many electrons. The energies of these electromagnetic particles are $\sim$ 10 MeV. There are many fewer muons with energies $\sim$ 1 GeV. Far from the core the shower particles are spread out in time by several μs. The tank is rather efficient in converting all the electromagnetic particles into Čerenkov light. Most muons deposit more than 240 MeV as they pass through the tank. The light is recorded by a FADC. Muons stand out in the FADC trace as large pulses on a low level continuum of the electromagnetic signal.

These properties have been experimentally verified by two full sized tanks installed at AGASA[23]. Figure 4 shows a trace recorded in a tank when a 10^{20} eV cosmic ray landed 1.7 km from the tank. The integrated signal corresponds to what would be produced by 36 muons passing vertically through the tank. Figure 5 shows the record of a 1.4$\times 10^{19}$ eV cosmic ray which landed 1.3 km from a tank. In this trace four individual muons can be resolved. Far from the core the muons arrive with a time spread which is half of that of the electromagnetic particles. The

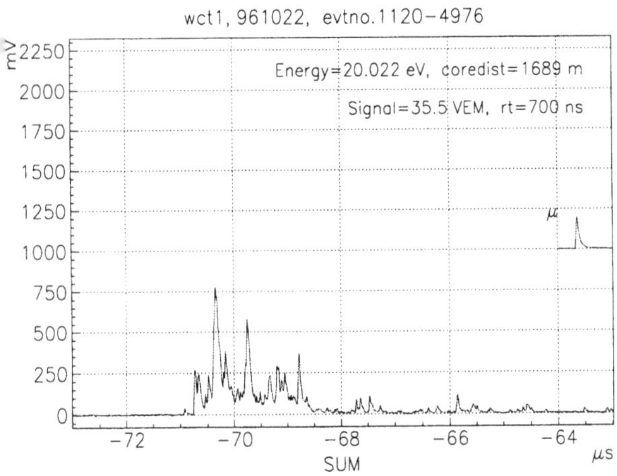

Figure 4. FADC trace of the signal recorded in a prototype Auger detector in the AGASA array. The signal was produced by a cosmic ray shower of energy 10^{20} eV at a distance of 1.7 km.

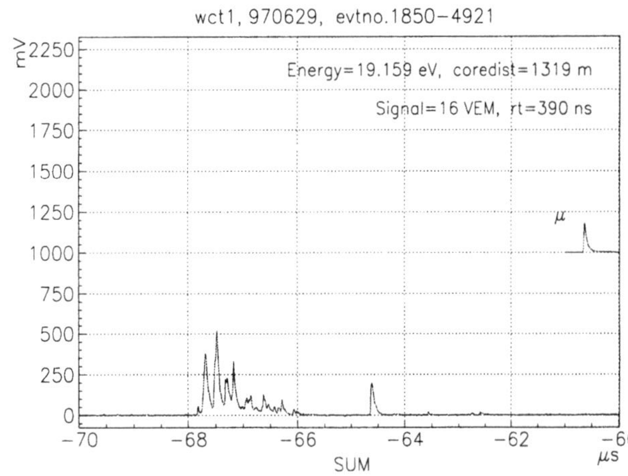

Figure 5. FADC trace of the signal recorded in a prototype Auger detector in the AGASA array. The signal was produced by a cosmic ray shower of energy 1.4×10^{19} eV at a distance of 1.3 km.

time for the tank signal to rise from 10% to 50% of its full height is a measure of the muon content as well. In Fig 6 we show the average pulse strength in equivalent muons for showers with energy $\geq 10^{18.5}$ eV as a function of distance between the tank and the shower core. The solid line corresponds to the expectation based on the Haverah Park experiment [2] which used water tanks as detectors. The agreement is excellent out to 1.5 km where the Haverah Park lateral distribution is well measured. The energy calibration of the AGASA experiment and the Haverah Park experiment are in good agreement.

The density of surface particles at a large fixed distance from the shower core for a vertical shower is proportional to the energy of the shower. This correspondence was first proposed by Hillas and has been confirmed by many simulations since Hillas's original prediction [24]. Application of this relation requires that the density observed on the ground for inclined showers be adjusted to what would have been observed had the shower been vertical. This involves corrections that can partly be measured and partly must be calculated.

Since the energy of a cosmic ray is one of the most important parameters to be measured, the Auger detector was designed to be a hybrid detector. The surface array is combined with a fluorescence detector (FD) [25]. By its very nature the FD is a natural addition to a ground array. Normally FD's must work in stereo; that is each shower must be observed by two FD's in order to have an accurate geometrical reconstruction and energy measurement. However, if a single FD is combined with a surface array and the relative time of the signals in the FD and surface array are measured, a single FD can reconstruct a shower with the same precision as if it had a stereo mate. In Fig. 7 we show the layout of the Auger detector in the reference design. At the threshold of 10^{19} eV every shower which triggers the surface array is observed by at least one of the three FD's. At 10^{20} eV the shower is usually observed by the ground array and at least two FD's, giving additional redundancy.

The basic features of the hybrid array have been experimentally verified with the HiRes prototype and the CASA-MIA surface detector [26]. Each technique working by itself has its strengths

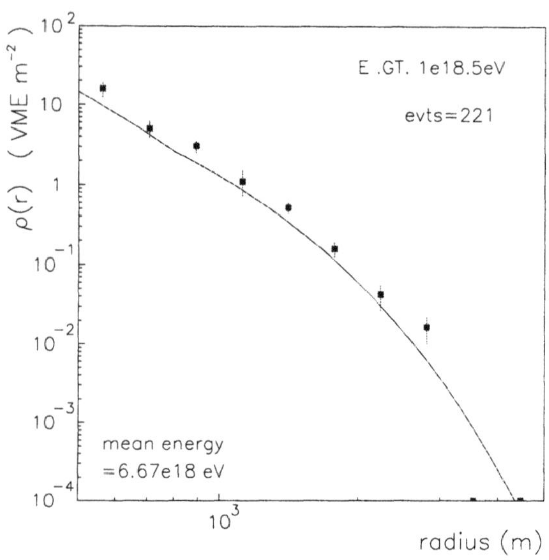

Figure 6. Lateral distribution of the average signal observed in a prototype tank at AGASA for showers with energies $\geq 5 \times 10^{18}$ eV. The solid line is the lateral distribution expected on the basis of the Haverah Park experiment.

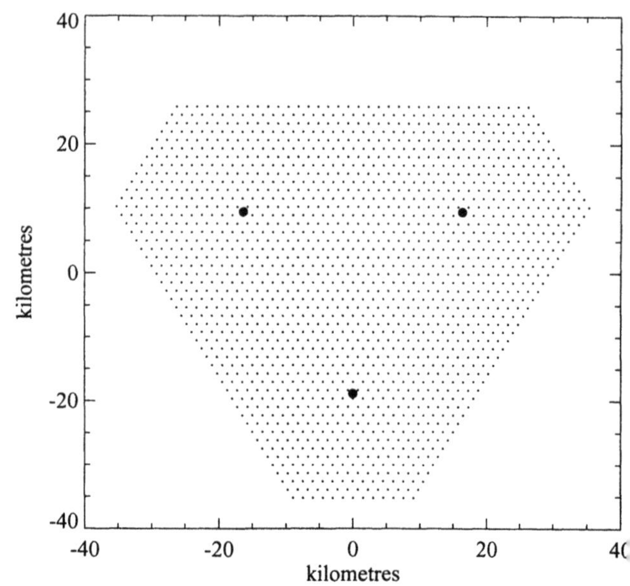

Figure 7. Reference design layout of the Auger detector. The circles indicate the location of the fluorescence detectors. Each fluorescence detector must measure showers out to a maximum distance of 20 km. The dots indicate the locations of the 1600 water tank detectors whose spacing is 1.5 km.

and weaknesses. Together they form the most powerful detector one can imagine. Simulations of one sort or another are needed in both cases. For the surface array the absolute energy must be based on shower simulations. Remarkably many seemingly different shower simulations give a relationship between the particle density at a fixed distance and the shower energy which varies by only 30% [27] and the relative energies are even more reliable. The surface array is easy to keep calibrated by the ever present single cosmic ray muons. The sensitivity of the surface array for a uniform grid is independent of where the shower lands. The fluorescence technique measures the ionization deposited in the atmosphere for the part of the shower seen, and hence is a measurement which is much more directly related to the shower energy. However to measure the energy spectrum means taking into account the aperture of the detector which varies with the night sky background, the atmospheric absorption which is much more important than in previous applications of the fluorescence technique (distances of up to 20 km must be seen instead of 3 km), and corrections for the direct and scattered Čerenkov light which can be significant (between 20% and 50%). With the exception of some geometrically favored showers, a prescribed shower shape is required for the analysis. An absolute calibration of the photomultipliers must be made and maintained. The aperture is a strong function of the energy, growing larger as the energy of the cosmic ray increases. The Auger group very quickly realized the advantages of combining the two techniques. In combination with the ground array the shower need be seen with only a single detector, and the combination permits the position of the core of the shower on the ground to be located, *independent* of the amplitudes of the signals in the tanks; only the knowledge of the relative time is required. This is important in the development of the reconstruction algorithms of the surface events where one in the beginning does not have a complete apriori knowledge of the lateral distributions in the shower plane.

An ideal detector should measure the energy, direction, and particle type of each detected cosmic ray. A real detector has finite angular and energy resolutions and at best can only separate the particle type on a statistical basis. Most of the events will be recorded with the surface array alone, as the fluorescence detector has a duty cycle of 10%. The energy resolution of surface array will have a standard deviation of 25% above 10^{19} eV essentially independent of the primary energy. This assumes that the cosmic rays come from a uniform mixture of all species between protons and iron and there is no knowledge of the species on an event by event basis. The angular resolution is defined as the half angle of a cone about the true direction containing 63% of the reconstructed events. It is 1.2° at 10^{19} eV falling to 0.7° at 10^{20} eV.

The composition sensitivity for the surface array uses the muon fraction measured in the water detectors combined with rise time of the signal from 10% to 50% of its full integral value. The power of these measurements is such that separation of a proton primary from an iron primary is about one standard deviation.

The hybrid events, in addition to allowing a cross calibration of the energy, permit a 1.5 σ separtion between protons and iron using the additional information of the nearly direct measurement of the depth in the atmosphere of the shower maximum. The energy resolution for the hybrid events will be about 15% and the angular resolution will be about 0.25°.

The performance of the Auger Observatory can be evaluated by calculating the event yield of the surface array for a specific model. We select a model where the source of the cosmic rays is due to metastable heavy relic particles, trapped in the halo of our galaxy [28] [29] superimposed with a universal distribution of sources. The former are not affected by the GZK cutoff while the latter will show a cutoff. This scenario is consistant with the most recent AGASA results [6]. In Fig 8 the observed distribution of events with resolution folded in is displayed. Only events with 5 tanks triggered and zenith angle $\leq$ 60° are accepted. Full efficiency is achieved at 10^{19} eV. The effect of the resolution is seen in the smoothing of the bump due to the pile up of the events due to the GZK cutoff. Expected in three years for both sites are 1400 events with energy $\geq$ 5 $\times$ 10^{19} eV and 260 events with energy $\geq$ 10^{20} eV. Because of the steep cosmic ray spectrum some 60,000 events will be recorded below 10^{19} eV while 18,000 events will be recorded above 10^{19} eV.

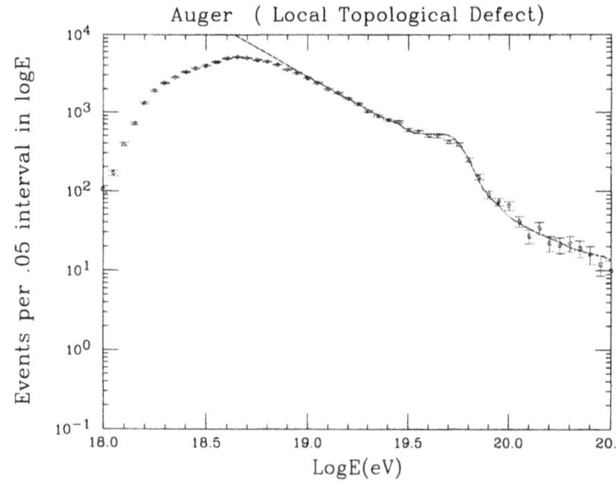

Figure 8. Plot of yield of events for three years of operation of the Auger observatory on the assumption that the cosmic rays come from a combination of heavy relics in the galactic halo and a universal distribution of extragalactic sources. The solid line is the yield expected for perfect resolution and full efficiency.

6. The Pièrre Auger Observatory as a Neutrino Detector.

Since this report is for the Neutrino98 conference we discuss the neutrino detection capabilities of the Auger observatory. A neutrino signature is the observation of nearly horizontal showers initiated deep in the atmosphere. As the surface detector consist of tanks with 1.2 meter vertical height, the area of the tank seen by horizontal particles is almost the same as for vertical particles. It is necessary to have measurements that can discriminate between the deep showers and those initiated high in the atmosphere by the ordinary cosmic ray particles. For a single Auger site a rough estimate of the target mass times solid angle can be made. Showers with zenith angles greater than 70°, which are initiated at altitudes less than 3 km, require the initiating particle to pass through some 3000 gm/cm^2 of atmosphere before interacting. Of the known particles only neutrinos will have this property. The solid angle for such events is 1 sr, the target volume is 9000 km^3 of air, which at a density of 0.001 gm/cm^2 gives a target of 9 km^3-sr (water equivalent). While such a mass is impressive, the requirement of a trigger of at least 5 tanks means that such apertures are achievable only for electron neutrinos with energy $\geq 10^{18}$ eV. Only the interaction of an electron neutrino places all its energy into a prompt shower. More detailed calculations have been made [30] [31] which give a mass×solid angle of 0.6, 10.0, 16.0 km^3-sr for 10^{17}, 10^{18}, and 10^{19} eV respectively. For these apertures an energy deposit of $\geq$ 1GeV is required in 5 or more tanks and evidence of electromagnetic energy is required in at least three tanks.

The background for the detection of neutrinos comes from near horizontal showers initiated high in the atmosphere by ordinary cosmic rays. Such showers have distinctive features. They consist entirely of muons; the electromagnetic component has died out. Even the occasional hard muon bremsstrahlung cannot mimic a deep shower. The muon showers are characterized by a very narrow time spread when they pass through a tank. By contrast a deep shower is characterized by an extended time spread. These two distributions are shown in Figs. 9 and 10. The large time spread of the signal in the tanks is a distinctive characteristic for a shower initiated deep in the atmosphere. An additional distinction is that the deeply initiated shower will have a significant electromagnetic component. This fact can be discerned from the pattern of energy deposit from

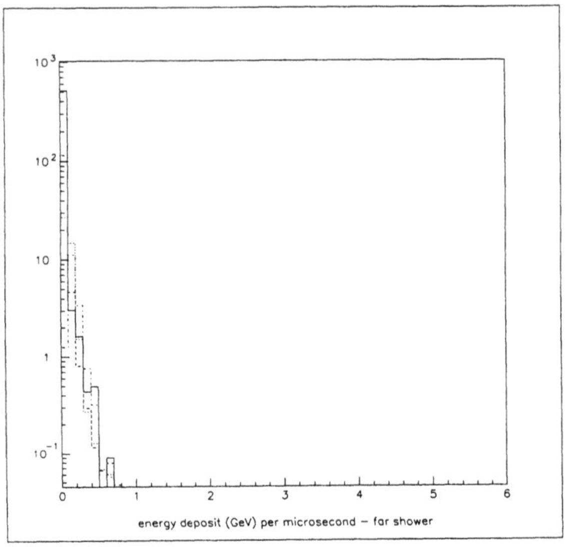

Figure 9. Energy deposit per unit time for a distant shower as a function of time. Ordinate: energy deposit in tank (GeV/μsec). Abcissa: Time (μsec). Curve with highest deposit is 0.5 km from shower axis. Successively lower curves are at 1.0 km, 1.5 km, and 2.0 km.

the FADC trace. The background rate of distant showers (which are of interest in themselves) is about 3000/ year.

Given the target mass and solid angle the sensitivity of the Auger observatory for neutrinos can be estimated. We use a cross section per nucleon given by [34]:

$$\sigma = 1.2(E/10^{18}\text{eV})^{0.4} \times 10^{-32}\ \text{cm}^2$$

We estimate the yearly rate in terms of the differential electron neutrino and anti-neutrino flux at an energy 10^{18} eV. The result is about the same for spectral indicies between 1.5 and 3.0. The sensitivity is 8×10^{24} I_0/year where I_0 is the neutrino flux at 10^{18}eV in units of (cm^2-sec-sr-GeV)$^{-1}$. An optimistic topological defect flux at an energy of 10^{18} eV is $I_0 = 10^{-24}$ which would produce 8 neutrino events per year.

While the neutrino sensitivity of the Auger ob-

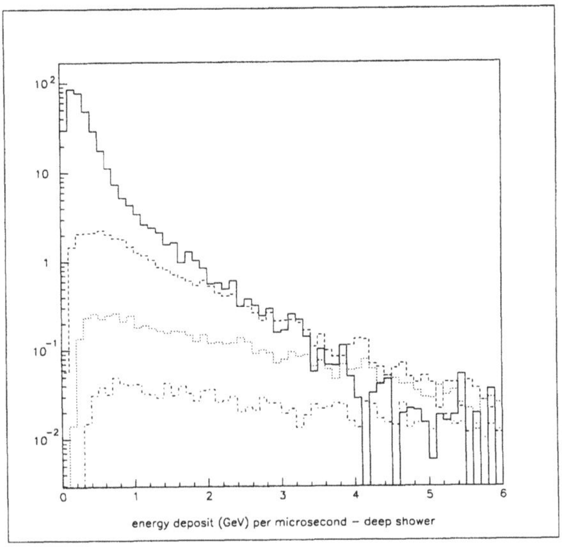

Figure 10. Energy deposit per unit time for a deep shower as a function of time. Ordinate: energy deposit in tank (GeV/μsec). Abcissa: Time (μsec). Curve with highest deposit is 0.5 km from shower axis. Successively lower curves are at 1.0 km, 1.5 km, and 2.0 km.

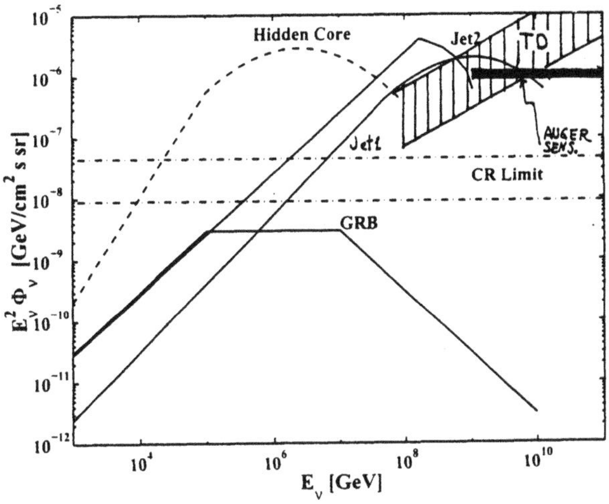

Figure 11. Various predictions of neutrino fluxes. The models Jet1 and Jet2 violate the limit of ref 32 which ties the neutrino flux to the observed cosmic ray fluxes. The hatched area covers the range of topological defect predictions [12]. The sensitivity of the Auger detector is indicated. The hidden core model is not constrained by the cosmic rays. The conclusions of ref 12 do not apply to the topological defect models.

servatory is significant, it remains marginal with respect to the predictions of the neutrino fluxes. Protheroe [32] has reviewed the various predictions of the neutrino fluxes from various sources at this conference. Many of them contradict an upper bound obtained by Waxman and Bahcall which relate the high energy cosmic ray flux to the neutrino flux [33]. One would not build an Auger observatory if the only possible signal were neutrinos. However Auger has a significant sensitivity and because of the rising neutrino interaction crossection one ton of target mass at 10^{18} eV is worth 1000 tons at 10^{12} eV. Further it is complementary to neutrino detectors that rely on the observation of upward going muons. At neutrino energies above 10^{16} eV the Earth becomes opaque. The Auger detector, whose target is the gossimer mass of air above it, is immune to this phenomenon and its sensitivity is only limited by the neutrino flux.

7. Present status of the Pièrre Auger Observatories

At the time of writing (September 1998)the project has been approved for a phased construction to begin in Argentina in calender year 1999. Construction in the north will be postponed for several years.

REFERENCES

1. Yoshida, S. and Dai H., 1998, J. Phys. **G 24**, 905; Sokolsky, P., P. Sommers, and B. R. Dawson, 1992 Physics Rep. **217**, 225; *Proceedings of the Paris Workshop on the Highest Energy Cosmic Rays*, 1992, Nucl. Phys. lo(Proc. Supp.) **B 28**, 213; *Proceedings of the International Symposium on Extremely High*

Energy Cosmic Rays: Astrophysics and Future Observations, 1996, ed M. Nagano, (Institute for Cosmic Ray Research, University of Tokyo); Swordy, S., rapporteur talk, 1994, *Proceedings of the 23rd International Cosmic Ray Conference, (Calgary)* 243; Watson, A. A., 1991, Nucl. Phys. (Proc. Supp.) **B 22**, 116; V. S. Berezinskiĭ, S. V. Bulanov, V. A. Dogiel, V. L. Ginzburg (editor) and V. S. Ptuskin, 1990, *Astrophysics of Cosmic Rays*, North Holland, Elsevier Science Publishers, The Netherlands.

2. Lawrence, M. A., et al., 1991, J. Phys. **G 17**, 773.
3. Afanasiev, B. N., et al., 1995, *Proceedings of the 24th International Cosmic Ray Conference (Rome)* **2**, 756.
4. Bird, D. J., et al., 1994, Ap. J. **424**, 491.
5. Yoshida S., et al., 1995, Astroparticle Physics **3**, 105.
6. Takeda, M., et al., 1998, Phys. Rev. Letters, **81**, 1163.
7. Bird D. J., et al., 1993, Phys. Rev. Letters, **71**, 3401.
8. Yoshida, S. and Dai H., 1998, J. Phys. **G 24**, 905.
9. Greisen, K., 1965, *Proceedings of the 9th International Cosmic Ray Conference (London)* **2**, 609.
10. Hillas, A. M., 1984, Ann. Rev. Astron. Astrophys. **22**, 425.
11. Biermann, P., 1997, J. Phys. **G 23**, 1.
12. Bhattacharjee, P., C. T. Hill, and D. N. Schramm, 1992, Phys. Rev. Letters **69**, 567.
13. Greisen, K., 1966, Phys. Rev. Letters **16**, 748.
14. Zatsepin, G. T. and V. A. Kuz'min, 1966, JETP Letters **4**, 78.
15. Kronberg, P. P., 1994, Rep. Prog. Phys. **57**, 325; Kronberg, P. P., 1994, Nature **370**, 179; Cole, P., Comments Astrophys. **16**, 1992, 45.
16. Hayashida, N., et al., 1994, Phys. Rev. Letters, **73**, 3491.
17. Bird, D. J., et al., 1995, Ap. J., **441**, 144; Elbert, J. W. and P. Sommers, 1995, Ap. J., **441**, 151.
18. Watson, A. A., 1997, University of Leeds, private communication.
19. Hayashida, N., et al., 1996, Phys. Rev. Letters **77**, 1000.
20. Hayashida, N., et al., 1998, preprint astro-ph/9807045.
21. The Pierre Auger Project, Design Report, 2nd Ed., November 1996, Fermilab; this report and technical (GAP) notes can be obtained from the world wide web at www-td-auger.fnal.gov:82.
22. Auger, P., et al., 1938, Comptes Rendus **206**, 1721; Auger, P., 1939, Rev. Mod. Phys. **11**, 288.
23. We thank Professor M. Nagano and N. Sakaki of the AGASA group and T. Kutter of the Univ. of Chicago for their cooperation in the construction and installation of these tanks.
24. Hillas, A. M., 1971, *Proceedings of the 12th International Cosmic Ray Conference (Hobart)*, **3**, 1001.
25. Baltrusaitis, R. M. et al., 1985, Nucl. Instr. Meth. **A240**, 410.
26. Fick B.F. et al., 1998, to be submitted to Nucl. Instr. Meth.
27. Knapp, J., Leeds University, private communication.
28. Hillas M., 1998, Nature, **395**,15.
29. Berezinsky, V., Kachelreiss, M., and Vilenkin, A., 1997 Phys. Rev. Lett. **79**, 4302.
30. Capelle, K. S. et al., 1998, Astropart. Phys. **8**, 321
31. Billoir, P., 1997, Auger GAP note 97-049; for availability see ref 21.
32. Protheroe, R., talk at this conference. References on the predicted neutrino fluxes are contained in these proceedings.
33. Waxman, E., and Bahcall, J., 1998, preprint, hep-ph/9807282, to be published, Phys. Rev. D.
34. Ralston, J. P., McKay, D. W., and Frichter, G. M., 1996, preprint, astro-ph/9606007.

Part 14

Conclusion

ELSEVIER

Nuclear Physics B (Proc. Suppl.) 77 (1999) 511–519

NUCLEAR PHYSICS B
PROCEEDINGS
SUPPLEMENTS

Beyond The Standard Model: This Time for Real

Frank Wilczek[a] *

[a]Institute for Advanced Study,
School of Natural Sciences, Olden Lane,
Princeton, New Jersey 08540

The value of the neutrino mass reported by the SuperK collaboration fits beautifully into the framework of gauge theory unification. Here I justify this claim, and review the other main reasons to believe in that framework. Supersymmetry and $SO(10)$ symmetry are important ingredients; nucleon instability is a dramatic consequence.

It has been a great privilege to attend this conference, which I am sure the future will regard as historic. I want to thank the organizers for making it in every way a very enjoyable experience, as well.

Undoubtedly it will take us, collectively, many years to do full justice to the wonderful discovery announced here, that neutrinos have non-zero mass. Many important tasks remain at the level of pure phenomenology, most obviously perhaps that of integrating the firm atmospheric oscillation results with the long-standing but still confusing solar neutrino anomalies, and the possible hints from LSND of a third distinct effect. However I am going to indulge myself by leaping over these vital issues, to discourse and speculate on the larger implications of the discovery for fundamental physics. Some of us have been hoping for many years to see results of this kind. Now that they are coming in, we look forward with both eagerness and trepidation to the confrontation of our dreams with reality. Let me remind you what's at stake.

1. A New Scale

It is important to realize that the degrees of freedom of the Standard Model permit neutrino masses. A minimal implementation of the construction requires an interaction of the type

$$\Delta\mathcal{L} \;=\; \kappa_{ij} L^{\alpha a i} L^{\beta b j} \epsilon_{\alpha\beta} \phi^\dagger_a \phi^\dagger_b + \text{h.c.}\,, \qquad (1)$$

where i and j are family indices; κ_{ij} is a symmetric matrix of coupling constants; the L fields are the left-handed doublets of leptons, with Greek spinor indices, early Roman weak $SU(2)$ indices, and middle Roman flavor indices; and finally ϕ is the Higgs doublet, with its weak $SU(2)$ index. Two-component notation has been used for the spinors, to emphasize that this way of forming mass terms, although different from what we are used to for quarks and charged leptons, is in some sense more elementary mathematically. $\Delta\mathcal{L}$ becomes a neutrino mass term when the ϕ field is replaced by its vacuum expectation value $\langle\phi^a\rangle = v\delta^a_1$.

Although this Eq. (1) is a possible interaction for the degrees of freedom in the Standard Model, it is usually considered "beyond" the Standard Model, for a very good reason. The new term differs from the terms traditionally included in the Standard Model in that the product of fields has mass dimension 5, so that the coefficient κ must have mass dimension -1. In the context of quantum field theory, it is a nonrenormalizable interaction. When one includes it in virtual particle loops, one will find amplitudes containing the dimensionless factors of the type $\kappa\Lambda$, where Λ is an ultraviolet cutoff. In this framework, therefore, one cannot accept $\Delta\mathcal{L}$ as an elementary interaction. It can only be understood within a larger theoretical context.

Given a numerical value for the neutrino mass, we can infer a scale beyond which $\Delta\mathcal{L}$ cannot be accurate, and degrees of freedom beyond the Standard Model must open up. To get oriented,

*Research supported in part by DOE grant DE-FG02-90ER40542. IASSNS-HEP98/79

PII S0920-5632(99)00488-0

let us momentarily pretend that κ is simply a number instead of a matrix, and that $m = 10^{-2}$ eV is the neutrino mass. Then, using $v = 250$ GeV for the vacuum expectation value, we calculate

$$1/M \equiv \kappa = m/v^2 = 1/(6 \times 10^{15}\ \text{GeV})\ . \qquad (2)$$

When energy and momenta of order M begin to circulate in loops the form of the interaction must be modified. Otherwise the dangerous factor $\kappa\Lambda$ will become larger than unity, inducing large and uncontrolled radiative corrections to all processes, and rendering the success of the Standard Model accidental.

Thus we trace the "absurdly small" value of the observed neutrino mass scale to an "absurdly large" fundamental mass scale. As I shall now discuss, this new scale provides, on the face of it, a wonderful confirmation of our best developed ideas for unification beyond the Standard Model.

Of course, experts will recognize that the foregoing argument is oversimplified; in due course, I shall revisit it in a more critical spirit.

2. Two Pillars of Unification

The standard model of particle physics is based upon the gauge groups SU(3)×SU(2)×U(1) of strong, electromagnetic and weak interactions acting on the quark and lepton multiplets as shown in Figure 1.

In this Figure I have depicted only one family (u,d,e,ν_e) of quarks and leptons; in reality there seem to be three families which are mere copies of one another as far as their interactions with the gauge bosons are concerned, but differ in mass. Actually in the Figure I have ignored masses altogether, and allowed myself the convenient fiction of pretending that the quarks and leptons have a definite chirality – right- or left-handed – as they would if they were massless. The more precise statement, of course, is that the gauge bosons couple to currents of definite chirality. The chirality is indicated by a subscript R or L. Finally the little number beside each multiplet is its assignment under the U(1) of hypercharge, which is the average of the electric charge of the multiplet.

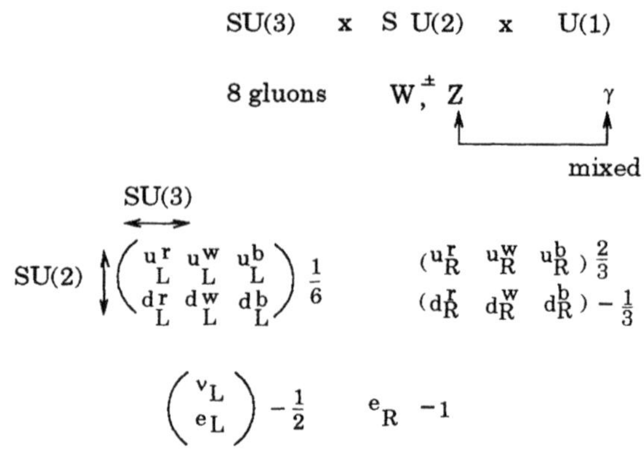

Figure 1. The gauge groups of the standard model, and the fermion multiplets with their hypercharges.

While little doubt can remain that the Standard Model is essentially correct, a glance at Figure 1 is enough to reveal that it is not a complete or final theory. To remove its imperfections, while building upon its solid success, is a worthy challenge.

There are two improvements on the Standard Model that are so deeply suggested in its structure, that I think it is perverse to deny them. Let me briefly recall these two pillars of unification:

3. Gauge Group and Fermion Unification

Given that the strong interactions are governed by transformations among three colors, and the weak by transformations between two others, what could be more natural than to embed both theories into a larger theory of transformations among all five colors [1]?

This idea has the additional attraction that an extra U(1) symmetry commuting with the strong SU(3) and weak SU(2) symmetries automatically appears, which we can attempt to identify with the remaining gauge symmetry of the standard model, that is hypercharge. For while in the separate SU(3) and SU(2) theories we must throw out the two gauge bosons which couple respectively to the color combinations R+W+B and G+P, in the SU(5) theory we only project out R+W+B+G+P, while the orthogonal combination (R+W+B)-$\frac{3}{2}$(G+P) remains.

Finally, the possibility of unified gauge symmetry breaking is plausible by analogy; after all, we know for sure that gauge symmetry breaking occurs in the electroweak sector.

Georgi and Glashow [2] showed how these ideas can be used to bring some order to the quark and lepton sector, and in particular to supply a satisfying explanation of the weird hypercharge assignments in the standard model. As shown in Figure 2, the five scattered SU(3)×SU(2)×U(1) multiplets get organized into just two representations of SU(5).

In making this unification it is necessary to allow transformations between (what were previously considered to be) particles and antiparticles of the same chirality, and also between quarks and leptons. It is convenient to work with left-handed fields only. Since the conjugate of a right-handed field is left-handed, we don't lose anything by doing so – though we must shed traditional prejudices about a rigorous distinction between matter and antimatter, since these get mixed up. Specifically, it will not be possible to declare that matter is what carries positive baryon and lepton number, since the unified theory does not conserve these quantum numbers.

As shown in Figure 2, there is one group of ten left-handed fermions that have all possible combinations of one unit of each of two different colors, and another group of five left-handed fermions that each carry just one negative unit of some color. These are the ten-dimensional antisymmetric tensor and the complex conjugate of the five-dimensional vector representation, commonly referred to as the "five-bar". In this way, *the structure of the standard model, with the particle assignments gleaned from decades of experimental effort and theoretical interpretation, is perfectly reproduced by a simple abstract set of rules for manipulating symmetrical symbols.* Thus for example the object RB in this Figure has just the strong, electromagnetic, and weak interactions we expect of the complex conjugate of the right-handed up-quark, without our having to instruct the theory further.

A most impressive, though simple, exercise is to work out the hypercharges of the objects in Figure 2 and checking against what you need in the Standard Model. These ugly ducklings of the Standard Model have matured into quite lovely swans.

SU(5): 5 colors RWBGP

10: 2 different color labels (antisymmetric tensor)

$$\begin{array}{llll} u_L: & RP, & WP, & BP \\ d_L: & RG, & WG, & BG \\ u^c_L: & RW, & WB, & BR \\ & (\bar{B}) & (\bar{R}) & (\bar{W}) \\ e^c_L: & GP & & \\ & (\,) & & \end{array} \qquad \begin{pmatrix} 0 & u^c & u^c & u & d \\ & 0 & u^c & u & d \\ & & 0 & u & d \\ & * & & 0 & e \\ & & & & 0 \end{pmatrix}$$

$\bar{5}$: 1 anticolor label

$$\begin{array}{llll} d^c_L: & \bar{R}, & \bar{W}, & \bar{B} \\ e_L: & \bar{P} & & \\ \nu_L: & \bar{G} & & \end{array} \qquad \begin{pmatrix} d^c & d^c & d^c & e & \nu \end{pmatrix}$$

$$\boxed{Y = -\tfrac{1}{3}\,(R+W+B) + \tfrac{1}{2}\,(G+P)}$$

Figure 2. Unification of fermions in SU(5)

4. Coupling Constant Unification

We have just seen that simple unification schemes are spectacularly successful at the level of classification. New questions arise when we consider dynamics.

Part of the power of gauge symmetry is that it fully dictates the interactions of the gauge bosons, once an overall coupling constant is specified. Thus if SU(5) or some higher symmetry were exact, then the fundamental strengths of the different color-changing interactions would have to be equal, as would the (properly normalized) hypercharge coupling strength. In reality the coupling strengths of the gauge bosons in SU(3)×SU(2)×U(1) are not observed to be equal, but rather follow the pattern $g_3 \gg g_2 > g_1$.

Fortunately, experience with QCD emphasizes that couplings "run"[3]. The physical mechanism of this effect is that in quantum field theory the vacuum must be regarded as a polarizable

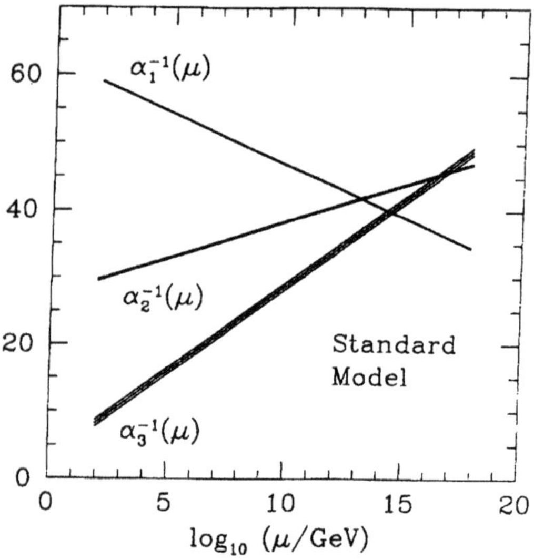

Figure 3. The failure of the running couplings, normalized according to SU(5) and extrapolated taking into account only the virtual exchange of the "known" particles of the standard model (including the top quark and Higgs boson) to meet. Note that only with fairly recent experiments [5], which greatly improved the precision of the determination of low-energy couplings, has the discrepancy become significant.

medium, since virtual particle-anti-particle pairs can screen charge. For charged gauge bosons, as arise in non-abelian theories, the paramagnetic (antiscreening) effect of their spin-spin interaction dominates, which leads to asymptotic freedom. As Georgi, Quinn, and Weinberg pointed out [4], if a gauge symmetry such as SU(5) is spontaneously broken at some very short distance then we should not expect that the effective couplings probed at much larger distances, such as are actually measured at practical accelerators, will be equal. Rather they will all have have been affected to a greater or lesser extent by vacuum screening and anti-screening, starting from a common value at the unification scale but then diverging from one another. The pattern $g_3 \gg g_2 > g_1$ is just what one should expect, since the antiscreening effect of gauge bosons is more pronounced for larger gauge groups.

The running of the couplings gives us a truly quantitative handle on the ideas of unification. To specify the relevant aspects of unification, one basically needs only to fix two parameters: the scale at which the couplings unite, (which is essentially the scale at which the unified symmetry breaks), and their common value when they unite. Given these, one calculates three outputs, the three *a priori* independent couplings for the gauge groups in SU(3)×SU(2)×U(1). Thus the framework is eminently falsifiable. The astonishing thing is, how close it comes to working (Figure 3).

The GQW calculation is remarkably successful in explaining the observed hierarchy $g_3 \gg g_2 > g_1$ of couplings and the approximate stability of the proton. In performing it, we assumed that the known and confidently expected particles of the standard model exhaust the spectrum up to the unification scale, and that the rules of quantum field theory could be extrapolated without alteration up to this mass scale – thirteen orders of magnitude beyond the domain they were designed to describe. It is a triumph for minimalism, both existential and conceptual.

On closer inspection, however, it is not quite good enough. Accurate modern measurements of the couplings show a small but definite discrepancy between the couplings, as appears in Figure 3. And heroic dedicated experiments to search for proton decay did not find it [6]; they currently exclude the minimal SU(5) prediction $\tau_p \sim 10^{31}$ yrs. by about two orders of magnitude.

If we just add particles in some haphazard way things will only get worse: minimal SU(5) nearly works, so a generic perturbation will be deleterious. Even if some *ad hoc* prescription could be made to work, that would be a disappointing outcome from what appeared to be one of our most precious, elegantly straightforward clues regarding physics well beyond the Standard Model.

Fortunately, there is a compelling escape from this impasse. That is the idea of supersymmetry [7]. Supersymmetry is certainly not a symmetry in nature: for example, there is certainly no bosonic particle with the mass and charge of the electron. However there are several reasons for thinking that supersymmetry might be spontaneously, and only relatively mildly broken, so that the superpartners are no more massive than ≈ 1 Tev. The most concrete arises in calculating radiative corrections to the (mass)^2 of the Higgs particle from diagrams of the type shown in Figure 4. One finds that they make an infinite, and also large, contribution. By this I mean that the divergence is quadratic in the ultraviolet cutoff. No ordinary symmetry will make its coefficient vanish. If we imagine that the unification scale provides the cutoff, we will find, generically, that the radiative correction to the (mass)^2 is much larger than the total value we need to match experiment. This is an ugly situation.

In a supersymmetric theory, if the supersymmetry is not too badly broken, it is possible to do better. For any set of virtual particles that might circulate in the loop there will be another graph with their supersymmetric partners circulating. If the partners were accurately degenerate, the contributions would cancel. Taking supersymmetry breaking into account, the threatened quadratic divergence will be cut off only at virtual momenta such that the difference in (mass)^2 between the virtual particle and its supersymmetric partner is negligible. Notice that we will be assured adequate cancellation if and only if supersymmetric partners are not too far split in mass – in the present context, if the splitting times the square root of the fine structure constant is not much greater than the weak scale.

The effect of low-energy supersymmetry on the running of the couplings was first considered long ago [8], in advance of the precise measurements of low-energy couplings or of the modern limits on nucleon decay. One might have feared that such a huge expansion of the theory, which essentially doubles the spectrum, would utterly destroy the approximate success of the minimal SU(5) calculation. This is not true, however. To a first approximation since supersymmetry is a space-time rather than an internal symmetry it does not affect the group-theoretic structure of the calculation.

Thus to a first approximation the absolute rate at which the couplings run with momentum is affected, but not the relative rates. The main effect is that the supersymmetric partners of the color gluons, the gluinos, weaken the asymptotic freedom of the strong interaction. Thus they tend to make its effective coupling decrease and approach the others more slowly. Thus their merger requires a longer lever arm, and the scale at which the couplings meet increases by an order of magnitude or so, to about 10^{16} Gev.

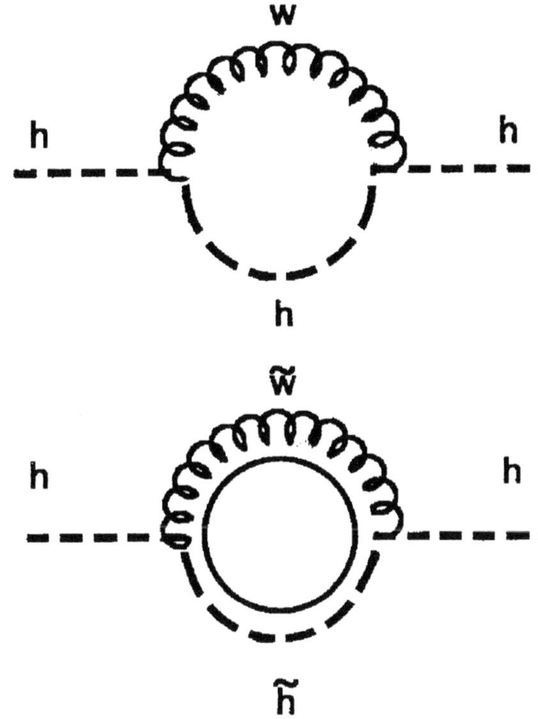

Figure 4. Contributions to the Higgs field self-energy. These graphs give contributions to the Higgs field self-energy which separately are formally quadratically divergent, but when both are included the divergence is removed. In models with broken supersymmetry a finite residual piece remains. If one is to obtain an adequately small finite contribution to the self-energy, the mass difference between Standard Model particles and their superpartners cannot be too great. This – and essentially only this – motivates the inclusion of virtual superpartner contributions in Figure 5 beginning at relatively low scales.

I want to emphasize that this very large new mass scale has emerged unforced from the internal logic of the Standard Model itself.

Its value is important in several ways. First, it explains why the exchange of gauge bosons that are in SU(5) but not in SU(3)×SU(2)×U(1) does not lead to catastrophically quick nucleon decay. Second, it brings us close to the Planck scale $M_{\rm Planck} \sim 10^{19}$ Gev at which exchange of gravitons competes quantitatively with the other interactions. Because $M_{\rm un.}$ is significantly smaller than the Planck mass, we need not be too nervous about the neglect of quantum gravity corrections to our calculation; but because it is not absurdly smaller, we can feel encouraged for the prospect of unification including both gravity and gauge forces.

Finally, as I shall be emphasizing, it can hardly be accidental that the unification scale found here is so close to the scale we previously gleaned from the neutrino mass.

There is another effect of low-energy supersymmetry on the running of the couplings, which although quantitatively small is of prime interest. There is an important exception to the general rule that adding supersymmetric partners does not immediately (at the one loop level) affect the *relative* rates at which the couplings run. That rule works for particles that come in complete SU(5) multiplets, such as the quarks and leptons, or for the supersymmetric partners of the gauge bosons, because they just renormalize the existing, dominant effect of the gauge bosons themselves. However there is one peculiar additional contribution, from the Higgs doublets. It affects only the weak SU(2) and hypercharge U(1) couplings. The net affect of doubling the number of Higgs fields (as, for slightly technical reasons, one must) and including their supersymmetric partners is a sixfold enhancement of the Higgs field contribution to the running of weak and hypercharge couplings. This causes a small, accurately calculable change in the unification of couplings calculation. From Figure 5 you see that it is a most welcome one. Indeed, in the minimal implementation of supersymmetric unification, it puts the running of couplings calculation right back on the money [9].

Since the running of the couplings with scale is logarithmic, the unification of couplings calculation is not terribly sensitive to the exact scale at which supersymmetry is broken, say between 100 Gev and 10 Tev. There have been attempts to push the calculation further, in order to address this question of the supersymmetry breaking scale, but there are many possibilities, and it is difficult to decide among them. An intriguing recent contribution is [10].

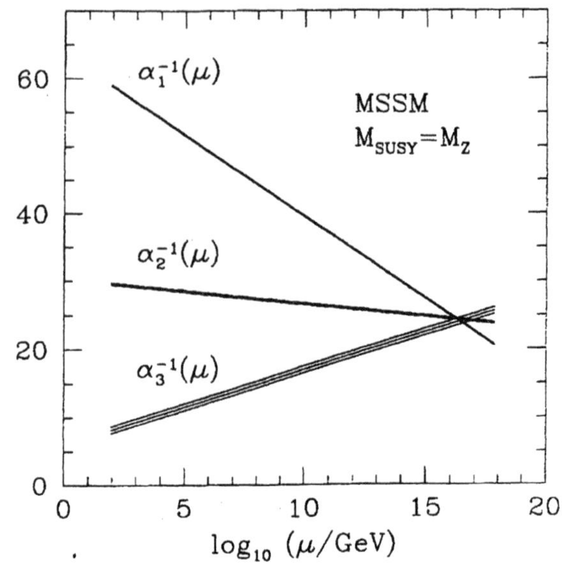

Figure 5. When the exchange of the virtual particles necessary to implement low-energy supersymmetry, a calculation along the lines of Figure 3 comes into adequate agreement with experiment.

5. SO(10), and a Third Pillar

There is a beautiful extension of SU(5) to the slightly larger group SO(10). With this extension, one can unite all the observed fermions of a family, plus one more, into a *single* multiplet [11]. The relevant representation for the fermions is a 16-dimensional spinor representation. Some of its features are depicted in Figure 6.

In addition to the conventional quarks and leptons the SO(10) spinor contains an additional particle, an SU(3)×SU(2)×U(1) singlet. (It is even an SU(5) singlet.) Usually when a theory predicts unobserved new particles they are an embarrassment. But these N particles – there are

SO(10): 5 bit register

($\pm\pm\pm\pm\pm$) : even # *of* −

10 :	$(++-\|+-)$	6	(u_L, d_L)
	$(+--\|++)$	3	u_L^c
	$(+++\|--)$	1	e_L^c
$\bar{5}$:	$(+--\|--)$	$\bar{3}$	d_L^c
	$(---\|+-)$	$\bar{2}$	(e_L, ν_L)
1 :	$(+++\|++)$	1	N_R

Figure 6. Unification of fermions in SO(10). The rule is that all possible combinations of 5 + and - signs occur, subject to the constraint that the total number of - signs is even. The SU(5) gauge bosons within SO(10) do not change the numbers of signs, and one sees the SU(5) multiplets emerging. However there are additional transformations in SO(10) but not in SU(5), which allow any fermion to be transformed into any other.

three of them, one for each family – are a notable exception. Indeed, they are central to the emerging connection between neutrino masses and unification [12].

Because the N^i are singlets, mass terms of the type

$$\Delta\mathcal{L}_N = \eta_{ij} N^{\alpha i} N^{\beta j} \epsilon_{\alpha\beta} \qquad (3)$$

with η_{ij} a symmetric coupling matrix, are consistent with $SU(3)\times SU(2)\times U(1)$ symmetry. This term of course greatly resembles the effective interaction responsible for neutrino masses, Eq. (1), but the difference is conceptually crucial. Because the Ns are Standard Model singlets the Higgs doublets that occurred in Eq. (1) need not appear here. A consequence is that the operators appearing in Eq. (3) have mass dimension 3, so that the η_{ij} must have mass dimension +1. This interaction therefore does not bring in any ultraviolet divergence problems.

What sets the scale for η? Although Eq. (3) is consistent with Standard Model gauge symmetries, or even $SU(5)$, it is not consistent with $SO(10)$. Indeed for the product of spinor 16 we have the decomposition $16\times16 = 10 + 120 + 126$, where only the 126 contains an $SU(5)$ singlet component. The most straightforward possibility for generating a term like Eq. (3) in the full theory is therefore to include a Higgs 126, and a Yukawa coupling of this to the 16s. If the appropriate components of the 126 acquire vacuum expectation values, Eq. (3) will emerge. The 126 is a five-index self-dual antisymmetric tensor under $SO(10)$, which may not be to everyone's taste. Alternatively, one can imagine that more complicated interactions, containing products of several simpler Higgs fields which condense, are responsible. These need not be fundamental interactions (they are, of course, non-renormalizable), but could arise through loop effects even in a renormalizable field theory.

At this level there are certainly many more options than constraints, so that without putting the discussion of N masses in a broader context, and making some guesses, one can't very specific or quantitatively precise. Nevertheless, I think it is fair to say that these general considerations strongly suggest that η is associated with breaking of unified symmetries down to the Standard Model. Thus, if the general framework is correct, the expected scale for its entries is set by the one we met in the unification of couplings calculation, i.e. $\eta \sim 10^{16}$ Gev.

The Ns communicate with the familiar fermions through the Yukawa interactions

$$\Delta\mathcal{L}_{N-L} = g^i_j \bar{N}_i L^{aj} \phi^\dagger_a + \text{h.c.}, \qquad (4)$$

using the previous notations but now, in this 'conventional' term, suppressing the Dirac spinor indices. These interactions are of precisely the type that generate masses for the quarks and charged leptons in the Standard Model. If N were otherwise massless, the effect of Eq. (4) would be to generate neutrino masses, of the same order as ordinary quark and lepton masses. In $SO(10)$, indeed, these masses would be related by simple Clebsch-Gordon and renormalization factors of order unity. Fortunately, as we have seen, N is far from massless.

Indeed, it is so massive that for purposes of low-energy physics we can and should integrate it out. This is easy to do. The effect of combining Eq. (3) and Eq. (4) and integrating out N is to generate

$$\Delta\mathcal{L}_{\text{eff.}} = g_i^k g_j^l (\eta^{-1})_{kl} L^{\alpha ai} L^{\beta bj} \epsilon_{\alpha\beta} \phi_a^\dagger \phi_b^\dagger + \text{h.c.} \quad (5)$$

Thus we arrive back at Eq. (1), with

$$\kappa_{ij} = g_i^k g_j^l (\eta^{-1})_{kl} . \quad (6)$$

This "seesaw" equation provides a much more precise version of the loose connection between unification scale and neutrino mass we discussed at the outset. There is much uncertainty in the details, since there is no reliable detailed theory for the g_i^k nor the ηs. But if g has an eigenvalue of order unity pointing toward the third family (as suggested by symmetry and the value of the top quark mass), and if we set the scale for η using the logic above, then we get close to 10^{-2} eV for the τ neutrino mass, as observed.

While at present it is less imposing than the others, this success promises to become the third pillar of unification.

The pattern of quark and charged lepton masses suggests that the other eigenvalues of g might be considerably smaller, thus generating a hierarchical pattern of neutrino masses. This is at least broadly consistent with proposed explanations of the solar neutrino anomalies, but will not readily accommodate the reported LSND results, nor neutrinos as cosmologically significant hot dark matter.

6. Summary and Prospect

A mass of approximately 10^{-2} eV for the heaviest neutrino fits beautifully into the framework of supersymmetric unification in SO(10). This sort of theory unifies the fermions in a particularly compelling way, with all the quarks and leptons in a generation fitting into a single multiplet, but requires the existence of new degrees of freedom, the Ns (one per family), which within the theory are predicted to be very heavy. The Ns themselves are not accessible, but they induce tiny masses for the observable neutrinos. Assuming supersymmetry is spontaneously and only mildly broken, this sort of theory also has impressive quantitative success in accounting for the disparate values of the gauge couplings of the Standard Model. Although I don't have time to discuss it here, one also finds here an attractive mechanism for understanding why the standard model Higgs field, unlike the other ingredients of the Standard Model, forms an incomplete multiplet of the unified symmetry [13].

In this talk I have taken a minimalist approach, extrapolating straight weak-coupling quantum field theory and gauge symmetry up to near- (but sub-) Planckian mass scales, using only degrees of freedom that the facts more or less directly require. This approach has the advantage of allowing us to make some simple, definite predictions.

General consequences of the minimalist framework are that the neutrino masses are Majorana and that there are no light sterile neutrinos. Also, it is hard to avoid a hierarchical pattern of neutrino masses. This makes it difficult to accommodate a cosmologically significant contribution of neutrino dark matter. These are eminently falsifiable assertions. Indeed, at this conference some have argued, implicitly or explicitly, that they already have been falsified. We shall see. If the minimalist framework really does break down, we will have learned a profound lesson.

The large mixing angle indicated by the atmospheric oscillation results, though by no means problematic, does come as something of a surprise. To do justice to experimental information at this level of detail, we must consider it in conjunction with the whole complex of questions around how unified symmetry is broken and how the pattern of quark and lepton masses is set. Some general considerations that guide this sort of phenomenology were discussed here by Professor Pati, and in rather different ways by Professors Langacker, Mohapatra, Ramond and Yanagida. In working on this subject with Babu and Pati, I have been pleasantly surprised at how well so many diverse facts can be fit together. But as yet no insight comparable to the "pillars" has emerged from thinking about the pattern of masses and mixings, and here one longs for a deeper, more compelling theory.

In any case, the acid test for this whole line of development is nucleon instability. Supersymmetric unification introduces new sources of nucleon instability that are precariously close to existing experimental limits. The large mixing indicated by the atmospheric neutrino oscillation results sharpens the problem from Higgsino exchange, because the dangerous Higgsino exchange is suppressed by the supposed smallness of its couplings to the light particles, and the straightforward relation of mass to coupling will be modified by mixing. Also, careful inclusion of the fields necessary to break the unified symmetry and generate neutrino masses brings to light additional potential sources of nucleon instability [14].

I hope and expect that at some future conference we will hear from SuperK – or their successors – reports of the other shoe dropping.

REFERENCES

1. J. Pati and A. Salam, Phys. Rev. Lett. **31**, 661-664 (1973).
2. H. Georgi and S. Glashow, Phys. Rev. Lett. **32**, 438 (1974).
3. D. Gross and F. Wilczek, Phys. Rev. Lett. **30**, 1343 (1973); H. D. Politzer, Phys. Rev. Lett. **30**, 1346 (1973).
4. H. Georgi, H. Quinn, and S. Weinberg, Phys. Rev. Lett. **33**, 451 (1974).
5. LEP Electroweak Working Group, preprint CERN-PPE/96-183 (Dec. 1996).
6. See for example G. Blewitt, *et al*, Phys. Rev. Lett. **55**, 2114 (1985), and the latest Particle Data Group compilations.
7. A very useful introduction and collection of basic papers on supersymmetry is S. Ferrara, *Supersymmetry* (2 vols.) (World Scientific, Singapore 1986). Another excellent standard reference is N.-P. Nilles, Phys. Reports **110**, 1 (1984).
8. S. Dimopoulos, S. Raby, and F. Wilczek, Phys. Rev. D **24**, 1681 (1981).
9. J. Ellis, S. Kelley, and D. Nanopoulos, Phys. Lett. **B260**, 131 (1991); U. Amaldi, W. de Boer, and H. Furstenau, Phys. Lett. **B260**, 447 (1991); for more recent analysis see P. Langacker and N. Polonsky, Phys. Rev. D **49**, 1454 (1994).
10. Barr, S.M., Preprint hep-ph/9806217.
11. H. Georgi, in *Particles and Fields – 1974*, ed. C. Carlson (AIP press, New York, 1975).
12. M. Gell-Mann, P. Ramond, and R. Slansky, in *Supergravity*, ed. P. van Neiuwenhuizen and D. Freedman (North Holland, Amsterdam, 1979), p. 315; T. Yanagida, Proc. of the Workshop on Unified Theory and Baryon Number in the Universe, eds. O. Sawada and A. Sugamoto (KEK, 1979).
13. S. Dimopoulos and F. Wilczek, in *The Unity of the Fundamental Interactions,* Proceedings of the 19th Course of the International School of Subnuclear Physics, Erice, Italy, 1981, edited by A. Zichichi (Plenum, New York, 1983); K.S. Babu and S.M. Barr, Phys. Rev. **D48** 5354 (1993).
14. K.S. Babu, J. Pati and F. Wilczek, "Unification, Neutrino Masses, and Nucleon Instability," to appear as IASSNS-HEP Preprint 98/80.

ELSEVIER

Nuclear Physics B (Proc. Suppl.) 77 (1999) 520–524

NUCLEAR PHYSICS B
PROCEEDINGS
SUPPLEMENTS

Comments

M. Koshiba
The University of Tokyo, Emeritus

It is indeed my great pleasure to have you here, old and new friends, in Takayama to attend this ν '98. Some people might, however, have noticed that we are missing two dear friends of ours missing this gathering. One is the late Professor David N. Schramm of the University of Chicago who died in a plane crash last December. Not only he was a great Cosmologist of our time but also his warm personality made us all his friends. I show his picture with his wife Judy at the time of the last conference at Toyama. (Photo.1) The other is the late Professor Teruhiro Suda of Kobe University who died in India almost exactly 5 years after the SN1987a outburst. He was the driving force of Kamiokandes and of the early phase of Super-Kamiokande. Here is a photo of our senior group members in which you see Prof. Suda at your right end. (Photo.2)

Let me propose to dedicate our one-minute-time here for recalling these two friends of ours.

Thank you very much.

Twenty years ago I had a problem. It is true that I could send the senior graduate students, ready for their thesis work, to our international collaboration experiments, DASP-DORIS and JADE-PETRA, for further training. The problem was; what can I do to entice younger students, up to 3 years in the graduate school and the undergraduates, to the experimental research work of this field? It was clear that: **We have to have some attractive experiment in the home country to keep them interested and to give them the training.**

It was December 1978 that H. Sugawara, Head of the theory group then at KEK, asked me by phone to think about an possible experiment to search for proton decays. The idea of imaging water Cherenkov detector deep underground immediately came to my mind because I discussed the possibilities of such an experiment, though in a primitive form, with Beppo Occhialini over beer glasses in Chicago in 1960. I prepared the drawing of the detector containing 3,000 tons of water and surrounded by photomultipliers over the entire surface and had it presented by my assistant of that time to the workshop at KEK.

Next month, January 1979, I learnt that a much larger, about 7,000 tons of water, experiment of the same design, is being seriously considered, with a budget of several million dollars, in the USA; IMB (Irvine-Michigan-Brookhaven). With an anticipated budget of 1 million dollars, including the excavation cost, how can we compete with them? There are two possible attitudes one can take depending on how much faith one puts on the SU(5) GUT model which predicted a life-time of about 10^{30} years with $p \rightarrow e^+ + \pi^0$ as the main decay mode. In this decay mode one expects about one half GeV electromagnetic cascade in one direction and another half GeV cascade in the opposite direction, thereby producing large number of Cherenkov photons in the back to back pattern. One 5-inch diameter PMT, off the shelf item of those days, per every m^2 of the surrounding surface would be ample enough to detect such an event and this was the attitude IMB had chosen.

If, however, one considers more seriously the possibilities of other types of GUT's, like SU (4) and/or SO(10), the important physics aims would include the determination of branching ratios of proton decays into various possible decay modes. One then has to improve the photon sensitivity as much as possible because other decay modes, like $p \rightarrow$ anti$-\nu + K^+$, produce much less Cherenkov

0920-5632/99/$ – see front matter
PII S0920-5632(99)00491-0

photons. This was the attitude I had taken in initiating Kamiokande; Kamioka-Nucleon-Decay-Experiment. Within the aforementioned budget how can we accomplish this task? I had a number of negotiations with the President, T. Hiruma, of HAMAMATSU PHOTONICS Co. with which our group had a good relation since the time of JADE. I finally succeeded to talk him into undertaking the development of 50-inch diameter PMT's in collaboration with our group. I assigned A. Suzuki and K. Arisaka to this job and they succeeded in one year to obtain good quality PMT's of this size. This was the reason why Kamiokande started, in July 1983, data-taking just about one year later than IMB. Now that we have, with one 50-inch PMT per every $m^2$of the surrounding surface, 16 times more sensitivity to the Cherenkov photons as compared to IMB and this was the reason why we could out-survive them. I learned a lesson. **'Insufficient funding can induce positive thinking'**. Since we decided to look for other, difficult to identify, decay modes of nucleon, we have to be very careful about the backgrounds and about the energy calibration. During the first three months of operation we found an event of $\mu^+ + 2\gamma$'s. The 2γ's nicely formed η mass. The total invariant mass was very close to that of proton. The total momentum-unbalance however was 275MeV/c which was a little too high for the Fermi momentum of proton in Oxygen nucleus. Is this a real proton decay event or a background event induced by the atmospheric neutrinos? We began a very serious study of atmospheric neutrinos which years later yielded the neutrino masses as reported in this conference. **'Today's backgrounds can yield tomorrow's signal,'** as contrasted to the saying 'today's signal is tomorrow's background' among some particle experimentalists.

As a means of energy calibration we looked at the energy spectrum of μ-e decay electrons and found that we could see clearly down to the electron energy of 12MeV below which the backgrounds dominated. Here is a nice possibility of observing the solar neutrinos by the recoil electrons from ν_e-e scattering in the water: Real-time-, directional- and spectral-observation of the solar neutrinos.

Besides the tremendous amount of backgrounds to be overcome, a calculation showed that we could expect at most one event every other day even if the Solar Standard Model is correct. This being such a nice possibility as to prompt me to propose a collaboration in the ICOBAN84 meeting at Park city, Utah, in January of 1984. Namely, besides reporting the preliminary results of proton decay search, I proposed two things: Proposal (1); an international collaboration experiment using Kamiokande on the feasibility of observing the solar neutrinos by means of ν_e-e scattering in the water. To this A. K. Mann showed an immediate interest and we formed the collaboration Kamiokande-II this year. Proposal (2); Super-Kamiokande of 50,000 tons of water with 4 times photon sensitivity as compared to Kamiokande as a real neutrino observatory at a cost of 100M$. This latter proposal could not attract anybody's interest even though I added we could call it JACK implying Japan-American-Collaboration at Kamioka. This Super-Kamiokande, however, was realized 12 years later in Japan.

In Kamiokande-II the American side was to provide a new set of electronics including TDC's for each PMT's while our group was to install the 4π-anti-counters completely surrounding the inner detector and to reduce the environmental backgrounds by many orders. After the strenuous effor.t of one and half year we could start data taking of the solar neutrinos in January of 1986.

About 14 months later near the end of February 1987, when we were beginning to prepare our first paper on the solar neutrinos, a supernova explosion was discovered in LMC. Since our detector has been working quietly and steadily with background level at 7.5MeV, the supernova neutrinos signal, time-bunched and of higher energies, was readily spotted in our data-tape. In the meantime Mont-Blanc experiment announced their SN neutrino signal, which were more than 4 hours ahead of our signal time. We had to be very careful and I ordered complete silence of all the collaborators until we finish all the possible checking. We disclosed our observation early in March and luckily our signal time was immediately confirmed by IMB. The first result of the solar neutrino observation by ν_e-e scattering was also published this year thereby establishing the birth of **'Observational Neutrino Astrophysics'**[1].

As of April 1, 1987, I retired from the University of Tokyo and I passed the spokesmanship of Kamiokande-II to Y. Totsuka but remained as a collaborator. The intensive study of the atmospheric neutrinos as background of proton decay search was beginning to reveal the first sign of something strange. Namely, among the totally confined events of clear single-ring events, the number of μ-events is not twice that of e-events while a simple physical argument leads to the number of ν_μ to that of ν_e to be two at low energies and larger still at higher energies[2]. Extensive, experimental and theoretical, studies of μ/e identification, neutron background, threshold effects of pseudo-elastic charged current interaction, etc., have been made and we published our second paper with a considerably better statistics[3]. Even outside of our Kamiokande collaboration some experimentalists started thinking about long-baseline neutrino oscillation experiment. The American collaborators shifted to SNO and Kamiokande-III started with the renewed electronics.

In June 1996 Super-Kamiokande started data taking. In June 1998 in this ν98 conference at Takayama we presented our Super-Kamiokande results on the solar neutrinos and on the atmospheric neutrino oscillation. **'The neutrinos do oscillate and thus have masses'**. In this year, 1998, KAMLAND led by A. Suzuki was fully approved. It is a 1,000 tons liquid scintillator ball to be installed in the old Kamiokande cave. It is to observe anti-ν's down to low energies and Kamioka experiments are now in the third generation. Let me now show you my personal guesses and wishes. Essentially the same guesses I have been showing since 1991, first at the Lohrmann-fest at DESY.

My Personal Guesses and Wishes

1) Taking the See-Saw mechanism, $m_{\nu_i} = m_{D_i}^p/N_i$, with p=1 rather than 2 and with m_D that of charged lepton, we obtain to within 20% error;

$m_e/m_{\nu e}=m_\mu/m_{\nu\mu}=m_\tau/m_{\nu\tau}$
$=N_e/m_e=N_\mu/m_\mu=N_\tau/m_\tau=3.7\text{x}10^{10}$,
$m_{\nu\tau}=4.8\text{x}10^{-2}$eV,
$m_{\nu\mu}=2.8\text{x}10^{-3}$eV,
$m_{\nu e}=1.4\text{x}10^{-5}$eV,
$N_e=2.0\text{x}10^{16}$eV,
$N_\mu=3.9\text{x}10^{18}$eV,
$N_\tau=6.6\text{x}10^{19}$eV.

Large mixing between ν_μ and ν_τ but small mixing between ν_e and ν_μ.

2) Now that the neutrinos have non-zero masses there occurs a very nice possibility of 'The total reflection of these neutrinos' even for 90^0 incident angle, at non-relativistic energies. Namely one can now hopefully work on the directional observation of 1.9K Cosmic Neutrino Background by implementing focussing mirror and/or Winston cone at very low temperature. Their detection, however, is a different matter and will require years of developmental work; single electron devices, phonon detector and/or Tera-Herz devices. The first thing to aim at would be to detect the same dipole anisotropy as observed in the Cosmic Microwave Background.

3) The τ-appearance experiment of $\nu_\mu - \nu_\tau$ oscillation is of vital importance. It is not easy but one of the three on-going plans, CERN to Gran-Sasso, FNAL to Sudan and KEK to Kamioka, will produce the results before the year 2010.

4) Now that the neutrinos can not responsible for the Dark Matter, it seems to me that the next likely candidate will be Axion of mass around 10^{-5}eV unless the on-going experiments kill it also. If this happens my guess for the Dark Matter is the other E(8) of the E(8)xE(8).

5) We now know that SU(5) is too small and for GUT we seem to have to take at least SUSY-SO(10) or SUSY-E(6). Unfortunately the theorists are not telling us experimentalists much about the predictions of these models.

6) During the 13 years, 1983 to 1996, the Imaging Water Cherenkov detector increased its size by a factor of 20. I should very much like to see a 1,000,000 tons Imaging Water Cherenkov detec-

tor constructed somewhere in the world by the year 2009.

REFERENCES

1. M.Koshiba; "Observational Neutrino Astrophysics", Physics Report, 220 Nos5&6, 1992, pp. 229-463
2. Phys. Lett., B205(1988)416
3. Phys. Lett., B280(1992)146

Photo 1

Photo 2

NUCLEAR PHYSICS B
PROCEEDINGS SUPPLEMENTS

ELSEVIER Nuclear Physics B (Proc. Suppl.) 77 (1999) 525–526

Concluding Words

G. Marx

Department of Atomic Physics, Eötvös University
Puskin utca 5, 1088 Budapest, Hungary

After concluding the successful Neutrino'98 Conference in Takayama, the International Neutrino Commitee has made plans for the future.

1. THE PAST

The first International Neutrino Conference was organized at the Lake Balaton in Hungary in 1972. At that conference Fred Reines reported about the first observed events of the $\bar{\nu}_e - e$ scattering at atomic reactor, B.C. Barish described plans for producing neutrinos at accelerator, and Ray Davis gave a progress report on the first solar neutrino observations. Several participants of Neutrino'72 are also now among us!

Since then over 25 years passed. Alltogether 18 international neutrino conferences were held; each of them offered new empirical evidences for three kinds of elusive neutrinos. We have witnessed the development of neutrino theory, concluding in the understanding of electroweak interactions. The Standard Model has become the coronation of the physics of the 20th century. The world acknowledged these achievements with a golden shower coming from Stockholm.

This is the first International Neutrino Conference at which Fred Reines, the pioneer of experimental neutrino physics, is not present. But his spirit is among us. On the occasion of his 80th birthday the International Neutrino Commitee greeted him by mail.

2. THE PRESENT

In Takayama, the Neutrino'98 Conference was well attended from all over the world. It offered sparkling new observations, presented by the physicists working at Superkamiokande and at other laboratories. Daring theoretical ideas were discussed in order to solve the new questions generated by these observations.

The 20th century is almost over. We can now firmly say that we see the deep interior of a star – our sun – where nuclear reactions are going on. From Superkamiokande one clearly sees the glowing center of the sun in ν_e light produced by the fusion fire. The variation of solar neutrino intensity due to the excentricity of earth's orbit has been observed with 84% confidence. This result – thanks to Homestake, GALLEX, SAGE and the Superkamiokande teams – is certainly one of the peak achievements of the physics in the 20th century, it will deserve a place in the schoolbooks of the 21st century.

3. THE FUTURE

The neutrino community leaves not only completed answers but burning questions as well for the 21st century. This is good news for the incoming generation of physicists. The main challenges are due to budget deficits:

Yes, the center of the sun can be seen, but its brightness seems to be certainly fainter than expected – this is a common finding of all the Neutrino Observatories. Is the ν_e light of the sun an alternating current? Or does the Standard Solar Model overestimate the solar neutrino luminosity? Let me quote from the novel "The Joshua Factor", written by Donald D. Clayton:

– *We all feel that the sun is telling us something important. The neutrino experiment is a means of communicating with the sun's center. There is always a sense of gratitude when a new truth is revailed. It's almost spiritual. It's as if the physicist were a prophet, going to the mountain wilderness in search of a message from god.*

PII S0920-5632(99)00492-2

And the message came from this god, the sun, and said: "You don't understand me."

The most challenging fact is, however, the deficit in the observed atmospheric ν_μ flux, discovered at Kamiokande. We know quantitavely that muons are generated by cosmic radiation bombarding our atmosphere. We know that energetic ν_μs do produce muons whith known cross section. But physicists observe less μ underground than expected, and this deficit seems to depend on whether the muon telescope looks upwards, downwards or sidewards. Do the ν_μs decay or oscillate underway? Anyway, it's hard to escape the conclusion that neutrinos possess rest mass. From the 21st century (from accelerator physicists) we await an answer to the question: what do the ν_μs have been transformed into? Into known particles or into new kinds of particles not yet detected (therefore called 'sterile' ones)? If one finds the answer, this will be the Great Leap for Humankind beyond the Standard Model. A New Century of Physics will begin.

Over 90% of gravitating matter, being present in the Universe, has not yet been seen. Is this dark matter composed (at least partially) by (conventional or 'sterile') neutrinos? It may be that the two last-mentioned deficits are closely connected. Neutrino oscillations (of ν_μ and ν_e) may indicate rest masses of neutrinos. The density profile of dark matter, observed astronomically because of its gravitational effect, can be explained by neutrinos possessing a rest mass of a few eV. The small anomalies in the decay spectrum of tritium are sensitive to the cosmic neutrino background. Is the reported tiny excess at the very end of the ^{3}H electron spectrum perhaps an indication for $\bar{\nu}_e$ induced β decay?

After sighting the neutrino glow of the solar center and the flash of Supernova 1987, a nearby astronomical object deserves quick attention: the earth. It has been suggested already at the very first neutrino conferences that the faint $\bar{\nu}_e$ glow from the interior of our planet brings valuable information about the amount of heavy elements, thus about the formation and history of earth. We have to rush developing neutrino geology before it's to late: the artificial $\bar{\nu}_e$ brightness of man-made reactors is going to overshine this faint natural $\bar{\nu}_e$ glow. Fortunately, promising attempts are in progress at Kamioka and Gran Sasso, to catch the last glimmer of this planet.

Well, these questions can be and will be answered. We may be sure that the upcoming International Neutrino Conferences will be as exciting as the previous ones were:

Neutrino'2000 will go to Sudbury in Canada where SNO (the Sudbury heavy-water Neutrino Observatory) will await for our visit. Our host will be Art B.McDonald, 1–613–7867546.

Neutrino'2002 will go to Munich, Germany, one of the spiritual centers of GALLEX. The conference will be hosted by Franz von Feilitzsch, 49–89–28912680.

Neutrino'2004 will go to the College de France in Paris. France is deeply involved in CHDS, BUGEY, GALLEX, NEMO, NOMAD, NUMU, CHOOZ, HELLAZ, LENCSE, ANTARES collaborations. The host will be Francois Vanucci, 33–1–44274638.

The vast majority of Neutrino Conferences happened in Asia and Europe, thus America awaits us. The first reactor neutrino experiment was designed in Los Alamos, for its semi-centenary there is a standing invitation to Santa Fe by Thomas J. Bowles, 1–505–6652676. New ν detectors will be operational at Fermilab in the next decade; the Soudan observatory is also within reach. There is an ivitation to have a conference at Fermilab by Adam Para, 1–630–8404343. Decisions will be made at the very next International Neutrino Conferences.

The International Neutrino Commitee, in the name of all of us, expresses its cordial thanks to our hosts in Takayama and Kamioka for the smooth and efficient organization, and its congratulations for the reported exciting experiments at Superkamiokande.

ELSEVIER

Nuclear Physics B (Proc. Suppl.) 77 (1999) 527–528

NUCLEAR PHYSICS B
PROCEEDINGS
SUPPLEMENTS

Neutrino 98

List of Contribution Papers

1. **C.H. Albright, K.S. Babu, S.M. Barr**, "Implications of a Minimal SO(10) Higgs Structure".
2. **V.B. Anikeev et al.**, "F_2, F_3 structure functions measurement in low Q^2 region with IHEP–JINR neutrino detector".
3. **R. Arnold et al. (NEMO Collaboration)**, "Double–β decay of ^{96}Zr and ^{94}Zr".
4. **A.S. Barabash et al.**, "2νββ decay of ^{100}Mo to the first 0^+ excited state in ^{100}Ru".
5. **I. Barnett et al.**, "Search for Charged Scalar Bosons in Muon Decay: A Precise Measurement of the Transverse Polarization of Positrons from the Decay of Polarized Muons".
6. **Kh.M. Beshtoev**, "Can the weak Interaction generates an enhancement of neutrino oscillations in matter".
7. **G.De Cataldo et al.**, "The NOE detector for a long baseline neutrino oscillation experiment".
8. **G. Conforto et al.**, "Solar Models and Neutrino Deficit".
9. **M. Daum et al.**, "The KARMEN Time Anomaly, Search for a Neutral Particle of Mass 33.9 MeV in Pion Decay".
10. **A. Geiser**, "Pseudo–Dirac neutrinos as a potential complete solution to the neutrino oscillation puzzle".
11. **A.N. Ivanov et al.**, "On the relaxation of the solar neutrino problem in the relativistic field theory model of the deuteron I".
12. **A.N. Ivanov et al.**, "On the relaxation of the solar neutrino problem in the relativistic field theory model of the deuteron II".
13. **Y. Koide**, "Universal Seesaw Mass Matrix Model and Neutrino Phenomenology".
14. **Q.Y. Liu and A.Yu. Smirnov**, "Neutrino Mass Spectrum with ν_μ–>ν_s Oscillations of Atmospheric Neutrinos".
15. **Q.Y. Liu, S.P. Mikeyev, A.Yu. Smirnov**, "Parametric Resonance in Oscillations of Atmospheric Neutrinos ?".
16. **M. Matsuda and M. Tanimoto**, "Natural Neutrino Mass Matrix".
17. **H. Minakata and O. Yasuda**, "Dark Matter Neutrinos Must Come with Degenerate Masses".
18. **L.A. Popeko**, "(ν,e)–Scattering and Search for Neutrino Magnetic Moment".
19. **P. Raychaudhuri**, "Solar Neutrino Flux Variations in Kamiokande Detector".
20. **P. Raychaudhuri**, "Variations of Solar Neutrino Flux in GALLEX and SAGE Detector".
21. **The Super–Kamiokande Collaboration**, "Measurement of a small atmospheric ν_μ/ν_e ratio".
22. **The Super–Kamiokande Collaboration**, "Study of the atmospheric neutrino flux in the multi–GeV energy range".
23. **The Super–Kamiokande Collaboration**, "Measurements of the Solar Neutrino Flux from Super–Kamiokande's First 300 Days".
24. **The Super–Kamiokande Collaboration**, "Search for Proton Decay via p–>$e^+\pi^0$ in a

Large Water Cherenkov Detector".
25. **M. Tanimoto**, "Indirect Search for CP Violation in Neutrino Oscillations".
26. **T. Totani**, "Electron Neutrino Mass Measurement by Supernova Neutrino Bursts and Implications for Hot Dark Matter".
27. **T. Totani et al.**, "Future Detection of Supernova Neutrino Burst and Explosion Mechanism".
28. **H.T. Wong and Jin Li**, "A Pilot Experiment with Reactor Neutrinos in Taiwan".
29. **O. Yasuda**, "Three flavor neutrino oscillation analysis of the SuperKamiokande atmospheric neutrino data".
30. **L. Berge et al.**, "Status of the EDELWEISS Experiment".
31. **M. Maris and S.T. Petcov**, "A Study of the Day-Night Effect for the Super-Kamiokande Detector: III. The Case of Transitions into Sterile Neutrino".
32. **S.T. Petcov**, "Diffractive-Like (or Parametric-Resonance-Like?) Enhancement of the Earth (Day-Night) Effect for Solar Neutrinos Crossing the Earth Core".
33. **A.N. Ivanov et al.**, "On the relaxation of the solar neutrino problem in the relativistic field theory model of the deuteron III".
34. **H.K. Tank**, "An Explanation for the Large Numbers in Astrophysics and some inshight into the nature of Fundamental Forces".
35. **H.K. Tank**, "Expressing energy-momentum four-vector of the special relativity in terms of wave-mechanics".
36. **Eduardo do Couto e Silva** (on behalf of NOMAD and TOSCA Collaborations), "Silicon detectors for neutrino oscillation experiments".

Nuclear Physics B (Proc. Suppl.) 77 (1999) 529–530

NUCLEAR PHYSICS B
PROCEEDINGS
SUPPLEMENTS

Neutrino 98

List of Poster Presentation

- Technical descriptions of Borexino — A. Ianni (Gran Sasso)
- Calculations of muon and hadronic fluxes from atmospheric neutrinos at high energies — L.V.Volkova(INR)
- Do not forget the physical Importance of the Supernova 87A signals recorded in the Kamiokande and IMB apparatus — H. Huzita (Padova)
- Atmospheric neutrinos in SNO — C. Waltham et al. (British Columbia)
- Silicon detectors for neutrino oscillation experiments — E. do Couto e Silva (CERN)
- NOE detector for a long baseline neutrino oscillation experiment — P. Spinelli (Bari)
- L3 cosmic experiment — R. Nahnhauer (DESY-Zeuthen)
- Some design, materials and construction features of SNO — E.D. Hallman (Laurentian)
- Development of Cryogenic Detectors for the Solar Neutrino Experiment GNO — M. Altmann (TU Muenchen)
- Improved calculation of atmospheric neutrinos and possibility of neutrino oscillations — S. Midorikawa (Aomori) et al.
- Extremely high energy particle astrophysics and the Telescope Array project — M. Sasaki (ICRR)
- F_2, F_3 structure function measurements in low Q^2 region with IHEP-JINR neutrino detector — V. Tumakov (Protvino) et al.
- Modern status of neutrino experiments at the underground neutrino laboratory of Kurchatov Institute near Krasnoyarsk nuclear reactor — Yu.Kozlov (Karchatov) et al.
- Neutrino Mass Spectrum with $\nu_\mu \to \nu_s$ Oscillations of Atmospheric Neutrinos/Parametric Resonance in Oscillations of Atmospheric Neutrinos? — Qiu-Yu Liu (SISSA)
- Universal Seesaw Mass Matrix Model and Neutrino Phenomenology — Y. Koide (Shizuoka)
- PeP-neutrino detector on 10 tons of metallic lithium — A. Kopylov (INR)
- Future Detection of Supernova Neutrino Bursts and Its Implications on Supernova Mechanism and Electron Neutrino Mass — T. Totani (Tokyo)
- Can long baseline experiments test models of neutrino mass matrix — M. Tanimoto (Ehime)
- Constraints of mixing angles from neutrino oscillation experiments and neutrinoless double beta decay — T. Fukuyama / K. Matsuda (Ritsumeikan)
- Current Status of the Solar Neutrino Problem with Superkamiokande — H. Minakata (Tokyo Metro.)
- MSW oscillations and flux-independent observables at SuperKamiokande and SNO — E. Lisi (Bari)

- Pseudo-Dirac neutrinos as a potential complete solution to the neutrino oscillation puzzle
  A. Geiser (CERN)
- Searching for Supernovae Using the SNO Detector
  J. Heise (British Columbia)
- An alternative solution for the solar neutrino problem
  Y. Tomozawa (Michigan)
- The KARMEN Time Anomaly, Search for a Neutral Particle of Mass 33.9 MeV in Pion Decay
  P.R. Kettle (PSI)
- Two neutrino double beta decay of ^{100}Mo to the first 0^+ excited state in ^{100}Ru/Double beta decay of ^{96}Zr and ^{94}Zr
  A. Barabash (ITEP)
- Can the weak interaction generate an enhancement of neutrino oscillations in matter?
  Kh. Beshtoev (Dubna)
- Constraints on Primordial Neutrino Decay in the Mass Range Near 0.01eV from Improved Limits to the IR Background
  S. Biller (Oxford)
- The Antares Project
  N. de Botton (Sacray) et al
- On the possibility to measure the neutrino magnetic moment down to 10^{-11} Bohr magnetor with artificial neutrino source
  I.R. Barabanov (INR)
- Analysis of neutrino-induced upward-going muons with Super-Kamiokande
  M. Yoshida (Osaka)/
  A. Habig (Boston)
- The International Supernova Early Alert Network
  A. Habig, K. Scholberg(Boston)/
  M. Vagins (UCI)
- Rn background in the Super-Kamiokande
  Y. Takeuchi (ICRR)
- LINAC calibration at Super-Kamiokande
  E. Blaufuss(LSU)/
  N. Sakurai (ICRR)
- HELLAZ
  P. Gorodetzky
  (PCC-College de France)

ELSEVIER

Nuclear Physics B (Proc. Suppl.) 77 (1999) 531–541

NUCLEAR PHYSICS B
PROCEEDINGS
SUPPLEMENTS

List of Participants

Albright, Carl H.
Theory Group , Northern Illinois Univ. Fermilab
albright@fnal.gov

Altmann, Michael
Physik Department E15, Technische Universitat Munchen
altmann@e15.physik.tu-muenchen.de

Annis, Patrio
Universite Libre de Bruxelles,
Patrio.Annis@cern.ch

Aoki, Ken-ichi
Physics Department, Kanazawa University
aoki@hep.s.kanazawa-u.ac.jp

Aoki, Shigeki
Div. of Sciences for Natural Environment, Faculty of Human Development, Kobe University
aoki@kobe-u.ac.jp

Arafune, Jiro
Institute for Cosmic Ray Research, Univ. of Tokyo
arafune@icrr.u-tokyo.ac.jp

Asaka, Takehiko
Institute for Cosmic Ray Research, University of Tokyo
asaka@icrr.u-tokyo.ac.jp

Bahcall, John N.
Institute for Advanced Study,
jnb@sns.ias.edu

Bakich, Andrew
School of Physics A28, University of Sydney
bakich@physics.usyd.edu.au

Baldo Ceolin, Milla
Physics Department, University of Padova
BALDOCEOLIN@PADOVA.INFN.IT

Bando, Masako
Aichi University,
bando@aichi-u.ac.jp

Barabanov, Igor
Institute for Nuclear Research of RAS, Senior Scientific Researcher
Barabanov@vaxmw.tower.ras.ru

Barabash, Alexander S.
Institute of Theoretical and Experimental Physics,
barabash@vxitep.itep.ru

Barbarino, Giancarlo
INFN and Dipartimento di scienze Fisiche Napoli, Complesso Universitario di Monte S.Angelo
BARBARINO@NA.INFN.IT

Barish, Barry C.
Caltech,
barish@ligo.caltech.edu

Barr, Giles
Division EP, CERN
giles.barr@cern.ch

Barszczak, Tomasz
University of California, Irvine
tomba@hepxvt.ps.uci.edu

Beier, Eugene W.
Department of Physics and Astronomy, University of Pennsylvania
geneb@upenn5.hep.upenn.edu

Bemporad, Carlo
Department of Physics, INFN and University of Pisa
Bemporad@axpia.pi.infn.it

Bergstrom, Lars
Dept of Physics, Stockholm University
lbe@physto.se

Bernstein, Robert H.
Fermilab,
rhbob@fnal.gov

Bilenky, Samoil
INFN, Torino Univ.
BILENKY@to.infn.it

Biller, Steven
Physics Dept., Oxford University
S.BILLER1@physics.oxford.ac.uk

Blaufuss, Erik
Louisiana State University,
blaufuss@suketto.icrr.u-tokyo.ac.jp

Booth, Norman E.
Department of Physics, University of Oxford
n.booth@physics.oxford.ac.uk

Bowles, Thomas J.
Los Alamos National Laboratory,
tjb@lanl.gov

Caldwell, David O.
University of California Santa Barbara,
caldwell@slac.stanford.edu

Camilleri, Leslie L.
CERN/PPE,
leslie.camilleri@cern.ch

Chen, Mark
Physics Department - Jadwin Hall, Princeton University
mcchen@princeton.edu

Chiba, Masami
Dept. of Phys., Faculty of Science, Tokyo Metropolitan Univ.
chiba@phys.metro-u.ac.jp

Chikira, Yuichi
Faculty of Science, Tokyo Institute of Technology
ychikira@th.phys.titech.ac.jp

Collar, Juan I.
U. Paris7/CERN, G.P.S.
collar@mail.cern.ch

Conforto, Gianni
University of Urbino,
CONFORTO@VXCERN.CERN.CH

Costa, Giovanni
Dept.of Physics, University of Padova
COSTA@PADOVA.INFN.IT

Cowen, Douglas F.
Department of Physics and Astronomy, University of Pennsylvania
cowen@dept.physics.upenn.edu

Cremonesi, Oliviero
INFN - Milano,
Oliviero.Cremonesi@mi.infn.it

Cribier, Michel
CEA/Saclay, DAPNIA/SPP
cribier@hep.saclay.cea.fr

Cronin, James W.
Enrico Fermi Institute, The University of Chicago
jwc@uchepa.uchicago.edu

Dalhed, Hollis E.
Lawrence Livermore National Laboratory, University of California
dalhed1@llnl.gov

de Botton, Nico
DAPNIA/SPP, CEA Saclay
botton@hep.saclay.cea.fr

Derbin, Alexander
St.Petersburg Nuclear Physics Inst.,
derbin@lnpi.spb.su

DiLella, Luigi
CERN/PPE,
dilella@axnd17.cern.ch

do Couto e Silva, Eduardo
CERN - EP Division,
EDUARDO.DO.COUTO.E.SILVA@CERN.CH

Doe, Peter J.
Nuclear Physics Lab., University of Washington
PDOE@U.WASHINGTON.EDU

Doki, Wataru
Niigata University,
doki@hep.sc.niigata-u.ac.jp

Domogatsky, Grigory V.
Institute for Nuclear Research, Russian Academy of Sciences
domogats@pcbai10.lpi.msk.su

Dore, Ubaldo
Dipartimento di Fisica, University 'la Sapienza' Roma
Ubaldo.dore@roma1.infn.it

Drexlin, Guido J.
Forschungszentrum Karlsruhe, Institut fuer Kernphysik 1
GUIDO@IK1.FZK.DE

Efremenko, Yuri
Research Association, University of Tennessee
efremenk@unix.cas.utk.edu

Ejiri, Hiroyasu
Research Center for Nuclear Physics
(RNCP), Osaka University
ejiri@rcnp.osaka-u.ac.jp

Elliott, Steven R.
Dept. of Physics, University of Washington
sre@u.washington.edu

Elsener, Konrad
CERN,
Konrad.Elsener@CERN.CH

Ereditato, Antonio
INFN Napoli, Italy,
antonio.ereditato@cern.ch

Etoh, Masayuki
Bubble Chamber Physics Lab., Tohoku University
etoh@awa.tohoku.ac.jp

Feldman, Gary J.
Department of Physics, Harvard University
feldman@physics.harvard.edu

Fernholz, Richard
Physics Department, Engineering Physicist Princeton University
richardf@Princeton.EDU

Fetscher, Wulf
Institute for Particle Physics, ETH Zuerich
fetscher@psi.ch

Fogli, Gian Luigi
Dipartimento di Fisica, University of Bari
Fogli@ba.infn.it

Foot, Robert T.
School of Physics, University of Melbourne
Foot@physics.unimelb.edu.au

Frekers, Dieter
Inst. of Nuclear Physics, University of Muenster
frekers@ikp.uni-muenster.de

Fujikawa, Kazuo
Department of Physics, University of Tokyo
fujikawa@hep-th.phys.s.u-tokyo.ac.jp

Fujiwara, Mamoru
Research Center for Nuclear Physics, Osaka University
fujiwara@rcnp.osaka-u.ac.jp

Fukuda, Yoshiyuki
Kamioka Observatory
Institute for Cosmic Ray Research, University of Tokyo
fukuda@suketto.icrr.u-tokyo.ac.jp

Fulgione, Walter
Istituto di Cosmogeofisica, corso Fiume 4
FULGIONE@cosmot.to.infn.it

Furuno, Koichiro
Research Center of Neutrino Science, Tohoku University
furuno@awa.tohoku.ac.jp

Fusaoka, Hideo
Aichi Medical University,
fusaoka@amugw.aichi-med-u.ac.jp

Futagami, Takahiro
Dept. of Physics, Faculty of Science, Tokyo Institute of Technology
futagami@hp.phys.titech.ac.jp

Gaillard, Jean-Marc
CERN/EP Division, Laboratoire Annecy-le-Vieux
Jean-Marc.Gaillard@cern.ch

Gaisser , Thomas K.
Bartol Research Institute, University of Delaware
gaisser@bartol.udel.edu

Gavrin, Vladimir N.
Institute for Nuclear Research of the, Russian Academy of Sciences
gavrin@adonis.iasnet.ru

Geiser, Achim
CERN/PPE Division,
Achim.Geiser@cern.ch

Giacomelli, Giorgio
Dipartimento di Fisica, University of Bologna
GIACOMELLI@BO.INFN.IT

Giunti, Carlo
INFN, Sezione di Torino,
GIUNTI@TO.INFN.IT

Glashow, Sheldon L.
Lyman Laboratory, Harvard University
glashow@physics.harvard.edu

Goldhaber, Maurice
Physics Department, Brookhaven National Laboratory
goldhaber@bnlarm.bnl.gov

Gomez-Cadenas, Juan J.
CERN/University of Valencia,
gomez@mail.cern.ch

Gorodetzky, Philippe
PCC-College de France,
Philippe.Gorodetzky@cern.ch

Gough, Douglas
Institute of Astronomy, University of Cambridge
douglas@ast.cam.ac.uk

Grant, Alan
CERN,
alan.grant@cern.ch

Gratta, Giorgio
Varian Physics Dept., Stanford University
gratta@hep.stanford.edu

Groom, Donald E.
Lawrence Berkeley National Laboratory,
deg@lbl.gov

Guyonnet, Jean-Louis
Institut de Recherches Subatomiques,
jean-louis.guyonnet@ires.in2p3.fr

Habig, Alec
Particle Astrophysics Group, Boston University
habig@budoe.bu.edu

Hagiwara, Kaoru
KEK Theory Group,
kaoru.hagiwara@kek.jp

Hahn, Richard L.
Brookhaven National Laboratory,
HAHN1@BNL.GOV

Haidt, Dieter
DESY,
dieter.haidt@desy.de

Hallman, Doug
Department of Physics and Astronomy, Laurentian University
edh@neutrino.phys.laurentian.ca

Halprin, Arthur
Department of Physics and Astronomy, University of the Delaware
halprin@udel.edu

Halzen, Francis L.
Physics Department, University of Wisconsin
halzen@pheno.physics.wisc.edu

Hamaguchi, Koichi
Dept. of Physics, Faculty of Science, Univ. of Tokyo
hama@hep-th.phys.s.u-tokyo.ac.jp

Hara, Toshio
Department of Physics, Faculty of Science, Kobe University
thara@kobe-u.ac.jp

Hara, Yasuo
Teikyo-Heisei University,
hara@cn.thu.ac.jp

Hasegawa, Takuya
Bubble Chamber Physics Lab., Tohoku University
hasegawa@awa.tohoku.ac.jp

Hashimoto, Michio
Department of Physics, Nagoya University
michioh@eken.phys.nagoya-u.ac.jp

Hata, Naoya
Institute for Advanced Study,
hata@ias.edu

Hatakeyama, Yutaka
Department of Physics, Tokai University
yutaka@tkikam.sp.u-tokai.ac.jp

Haxton, Wick
NINT, University of Washington
haxton@phys.washington.edu

Hayato, Yoshinari
National Laboratory for High Energy Accelerator Research Organization, (KEK)
hayato@neutrino.kek.jp

Heinz, Richard M.
Physics Dept., Indiana University
HEINZ@INDIANA.EDU

Heise, Jaret
University of British Columbia,
jaret@physics.ubc.ca

Helmer, Rich
TRIUMF
Sudbury Neutrino Observatory, Creighton Mine
helmer@surf.sno.laurentian.ca

Hidaka, Keisho
Department of Physics, Tokyo Gakugei University
hidaka@u-gakugei.ac.jp

Hill, James
SUNY Stony Brook/ JSPS (KEK), 62 Experimental Group
jimhill@mochi.icrr.u-tokyo.ac.jp

Hime, Andrew
Physics Division, Los Alamos National Laboratory
ahime@lanl.gov

Hisano, Junji
KEK (Theory Group),
hisano@theory.kek.jp

Holzschuh, Eugen
Physik-Institut, University of Zurich
Holzschuh@uzph3.physik.unizh.ch

Honda, Morihiro
Institute for Cosmic Ray Research,
mhonda@icrr.u-tokyo.ac.jp

Huzita, Humiaki
INFN sezione di Padova, c/o Physic Dept., Univ. of Padova
Huzita@Padova.infn.it

Ianni, Aldo
Laboratori Nazionali del Gran Sasso
INFN, Physics Department University of L'Aquila
Aldo.Ianni@lngs.infn.it

Inagaki, Takahiro
KEK, High Energy Accelerator Research Organization
inagaki@neutrino.kek.jp

Inoue, Kunio
Kamioka Observatory
Institute for Cosmic Ray Research, University of Tokyo
inoue@icrr.u-tokyo.ac.jp

Ishino, Hirokazu
Kamioka Observatory
Institute for Cosmic Ray Research, University of Tokyo
ishino@suketto.icrr.u-tokyo.ac.jp

Itow, Yoshitaka
Kamioka Observatory
Institute for Cosmic Ray Research, University of Tokyo
itow@icrr.u-tokyo.ac.jp

Iwamoto, Toshiyuki
Bubble Chamber Physics Lab., Tohoku University
iwamoto@awa.tohoku.ac.jp

Jonkmans, Guy
Institut de Physique, University of Neuchatel
Guy.jonkmans@iph.unine.ch

Jung, Chang Kee
Dept. of Physics and Astronomy, The State University of New York at Stony Brook
alpinist@sbhep.physics.sunysb.edu

Kabe, Seiji
National Laboratory for High Energy Accelerator Research Organization (KEK),
seiji.kabe@kek.jp

Kajita, Takaaki
Kamioka Observatory
Institute for Cosmic Ray Research, University of Tokyo
kajita@icrr.u-tokyo.ac.jp

Kameda, Jun
Kamioka Observatory
Institute for Cosmic Ray Research, University of Tokyo
kameda@suketto.icrr.u-tokyo.ac.jp

Kaneyuki, Kenji
Dept. of Physics, Faculty of Science, Tokyo Institute of Technology
kaneyuki@hp.phys.titech.ac.jp

Karle, Albrecht
High Energy Physics, University of Wisconsin - Madison
karle@alizarin.physics.wisc.edu

Kasahara, Katsuaki
Shibaura Institute of Technology, Department of Systems Engineering
kasahara@icrr.u-tokyo.ac.jp

Kawasaki, Masahiro
Institute for Cosmic Ray Research, University of Tokyo
kawasaki@ctsun2.icrr.u-tokyo.ac.jp

Kayser, Boris
Physics Division, National Science Foundation
bkayser@nsf.gov

Kearns, E.
Dept. of Physics, Boston Univ.

Kettle, Peter-Raymond
Paul Scherrer Institute (PSI), Research Dept. F1, Nuclear and Particle Physics
Peter-Raymond.Kettle@psi.ch

Kibayashi, Atsuko
University of Hawaii,
atsuko@suketto.icrr.u-tokyo.ac.jp

Kielczewska, Danuta M.
Institute of Experimental Physics, Warsaw University/ University of California at Irvine
danka@fuw.edu.pl

Kiers, Ken
High Energy Theory
Department of Physics, Brookhaven National Laboratory
kiers@bnl.gov

Kiko, Juergen
Max-Planck-Institut f. Kernphysik,
kiko@kosmo.mpi-hd.mpg.de

Kim, Byung-Kyu
Louisiana State University,
bkkim@suketto.icrr.u-tokyo.ac.jp

Kim, Chung W.
Korea Institute for Advanced Study and, Johns Hopkins
cwkim@kias.kaist.ac.kr

Kim, Soo-Bong
Dept. of Physics, Seoul National University
sbkim@neutrino.kek.jp

Kirsten, Till A.
Max-Planck-Inst.flr Kernphysik,
KIRST@KOSMO.MPI-HD.MPG.DE

Kitagaki, Toshio
Tohoku Gakuin University,

Klapdor-Kleingrothaus, Hans Volker
Max Planck Institut fuer Kernphysik,
Klapdor@enull.mpi-hd.mpg.de

Kobayashi, Kazuyoshi
Kamioka Observatory
Institute for Cosmic Ray Research , University of Tokyo
kenkou@icrkm4.icrr.u-tokyo.ac.jp

Kobayashi, Yuichi
Kamioka Observatory
Institute for Cosmic Ray Research, University of Tokyo
yuichi@suketto.icrr.u-tokyo.ac.jp

Koide, Yoshio
Dept. of Physics, University of Shizuoka
koide@u-shizuoka-ken.ac.jp

Koike, Masafumi
Institute for Cosmic Ray Research, University of Tokyo
koike@icrr.u-tokyo.ac.jp

Konuma, Michiji
Faculty of Environmental and Information Studies, Musashi Institute of Technology
konuma@yc.musashi-tech.ac.jp

Kopylov, Anatoli V.
Institute for Nuclear Research of the, Russian Academy of Sciences
kopylov@al20.inr.troitsk.ru

Koshiba, Masatoshi
The University of Tokyo,
mkoshiba@ap-soft-tech.co.jp

Koshio, Yusuke
Kamioka Observatory
Institute for Cosmic Ray Research, University of Tokyo
koshio@icrr.u-tokyo.ac.jp

Kotchetov, Oleg
Laboratory Nuclear Problems, JINR - DUBNA
kochet@nusun.jinr.ru

Kozlov, Iouri
Inst. of General and Nuclear Phys. of RRC, Kurchatov Institute
kozlov@dnuc.polyn.kiae.su

Krivosheina, Irina V.
Max-Planck-Institut fuer, Kernphysik,
irina@enull.mpi-hd.mpg.de

Kropp, William R.
Dept.of Physick and Astronomy, University of California
wkropp@uci.edu

Kuo, Tzee-Ke
Physics Dept., Purdue Univ.
TKKUO@PHYSICS.PURDUE.EDU

Lande, Kenneth
Physics Department, University of Pennsylvania
klande@sas.upenn.edu

Lang, Karol
Department of Physics, University of Texas at Austin
lang@hep.utexas.edu

Langacker, Paul
Department of Physics and Astronomy, University of Pennsylvania
pgl@langacker.hep.upenn.edu

Lanou, Robert E.
Department of Physics, Brown University
lanou@brownvm.brown.edu

Law, Jimmy
Dept. of Physics, University of Guelph
jlaw@uoguelph.ca

Leung, Chung N.
Department of Physics & Astronomy, University of Delaware
leung@physics.udel.edu

Linssen, Lucie
CERN, EP Division
Lucie.Linssen@cern.ch

Lipari, Paolo
INFN sez. Roma, and Dipartimento di Fisica, Universita' di Roma I
paolo.lipari@roma1.infn.it

Lisi, Eligio
Dip. di Fisica and INFN, Universita di Bari
lisi@ba.infn.it

Liu, Qui-Yu
Scuola Internationel Superiore di Studi Avonzati (SISSA),
qyliu@sissa.it

Lobashev, Vladimir M.
Institute for Nuclear Research of the Russian, Academy of Sciences
lobashev@al20.inr.troitsk.ru

Loh, Eugene C.
Dept. of Physics, High Energy Astrophysics Inst., The University of Utah
loh@mail.physics.utah.edu

Lundberg, Byron
Fermilab,
lundberg@FNAL.GOV

Maekawa, Nobuhiro
Dept. of Physics, Kyoto University
maekawa@gauge.scphys.kyoto-u.ac.jp

Maki, Ziro
Phys. Dept., Faculty of Science and, Technology, Kinki University
ziromaki@phys.kindai.ac.jp

Martens, Kai
Dept. of Physics and Astronomy, SUNY at Stony Brook
kai@icrr.u-tokyo.ac.jp

Marx, George
Department of Atomic Physics, Eotvos University
atom@hercules.elte.hu

Matsuda, Atsushi
Institution of Cosmic Ray Research, University of Tokyo
matsuda@icrr.u-tokyo.ac.jp

Matsuda, Koichi
Ritsumeikan University,
sph30101@se.ritsumei.ac.jp

Matsuda, Satoshi
FIHS, Dept. of Fundamental Sciences, Kyoto Univ.
matsuda@phys.h.kyoto-u.ac.jp

Mauger, Christopher
State University of New York at Stony Brook,
cmauger@suketto.icrr.u-tokyo.ac.jp

McDonald, Arthur B.
Director, Sudbury Neutrino Observatory Institute, Stirling Hall, Queen's University
mcdonald@sno.phy.queensu.ca

McGrew, Clark
Physics Dept., State University of New York at Stony Brook
clark.McGrew@sunysb.edu

McKee, Shawn
Physics Department, University of Michigan
smckee@umich.edu

Melzer, Oliver
NIKHEF
Physics Department, Nagoya University
Oliver.Melzer@cern.ch

Meyer, Hinrich O.
University of Wuppertal,
meyer@wpos7.physik.uni-wuppertal.de

Midorikawa, Shoichi
Faculty of Engineering, Aomori University
midori@aomori-u.ac.jp

Migliozzi, Pasquale
CERN, Division EP
pasquale.migliozzi@cern.ch

Mimura, Yukihiro
Institute for Cosmic Ray Research, University of Tokyo
mimura@icrr.u-tokyo.ac.jp

Minakata, Hisakazu
Dept. of Physics, Faculty of Science, Tokyo Metropolitan Univ.
minakata@phys.metro-u.ac.jp

Mine, Shunichi
National Laboratory for High Energy Accelerator Research Organization, (KEK)

Miura, Makoto
Kamioka Observatory
Institute for Cosmic Ray Research, University of Tokyo
miuram@suketto.icrr.u-tokyo.ac.jp

Miyano, Kazumasa
Department of Physics, Niigata University
miyano@hep.sc.niigata-u.ac.jp

Mizutani, Kohei
Department of Physics, Saitama University
mizutani@cr.phy.saitama-u.ac.jp

Mohapatra, Rabindra N.
Dept. of Physics, Univ. of Maryland
RMOHAPATRA@UMDHEP.UMD.EDU

Monacelli, Piero
L'Aquila Univ.-INFN Gran Sasso Laboratory, Dipartimento di Fisica Universita
Piero.Monacelli@lngs.infn.it

Montanet, Francois
CPPM Marseille, IN2P3/CNRS - Univ. Mediterranee
montanet@cppm.in2p3.fr

Morales, Angel
Laboratory of Nuclear and High Energy Physics, Facultad de Ciencias, University of Zaragoza
amorales@posta.unizar.es

Morales, Julio
Laboratorio de Fisica Nuclear Y Altas Energias, Facultad de Ciencias, University of Zaragoza
Julio.Morales@posta.unizar.es

Moroi, Takeo
Theoretical Physics Group, Lawrence Berkeley National Laboratory
moroi@thsrv.lbl.gov

Moscoso, Luciano
DAPNIA/SPP, CEA/Saclay
moscoso@hep.saclay.cea.fr

Mourao, Ana Maria
Dep. of Physics, CENTRA/IST LISBON-PORTUGAL
amourao@calvin.ist.utl.pt

Muciaccia, Maria-Teresa
Dip. Fisica, Bari University
MUCIACCIA@BARI.INFN.IT

Mufson, Stuart Lee
Astronomy Department, Indiana University
mufson@mimosa.astro.indiana.edu

Muraki, Yasushi
Solar-Terrestrial Environment Lab., Nagoya Univ.
muraki@stelab.nagoya-u.ac.jp

Murayama, Akihiro
Faculty of Education, Shizuoka University
murayama@ed.shizuoka.ac.jp

Murayama, Hitoshi
Assistant Professor of Physics, University of California
murayama@thsrv.lbl.gov

Nagashima, Yorikiyo
Physics Department, Faculty of Science, Osaka University
naga@hep.sci.osaka-u.ac.jp

Nahnhauer, Rolf
DESY - ZEUTHEN,
nahnhaue@ifh.de

Nakahata, Masayuki
Kamioka Observatory
Institute for Cosmic Ray Research, University of Tokyo
nakahata@suketto.icrr.u-tokyo.ac.jp

Nakamura, Kenzo
National Laboratory for High Energy Accelerator Research Organization (KEK),
KENZO.NAKAMURA@KEK.JP

Nakamura, Mituhiro
F-lab, Department of Physics, Nagoya University
nakamura@flab.phys.nagoya-u.ac.jp

Nambu, Yoichiro
E. Fermi Institute, University of Chicago
nambu@yukawa.uchicago.edu

Nishijima, Kyoshi
Department of Physics, Tokai University
kyoshi@tkikam.sp.u-tokai.ac.jp

Nishikawa, Koichiro
High Energy Accelerator Research Organization, (KEK)
nishikaw@kekvax.kek.jp

Nishiura, Hiroyuki
Junior College, Osaka Institute of Technology
nishiura@jc.oit.ac.jp

Niu, Kiyoshi
CERN, Nagoya University
kiyoshi.Niu@cern.ch

Niwa, Kimio
NAGOYA UNIVERSITY,
niwa@kekvax.kek.jp

Nomura, Daisuke
Dept. of Physics, Faculty of Science, Univ. of Tokyo
daisuke@hep-th.phys.s.u-tokyo.ac.jp

Nomura, Yasunori
Dept. of Physics, Faculty of Science, Univ. of Tokyo
yasunori@hep-th.phys.s.u-tokyo.ac.jp

Nunokawa, Hiroshi
Instituto de Fisica Gleb Wataghin, Universidade Estadual de Campinas-UNICAMP
nunokawa@ifi.unicamp.br

Obayashi, Yoshihisa
Kamioka Observatory
Institute for Cosmic Ray Research, University of Tokyo
ooba@suketto.icrr.u-tokyo.ac.jp

Oberauer, Lothar
Borexino Collaboration, Technische Universitat Munchen
oberauer@e15.physik.tu-muenchen.de

Ochi, Nobuaki
Department of Physics, Okayama University
ochi@fphy.hep.okayama-u.ac.jp

Ogawa, Hiroshi
Bubble Chamber Physics Lab., Tohoku University
ogawa@awa.tohoku.ac.jp

Ohashi, Akiko
Department of Physics, Okayama University
ohashi@fphy.hep.okayama-u.ac.jp

Okada, Atsushi
Institute for Cosmic Ray Research, University of Tokyo
okada@icrr.u-tokyo.ac.jp

Okei, Kazuhide
Department of Physics, Okayama University
okei@fphy.hep.okayama-u.ac.jp

Okumura, Kimihiro
Kamioka Observatory
Institute for Cosmic Ray Research, University of Tokyo
okumura@icrr.u-tokyo.ac.jp

Ollerhead, Robin
Department of Physics, University of Guelph
rwo@physics.uoguelph.ca

Oyama, Yuichi
National Laboratory for High Energy Accelerator Research Organization , (KEK)
oyamay@kekvax.kek.jp

Pakvasa, Sandip
University of Hawaii,
pakvasa@uhheph.phys.hawaii.edu

Palladino, Vittorio
CERN/EP Division, Univ. of Naples
Vittorio.Palladino@cern.ch

Panman, Jaap
CERN/EP,
Jaap.Panman@cern.ch

Para, Adam
Fermilab,
para@fnal.gov

Parke, Stephen
Theoretical Physics Department, Fermi National Accelerator Laboratory
parke@fnal.gov

Pati, Jogesh C.
Physics Department, Univ.of Maryland
pati@umdhep.umd.edu

Peltoniemi, Juha T.
Department of Physics, University of Helsinki
juha@neutrino.pc.helsinki.fi

Perez, Patrice
DAPNIA/SPP
Bat.141, CEA-Saclay
perez@hep.saclay.cea.fr

Pesen, Erhan
CERN, EP Division
Erhan.Pesen@cern.ch

Petcov, Serguey T.
SISSA/INFN - Trieste, Italy and Bulgarian, Academy of Sciences, Sofia, Bulgaria
PETCOV@HE.SISSA.IT

Peterson, Earl A.
School of Physics and Astronomy, University of Minnesota
EAP@MNHEP1.HEP.UMN.EDU

Piepke, Andreas
Caltech, Norman Bridge Lab. of Physics
andreas@citnp1.caltech.edu

Pietropaolo, Francesco
INFN Padova,
francesco.pietropaolo@cern.ch

Pietschmann, Herbert V.R.
Inst. Theor. Phys., Universitaet Wien
piet@pap.univie.ac.at

Piquemal, Fubrice
CEN Bordeaux-Gradignan, CNRS-IN2P3 Bordeaux Universite
piquemal@cenbg.in2p3.fr

Popeko, Ludvig
St-Petersburg Nuclear Physics, Institute RAS
popeko@hep486.pnpi.spb.ru

Price, LeRoy R.
Dept. of Physics and Astronomy, University of California, Irvine
price@leroy.ps.uci.edu

Protheroe, Raymond J.
Department of Physics and Mathematical Physics, The University of Adelaide
rprother@physics.adelaide.edu.au

Raffelt, Georg G.
Max-Planck-Institut fur Physik,
raffelt@mppmu.mpg.de

Raghavan, Raju S.
Bell Laboratories,
raju@physics.lucent.com

Ramanamurthy, Poolla V.
S.T.E. Laboratory, Nagoya University
murthy@stelab.nagoya-u.ac.jp

Ramond, Pierre M.
Department of Physics, Institute for Fundamental Theory, Univ. of Florida
ramond@phys.ufl.edu

Ranucci, Gioacchino
Instituto Nazionale di Fisica Nucleare
Sezione di Milano,
Ranucci@mi.infn.it

Robertson, R.G. Hamish
Department of Physics, University of Washington
rghr@u.washington.edu

Roe, Byron P.
Dept. Physics, Randall Laboratory, University of Michigan
Byronroe@UMICH.edu

Ronga, Francesco
INFN Laboratori Nazionali di Frascati,
RONGA@axlnf2.lnf.infn.it

Roos, Matts
High Energy Physics, University of Helsinki
Matts.Roos@Helsinki.fi

Rosen, Peter S.
U.S. Department of Energy,
Peter.Rosen@mailgw.er.doe.gov

Rowley, Keith J.
Chemistry Department-555, Brookhaven National Laboratory
rowley@glx.chm.bnl.gov

Rubbia, Andre
CERN/ETHZ,
andre.rubbia@cern.ch

Rubinstein, Hector
Theoretical Physics, Uppsala University
rub@physto.se

Saavedra, Oscar
Department of Physics, Universita' di Torino, Dipartimento di Fisica Generale
saavedra@to.infn.it

Sacquin, Yves
COMMISSARIAT A L'ENERGIE A TOMIQUE, DAPNIA/SPP - CEA/SACLAY
sacquin@hep.saclay.cea.fr

Sadoulet, Bernard
Department of Physics, University of California, UC Berkeley
Center for Particle Astrophysics
sadoulet@physics.berkeley.edu

Saitta, Biagio
EP Division, CERN, University of Cagliari
biagio.saitta@cern.ch

Saji, Choji
Department of Physics, Niigata University
saji@hep.sc.niigata-u.ac.jp

Sakai, Atsushi
KEK, High Energy Accelerator Research Organization
sakai@neutrino.kek.jp

Sakuda, Makoto
National Laboratory for High Energy Physics (KEK),
sakuda@kekvax.kek.jp

Sakurai, Nobuyuki
Kamioka Observatory
Institute for Cosmic Ray Research, University of Tokyo
sakurai@sukettoe.icrr.u-tokyo.ac.jp

Sanda, Ichiro
Dept. of Physics, Faculty of Science, Nagoya Univ.
sanda@eken.phys.nagoya-u.ac.jp

Santacesaria, Roberta
Universita' di Roma Dipartimento di Fisica dell', INFN
Roberta.Santacesaria@cern.ch

Sasaki, Makoto
Institute for Cosmic Ray Research, University of Tokyo
sasakim@icrr.u-tokyo.ac.jp

Sato, Joe
Dept. of Physics, Faculty of Science, Univ. of Tokyo
joe@hep-th.phys.s.u-tokyo.ac.jp

Sato, Osamu
Department of Physics, Nagoya University
sato@flab.phys.nagoya-u.ac.jp

Schmitz, Norbert
Max-Planck-Institut Fuer Physik,
nschmitz@mppmu.mpg.de

Schneps, Jacob
Department of Physics, Tufts University
JACOB@TUHEP3.PHY.TUFTS.EDU

Scholberg, Kate
Boston University,
kate@suketto.icrr.u-tokyo.ac.jp

Shafi, Qaisar
Bartol Research Institute, University of Delaware
Shafi@bartol.udel.edu

Shiga, Kunio
Physics Department, Faculty of International Studies of Culture, Kyusyu Sangyo University
shiga@phys.kyusan-u.ac.jp

Shiozawa, Masato
Kamioka Observatory
Institute for Cosmic Ray Research, University of Tokyo
masato@icrkm4.icrr.u-tokyo.ac.jp

Shirai, Junpei
Research Center for Neutrino Science, Tohoku University
shirai@awa.tohoku.ac.jp

Smirnov, Alexei Yu
International Center for Theoretical Physics, (ICTP)
smirnov@ictp.trieste.it

Smy, Michael
Department of Physics and Astronomy, University of California, Irvine
smy@suketto.icrr.u-tokyo.ac.jp

Snellman, Hakan
Department of Theoretical Physics, Royal Institute of Technology
snell@theophys.kth.se

Sobel, Henry W.
Department of Physics and Astronomy, University of California Irvine
SOBEL@MASTER.PS.UCI.EDU

Spentzouris, Panagiotis G.
Fermi National Accelerator Laboratory,
spentz@hecate.fnal.gov

Spinelli, Paolo
Department of Physics and INFN, Bari University
Paolo.Spinelli@ba.infn.it

Spiro, Michel
CEA Saclay,
spiro@hep.saclay.cea.fr

Stiegler, Ulrich
CERN/PPE,
Ulrich.Stiegler@cern.ch

Stone, James
Dept. of Physics, Boston Univ.
stone@superk.bu.edu

Strolin, Paolo
EP Division, Univ. of Napoli and CERN
Paolo.Strolin@CERN.CH

Suekane, Fumihiko
Department of Physics, Tohoku University
suekane@awa.tohoku.ac.jp

Suematsu, Daijiro
Department of Physics, Kanazawa University
suematsu@hep.s.kanazawa-u.ac.jp

Sugawara, Toru
Department of Physics, Nagoya University
sugawara@eken.phys.nagoya-u.ac.jp

Sugimoto, Takashi
Dept. of Physics, Faculty of Science, Univ. of Tokyo
sugi@hep-th.phys.s.u-tokyo.ac.jp

Sugiura, Masaki
Department of Physics, Nagoya University
sugiura@eken.phys.nagoya-u.ac.jp

Sulak, Larry R.
Dept. of Physics, Boston Univ.
sulak@Bu.phyk.bu.edu

Suzuki, Atsumu
Faculty of Science, Department of Physics, Kobe University
suzuki@phys01.phys.kobe-u.ac.jp

Suzuki, Atsuto
Research Center for Neutrino Science, Graduate School of Science, Tohoku Univ.
suzukia@awa.tohoku.ac.jp

Suzuki, Yoichiro
Kamioka Observatory
Institute for Cosmic Ray Research, University of Tokyo
suzuki@suketto.icrr.u-tokyo.ac.jp

Takasugi, Eiichi
Dept. of Physics, RCNP/Osaka University
takasugi@phys.wani.osaka-u.ac.jp

Takayama, Jun
B+P Company, Sony Corporation
akiyo@vid.cpg.sony.co.jp

Takeda, Gyo
Tohoku Gakuin University,

Takeuchi, Hideki
Kamioka Observatory
Institute for Cosmic Ray Research, University of Tokyo
hideki@suketto.icrr.u-tokyo.ac.jp

Takeuchi, Yasuo
Kamioka Observatory
Institute for Cosmic Ray Research, University of Tokyo
takeuchi@icrr.u-tokyo.ac.jp

Takita, Masato
Faculty of Science, Osaka University
takita@oskjcc.hep.sci.osaka-u.ac.jp

Tamae, Kyoko
Tohoku University,
tamae@awa.tohoku.ac.jp

Tanimoto, Morimitsu
Faculty of Education, Ehime University
tanimoto@edserv.ed.ehime-u.ac.jp

Tao, Charling
PCC College de France,
Charling.Tao@cdf.in2p3.fr

Tartaglia, Roberto
Laboratori Nazionali del Gran Sasso, INFN
tartaglia@lngs.infn.it

Tasaka, Shigeki
Faculty of General Education, Gifu Univ.
tasaka@physics.physics.gifu-u.ac.jp

Tomozawa, Yukio
Physics Department
Randall Laboratory Physics, University of Michigan
tomozawa@umich.edu

Totani, Tomonori
Dept. Phys., University of Tokyo
totani@utaphp2.phys.s.u-tokyo.ac.jp

Totsuka, Yoji
Kamioka Observatory
Institute for Cosmic Ray Research, University of Tokyo
totsuka@icrr.u-tokyo.ac.jp

Toumakov, Vladimir
Institute for High Energy Physics, Russia & High Energy Accelerator Research Organization
tumakov@pnahp.kek.jp

Vagins, Mark
University of California, Irvine
VAGINS@DANKA.PS.UCI.EDU

van Dantzig, Rene
NIKHEF,
rvd@nikhef.nl

Vannucci, Francois
LPNHE, Univ. Paris 7
Vannucci@IN2P3.FR

Viollier, Raoul D.
Department of Physics, University of Cape Town
VIOLLIER@PHYSCI.UCT.AC.ZA

Viren, Brett
State University of New York (SUNY), at Stony Brook
bviren@superk.physics.sunysb.edu

Vitale, Sandro
Dipartimento di Fisica, Universita' di Genova, INFN Genova
VITALE@GE.INFN.IT

Volkas, Raymond R.
School of Physics, The University of Melbourne
r.volkas@physics.unimelb.edu.au

Volkova, Lioudmila
Inst. for Nuclear Research of Russian Academy, of Sciences
volkova@inr.npd.ac.ru

Wahl, Heinrich
CERN,
wahl@mail.cern.ch

Walter, Christopher
Boston University,
Walter@Budoe.Bu.edu

Waltham, Chris
Department of Physics and Astronomy, University of British Columbia
waltham@physics.ubc.ca

Wark, David L.
Oxford University,
d.wark1@physics.ox.ac.uk

Watanabe, Yasushi
Faculty of Science, Tokyo Institute of Technology
watanabe@hp.phys.titech.ac.jp

Weinheimer, Christian
Inst. F. Physics, University of Mainz
christian.weinheimer@uni-mainz.de

White, D. Hywel
Physics Division, Los Alamos National Laboratory H846
white@lanl.gov

Wilczek, Frank
Institute for Advanced Study,
wilczek@sns.ias.edu

Winter, Klaus H.
CERN-PPE, Humboldt University Berlin
Klaus.Winter@CERN.ch

Wojcicki, Stanley G.
Stanford University,
sgweg@slac.stanford.edu

Wold, Donald
University of Arkansas, Little Rock,
DCWOLD@UALR.EDU

Wong, Henry Tsz-King
Institute of Physics, Academia Sinica
htwong@phys.sinica.edu.tw

Yamada, Sakue
Inst. of Particle and Nuclear Studies, High Energy Accelerator Research Organization
yamada@insux1.tanashi.kek.jp

Yamada, Shuei
Kamioka Observatory
Institute for Cosmic Ray Research, University of Tokyo
shuei@icrkm4.icrr.u-tokyo.ac.jp

Yamaguchi, Yoshio
Former - President, IUPAP

Yanagida, Tsutomu
Dept. of Physics, Faculty of Science, Univ. of Tokyo
yanagida@hep-th.phys.s.u-tokyo.ac.jp

Yanagisawa, Chiaki
Dept. of Physics and Astronomy, State University of New York at Stony Brook
chiaki@sbhep.physics.sunysb.edu

Yasuda, Osamu
Department of Physics, Tokyo Metropolitan University
yasuda@phys.metro-u.ac.jp

Yoshida, Makoto
Department of Physics, Faculty of Science, Osaka University
yoshida@hep.sci.osaka-u.ac.jp

Yoshimura, Motohiko
Department of Physics, Tohoku University
yoshim@tuhep.phys.tohoku.ac.jp

Yoshioka, Koichi
Dept. of Physics, Kyoto University
yoshioka@gauge.scphys.kyoto-u.ac.jp

Young, Kenneth
Department of Physics, University of Washington
young@phys.washington.edu

Yuda, Toshinori
Institute for Cosmic Ray Research, University of Tokyo
yuda@icrr.u-tokyo.ac.jp

Zaimidoroga, Oleg
Institute for Nuclear Research, Dubna,
zaimidor@cv.jinr.ru

Zdesenko, Yuri G.
Kiev Institute for Nuclear Research,, National Academy of Sciences, Ukraine
interdep@kinr.kiev.ua

Zeitnitz, Bernhard R.
Universitat Karlsruhe,
Zeitnitz@ik1.fzk.de

AUTHOR INDEX

GENERAL INFORMATION

Nuclear Physics B – Proceedings Supplements (PS) is the premier publication outlet for the proceedings of key conferences on high-energy physics and related areas. The series covers both large international conferences and more specialized topical meetings. Under the guidance of the Editorial Board, the newest discoveries and the latest developments, reported at carefully selected meetings, are published covering experimental as well as theoretical particle physics, hadronic physics, cosmology, astrophysics and gravitation, field theory and statistical systems.

Note to Confrence Organizers

Organizers of upcoming meetings who are interested in exploring the possibilities of *Nuclear Physics B – Proceedings Supplements* as the publication outlet of their proceedings are invited to send full details of the planned conference to:

Drs. Carl Schwarz
Publishing Editor
Elsevier Science B.V., P.O. Box 103, 1000 AC Amsterdam, The Netherlands
Telephone: +31 20 485 2355; Telefax: +31 20 485 2580; Telex: 10704.
E-mail: C.SCHWARZ@ELSEVIER.NL

Proceedings will be produced from camera-ready copy and will be published within three months of the Publisher's receipt of the final and complete typescript. Attractive bulk-order arrangements will be offered to conference organizers for participants' copies.

Subscription information

In 1999, eleven volumes of *Proceedings Supplements* will be published, comprising ca. 12 individual Proceedings.

Subscription prices are available upon request from the Publisher. A combined subscription to Nuclear Physics A volumes 645-661, Nuclear Physics B volumes 536-563, and Nuclear Physics B – Proceedings Supplements volumes 69-79 is available at a reduced rate. Single copies of proceedings published in the *Proceedings Supplements* can be ordered directly from the Publisher.

Subscriptions are accepted on a prepaid basis only and are entered on a calendar year basis. Issues are sent by surface mail except to the following countries where air delivery via SAL is ensured: Argentina, Australia, Brazil, Canada, PR China, Hong Kong, India, Israel, Japan, Malaysia, Mexico, New Zealand, Pakistan, Singapore, South Africa, South Korea, Taiwan, Thailand, USA. For all other countries airmail rates are available upon request.

www.ingramcontent.com/pod-product-compliance
Ingram Content Group UK Ltd.
Pitfield, Milton Keynes, MK11 3LW, UK
UKHW050949290926
15470UKWH00005B/1730

9 780444 502896